The Essays and Articles of Alfred North Whitehead, 1886–1916

The Edinburgh Critical Edition of the Complete Works of Alfred North Whitehead

The complete, collected critical edition of the unpublished and published manuscripts of British-American philosopher Alfred North Whitehead

This Critical Edition brings together for the first time, in a series of critically edited volumes, the complete, collected published works and previously unpublished lectures, papers and correspondence of Alfred North Whitehead, one of the 20th century's most original and significant philosophers.

Newly discovered materials, long thought lost or destroyed, including illustrations, equations and chalkboard diagrams, which Whitehead used in classroom settings, have never before been seen by contemporary scholars. A projected six volumes of unpublished material will illuminate many factors influencing the development of Whitehead's initial and later thought, as well as elucidate, in considerably greater detail than ever before, many of the principal concepts later set out in his body of published philosophical reflection.

New critical editions of his famous lecture series at Harvard University, such as *Science and the Modern World, Symbolism* and *The Function of Reason*, his world-renowned Gifford lecture series at Edinburgh University in 1928, *Process and Reality*, together with new editions of justifiably famous collections of later essays and public lectures, such as *Adventures of Ideas* and *Modes of Thought*, will round out this ambitious and authoritative Critical Edition.

Available
The Harvard Lectures of Alfred North Whitehead, 1924–1925: Philosophical Presuppositions of Science edited by Paul A. Bogaard and Jason Bell

The Harvard Lectures of Alfred North Whitehead, 1925–1927: General Metaphysical Problems of Science edited by Brian G. Henning, Joseph Petek and George Lucas

The Collected Essays and Articles of Alfred North Whitehead, 1886–1916 edited by Brian G. Henning, Joseph Petek and Robert J. Valenza

The Collected Essays and Articles of Alfred North Whitehead, 1917–1942 edited by Brian G. Henning and Joseph Petek

Visit the series page at edinburghuniversitypress.com/series-the-edinburgh-critical-edition-of-the-complete-works-of-alfred-north-whitehead.html

The Essays and Articles of Alfred North Whitehead, 1886–1916

Edited by Brian G. Henning, Joseph Petek and Robert J. Valenza

EDINBURGH
University Press

Edinburgh University Press is one of the leading university presses in the UK. We publish academic books and journals in our selected subject areas across the humanities and social sciences, combining cutting-edge scholarship with high editorial and production values to produce academic works of lasting importance. For more information visit our website: edinburghuniversitypress.com

Edinburgh University Press Ltd
13 Infirmary Street
Edinburgh EH1 1LT

Typeset in Garamond and Cambria by R. J. Footring Ltd, Derby, UK, and printed and bound in Great Britain

A CIP record for this book is available from the British Library

ISBN 978 1 3995 3120 7 (hardback)
ISBN 978 1 3995 3122 1 (webready PDF)
ISBN 978 1 3995 3123 8 (epub)

EU Authorised Representative:
Easy Access System Europe
Mustamäe tee 50, 10621 Tallinn, Estonia
gpsr.requests@easproject.com

MIX
Paper | Supporting responsible forestry
FSC
www.fsc.org
FSC® C018072

This book is printed using paper from well-managed forests, recycling and other controlled sources

This book is in honour of Donald Phillip Verene
by Frederick and Nancy Marcus

Contents

General introduction

Brian G. Henning, *General Editor*
Joseph Petek, *Executive Editor*

Alfred North Whitehead, a British-born mathematician and philosopher who attained widespread acclaim in America in the first half of the twentieth century, is perhaps best known among the wider public for his famous saying that the European philosophical tradition 'consists of a series of footnotes to Plato'.[1]

A student of mathematics at Trinity College, Cambridge, in the 1880s, Whitehead was elected a Fellow of the College in 1884, where he mentored the mathematical studies of such notable figures as John Maynard Keynes and Bertrand Russell, and was also a member of one of the most elite societies in the English-speaking university world at the time, the famed 'Cambridge Apostles'.[2]

In 1910, Whitehead resigned his fellowship at Trinity College and moved to London, where (after a year spent in research and writing) he obtained a lectureship in mathematics at University College London. In 1914, he began lecturing at the newly organised Imperial College of Science and Technology (now Imperial College London) before being elected Dean of the Faculty of Science at the University of London in 1918 and Chairman of its Academic Council in 1920.[3]

At the beginning of the twentieth century, Whitehead began a collaboration with his student and protégé, Bertrand Russell, that lasted well over a decade and resulted in the publication (in 1910, 1912 and 1913) of the three-volume *Principia Mathematica*, a monumental study intended to establish the foundations of mathematics in formal logic. Subsequently, in 1922, Whitehead himself concluded two-decades of study of the geometry of space-time in *The Principle of Relativity*, a path-breaking work of theoretical cosmology explicitly intended to pose an alternative to Albert Einstein's General Theory. Other works spanning Whitehead's academic career in England include *A Treatise on Universal Algebra* (1898), 'On mathematical concepts of the material world' (1906), *An Introduction to Mathematics* (1911), *The Organisation of Thought* (1917), *An Enquiry Concerning the Principles of Natural Knowledge* (1919) and a significant work on the philosophy of science, *The Concept of Nature* (1920). These works cemented Whitehead's reputation as one of the leading figures in the early development of analytic philosophy, alongside Russell, G. E. Moore and Ludwig Wittgenstein.

---

1. Alfred North Whitehead, *Process and Reality* (New York: Free Press, 1978 [1929]), p. 39.
2. For more on Whitehead and the Apostles, see Victor Lowe, *Alfred North Whitehead: The Man and His Work, Volume I: 1861–1910* (Baltimore: Johns Hopkins University Press, 1985), pp. 111–145.
3. Both University College London and the Imperial College of Science and Technology were member colleges of the University of London.

In 1924, at the age of sixty-three, Whitehead retired from his position at the University of London and sailed to America to accept a position at Harvard University, where he would teach until his retirement in 1937. He would live the rest of his life in Cambridge, Massachusetts.

Upon his arrival in the United States, and with inspiration allegedly derived from the uniquely American philosophy of 'experience' (particularly that of William James and John Dewey), Whitehead undertook a series of works on far broader philosophical themes. His later work encompasses the history and philosophy of science, the sociological development of civilisation, the philosophy of history, philosophical reflections on education and, most famously, an ambitious attempt to develop a descriptive metaphysical system commensurate with early twentieth-century relativistic cosmology and quantum mechanics, which he called the 'philosophy of organism' and which scholars now commonly refer to as 'process philosophy'. Works during this American period of his career include *Science and the Modern World* (1925), *Religion in the Making* (1926), *Symbolism* (1927), *The Function of Reason* (1929) and his best-known work, *Process and Reality* (1929), based upon his Gifford lectures at the University of Edinburgh delivered in June 1928. In addition to subsequent books, such as *Adventures of Ideas* (1933) and *Modes of Thought* (1938), Whitehead also published a number of lectures and essays on diverse topics during this period, many of which he later published in collections such as *The Aims of Education* (1929) and *Essays in Science and Philosophy* (1947).

Whitehead's stature as a pivotal figure in American philosophy, his significant contributions to the rise of analytic philosophy, and the recent and worldwide renewal of interest in the constructive contribution of his seminal ideas to a number of important contemporary metaphysical and cosmological problems in philosophy argue for the appropriateness and timeliness of a complete critical edition of his works – an edition intended to rank alongside critical collected editions already produced or in progress on the works of his eminent predecessors and contemporaries, including William James, John Dewey, Josiah Royce, George Santayana and C. S. Peirce, as well as Einstein, Russell and Wittgenstein.

The need for a complete critical edition of the unpublished and published manuscripts of Whitehead is compelling. Beginning around 2010, an enormous body of unpublished material steadily began to come to light, thus helping to dispel the myth that all of the Whiteheadian *Nachlass* had been lost or destroyed.[1] The new discoveries included copious and detailed lecture notes taken by both students and faculty colleagues attending Whitehead's classes. These lecture notes were recorded by philosophers who became quite accomplished in their own right, such as W. V. O. Quine, Charles Hartshorne, Raphael Demos, Susanne Langer and Paul Weiss.

Despite the fact that much of his work undertaken in America was written comparatively recently (that is, during the early part of the last century), scholars

1. For more on this see Brian G. Henning, 'Preface: A Brief History of the Critical Edition of Whitehead', in Brian G. Henning and Joseph Petek (eds), *Whitehead at Harvard, 1924–1925* (Edinburgh: Edinburgh University Press, 2020); Brian G. Henning, 'On the Recently Discovered Whitehead Papers', Whitehead Research Project, 14 January 2019, <https://whiteheadresearch.org/2019/01/14/on-the-recently-discovered-whitehead-papers> (accessed January 2025); and Joseph Petek, *Unearthing the Unknown Whitehead* (Lanham: Lexington Press, 2022), pp. 8–9.

at present do not have a clear sense of what motivated Whitehead or what led to the development of his thought. Part of the reason for this has to do with the fact that his personal correspondence with colleagues – along with the lecture notes of his students and colleagues which were not either discarded, donated or destroyed following his death – had been scattered in libraries and archives across the United States, Canada and the UK.

The editors of the Critical Edition of Whitehead have invested considerable time and effort locating, collecting and cataloguing a substantial body of these extant unpublished materials. The newly discovered materials in particular shed considerable light on the factors influencing the development of Whitehead's later philosophy, and also often elucidate – in considerably greater detail than ever before – many of the principal concepts encompassed within that body of philosophical reflection (e.g. through the apparent use by Whitehead himself of illustrations, equations and blackboard diagrams in classroom settings which have never before been seen by contemporary scholars).

Whitehead's published works, meanwhile, remain in disarray: some are out of print, while others remain in print from multiple sources, replete with inconsistencies and extensive uncorrected typographical and textual errors. Many lack more than a cursory index. Apart from a newly reprinted edition of Whitehead and Russell's *Principia Mathematica*, the sole exception was the re-issue (in 1978) of a so-called 'Corrected Edition' of *Process and Reality*. This proved to be a helpful interim work but it did not fully adhere to the rigorous protocols since established for scholarly editions generally.[1] Finally, while multiple copyrights for extant works are claimed by several publishers, in point of fact all of Whitehead's materials (published and unpublished) continue to remain legally in the sole possession of the Whitehead family estate, whose sole executor[2] has granted full permission for the publication of both unpublished and published works to the Whitehead Research Project.

When finally completed, the Edinburgh Critical Edition of the Complete Works of Alfred North Whitehead will encompass the entire collected works of the author, published and unpublished. Furthermore, by organising, archiving, digitising, transcribing, editing and indexing all of his writings, the full breadth and significance of Whitehead's diverse work will, perhaps for the first time, come fully into view. Not only will this make available materials that have never been published, but it will also improve the scholarly understanding of previously published items. Perhaps the most useful and exciting aspect of this project to many scholars, however (beyond the publication of newly discovered material), will be the creation of a fully searchable electronic archive of these primary source materials.[3] In these ways, this Critical Edition should facilitate and energise the scholarly study of one of the twentieth century's most original and influential intellectual figures.

1. See for example Mary-Jo Kline and Susan Holbrook Perdue, *A Guide to Documentary Editing*, 3rd edition (Charlottesville: University of Virginia Press, 2008).
2. George W. Whitehead (a grandson).
3. Digital scans of many items can be viewed online at <https://wrl.whiteheadresearch.org>.

Chronology for Alfred North Whitehead

1861 15 February. Birth as the last of four children.
1875 September. Whitehead leaves home to attend Sherborne School, a boarding school for boys (through to 23 June 1880).
1877 June. Whitehead's grandfather dies.
1880 October. Whitehead attends Trinity College, Cambridge, as a student (through to October 1884).
1884 17 May. Election to Cambridge Apostles.
1884 9 October. Whitehead elected a Fellow of Trinity College, Cambridge, and begins teaching mathematics there.
1888 First two papers published.
1888 Begins afternoon teaching at Girton College, Cambridge (then a women's college), probably for no pay, and continues until 1913.
1890 16 December. Marriage to Evelyn Wade.
1891 31 December. Birth of T. North Whitehead.
1894 23 February. Birth of Jessie Whitehead.
1895 October. Bertrand Russell elected a Fellow of Trinity College.
1898 March. Whitehead's father dies.
1898 27 November. Birth of Eric Whitehead.
1899 Whitehead moves from Cambridge to Grantchester.
1901 Whitehead is appointed to the Council of Newnham College, Cambridge, although he never teaches there.
1901 Bertrand and Alys Russell move in with the Whiteheads and live with them for about a year. Russell and Whitehead begin collaboration on the *Principia.*
1903 13 February. Whitehead is elected a Fellow of the Royal Society.
1907 January. Whitehead's brother-in-law, John Branch (married to his sister Shirley), kills himself. Whitehead visits to help put affairs in order.
1907 April. Whitehead moves back to Cambridge.
1907 May. Whitehead becomes Chairman of the Cambridge branch of the Men's League for Women's Suffrage.
1910 April. Whitehead resigns his Trinity lectureship and soon moves to Chelsea, London. He and his wife also buy a country home in Lockeridge, Wiltshire.
1910 December. First volume of the *Principia* is published.
1911 July. Whitehead is appointed to a teaching position at University College London.
1912 Whitehead is elected President for a year of both the South-Eastern Mathematical Association and the London branch of the Mathematical Association.
1914 10 July. Whitehead is appointed to a professorship at Imperial College London.

1914 15 July. Whitehead resigns from University College London.
1914 31 July. Gertrude Stein and Alice Toklas stay with the Whiteheads for six weeks.
1914 1 September. Whitehead begins teaching at Imperial College London.
1915 Whitehead joins the Aristotelian Society, in which he is active through to his departure for America in 1924 (serves as President, 1922–3).
1915 Whitehead is elected President of the Mathematical Association and serves for two years.
1916 11 July. Russell is dismissed from his Trinity lectureship due to his anti-war views.
1917 May. Eric Whitehead is enlisted in the Royal Flying Corps.
1917 The Whiteheads sell their country home in Lockeridge. They buy another in Oxted, Surrey.
1918 13 March. Eric Whitehead is killed in action.
1918 6 May. Russell is imprisoned for his anti-war views until 11 September of the same year.
1918 31 October. Whitehead is elected Dean of the Faculty of Science at the University of London and holds the post for four years.
1919 18 October. Whitehead delivers the first of his Tarner lectures (he gives seven in total, through to 29 November that year).
1920 Whitehead is appointed Chairman of the Academic (Leadership) Council of the Senate of the University of London and holds the post for four years.
1920 July. Whitehead is appointed Chairman of the Delegacy for Goldsmiths' College, London, and holds the post until he leaves for Harvard in 1924.
1920 3 November. T. North Whitehead marries Margaret Schuster.
1921 Eric Alfred North Whitehead is born, T. North's son and Whitehead's grandson.
1922 April. Whitehead accepts first invitation to cross the Atlantic to America for a conference at Bryn Mawr College in honour of mathematics professor Charlotte Angas Scott.
1924 2 February. Whitehead's mother dies.
1924 24 February. Whitehead accepts a five-year appointment at Harvard.
1924 1 April. Whitehead agrees to give the Lowell lectures, responding to an invitation of 18 March.
1924 16 August. Whitehead leaves England for Cambridge, Massachusetts.
1924 27 August. Whitehead arrives in Cambridge.
1924 25 September. Whitehead gives his first lectures at Harvard.
1924 4 October. Whitehead moves into a new flat in Radnor Hall (now part of the Memorial Drive Apartments Historic District), overlooking the Charles River.
1925 2 February. Whitehead delivers the first of his Lowell lectures, which he proceeds to give through to 2 March, every Monday and Thursday at 5 p.m., with the exception of 23 February (these would become *Science and the Modern World*).
1925 5 April. Whitehead delivers a lecture on 'Religion and science' at Phillips Brooks House at Harvard.

1925 14 April. Whitehead gives a lecture on 'Mathematics as an element in the history of thought' to the Mathematical Society at Brown University, Providence, Rhode Island.

1925 29 June. Jessie Whitehead arrives in Cambridge, Massachusetts, and begins working in the Harvard library on 20 July.

1925 30 December. Whitehead attends his first meeting of the American Philosophical Association and gives a paper on the subject of time.

1926 February. Whitehead delivers a second series of Lowell lectures (these would become *Religion in the Making*).

1926 February. The 'Committee of Four' begins meeting (Whitehead, L. J. Henderson, John Livingston Lowes, Charles P. Curtis). Together they choose four to eight PhD students per year for the Society of Fellows for a term of three years each with an option for three more (among these Junior Fellows was Willard van Orman Quine, chosen in the first batch).

1926 During Harvard's spring break, Whitehead delivers lectures at McGill University (Montreal, Canada), the University of Michigan and the University of Illinois.

1926 Whitehead delivers lectures at Dartmouth College.

1926 September. Whitehead attends the Sixth International Congress of Philosophy at Harvard and gives a paper entitled 'Time'.

1927 27 February. Whitehead accepts an invitation to deliver the Gifford lectures for 1928.

1927 April. Whitehead delivers the Barbour-Page lectures at the University of Virginia (these would become *Symbolism: Its Meaning and Effect*).

1928 1 June. Whitehead delivers the first of the Gifford lectures (which he proceeds to give through to 22 June).

1929 March. Whitehead delivers the Vanuxem lectures at Princeton University, New Jersey (these would become *The Function of Reason*).

1929 Whitehead delivers a series of four Mary Flexner Lectures at Bryn Mawr College, which would become chapters in *Adventures of Ideas*.

1930 Whitehead attends the Seventh International Congress of Philosophy in Oxford.

1931 14 February. Whitehead attends a symposium for his seventieth birthday, arranged by James Woods.

1931 February. T. North Whitehead joins a human relations study group at Harvard Business School and moves to Cambridge, Massachusetts.

1931 Whitehead delivers an address at the Harvard Business School.

1931 29 December. Whitehead delivers his presidential address to the eastern division of the American Philosophical Association at New Haven.

1931 Whitehead is elected a Fellow of the British Academy.

1932 March. Whitehead delivers the Davies Lecture in Philosophy at the Institute of Arts and Sciences, Columbia University, which would become a chapter in *Adventures of Ideas*.

1933 October. Whitehead delivers two lectures at the University of Chicago, which would become *Nature and Life* and, later, two chapters in *Modes of Thought*.

1935 25 October. Whitehead gives a short address to graduate students of the

Harvard and Radcliffe Philosophical Departments, which would be adapted into the epilogue of *Modes of Thought*.

1937 7 May. Whitehead delivers his final Harvard lecture.

1937 29 October. Whitehead delivers the first of a series of six inaugural Calkins lectures at Wellesley College, which would become the first six chapters of *Modes of Thought*.

1941 15 February. Whitehead attends a party for his eightieth birthday, organised by his Harvard colleagues.

1945 1 January. Whitehead is awarded the Order of Merit.

1947 30 December. Whitehead dies from a cerebral haemorrhage.

Published Works of Alfred North Whitehead

1886 10 February. 'A Celebrity at Home: The Clerk of the Weather'. *Cambridge Review*, vol. 7, no. 167, pp. 202–203.

1886 12 May. 'Davy Jones'. *Cambridge Review*, vol. 7, no. 174, pp. 311–312.

1888 February. 'On the Motion of Viscous Incompressible Fluids'. *Quarterly Journal of Pure and Applied Mathematics*, vol. 23, no. 89, pp. 78–93.

1888 6 March. 'A Visitation'. *Cambridge Fortnightly*, vol. 1, no. 4, pp. 81–83.

1888 June. 'Second Approximations to Viscous Fluid Motion'. *Quarterly Journal of Pure and Applied Mathematics*, vol. 23, no. 90, pp. 143–152.

1891 20 February. 'The Fens as Seen from Skates'. *Cambridge Review*, vol. 12, no. 297, pp. 212–213. (Attribution to Whitehead not wholly certain.)

1896 14 May. 'On Ideals: With Reference to the Controversy Concerning the Admission of Women to Degrees in the University'. *Cambridge Review*, vol. 17, pp. 310–311.

1898 10 March. 'The Geodesic Geometry of Surfaces in Non-Euclidean Space'. *Proceedings of the London Mathematical Society*, vol. 29, pp. 275–324.

1898 *A Treatise on Universal Algebra*. Cambridge University Press.

1899 2 February. 'Sets of Operations in Relation to Groups of Finite Order' (abstract). *Proceedings of the Royal Society of London*, vol. 64, pp. 319–320.

1901 April. 'Memoir on the Algebra of Symbolic Logic (Part I)'. *American Journal of Mathematics*, vol. 23, no. 2, pp. 139–165.

1901 October. 'Memoir on the Algebra of Symbolic Logic (Part II)'. *American Journal of Mathematics*, vol. 23, no. 4, pp. 297–316.

1902 October. 'On Cardinal Numbers'. *American Journal of Mathematics*, vol. 24, no. 4, pp. 367–394.

1903 April. 'The Logic of Relations, Logical Substitution Groups, and Cardinal Numbers'. *American Journal of Mathematics*, vol. 25, no. 2, pp. 157–178.

1903 14 May. 'The University Library'. *Cambridge Review*, vol. 24, p. 295.

1904 January. 'Theorems on Cardinal Numbers'. *American Journal of Mathematics*, vol. 26, no 1, pp. 31–32.

1905 November. 'Note'. *Revue de Metaphysique et de Morale*, vol. 13, no. 6, pp. 916–917.

1906 May. 'On Mathematical Concepts of the Material World'. *Philosophical Transactions of the Royal Society of London*, vol. 205, pp. 465–525.

1906 *The Axioms of Projective Geometry*. Cambridge University Press.

1906 5 November. 'Liberty and the Enfranchisement of Women'. Cambridge Women's Suffrage Association, pamphlet.

1907 March. *The Axioms of Descriptive Geometry*. Cambridge University Press.

1910 With Bertrand Russell. 'Non-Euclidean Geometry'. *Encyclopaedia Britannica*, 11th edition, vol. 11, pp. 724–730.

1910 'Axioms of Geometry'. *Encyclopaedia Britannica* (11th edition), vol. 11, pp. 730–736.

1910 October. 'The Philosophy of Mathematics'. *Science Progress in the Twentieth Century*, vol. 5, no. 18, pp. 234–239.

1910 December. With Bertrand Russell. *Principia Mathematica, Volume I*. Cambridge University Press.

1911 *An Introduction to Mathematics*. Williams and Norgate, Henry Holt.

1911 'Mathematics'. *Encyclopaedia Britannica* (11th edition), vol. 17, pp. 878–883.

1911 December. 'The Place of Mathematics in a Liberal Education'. *Journal of the Association of Teachers of Mathematics for the Southeastern Part of England*, vol. 1, no. 1, pp. 2–12.

1912 With Bertrand Russell. *Principia Mathematica, Volume II*. Cambridge University Press.

1913 With Bertrand Russell. *Principia Mathematica, Volume III*. Cambridge University Press.

1913 'The Principles of Mathematics in Relation to Elementary Teaching'. *Proceedings of the Fifth International Congress of Mathematicians, Volume II*, Cambridge University Press, pp. 449–454. Also contemporaneously published in *L'Enseignment Mathematique*.

1913 March. 'Presidential Address to the London Branch of the Mathematical Association'. *Mathematical Gazette*, vol. 7, no. 104, pp. 87–94.

1914 'Report of the Council of the Royal Society of London'. *Year-Book of the Royal Society*, pp. 177–87.

1915 'Report of the Council of the Royal Society of London'. *Year-Book of the Royal Society*, pp. 176–85.

1916 'Space, Time, and Relativity'. *Proceedings of the Aristotelian Society*, vol. 16, pp. 104–129.

1916 January. 'The Aims of Education: A Plea for Reform'. *Mathematical Gazette*, vol. 8, no. 121, pp. 191–203.

1916 May. 'La Theorie Relationniste de L'Espace'. *Revue de Métaphysique et de la Morale*, vol. 23, no. 3, pp. 423–454.

1917 'The Organisation of Thought'. *Report of the 86th Meeting of the British Association for the Advancement of Science*, pp. 355–365. Also contemporaneously published with slight differences in six other places (see p. 452 for list).

1917 *The Organisation of Thought*. Williams and Norgate, Greenwood Press.

1917 January. With A. W. Siddons. 'Letter to the Editor'. *Mathematical Gazette*, vol. 9, no. 127, p. 14.

1917 March. 'Technical Education and Its Relation to Science and Literature'. *Mathematical Gazette*, vol. 9, no. 128, pp. 20–33. Also contemporaneously published in *Technical Journal*, vol. 10 (January), pp. 59–74.

1917 March. A Polytechnic in War-Time. *Borough Polytechnic News*, vol. 27, no. 3, pp. 27–29.

1918 1 May. 'Graphical Solution for High-Angle Fire'. *Proceedings of the Royal Society*, Series A, vol. 94, no. 661, pp. 301–307.

1919 *An Address . . . to the Students and Parents, On Founder's Day . . .* [Stanley Technical Trade School, South Norwood, London]. Coventry and Son.

1919 With Oliver Lodge, J. W. Nicholson, Henry Head, Adrian Stephen and H. Wildon Carr. 'Symposium. Time, Space and Material: Are They, and If So in What Sense, the Ultimate Data of Science?' *Proceedings of the Aristotelian Society, Supplementary Volumes*, vol. 2, *Problems of Science and Philosophy*, pp. 44–108.

1919 *An Enquiry Concerning the Principles of Natural Knowledge*. Cambridge University Press.

1919 15 November. 'A Revolution in Science'. *The Nation*, vol. 26, pp. 232–233.

1920 *The Concept of Nature*. Cambridge University Press.

1920 'Discussion upon Fundamental Principles of Education'. In *Report of the Eighty-Seventh Meeting of the British Association for the Advancement of Science*, p. 361. John Murray.

1920 12 February. 'Einstein's Theory: An Alternative Suggestion'. *Times Educational Supplement*, p. 83.

1921 'Science in General Education'. In *Second Congress of the Universities of the Empire*, pp. 31–39. G. Bell and Sons.

1921 With many others. *Report of the Committee Appointed by the Prime Minister to Inquire into the Position of Classics in the Educational System of the United Kingdom*. His Majesty's Stationery Office.

1922 20 February. With H. W. Carr, T. P. Nunn and Dorothy Wrinch. 'Discussion: The Idealistic Interpretation of Einstein's Theory'. *Proceedings of the Aristotelian Society*, vol. 22, pp. 123–138.

1922 18 April. 'Some Principles of Physical Science', presentation at Bryn Mawr College conference put in print as a booklet and also published later that year as a chapter in *The Principle of Relativity*.

1922 16 July. 'The Philosophical Aspects of the Principle of Relativity'. *Proceedings of the Aristotelian Society*, vol. 22, pp. 215–223.

1922 *The Principle of Relativity, with Applications to Physical Science*. Cambridge University Press.

1922 *The Rhythm of Education*. London: Christophers. Also published in an abbreviated form in the *Bulletin of the American Association of University Professors*, vol. 9, no. 7 November 1923, pp. 17–19.

1923 'The First Physical Synthesis'. In *Science and Civilization*, pp. 161–178. Oxford University Press.

1923 'Uniformity and Contingency'. *Proceedings of the Aristotelian Society*, vol. 23, pp. 1–18.

1923 January. 'The Place of Classics in Education'. *Hibbert Journal*, vol. 21, no. 2, pp. 248–261.

1923 17 February. 'Letter to the Editor: Reply to Review of *The Principle of Relativity*'. *New Statesman*, vo. 20, p. 568.

1923 July. 'The Rhythmic Claims of Freedom and Discipline'. *Hibbert Journal*, vol. 21, no. 4, pp. 657–668.

1923 With H. W. Carr and R. A. Sampson. 'Symposium. The Problem of Simultaneity: Is There a Paradox in the Principle of Relativity in Regard to the Relation of Time Measured and Time Lived?' *Aristotelian Society, Supplementary Volumes*, vol. 3, *Relativity, Logic and Mysticism*, pp. 15–41.

1925 April. 'The Importance of Friendly Relations Between England and the United States'. *Phillips Bulletin*, vol. 19, no. 3, pp. 15–18.
1925 August. 'Religion and Science'. *Atlantic Monthly*, vol. 136, pp. 200–207.
1925 June. 'Remarks at Wellesley College Semi-Centennial Celebration'. *Wellesley Alumnae Magazine*, pp. 234–235.
1925 October. *Science and the Modern World.* Macmillan.
1926 September. *Religion in the Making.* Macmillan and Cambridge University Press.
1926 January. '*Principia Mathematica* – To the Editor of *Mind*'. *Mind*, vol. 35, no. 137, p. 130.
1926 August. 'The Education of an Englishman'. *Atlantic Monthly*, vol. 138, pp. 192–198.
1927 'Time'. In *Proceedings of the Sixth International Congress of Philosophy*, pp. 59–64. Longmans, Green and Co.
1927 June. 'England and the Narrow Seas'. *Atlantic Monthly*, vol. 139, pp. 791–798.
1927 November. *Symbolism: Its Meaning and Effect.* Macmillan.
1928 May. 'Universities and Their Function'. *Atlantic Monthly*, vol. 141, pp. 638–644.
1929 *Process and Reality.* Macmillan and Cambridge University Press.
1929 *The Aims of Education and Other Essays.* Macmillan, Williams and Norgate.
1929 *The Function of Reason.* Princeton University Press.
1930 January. 'An Address Delivered at the Celebration of the Fiftieth Anniversary of the Founding of Radcliffe College'. *Radcliffe Quarterly*, vol. 14, no. 1, pp. 1–5.
1930 'Prefatory Note'. In Susanne K. Langer, *The Practice of Philosophy*, p. vii. Henry Holt and Co.
1931 'On Foresight'. Introduction, in Wallace Brett Donham, *Business Adrift*, pp. xi–xxix. McGraw-Hill.
1931 'Objects and Subjects'. *Proceedings and Addresses of the American Philosophical Association*, vol. 5, pp. 130–146. It was later published as chapter XVI of *Adventures of Ideas.*
1932 'Symposium in Honor of the Seventieth Birthday of Alfred North Whitehead', speech included in a booklet of symposium proceedings printed by Harvard University Press, pp. 22–29.
1933 *Adventures of Ideas.* Macmillan and Cambridge University Press.
1933 July. 'The Study of the Past – Its Uses and Its Dangers'. *Harvard Business Review*, vol. 11, no. 4, pp. 436–444.
1934 'Foreword'. In *The Farther Shore: An Anthology of World Opinion on the Immortality of the Soul.* Houghton Mifflin.
1934 *Nature and Life.* University of Chicago Press. (This book, which comprises two essays, was republished in its entirety as Part 3 of *Modes of Thought*, 1938.)
1934 'Philosophy of Life'. In Dagobert D. Runes (ed.), *Twentieth Century Philosophy: Living Schools of Thought*, pp. 131–144. Philosophical Library. This was later published as chapter VIII of *Modes of Thought.*

1934 'Foreword'. In Willard van Orman Quine, *A System of Logistic*, pp. ix–x. Harvard University Press.
1934 July. 'Indication, Classes, Numbers, Validation'. *Mind*, vol. 43, no. 171, pp. 281–297.
1934 October. 'Corrigenda. Indication, Classes, Numbers, Validation'. *Mind*, vol. 43, no. 172, p. 543.
1935 With Walter E. Clark, George H. Chase and Ralph Barton Perry. 'Minute on the Life and Services of Professor James Haughton Woods'. *Harvard University Gazette*, vol. 30, pp. 153–155.
1935 'The Aim of Philosophy'. *Harvard Alumni Bulletin*, vol. 38, pp. 234–235. This was later published as the Epilogue in *Modes of Thought*.
1935 'Eulogy of Bernard Bosanquet'. In *Bernard Bosanquet and His Friends: Letters Illustrating the Sources and the Development of His Philosophical Opinions*, p. 316. George Allen and Unwin.
1936 June. 'Memories'. *Atlantic Monthly*, vol. 157, no. 6, pp. 672–679.
1936 September. 'Harvard: The Future'. *Atlantic Monthly*, vol. 158, no. 3, pp. 260–270.
1937 'Remarks'. *Philosophical Review*, vol. 46, pp. 178–186.
1937 March. 'Analysis of Meaning'. *Philosophical Review*, vol. 46, no. 2, pp. 178–186.
1938 *Modes of Thought*. Macmillan and Cambridge University Press.
1939 March. 'An Appeal to Sanity'. *Atlantic Monthly*, vol. 163, no. 3, pp. 309–320.
1939 'John Dewey and His Influence'. In Paul Arthur Schilpp (ed.), *The Philosophy of John Dewey*, pp. 477–479. Northwestern University Press.
1940 'Aspects of Freedom'. In Ruth Nanda Anshen (ed.), *Freedom: Its Meaning*, pp. 42–67. Harcourt, Brace.
1940 24 December. 'The Issue: Freedom'. *Boston Daily Globe*.
1941 'Mathematics and the Good'. In Paul Arthur Schilpp (ed.), *The Philosophy of Alfred North Whitehead*, pp. 666–681. Northwestern University Press.
1941 'Immortality'. In Paul Arthur Schilpp (ed.), *The Philosophy of Alfred North Whitehead*, pp. 682–700. Northwestern University Press.
1941 'Autobiographical Notes'. In Paul Arthur Schilpp (ed.), pp. 3–14. *The Philosophy of Alfred North Whitehead*. Northwestern University Press.
1942 February. 'The Problem of Reconstruction'. *Atlantic Monthly*, vol. 169, no. 2, pp. 172–175.
1942 October. 'Statesmanship and Specialized Learning'. *Proceedings of the American Academy of Arts and Sciences*, vol. 75, no. 1, pp. 1–5.
1945 Preface to 'The Organization of a Story and a Tale', by William Morgan. *Journal of American Folklore*, vol. 58, no. 229, p. 169.
1947 *Essays in Science and Philosophy*. Philosophical Library.

Abbreviations of Whitehead Titles

AE	*The Aims of Education and Other Essays* (1929)
AI	*Adventures of Ideas* (1933)
CN	*The Concept of Nature* (1920)
EA1	*The Essays and Articles of Alfred North Whitehead, 1886–1916* (2025)
EA2	*The Essays and Articles of Alfred North Whitehead, 1917–1942* (2025)
ESP	*Essays in Science and Philosophy* (1947)
FR	*The Function of Reason* (1929)
HL1	*The Harvard Lectures of Alfred North Whitehead, 1924–1925: Philosophical Presuppositions of Science* (2017)
HL2	*The Harvard Lectures of Alfred North Whitehead, 1925–1927: General Metaphysical Problems of Science* (2021)
IM	*An Introduction to Mathematics* (1911)
MT	*Modes of Thought* (1938)
OT	*The Organisation of Thought, Educational and Scientific* (1917)
PM	*Principia Mathematica* (1910–1913)
PNK	*An Enquiry Concerning the Principles of Natural Knowledge* (1919)
PR	*Process and Reality: An Essay in Cosmology* (1929)
R	*The Principle of Relativity: With Applications to Physical Science* (1922)
RM	*Religion in the Making* (1926)
S	*Symbolism: Its Meaning and Effect* (1927)
SMW	*Science and the Modern World* (1925)
UA	*Universal Algebra* (1898)

Introduction to *The Essays and Articles of Alfred North Whitehead, 1886–1916*

Brian G. Henning, *General Editor*
Joseph Petek, *Executive Editor*

This volume of the Edinburgh Critical Edition of the Complete Works of Alfred North Whitehead presents, for the first time, critically edited versions of all of Whitehead's essays and articles published between 1886 and 1916.[1] Besides presenting one article for the first time,[2] it collects numerous Whitehead publications that have either been mostly forgotten or were not particularly accessible, some having never been republished since their first appearance. Through the chronological ordering of these texts, readers will be able to trace more easily the line of development of Whitehead's thought than has previously been possible. This volume, EA1, covers covers Whitehead's initial career as a mathematician and his first forays into physics and philosophy.

In addition to correcting numerous errors,[3] this volume provides editorial introductions to each individual essay or article, adds explanatory footnotes for people, places and events which Whitehead mentions, and provides a complete record of differences between all the versions of each text that appeared in Whitehead's lifetime, with the aim of making this collection the definitive versions of Whitehead's essays and articles.

Text selection

Our first task in assembling the two volumes of Whitehead's essays and articles was to decide which texts to include and which to exclude, that is, to decide which texts count as 'essays' or 'articles'. This may sound like a straightforward task, and yet for any set of criteria there are edge cases. For instance, are brief letters to the editor 'articles'? Do unpublished obituaries count as 'essays'? What about forewords by

1. The Editors searched mightily for a way of dividing Whitehead's essays into two intellectual units based on some philosophical basis, such as before London and after, before PM and after, and so on. Unfortunately, every philosophically satisfying division made the two volumes hopelessly lopsided in terms of length, a reasonable concern of our publisher. In the end, the division used here, with 1916 and earlier in EA1, and 1917 and later in EA2, was arrived at solely with overall volume length in mind. This is not philosophically satisfying, but at times practical matters of space win out.
2. Namely, 'Sets of Operations in Relation to Groups of Finite Order' (p. 87), which was submitted to the *Proceedings of the Royal Society of London* but rejected.
3. In the case of the mathematical articles, the number of errors was sometimes in the dozens.

Whitehead for books written by other people? While the aim of the Critical Edition of Whitehead is to publish all of Whitehead's writings, not all of them belong in the two *Essays and Articles* volumes. Some will appear in volumes of Whitehead's correspondence and others in a volume of *Papers*, which contains more miscellaneous writings.

In the end, we decided upon two criteria for inclusion.

First, the text must in some sense be intended for the public sphere, rather than intended only for private consumption. Of course this includes any of Whitehead's texts that were *published* in his lifetime, but also unpublished articles which were clearly intended for publication,[1] and any public addresses he delivered (published or unpublished), since these were by definition not private matters. On the other hand, it excludes purely private or internal writings, however formal they may be, including referee reports on scholarly articles or addresses delivered to internal administrative bodies.[2]

Second, the text should primarily concern and be expressive of Whitehead's own work and thought, rather than be directed chiefly towards another person. Under this criterion we have mainly excluded obituaries and Whitehead's forewords to books written by others. Whitehead's forewords are invariably short and lack any real interest absent the books within which they appear. These sorts of materials will in due course appear in the volume of *Papers*.

On the inclusion of essays from Whitehead's books

Additionally, we found ourselves faced with the decision of which of Whitehead's books should be 'split up', with their contents included in these *Essays and Articles* volumes. While some of Whitehead's books are cohesive wholes, either written all at once or delivered as a single set of lectures (or nearly so),[3] others were looser collections of essays or addresses. The short answer to this problem is that the Editors, in consultation with the Editorial Advisory Board, decided to split up Whitehead's books *The Organisation of Thought, Educational and Scientific* (1917, abbreviated here as 'OT'), *The Aims of Education and Other Essays* (1929, abbreviated as 'AE') and *Essays in Science and Philosophy* (1947, abbreviated as 'ESP').

It was a particularly easy decision to break up the contents of OT and AE because (1) they shared so much material in common and (2) they were compilations of essays and articles delivered over fairly long periods of time.[4] OT comprises eight chapters, while AE has ten, with six chapters appearing in both. In addition to the simple fact that each of the twelve distinct essays that appeared in these two books were all written at different times for different occasions, it would not have made sense for a critical edition of Whitehead's works to edit these books separately when so much of the material is redundant.

1. For example, his 1899 'Sets of Operations in Relation to Groups of Finite Order' (p. 87), submitted to the Royal Society for publication.
2. An example of the latter is an 1899 address to the Council of Trinity College concerning the deanship.
3. For example, *Science and the Modern World* and *Modes of Thought* both included chapters that were added subsequent to the lecture series upon which they were primarily based.
4. The essays in OT spanned 1912–1917, while those in AE spanned 1913–1928.

While ESP does not have the same redundancy problem, it is even less cohesive than either OT or AE – its chief aim having seemingly been to collect in a book nearly all of Whitehead's essays since 1910 that had *not* previously been published in OT or AE. There was simply no virtue in keeping this particular set of Whitehead's writings cordoned off as a distinct entity. This decision was further supported by ESP's problematic publication history.

Overlooking the manifestly sloppy editing,[1] the especially strange thing about ESP is that another book by Whitehead, entitled *Science and Philosophy*, appeared in 1948, just a year after *Essays in Science and Philosophy*, and it was from the same publisher (Philosophical Library). What was the difference between the two books? Very little. Besides different typesetting, *Science and Philosophy* omitted two chapters from ESP[2] and did not include an index. Otherwise, it was the same book. Moreover, in the 'Acknowledgements' section, the original publishers are thanked for allowing the reprint of the two articles which *Science and Philosophy* had actually dropped. In fact, it is not even clear whether Philosophical Library had copied the acknowledgements from ESP and forgotten to omit mention of these two articles, or – in its rush to put out a new edition of ESP to capitalise on renewed interest in Whitehead's work subsequent to his death in late 1947 – somehow accidentally omitted the two chapters themselves.[3] Our inquiries to Philosophical Library yielded no answers, and it may remain forever unclear exactly how and why this second, nearly identical collection of Whitehead's essays came to be.

Text order

While a thematic arrangement of Whitehead's writings is perhaps *feasible*, it quickly became clear to us that a chronological ordering made the most sense. A key advantage to a chronological ordering is that it allows readers to easily trace the development of Whitehead's thought in a way that would have been difficult until now. Many of these essays had been included within a single book despite being written many years apart, while others had been largely forgotten or were not readily

1. One reviewer amusingly noted: 'Misleadingly heralded as a "new book" by Professor Whitehead, this is in fact a collection of previously printed papers, ranging in date of initial publication from 1911 to 1941, whose novelty consists chiefly in the occurrence, in this reprinting, of typographical errors happily absent from the original. The reader who here encounters for the first time the essay on "Mathematics and the Good" will be relieved, I am sure, to know that the "deflections" of the author referred to on page 102 were "reflections" in the volume from which this imperfect copy is derived' (Arthur E. Murphy, 'Review: *Essays in Science and Philosophy*', *Philosophical Review*, vol. 56 (1947), pp. 709–711). Further, the chapter entitled 'Indication, Classes, Numbers, Validation' had originally appeared in the journal *Mind* with numerous errors, necessitating the publication of corrigenda in order to render it intelligible, but when republished in ESP thirteen years later, the corrigenda were not taken into account, so that the ESP version repeated all of the same errors (see the editorial introduction to 'Indication, Classes, Numbers, Validation', EA2, p. 277).
2. These being 'Non-Euclidean Geometry' (p. 275) and 'Indication, Classes, Numbers, Validation' (EA2, p. 277).
3. This latter may seem quite implausible, but it is not outside the realm of possibility, especially when one considers that this same publisher had also accidentally omitted a chapter from the table of contents in ESP ('Uniformity and Contingency').

accessible, all of which made it challenging to get a bird's-eye view of the development of Whitehead's thought.

In applying a chronological order to the source texts, we then had to decide whether the controlling date for this order would be the date of *publication* or the date of *composition* or *delivery* (if an address). This question was complicated by the fact that some texts were never published and hence have no publication date, while for others we simply do not know the exact date of their composition or delivery. In the end, for those pieces for which dates are known, texts are ordered *primarily* by initial publication date and *secondarily* by date of composition or delivery.

Text versions

The next thorny question we had to answer in editing these volumes was which version of each text – where more than one exists – would serve as the 'base text' for editing. We further needed to decide to what extent we would track and/or integrate changes made in alternative versions.

It was decided early on that we would not concern ourselves with versions of these texts that appeared after Whitehead's death and thus cannot have been made with his involvement. We are interested in how *Whitehead himself* altered things from one version to the next; it is ultimately irrelevant if the 1967 Free Press edition of AE made small changes to the original 1929 version, introduced by a later editor.

With this understood, whenever possible, the earliest available or 'most original' version is used as the 'base text' relative to which edits are indicated. When specific factors relevant to a given piece require the use of something other than the earliest variant, this is explained and justified in the introduction to the piece. For clarity, the available versions of each essay are in the heading of each piece, with the base text set in bold.

There are a number of reasons for privileging the most original version. One practical reason is that Whitehead's essays and articles tended to be trimmed rather than expanded in later versions. His 1916 article 'Space, Time, and Relativity' is an extreme example of this, in which half the essay (about 4,000 words) was cut for republication in AE. More commonly, the later versions of other essays and articles that appeared in books would often eliminate event-specific preambles and references. For example, the first two paragraphs of the 1911 article 'The Place of Elementary Mathematics in a Liberal Education' were dropped for the reprinting in ESP in 1947. These paragraphs referred, in part, to the formation of the teachers' association of which Whitehead had been elected the first President and to which he was making his address. It is perfectly understandable why such paragraphs would be cut for inclusion in a book: readers would not be members of this association and so these portions might be considered largely irrelevant to them. But in a critically edited volume of Whitehead's collected papers, there is every reason to include such material. It gives the essays a further context that later versions obscure.

Perhaps the most important reason for privileging the most original version has to do with the chronological ordering of the texts. Since these essays and articles are being presented in chronological order in the hopes of revealing a clearer line of development in Whitehead's thinking, it is preferable to present the original version of each as reflecting Whitehead's thinking *at that time*. Later versions of a text may actually contain substantive changes that *obscure* the development of Whitehead's thinking, that is, are *not authentic to his thinking at the time of original composition*.

For instance, in 'Space, Time, and Relativity', for the AE version, which appeared more than a decade after the paper was first published, he changed a reference to 'individual experience' to 'individual *conscious* experience'.[1] This is clearly a significant change, but if adopted silently, it would appear that this distinction had occurred to Whitehead at the time of writing, and not more than a decade later. By selecting the earlier text as the base text and then indicating to the reader where and when Whitehead made substantive changes, it is easier to see the development of his thinking. See the next section, on the editorial handling of the texts, for more on this issue.

Given these considerations, we have chosen the most original version of each essay or article as our base text for critical editing, unless there were clear mitigating factors that demanded an alternative. Such exceptions are discussed in the editorial introductions to the individual essays and articles in question.[2]

Editorial handling of the texts

In privileging the most original version of each essay and article, we decided not to standardise the style of punctuation across all texts. Instead, each has been reproduced exactly as it first appeared, with the addition of a record of the original pagination between vertical lines (e.g. |63|).[3] We have corrected *errors and inconsistencies* (e.g. in capitalisation) but such changes to the base text are always footnoted.

The only *silent standardisation* we have undertaken is (1) to change underlining to italics[4] and (2) to remove credit lines and contextual information appearing at the beginning or end of essays and articles, moving it to the editorial introductions if necessary. For instance, we have removed all of Whitehead's credit lines, as in this edition they are wholly redundant, while information appearing with the title that gives context for the essay or article[5] is discussed in the introduction to the item.

As for changes made in alternative (and usually later) versions of the text, these have been divided into two classes, namely *significant* and *insignificant*.

We considered a change *significant* if it could affect meaning. This usually involved the change, subtraction or addition of words or sentences. Such changes are tracked

1. Emphasis ours. See p. 382.
2. For example, for 'Technical Education and Its Relation to Science and Literature' (EA2, p. 3), a copy of the journal in which it was originally published could not be found; for 'The Organisation of Thought' (p. 452), seven different versions were published in close proximity to one another, and we chose the BAAS version as the source text because it was the organisation at which Whitehead originally presented the address, and thus it might be considered the most authoritative.
3. Some essays and articles were originally printed in multiple columns per page; in these instances letters have been appended to the page numbers (for example, |312a| and |312b|).
4. Underlining occurred only in previously unpublished essays and articles that were either handwritten or done on typewriters, which were unable to produce italics.
5. For example, Whitehead's 'Uniformity and Contingency' (EA2, p. 132), as it originally appeared in the *Proceedings of the Aristotelian Society*, contained a note just above the title which specified that it had been read at a 'Meeting of the Aristotelian Society at 21, Gower Street, London, W.C.1, on November 6th, 1922, at 8 P.M.'

through footnotes inserted at the appropriate places, so that readers may easily and immediately see the ways in which a text was significantly altered between versions.

We considered a change *insignificant* if we judged that it had no significant bearing on meaning. Examples of such changes include the addition or subtraction of punctuation, a change from American to British spelling or vice versa, changes in capitalisation and the introduction of new errors. In keeping with a critical edition, even these sorts of changes have been exhaustively tracked in the back-of-book record, but are not otherwise noted, so as not to overburden texts with footnotes.

Some of the essays and articles in this volume had never previously been published, and these have been handled slightly differently. Whitehead sometimes made corrections, either typed or by hand, throughout unpublished manuscripts, in some cases across multiple drafts. We have endeavoured to edit these pieces into their 'final' form, that is, as Whitehead would have wished them to appear had they been published in his lifetime. As such, rather than footnoting all of Whitehead's changes – which would have hopelessly cluttered these texts – we have adopted most corrections silently, unless they have a possible effect on meaning. Again, the back-of-book record tracks all changes, including differences between drafts.

Finally, *contextual notes* have been added to provide the reader with useful information about people, books or events to which Whitehead makes reference. Whitehead's own footnotes – which are few – have been marked with a dagger (†) to distinguish them from footnotes added by the editors.

On editorial introductions to individual essays and articles

Editorial introductions precede each essay and article. A central editorial philosophy of the Critical Edition of Whitehead is that editorial influence should be minimised to the greatest extent possible. Our goal is to provide clear, readable texts that may serve as a basis for scholars' own investigations. Given this, the purpose of editorial introductions is *not* to explain the significance of a text or to interpret its meaning. There are four things that we have aimed to provide in the editorial introductions:

- a listing of all versions of the text that appeared within Whitehead's lifetime;
- an overview of the types of changes that were made between different versions of the text;
- historical and biographical context of the writing or delivery of the text;
- a précis that endeavours to avoid or keep at an austere minimum the imposition of interpretation of the text.

(1) List of versions

The editorial introduction for each text begins with a listing of all versions which appeared within Whitehead's lifetime and **bolds** the text which has been used as a base for editing.

(2) Overview of changes

In many cases, the changes made between different versions of an essay or article were fairly minor, such as changes from British to American spelling, or the correction (or addition!) of small typographical errors. These differences between versions have all been exhaustively catalogued, either in footnotes to the text (when they are deemed substantive and may affect meaning) or in the back-of-book record (when they are deemed not to affect meaning). However, in each of our introductions we do endeavour to provide a general overview of the changes made from one version to the next.

(3) Historical and biographical context

Context is often key to understanding a text. Where applicable and available, relevant historical and personal context has been provided. For instance, if the text was originally an address, then on what date, in what location and at what occasion was it delivered? Had Whitehead recently been appointed to any new professional positions? Had his living situation changed? What larger world events or social upheavals – such as world wars, or voting rights movements, or a change in the teaching of geometry in the UK – might have been affecting or motivating Whitehead's writing? And what personal events and relationships in Whitehead's life – such as the death of his son, Eric, or his relationships with some of the great thinkers of the twentieth century – might have done the same?

(4) Précis of the content

The final purpose of the introductions to individual texts found in this volume is to concisely introduce the content of those texts in a way that makes them more accessible to a wider audience. A core concern is that such introductory material not be interpretive. To this end, we have tended towards brevity and caution. Not only can it be deceptively difficult to write an introduction that does not unconsciously import twenty-first-century biases and assumptions, but also this volume's purpose is not to *explain* Whitehead's thinking to its readers but, rather, to provide definitive and corrected versions of Whitehead's writings so that readers may come to their own conclusions. Readers are here for Whitehead's writings, not the writings of his editors, and so those editors should intervene only so far as is necessary for Whitehead to be understood, and no further. However, with all of this being said, there is, at times, a self-evident difference in difficulty between Whitehead's various texts. Some essays and articles are readily comprehensible to a general audience, while others require advanced, specialised training to comprehend.

For instance, as a general rule, Whitehead's writings on education are not especially difficult for a general audience. Any person who has attended several years of schooling will immediately grasp what Whitehead means when he says that 'Vague generalities are worse than useless, and if we attempt to embody abstractions in short, precise formulæ, the pupils will simply learn them by heart as empty sounds'.[1] For who among us has not, at some point in our education, had to memorise some arcane bit of knowledge which we felt certain would have no applicability to our lives beyond the exam which required its memorisation? For essays

1. 'The Place of Elementary Mathematics in a Liberal Education', p. 344.

and articles of this nature, an exhaustive introduction and summary of the content would be a needless redundancy.

On the other hand, many of Whitehead's essays and articles on mathematics or physics approach complete incomprehensibility for a general audience. For example, when Whitehead begins giving scores of equations on the motion of viscous incompressible fluids, readers without advanced training in mathematics are very unlikely to be able to follow him. In such cases, when the knowledge required to understand one of Whitehead's texts is more specialised, our introductions to the content have tended to be more extensive. The goal in these cases where an essay or article may be inaccessible to a large swath of potential readers is for the introduction to provide at least a general idea of its themes, goals and importance, even if they are unable to decipher the text itself.

Some further notes and special cautions are in order regarding the essentially mathematical or scientific papers in this volume. The complete accuracy of publications heavy in technical usages, both terminology and especially symbolism, is historically more challenging than otherwise. The final typesetting is unlikely to have had the benefit of anyone with more than a nodding acquaintance with the syntactical conventions of the notation and even less so to have had the benefit of someone guided by the semantic sense of the work. Until technology was sufficient to eliminate virtually all the manual elements of this process, the accuracy of the resultant publication depended not only on the accuracy of the authors and their reviewers, but also on the diligent review of the near-final product before it went to press. In Whitehead's case, we have abundant evidence in his publications to suggest that he often neglected this last step. Consequently, errors appeared in his works that were sometimes egregious, violating the most basic syntactical elements or any semantic plausibility. Hence in the critical versions of the papers included in this volume, the reader needs to be aware that while the editors have invested vast effort in eliminating typographical errors, given the number of them, it is almost certain that some persist. Moreover, in many cases, Whitehead's compact style does not admit a reproduction of his myriad complex calculations. Where these were reproducible, we found excellent agreement between intent and execution, but sometimes only the plausibility of a result could be checked due to inevitable resource constraints. Finally, quite a few of the essays and articles appeared at a time of historical ferment, especially in connection with the foundations of mathematics. Under such circumstances, there are also deep oversights in some of these papers that are not subject to local correction.[1] Such issues, however, would seem to enhance the historical value of the works more so than to detract from their mathematical value.

Even in our present highly automated electronic culture, mathematical and scientific articles still admit oversights, and almost any researcher relying on refereed publications has learned to beware: the deep and long-term consistency of decidedly technical results is ultimately the grounds for their acceptance, and this can take

1. For example, in his discussions of cardinality, Whitehead speaks freely of the set of all sets, a construction that would soon be deemed invalid by virtue of Russell's Paradox and the ensuing rise of formal set theory. This and other related problems are covered in the individual introductions to the corresponding essays and articles.

decades or sometimes centuries. While the results included in the mathematical and scientific essays and articles edited and reproduced here seem unlikely to serve as the foundation for further research in mathematics or science – as opposed to history or philosophy – one should, nonetheless, consider the issues raised above when engaging with these texts.

In summation, we have attempted to make our editorial introductions to the content of individual essays and articles as brief as possible in order that Whitehead can speak for himself to the maximum extent without being filtered through the medium of the editors.

In most cases, editorial introductions are jointly authored by the editors. The exception is one introduction written by Ronny Desmet.[1] Further, each introduction was reviewed by at least one member of the Whitehead Research Project Editorial Advisory Board.[2]

Acknowledgements

This book is in honour of Donald Phillip Verene by Frederick and Nancy Marcus.

The Whitehead Research Project Editorial Advisory Board has provided invaluable editorial advice throughout the long process of editing this book. Moreover, two board members – Nancy Frankenberry and George R. Lucas, Jr. – have provided significant financial support. Other financial supporters included John Buchanan, an anonymous foundation and Charles R. Arterburn. Without their contributions, this Critical Edition would not have been possible.

We have received institutional support from Gonzaga University, which became the new home of the Critical Edition of Whitehead in mid-2021, and of the Whitehead Research Project in October 2023. Prior to these moves, the Project had been supported since its inception in 2006 by Claremont School of Theology and the Center for Process Studies. We also want to thank Roland Faber, who founded and led the Whitehead Research Project for many years prior to his retirement in July 2023.

Many of the essays and articles in this book were initially transcribed by Jenna Petsche, an Editorial Assistant for the Project from 2019 to 2021.

We were helped by a number of archivists who went beyond the call in researching our questions about these various essays and articles. We particularly want to thank Jim Stimpert, Special Collections Senior Reference Archivist at Johns Hopkins University (now retired), for helping with questions about the Victor Lowe collection for more than a decade, and Steven Gross, the Chair of the Philosophy Department at Johns Hopkins, for helping us to find graduate students to scan large portions of Lowe's papers for us.

Finally, we are grateful to Joan Caywood for her exceptional skill in indexing the volume.

1. 'La Théorie Relationniste de L'Espace (The Relational Theory of Space)', p. 401.
2. Board members are listed at the front of this volume; short biographies can be found at <https://whiteheadresearch.org/initial-aims/editorial-advisory-board/> (accessed January 2025).

A Celebrity at Home: The Clerk of the Weather

1886 *Cambridge Review*, vol. 7, no. 167 (February), pp. 202–203

This playful essay appeared five days before Whitehead's twenty-fifth birthday and less than two years after he was elected a Fellow of Trinity College, Cambridge (on 9 October 1884). Like 'Davy Jones', which was published just a few months later, in May 1886, and 'On Ideals', published ten years later, in 1896, this essay was simply signed 'W.', making its authorship less than completely certain. Whitehead's daughter, Jessie, was confident of her father's authorship of the first two essays from the 1880s, but could only say that 'On Ideals' was 'probably' authored by her father.[1] Additional confidence that Whitehead was the author of this essay may be derived from the fact that all three of these essays appeared in the official bibliography of Whitehead's works constructed by Lowe for the 'Library of Living Philosophers' volume dedicated to Whitehead's work.[2] It is reasonable to infer that Whitehead, who was to some degree involved with the production of the volume, would have reviewed and approved the bibliography. Assuming this piece is by Whitehead, it may in fact be his first published essay. Finally, stylistically, this essay (and 'Davy Jones') are similar to the 1888 essay 'A Visitation', which is signed 'A. N. W.' All three essays make use of the rhetorical device of a fictional conversation.

This essay depicts a fictional conversation between the narrator and the Clerk of the Weather, 'an imaginary functionary facetiously supposed to direct the weather'.[3] The Clerk is upfront about his status as an imaginary being, lamenting 'If I were only real . . . . It is a horrible thing to exist only in other people's thoughts'.

I was never on speaking terms with the man in the moon; in fact, as far as I am aware, he never took the slightest notice of me. This neglect would have wounded my self-love, if he had been in the habit of unbending to others; but as he treated all with the same proud disdain, I never troubled my head about it.

But the Clerk of the Weather was a very different sort of person, fond of his chat. At one time I knew him pretty intimately. He was about five foot eight in height, past

1. Victor Lowe, *Alfred North Whitehead: The Man and His Work, Volume I: 1861–1910* (Baltimore: Johns Hopkins University Press, 1985),pp. 214–215.
2. Paul Schilpp (ed.), *The Philosophy of Alfred North Whitehead* (Evanston: Northwestern University Press, 1941), p. 705.
3. Thomas Davidson, *Chambers's Twentieth Century Dictionary of the English Language* (London: W. and R. Chambers, 1907), p. 175. According to the *Oxford English Dictionary*, the first appearance of 'Clerk of the Weather' was in Basil Hall's *Fragments of Voyages and Travels, Including Anecdotes of a Naval Life: Chiefly for the Use of Young Persons, Volume III* (Edinburgh: Robert Cadell, 1831), p. 296, 'Owing to their interest with the clerk of the weather office' (*Oxford English Dictionary*, s.v. 'Clerk of the Weather (*n.*)', June 2024, <https://doi.org/10.1093/OED/5518823350> [accessed January 2025]).

middle age, dressed in black, and in his happier days could appreciate a joke. With all this there was a vein of sadness running through him, and his temper was at times uncertain.

Like many other worthy folk, he has been thrown out of employment by the progress of science. He never could master the new ideas, and so three years ago he had to resign owing to his scandalous mismanagement of the summer weather.[1] Since this event his temper has become soured. When I am alone he often pours out his complaints to me. Poor old man, like others who have seen better days, there are few who will attend to him now.

"If I were only real," he once said to me, "I think I could stand it. Take yourself for instance: suppose your present employment failed you, there are hundreds of other ways in which you could live – you might not like your new occupation; but I say," he continued, raising his voice excitedly, "you could do it – you might be a ploughman, bishop, fisherman, or what not. Never mind the unpleasantness; there you would be, flesh and blood. But," – here he relapsed into a melancholy calm, – "I am the Clerk of the Weather, or I am nothing. Now I am no longer the Clerk of the Weather, for those Kew[2] people have made me resign, so what am I?" This was a question which I did not feel competent to answer off-hand, so I let him run on. "Yes, for a long time past I had been growing uneasy about the future, and my post never was a comfortable one. It is a horrible thing to exist only in other people's thoughts. You are absolutely at their mercy; and no amount of reasoning can make the situation the more pleasant. I go wandering about from mind to mind, like the soul of a departed Positivist; and I can tell you, that at times I have hard enough work to find a resting-place. Children are my best friends. Sometimes when a child is not allowed, because of the rain, to run out of doors and trundle its hoop or spin its top, its mother tells it to ask me to take away the clouds; and then it will stand at the window, and wonder who I am, and whether I would mind being spoken to. So if I am able, I will just clear the weather up a bit; and perhaps the child will think of me afterwards as a kindly old gentleman who helps good boys and girls. But, O my friend," – and here he seemed to sob and I fancied I felt a moist pressure on my hand, but perhaps it was only a drop of rain and a slight breath of air rustling the trees, – "there are few who think of me as you do; they mostly forget me when they go to school."

The Clerk is a thorough Englishman, and is all the world like a retired professional man, as indeed he is. When in office, he would cross the channel once in six weeks or so, just to fix up the weather on the other side; but he never took kindly to foreign countries. "It is extraordinary how little is doing there," he once said when

1. It is unclear whether 'three years ago' specifically was chosen arbitrarily by Whitehead for dramatic effect or referred to actual weather events around this time. Historical records of weather events in the UK report: (1) heavy rains and flooding from September 1882 to February 1883, (2) heavy snowfall between 4 and 8 December 1882, leading to metre-high snow drifts, (3) a drought in the spring of 1883, with rainfall only 60–70% of normal, (4) dramatically colder weather beginning on the 5/6 March 1883, in the top five coldest months since 1659, (5) the eruption of Krakatoa in late August 1883, which led to a decrease in world temperatures of about half a degree Celsius and dust veils creating spectacular sunrises and sunsets (the sound of the eruption is said to have been heard thousands of kilometres away) (Martin Rowley, 'Weather in History 1850 to 1899 AD', Weatherwebdotnet, <https://premium.weatherweb.net/weather-in-history-1850-to-1899-ad/> [accessed January 2025]).
2. The Kew Observatory was founded by George III in 1769 and is now known as the King's Observatory.

chatting of them; "No life whatever among them: and if anything does happen, like enough it is an earthquake or some such piece of quackery, which bears about the same relation to weather, real genuine English weather, as brass does to gold." He was never tired of railing against the sun, a body with which he was at constant feud, all the more bitter because they had to work together. It seems that he was under contract to deliver down on the earth all the water drawn up by the sun. "Talking about the present over-production in trade," – he would say, – "Why it is nothing to what I have seen. There is the sun, who would go on draw, draw, and there was nothing that would stop him. A pretty job I have had at times to get it all down again. Things too grew harder to manage as years went on. In old times, if it came to the worst, the bishops would get out their images and holy water, and I sometimes managed to sneak out of the bargain. But now the men of science with their talk about laws of nature and rain gauges have stopped all those practices; and a hard time I had of it in my old age."

Nothing would stir him so quickly as an allusion to the new |203| system of weather prediction; if my memory does not deceive me, "tomfoolery" was the exact expression he once applied to it. "In my time," he said, "I did things how I liked and when I liked. To be sure, there was the sun's water to be got down, but if I was pretty well advanced with that business, I did as I pleased. I never made up my mind five minutes beforehand. How should I? For, as I have often told you, I never had a mind of my own, but have always lived in those of other people."

He has not been to see me lately. The fact is, that at our last meeting I offended him deeply. I had asked him if the moon in old times had any connection with the weather. "Oh yes," he said, "I let it meddle a bit. It was not like the sun, for I could always stop it whenever I chose. I used to reckon," he continued, chuckling as he spoke of it, "that all the water it drew up in a year could be sent down again in dew in the course of one night. But it was a kindness to humour it, so I now and then allowed it to melt a small cloud or two, on nights when it was full." I then asked him if there was any foundation for the belief that the weather changed when the new moon was on its back. "None whatever," he said laughing, "but the moon thought so, and I never denied it, though it was all nonsense." Now, unfortunately, during the same morning I had been teaching a dull boy the rudiments of meteorology, so I am afraid my temper was rather ruffled, and I answered testily, "It seems to me that there was a good deal of nonsense in your time." In a moment the quick-tempered old fellow lightened up with anger, and thundered out of my presence before I could stop him.

I have not seen him since, and I am afraid he is not getting on very well, for he has but few friends, and is rarely spoken of. So I thought it well to set down on paper what I can remember of him, as it would be a pity if the old gentlemen were quite forgotten.

Davy Jones

1886 *Cambridge Review*, vol. 7, no. 174 (May), pp. 311–312

Like 'Clerk of the Weather', which was published a few months earlier in February and 'On Ideals', published ten years later in 1896, this imaginative essay was simply signed 'W.', making its authorship less than completely certain. However, Whitehead's daughter, Jessie, was confident of her father's authorship of this essay,[1]and in addition to other evidence that Whitehead is the author,[2] its mention of the 'fish torpedo', otherwise known as the 'Whitehead torpedo' – coupled with the ensuing fictional conversation – seems to strongly suggest that the piece is his.[3]

This essay depicts a conversation between the narrator and the fictional character Davy Jones, an evil spirit of the sea, said to bring disasters to ships, drowning sailors and forever holding them in 'Davy Jones's locker,' that is, the bottom of the sea.[4]

One fine summer's afternoon I was walking on the shore near the North Foreland.[5] There was not a breath of air; and the sea had that glassy appearance which betokens a perfect calm. I happened by chance to come across Davy Jones, slouching idly under the cliff. We had met before, so he greeted me pleasantly enough. His appearance is not prepossessing; this is chiefly due to the fact that he never looks frankly into one's face, but continually glances up from the corners of his eyes in a way that inspires distrust. Otherwise he is a tall, well-built fellow, with a slight stoop forward, bronzed and weatherbeaten. He has always been very civil to me, so I will not say a word against him; though it must be confessed that his character is not good. On the afternoon in question we entered into a little conversation. "It is a slack time with me now, so I have taken a day off," said he in explanation of his presence on shore, and at the same time he glanced with an air of careless abstraction at the advancing tide, as it gradually surrounded two children at play. "Excuse me, my friend, for one moment," I replied, as I ran off to warn the little fellows of their danger. This did not take me long to accomplish, and in a few seconds I was

---

1. Victor Lowe, *Alfred North Whitehead: The Man and His Work, Volume I: 1861–1910* (Baltimore: Johns Hopkins University Press, 1985), pp. 214–215.
2. See the editorial introduction to 'A Celebrity at Home: The Clerk of the Weather', p. 1.
3. See p. 6n3.
4. The origins of 'Davy Jones' as a spirit/devil of the sea are unclear, but its use dates back to at least the early 1700s.
5. North Foreland is a chalk headland on the Kent coast of south-east England and forms the eastern end of the Isle of Thanet, where Whitehead was born.

back again by his side, panting for breath. He seemed rather annoyed at my conduct, and muttered some unkindly wishes before replying to the inquiries with which I restarted the con-|312a|versation. I am fond of gathering curious information from eccentric people; so, with the view of ultimately drawing him out, I had asked after his affairs. "Well," he said, "business may perhaps be a little brisker now than it was a few years back; but there has been nothing of any real consequence since the Flood." It will be seen from this answer that Davy is one of those men whose character has been ruined by too great prosperity in early life. When he was quite a young man the Noachian Deluge[1] turned his head. He did not recognise that the riches, then acquired, should be reckoned rather as capital than as income. Accordingly, after a few years of reckless profusion, he was reduced to a precarious livelihood on chance fishermen. The escape of the Ark and its freight was at the time a great annoyance to him; though he has since owned that it was a blessing in disguise. If he had succeeded in accomplishing the entire destruction of mankind he must have literally starved in later years. I took up the first part of his reply, disregarding the allusion to the Flood, a topic on which he is, in truth, apt to become rather wearisome, "Yes, the expansion of trade doubtless has done you a good turn." In reply to this, he went off into enthusiastic praise of modern commerce. "Such dash, such enterprise; none of that petty calculation which, though it may save a life or two, is the sure sign of a small mind."

Now that he had warmed himself up, there was no difficulty in inducing him to launch out in a series of reminiscences. His estimates of people, with whom he had come into contact, were curiously different from those formed by the rest of the world. "Yes," he said, in answer to a question on the point, "I watched Ulysses' voyage with some interest, and I will not deny but that I took a pleasure in it; for I managed to pick up a few trifles, but he was a mean-spirited, tricky fellow, always baulking just expectations.[2] As for Æneas, he was my pet aversion. What did he want with ships that changed into fish, or gods, I forget which, and were attended by water-nymphs?[3] They were worse than life-boats." As he said this, Davy took up a flint-stone and jerked it towards the neighbouring boat-house, in which was kept one of the last-named offending vessels; and he seemed pleased when it made an ugly mark on the fresh paint. He then continued in reference to Æneas—"He and

---

1. This is another name for the flood experienced by Noah and his family as depicted in the Old Testament, Genesis 5–10, referenced a few sentences above.
2. Homer's *Iliad* follows Odysseus (otherwise known as Ulysses) on his ten-year return voyage from the Trojan war. Though his entire crew was killed, Odysseus survived against all odds, apparently much to the annoyance of Davy Jones.
3. Aeneas is a figure in Greek and Roman mythology and a character in Homer's *Iliad* and Virgil's *Aeneid*. The reference here is to a passage in the *Aeneid* in which the Trojan ships transform into 'sea-nymphs':

Et sua quaeque	And all at once,
continuo puppes abrumpunt vincula ripis	each vessel snapping her cables free of the bank,
delphinumque modo demersis aequora rostris	they dive like dolphins, plunging headlong beaks
ima petunt. Hinc virgineae (mirabile monstrum)	to the bottom's depths, then up they surface,
quot prius aeratae steterant ad litora prorae	turned into lovely virgins – wondrous omen –
reddunt se totidem facies pontoque feruntur.	each a sea-nymph sweeping out to sea.

Virgil, *The Aeneid*, translated by Robert Fagles (London: Penguin Books, 2010), book 9, lines 117–122.

his crew were a wretched set,—except Palinurus;[1] I liked him." He declared that the greatest disappointment, he ever suffered, was occasioned by the loss, when they were almost in his grasp, of Cæsar and all his fortunes; and that even now he could never think of that miserable episode without a pang.

Davy Jones is a man with some enthusiasm for his profession. Before the present influx of business, which leaves him but little spare time, he tried some interesting experiments. For instance, at special request, he once endeavoured to drown the wandering Jew;[2] but the attempt, he said, was perfectly futile, for the old gentleman showed a power of floatation which would have been surprising even in the case of a younger man.

The conversation was now interrupted for a few minutes by the passage round the Foreland of an emigrant ship, which Davy watched with intense eagerness, whistling for the wind. This manner of his made me feel rather uncomfortable; so, by way of diverting our thoughts, I asked him some questions relative to naval battles. To my surprise, I found that he was without interest in them. "It is a miserable way," he explained, "of obtaining men—dead before they reach me; as far as I am concerned they might just as well have died on shore. Two men on a raft give me more innocent amusement than ten battles." I then explained to him the principle of the fish-torpedo, |312b| which he had not heard of before. The invention pleased him mightily; and he congratulated me warmly on my relationship, as he supposed, to the inventor,[3] assuring me with great politeness that his interest in the next sea-fight would be vastly heightened by the effective use of the weapon.

We had now been chatting for some time; and I rose to my feet stretching myself, preparatory to walking away. "Are there any new books worth looking at?" he enquired, for he is a voracious reader of nautical stories. I mentioned Louis Stevenson's book, "The Treasure Island,"[4] as a work likely to interest him. "Many drowned?" he asked. "A first mate," I replied. "Well," he said, "that is better than nothing; I will look at it." He once himself started to write a book of adventures; but could not resist the temptation of submerging the whole shipload in the second chapter—where, in ordinary novels of this kind, the hero is merely sea-sick—and this brought the work to an abrupt conclusion.

1. Palinurus, in Roman mythology and especially Virgil's *Aeneid*, is the coxswain of Aeneas's ship, or the person in charge of navigation and steering. At the end of book 5 of the *Aeneid*, he is chosen by the gods to be sacrificed in order to guarantee the safe passage of the Trojans, and drowns after falling overboard.
2. The 'wandering Jew' is said to be a man who taunted Jesus on his way to the crucifixion and was cursed to wander the earth until Jesus' second coming.
3. This is a significant detail as far as ascribing the authorship of the essay to Whitehead. The 'fish torpedo' was the world's first self-propelled torpedo, invented by the English engineer Robert Whitehead (1823–1905) in 1866, and was also known as the 'Whitehead torpedo'. Though Robert Whitehead and Alfred North Whitehead shared the same surname, they were not related; nevertheless, the essay's reference to the torpedo's inventor might be due to this shared surname.
4. *Treasure Island* was originally published in serialised form in *Young Folks* in 1881 and 1882. *Treasure Island* was first published as a book by Cassel and Co. in 1883, just a few years before the present piece.

As a parting shot, while taking my leave, I asked him what he thought of Plimsoll[1] and Chamberlain.[2] "They are more offensive to me," he answered warmly, "even than Jonah's whale." And with this emphatic expression of disapproval, he walked away.

1. Samuel Plimsoll (1824–1898) was an English politician remembered for devising the Plimsoll line, 'an internationally agreed-upon reference mark located on a ship's hull that indicates the maximum depth a vessel may safely immerse when loaded with cargo.' He was a member of the British Parliament who was concerned with the loss of ships and crews due to vessel overloading. See Haley Mummert, 'The History of the Plimsoll Line', TRG, 6 May 2020, <https://traderiskguaranty.com/trgpeak/history-plimsoll-line/> (accessed January 2025).
2. Joseph Chamberlain (1836–1914) was a British statesman and the father of Neville Chamberlain. From 1880 to 1885, he was President of the British Board of Trade. In May 1884, two years before the publication of this essay, Chamberlain had argued strongly for the passing of the Merchant Shipping Bill, which aimed to promote the security of life and property at sea. Chamberlain said that 'I proposed to make it impossible for any man to make a profit by the loss of his ship, and of the crew that sailed in her. I proposed to make it impossible for any man to contract himself out of the liability for the negligence of himself or the persons whom he employed. I proposed that the Employers' Liability Act should also be applied to the sea service'. See Alexander Mackintosh, *Joseph Chamberlain: An Honest Biography* (London: Hodder and Stoughton, 1914), p. 72. The measure was defeated by the shipowners and their underwriters, who accused Chamberlain of 'wanton interference with trade interests' and 'libellous attacks on an honourable class' (ibid.).

On the Motion of Viscous Incompressible Fluids

1888 *Quarterly Journal of Pure and Applied Mathematics*, vol. 23, no. 89 (February), pp. 78–93

This article lies in the domain of applied mathematics, and specifically in fluid dynamics. No new major concepts are deployed, but Whitehead succeeds in using his virtuoso computational skills to extend the analysis of the motion of fluids into new territories of forbidding complexity. Here we shall introduce his background assumptions, general approach and his most important results.

The key terms in Whitehead's title are easily understood. The concept of a fluid comprehends both gases and liquids, but a mass in a gaseous state responds to increased pressure on its container by significant compression. Hence its density (mass per unit volume) increases. Liquids under similar circumstances, in contrast, essentially maintain their density under increasing pressure, hence the term incompressible, and this indeed is the fundamental distinction between these two states of matter. So, Whitehead is speaking of the motion of liquids in this work. Viscosity is the tendency of the flow of one layer of a fluid to be transferred to an adjacent layer, and in some ways this may be thought of as a kind of friction. For liquids, we may think of viscosity more intuitively as thickness, in the sense that we would say blood is thicker than water, and molasses is thicker than blood. In summary, then, Whitehead is addressing liquids for which a flow in any given region induces a flow in adjacent regions. All of this can, of course, be suitably quantified, and that brings us to Whitehead's approach.

While we may see any fluid as a distribution of particles subject to the classical laws of motion, the number of particles and their interactions makes such an approach computationally intractable. Accordingly, Leonhard Euler (1707–1783) introduced the standard mathematical framework for fluid dynamics as the study of flow velocity and medium density at any point within the distribution. Whitehead's calculations are made within this framework; his main goal is to extend the range and accuracy of earlier analyses by applying correspondingly more complex methods, as we shall next explain.

We may recall from elementary physics that there is a simple formula showing that the period of a pendulum is proportional to the square root of its length. We may be much less likely to remember that the simplicity of this relationship depends on the approximation of the sine of an angle, θ (measured in radians), to the angle θ itself – an approximation which is fairly accurate only for small angles. If one wanted to improve the accuracy of the period, one way to proceed would be to use a better approximation of the sine, which is to say, to use higher powers of θ in the approximation. Indeed, this leads to more accurate but distinctly more complicated formulas. Briefly put, what Whitehead is doing in this paper is directly analogous: by including terms of the second degree (that is, squares and products of the velocity components), he refines a previously known approximation of the motion of a fluid that had been derived using only the first powers of these quantities.

The general analysis for this refinement is carried out in Section 2, with a simplification introduced in Section 2a that is valid for all subsequent applications. The heart of the article is Section 3, in which Whitehead applies his results to the case of a rotating sphere imagined to be immersed in an infinite lake. The key result is the set of equations (p. 18) for the flow velocity at any point of the lake, and these include terms involving the square of the magnitude of the angular velocity. Shifting to polar coordinates, Whitehead is then able to give an analytic explanation of the flow of the lake that is consistent with that conjectured earlier by George Stokes (1819–1903). Section 4 shows that the standard approximation of the motion of a liquid in a pipe remains unchanged under Whitehead's more sophisticated treatment, and Section 5 promises a second piece to address a non-rotating sphere moving through an infinite lake, a work that is subsequently included in this volume.[1]

A method of approximation.

1. The object of this paper is to point out a method of approximation to the solutions of problems in viscous fluid motion when the squares and products of the velocities are retained. It will easily be seen that the method is perfectly general and could be applied to find a solution correct as far as terms involving the $(n+1)^{\text{th}}$ powers of the velocities, when a solution involving their n^{th} powers is known. For the sake of simplicity we confine ourselves to the first step in the approximation.

It is true that, since the general equations of motion of viscous fluids can only be depended upon for small velocities, the ordinary solutions with the squares and products of the velocities neglected give in general all that can be safely concluded from them. But solutions in which the approximation has been carried a step further, to include the squares and products of the velocities, may find some use by enabling the limits of the applicability of the equations of motion to be tested from a comparison of their results with experiments: further they will shew[2] the kind of motion to be expected in various cases, although the numerical results may be inaccurate.

|79| 2. The fluid is homogeneous and incompressible; its density is ρ, its coefficient of viscosity μ. Let p be the pressure at any point of fluid, u, v, w the resolved parts of the velocity at the point along three rectangular axes, x, y, z the coordinates of the point referred to those axes. Also let ξ, η, ζ be the components of the 'angular velocity' at the point, so that

$$\xi = \tfrac{1}{2}\left(\frac{dw}{dy} - \frac{dv}{dz}\right),$$

with similar expressions for η, ζ.

Let P denote the expression

$$\frac{p}{\rho} + \tfrac{1}{2}\left(u^2 + v^2 + w^2\right).$$

---

1. See 'Second Approximations to Viscous Fluid Motion', p. 25.
2. 'Shew' is an archaic spelling of 'show', and it appears that 'show' had already become the more standard spelling by the time this article was published.

The general equations of motion may now be written

$$\left.\frac{du}{dt}-2(v\xi-w\eta)=-\frac{dP}{dx}+\frac{\mu}{\rho}\left[\frac{d^2u}{dx^2}+\frac{d^2u}{dy^2}+\frac{d^2u}{dz^2}\right]\right\} \quad \ldots\ldots\ldots\ldots (A),$$

with two similar equations.

Here $\frac{d}{dt}$ signifies differentiation with respect to the time without change of position in space.

The equation of continuity gives

$$\frac{du}{dx}+\frac{dv}{dy}+\frac{dw}{dz}=0 \quad \ldots\ldots\ldots\ldots\ldots\ldots\ldots (B).$$

Now suppose the steady motion of the fluid is required when subject to given boundary conditions.

Let u_0, v_0, w_0 be the values of u, v, w giving the solution of the problem, when the squares and products of the velocities are neglected; let P_0 be the corresponding expression for P under the same conditions. Then these expressions satisfy the equations[1]

$$\left.\begin{array}{c}\frac{dP_0}{dx}=\frac{\mu}{\rho}\left[\frac{d^2u_0}{dx^2}+\frac{d^2u_0}{dy^2}+\frac{d^2u_0}{dz^2}\right]\\ \text{with two similar equations}\end{array}\right\},$$

and

$$\frac{du_0}{dx}+\frac{dv_0}{dy}+\frac{dw_0}{dz}=0.$$

Also u_0, v_0, w_0 satisfy certain conditions at the boundaries.

To carry the approximation a step further, let u_0+u_1, v_0+v_1, w_0+w_1, P_0+P_1 be the expressions for u, v, w, P which give the solution of the problem when the squares and products |80| of the velocities are retained: then u_1, v_1, w_1, P_1 are expressions involving these squares and products only. Substitute these expressions in the equations (A) and neglect terms involving cubes and higher powers of the velocities: the equations become

$$\left.\begin{array}{c}\frac{dP_1}{dx}-2(v_0\zeta_0-w_0\eta_0)=\frac{\mu}{\rho}\left[\frac{d^2u_1}{dx^2}+\frac{d^2u_1}{dy^2}+\frac{d^2u_1}{dz^2}\right]\\ \text{with two similar equations}\end{array}\right\} \quad \ldots\ldots\ldots\ldots (C);$$

also u_1, v_1, w_1 satisfy the equation of continuity,

$$\frac{du_1}{dx}+\frac{dv_1}{dy}+\frac{dw_1}{dz}=0.$$

Differentiate the first of equations (C) with respect to x, the second with respect to y, the third with respect to z; add the results: we obtain, remembering the equation of continuity satisfied by u_1, v_1, w_1, the equation

$$\frac{d^2P_1}{dx^2}+\frac{d^2P_1}{dy^2}+\frac{d^2P_1}{dz^2}=2\frac{d(v_0\zeta_0-w_0\eta_0)}{dx}+2\frac{d(w_0\xi_0-u_0\zeta_0)}{dy}+2\frac{d(u_0\eta_0-v_0\xi_0)}{dz}.$$

The right-hand side of this equation is a known expression, since u_0, v_0, w_0 are known by hypothesis; therefore this is a differential equation to find P_1.

1. The bracketed expression below was erroneously rendered as $\left[\frac{d^2u_0}{dx^2}+\frac{d^2v_0}{dy^2}+\frac{d^2w_0}{dz^2}\right]$.

For the sake of brevity put

$$A_0 = v_0\zeta_0 - w_0\eta_0,\ B_0 = w_0\xi_0 - u_0\zeta_0,\ C_0 = u_0\eta_0 - v_0\xi_0,$$

also let

$$K_0 = \frac{dA_0}{dx} + \frac{dB_0}{dy} + \frac{dC_0}{dz}.$$

Then

$$\frac{d^2P_1}{dx^2} + \frac{d^2P_1}{dy^2} + \frac{d^2P_1}{dz^2} = 2K_0$$

is the equation for P_1.

The solution of this equation is known: let x', y', z' be the coordinates of a point Q in the fluid; let $d\tau'$ denote an element of volume $dx'dy'dz'$; and let dashed letters, such as K_0', denote the values of the other expressions at Q. Let σ denote the distance between the points (x,y,z) and (x',y',z'), [1] so that

$$\sigma^2 = (x-x')^2 + (y-y')^2 + (z-z')^2.$$

Then

$$P_1 = -\frac{1}{2\pi}\iiint K_0' \frac{1}{\sigma} d\tau' + \Pi_1,$$

|81| where Π_1 satisfies Laplace's equation

$$\frac{d^2\Pi_1}{dx^2} + \frac{d^2\Pi_1}{dy^2} + \frac{d^2\Pi_1}{dz^2} = 0,$$

and will be called the 'Laplacian' term. The triple integration is to include the whole fluid.

Equations (C) may be considered as three equations to find u_1, v_1, w_1 and solved in the same way as that to find P_1. With the notion just explained, we obtain

$$\frac{\mu u_1}{\rho} = -\frac{1}{4\pi}\iiint \frac{dP_1'}{dx'}\frac{1}{\sigma} d\tau' + \frac{1}{2\pi}\iiint A_0' \frac{1}{\sigma} d\tau' + \frac{\mu U_1}{\rho},$$

with two similar equations, where U_1, V_1, W_1 are Laplacian terms, and the triple integrations are to be taken to include the whole fluid.

The conditions to be satisfied by u_1, v_1, w_1 are

$$\frac{du_1}{dx} + \frac{dv_1}{dy} + \frac{dw_1}{dz} = 0,$$

at each point of the fluid, and the boundary conditions: the Laplacian terms Π_1, U_1, V_1, W_1 must be chosen to satisfy these conditions, if possible.

The equation of continuity can always be satisfied: for substituting for u_1, v_1, w_1 their values as found above, the equation becomes

$$\frac{\mu}{\rho}\left[\frac{dU_1}{dx} + \frac{dV_1}{dy} + \frac{dW_1}{dz}\right]$$

$$= \frac{1}{4\pi}\frac{d}{dx}\iiint \frac{dP_1'}{dx'}\frac{1}{\sigma} d\tau' + \text{ two similar terms in } y \text{ and } z$$

$$- \frac{1}{2\pi}\frac{d}{dx}\iiint A_0' \frac{1}{\sigma} d\tau' + \text{ two similar terms in } y \text{ and } z.$$

1. Commas have been inserted in the two preceding ordered triples to avoid confusion with a product.

Now suppose that the fluid, if it extends to infinity, is at rest there. Let ds' be an element of the boundary surface; and dn' an element of the normal to the boundary at ds', drawn positively into the fluid; let its direction cosines be l', m', n'. Integrate by parts and use the formulæ

$$\frac{d}{dx}\frac{1}{\sigma}+\frac{d}{dx'}\frac{1}{\sigma}=0, \quad \&c.$$

|82| Then

$$\frac{\mu}{\rho}\left[\frac{dU_1}{dx}+\frac{dV_1}{dy}+\frac{dW_1}{dz}\right]$$
$$=\frac{1}{4\pi}\iint\frac{dP_1'}{dn'}\frac{1}{\sigma}ds'-\frac{1}{2\pi}\iint[l'A_0'+m'B_0'+n'C_0']\frac{1}{\sigma}ds'$$
$$+\frac{1}{4\pi}\iiint\nabla^2P_1'\frac{1}{\sigma}d\tau'-\frac{1}{2\pi}\iiint\left[\frac{dA_0'}{dx'}+\frac{dB_0'}{dy'}+\frac{dC_0'}{dz'}\right]\frac{1}{\sigma}d\tau',$$

where ∇^2P_1' stands for

$$\left[\frac{d^2}{dx'^2}+\frac{d^2}{dy'^2}+\frac{d^2}{dz'^2}\right]P_1'.$$

But

$$\nabla^2P_1'=2\left[\frac{dA_0'}{dx'}+\frac{dB_0'}{dy'}+\frac{dC_0'}{dz'}\right];$$

hence the triple integrals in the above equations vanish.

The equation becomes therefore

$$\frac{\mu}{\rho}\left[\frac{dU_1}{dx}+\frac{dV_1}{dy}+\frac{dW_1}{dz}\right]=\frac{1}{4\pi}\iint\left[\frac{dP_1'}{dn'}-2\{l'A_0'+m'B_0'+n'C_0'\}\right]\frac{1}{\sigma}ds'.$$

The limits of integration of the right-hand side are to include all the boundary surface of the fluid. Now the left-hand side of this equation must satisfy Laplace's equation, because U_1, V_1, W_1 do so; therefore the right-hand side must be a function of x, y, z satisfying it also at any point in the fluid. This is the case, since there is only a surface integral involved: accordingly it is always possible to choose U_1, V_1, W_1, and Π_1, so as to satisfy this equation.

The remaining arbitrary parts of U_1, V_1, W_1, Π_1 must be chosen so as to satisfy, if possible, the boundary conditions.

For the sake of example let us assume as one of the boundary conditions, that there is no slipping of the fluid over the bounding surfaces. Suppose also that u_0, v_0, w_0 have been determined so that at each point of the boundary surface they are the components of the velocity of the surface. Then at every point of the bounding surface $u_1=v_1=w_1=0$.

|83| 2a.[1] The work of approximation may be materially shortened in the one important case. In all the special examples of viscous fluid motion solved to a first approximation P is constant, and

$$\nabla^2u_0=0, \qquad \nabla^2v_0=0, \qquad \nabla^2w_0=0.$$

1. The section has been renumbered here 2a instead of 2, since the original already contained a section 2, above.

It follows from these three equations that ξ_0, η_0, ζ_0 are derivable by differentiation from some function ψ; so that

$$\xi_0 = \frac{d\psi}{dx}, \qquad \eta_0 = \frac{d\psi}{dy}, \qquad \zeta_0 = \frac{d\psi}{dz}.$$

Now ξ_0, η_0, ζ_0 satisfy the equation

$$\frac{d\xi_0}{dx} + \frac{d\eta_0}{dy} + \frac{d\zeta_0}{dz} = 0;$$

therefore ψ satisfies Laplace's equation. Also, by simple differentiation, we can prove that

$$\xi_0^2 = \frac{d}{dx}\left(\psi \frac{d\psi}{dx}\right) - \psi \frac{d^2\psi}{dx^2}$$

$$= \tfrac{1}{2}\frac{d^2}{dx^2}\psi^2 - \psi\frac{d^2\psi}{dx^2};$$

with similar expressions for η_0^2 and ζ_0^2 respectively. Add these three expressions for ξ_0^2, η_0^2, ζ_0^2, and remember that ψ satisfies Laplace's equation; the result is the equation

$$\xi_0^2 + \eta_0^2 + \zeta_0^2 = \tfrac{1}{2}\left[\frac{d^2}{dx^2} + \frac{d^2}{dy^2} + \frac{d^2}{dz^2}\right]\psi^2.$$

But in this special case $K_0 = 2\left(\xi_0^2 + \eta_0^2 + \zeta_0^2\right)$; therefore we obtain finally

$$K_0 = \nabla^2\psi^2.$$

The differential equation to find P_1 becomes

$$\left[\frac{d^2}{dx^2} + \frac{d^2}{dy^2} + \frac{d^2}{dz^2}\right]\left(P_1 - 2\psi^2\right) = 0.$$

The solution of this equation is

$$P_1 = 2\psi^2 + \Pi_1 \quad \ldots\ldots\ldots\ldots\ldots\ldots\ldots\ldots\ldots (D),$$

where Π_1 is the Laplacian term involving constants as yet undetermined.

It is interesting to note the expression for the pressure at any point (p), which is immediately obtained by the aid of equation (D); it is

$$p = p_0 - \tfrac{1}{2}\rho\left(u_0^2 + v_0^2 + w_0^2\right) + 2\rho\psi^2 + \rho\Pi_1.$$

|84| *Special Case.*

3. A sphere of radius a is rotating with uniform angular velocity ω about a diameter, and is surrounded by a viscous incompressible fluid which extends to infinity and is at rest there; there is no slipping of the fluid over the surface of the sphere.

To find the motion of the fluid so as to include terms involving ω^2 take the centre of the sphere as origin of coordinates, and the axis of rotation as axis of z. Then the usual solution gives

$$u_0 = -a^3\omega\frac{y}{r^3}, \quad v_0 = a^3\omega\frac{x}{r^3}, \quad w_0 = 0.$$

By differentiating we find

$$\xi_0 = \frac{\omega a^3}{2}\frac{d^2}{dx\,dz}\frac{1}{r}, \quad \eta_0 = \frac{\omega a^3}{2}\frac{d^2}{dy\,dz}\frac{1}{r}, \quad \zeta_0 = \frac{\omega a^3}{2}\frac{d^2}{dz^2}\frac{1}{r}.$$

Hence, from inspection,

$$\psi = \frac{\omega a^3}{2}\frac{d}{dz}\frac{1}{r},$$

then equation (D) gives

$$P_1 = \frac{\omega^2 a^6}{2}\left\{\frac{d}{dz}\frac{1}{r}\right\}^2 + \Pi_1;$$

and from the symmetry of the motion it can be seen that

$$\Pi_1 = \sum p_n \frac{d^n}{dz^n}\frac{1}{r},$$

where the p's are constants to be determined.

To find u_1, v_1, w_1, we have

$$\frac{\mu u_1}{\rho} = -\frac{1}{4\pi}\iiint \frac{dP_1'}{dx'}\frac{1}{\sigma}d\tau' + \frac{1}{2\pi}\iiint A_3'\frac{1}{\sigma}d\tau' + \frac{\mu U_1}{\rho},$$

with similar equations for v_1, w_1.

For the calculation of the volume integrals depending on P_1, let M denote the point (x, y, z), Q the point (x', y', z'),[1] O the centre of the sphere. The length of OM is r, of OQ is r'; let the angle MOQ be χ.

Then if $r' < r$,

$$\frac{1}{\sigma} = \frac{1}{r} + \sum \frac{r'^n}{r^{n+1}} P_n(\cos\chi),$$

|85| if $r' > r$,

$$\frac{1}{\sigma} = \frac{1}{r'} + \sum \frac{r^n}{r'^{n+1}} P_n(\cos\chi),$$

where $P_n(\cos\chi)$ is a zonal harmonic with OM as axis.

First consider the term involving as a factor the arbitrary constant p_n. Let $d\omega'$ be an element of solid angle contained by a slender cone described with O as vertex, OQ as axis. Then, integrating so as to include all angular space, we have

$$\iint \frac{d^{n+1}}{dz'^n dx'}\left(\frac{1}{r'}\right)\cdot P_m(\cos\chi)\,d\omega' = 0,$$

when m is not equal to $n + 1$, and also

$$\iint \frac{d^{n+1}}{dz'^n dx'}\left(\frac{1}{r'}\right)\cdot P_{n+1}(\cos\chi)\,d\omega' = \frac{4\pi}{2n+3}\frac{r^{n+2}}{r'^{n+2}}\frac{d^{n+1}}{dz^n dx}\frac{1}{r}.$$

In the triple integral which we have to evaluate, $d\tau'$ may be written $r'^2 dr' d\omega'$, and the integration for angular space has just been accomplished. The integration with respect to r' must be separated into two parts; the first part is from $r' = a$ to $r' = r$, when the first formula for $\frac{1}{\sigma}$ must be used; the second part is from $r' = r$ to $r' = \infty$, when the second formula for $\frac{1}{\sigma}$ is to be used.

Then selecting from the expression $\iiint \frac{dP_1'}{dx'}\frac{1}{\sigma}d\tau'$ the term involving p_n, and performing the integrations round all angular space, there is obtained

1. Commas have been inserted in the two preceding ordered triples to avoid confusion with a product.

$$p_n\frac{4\pi}{2n+3}\frac{d^{n+1}}{dx\,dz^n}\left(\frac{1}{r}\right)\cdot\left[\int_a^r\frac{r'^{n+1}}{r^{n+2}}\frac{r^{n+2}}{r'^{n+2}}r'^2dr'+\int_r^\infty\frac{r^{n+1}}{r'^{n+2}}\frac{r^{n+2}}{r'^{n+2}}r'^2dr'=2\right],$$

that is
$$p_n\frac{4\pi}{2n+3}\frac{d^{n+1}}{dx\,dz^n}\left(\frac{1}{r}\right)\cdot\left[\int_a^r r'dr'+r^{2n+3}\int_r^\infty\frac{1}{r'^{2n+2}}dr'\right],$$

that is
$$p_n\frac{2\pi}{2n+3}\left[\frac{2n+3}{2n+1}r^2-a^2\right]\frac{d^{n+1}}{dx\,dz^n}\frac{1}{r},$$

that is
$$2\pi p_n\left[\frac{r^2}{2n+1}-\frac{a^2}{2n+3}\right]\frac{d^{n+1}}{dx\,dz^n}\frac{1}{r}.$$

|86| Consider in the next place the term

$$\iiint\frac{d}{dx'}\left\{\frac{d}{dz'}\frac{1}{r'}\right\}^2\frac{1}{\sigma}d\tau'.$$

By simple differentiation

$$\frac{d}{dx'}\left\{\frac{d}{dz'}\frac{1}{r'}\right\}^2=\tfrac{2}{5}\frac{1}{r'}\frac{d^3}{dx'dz'^2}\frac{1}{r'}+\tfrac{6}{5}\frac{1}{r'^3}\frac{d}{dx'}\frac{1}{r'}.$$

These two terms can be integrated separately by exactly the same method as that just used, the only difference between the cases is that instead of p_n the numerical factors must be substituted, that n is 2 for the first integral and 1 for the second, and that, under the integral signs with respect to r', extra factors, which are $\frac{1}{r'}$ for the first term and $\frac{1}{r'^3}$ for the second term, must be substituted.

The first term becomes

$$\tfrac{2}{5}\frac{4\pi}{7}\frac{d^3}{dx\,dz^2}\left(\frac{1}{r}\right)\cdot\left[\int_a^r\frac{1}{r'}r'dr'+r^7\int_r^\infty\frac{1}{r'}\frac{1}{r'^6}dr'\right],$$

that is
$$\frac{8\pi}{5}\left[\tfrac{1}{6}r-\tfrac{1}{7}a\right]\frac{d^3}{dx\,dz^2}\frac{1}{r};$$

the second term becomes

$$\frac{24\pi}{5}\left[\tfrac{1}{3}\frac{1}{a}-\tfrac{1}{4}\frac{1}{r}\right]\frac{d}{dx}\frac{1}{r}.$$

Now the expression for P_1 is

$$\frac{\omega^2a^6}{2}\left\{\frac{d}{dz}\frac{1}{r}\right\}^2+\sum p_n\frac{d^n}{dz^n}\frac{1}{r};$$

combining these results, we find that

$$\frac{1}{4\pi}\iiint\frac{dP_1'}{dx'}\frac{1}{\sigma}d\tau'=\frac{3\omega^2a^6}{5}\left[\tfrac{1}{3}\frac{1}{a}-\tfrac{1}{4}\frac{1}{r}\right]\frac{d}{dx}\frac{1}{r}$$

$$+\frac{\omega^2a^6}{5}\left[\tfrac{1}{6}r-\tfrac{1}{7}a\right]\frac{d^3}{dx\,dz^2}\frac{1}{r}+\sum\tfrac{1}{2}p_n\left[\frac{r^2}{2n+1}-\frac{a^2}{2n+3}\right]\frac{d^{n+1}}{dx\,dz^n}\frac{1}{r}.$$

To find the corresponding expression for the y-equation, we change $\frac{d}{dx}$ into $\frac{d}{dy}$ throughout the equation just obtained.

|87| To find $\frac{1}{4\pi}\iiint \frac{dP_1'}{dz'}\frac{1}{\sigma}d\tau'$, we have, by simple differentiation,

$$\frac{d}{dz}\left\{\frac{d}{dz}\frac{1}{r}\right\}^2 = \tfrac{2}{5}\frac{1}{r}\frac{d^3}{dz^3}\frac{1}{r} + \tfrac{8}{5}\frac{1}{r^3}\frac{d}{dz}\frac{1}{r}.$$

Now the methods of integration to be used for these integrals are exactly the same as those already employed, therefore by comparing the numerical coefficients, we find at once

$$\frac{1}{4\pi}\iiint \frac{dP_1'}{dz'}\frac{1}{\sigma}d\tau' = \frac{4\omega^2a^6}{5}\left[\tfrac{1}{3}\frac{1}{a} - \tfrac{1}{4}\frac{1}{r}\right]\frac{d}{dz}\frac{1}{r}$$
$$+\frac{\omega^2a^6}{5}\left[\tfrac{1}{6}r - \tfrac{1}{7}a\right]\frac{d^3}{dx\,dz^2}\frac{1}{r} + \sum \tfrac{1}{2}p_n\left[\frac{r^2}{2n+1} - \frac{a^2}{2n+3}\right]\frac{d^{n+1}}{dz^{n+1}}\frac{1}{r}.$$

We have now to calculate the integrals

$$\iiint A_0'\frac{1}{\sigma}d\tau',\ \iiint B_0'\frac{1}{\sigma}d\tau',\ \iiint C_0'\frac{1}{\sigma}d\tau';$$

where[1]

$$A_0 = v_0\zeta_0 - w_0\eta_0 = \tfrac{1}{2}\omega^2a^6\frac{x}{r^3}\frac{d^2}{dz^2}\frac{1}{r},$$

$$B_0 = w_0\xi_0 - u_0\zeta_0 = \tfrac{1}{2}\omega^2a^6\frac{y}{r^3}\frac{d^2}{dz^2}\frac{1}{r},$$

$$C_0 = u_0\eta_0 - v_0\xi_0 = \tfrac{1}{2}\omega^2a^6\left[\frac{y}{r^3}\frac{d^2}{dy\,dz}\frac{1}{r^3} + \frac{x}{r^3}\frac{d^2}{dx\,dz}\frac{1}{r}\right].$$

By simple differentiation

$$\frac{x}{r^3}\frac{d^2}{dz^2}\frac{1}{r} = \tfrac{2}{5}\frac{1}{r^3}\frac{d}{dx}\frac{1}{r} - \tfrac{1}{5}\frac{1}{r}\frac{d^3}{dx\,dz^2}\frac{1}{r},$$

$$\frac{y}{r^3}\frac{d^2}{dz^2}\frac{1}{r} = \tfrac{2}{5}\frac{1}{r^3}\frac{d}{dy}\frac{1}{r} - \tfrac{1}{5}\frac{1}{r}\frac{d^3}{dy\,dz^2}\frac{1}{r},$$

$$\frac{y}{r^3}\frac{d^2}{dy\,dz}\frac{1}{r} + \frac{x}{r^3}\frac{d^2}{dx\,dz}\frac{1}{r} = -\tfrac{6}{5}\frac{1}{r^3}\frac{d}{dz}\frac{1}{r} + \tfrac{1}{5}\frac{1}{r}\frac{d^3}{dz^3}\frac{1}{r}.$$

It will be seen by a reference to what precedes that all the integrals which we want have already been evaluated; so |88| the results can be stated at once as follows:[2]

$$\frac{1}{2\pi}\iiint A_0'\frac{1}{\sigma}d\tau' = \frac{2\omega^2a^6}{5}\left[\tfrac{1}{3}\frac{1}{a} - \tfrac{1}{4}\frac{1}{r}\right]\frac{d}{dx}\frac{1}{r} - \frac{\omega^2a^6}{5}\left[\tfrac{1}{6}r - \tfrac{1}{7}a\right]\frac{d^3}{dx\,dz^2}\frac{1}{r},$$

$$\frac{1}{2\pi}\iiint B_0'\frac{1}{\sigma}d\tau' = \frac{2\omega^2a^6}{5}\left[\tfrac{1}{3}\frac{1}{a} - \tfrac{1}{4}\frac{1}{r}\right]\frac{d}{dy}\frac{1}{r} - \frac{\omega^2a^6}{5}\left[\tfrac{1}{6}r - \tfrac{1}{7}a\right]\frac{d^3}{dy\,dz^2}\frac{1}{r},$$

$$\frac{1}{2\pi}\iiint C_0'\frac{1}{\sigma}d\tau' = \frac{6\omega^2a^6}{5}\left[\tfrac{1}{3}\frac{1}{a} - \tfrac{1}{4}\frac{1}{r}\right]\frac{d}{dz}\frac{1}{r} - \frac{\omega^2a^6}{5}\left[\tfrac{1}{6}r - \tfrac{1}{7}a\right]\frac{d^3}{dz^3}\frac{1}{r}.$$

---

1. The bracketed expression below erroneously had $\frac{x}{r}$ instead of $\frac{x}{r^3}$.

2. The third equation below erroneously had $\frac{d^3}{dy\,dz^2}$ instead of $\frac{d^3}{dz^3}$.

All the integrals which are wanted in the expressions to find u_1, v_1, w_1 have now been calculated. Besides these integrals, terms U_1, V_1, W_1 occur in the expressions giving u_1, v_1, w_1 : they must each satisfy Laplace's equation, but are as yet otherwise undetermined; though the conditions necessary for their determination have been found and stated in the general explanation of the method, namely that u_1, v_1, w_1 must be zero at the surface of the sphere. Without more ado we may assume U_1, V_1, W_1 to be given by the expressions

$$\frac{\mu U_1}{\rho}=\frac{\omega^2 a^6}{5}\sum \beta_n \frac{d^{n+1}}{dx\,dz^n}\frac{1}{r},$$

$$\frac{\mu V_1}{\rho}=\frac{\omega^2 a^6}{5}\sum \gamma_n \frac{d^{n+1}}{dy\,dz^n}\frac{1}{r},$$

$$\frac{\mu W_1}{\rho}=\frac{\omega^2 a^6}{5}\sum \delta_n \frac{d^{n+1}}{dz^{n+1}}\frac{1}{r};$$

where the β's, γ's, δ's are arbitrary constants to be determined.

Furthermore, the terms satisfying Laplace's equation which occur in the integrals just evaluated may be absorbed for the sake of shortness into the corresponding terms in U_1, V_1, W_1 when forming the expressions for u_1, v_1, w_1. These expressions |89| now are

$$\frac{\mu u_1}{\rho}=\left[\frac{\omega^2 a^6}{5}\left(\beta_0+\tfrac{1}{4}\frac{1}{r}\right)-\tfrac{1}{2}p_0 r^2\right]\frac{d}{dx}\frac{1}{r}$$
$$+\left[\frac{\omega^2 a^6}{5}\left(\beta_2+\tfrac{1}{3}r\right)-\tfrac{1}{10}p_2 r^2\right]\frac{d^3}{dx\,dz^2}\frac{1}{r}+\sum \tfrac{1}{2}p\frac{r^2}{2n+1}\frac{d^{n+1}}{dx\,dz^n}\frac{1}{r}$$
$$+\frac{\omega^2 a^6}{5}\sum \beta_n \frac{d^{n+1}}{dx\,dz^n}\frac{1}{r};$$

$\frac{\mu v_1}{\rho}$ = a similar expression found by changing *x* into *y*;

$$\frac{\mu w_1}{\rho}=\left[\frac{\omega^2 a^6}{5}\left(\delta_0+\tfrac{1}{2}\frac{1}{r}\right)-\tfrac{1}{2}p_0 r^2\right]\frac{d}{dz}\frac{1}{r}$$
$$+\left[\frac{\omega^2 a^6}{5}\left(\beta_2+\tfrac{1}{3}r\right)-\tfrac{1}{10}p_2 r^2\right]\frac{d^3}{dz^3}\frac{1}{r}+\tfrac{1}{2}p_n\frac{r^2}{2n+1}\frac{d^{n+1}}{dz^{n+1}}\frac{1}{r}$$
$$+\frac{\omega^2 a^6}{5}\sum \beta_n \frac{d^{n+1}}{dz^{n+1}}\frac{1}{r}.$$

Since u_1, v_1, w_1 must vanish at an infinite distance, $p_0=0$. Also u_1, v_1, w_1 must vanish when $r=a$; determining the β's, γ's, δ's by this condition, we find that

$$\frac{\mu u_1}{\rho}=-\frac{\omega^2 a^6}{20}\left(\frac{1}{a}-\frac{1}{r}\right)\frac{d}{dx}\frac{1}{r}-\left[\frac{\omega^2 a^6}{15}(r-a)+\frac{p_2}{10}\left(r^2-a^2\right)\right]\frac{d^3}{dx\,dz^2}\frac{1}{r}$$

+ terms which will be proved to vanish, depending on p_3, p_4, &c.;

$\frac{\mu v_1}{\rho}$ = a similar expression;

$$\frac{\mu w_1}{\rho}=-\frac{\omega^2 a^6}{10}\left(\frac{1}{a}-\frac{1}{r}\right)\frac{d}{dz}\frac{1}{r}-\left[\frac{\omega^2 a^6}{15}(r-a)+\frac{p_2}{10}\left(r^2-a^2\right)\right]\frac{d^3}{dz^3}\frac{1}{r}$$

+ terms which will be proved to vanish.

To determine the p's, we notice that u_1, v_1, w_1 must satisfy the equation of continuity; from this it can be easily seen to follow that all the p's are zero, with the exception of p_2.

Now substituting $\frac{\mu u_1}{\rho}$, $\frac{\mu v_1}{\rho}$, $\frac{\mu w_1}{\rho}$ in the equation of continuity, and putting all the p's except p_2 zero, all the terms vanish identically except

$$-\frac{\omega^2 a^5}{20}\left[\frac{d^2}{dx^2}\frac{1}{r}+\frac{d^2}{dy^2}\frac{1}{r}-2\frac{d^2}{dz^2}\frac{1}{r}\right]-\tfrac{1}{5}p_2\left[x\frac{d}{dx}+y\frac{d}{dy}+z\frac{d}{dz}\right]\frac{d^2}{dz^2}\frac{1}{r},$$

|90| that is,

$$\frac{3\omega^2 a^5}{20}\frac{d^2}{dz^2}\frac{1}{r}+\tfrac{3}{5}p_2\frac{d^2}{dz^2}\frac{1}{r}.$$

Hence, if p_2 be taken equal to $-\frac{1}{4}\omega^2 a^5$ the equation of continuity is satisfied.

The motion is now completely determined; the complete expressions for the velocities to our order of approximation are:[1]

$$u=\omega a^3\frac{d}{dy}\frac{1}{r}-\frac{\rho\omega^2 a^6}{20\mu}\left(\frac{1}{a}-\frac{1}{r}\right)\frac{d}{dx}\frac{1}{r}-\frac{\rho\omega^2 a^6}{5\mu}\left[\frac{r-a}{3}-\frac{r^2-a^2}{8a}\right]\frac{d^3}{dx\,dz^2}\frac{1}{r},$$

$$v=-\omega a^3\frac{d}{dx}\frac{1}{r}-\frac{\rho\omega^2 a^6}{20\mu}\left(\frac{1}{a}-\frac{1}{r}\right)\frac{d}{dy}\frac{1}{r}-\frac{\rho\omega^2 a^6}{5\mu}\left[\frac{r-a}{3}-\frac{r^2-a^2}{8a}\right]\frac{d^3}{dx\,dz^2}\frac{1}{r},$$

$$w=\frac{\rho\omega^2 a^6}{10\mu}\left(\frac{1}{a}-\frac{1}{r}\right)\frac{d}{dz}\frac{1}{r}-\frac{\rho\omega^2 a^6}{5\mu}\left[\frac{r-a}{3}-\frac{r^2-a^2}{8a}\right]\frac{d^3}{dz^3}\frac{1}{r}.$$

The type of motion indicated by these expressions is more easily recognised by transferring to polar coordinates. Let r, θ, ϕ be the coordinates of any point M in the fluid, so that θ is the angle which the radius vector r makes with the axis of rotation, and ϕ is the angle which the plane through the axis and the point M makes with the fixed plane of zx, where it is to be remembered that the axis of rotation is the axis of z. Let R, Θ, Φ be the three component velocities resolved along r, perpendicular to r in the plane ZOM so as to increase θ when positive, and perpendicular to the plane ZOM so as to increase ϕ when positive, respectively.

Then

$$R=u\frac{x}{r}+v\frac{y}{r}+w\frac{z}{r}=-\frac{\rho\omega^2 a^6}{10\mu}\frac{1}{r^2}\left(\frac{1}{a}-\frac{1}{r}\right)P_2(\cos\theta)$$
$$+\frac{6\rho\omega^2 a^6}{5\mu}\frac{1}{r^4}\left[\frac{r-a}{3}-\frac{r^2-a^2}{8a}\right]P_2(\cos\theta)$$
$$=-\tfrac{1}{4}\frac{\rho\omega^2 a^3}{\mu}\frac{a^4}{r^2}\left(\frac{1}{a}-\frac{1}{r}\right)^2 P_2(\cos\theta),$$

where $P_2(\cos\theta)$ is a zonal harmonic of the second order.

1. The second equation below erroneously had $v=-\omega^2 a^3$ instead of $v=-\omega a^3$.

|91| And

$$\Theta = (u\cos\phi + v\sin\phi)\cos\theta - w\sin\theta$$
$$= \tfrac{3}{20}\frac{\rho\omega^2 a^6}{\mu}\frac{1}{r^2}\left[\frac{1}{a}-\frac{1}{r}\right]\sin\theta\cos\theta$$
$$+\tfrac{6}{5}\frac{\omega^2 a^6}{\mu}\frac{1}{r^4}\left[\frac{r-a}{3}-\frac{r^2-a^2}{8a}\right]\sin\theta\cos\theta$$
$$= \tfrac{1}{4}\frac{\rho\omega^2 a^3}{\mu}\frac{a^4}{r^3}\left[\frac{1}{a}-\frac{1}{r}\right]\sin\theta\cos\theta,$$
$$\Phi = \frac{\omega a^3\sin\theta}{r^2}.$$

The character of the fluid motion can now be easily described; near the axis of z, R is negative and the fluid is moving towards the sphere; when $\cos\theta = \pm\frac{1}{\sqrt{3}}$, $R = 0$; when $\cos\theta$ lies between the limits $\pm\frac{1}{\sqrt{3}}$, R is positive; therefore in the equatoreal regions the fluid is moving away from the sphere. Also, from the consideration of Θ, we see that there is a motion from the polar regions towards the equatoreal regions.

Stokes, in his paper 'On the Friction of Fluids in Motion,'[1†] mentions that some sort of motion like this is necessary; his words are:—'It may be shewn from the general equations that in this case a permanent motion in annuli is impossible, and that, whatever may be the law of friction between the solid sphere and the fluids. Hence it appears that it is necessary to suppose that the particles move in planes passing through the axis of rotation, while they, at the same time, move round it. In fact, it is easy to see that from the excess of centrifugal force in the neighbourhood of the equator of the revolving sphere, the particles in that part will recede from the sphere, and approach it again in the neighbourhood of the poles, and this circulating motion will be combined with a motion about the axis.' Siemens also in his theory of the Conservation of Solar Energy[2] assumed that this sort of motion exists in matter diffused throughout space, being due to the sun's rotation. Curiously enough M. Faye, in his criticism on Siemens' theory, appears to have denied that such a motive is possible, cf. Siemens' collection of papers and discussions of this subject.

|92| 4. *Special case.* The flow in a pipe.

The fluid is flowing steadily through a straight pipe of circular cross section, and there is no slipping at the boundary surface. We proceed to shew that the usual solution is absolutely accurate. This problem is best solved by using cylindrical coordinates throughout. The case of the sphere might also have been treated in a[3] similar way with polar coordinates, and by the aid of a little guess-work in the solution of some of the partial differential equations the work can be made somewhat shorter than that given above; but the dependence of the solution on the general theory is

1. †cf. Collected Papers, Vol. I., p. 103. [The full citation is: George Gabriel Stokes, 'On the Theories of the Internal Friction of Fluids in Motion, and of the Equilibrium and Motion of Elastic Solids', in George Gabriel Stokes, *Mathematical and Physical Papers, Volume I* (Cambridge: Cambridge University Press, 1880), p. 103.]
2. C. William Siemens, *On the Conservation of Solar Energy: A Collection of Papers and Discussions* (London: Macmillan, 1883).
3. The original omitted 'a'.

not so clearly exemplified as by the work actually given. In the present problem take the axis of the pipe as axis of z, and let the usual cylindrical coordinates of any point be z, r, θ. Let the corresponding resolved parts of the velocities at any point of the fluid be w, R, Θ, and the corresponding resolved parts of the 'angular velocity' be ζ, R_1, Θ_1.

For the sake of brevity we will at once state the evident simplifications which hold for this problem of the straight pipe; we have

$$R=0, \quad \Theta=0;$$

also none of the remaining expressions for the velocity or angular velocity involve θ or z.

From this we deduce

$$R_1=0, \quad \Theta_1=-\tfrac{1}{2}\frac{dw}{dr}, \quad \zeta=0.$$

The equation of continuity is satisfied by means of these assumptions, for w does not contain z.

The equations[1] of steady motion become

$$w\frac{dw}{dr}=\frac{dP}{dr}$$

$$0=\frac{dP}{dz}+\frac{\mu}{\rho}\left[\frac{d^2w}{dr^2}+\frac{1}{r}\frac{dw}{dr}\right].$$

The solution of these equations, found by neglecting the square of the velocity, is given by Stokes in his paper on Fluid Friction,[2†] it is

$$w=\tfrac{1}{4}V\frac{r^2-a^2}{a^2},$$

|93| where we have assumed that there is no slipping of the fluid over the boundary surface of the pipe.

Substituting this solution in the equations of motion and retaining the squares of small quantities, we have[3]

$$\frac{dP}{dr}=\tfrac{1}{8}V^2\frac{r\left(r^2-a^2\right)}{a^4}, \quad \frac{dP}{dz}=-\tfrac{1}{4}\frac{\mu V}{\rho a^2}.$$

Both these equations can be satisfied by assuming that

$$P=-\tfrac{1}{4}\frac{\mu V}{\rho a^2}z+\tfrac{1}{16}\frac{V^2}{a^2}\left(\tfrac{1}{2}r^4-a^2r^2\right)+P_0,$$

where P_0 is some arbitrary constant to be determined by the end conditions of the pipe.

The equations of motion have therefore been proved to be satisfied exactly by this solution, and not merely approximately for the first powers of the velocity only.

It is interesting to notice that the extra term involving V^2 which we have added

1. The original had 'equation'.
2. † cf. Collected Papers, Vol. I., p. 105.
3. The first equation below erroneously had $\frac{r\left(r^2-a\right)}{a^4}$ instead of $\frac{r\left(r^2-a^2\right)}{a^4}$.

to P does not affect the expression for the pressure at any point; for, if p stands[1] for the pressure,

$$p = \rho P - \tfrac{1}{2}\rho w^2$$

$$= -\tfrac{1}{4}\frac{\mu V}{a}z + \frac{\rho V^2}{a^2}\left(\tfrac{1}{32}r^4 - \tfrac{1}{16}a^2r^2\right) + \text{const.} - \tfrac{1}{32}V^2\left(\frac{r^2 - a^2}{a^2}\right)^2$$

$$= -\tfrac{1}{4}\frac{\mu V}{a}z + \text{const.} = -\tfrac{1}{4}\frac{\mu V}{a}z + p_0,$$

which is the same expression as is found by neglecting V^2.

5. Another special case which can be dealt with by this method is that of a sphere moving uniformly forward without rotation in a fluid at rest at infinity. This problem will be treated in another communication. The expressions for the velocity are somewhat longer than those found above for the rotating sphere.

1. The original had 'stand'.

A Visitation

1888 *Cambridge Fortnightly*, vol. 1, no. 4 (6 March), pp. 81–83

The Cambridge Fortnightly *ran for a year, with this essay appearing in the fourth of five issues. Unlike the two essays from 1886, 'Clerk of the Weather' and 'Davy Jones', which were simply signed 'W.', this essay is signed 'A. N. W.', making its attribution to Whitehead more secure. Further, like those other two essays, the entry for this essay appears in Victor Lowe's 1941 bibliography for the Library of Living Philosophers volume on Whitehead's work, which Whitehead likely reviewed and approved, and makes use of the same rhetorical device of a fictional conversation.*[1] *Additionally, in his biography of Whitehead, Lowe notes that* Fortnightly *was edited by two friends and fellow Cambridge Apostles, Roger Fry and Nathaniel Wedd.*[2]

This brief and casual essay begins by using the occasion of happy families visiting Cambridge University with their sons who are students to imagine what it would be like if the saintly founders of the University were given a tour of their own college five centuries after its founding.

During the May Term it is very pleasant to watch family parties as they stroll through a college: in front walk the father and mother full of secret pride in their son, who is following, well-dressed and clean, in company with his two sisters and a cousin. The young folk are all laughing gaily at some remark just made by the young man who seems to be pointing towards the dean's window. They linger for a moment at the gateway, give a last look round, and then disappear, to be succeeded by another party and yet another, all of the same type, happy and careless. Sometimes the father is absent, perhaps he cannot leave his business, or perhaps he is calling alone on some old college friend, now a middle-aged tutor who would object to the intrusion of ladies. But it may be that he is absent for another reason, and then the son will be walking by the side of his mother, who is still wearing her weeds,[3] while the three young ladies from behind occasionally interject between them a remark, which they turn round to answer.

But we also receive visitors of another sort: a band of excursionists will occasionally troop through the court, serious and middle-class. If they are accompanied by undergraduates, vigorously acting as guides, then we know that their seriousness is partly acquired and is the result of attendance at courses of "Extension Lectures."

1. See the introduction to 'A Celebrity at Home: the Clerk of the Weather', p. 1.
2. Victor Lowe, *Alfred North Whitehead: The Man and His Work, Volume I: 1861–1910* (Baltimore: Johns Hopkins University Press, 1985), p. 165.
3. A 'widow's weeds' refers to the black dress and veil traditionally worn by a widow while mourning the death of her husband.

Next to watching the enjoyment of others as they pass by, comes the pleasure of rambling alone on some summer afternoon through the quads as they stand silent and empty, while the men are away boating and cricketing. These buildings remind me of the people who will only talk to those who visit them alone and thoughtful.

But sometimes I have visitors of my own to attend to; it is the most amusing task possible to conduct in imagination a founder round his own college. A short time ago I took the united guilds of the Corpus Christi and of the Blessed Virgin Mary round the college which they had founded,[1] and will end these reflections by retailing my experience. On these occasions I generally meet my guests outside the college gates; we adopted the usual plan in this instance and had a good look at the exterior before proceeding. They were very content with its appearance and highly pleased at recognizing the emblems of |82| the four Evangelists at the top of the neighbouring church tower: they took it as a happy augury that the college and its successive masters had never ceased to rely on the prayers of the saints as the source of all prosperity. We next went inside: the good men were a little startled at first by the changed appearance of the porter, but I soon made them to understand that outward habiliments must naturally change in the course of five hundred long years. Indeed they made rather too free a use of this idea after I had explained it to them, for they were delighted when we met one of the fellows in flannels and a boating blazer entering the college with a lawn tennis racket in his hand: and deemed him to be the master or at least a priest fresh from some festal service in honour of their peculiar saint. Before I had had time to explain to them their mistake, and in truth it would have been no easy matter, he had caught us up as we were sauntering in a large group towards the chapel, staring about us as we went. My friend, whom I introduced, was very pleased to meet with the members of the guild: he felt that he had deserved well of them, for he had stoutly defended the ecclesiastical character of the college when the new statutes were under discussion, and had written an excellent article in a magazine protesting against disregard of the wishes of pious founders. He explained to them his feelings on these disputed points, and they in return gave praise to the Blessed Virgin that such men should still inhabit their college. The visitors were now in high, good humour, and my secret anxiety as to the result of this interview was beginning to abate. As we neared the chapel my friend very wisely tried to prepare them for the inevitable shock which they must experience on entering the building: "You will find," said he, "that in some points we have deviated from the old paths. Religion is now based more directly upon the study of the Holy Scriptures than it was perhaps in your times." The master of the guild, in almost my own words, answered that he had bethought him how that there must be many a change in the course of five hundred long years, and that with the approval of His Holiness the Pope even the Scriptures might be sparingly read. "But 'tis a great change," he mused, as he slowly mounted the chapel steps. My friend's embarrassed reply was cut short by the commotion among the other members of the guild consequent on their failure to |83| discover any holy water.

1. Whitehead is referring to 'The College of Corpus Christi and the Blessed Virgin Mary' – more commonly known simply as 'Corpus Christi College, Cambridge', or just 'Corpus' – a constituent college of the University of Cambridge.

"Has his Holiness allowed even that to be removed?" cried the master in astonishment, "by St. Mary things are changed indeed!" "The fact is," was the reply, "we have discovered that the authority of the Pope himself is not based on any sound scriptural warrant." "You do not obey the Pope?" gasped out the master. "But the patron saint of our college—tell me—you worship her?" "To tell the truth we no longer—," when the answer had reached this length, I heard a faint "Oh!" ejaculated by my side, and on looking round I found that the feelings of the old man had been too much for him, and he had vanished from mortal view. The rest of the guild were also gone, and we men were left staring at each other blankly and foolishly. "With reference to the intentions of pious founders—," I began reflectively—"Ah!" said my friend, and he too turned on his heel and departed from me. I was now left alone and slowly wended my way home, reflecting that if ghosts are entrusted, as some deem, with visitorial powers, the couch of many a college fellow must be infested with nightly horrors.

Second Approximations to Viscous Fluid Motion

1888 *Quarterly Journal of Pure and Applied Mathematics*, vol. 23, no. 90 (June), pp. 143–152

'Second Approximations' was also published in 1888 and is a natural extension of 'Incompressible Fluids'. The domain is therefore again fluid dynamics and the reader might do well to review the second, third and fourth paragraphs of our introduction to the former work. In this introduction, we shall briefly review the prior article, describe the new specific problem at hand, identify the key differences in Whitehead's assumptions and general approach between the two papers, and then summarise his principal result for this second article.

'Incompressible Fluids' develops a mathematical framework for the equations of motion that govern the flow of an incompressible, viscous fluid with the associated analysis expanded to include second-order terms of the velocity vector associated with the flow. The principal application of the previous paper is to the motion induced in an infinite lake by a rotating sphere. The present paper inherits the full framework of the previous analysis but addresses the specific problem of a non-rotating sphere moving through the lake at a uniform velocity. Early on, Whitehead notes that by a simple change of coordinate systems, this is equivalent to the problem of how a stationary sphere would affect the flow of a body of fluid moving past it at an otherwise constant velocity. Thus, we may now think of a stationary sphere in an infinite river, and Whitehead chooses this equivalent characterisation of the problem as mathematically more tractable.

The ideas of incompressibility and viscosity are, of course, still in play, but Whitehead also makes explicit here an assumption that is implicit in 'Incompressible Fluids' and seems at first too obvious to mention: namely, that the second-order contributions to the velocity (hence also the angular velocity) must remain finite at every point in the flow. (See the second paragraph of Section 3 on p. 28.) The reason for this explicit condition appears later in the paper, when it is used to impose decisive conditions on variables and on the ultimate interpretation of his results (pp. 31–34). We shall say a little more about this below.

The approach in 'Second Approximations' is to treat the problem in terms of differential equations rather than by the numerous triple integrals calculated in its antecedent. Whitehead explains this choice at the end of his introduction, and indeed his methods are well suited to the problem at hand (p. 26). The upshot of his analysis is that 'no second approximation to the solution of the equations of motion is possible in this problem consistent with the "angular velocity" of the fluid being finite at each point of space' (p. 34). To understand the conclusion he draws from this, we shall need to say a few words about aspects of mathematical models in general and to revisit the finiteness condition introduced above.

The correspondence between a mathematical model and a physical system is bi-directional. For example, the spherical shape of a floating soap bubble suggests that the most efficient way to enclose a given volume is with a sphere. Subsequently, once this is proven mathematically, it then 'explains' the shape of a soap bubble. Problems with infinity in mathematical models are likewise connected to the behaviour of physical systems. Newton's inverse-square law in his theory of universal gravitation suggests an infinite force whenever two bodies, such as a hand and a rock, make contact. Of course, no such force occurs in nature –we do throw stones – and this raises two possible implications: either Newton's theory is wrong, or ordinary things don't really touch. In this example, both possibilities are true; universal gravitation has been superseded by general relativity, while quantum theory tells us, via the Pauli exclusion principle, that ordinary things keep their distance at microscopic scales. (This is not to say that quantum theory and general relativity have yet been reconciled.)

Whitehead's final words in 'Second Approximations' speak to the missing second-order solution as being consistent with the observation of dead water behind a spherical body moving in an otherwise still fluid. Thus, he acknowledges that in this case an unexpected analytic outcome is supported by physical evidence, but he carefully makes no claim that his mathematical analysis is yet sufficient to compel this physical state, leaving matters in a delightful theoretical–empirical equilibrium.

A sphere moving steadily in a straight line.

1. A homogeneous incompressible viscous fluid is flowing steadily past a fixed spherical obstacle, the motion of the fluid is the same at all points at a great distance from the sphere and will be denoted by V. If we impress on the sphere and on each element of the fluid a velocity V, so as to reduce to rest the fluid at a great distance from the sphere, we arrive at the case of a sphere moving uniformly forward with velocity V and disturbing a fluid |144| otherwise at rest. These are two methods of stating what is really the same problem; we will choose the first form of statement and consider the sphere as fixed.

We have to find the motion of the fluid when terms involving the square of V are retained. The general theory of our method of approximation was explained in the previous paper in this Journal; the work of this paper will explain itself as it proceeds.

It will be found convenient, because of the finite motion of the fluid at an infinite distance, to avoid triple integrals; our procedure therefore will be slightly different from that adopted in the paper referred to.

2. Let the radius of the sphere be a, the coefficient of viscosity of the fluid μ, the density of the fluid ρ. The fluid in contact with the surface of the sphere is at rest relatively to the surface. Let the origin of coordinates be at the centre of the sphere, and let the axis of z be drawn with its positive direction opposite to the direction of motion of the fluid at a great distance from the sphere. The polar coordinates of any point Q in the fluid are r, θ, ϕ, where r is the distance of Q from the centre O of the

sphere, θ is the angle ZOQ, and ϕ is the angle that the plane ZOQ makes with the plane of zx. Let the corresponding resolved parts of the velocity of the fluid at Q be u, v, w; so that u is the velocity along r, v is the velocity in the plane ZOQ perpendicular to OQ, and w is the velocity perpendicular to the plane ZOQ. Let the corresponding 'angular velocities' of the fluid at Q be ξ, η, ζ. Then the square of the velocity of the fluid at Q is $\left(u^2+v^2+w^2\right)$; and as in the previous paper P denotes the expression

$$\frac{p}{\rho}+\frac{1}{2}\frac{u^2+v^2+w^2}{\rho},$$

p being the pressure at Q.

The equations of motion transferred to polar coordinates are

$$\frac{du}{dt}-2(v\zeta-w\eta)=-\frac{dP}{dr}-\frac{2\mu}{\rho}\left(\frac{1}{r}\frac{d\zeta}{d\theta}+\frac{\zeta\cot\theta}{r}-\frac{1}{r\sin\theta}\frac{d\eta}{d\phi}\right),$$

$$\frac{dv}{dt}-2(w\xi-u\zeta)=-\frac{1}{r}\frac{dP}{d\theta}-\frac{2\mu}{\rho}\left(\frac{1}{r\sin\theta}\frac{d\xi}{d\phi}-\frac{d\zeta}{dr}-\frac{\zeta}{r}\right),$$

$$\frac{dw}{dt}-2(u\eta-v\xi)=-\frac{1}{r\sin\theta}\frac{dP}{d\phi}-\frac{2\mu}{\rho}\left(\frac{d\eta}{dr}+\frac{\eta}{r}-\frac{1}{r}\frac{d\xi}{d\theta}\right).$$

|145| Also, ξ, η, ζ are found from u, v, w by means of the following equations:

$$2\xi=\frac{1}{r}\frac{dw}{d\theta}+\frac{w\cot\theta}{r}-\frac{1}{r\sin\theta}\frac{dv}{d\phi},$$

$$2\eta=\frac{1}{r\sin\theta}\frac{du}{d\phi}-\frac{dw}{dr}-\frac{w}{r},$$

$$2\zeta=\frac{dv}{dr}+\frac{v}{r}-\frac{1}{r}\frac{du}{d\theta}.$$

Now from the symmetry of the motion we see that w is everywhere zero in the case of the fluid motion now being considered, and also that ϕ does not enter into the expressions for u or v. It follows then from the equations just written down that ξ, η are also in this case everywhere zero. The equations of steady motion become with these limitations

$$\left.\begin{aligned}2v\zeta&=\frac{dP}{dr}+\frac{2\mu}{\rho}\left(\frac{1}{r}\frac{d\zeta}{d\theta}+\frac{\zeta\cot\theta}{r}\right)\\-2u\zeta&=\frac{1}{r}\frac{dP}{d\theta}-\frac{2\mu}{\rho}\left(\frac{d\zeta}{dr}+\frac{\zeta}{r}\right)\end{aligned}\right\}\ldots\ldots\ldots\ldots\ldots(A).$$

The equation of continuity satisfied by u and v is

$$\frac{du}{dr}+\frac{2u}{r}+\frac{1}{r}\frac{dv}{d\theta}+\frac{v\cot\theta}{r}=0\ldots\ldots\ldots\ldots\ldots(B).$$

The equation connecting ζ with u and v is

$$2\zeta=\frac{dv}{dr}+\frac{v}{r}-\frac{1}{r}\frac{du}{d\theta}\ldots\ldots\ldots\ldots\ldots(C).$$

A solution of equations (*A*) and (*B*), the squares of the velocity being omitted, has been given for the problem under consideration, cf. Lamb's 'Motion of Fluids,' p. 226.[1] The expressions there found for u, v are

$$\left.\begin{aligned} u &= V\left[-\tfrac{1}{2}\frac{a^3}{r^3}+\tfrac{3}{2}\frac{a}{r}-1\right]\cos\theta \\ v &= V\left[-\tfrac{1}{4}\frac{a^3}{r^3}+\tfrac{3}{4}\frac{a}{r}+1\right]\sin\theta \end{aligned}\right\} \quad\ldots\ldots\ldots\ldots\ldots\ldots\ldots(D).$$

Substituting these expressions for u, v in equation (C), we find

$$\zeta = \tfrac{3}{4}\frac{Va}{r^2}\sin\theta.$$

|146| In order to find P to this order of approximation, we substitute this value of ζ in equations (*A*) and neglect their left-hand sides; then

$$\left.\begin{aligned} \frac{dP}{dr} &= -3\frac{\mu Va}{\rho}\frac{\cos\theta}{r^3} \\ \frac{dP}{d\theta} &= -\tfrac{3}{2}\frac{\mu Va}{\rho}\frac{\sin\theta}{r^2} \end{aligned}\right\}.$$

It follows that

$$P = \tfrac{3}{2}\frac{\mu Va}{\rho}\frac{\cos\theta}{r^2} + \text{constant}.$$

We may notice in passing the fact that to this order of approximation the equation just found may be written

$$p = \tfrac{3}{2}\mu Va\frac{\cos\theta}{r^2} + \text{constant};$$

which is an expression for the pressure containing the first power of the velocity.

3. We now proceed to find a solution in which terms involving the square of V are retained. Let the expressions for the resolved parts of the velocity, namely for u, v, and that for P, which satisfy the conditions of the problem, be written u_0+u_1, v_0+v_1, P_0+P_1; where u_0, v_0, P_0 are the expressions found above for u, v, P, which involve only the first powers of V, and u_1, v_1, P_1 involve V^2 as a factor. Also ζ will then split up into two parts $\zeta_0+\zeta_1$, where ζ_0 has been found above and ζ_1 depends on V^2.

The boundary conditions which u_1, v_1 are to satisfy, if possible, are that (1) they are to vanish at the surface of the sphere, (2) they are to vanish at an infinite distance from the sphere; they must also (though this is not a 'boundary' condition) be finite at each point of space occupied by the fluid. If u_1, v_1 satisfy these conditions ζ_1 must vanish at infinity; we also introduce the condition respecting it, that it be finite at each point of space occupied by the fluid. The boundary condition to be satisfied by P_1 is that it is to be constant at infinity, but an arbitrary constant has

1. Horace Lamb, *A Treatise on the Mathematical Theory of the Motion of Fluids* (Cambridge: Cambridge University Press, 1879).

already been introduced in P_0; therefore, without loss of generality, we may say that P_1 is to vanish at infinity.

We have now to satisfy equations (*A*), (*B*), (*C*). Since u_0, v_0, ζ_0 satisfy equations (*B*), (*C*) when substituted in them |147| for u, v, ζ respectively, and also since u_0+u_1, v_0+v_1, $\zeta_0+\zeta_1$ have to satisfy the same equation, it follows that u_1, v_1, ζ_1 satisfy these equations. Again, with reference to the equations (*A*), the terms in it involving only the first power of *V* satisfy them by themselves, and we are neglecting the cubes of *V*; therefore, on the left-hand sides of equations (A), u_0, v_0, ζ_0 may be substituted for u, v, ζ, and on the right-hand sides ζ_1, P_1 for ζ, P. Making these substitutions and multiplying out the products $u_0\zeta_0$, $v_0\zeta_0$ on the left-hand sides, equations (*A*) become

$$\left.\begin{aligned}\tfrac{3}{2}V^2\left[-\tfrac{1}{4}\frac{a^4}{r^5}-\tfrac{3}{4}\frac{a^2}{r^3}+\frac{a}{r^2}\right]\sin^2\theta&=\frac{dP_1}{dr}+2\frac{\mu}{\rho}\left[\frac{1}{r}\frac{d\zeta_1}{d\theta}+\frac{\zeta_1\cot\theta}{r}\right]\\ -\tfrac{3}{2}V^2\left[-\tfrac{1}{2}\frac{a^4}{r^5}+\tfrac{3}{2}\frac{a^2}{r^3}-\frac{a}{r^2}\right]\sin\theta\cos\theta&=\frac{dP_1}{rd\theta}-2\frac{\mu}{\rho}\left[\frac{d\zeta_1}{dr}+\frac{\zeta_1}{r}\right]\end{aligned}\right\}\cdots\cdots\cdots(E).$$

By simple differentiation and addition of these equations so as to eliminate ζ_1 we find

$$\frac{d^2P_1}{dr^2}+\frac{2}{r}\frac{dP_1}{dr}+\frac{1}{r^2}\frac{d^2P_1}{d\theta^2}+\frac{\cot\theta}{r}\frac{dP_1}{d\theta}=f(r)+\phi(r)P_2(\cos\theta),$$

where $\qquad P_2(\cos\theta)$ stands for $\tfrac{1}{2}(3\cos^2\theta-1)$,

$$f(r)\text{ for }\tfrac{3}{4}V^2\left[\frac{a^4}{r^6}+\frac{a^2}{r^4}\right],$$

$$\phi(r)\text{ for }3V^2\left[\tfrac{1}{4}\frac{a^4}{r^6}-\tfrac{7}{4}\frac{a^2}{r^4}+\frac{a}{r^3}\right].$$

To find a particular integral of this equation put

$$P_1=x+yP_2(\cos\theta),$$

x and *y* being functions of *r* only. Then remembering that the equation above is really Poisson's equation, the particular integral is found by choosing particular integrals of the equations

$$\frac{d^2x}{dr^2}+\frac{2}{r}\frac{dx}{dr}=f(r)=\tfrac{3}{4}V^2\left[\frac{a^4}{r^6}+\frac{a^2}{r^4}\right],$$

$$\frac{d^2y}{dr^2}+\frac{2}{r}\frac{dy}{dr}-6\frac{y}{r^2}=\phi(r)=3V^2\left[\tfrac{1}{4}\frac{a^4}{r^6}-\tfrac{7}{4}\frac{a^2}{r^4}+\frac{a}{r^3}\right].$$

|148| These equations will be found to be satisfied by[1]

1. In the first equation below, the original erroneously had $\frac{a^4}{r^2}$ instead of $\frac{a^4}{r^4}$. Also, in the second equation below, the original had $\frac{2\ 1}{1\ 6}$ instead of $\frac{21}{16}$. This space in the middle of the line was likely a

$$x = V^2\left[\tfrac{1}{16}\frac{a^4}{r^4} + \tfrac{3}{8}\frac{a^2}{r^2}\right],$$

$$y = V^2\left[\tfrac{1}{8}\frac{a^4}{r^4} + \tfrac{21}{16}\frac{a^2}{r^2} - \tfrac{1}{2}\frac{a}{r}\right].$$

The complementary function which satisfies Laplace's equation is shown by the symmetry of the motion to depend on zonal harmonics only; we will write it

$$\Sigma p_n \frac{a^{n+1}}{r^{n+1}} P_n(\cos\theta),$$

where p_0, p_1, &c. are undetermined constants. Then grouping together the coefficients of each of the functions $P_n(\cos\theta)$, the final solution of the equation is

$$P_1 = V^2\left[\tfrac{1}{16}\frac{a^4}{r^4} + \tfrac{3}{8}\frac{a^2}{r^2} + p_0\frac{a}{r}\right]$$
$$+V^2\left[\tfrac{1}{8}\frac{a^4}{r^4} + p_2\frac{a^3}{r^3} + \tfrac{21}{16}\frac{a^2}{r^2} - \tfrac{1}{2}\frac{a}{r}\right]P_2(\cos\theta) + V^2\Sigma p_n \frac{a^{n+1}}{r^{n+1}}P_n(\cos\theta).$$

Substitute this value of P_1 in equations (E), and remember that $3\sin^2\theta = 2\{1 - P_2(\cos\theta)\}$. The equations to determine ζ_1 become

$$\frac{2\mu}{\rho}\left[\frac{d\zeta_1}{rd\theta} + \frac{\zeta_1\cot\theta}{r}\right] = (p_0+1)V^2\frac{a}{r^2}$$
$$+V^2\left[\tfrac{3}{4}\frac{a^4}{r^5} + 3p_2\frac{a^3}{r^4} + \tfrac{27}{8}\frac{a^2}{r^3} - \tfrac{3}{2}\frac{a}{r^2}\right]P_2(\cos\theta)$$
$$+V^2\Sigma(n+1)p_n\frac{a^{n+1}}{r^{n+2}}P_2(\cos\theta),$$

$$\frac{2\mu}{\rho}\left[\frac{d\zeta_1}{dr} + \frac{\zeta_1}{r}\right] = -V^2\left[\tfrac{9}{8}\frac{a^4}{r^5} + 3p_2\frac{a^3}{r^4} + \tfrac{27}{16}\frac{a^2}{r^3}\right]\sin\theta\cos\theta$$
$$+V^2\Sigma p_n\frac{a^{n+1}}{r^{n+2}}\frac{d}{d\theta}P_n(\cos\theta).$$

The first of these two equations is a differential equation with θ as the independent variable. A particular integral can be easily found by noting the relations

$$\frac{d\cot\theta}{d\theta} + \cot^2\theta = -1,\quad \left[\frac{d}{d\theta} + \cot\theta\right]\frac{d}{d\theta}P_n(\cos\theta) = -n(n+1)P_n(\cos\theta).$$

|149| The particular integral is then found to be

typesetting limitation rather than an actual error. The compositor would have had a complete set of single-numeral fractions as single pieces of type and would juxtapose them to make up those which involved more than one numeral, top or bottom. This would leave a short letter-space between the two. We have closed up this space here, and have done so silently for subsequent instances (they are still recorded in the back-of-book record).

$$\frac{2\mu}{\rho}\zeta_1 = -(p_0+1)V^2\frac{a}{r}\cot\theta$$

$$+V^2\left[\tfrac{3}{8}\frac{a^4}{r^4}+\tfrac{3}{2}p_2\frac{a^3}{r^3}+\tfrac{27}{16}\frac{a^2}{r^2}-\tfrac{3}{4}\frac{a}{r}\right]\sin\theta\cos\theta$$

$$-V^2\Sigma\frac{1}{n}p_n\frac{a^{n+1}}{r^{n+1}}\frac{d}{d\theta}P_n(\cos\theta).$$

This is also a particular integral of the second equation. The complementary function which satisfies both equations is $\frac{A}{r}\operatorname{cosec}\theta$.

But ζ_1 is to be finite at all points of space occupied by the fluid, and therefore at points along the axis of z where $\theta=0$. To fulfil this condition we must put $p_0+1=0$, $A=0$. Therefore our equation for ζ_1 is

$$\frac{2\mu}{\rho}\zeta_1 = V^2\left[\tfrac{3}{8}\frac{a^4}{r^4}+\tfrac{3}{2}p_2\frac{a^3}{r^3}+\tfrac{27}{16}\frac{a^2}{r^2}-\tfrac{3}{4}\frac{a}{r}\right]\sin\theta\cos\theta - V^2\Sigma\frac{1}{n}p_n\frac{a^{n+1}}{r^{n+1}}\frac{d}{d\theta}P_n(\cos\theta).$$

We have now to find the radial and transverse velocities u_1, v_1 by the aid of the equations (*B*), (*C*); namely

$$\frac{du_1}{dr}+\frac{2u_1}{r}+\frac{1}{r}\frac{dv_1}{d\theta}+\frac{v_1\cot\theta}{r}=0 \quad\text{.......................}(B),$$

$$\frac{dv_1}{dr}+\frac{v_1}{r}-\frac{1}{r}\frac{du_1}{d\theta}=2\zeta_1 \quad\text{.........................}(C).$$

To satisfy these equations we substitute

$$u_1 = xP_2(\cos\theta)+\Sigma x_nP_n(\cos\theta),$$

$$v_1 = y\sin\theta\cos\theta+\Sigma y_n\frac{d}{d\theta}P_n(\cos\theta),$$

where the x's and y's are functions of r only. Then remembering the differential equation satisfied by $P_n(\cos\theta)$, we see that equation (*B*) can be satisfied by assuming that x, y satisfy the equation

$$\frac{1}{r}\frac{d}{dr}(r^2x)+2y=0 \quad\text{............................}(F),$$

and that x_n, y_n satisfy the equation

$$\frac{1}{r}\frac{d}{dr}(r^2x_n)-n(n+1)y_n=0 \quad\text{.......................}(G).$$

|150| Also equation (*C*) can be satisfied by assuming that

$$\frac{d}{dr}(ry)+3x=\frac{\rho V^2}{\mu}\left[\tfrac{3}{8}\frac{a^4}{r^3}+\tfrac{3}{2}p_2\frac{a^3}{r^2}+\tfrac{27}{16}\frac{a^2}{r}-\tfrac{3}{4}a\right] \quad\text{...............}(H),$$

and that

$$\frac{d}{dr}(ry_n)-x_n=-\frac{1}{n}p_n\frac{\rho V^2}{\mu}\frac{a^{n+1}}{r^n} \quad\text{.......................}(K).$$

Equation (H) can be written for the sake of conciseness in the form

$$\frac{d}{dr}(ry)+3x=\Sigma\frac{\lambda}{r^{\alpha}}.$$

Eliminating y between this and equation (F), we find

$$\frac{d^2}{dr^2}\left(r^2x\right)-6x=-2\Sigma\frac{\lambda}{r^{\alpha}}.$$

The solution of this equation is

$$x=\frac{A}{r^4}+Br-2\Sigma\frac{\lambda}{(\alpha-4)(\alpha+1)}\frac{1}{r^{\alpha}},$$

where we notice that none of the special values of α included under the sign Σ are 4 or -1. But x must not be infinite at a great distance from the sphere; this condition gives B=0. Replace A by $\frac{\rho V^2}{\mu}ca^5$, and put the special numerical values for λ, α; then x may be written

$$x=\frac{\rho V^2}{\mu}\left[c\frac{a^5}{r^4}+\tfrac{3}{16}\frac{a^4}{r^3}+\tfrac{1}{2}p_2\frac{a^3}{r^2}+\tfrac{9}{16}\frac{a^2}{r}-\tfrac{3}{8}a\right],$$

where c, p_2 are arbitrary constants which must be determined.

Equation (F) now determines y by mere differentiation, giving

$$y=\frac{\rho V^2}{\mu}\left[c\frac{a^5}{r^4}+\tfrac{3}{32}\frac{a^4}{r^3}-\tfrac{9}{32}\frac{a^2}{r}+\tfrac{3}{8}a\right].$$

Equations (G) and (K) determine x_n and y_n.

Eliminating y_n, we find

$$\frac{d^2}{dr^2}\left(r^2x_n\right)-n(n+1)x_n=-(n+1)p_n\frac{\rho V^2}{\mu}\frac{a^{n+1}}{r^n}.$$

The complementary function of this equation is

$$\lambda_n\frac{\rho V^2}{\mu}\frac{a^{n+3}}{r^{n+2}}+Br^{n-1}.$$

|151| But x_n is to vanish at an infinite distance from the sphere; therefore we may at once put B=0.

The solution of the equation may now be written

$$x_n=\frac{\rho V^2}{\mu}\left[\lambda_n\frac{a^{n+3}}{r^{n+2}}+\frac{n+1}{2(2n+1)}p_n\frac{a^{n+1}}{r^n}\right].$$

Equation (G) gives y_n by means of differentiation only, when the expression for x_n is known; we deduce

$$y_n=-\frac{\rho V^2}{\mu}\left[\frac{\lambda_n}{n+1}\frac{a^{n+3}}{r^{n+2}}+\frac{n-2}{2n(2n+1)}p_n\frac{a^{n+1}}{r}\right].$$

The arbitrary constants which occur in x and y, and also in each pair of the set x_n, y_n, must be determined by the condition that all the x's and y's are to vanish at the surface of the sphere, that is, when $r=a$.

The condition $x=0$, when $r=a$, gives

$$c+\tfrac{1}{2}p_2+\tfrac{3}{8}=0.$$

The condition $y=0$, when $r=a$, gives

$$c+\tfrac{3}{16}=0.$$

Hence $\qquad c=-\tfrac{3}{16},\ p_2=-\tfrac{3}{8}.$

The condition $x_n=0$, when $r=a$, gives

$$\lambda_n+\frac{n+1}{2(2n+1)}p_n=0.$$

The condition $y_n=0$, when $r=a$, gives

$$\frac{\lambda_n}{n+1}+\frac{n-1}{2n(2n+1)}p_n=0.$$

These equations involve $\lambda_n=0$, $p_n=0$; and therefore $x_n=0$, $y_n=0$ for all values of *r*.

We have now completely determined values of u_1 and v_1, which satisfy equations (*B*) and (*C*), namely

$$u_1=-\tfrac{3}{16}\frac{\rho V^2}{\mu}\left[\frac{a^5}{r^4}-\frac{a^4}{r^3}+\frac{a^3}{r^2}-3\frac{a^2}{r}+2a\right]P_2(\cos\theta)$$

$$=-\tfrac{3}{16}\frac{\rho V^2 a}{\mu}\left(\frac{a}{r}-1\right)^2\left(\frac{a^2}{r^2}+\frac{a}{r}+2\right)P_2(\cos\theta),$$

$$v_1=-\tfrac{3}{16}\frac{\rho V^2}{\mu}\left[\frac{a^5}{r^4}-\tfrac{1}{2}\frac{a^4}{r^3}+\tfrac{3}{2}\frac{a^2}{r}-2a\right]\sin\theta\cos\theta$$

$$=-\tfrac{3}{16}\frac{\rho V^2 a}{\mu}\left(\frac{a}{r}-1\right)\left(\frac{a^3}{r^3}+\tfrac{1}{2}\frac{a^2}{r^2}+\tfrac{1}{2}\frac{a}{r}+2\right)\sin\theta\cos\theta.$$

|152| These expressions for u_1 and v_1 will not serve as the required solution of the problem. They do not vanish at an infinite distance from the sphere, but become

$$u_1=-\tfrac{3}{8}\frac{\rho V^2 a}{\mu}P_2(\cos\theta),$$

$$v_1=\tfrac{3}{8}\frac{\rho V^2 a}{\mu}\sin\theta\cos\theta.$$

Also by reason of the presence of these terms which do not vanish at infinity the energy dissipated per unit of time would be infinite if this solution stood.

Can complementary functions be added to our solutions of the equations (*B*) and (*C*) which would cut out these terms? That is to say, a solution is to be found which satisfies equation (B) and equation (C) with its right-hand side, i.e. ζ, put equal to zero. In other words, we are to find an irrotational motion which vanishes at the surface of the sphere $r=a$, and at an infinite distance from the sphere becomes

$$u = \tfrac{3}{8}\frac{\rho V^2 a}{\mu} P_2(\cos\theta),$$

$$v = -\tfrac{3}{8}\frac{\rho V^2 a}{\mu} \sin\theta\cos\theta.$$

But this is obviously impossible since these expressions which are to hold at a great distance from the sphere are not themselves derivable from a potential function. Hence no second approximation to the solution of the equations of motion is possible in this problem consistent with the 'angular velocity' of the fluid being finite at each point of space. But the angular velocity becoming infinite at all points on a certain surface is just the condition which is required for the formation of surfaces of discontinuity and of dead water. Thus our analysis, when carried to a second approximation, points towards—though it does not actually prove—the very state of things which occurs in nature, namely the formation of a mass of dead water behind the moving body.

The Fens as Seen from Skates

1891 *Cambridge Review*, vol. 12, no. 297 (February), pp. 212–213

This essay appeared five days after Whitehead's thirtieth birthday and a little more than two months after his marriage to Evelyn Wade. Like 'The Clerk of the Weather' and 'Davy Jones' (both from 1886), this essay is simply signed 'W.' However, unlike those essays, 'The Fens' does not appear among the list of Whitehead's works in the official 1941 bibliography of Whitehead's works by Victor Lowe in the Library of Living Philosophers volume that Whitehead was likely to have reviewed.[1] *However, 'The Fens' is attributed to Whitehead by Lowe in a 1977* Process Studies *article.*[2] *Despite this attribution, eight years later (1985) Lowe did not include 'The Fens' in his biography and its list of Whitehead's works, despite including those earlier (1886) articles. The omission of 'The Fens' goes unmentioned by Lowe. In the end, the authorship of this essay remains uncertain and scholars should hesitate to attribute it to Whitehead without proper qualification and context.*

This essay captures the author's remembrances of what it is like to experience the Fens while on ice skates. A twenty-first-century author describes them as follows:

> *The little that remains of the fenlands of East Anglia provide only a glimpse of what this desolate marsh area was like for centuries. Since the age of the Romans, the fens were a land of tufts of solid ground rising above shallow water, reed-beds, pools and mires. The rivers of the Welland, Witham, Glen, Nene, Cam, Great Ouse, and Little Ouse made of the fens a vast inland waterway navigable only in shallow-bottomed boats, the forerunners of today's punts. . . . The Saxons founded a series of isolated monasteries on islands in the fens. Ely is one such island, and its name is a reminder of the area's rich sea life; Ely translates as 'island of eels'.*
>
> *Elizabeth I was interested in draining the fens to provide for agriculture, but it was left to the Duke of Bedford in the 17th century to take matters in hand. . . . The Great Fen reclamation drained over 40,000 acres of land but had the unexpected consequence of causing the neighbouring farmland to sink. This happened when the peat topsoil dried out and settled. In some cases land sunk up to 20 feet below the level of the cuts. Today the old fens survive in isolated pockets.*[3]

---

1. Paul Schilpp (ed.), *The Philosophy of Alfred North Whitehead* (Evanston: Northwestern University Press, 1941), p. 705.
2. In 1977 the journal *Process Studies* published what was then a newly discovered essay by Whitehead called 'Liberty and the Enfranchisement of Women'. The first footnote attached to the article by the editor says 'We are indebted to Victor Lowe for calling this item to our attention. During recent years Lowe has learned of several previously unidentified works of Whitehead besides this one: . . . "The Fens as Seen from Skates,"' (Alfred North Whitehead, 'Liberty and the Enfranchisement of Women', *Process Studies*, vol. 7, no. 1 (spring 1977), p. 39).
3. David Ross, 'The Cambridgeshire Fens', Britain Express website, <https://www.britainexpress.com/counties/cambridgeshire/fens.htm> (accessed January 20205).

If it has done nothing else, this winter has given us a haul over our fathers and grandfathers, who have of recent years had the monopoly of yarns about "the days when they skated to Ely," or "when all the fens were frozen;" but one is inclined to think that they made more use of such frosts and of the fen ice than we do in our day: a fact in great measure due no doubt to the growth of figure skating and the use of acmes, the first of which is an entirely different pursuit to "running," and the second the most tiring weapon with which to tackle long distances, and the obstructions one is bound to meet in a day's going. But the fen country is as full of interest—physical, historical, and architectural—as ever. The great |213| length and the octagon at Ely,[1] the magnificent Norman at Peterborough,[2] the romance (and the crow tribe) which circles round the now solitary tower of Crowland,[3] or the splendid west front of Thorney,[4] still reign and spread their influence across the icebound dykes and drains; and here and there, from many a fine tower, still chimes a merry peal of a Sunday night, till all the air is full of the cheerful music, or down the long silent reaches of the Hundred Foot[5] comes on the wind no sound but that of the ring of the skates of a market gang two miles away upon the ice.

And the fenmen are not one whit less cheerful or less old-fashioned than they were fifty years ago, although big engines have to a great extent replaced the groups of windmills which used to dot the country; although under their pumping the fen land has in many places sunk as much as six feet in the past twenty years, and although ague[6] is one of those evils of former days on which the oldest inhabitant will dilate to a Saturday night gathering round the big peat fire. Yet still they talk of "buskins" and "pattens"; still their greatest enemy is the flood; still the ice becomes their highway to market or to church.

And it is seldom a winter passes but the drains and dykes present a goodly reach of ice to the fen runners; for, as the meteorologist[7] knows, the thermometer at Cambridge reads two or three degrees lower than any other place whenever it gets a chance, and away north-eastward in the Bedford Level[8] the mercury will show, as a rule, even a lower temperature; and so it is that a fortnight or so of frost is little recked of, but is rather a matter of course; an event to be expected each winter by the fenmen, and with cheery faces young and old strap on their pattens, nor grumble a bit to see old Jack Frost again. And if his grip become too firm, distress begins to show his ugly face in some lonely fen-bound cottages, why then the parish has a fine

1. The fourteenth-century Octagon at Ely Cathedral.
2. The tower of the Peterborough Cathedral, which was rebuilt in the Norman style in the twelfth century.
3. The tower of the twelfth-century Crowland Abbey.
4. Thorney Abbey, on the island of Thorney in the Fens, was originally completed in the early twelfth century, and later restored in 1638 as the Church of St Mary and St Botolph.
5. Known today as the 'Ouse Washes', this is a 'floodplain known for more than three and a half centuries as the "Hundred Foot Washes" . . . a long, narrow strip of land through which the waters of the Great Ouse River flow in various man-made channels en-route to the sea' (Peter Cox, 'My Ouse Washes history and management research project', <http://www.ousewashes.info/>, accessed January 2025).
6. 'Ague' here likely means 'a fit of shivering'.
7. Spelling corrected from 'meterologist'.
8. The town of Bedford, which lies west by south-west of Cambridge, is the historic county town of Bedfordshire and was founded at a ford on the River Great Ouse. The Bedford Level experiment was carried out along the Old Bedford River in the mid-1800s to measure the curvature of the Earth.

day's outing—a kind of grand regatta—and all sorts and conditions turn to and race together on the well-swept course; and the handicapping produces plenty of good sport, even among the veterans who have "most done wi' running;" and home the victors go, laden with loaves of bread, and the vanquished too, for that matter, so long as they are in need, and for a fortnight afterwards things cannot go wrong, even in the poorest cottage, and many's the yarn spun of that day's races and the champions it brought forth.

Thus every village has its running team, which is in just as great honour as its cricket eleven, and just as much subject of discussion: thus upon the ice everyone is your fellow, and is pleased to pull up and inquire as to the distance you have run that morning, your destination, or things in general—from the veriest urchin to the old skating champion, from the lock-keeper to the parson; and, should you be going the same way, they will fall in behind you and swing away, step for step with you, for many a mile, and very pleasant often do they make their company. Nowhere but at sea does one feel such interest in those one passes. When you first see each other six miles off, it is wonderful with what an old-friend sort of greeting it is you meet, or with what regret you bid your man farewell, and hear the skates singing further and further away up the Long Reach,[1] till once again their owner is a small speck miles away in the eye of the setting sun. And with darkness comes many a tumble in cat-ice, or open hole, and perhaps a wish or two for one's snug rooms at Cambridge. But all is made up for by the warm welcome you never fail to get at the Inn, and a supper of homemade sausages and a yarn from the landlord over his grog must cure all weariness.

1. This may be referring to the village of Reach on the edge of the fenland in east Cambridgeshire at the north end of Devil's Dyke.

On Ideals: With Reference to the Controversy Concerning the Admission of Women to Degrees in the University

1896 *Cambridge Review*, vol. 17 (14 May), pp. 310–311.

Like 'Clerk of the Weather' and 'Davy Jones,' this essay was simply signed 'W.', making its authorship less than completely certain. Whitehead's daughter, Jessie, was confident of her father's authorship of the two aforementioned essays, but could say of this one only that it was 'probably' by her father.[1] *Whitehead's biographer, Victor Lowe, is more confident, writing that 'Its content, I think, makes the probability that some other "W" wrote it almost nil, so I rule that out.'*[2] *The topic is well described by the subtitle. As the author notes in the essay, the university already had for more than two decades (since 1869) maintained a residential college, Girton College, to educate women. Whitehead himself had taught at Girton since 1888, likely for no pay, a practice he continued until 1913.*[3] *Lowe notes that five years after this letter appears (in 1901) Whitehead became associated with Cambridge's second women's college, Newnham College, and served on its Council from 1907 until 1911.*[4]

Despite Cambridge being one of the first universities to create a college for women, women were not eligible for a degree. The author of this essay is joining what was at this time a vigorous debate as to whether to change this policy. The debate came to a head a year and a week after the essay appeared, on 21 May 1897, when a historic vote was taken on whether to grant women degrees. The Senate vote was 1,707 to 661 against.[5] *A celebratory 'riot' broke out among men, with some students burning effigies of female students. Cambridge was the last British university to grant women degrees, not doing so until 1948.*[6]

In this controversy, there is heard, as it were in an undercurrent, a murmur of ideals. Reference is made to the ideal of education, to the ideal of womanhood, to the ideal of things in general.

1. Victor Lowe, *Alfred North Whitehead: The Man and His Work, Volume I: 1861–1910* (Baltimore: Johns Hopkins University Press, 1985), p. 215.
2. Lowe, *Alfred North Whitehead, Volume I*, p. 215.
3. See Lowe, *Alfred North Whitehead, Volume I*, pp. 160 and 217. As Lowe notes, 'Since he [Whitehead] moved to London in 1910, the implication is that for three years he took the train to Cambridge at least one day a week to teach women' (p. 217).
4. Lowe, *Alfred North Whitehead, Volume I*, p. 217.
5. Lowe, *Alfred North Whitehead, Volume I*, p. 217.
6. Stuart Roberts, 'The Rising Tide: Women at Cambridge', University of Cambridge, <https://www.cam.ac.uk/stories/the-rising-tide> (accessed January 2025).

But one side alone seems to use the word – to claim to guide and justify its action on ideal and elevated grounds.

Those who propose a separate, an "Imperial University", for Women, desire it, they say, because women will there be free to follow an ideal course of education, based on consideration of their truest and highest interests. To |310b| these advocates of an Imperial University there appears nothing ideal in the persistent, irritating, and stupid clinging of women to the University of Cambridge – the desire to have the same degrees as men, to tie themselves down to a course of study and to a system of education which no one has ever attempted to call ideal, or even satisfactory.

In various tones, sometimes impatiently, but for the most part kindly and politely, those who oppose the admission of women to the University say:– "We speak in your highest, your most sacred interests, when we tell you that it is much better for you not to be admitted to the University. Please start a nice University of your own somewhere else. Never mind about having no money, no teachers, no standards; you will have freedom there to follow a great ideal. We are not free here: we are hampered by traditions, which make our system clumsy, which make us move slowly and fall into terrible mistakes, and we are afraid we are likely to make more mistakes in the future. We should like to see your education unfettered. We will not cast you off entirely; we will help you; you shall have our advice in determining what is most suitable for women's education; and we think it probable you will be a great success.

"Besides, to admit you to the University would be most inconvenient to us in many ways;–most injudicious, and probably very dangerous. Please, please go away."

To this the other side replies:–

"We, too, have our ideal. In clinging to you with all your faults, which you magnanimously point out, and which we quite recognize, in desiring to be admitted as members of your University, we aspire to share your great ideal, your high and imperishable standards of knowledge, of scholarship, and of research, without which no higher education, whether of men or women, is possible. We hope and believe that the highest knowledge, the truest scholarship, like the highest and divinest virtues, – Faith, Hope, and Charity – are the same for men and women; and that the admission of this by the University will have no tendency to confuse the sexes, or to interfere with the sanctity of domestic life.

"We hoped that during the twenty years we have been within your gates, during which time you have been most generous to us, that though we may have offended many prejudices, which we fear is always the lot of those who desire new things, yet that we had never seriously disturbed you, or caused you serious inconvenience, or lowered in any way the tone of University life, or its intellectual standards.

"We have regarded with reverence the sacred trust which you hand down through the generations to keep pure and undefiled the love of knowledge and the energy of research. We have breathed with awe the atmosphere peculiar to you as an ancient seat of learning; we have trembled with sensitiveness to your spirit and the finer breath with which you inspire and animate education. We feel that it is alive with the spirit and souls of the thousands of patient and single-minded scholars and students who have worked within your walls and under your protection, spending laborious days, perhaps in what the common sense of all ages would brand as futilities. From the days when it seemed to scholars important to determine how many angels could stand on the point of a needle, to these days when great minds are occupied in measuring by thousands the legs of crabs, the history of thought shows that it has always been your peculiar glory and privilege to encourage what seems to the worldly utilitarian, and therefore short-sighted, eye, a waste of time resulting in no achievement, no gain to the human race.

|311a| "We see that your blind instinct for knowledge, your love of learning as an end in itself, is what makes your influence so unique, so invaluable, your spirit so priceless in animating the higher education of the country. Without you where would be the standards, the ideals of education in a country? We do not fear that there will ever be, in our national life, an unhealthy want of that common sense which desires to subordinate knowledge as a means to the end of practical life. That is the growth of a day, and the possession of every school board and training college, but to you belongs alone the sense of the sacredness of knowledge and of the intellectual life which has grown up and been cherished within your walls throughout the ages, a cause which can count by hundreds its heroes and its martyrs among your scholars.

"It is from this spirit, you tell us, we should cut ourselves off, and rejoice that we do so.

"But it is not only a vague sentiment of reverence and enthusiasm for your spirit, that makes us regard with despair a separation from you, but the consciousness that without it, all higher education is but a name, and all secondary education worthless. We know that in the past the absence in our education of all standards, of all ideals lowered it to be the creature of circumstance, the sport of prejudice and caprice.

"One of the strongest opponents of the admission of women to the University – Prof. Marshall – says in his pamphlet on the subject: 'A generation ago there was much that was unsatisfactory in the position of women in England. The narrow and cruel customs which had debarred women, save of the working classes, from suitable methods of earning a livelihood were partly the cause and partly the effect of radical defects in their education. Few schoolmistresses even of expensive schools had been taught sufficiently well to be able to teach well. Their aims were often fairly high, but they were generally compelled to give their chief energies to imparting accomplishments: that is, forms of knowledge and art which ripened rapidly and were easily turned to account in general society, but did not develop thoroughly the best faculties of the learner. Girls often worked longer hours than their brothers, but their methods were bad: they made a business of trifles . . .' We would wish to speak with every respect of the enlightened ladies and gentlemen who will form the Council of the Imperial University: yet they are but human, and what is to insure that their standard of what women ought to be, their idea as to what are the highest and best interests of womanhood (which, it is implied, are the sole considerations worthy to be the basis of the scheme of education of an Imperial University) are truer, worthier, less fluctuating than of old, less coloured and blown about by every breath of prejudice? How can we be sure that they too may not incline towards the forms of knowledge that ripen quickly and are easily turned to account?

"We distrust a scheme of education based on considerations of our highest interests: we cling to those traditions of learning and knowledge which you say will merely hamper us. The mistakes you make with these are much less deadly, less dangerous than those we shall make without them.

"The highest interests of women will not probably be closer to the hearts of the Council of the Imperial University than were the highest interests of the whole human race to the hearts of the political enthusiasts of the French revolution. Both glory in their freedom to pursue great ideals, unhampered by the traditions of the past. Do you think that the Council of the Imperial |311b| University is likely to be more successful in framing a worthy and stable system of education than were those politicians in framing their ideal social and political constitutions? It is true they had the welfare of the whole human race to consider as the basis of their schemes: the Imperial University confines itself to considering the highest interests of only a section of humanity. But the future of that section is as obscure, its development as dependent on the working of the blind forces of nature, as little within the range of definite prophecy, as the future of the human race as a whole.

"There is, we feel, one reason which, though it might sadden us unfathomably, we should accept without a murmur, as inevitably closing against us the doors of the University. If you feel that it is proved beyond a doubt that our admission would lower the ideals and the tone, the soul and spirit of the University, if you think that during the twenty years we have been within your gates, we have shown, beyond a doubt, that we have no love of knowledge, no desire to pursue truth for its own sake, no capacities and no aspiration, that in admitting us you are traitors to your sacred and inherited trust to preserve untainted by the world, and undegraded by the desire of mere achievement, the love and pursuit of knowledge. If this is really so, then in the highest interests of humanity, call us not even *Hon. B.A.*, or *E.B.A.*, but banish us at once far from the precincts of the University.

"But we beg you very patiently, very seriously, very impartially, to consider if this be really so. And if, as we hope and believe, you find that experience does not prove it, but the contrary, and that it is on the possible dangers of the future that your anxious eyes are fixed, will you not consider whether it is not possible to guard in detail against the dangers as they arise, without denying us a share in the priceless benefits of your intellectual life and traditions."

So then, it appears that in this controversy both sides have ideals. If I seem to have laid undue stress on the ideals of one side, it is, I suppose, because these ideals were likely to attract small notice, being old and dim and obscure, compared to the brilliant and brand-new ones of the Imperial University for Women.

The Geodesic Geometry of Surfaces in Non-Euclidean Space

1898 *Proceedings of the London Mathematical Society*, vol. 29, pp. 275–324

A note just after the title of this paper specifies that it was 'Received and read March 10th, 1898'. Before its publication in the Proceedings, some material was added by Whitehead (p. 45, p. 74 and p. 81), as was an appendix by William Burnside[1] *(pp. 84–86), at Whitehead's invitation.*

The geometry of geodesics, as Whitehead often speaks of it in this paper (p. 44), lies within what one would call differential geometry *in the present day. Its most salient feature is its contrast with* synthetic geometry. *In the latter, one deals with axioms and postulates independently of any coordinate system, and this subject is much wider than the Euclidean geometry studied in secondary schools.*[2] *Differential geometry, in contrast, requires a coordinate system, calculus and other elements of more advanced mathematics, as described below.*

The shortest path between two points on a surface is always a geodesic, but the converse is not quite true: in general, we can say only that a geodesic is the shortest path between points that are arbitrarily close. The more intrinsic property is that the acceleration *of the path (see below) should be at right angles to the surface under investigation. So, for example, on a sphere, having selected two antipodal points (that is, a north pole and a south pole), such a curve could be a meridian line or the intersection of any plane containing the centre of the sphere with the sphere itself; one calls arcs of this kind great circles. Here we see clearly that given two nearby points, A and B, on the sphere – for instance, on its equator as defined by the selected poles – we have two geodesics connecting A and B, both along the equator: the shorter, more direct one, and the longer one that goes the other way around. The key here is that if we look at these great circles in ordinary three-dimensional space, they are not ordinary lines, but their defect is revealed in a turning – or, more technically, in an acceleration – everywhere at right angles to the surface of the sphere. In consequence, as paths* on the sphere, *they have no observable component of acceleration. If we now identify points with particles and accelerations with forces, a geodesic represents motion without any impinging force and this, critically, is exactly the interpretation taken in general relativity as far as we conceive gravity to be a distortion of Euclidean space by the distribution of masses in the universe. Hence the notion of a geodesic is truly of universal importance.*

1. William Burnside (1852–1927) was an English mathematician best known for his theories of finite groups. The year after this paper was published, Burnside would serve as a referee for Whitehead's paper 'Sets of Operations in Relation to Groups of Finite Order' and recommend against its publication (see p. 88).
2. Of note are the studies of projective and non-Euclidean synthetic geometries, to which Whitehead was no stranger. See, for instance, his encyclopaedia article 'Axioms of Geometry', p. 299.

The geometry of the sphere typifies one species of non-Euclidean geometry, called elliptic geometry. *Here we can see that the parallel postulate does not hold because any two distinct great circles intersect at two points. The other species is called* hyperbolic geometry; *in this case the parallel postulate fails in the complementary sense: given any straight line (geodesic) and a point* not *on that line, there are infinitely many distinct lines through said point that do not intersect the original line. The most naively accessible model for hyperbolic geometry is the* pseudosphere, *a radially symmetric surface that resembles the bell end of a musical instrument in the brass family.*

In investigating the geodesics of a surface, Whitehead follows and extends the work of Bolyai (on his p. 45) and Darboux (pp. 47, 58) to both elliptic and hyperbolic geometry. The work leans heavily on Whitehead's previous work in Universal Algebra *(UA), which he references throughout. UA is in turn based largely on Hermann Grassmann's 'Calculus of Extension', which we might speak of in the present day as the notion of a* Grassmann *or* exterior algebra. *Exterior algebras formalise the concept of oriented volume and thereby unify the classical theorems of multivariable calculus. The complementary (and less forbidding) notion of an* inner product *formalises the concepts of length and angle. Whitehead summarises both ideas in the first section of his paper (beginning on p. 47) in terms of the fundamental definitions and transformations.*

The emphasis on these formalisms is both powerful and revealing. They reduce much of the analysis to a computational system just as the identities learned in first-year algebra render the general rules for calculation that apply to the arithmetic of real numbers. The significance of Whitehead's summary is that he gives himself and others the framework in which to work via logical transformations that detach the calculations from their semantic weight. (This is both a liberation and a curse, with the latter aspect well reflected in the unforgiving attitude of many a beginning algebra student.) Supplementing this development with some formidable computations using comparatively elementary calculus, Whitehead obtains his stated objectives in extending previous work uniformly to both species of non-Euclidean geometry. Late in the paper, he looks at specific examples that both confirm and illustrate his calculations in a way reminiscent of his papers on fluid dynamics and very much in accord with the general principle of Cartesian synthesis as a necessary step following deduction in the acquisition of knowledge.

While differential geometry is not so austere as logic, formal set theory or even abstract algebra, the shift in emphasis here away from synthetic geometry and towards the formalisms mentioned above is in accord with Whitehead's early work on naive set theory and later work on logicism. His embrace of formal relations here suggests a trajectory in the direction of twentieth-century mathematics, with its enormous emphasis on informal axiomatic theories, a trajectory that was perhaps deflected by his absorption into the foundational work with Russell and his later discovery of and immersion in process metaphysics. Nonetheless, he did return to the foundational issues of analytic geometry and to a number of extremely technical formalisms in his book The Principle of Relativity: With Applications to Physical Science *(1922), which presented an alternative approach to Einstein's theory of gravity.*

Introduction.

The present paper is concerned with the properties of geodesic lines drawn on surfaces in elliptic and in hyperbolic space. There are two forms of elliptic space, which have been named by Burnside the "single" and the "double" forms. They have also been named the "polar" and "antipodal" forms. The antipodal (or double) form has also been called "spherical." In this paper the polar form of elliptic space will be the only one considered. Thus, both in the elliptic and the hyperbolic geometry of this paper, two straight lines in a plane intersect in one point only; and, in hyperbolic geometry, this point may lie in antispace (also called "ideal space").

Now the relations between the properties of geodesics on surfaces and non-Euclidean geometry, as far as they have hitherto been investigated, to my knowledge, are as follows:

It has been proved by Beltrami[1] that the "geodesic geometry" of surfaces of constant curvature in *Euclidean* space is the same as the geometry of straight lines in planes in elliptic or in hyperbolic space, according as the curvature of the surface is positive or negative.

The geometry of great circles on a sphere of radius ρ in elliptic space of "space-constant" γ is the same as the geometry of straight lines in planes in elliptic space of space-constant $\gamma \sin\frac{\rho}{\gamma}$.

The geometry of great circles on a sphere of radius ρ in hyperbolic space of "space-constant" γ is the same as the geometry of straight lines in planes in elliptic space of space-constant $\gamma \sinh\frac{\rho}{\gamma}$.

The geometry of geodesics (that is, lines of equal distance), on a surface of equal distance, σ, from a plane in hyperbolic space of space-constant γ, is the same as that of straight lines in planes in hyperbolic space of space-constant $\gamma \cosh\frac{\sigma}{\gamma}$.

Finally, the geometry of geodesics (that is, limit-lines), on a limit-surface in hyperbolic space—which may be conceived either as a sphere of infinite radius or as a surface of equal, but infinite, distance |276| from a plane—is the same as that of straight lines in planes in Euclidean space.

The preceding four propositions are due directly, or almost directly, to John Bolyai, though, of course, he only directly treats of hyperbolic space. I have given demonstrations of them elsewhere.[2†]

From the popularization of Beltrami's results by Helmholtz, and from the unfortunate adoption of the name "radius of space curvature" for γ (here called the space-constant), many philosophers, and it may be suspected from their language, many mathematicians, have been misled into the belief that some peculiar property of flatness is to be ascribed to Euclidean space, in that planes of other sorts of space can be represented as surfaces in it. This idea is sufficiently refuted, at least, as regards hyperbolic space, by Bolyai's theorem respecting the geodesic geometry of limit-surfaces. For a Euclidean plane can thereby be represented by a surface in hyperbolic space.

---

1. Eugenio Beltrami (1835–1900) was an Italian mathematician best known for his work on differential geometry.
2. † *Cf. A Treatise on Universal Algebra*, § 262.

[*Added*, May, 1898.[1] It is to be noticed that the comparison of the lengths of the space-constants of different "spaces" is nonsense. A space-constant can only be said to be of length γ in comparison with the length of some arbitrary unit straight line *in that space*. No comparison exists between lines in different spaces. But we may compare space-constants when we are really, as above, discussing the geodesic geometry of surfaces in a space of three dimensions; for, in this case, the test of congruence can be applied so as to compare the length of a geodesic on one surface with the length of any other line in the complete space.

All non-Euclidean geometry can be interpreted as geodesic geometry (allowing space of four dimensions). The preceding argument of this preface is directed against the assumption that this is the necessary interpretation, the refutation being based on the fact that Euclidean geometry can also be interpreted as geodesic geometry in a non-Euclidean space.

In the application of geometry to the interpretation of phenomena the idea of a fourth dimension is useless, as we have no such intuition. But the comparison of the space-constant γ to our empirically given units, such as a mile or the earth's diameter, is the fundamental problem of applied geometry. In pure geometry a disembodied mind is contemplating a possible idea of space. To such a mind there are only three kinds of geometry: namely hyperbolic, Euclidean, elliptic (polar and antipodal). But in applied geometry there is the mini-|277|mum length which can be perceived—some small fraction of an inch, —and a maximum length—namely, the distance of the furthest fixed star. These are empirically given lengths, and the properties of space in relation to our experience will be different according to the ratio of the space-constant to some intermediate empirically given standard length, such as the Earth's diameter.

It has been stated that any value of the space-constant may be assumed, since the laws of nature can be altered to suit. This is untrue if the possibility of measuring lines apart from any assumption of a special geometry is allowed. Thus the ratio of the circumference of a circle of radius *r* to its diameter is $\pi\gamma \sin {}^{r}\!/\!_{\gamma}$, π, and $\pi\gamma \sinh {}^{r}\!/\!_{\gamma}$ in the three geometries respectively: accordingly, it is possible to determine γ by actual measurement without any dependence on the more hypothetical laws of nature. The practical accuracy which can be thus attained is not to the point; the possibility of the experiment proves that γ is not arbitrary.]

It is the object of this paper to extend and complete Bolyai's theorem by investigating the properties of the general class of surfaces in any non-Euclidean space, elliptic or hyperbolic, which are such that their geodesic geometry is that of straight lines in a Euclidean plane. Such surfaces are proved to be real in elliptic as well as in hyperbolic space, and their general equations are found for the case when they are surfaces of revolution. In hyperbolic space, Bolyai's limit-surfaces are shown to be a particular case of such surfaces of revolution. The surfaces fall into two main types, of which figures of meridian sections are given; the limit-surfaces form a transition case between these types.

---

1. That is, subsequent to the paper's original delivery on 10 March 1898.

In elliptic space there is only one type of such a surface of revolution; but it may have a finite or an infinite number of sheets according as a certain number is rational or irrational. The simplest surface of this type is one-sheeted, and a figure of it is given. Its equation, referred to the most convenient rectangular axes, is

$$3\left(\xi^2+\xi_3^2\right)^2\xi_3^2=3\left(\xi^2+\xi_3^2\right)\left(\xi^2+\xi_3^2\right)\left(\xi_1^2+\xi_2^2\right)+\left(2\xi^2-\xi_3^2\right)\left(\xi_1^2+\xi_2^2\right)^3.$$

The same principles, only more laborious in their application, would enable the problem to be solved of the discovery in any kind of space of surfaces with their "geodesic" geometry identical with that of planes in any other kind of space. The distinguishing characteristic of surfaces is found, but I have not worked out the problem in detail.

|278| As a preliminary to these investigations some general theorems are proved. If a family of geodesics and its orthogonal curves be traced on a surface in Euclidean geometry, Gauss[1] has proved that the formula for an element of arc becomes

$$(\delta\sigma)^2=(\delta\rho)^2+\Pi^2(\delta\phi)^2,$$

where the curves $\phi=$ constant are the geodesics, and the curves $\rho=$ constant are the orthogonal family. Such curvilinear coordinates are called in this paper "semi-geodesic orthogonal coordinates." Polar geodesic coordinates form a particular case of such coordinates. It is shown in this paper that all the well-known general properties of semi-geodesic orthogonal coordinates also hold for non-Euclidean space. In particular, the important equation

$$\frac{\partial^2\Pi}{\partial\rho^2}+\frac{\Pi}{\rho_1\rho_2}=0,$$

where $\frac{1}{\rho_1}$ and $\frac{1}{\rho_2}$ are the principal measures of curvature of the surface at the point, becomes in elliptic space

$$\frac{\partial^2\Pi}{\partial\rho^2}+\left(\frac{1}{\gamma^2}+\frac{1}{\rho_1\rho_2}\right)\Pi=0;$$

and in hyperbolic space

$$\frac{\partial^2\Pi}{\partial\rho^2}+\left(\frac{1}{\rho_1\rho_2}-\frac{1}{\gamma^2}\right)\Pi=0.$$

In this connexion it is necessary to note that in non-Euclidean geometry a distinction has to be drawn between the inverse of the measure of curvature of a curve, which may be called the radius of curvature, and the radius of the circle through three consecutive points on the curve, which may be called the radius of the circle of curvature. If $\delta\sigma$ be a small arc, and $\delta\varepsilon$ the angle between the tangents at its extremities, the radius of curvature is the limit of $\delta\sigma/\delta\varepsilon$ Also, if this be called κ, and μ be the radius of the circle of curvature, then it is easy to prove[2†] that

$$\kappa=\gamma\,\frac{\tan}{\tanh}\,\frac{\mu}{\gamma},$$

according as the space is elliptic or hyperbolic.

1. Johann Carl Friedrich Gauss (1777–1855) was a German mathematician, astronomer and physicist.
2. † *Cf. Universal Algebra*, § 288 (4).

|279| The condition that a curve drawn on a surface in non-Euclidean space may be a geodesic is investigated, and this condition is put into different forms. Thus, if any point on the surface be denoted by curvilinear coordinates θ and ϕ, the condition, that the curve

$$\theta = f(r), \quad \phi = F(\tau)$$

is a geodesic, is found in the form of a differential equation involving, $\theta, \phi, \dot{\theta}, \dot{\phi}, \ddot{\theta}, \ddot{\phi}$: this equation is the same equation as that which holds for Euclidean space. In particular, if ρ and ϕ be semi-geodesic orthogonal coordinates, and ρ be taken as the independent variable, the simplified form of the differential equation of a geodesic, involving ρ, ϕ, $\frac{d\phi}{d\rho}, \frac{d^2\phi}{d\rho^2}$, is found. This is also the same as the corresponding equation for Euclidean space.

Hence the proof given by Darboux[1] for Beltrami's theorem on the geodesic representation of surfaces on planes is found to hold for non-Euclidean space; thus the theorem is extended to non-Euclidean space. The theorem is as follows:

Surfaces of constant curvature are the only surfaces for which geodesics can be transformed into straight lines when the surface is represented point for point on a plane. Surfaces of constant curvature are here named elliptic, parabolic, or hyperbolic, according as in elliptic geometry $\frac{1}{\rho_1\rho_2} + \frac{1}{\gamma^2}$ is positive, zero, or negative, and according as in hyperbolic geometry $\frac{1}{\rho_1\rho_2} - \frac{1}{\gamma^2}$ is positive, zero, or negative.

The "geodesic" geometry of an elliptic surface of constant curvature would be that of a plane in elliptic geometry; of a hyperbolic surface of constant curvature would be that of a plane in hyperbolic geometry. But these cases are not here considered in detail. The case of parabolic surfaces of constant curvature is considered in detail, with the results already mentioned.

The elementary theorems respecting the curvature of surfaces in non-Euclidean geometry, and the application of Gauss' curvilinear coordinates to such surfaces, are given by Darboux, *Théorie générale des Surfaces*, Livre vii., chapter xiv.[2] I have also developed them by the same methods and notation as are used in this paper, in *A Treatise on Universal Algebra*, Book vi., chapter vii.

The theorems in this paper are proved by the use of Grassmann's[3] "Calculus of Extension." A sketch of the method of this Calculus, |280| especially in relation to non-Euclidean space, has been given by Buchheim[4] in the *Proceedings* of this Society; *cf.* Vol. xv.[5] The Calculus is also investigated in detail in the previously mentioned volume (*Universal Algebra*), and I have ventured to shorten my paper by reference to it for results assumed.

1. Jean-Gaston Darboux (1842–1917) was a French mathematician.
2. Jean-Gaston Darboux, *Leçons sur la théorie générale des surfaces et les applications géométriques du calcul infinitésimal, troisième partie* [*Lessons on the General Theory of Surfaces and Geometric Applications of Infinitesimal Calculus, Third Part*] (Paris: Gauthier-Villars, 1894).
3. Hermann Günther Grassmann (1809–1877) was a German linguist, physicist and mathematician.
4. Arthur Buchheim (1859–1888) was a British mathematician.
5. Arthur Buchheim, 'On the Theory of Screws in Elliptic Space', *Proceedings of the London Mathematical Society*, vol. 15 (1883), pp. 83–98.

The brevity and facility of reasoning gained by the use of Grassmann's methods are very great. They can be immediately transformed into ordinary equations; but, by their use, the theorems of non-Euclidean geometry can be proved analytically even more easily than those of Euclidean geometry by ordinary methods.

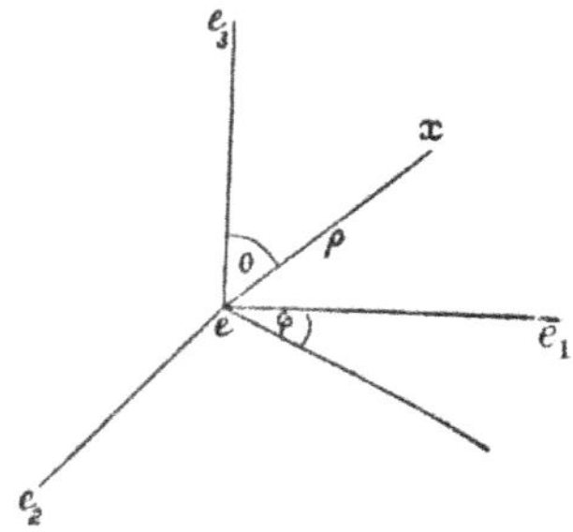

1. Let the small italic letters represent points. Thus, let e be the origin of coordinates, and ee_1, ee_2, ee_3 be three mutually rectangular axes. Also, let e, e_1, e_2, e_3 be a set of four quadrantal points; so that, if γ be the "space-constant," the distance between any two of these points is $\frac{1}{2}\pi\gamma$, being half the complete length of a straight line. Let x be any point, let the length ex be ρ, the angle xee_3 be θ, and the angle between the planes e_1ee_3 and xee_3 be ϕ. Thus ρ, θ, ϕ are analogous to the polar coordinates of Euclidean space. Also it is well known that, if we assume ξ, ξ_1, ξ_2, ξ_3, so that

$$\frac{\xi}{\cos\frac{\rho}{\gamma}}=\frac{\xi_1}{\sin\frac{\rho}{\gamma}\sin\theta\cos\phi}=\frac{\xi_2}{\sin\frac{\rho}{\gamma}\sin\theta\sin\phi}=\frac{\xi_3}{\sin\frac{\rho}{\gamma}\cos\theta},$$

then ξ, ξ_1, ξ_2, ξ_3 have analogous properties partly to rectangular Cartesian coordinates, and partly to quadriplanar coordinates in Euclidean space. All the equations can be made homogeneous, and equations of the first degree represent planes. When the above proportions are turned into equations, so that

$$\xi=\cos\rho/\gamma,\ \text{\&c.},$$

then ξ, ξ_1, ξ_2, ξ_3 will be called *the* rectangular coordinates of the point x. In this case they satisfy the equation

$$\xi^2+\xi_1^2+\xi_2^2+\xi_3^2=1.$$

|281| The symbols which represent points are combined according to the laws of Grassmann's "Calculus of Extension." Thus, if x and y be any two points,

$$xy=-yx \ \text{ and } \ xx=0$$

Also xy is called a "linear element" or a "force." Again, if z and u be two other points,

$$xyz=yzx=zxy=-yxz=\text{\&c.}$$

Also,
$$xxy=0.$$

A product of three points, such as xyz, is called a "planar element." Again,

$$(xyzu)=-(uxyz)=(zuxy)=\text{\&c.};$$

such a product is treated as an ordinary algebraic quantity, a mere number.

The relation of the point x to the points e, e_1, e_2, e_3 is expressed by the equation

$$x=\xi e+\xi_1e_1+\xi_2e_2+\xi_3e_3,$$

where ξ, ξ_1, ξ_2, ξ_3 are the quantities already defined by reference to the geometrical relations of x and the axes ee_1, ee_2, ee_3. Then the products xy, xyz, $(xyzu)$

can be expressed in terms of the products of e, e_1, e_2, e_3, taken respectively two together, three together, and four together. But the mention of $(ee_1e_2e_3)$ is avoided by the convention that in elliptic geometry

$$(ee_1e_2e_3)=1,$$

and in hyperbolic geometry

$$(ee_1e_2e_3)=\sqrt{-1}.$$

Thus, if

$$x=\xi e+\xi_1e_1+\xi_2e_2+\xi_3e_3,$$
$$y=\eta e+\eta_1e_1+\eta_2e_2+\eta_3e_3,$$
$$z=\zeta e+\zeta_1e_1+\zeta_2e_2+\zeta_3e_3,$$
$$u=\nu e+\nu_1e_1+\nu_2e_2+\nu_3e_3,$$

then

$$xy=(\xi\eta_1-\xi_1\eta)ee_1+(\xi\eta_2-\xi_2\eta)ee_2+\text{\&c.},$$

$$xyz=\begin{vmatrix}\xi, & \xi_1, & \xi_2\\ \eta, & \eta_1, & \eta_2\\ \zeta, & \zeta_1, & \zeta_2\end{vmatrix}ee_1e_2+\text{\&c.},$$

|282| and

$$(xyzu)=\begin{vmatrix}\xi, & \xi_1, & \xi_2, & \xi_3\\ \eta, & \eta_1, & \eta_2, & \eta_3\\ \zeta & \zeta_1, & \zeta_2, & \zeta_3\\ \nu & \nu_1, & \nu_2, & \nu_3\end{vmatrix}.$$

Accordingly $xyz=0$ denotes that x, y, and z are collinear, and $(xyzu)=0$ denotes that x, y, z, u are coplanar.

Also in elliptic geometry

$$(x|y)=\xi\eta+\xi_1\eta_1+\xi_2\eta_2+\xi_3\eta_3,$$

and in hyperbolic geometry

$$(x|y)=\xi\eta-\xi_1\eta_1-\xi_2\eta_2-\xi_3\eta_3.$$

Furthermore, the following formulæ always hold:—

$$(xy|zu)=(x|z)(y|u)-(x|u)(y|z),$$

$$(xyz|uvw)=\begin{vmatrix}(x|u), & (x|v), & (x|w)\\ (y|u), & (y|v), & (y|w)\\ (z|u), & (z|v), & (z|w)\end{vmatrix}.$$

The most important special cases of these formulæ arise in the evaluation of the forms $(xy|xy)$ and $(xyz|xyz)$.

If ξ, ξ_1, ξ_2, ξ_3 are the coordinates of the point x, then $(x|x)=1$. The point x is said to be at unit intensity. But, if ξ, ξ_1, ξ_2, ξ_3 are only proportionals to these coordinates, then $+\sqrt{(x|x)}$ is called the intensity of the point x. It will usually be convenient to work with points at unit intensity.

The polar plane of a point x with respect to the absolute is the planar element $|x$, the pole of a planar element P is $|P$, the polar line of the linear element xy is $|xy$. Also $\|x=-x$, $\|P=-P$, $\|xy=xy$, and $|xy=|x\cdot|y$.

These formulæ, and others, are more particularly explained in the work (*Universal Algebra*, Books IV. and VI.) already referred to. But the preceding explanations will enable any reader unacquainted with the calculus of extension to translate the reasoning into the language of ordinary algebra.

Hence, with the notation of the calculus of extension as applied to this subject (*cf. Universal Algebra*, Book IV., chapter iii., also §§204, 210), the distance $\overline{xy}$ between two points x and y is given by

$$\cos\frac{\overline{xy}}{\gamma}=\frac{(x|y)}{\sqrt{(x|x)(y|y)}},\quad \sin\frac{\overline{xy}}{\gamma}=\sqrt{\frac{(xy|xy)}{(x|x)(y|y)}}.$$

|283| 2. Now (*cf. Universal Algebra*, Book VI., chapter vii.) let x be any unit point on a surface, so that throughout the subsequent reasoning we assume $(x|x)=1$. Also let the position of x on the surface be defined, according to Gauss' method, by two curvilinear coordinates θ and ϕ, which are not to be identified with the θ and ϕ of the previous explanations respecting rectangular axes. Thus the equations $\theta=\text{constant}$ and $\phi=\text{constant}$ represent two families of curves traced on the surface, and the ordinary algebraic quantities which define x are functions of θ and ϕ.

Let x_1 stand for $\frac{\partial x}{\partial\theta}$, x_{11} for $\frac{\partial^2 x}{\partial\theta^2}$, x_2 for $\frac{\partial x}{\partial\phi}$, x_{12} for $\frac{\partial^2 x}{\partial\theta\,\partial\phi}$, x_{22} for $\frac{\partial^2 x}{\partial\phi^2}$, and so on for partial differential coefficients of a higher order. Thus, if x be written in the form

$$x=\xi e+\xi_1 e_1+\xi_2 e_2+\xi_3 e_3,$$

$$x_1=\frac{\partial\xi}{\partial\theta}e+\frac{\partial\xi_1}{\partial\theta}e_1+\frac{\partial\xi_2}{\partial\theta}e_2+\frac{\partial\xi_3}{\partial\theta}e_3,$$

and similarly for x_{11}, and so on. Then, by differentiating the equation

$$(x|x)=1,$$

which is to hold for all positions of x on the surface, we find

$$(x|x_1)=0=(x|x_2),$$

also

$$(x_1|x_1)+(x|x_{11})=0,$$

$$(x_1|x_2)+(x|x_{12})=0,$$

and so on.

The tangent plane at x is represented by the planar element xx_1x_2 [*cf. Universal Algebra*, Book IV., chapters i. and ii., and §293 (5)]; the normal at x is represented by the linear element $|x_1x_2$; the tangent lines to the curves $\phi=\text{constant}$ and $\theta=\text{constant}$ through x are xx_1 and xx_2 respectively.

The angle between these tangent lines [*cf. Universal Algebra*, §295 (2)] is

$$\cos^{-1}\frac{(x_1|x_2)}{\sqrt{(x_1|x_1)(x_2|x_2)}}.$$

The length $(\delta\sigma)$ of the element of arc joining the points x and $x+\delta x$ on the surface, where $x+\delta x$ corresponds to the values $\theta+\delta\theta$ and |284| $\phi+\delta\phi$ of the coordinates, is given (*cf. Universal Algebra*, §293) by

$$\frac{(\delta\sigma)^2}{\gamma^2}=(x_1|x_1)(\delta\theta)^2+2(x_1|x_2)\delta\theta\,\delta\phi+(x_2|x_2)(\delta\phi)^2.$$

Thus $(x_1|x_1)$, $(x_1|x_2)$, and $(x_2|x_2)$ are the quantities which in Gauss' theory of surfaces are usually denominated by E, F, and G, respectively. Also

$$H^2=EG-F^2=(x_1|x_1)(x_2|x_2)-(x_1|x_2)^2=(x_1x_2|x_1x_2).$$

The advantages of the forms $(x_1|x_1)$, &c., are that they are nearly as short as the single letters E, F, G, and at the same time they express completely the rules for their formation and transformation.

Also, if $\frac{1}{\rho_1}$ and $\frac{1}{\rho_2}$ be the principal *measures of curvature* of the surface at x, then [*cf. Universal Algebra*, §294 (3)]

$$\frac{\gamma^2\left\{(x_1|x_1)(x_2|x_2)-(x_1|x_2)^2\right\}^2}{\rho_1\rho_2}=\Delta_1-\Delta_2+\left\{(x_1|x_1)(x_2|x_2)-(x_1|x_2)^2\right\}$$
$$\times\left\{(x_1|x_2)^2-(x_1|x_1)(x_2|x_2)+(x_1|x_2)_{12}-\tfrac{1}{2}(x_2|x_2)_{11}-\tfrac{1}{2}(x_1|x_1)_{22}\right\}, \tag{1}$$

where

$$(x_1|x_2)_{12}=\frac{\partial^2}{\partial\theta\,\partial\phi}(x_1|x_2),$$

with similar meaning for the other terms of similar form; Δ_1 and Δ_2 stand for the determinants

$$\Delta_1=\begin{vmatrix}(x_1|x_1), & (x_1|x_2), & (x_1|x_2)_2-\tfrac{1}{2}(x_2|x_2)_1\\ (x_1|x_2), & (x_2|x_2), & \tfrac{1}{2}(x_2|x_2)_2\\ \tfrac{1}{2}(x_1|x_1)_1, & (x_1|x_2)_1-\tfrac{1}{2}(x_1|x_1)_2, & 0\end{vmatrix},$$

$$\Delta_2=\begin{vmatrix}(x_1|x_1), & (x_1|x_2), & \tfrac{1}{2}(x_1|x_1)_2\\ (x_1|x_2), & (x_2|x_2), & \tfrac{1}{2}(x_2|x_2)_1\\ \tfrac{1}{2}(x_1|x_1)_2, & \tfrac{1}{2}(x_2|x_2)_1, & 0\end{vmatrix}.$$

3. The equations $$\theta=f(\tau),\quad \phi=\mathrm{F}(\tau),$$

where τ is some independent variable, and f and F are functional signs, define a curve on the surface.

Let the point (at unit intensity) corresponding to the value $r+\delta r$ |285| of the variable be denoted by

$$x+\dot{x}\delta\tau+\tfrac{1}{2}\ddot{x}\delta\tau^2+\ldots;$$

then $$\dot{x} = \frac{dx}{d\tau}, \; \ddot{x}\frac{d^2x}{d\tau^2},$$

and so on. Hence

$$\dot{x} = x_1\dot{\theta} + x_2\dot{\phi},$$
$$\ddot{x} = x_1\ddot{\theta} + x_2\ddot{\phi} + x_{11}\dot{\theta}^2 + 2x_{12}\dot{\theta}\dot{\varphi} + x_{22}\dot{\phi}^2.$$

Also the length of an element of arc $(\delta\sigma)$ on the curve is given by

$$\left(\frac{\delta\sigma}{\gamma}\right)^2 = (\dot{x}|\dot{x})\delta\tau^2.$$

Thus, if τ denote the length of the arc of the curve measured from a fixed point, then in this special case $$(\dot{x}|\dot{x}) = \frac{1}{\gamma^2},$$

since $$\delta\tau = \delta\sigma.$$

The osculating plane of the curve is [*cf. Universal Algebra*, §287 (6)] $x\dot{x}\ddot{x}$; the tangent line is $x\dot{x}$.

Let ω be the angle between the normal section at x containing the tangent line $x\dot{x}$ and the osculating plane of the curve. Then the angle between the tangent plane at x and the osculating plane at x is $\frac{1}{2}\pi - \omega$. Thus (*cf. Universal Algebra*, §221)

$$\sin\omega = \frac{(x\dot{x}\ddot{x}|xx_1x_2)}{(x\dot{x}\ddot{x}|x\dot{x}\ddot{x})(xx_1x_2|xx_1x_2)}.$$

Now, if $\frac{1}{\kappa}$ be the measure of curvature of the curve, that is $\frac{\delta\varepsilon}{\delta\sigma}$, where $\delta\varepsilon$ is the angle of contingence, it is easily proved [*cf. Universal Algebra*, §288 (1)] that

$$\frac{\gamma^2}{\kappa^2} = \frac{(x\dot{x}\ddot{x}|x\dot{x}\ddot{x})}{(\dot{x}|\dot{x})^3}.$$

Also $$(xx_1x_2|xx_1x_2) = (x_1x_2|x_1x_2),$$

since $$(x|x_1) = 0 = (x|x_2);$$

|286| and similarly $$(x\dot{x}\ddot{x}|xx_1x_2) = (\dot{x}\ddot{x}|x_1x_2).$$

Thus $$\sin\omega = \frac{\kappa(\dot{x}\ddot{x}|x_1x_2)}{\gamma\sqrt{(\dot{x}|\dot{x})^3(x_1x_2|x_1x_2)}}.$$

Let $\frac{\sin\omega}{\kappa}$ be defined as the geodesic curvature of the curve, and let it be written $\frac{1}{\mu_g}$. Then

$$\frac{1}{\mu_g} = \frac{(\dot{x}\ddot{x}|x_1x_2)}{\gamma\sqrt{\{(\dot{x}|\dot{x})^3(x_1x_2|x_1x_2)\}}}.$$

The condition for a geodesic is that

$$(\dot{x}\ddot{x}|x_1x_2) = 0 \tag{2}$$

may hold at every point of the curve. For this condition secures that the osculating planes of the curve are normal planes of the surface; also, since the geometry of infinitely small figures becomes ultimately that of Euclidean space, the condition is thereby secured that the length of each element of arc of the curve is a minimum distance (on the surface) between its extremities.

This equation can be put into another form, which explicitly relates θ and ϕ and their differential coefficients with respect to τ. For

$$\dot{x} = x_1\dot{\theta} + x_2\dot{\phi},$$

$$\ddot{x} = x_1\ddot{\theta} + x_2\ddot{\phi} + x_{11}\dot{\theta}^2 + 2x_{12}\dot{\theta}\dot{\phi} + x_{22}\dot{\phi}^2;$$

hence
$$\dot{x}\ddot{x} = \left(\dot{\theta}\ddot{\phi} - \ddot{\theta}\dot{\phi}\right)x_1x_2 + x_1x_{11}\dot{\theta}^3 + \left(2x_1x_{12} + x_2x_{11}\right)\dot{\theta}^2\dot{\phi} + \left(2x_2x_{12} + x_1x_{22}\right)\dot{\theta}\dot{\phi}^2 + x_2x_{22}\dot{\phi}^3.$$

Hence the equation
$$\left(\dot{x}\ddot{x}\,|\,x_1x_2\right) = 0$$
becomes

$$\left(\dot{\theta}\ddot{\phi} - \ddot{\theta}\dot{\phi}\right)\left(x_1x_2\,|\,x_1x_2\right) + \dot{\theta}^3\left(x_1x_{11}\,|\,x_1x_2\right) + \left\{2\left(x_1x_{12}\,|\,x_1x_2\right) + \left(x_2x_{11}\,|\,x_1x_2\right)\right\}\dot{\theta}^2\dot{\phi} + \left\{2\left(x_2x_{12}\,|\,x_1x_2\right) + \left(x_1x_{22}\,|\,x_1x_2\right)\right\}\dot{\theta}\dot{\phi}^2 + \left(x_2x_{22}\,|\,x_1x_2\right)\dot{\phi}^3 = 0.$$

The coefficients can easily be expressed in terms of the differential coefficients of $\left(x_1|x_1\right)$, $\left(x_1|x_2\right)$, $\left(x_2|x_2\right)$. For

$$\left(x_1x_{11}\,|\,x_1x_2\right) = \left(x_1|x_1\right)\left(x_{11}|x_2\right) - \left(x_1|x_2\right)\left(x_1|x_{11}\right)$$
$$= \left(x_1|x_1\right)\left(x_1|x_2\right)_1 - \tfrac{1}{2}\left(x_1|x_1\right)\left(x_1|x_1\right)_2 - \tfrac{1}{2}\left(x_1|x_2\right)\left(x_1|x_1\right)_1,$$

|287| and $2\left(x_1x_{12}\,|\,x_1x_2\right) + \left(x_2x_{11}\,|\,x_1x_2\right)$

$$= \tfrac{1}{2}\left(x_1x_2\,|\,x_1x_2\right)_1 + \tfrac{1}{2}\left(x_1|x_1\right)\left(x_2|x_2\right)_1 - \tfrac{1}{2}\left(x_1|x_2\right)\left(x_1|x_1\right)_2$$

with similar equations for the coefficients of $\dot{\theta}\dot{\phi}^2$ and of $\dot{\phi}^3$.

Hence equation (2) can be written in the form

$$\begin{aligned}
2\left(\dot{\theta}\ddot{\phi} - \ddot{\theta}\dot{\phi}\right)&\left(x_1x_2\,|\,x_1x_2\right) \\
= &\left\{\left(x_1|x_1\right)\left(x_1|x_1\right)_2 + \left(x_1|x_2\right)\left(x_1|x_1\right)_1 - 2\left(x_1|x_1\right)\left(x_1|x_2\right)_1\right\}\dot{\theta}^3 \\
&+\left\{3\left(x_1|x_2\right)\left(x_1|x_1\right)_2 - 3\left(x_1|x_1\right)\left(x_2|x_2\right)_1 + \left(x_1x_2\,|\,x_1x_2\right)_1\right\}\dot{\theta}^2\dot{\phi} \\
&-\left\{3\left(x_1|x_2\right)\left(x_2|x_2\right)_1 - 3\left(x_2|x_2\right)\left(x_1|x_1\right)_2 + \left(x_1x_2\,|\,x_1x_2\right)_2\right\}\dot{\theta}\dot{\phi}^2 \\
&-\left\{\left(x_2|x_2\right)\left(x_2|x_2\right)_1 + \left(x_1|x_2\right)\left(x_2|x_2\right)_2 - 2\left(x_2|x_2\right)\left(x_1|x_2\right)_2\right\}\dot{\phi}^3. \qquad (3)
\end{aligned}$$

It is easy to verify that this equation for geodesics on surfaces in elliptic space is the same as that for Euclidean space given by Darboux in his *Leçons sur la Théorie générale des Surfaces*, LIVRE V., chapter iv., §314.[1] It is obvious that equation (3) must

1. Jean-Gaston Darboux, *Leçons sur la théorie générale des surfaces et les applications géométriques du calcul infinitésimal, deuxième partie* [*Lessons on the General Theory of Surfaces and Geometric Applications of Infinitesimal Calculus, Second Part*] (Paris: Gauthier-Villars, 1889).

be the same in Euclidean and in non-Euclidean geometry; for, in each case, it is the immediate expression of the fact that $E\delta\theta^2 + 2F\delta\theta\delta\phi + G\delta\phi^2$ is a minimum. Equation (3) is practically an interpretation of equation (2) in terms of ordinary algebra, and is, therefore, useful when it is required to integrate it or to discuss it in any way in connexion with the theory of differential equations.

The equation of a curve on a surface can be put into the form

$$\psi(\theta, \phi) = \text{constant},$$

thus avoiding the use of an auxiliary variable τ. The preceding formula can be transformed into this notation. For

$$\dot{x} = x_1\dot{\theta} + x_2\dot{\phi},$$

where

$$\frac{\partial\psi}{\partial\theta}\dot{\theta} + \frac{\partial\psi}{\partial\phi}\dot{\phi} = 0.$$

Now let $\frac{\partial\psi}{\partial\theta}, \frac{\partial\psi}{\partial\phi}$ be denoted by ψ_1 and ψ_2 respectively. Then

$$\frac{\dot{\theta}}{\psi_2} = \frac{\dot{\phi}}{-\psi_1} = \frac{\dot{\sigma}}{\gamma\sqrt{(x_1|x_1)\psi_2^2 - 2(x_1|x_2)\psi_1\psi_2 + (x_1|x_1)\psi_1^2}},$$

where $\dot{\sigma}$ stands for $\frac{d\sigma}{d\tau}$.

|288| Now, following Beltrami's notation for Euclidean space, put

$$\{(x_1|x_1)\psi_2^2 - 2(x_1|x_2)\psi_1\psi_2 + (x_2|x_2)\psi_1^2\}/(x_1x_2|x_1x_2) = \Delta\psi.$$

Hence

$$\frac{\dot{\theta}}{\psi_2} = \frac{\dot{\phi}}{-\psi_1} = \frac{\dot{\sigma}}{\gamma\{(x_1x_2|x_1x_2)\Delta\psi\}^{\frac{1}{2}}} = \frac{\dot{\sigma}}{\lambda}, \text{ say,}$$

where

$$\lambda = \gamma\{(x_1x_2|x_1x_2)\Delta\psi\}^{\frac{1}{2}}.$$

Hence, choosing τ so that $\dot{\sigma} = 1$,

$$\dot{x} = \frac{1}{\lambda}(x_1\psi_2 - x_2\psi_1) = \frac{1}{\lambda}\left(\psi_2\frac{\partial}{\partial\theta} - \psi_1\frac{\partial}{\partial\phi}\right)x.$$

Now let δ stand for the symbolic operator

$$\left(\psi_2\frac{\partial}{\partial\theta} - \psi_1\frac{\partial}{\partial\phi}\right);$$

also let

$$\delta^2 = \left(\psi_2\frac{\partial}{\partial\theta} - \psi_1\frac{\partial}{\partial\phi}\right)\left(\psi_2\frac{\partial}{\partial\theta} - \psi_1\frac{\partial}{\partial\phi}\right).$$

Then

$$\dot{x} = \frac{1}{\lambda}\delta x;$$

also

$$(\dot{x}|\dot{x}) = \frac{1}{\gamma^2} = \frac{1}{\lambda^2}(\delta x|\delta x).$$

Similarly,

$$\ddot{x} = \frac{1}{\lambda}\delta\frac{1}{\lambda}\delta x + \frac{1}{\lambda^2}\delta^2 x.$$

Hence $$\dot{x}\ddot{x}=\frac{1}{\lambda^3}\delta x\delta^2 x.$$

Thus $$\frac{\gamma}{\kappa}=\frac{\left(x\delta x\delta^2 x\,|\,x\delta x\delta^2 x\right)^{\frac{1}{2}}}{\left\{\left(x_1x_2\,|\,x_1x_2\right)\Delta\psi\right\}^{\frac{3}{2}}};$$

also $$\frac{\gamma}{\mu_g}=\frac{\left(\delta x\delta^2 x\,|\,x_1x_2\right)}{\left(x_1x_2\,|\,x_1x_2\right)\left(\Delta\psi\right)^{\frac{3}{2}}}.$$

The condition for a geodesic is given by

$$\left(\delta x\delta^2 x\,|\,x_1x_2\right)=0. \tag{4}$$

Thus, in the special case in which $\psi=\phi$, then $\psi_2=1$ and $\psi_1=0$; hence δ becomes $\frac{\partial}{\partial\theta}$. Thus the condition for $\phi=\text{constant}$ being a geodesic is that $\left(x_1x_{11}\,|\,x_1x_2\right)=0$ at all points of it.

|289| 4. Now assume that the curves $\phi=\text{constant}$ are a family of geodesics, and that the curves $\theta=\text{constant}$ are the family of curves cutting the geodesics at right angles.

Then the condition that the two families are everywhere orthogonal is

$$\left(x_1\,|\,x_2\right)=0.$$

The condition that the family $\phi=\text{constant}$ is geodesic is

$$\left(x_1x_{11}\,|\,x_1x_2\right)=0;$$

that is, $$\left(x_1\,|\,x_1\right)\left(x_{11}\,|\,x_2\right)-\left(x_1\,|\,x_2\right)\left(x_1\,|\,x_{11}\right)=0.$$

Hence, using the condition of orthogonality, we find

$$\left(x_{11}\,|\,x_2\right)=0.$$

But, by differentiating $\left(x_1\,|\,x_2\right)=0$ partially with respect to ϕ, we find

$$\left(x_{11}\,|\,x_2\right)+\left(x_1\,|\,x_{12}\right)=0.$$

Hence $$\left(x_1\,|\,x_{12}\right)=0.$$

But $$\left(x_2\,|\,x_{12}\right)=\tfrac{1}{2}\left(x_1\,|\,x_1\right)_2.$$

Hence $$\left(x_1\,|\,x_1\right)_2=0;$$

that is to say, $\left(x_1\,|\,x_1\right)$ is independent of ϕ. Now

$$\frac{\delta\sigma^2}{\gamma}=\left(x_1\,|\,x_1\right)\delta\theta^2+\left(x_2\,|\,x_2\right)\delta\phi^2.$$

But, since $\left(x_1\,|\,x_1\right)$ is independent of ϕ, we may write

$$\int\left(x_1\,|\,x_1\right)d\theta=\frac{\rho}{\gamma},$$

and use ρ and ϕ as coordinates instead of θ and ϕ. The curves $\rho=\text{constant}$ and $\phi=\text{constant}$ are the same as the curves $\theta=\text{constant}$ and $\phi=\text{constant}$. Also let the subscript 1 denote for the future partial differentiation with respect to ρ. Then

$$\delta\sigma^2 = \delta\rho^2 + \gamma^2(x_2|x_2)\delta\phi^2.$$

Thus, if a family of geodesics be cut orthogonally, the distance between any two of the orthogonal curves measured along an orthogonal geodesic is the same for that pair of curves whatever geodesic be chosen. For consider the pair of curves $\rho = \rho_1$ and $\rho = \rho_2$; and let |290| the geodesic be $\phi = \alpha$. Then an element of arc along the geodesic is

$$\delta\sigma = \delta\rho.$$

Then $$\sigma = \rho_2 \sim \rho_1.$$

Let $\rho_2 \sim \rho_1$ be called the "geodesic distance between the curves."

5. Conversely, assume that the curves $\phi = \text{constant}$ are a family of geodesics, and that the length of the arc of any geodesic $\phi = \alpha$, between any given pair of curves $\theta = \theta_1$ and $\theta = \theta_2$, is the same for all values of α. It is required to prove that the two families are orthogonal, if any particular curve $\theta = \theta_0$ is orthogonal to all the curves $\phi = \text{constant}$. The conditions are

$$(x_1 x_{11}|x_1 x_2) = 0,$$

and that $(x_1|x_1)$ is independent of ϕ. For then the element of the arc of $\phi = \alpha$ is given by

$$\delta\sigma = \gamma\sqrt{(x_1|x_1)}\delta\theta,$$

and is independent of α.

It follows from the second condition, by differentiating with respect to ϕ, that

$$(x_1|x_{12}) = 0.$$

Hence $$(x_1|x_2)_1 = (x_{11}|x_2) + (x_1|x_{12}) = (x_{11}|x_2).$$

Again, the first condition becomes, after performing the multiplications indicated,

$$(x_1|x_1)(x_2|x_{11}) - (x_1|x_2)(x_1|x_{11}) = 0;$$

that is, $$(x_1|x_1)(x_1|x_2)_1 - \tfrac{1}{2}(x_1|x_2)(x_1|x_1)_1 = 0.$$

Hence, $$2\frac{(x_1|x_2)_1}{(x_1|x_2)} = \frac{(x_1|x_1)_1}{(x_1|x_1)}.$$

Hence, by integrating, we find

$$(x_1|x_2)^2 = \Phi(x_1|x_1);$$

where Φ is a function of ϕ only. But, by hypothesis, when $\theta = \theta_0$,

$$(x_1|x_2) = 0,$$

for all values of ϕ. Hence

$$\Phi(x_1|x_1)_{\theta=\theta_0} = 0,$$

for all values of ϕ. But $(x_1|x_1)$ cannot vanish, for an element of the arc of any geodesic $\phi = \alpha$ at the point (θ_0, α) is

$$\delta\sigma = \gamma(x_1|x_1)\delta\theta.$$

|291| Thus $\Phi = 0$ for all values of ϕ. Hence

$$(x_1|x_2) = 0$$

at all points of the surface.

6. Collecting results, we see that, if the coordinate curves are geodesics and their orthogonals, and if the length ρ of the geodesic arc from a given orthogonal be taken as one coordinate, then

$$(x|x_2) = \frac{1}{\gamma^2}, \quad (x_1|x_2) = 0.$$

To these must be added the general condition that x is a unit point, namely,

$$(x|x) = 1.$$

Also all equations hold which can be deduced from these by partial differentiations with respect to ρ or ϕ.

Any element of arc $\delta\sigma$ takes the form

$$(\delta\dot{\sigma})^2 = (\delta\rho)^2 + \gamma^2(x_2|x_2)(\delta\phi)^2 = \delta\rho^2 + \gamma^2\Pi^2(\delta\phi)^2, \tag{5}$$

where Π^2 is written for $(x_2|x_2)$. Let this type of coordinates be called "semi-geodesic orthogonal coordinates."

7. A special form of such coordinates can always be found as follows:—Let the family of geodesics be the geodesics issuing from any point c on the surface; let ρ be the arc of the geodesic through the point x, measured from c, and let ϕ be the angle which that geodesic makes at c with a given geodesic of the family.

Let such coordinates be called geodesic polar coordinates; and let the curves $\rho = \text{constant}$ be called "geodesic circles centre c," and the curves $\phi = \text{constant}$ be called the "geodesic radii from c."

It is easily seen that the orthogonal family to the geodesic radii is the family of geodesic circles. For, by §5, we have only to prove that one of the geodesic circles is orthogonal to the radii. But it is obvious that a geodesic circle with an infinitely small radius, ρ, is orthogonal, since the figure becomes ultimately a plane figure.

Since geodesic circles with small radii are ultimately plane, and since the properties of small figures are ultimately the same as those of figures in Euclidean space, it follows that an element of arc of the circle $\rho = \alpha$, when α is small, takes the form $\rho\delta\phi$. Hence the limit of Π, when ρ is small, is ρ/γ. This result is only necessarily true for polar geodesic coordinates.

|292| 8. The formula (1) in §2 for $\dfrac{1}{\rho_1\rho_2}$ becomes, using the semi-geodesic orthogonal coordinates of §6 and simplifying,

$$\frac{(x_2|x_2)^2}{\gamma^2\rho_1\rho_2} = \Delta_1 - \Delta_2 + \frac{(x_2|x_2)}{\gamma^2}\left\{-\frac{(x_2|x_2)}{\gamma^2} - \tfrac{1}{2}(x_2|x_2)_{11}\right\},$$

where $$\Delta_1 = 0, \quad \Delta_2 = -\tfrac{1}{4}\frac{(x_2|x_2)_1^2}{\gamma^2}.$$

Hence $$\frac{(x_2|x_2)^2}{\rho_1\rho_2} = \tfrac{1}{4}(x_2|x_2)_1 - \frac{(x_2|x_2)^2}{\gamma^2} - \tfrac{1}{2}(x_2|x_2)(x_2|x_2)_{11}.$$

Now $$\Pi^2 = (x_2|x_2).$$

Therefore, substituting and reducing, we find

$$\frac{\partial^2\Pi}{\partial\rho^2} + \left(\frac{1}{\gamma^2} + \frac{1}{\rho_1\rho_2}\right)\Pi = 0. \tag{6}$$

This is the analogue in elliptic geometry of Gauss's well-known equation in Euclidean geometry for the measure of curvature at any point of a surface in terms of semi-geodesic orthogonal coordinates (*cf.* Salmon's *Solid Geometry*, 3rd ed.,[1] §329, and Darboux, §524. Darboux also considers for Euclidean space semi-geodesic orthogonal coordinates, Livre v., chapter v.[2]). If γ be made infinite, the equation reduces to the Euclidean form, as it ought to do.

9. Now, transforming equation (3) into the special form which it assumes when the semi-geodesic orthogonal coordinates (ρ, ϕ) are substituted, we obtain as the equation satisfied by geodesics, after division by $2\Pi^2/\gamma^4$,

$$(\dot\rho\ddot\phi - \ddot\rho\dot\phi) = -\frac{2}{\Pi}\frac{\partial\Pi}{\partial\rho}\dot\rho^2\dot\phi - \frac{1}{\Pi}\frac{\partial\Pi}{\partial\phi}\dot\rho\dot\phi^2 - \Pi\frac{\partial\Pi}{\partial\rho}\dot\phi^3. \tag{7}$$

Also ρ can be taken instead of τ as the independent variable. Hence

$$\dot\rho = 1, \quad \ddot\rho = 0.$$

Thus the equation (7) becomes

$$\ddot\phi = -\frac{2}{\Pi}\frac{\partial\Pi}{\partial\rho}\dot\phi - \frac{1}{\Pi}\frac{\partial\Pi}{\partial\phi}\dot\phi^2 - \Pi\frac{\partial\Pi}{\partial\rho}\dot\phi^3. \tag{8}$$

This is the same as the equation for Euclidean geometry given by Darboux, §598.

|293| 10. Beltrami's theorem, that the surfaces of constant curvature are the only surfaces for which geodesics are transformed into straight lines when the surface is represented on a plane, can now be proved by exactly the same method as is employed by Darboux (*cf.* §598) to prove it for Euclidean space.

The following proof is substantially taken from Darboux:—

Let the point e_0 on the plane correspond to the point *e* on the surface. Then the straight lines through e_0 correspond to the geodesics on the surface through *e*. Also, if ρ_0 and θ be the polar coordinates of any point on the plane, and ρ and ϕ the polar geodesic coordinates of the corresponding point on the surface, we must have

$$\rho_0 = f(\rho, \phi), \quad \theta = F(\phi),$$

where θ and ϕ are measured respectively from a corresponding pair formed by a straight line through e_0 and a geodesic through *e*. But the general equation of a straight line is

1. George Salmon, *A Treatise on the Analytic Geometry of Three Dimensions*, 3rd edition (Dublin: Hodges, Foster, & Co., 1874).
2. Darboux, *Leçons, deuxième partie.*

$$\alpha \sin\frac{\rho_0}{\gamma}\cos\theta + \beta \sin\frac{\rho_0}{\gamma}\sin\theta + \delta = 0.$$

Hence $$\alpha + \beta\mu + \delta\sigma = 0,$$

where μ is a function of ϕ only and σ is a function of ρ and ϕ, is the general integral for geodesics on the surface.

But the differential equation (8) can be written, substituting μ for ϕ as the coordinate,

$$\ddot{\mu} = -\frac{2}{\Pi}\frac{\partial\Pi}{\partial\rho}\dot{\mu} - \frac{1}{\Pi}\frac{\partial\Pi}{\partial\mu}\dot{\mu}^2 - \Pi\frac{\partial\Pi}{\partial\rho}\dot{\mu}^3.$$

But, differentiating the integral form, we have

$$\beta\dot{\mu} - \delta\left(\sigma_1 + \sigma_2\dot{\mu}\right) = 0,$$

$$\beta\ddot{\mu} + \delta\left(\sigma_{11} + 2\sigma_{12}\dot{\mu} + \sigma_{22}\dot{\mu}^2 + \sigma_2\ddot{\mu}\right) = 0.$$

Eliminating β and δ,

$$\ddot{\mu}\left(\sigma_1 + \sigma_2\dot{\mu}\right) = \dot{\mu}\left(\sigma_{11} + 2\sigma_{12}\dot{\mu} + \sigma_{22}\dot{\mu}^2 + \sigma_2\ddot{\mu}\right).$$

Hence $$\ddot{\mu} = \frac{\sigma_{11}}{\sigma_1}\dot{\mu} + \frac{2\sigma_{12}}{\sigma_1}\dot{\mu}^2 + \frac{\sigma_{22}}{\sigma_1}\dot{\mu}^3.$$

Accordingly, by comparison,

$$\frac{\sigma_{11}}{\sigma_1} = -\frac{2}{\Pi}\frac{\partial\Pi}{\partial\rho}, \quad \frac{2\sigma_{12}}{\sigma_1} = -\frac{1}{\Pi}\frac{\partial\Pi}{\partial\mu}, \quad \frac{\sigma_{22}}{\sigma_1} = -\Pi\frac{\partial\Pi}{\partial\rho}. \tag{9}$$

|294| From the first two of the equations (9) we deduce

$$\Pi^2\sigma_1 = \frac{1}{\Phi^3}, \quad \Pi\sigma_1^2 = \frac{1}{\mathrm{P}^3},$$

where Φ and P are arbitrary functions of μ and ρ respectively. Hence

$$\Pi = \frac{\mathrm{P}}{\Phi^2}, \quad \sigma_1 = \frac{\Phi}{\mathrm{P}^2}.$$

Thus $$\sigma = \Phi\int\frac{d\rho}{\mathrm{P}^2} + \mathrm{M},$$

where M is another arbitrary function of μ only. Hence, from the third of equations (9),

$$\Phi_{22}\int\frac{d\rho}{\mathrm{P}^2} + \mu_{22} = -\frac{\Phi}{\mathrm{P}^2}\frac{\mathrm{P}}{\Phi^2}\frac{\mathrm{P}_1}{\Phi^2} = -\frac{\mathrm{P}_1}{\mathrm{P}\Phi^3}.$$

Differentiating both sides with respect to ρ,

$$\frac{\Phi_{22}}{\mathrm{P}^2} = -\frac{1}{\Phi^3}\frac{d}{d\rho}\left(\frac{\mathrm{P}_1}{\mathrm{P}}\right).$$

Hence $$\Phi^3\Phi_{22} = -\mathrm{P}^2\frac{d}{d\rho}\left(\frac{\mathrm{P}_1}{\mathrm{P}}\right).$$

But the left-hand side is a function of μ only, and the right-hand side is a function of ρ only. Hence both sides are constant. Thus

$$\Phi^3\Phi_{22} = -\varepsilon, \quad P^2 \frac{d}{d\rho}\left(\frac{P_1}{P}\right) = \varepsilon.$$

The second equation can be written

$$\frac{P_1}{P}\frac{d}{d\rho}\left(\frac{P_1}{P}\right) = \frac{\varepsilon P_1}{P^3}.$$

Hence, by integration,
$$\tfrac{1}{2}\frac{P_1^2}{P^2} = -\frac{\varepsilon}{2P^2} + \tfrac{1}{2}\eta,$$

where η is constant. Hence

$$P_1^2 = -\varepsilon + \eta P^2.$$

Therefore
$$P_1 P_{11} = \eta P P_1, \quad P_{11} = \eta P.$$

Now, from equation (6),

$$\frac{1}{\Pi}\frac{\partial^2 \Pi}{\partial \rho^2} = -\left(\frac{1}{\gamma^2} + \frac{1}{\rho_1 \rho_2}\right).$$

But
$$\Pi = \frac{P}{\Phi^2};$$

|295| therefore
$$\frac{1}{\Pi}\frac{\partial^2 \Pi}{\partial \rho^2} = \frac{P_{11}}{P} = \eta.$$

Hence
$$\frac{1}{\rho_1 \rho_2} = -\eta - \frac{1}{\gamma^2}.$$

Thus the only surfaces for which the geodesics correspond to straight lines are surfaces of constant curvature.

11. There are three types of surfaces of constant curvature, according as $\frac{1}{\rho_1\rho_2} + \frac{1}{\gamma^2}$ is positive, negative, or zero.

Let
$$\frac{1}{\rho_1 \rho_2} + \frac{1}{\gamma^2} = \frac{1}{\varepsilon^2}, \quad -\frac{1}{\varepsilon^2}, \text{ or } 0,$$

in the three cases.

$$\left.\begin{aligned} &\text{In case (1)} \quad && \Pi = \sigma \sin\left(\frac{\rho}{\varepsilon} - \tau\right) \\ &\text{in case (2)} && \Pi = \sigma \sinh\left(\frac{\rho}{\varepsilon} - \tau\right) \\ &\text{in case (3)} && \Pi = \sigma\rho + \tau \end{aligned}\right\}, \qquad (10)$$

where, if semi-geodesic orthogonal coordinates be used, σ and τ are functions of ϕ only.

Let the surfaces of type (1) be called "elliptic surfaces of constant curvature," of type (2) "hyperbolic surfaces of constant curvature," of type (3) "parabolic surfaces of constant curvature."

12. If geodesic polar coordinates be used, it follows from §7, by making ρ small, that $\tau = 0$ and $\sigma = \varepsilon/\gamma$ for the elliptic and hyperbolic surfaces, and $\tau = 0$, $\sigma = \frac{1}{\gamma}$ for the parabolic surfaces.

Hence, with these coordinates,

$$\Pi = \frac{\varepsilon}{\gamma}\sin\frac{\rho}{\varepsilon} \text{ for elliptic surfaces,}$$

$$\Pi = \frac{\varepsilon}{\gamma}\sinh\frac{\rho}{\varepsilon} \text{ for hyperbolic surfaces,}$$

$$\Pi = \frac{\rho}{\gamma} \text{ for parabolic surfaces.}$$

Conversely, if the limit of Π be $\frac{\sigma\rho}{\varepsilon}$, when ρ is small, then the curves $\rho = \text{constant}$ are geodesic circles.

|296| For, if $\delta\sigma_2$ be an element of the arc of one of these curves, for which ρ is a small constant,

$$\delta\sigma_2 = \frac{\gamma\sigma}{\varepsilon}\rho\delta\phi.$$

Now write $\psi = \frac{\gamma}{\varepsilon}\int\sigma d\phi$, remembering that σ is a function of ϕ only. Then

$$[\sigma_2]_{\psi_1}^{\psi_2} = \rho(\psi_2 - \psi_1).$$

Hence the length of arc $[\sigma_2]_{\psi_1}^{\psi_2}$ is very small, when ρ is small. Thus the limit of the curve $\rho = \text{constant}$ must be an evanescent curve round a point. Hence ρ and ψ are polar geodesic coordinates.

13. The only surfaces in elliptic space on which two orthogonal families of geodesic curves can be drawn are parabolic surfaces of constant curvature. For, by the collected results of §6, if we take these orthogonal geodesics as curvilinear coordinate curves, it follows that we may write

$$(x_1|x_1) = \frac{1}{\gamma^2} = (x_2|x_2), \text{ and } (x_1|x_2) = 0. \tag{11}$$

Here the orthogonal curves are the families $\xi = \text{constant}$ and $\eta = \text{constant}$, and the element of arc $\delta\sigma$ takes the form

$$(\delta\sigma)^2 = (\delta\xi)^2 + (\delta\eta)^2;$$

also the suffix 1 denotes partial differentiation with respect to ξ, and the suffix 2 with respect to η.

It follows from equations (11) that

$$(x_1|x_1)_1 = 0 = (x_1|x_1)_2 = (x_2|x_2)_1 = (x_2|x_2)_2 = (x_1|x_2)_1 = (x_1|x_2)_2.$$

Hence the equation (1) in §2 for $\frac{1}{\rho_1\rho_2}$ becomes

$$\frac{1}{\rho_1\rho_2} = -\frac{1}{\gamma^2}.$$

This proves the proposition.

Conversely,[1] two orthogonal families of geodesics can always be drawn on any parabolic surface of constant curvature so as to include any given geodesic as a member of one of the families. For, from each point of the given geodesic, draw the geodesics cutting the given geodesic at right angles. Let the length of the geodesic arc drawn in this manner to any point x be ρ. Let the family of geodesics so |297| drawn be denoted by the equation $\phi = \text{constant}$. Then (by §5 above) ρ and ϕ are semi-geodesic orthogonal coordinates; and, of the curves $\rho = \text{constant}$, on curve, namely, $\rho = 0$, is a geodesic. We will now prove that all the curves $\rho = \text{constant}$ are geodesics.

For, from §§6 and 11, we have

$$(x_1|x_1) = \frac{1}{\gamma^2}, \quad (x_1|x_2) = 0, \quad (x_2|x_2) = (\sigma\rho + \tau)^2,$$

where σ and τ are functions of ϕ only. Also, by §3, equation (4), the condition that the curves $\rho = \text{constant}$ may be geodesics is that $(x_2 x_{22}|x_1 x_2)$ should vanish at each point of the surface. Now

$$\begin{aligned}(x_2 x_{22}|x_1 x_2) &= (x_1|x_2)(x_2|x_{22}) - (x_2|x_2)(x_1|x_{22}) \\ &= -(x_2|x_2)(x_1|x_{22}).\end{aligned}$$

But $$(x_{12}|x_2) + (x_1|x_{22}) = (x_1|x_2)_2 = 0;$$

and $$(x_{12}|x_2) = \tfrac{1}{2}(x_2|x_2)_1.$$

Hence $$(x_1|x_{22}) = -\tfrac{1}{2}(x_2|x_2)_1.$$

Therefore $$\begin{aligned}(x_2 x_{22}|x_1 x_2) &= \tfrac{1}{2}(x_2|x_2)(x_2|x_2)_1 \\ &= \sigma(\sigma\rho + \tau)^3.\end{aligned}$$

But, when $\rho = 0$, by hypothesis,

$$(x_2 x_{22}|x_1 x_2) = 0.$$

Hence $$\sigma\tau^3 = 0.$$

Now, if $\tau = 0$, the curve $\rho = 0$ is, by §12, an evanescent circle, and is, therefore, not a geodesic, contrary to hypothesis.

Hence, $\sigma = 0$; and, therefore, at all points of the surface

$$(x_2 x_{22}|x_1 x_2) = 0.$$

Thus, the family of curves $\rho = \text{constant}$ are geodesics.

Thus, on parabolic surfaces of constant curvature, and only on such surfaces, geodesic orthogonal coordinates are possible; and the coordinate curves can be chosen so as to include any assigned geodesic.

14. On any parabolic surface of constant curvature let geodesic orthogonal coordinates be chosen. Let the curves $\xi = \text{constant}$ denote one family of geodesics, and the curves $\eta = \text{constant}$ denote the orthogonal family of geodesics. Also choose

1. This paragraph was numbered 14, a numbering error since there is also a section 14 below. Since this misnumbered section is never referenced, we have simply deleted the number rather than renumbered subsequent sections.

ξ and η so that $\delta\xi$ denotes an element of arc of a curve $\eta=$constant, and $\delta\eta$ denotes an |298| element of arc of a curve $\xi=$constant. Then, by §6,

$$(x_1|x_1)=\frac{1}{\gamma^2}=(x_2|x_2),$$

and

$$(x_1|x_2)=0.$$

Hence[1]

$$(\delta\sigma)^2=(\delta\xi)^2+(\delta\eta)^2.$$

The identity of the geodesic geometry of parabolic surfaces of constant curvature with the geometry of straight lines in a Euclidean plane follows immediately from this equation. It may be convenient to some readers to illustrate this identity by the following investigation, occupying the remainder of this article (14).

Let any geodesic be given by

$$\xi=f(\tau),\quad \eta=\phi(\tau);$$

and let differentiation with respect to τ be denoted by $\dot{\xi}$, $\dot{\eta}$, and so on. The condition for a geodesic has been given by equation (3) of §2. This now takes the special form

$$\dot{\xi}\ddot{\eta}-\ddot{\xi}\dot{\eta}=0.$$

Integrating the form

$$\frac{\ddot{\xi}}{\dot{\xi}}-\frac{\ddot{\eta}}{\dot{\eta}}=0,$$

we have

$$\dot{\eta}=\alpha\dot{\xi}. \tag{12}$$

Integrating again,

$$\eta=\alpha\xi+\beta. \tag{13}$$

This is the general equation of a geodesic in terms of orthogonal curvilinear coordinates. The tangent line at any point x of the geodesic (13) is $x\dot{x}$. But

$$x\dot{x}=x(x_1\dot{\xi}+x_2\dot{\eta})=x(x_1+\alpha x_2)\dot{\xi}.$$

Hence $x(x_1+\alpha x_2)$ is the tangent line. The tangent line to any other geodesic

$$\eta=\alpha'\xi+\beta',$$

through the same point x, is $x(x_1+\alpha' x_2)$. The angle θ between these lines is given by

$$\cos\theta=\frac{\{x(x_1+\alpha x_2)|x(x_1+\alpha' x_2)\}}{\sqrt{\{x(x_1+\alpha x_2)|x(x_1+\alpha x_2)\}\{x(x_1+\alpha' x_2)|x(x_1+\alpha' x_2)\}}}$$

$$=\frac{1+\alpha\alpha'}{\sqrt{(1+\alpha^2)(1+\alpha'^2)}};$$

|299| that is,

$$\tan\theta=\frac{\alpha\sim\alpha'}{1+\alpha\alpha'}.$$

1. The following equation was numbered (11), a numbering error since there is already an equation (11). Since this misnumbered equation is never referenced, we have simply deleted the number rather than renumber all subsequent equations.

Thus the angles that the geodesics severally make with $\eta = 0$ at their points of intersection with it are $\tan^{-1}\alpha$ and $\tan^{-1}\alpha'$.

15. We will now find the equation of the general type of those parabolic surfaces of constant curvature which are also surfaces of revolution. By so doing we also prove the important proposition that parabolic surfaces of constant curvature in elliptic space are real surfaces.

Let ee_3 be the axis of revolution, and let e, e_1, e_2, e_3 be a set of unit-normal (or quadrantal) points e_3 [*cf. Universal Algebra*, §233 (1) and §245]. Then the three lines ee_1, ee_2, ee_3 form a system of three axes mutually at right angles, origin e.

Any unit point $$x = \xi e + \xi_1 e_1 + \xi_2 e_2 + \xi_3 e_3$$

satisfies the condition $$(x|x) = 1,$$

that is, $$\xi^2 + \xi_1^2 + \xi_2^2 + \xi_3^2 = 1;$$

and ξ, ξ_1, ξ_2, ξ_3 are the ordinary rectangular coordinates of elliptic geometry.

The meridian curves of the surface of revolution lie in planes through ee_3, and are geodesics of the surface. The parallels of latitude are curves (not geodesics) which cut the meridians orthogonally. Take these two families of curves to define a system of semi-geodesic orthogonal coordinates. In accordance with the notation of §6, let a point x on the surface be defined by ρ, which is the geodesic arc measured from some fixed parallel of latitude, and by ϕ, which is the angle between the plane ee_3x and the plane ee_3e_1.

Then, from §11, equation (10), we have

$$\Pi = \alpha\rho + \beta, \tag{14}$$

where α and β are constants, since the surface is one of revolution.

Also, since the surface is one of revolution, we may assume

$$x = \xi e + \mu(e_1 \cos\phi + e_2 \sin\phi) + \xi_3 e_3, \tag{15}$$

where ξ, μ, and ξ_3 are functions of ρ only.

Also, since $$(x|x) = 1,$$

we have $$\xi^2 + \mu^2 + \xi_3^2 = 1. \tag{16}$$

|300| Since $$(x_1|x_1) = \frac{1}{\gamma^2},$$

we have $$\left(\frac{d\xi}{d\rho}\right)^2 + \left(\frac{d\mu}{d\rho}\right)^2 + \left(\frac{d\xi_3}{d\rho}\right)^2 = \frac{1}{\gamma^2}. \tag{17}$$

Since $$(x_2|x_2) = \Pi^2 = (\alpha\rho + \beta)^2,$$

we have $$\mu^2 = (\alpha\rho + \beta)^2;$$

that is, $$\mu = \alpha\rho + \beta. \tag{18}$$

The condition $(x_1|x_2) = 0$ is satisfied identically.

From (17) and (18) we find

$$\left(\frac{d\xi}{d\rho}\right)^2+\left(\frac{d\xi_3}{d\rho}\right)^2=\frac{1}{\gamma^2}-\alpha^2.$$

Hence for a real surface we must have $\alpha^2<\frac{1}{\gamma^2}$. Put

$$\alpha=\frac{\cos\delta}{\gamma}.$$

Then
$$\left(\frac{d\xi}{d\rho}\right)^2+\left(\frac{d\xi_3}{d\rho}\right)^2=\frac{\sin^2\delta}{\gamma^2}. \tag{19}$$

Also equations (16) and (18) give

$$\xi^2+\xi_3^2=1-(\alpha\rho+\beta)^2=\sigma^2,$$

say. Assume[1]
$$\xi=\sigma\cos\varpi,\quad \xi_3=\sigma\sin\varpi,$$

where ϖ is a function of ρ only. Hence equation (19) becomes

$$\sigma^2\left(\frac{d\varpi}{d\rho}\right)^2+\left(\frac{d\sigma}{d\rho}\right)^2=\frac{\sin^2\delta}{\gamma^2}.$$

Thus[2]
$$\frac{d\varpi}{d\rho}=\frac{\left\{\frac{\sin^2\delta}{\gamma^2}-\left(\frac{d\sigma}{d\rho}\right)^2\right\}^{\frac{1}{2}}}{\sigma}.$$

Now, put
$$\alpha\rho+\beta=\tau,$$

so that
$$\sigma^2=1-\tau^2.$$

Then
$$\varpi=\sec\delta\int\frac{(\sin^2\delta-\tau^2)^{\frac{1}{2}}}{1-\tau^2}d\tau.$$

The argument of the integral is rationalized by the assumption

$$\sin^2\delta-\tau^2=\tau^2\zeta^2.$$

|301| Finally, we find

$$\varpi-\varepsilon=\tan^{-1}\frac{\left\{\gamma^2\sin^2\delta-(\rho\cos\delta+\kappa)^2\right\}^{\frac{1}{2}}}{(\rho\cos\delta+\kappa)\cos\delta}-\sec\delta\tan^{-1}\frac{\left\{\gamma^2\sin^2\delta-(\rho\cos\delta+\kappa)^2\right\}^{\frac{1}{2}}}{\rho\cos\delta+\kappa}, \tag{20}$$

where we have written
$$\gamma\beta=\kappa,$$

and ε is an arbitrary constant.

1. The following equation was numbered (20), a numbering error since there is another equation (20) below. Since this misnumbered equation is never referenced, we have simply deleted the number rather than renumber all subsequent equations.
2. For the following equation, the original erroneously omitted the '$\frac{1}{2}$'.

Thus the required surface is given by

$$\left.\begin{aligned} x &= \xi e + \frac{\rho\cos\delta+\kappa}{\gamma}(e_1\cos\phi+e_2\sin\phi)+\xi_3 e_3, \\ \text{where}\qquad \xi &= \left\{1-\frac{(\rho\cos\delta+\kappa)^2}{\gamma^2}\right\}^{\frac{1}{2}}\cos\varpi, \\ \xi_3 &= \left\{1-\frac{(\rho\cos\delta+\kappa)^2}{\gamma^2}\right\}^{\frac{1}{2}}\sin\varpi, \end{aligned}\right\} \qquad (21)$$

and ϖ is given by equation (20).

If we measure ρ from the parallel of latitude which in equations (21) is denoted by

$$\rho = -\kappa\sec\delta,$$

the equations become

$$\left.\begin{aligned} x &= \xi e + \frac{\rho\cos\delta}{\gamma}(e_1\cos\phi+e_2\sin\phi)+\xi_3 e_3, \\ \xi &= \left\{1-\frac{\rho^2\cos^2\delta}{\gamma^2}\right\}^{\frac{1}{2}}\cos\varpi, \\ \xi_3 &= \left\{1-\frac{\rho^2\cos^2\delta}{\gamma^2}\right\}^{\frac{1}{2}}\sin\varpi, \\ \varpi-\varepsilon &= \tan^{-1}\frac{\{\gamma^2\sin^2\delta-\rho^2\cos^2\delta\}^{\frac{1}{2}}}{\rho\cos^2\delta}-\sec\delta\tan^{-1}\frac{\{\gamma^2\sin^2\delta-\rho^2\cos^2\delta\}^{\frac{1}{2}}}{\rho\cos\delta}. \end{aligned}\right\} \qquad (22)$$

The equation of the surface can also be put in the form

$$\tan^{-1}\frac{\xi_3}{\xi}-\varepsilon = \tan^{-1}\frac{\{\sin^2\delta-\xi_1^2-\xi_2^2\}}{(\xi_1^2+\xi_2^2)^{\frac{1}{2}}\cos\delta}-\sec\delta\tan^{-1}\frac{\{\sin^2\delta-\xi_1^2-\xi_2^2\}^{\frac{1}{2}}}{(\xi_1^2+\xi_2^2)^{\frac{1}{2}}}. \qquad (23)$$

The equations (22) and (23) can be conceived in two different ways, which are not quite identical in their results. Equation (23) may be |302| taken as the fundamental equation; and ρ must then be conceived as standing merely for the analytical expression

$$\frac{\sec\delta}{\gamma}\{\xi_1^2+\xi_2^2\}^{\frac{1}{2}}.$$

Then the equations (22), together with the equations

$$\xi_1 = \frac{\rho\cos\delta}{\gamma}\cos\phi,\quad \xi_2 = \frac{\rho\cos\delta}{\gamma}\sin\phi,$$

define the coordinates ξ, ξ_1, ξ_2, ξ_3 in terms of the auxiliary variables ρ and ϕ. The surface will be found to consist of many sheets, and the geometrical meaning of ρ must be varied to make it appropriate to each particular sheet.

The second method of conceiving equations (22) and (23) is the one by which they have been attained; namely, ρ is conceived as the arc of the meridian geodesic measured from a given parallel of latitude. This method is appropriate for studying the properties of any one particular sheet of the surface; and, since the different sheets are merely repetitions of each other, which arise owning to the multiple values of the inverse tangents in equation (23), this will be the convenient method for us to continue to adhere to for the present. It will save confusion if, during our adherence to this geometrical definition of ρ, the surface is called "the sheet." The change of name is appropriate; for we are considering one sheet of the surface obtained by the first point of view.

16. The parallel of latitude $\rho = 0$ is an evanescent circle surrounding and coinciding with some point on the axis ee_3. Let this point be called "the conical point of the corresponding sheet," and let the sheet be considered as consisting of all that part of the complete surface denoted by equation (23) which is represented by equations (22) when the ambiguities of the inverse tangents are so determined that ρ represents the length of the meridian geodesic drawn from their conical point.

Firstly, let ρ be assumed to be positive, and let λ and μ be positive acute angles such that

$$\cot\lambda = \frac{\{\gamma^2\sin^2\delta - \rho^2\cos^2\delta\}^{\frac{1}{2}}}{\rho\cos^2\delta}, \quad \cot\mu = \frac{\{\gamma^2\sin^2\delta - \rho^2\cos^2\delta\}^{\frac{1}{2}}}{\rho\cos\delta}.$$

Then the most general expression of the last of equations (22), without any determination of the ambiguities, is

$$\varpi - \varepsilon = \left(n\pi + \tfrac{1}{2}\pi - \lambda\right) - \sec\delta\left(m\pi + \tfrac{1}{2}\pi - \mu\right),$$

|303| where m and n are any two integers, positive or negative. Now let

$$\varepsilon = \tfrac{1}{2}\pi\left(\sec\delta - 1\right). \tag{24}$$

Then the equation for ϖ becomes

$$\varpi = \left(n - m\right)\pi - 2m\varepsilon + \mu\sec\delta - \lambda. \qquad (25)^{1}$$

The only equations in which ϖ is used are the following:—

$$\left.\begin{aligned} \xi &= \left\{1 - \frac{\rho^2\cos^2\delta}{\gamma^2}\right\}^{\frac{1}{2}}\cos\varpi = \left(-1\right)^{n-m}\left\{1 - \frac{\rho^2\cos^2\delta}{\gamma^2}\right\}^{\frac{1}{2}}\cos\{-2m\varepsilon + \mu\sec\delta - \lambda\} \\ \xi_3 &= \left\{1 - \frac{\rho^2\cos^2\delta}{\gamma^2}\right\}^{\frac{1}{2}}\sin\varpi = \left(-1\right)^{n-m}\left\{1 - \frac{\rho^2\cos^2\delta}{\gamma^2}\right\}^{\frac{1}{2}}\sin\{-2m\varepsilon + \mu\sec\delta - \lambda\} \end{aligned}\right\}. \tag{26}$$

Secondly, let ρ be assumed to be negative and equal to $-\rho'$, and let λ' and μ' be positive acute angles such that

$$\cot\lambda' = \frac{\{\gamma^2\sin^2\delta - \rho'^2\cos^2\delta\}^{\frac{1}{2}}}{\rho'\cos^2\delta}, \quad \cot\mu' = \frac{\{\gamma^2\sin^2\delta - \rho'^2\cos^2\delta\}^{\frac{1}{2}}}{\rho'\cos\delta}.$$

Then λ' and μ' are respectively the same functions of ρ' as λ and μ are of ρ.

Hence the last of equations (22) becomes, when ρ is negative,[2]

$$\varpi - \varepsilon = \left(n'\pi + \tfrac{1}{2}\pi + \lambda'\right) - \sec\delta\left(m'\pi + \tfrac{1}{2}\pi + \mu'\right),$$

---

1. The original erroneously had '$\varpi = \left(n - m\right)\pi - 2m\varepsilon - \mu\sec\delta - \lambda$'.
2. For the following equation, the original erroneously had 'e' instead of 'ε'.

where m' and n' are any integers, positive or negative. Hence, using equation (24), this equation becomes

$$\varpi = (n'-m')\pi - 2m'\varepsilon - \mu'\sec\delta + \lambda'. \tag{27}$$

Substituting in the other equations (22), we find

$$\left.\begin{aligned} \xi &= \left\{1-\frac{\rho^2\cos^2\delta}{\gamma^2}\right\}^{\frac{1}{2}}\cos\varpi = (-1)^{n'-m'}\left\{1-\frac{\rho'^2\cos^2\delta}{\gamma^2}\right\}^{\frac{1}{2}}\cos\{2m'\varepsilon+\mu'\sec\delta-\lambda'\} \\ \xi_3 &= \left\{1-\frac{\rho'^2\cos^2\delta}{\gamma^2}\right\}^{\frac{1}{2}}\cos\varpi = (-1)^{n'-m'+1}\left\{1-\frac{\rho'^2\cos^2\delta}{\gamma^2}\right\}^{\frac{1}{2}}\sin\{2m'\varepsilon+\mu'\sec\delta-\lambda'\} \end{aligned}\right\}. \tag{28}$$

The factor $(-1)^{n-m}$ in equations (26) and $(-1)^{n'-m'}$ in equations (28) may be omitted. For the points[1] $(\pm\xi, \pm\xi_3, \xi_1, \xi_2)$, that is to say, the points $(\xi, \xi_3, \pm\xi_1, \pm\xi_2)$, are both on the surface if one of them is.

17. Now, when ρ is small, so that ρ^2 is neglected,

$$\lambda = \frac{\rho\cos^2\delta}{\gamma\sin\delta}, \quad \mu = \frac{\rho\cos\delta}{\sin\delta}, \quad \mu\sec\delta - \lambda = \frac{\rho\sin\delta}{\gamma}.$$

|304| Hence equations (26) become

$$\left.\begin{aligned} \xi &= \quad\cos 2m\varepsilon + \frac{\rho\sin\delta}{\gamma}\sin 2m\varepsilon \\ \xi_3 &= -\sin 2m\varepsilon + \frac{\rho\sin\delta}{\gamma}\cos 2m\varepsilon \end{aligned}\right\}. \tag{29}$$

Similarly, when ρ' is small, equations (28) become

$$\left.\begin{aligned} \xi &= \quad\cos 2m'\varepsilon + \frac{\rho'\sin\delta}{\gamma}\sin 2m'\varepsilon \\ \xi_3 &= -\sin 2m'\varepsilon + \frac{\rho'\sin\delta}{\gamma}\cos 2m'\varepsilon \end{aligned}\right\}. \tag{30}$$

Hence the series of conical points on ee_3 are given by

$$\xi = \cos 2m\varepsilon, \quad \xi_3 = -\sin 2m\varepsilon \,(m = 0, \pm 1, \pm 2, \ldots).$$

Let these points be denoted by $d_0, d_1, d_{-1}, d_2, d_{-2}, \ldots$, so that

$$d_m = e\cos 2m\varepsilon - e_3\sin 2m\varepsilon, \quad d_{-m} = e\cos 2m\varepsilon + e_3\sin 2m\varepsilon;$$

also d_0 is the point e. The distance between any two consecutive conical points, such as d_m and d_{m+1} is κ, where

$$\cos\frac{\kappa}{\gamma} = \frac{(d_m|d_{m+1})}{\sqrt{(d_m|d_m)(d_{m+1}|d_{m+1})}} = (d_m|d_{m+1})$$

$$= \cos 2m\varepsilon\cos\overline{2m+2}\varepsilon + \sin 2m\varepsilon\sin\overline{2m+2}\varepsilon = \cos 2\varepsilon.$$

Hence $\kappa = 2\gamma\varepsilon$, where, since the whole line is of length $\pi\gamma$, a number of complete circuits of the line may be included. If ε/π be a rational number, that is, if $\cos\delta$ be a rational number, there are only a finite number of conical points to the surface. But,

1. For the following expression, the original erroneously had 'ξ_1' instead of 'ξ', and was missing a comma.

if $\cos\delta$ be an irrational number, there are an infinite number of conical points to the surface.

18. Let us now study the sheet of which *e*, that is d_0, is the conical point.
Putting $m=0$, equations (26) become

$$\left.\begin{aligned}\xi &= \left\{1-\frac{\rho^2\cos\delta}{\gamma^2}\right\}^{\frac{1}{2}}\cos\{\mu\sec\delta-\lambda\}\\ \xi_3 &= \left\{1-\frac{\rho^2\cos\delta}{\gamma^2}\right\}^{\frac{1}{2}}\sin\{\mu\sec\delta-\lambda\}\end{aligned}\right\}, \tag{31}$$

|305| where λ and μ are positive acute angles given by

$$\tan\lambda=\frac{\rho\cos^2\delta}{\{\gamma^2\sin^2\delta-\rho^2\cos^2\delta\}^{\frac{1}{2}}},\quad \tan\mu=\frac{\rho\cos\delta}{\{\gamma^2\sin^2\delta-\rho^2\cos^2\delta\}^{\frac{1}{2}}}.$$

Also, when ρ is small and ρ^2 is neglected, equations (29) become

$$\xi=1,\quad \xi_3=\frac{\rho\sin\delta}{\gamma}. \tag{32}$$

Also putting $m'=0$, equations (28) become

$$\left.\begin{aligned}\xi &= \left\{1-\frac{\rho'^2\cos^2\delta}{\gamma^2}\right\}^{\frac{1}{2}}\cos\{\mu'\sec\delta-\lambda'\}\\ \xi_3 &= \left\{1-\frac{\rho'^2\cos^2\delta}{\gamma^2}\right\}^{\frac{1}{2}}\sin\{\mu'\sec\delta-\lambda'\}\end{aligned}\right\}, \tag{33}$$

where λ' and μ' are positive acute angles given by

$$\tan\lambda'=\frac{\rho'\cos^2\delta}{\{\gamma^2\sin^2\delta-\rho'^2\cos^2\delta\}^{\frac{1}{2}}},\quad \tan\mu'=\frac{\rho'\cos\delta}{\{\gamma^2\sin^2\delta-\rho'^2\cos^2\delta\}^{\frac{1}{2}}}.$$

Also, when ρ' is small and ρ'^2 is neglected, equations (30) become

$$\xi=1,\quad \xi_3=-\frac{\rho'\sin\delta}{\gamma}. \qquad (34)^1$$

Hence, from equations (32), (34), and the first of equations (22), a point *x* on the sheet in the neighbourhood of *e* is

$$x=e+\frac{\rho\cos\delta}{\gamma}(e_1\cos\phi+e_2\sin\phi)+\frac{\rho\sin\delta}{\gamma}e_3; \tag{35}$$

and this equation holds whether ρ be positive or negative. Hence, in the neighbourhood of *e*,

$$x_1=\frac{\cos\delta}{\gamma}(e_1\cos\phi+e_2\sin\phi)+\frac{\sin\delta}{\gamma}e_3. \tag{36}$$

Hence (*cf. Universal Algebra*, §293) the tangent at *e* to the meridian geodesic in the plane ϕ is the line ex_1, that is,

---

1. This equation erroneously had 'γ^2' instead of 'γ'.

$$\frac{\cos\delta}{\gamma}\left(ee_1\cos\phi+ee_2\sin\phi\right)+\frac{\sin\delta}{\gamma}ee_3.$$

Now the angle χ between ex_1 and ee_3 is given by

$$\cos\chi=\frac{\left(ex_1|ee_3\right)}{\sqrt{\left(ex_1|ex_1\right)\left(ee_3|ee_3\right)}}=\sin\delta,$$

since $$\left(ee_1|ee_3\right)=0=\left(ee_2|ee_3\right),\ \text{and}\ \left(ee_3|ee_3\right)=1.$$

Hence $$\chi=\tfrac{1}{2}\pi-\delta.$$

|306| Hence the tangents at *e* to the meridian geodesics form a cone of revolution of semi-vertical angle $\frac{1}{2}\pi-\delta$.

It is now easy to state the relation of the semi-geodesic coordinates ρ and ϕ to polar geodesic coordinates ρ and θ. For the two geodesics ϕ and $\phi+\delta\phi$ make an angle $\delta\theta$ with each other at their point of intersection *e*. Now let *p* and p' be

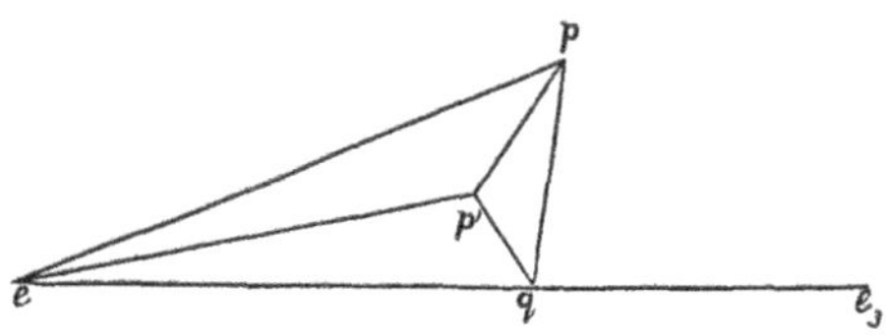

two points respectively on these geodesics, and at the same small geodesic distance ρ from *e*. Draw the perpendiculars *pq* and *p'q* to ee_3. Then, remembering that the properties of small figures are ultimately those of Euclidean space,

$$\overline{pp'}=\overline{ep}\cdot\delta\theta=\rho\delta\theta.$$

Also $$\overline{pp'}=\overline{pq}\cdot\delta\phi=\rho\cos\delta\,\delta\phi.$$

Hence $$\delta\theta=\cos\delta\,\delta\phi,\quad \theta=\phi\cos\delta.$$

Hence, if $\delta\sigma$ be an element of arc at any point of the surface,

$$\begin{aligned}(\delta\sigma)^2&=(\delta\rho)^2+\rho^2\cos^2\delta(\delta\phi)^2\\&=(\delta\rho)^2+\rho^2(\delta\theta)^2;\end{aligned}$$

and this agrees with §12 and with the formula of Euclidean plane geometry which holds for the geodesic geometry of this sheet.

19. Recurring to equations (31), note that, as ρ increases from 0 to $\gamma\tan\delta$, λ and μ increase from 0 to $\frac{1}{2}\pi$; and that, when $\rho>\gamma\tan\delta$, λ and μ are imaginary. Hence, for ρ positive, the sheet is real only when $\rho<\gamma\tan\delta$.

Thus, as ρ increases from 0 to $\gamma\tan\delta$, ξ decreases from 1 to $\cos\delta\cos\varepsilon$, ξ_3 increases from 0 to $\cos\delta\sin\varepsilon$.

The plane of the parallel of latitude through *x* is perpendicular to ee_3, and is, therefore, the plane xe_1e_2.

Hence the centre of the circular parallel of latitude is the point $xe_1e_2\cdot ee_3$, that is, $\left(xe_1e_2e_3\right)e-\left(xe_1e_2e\right)e_3$, that is, $\xi e+\xi_3e_3$.

Hence, as ρ increases from 0 to $\gamma\tan\delta$, the centres of the successful parallels of latitude of the sheet occupy successive positions on the line ee_3 between the point e and the point $e\cos\varepsilon + e_3\sin\varepsilon$, which is the centre of the extreme parallel of latitude on the sheet. Let $e\cos\varepsilon + e_3\sin\varepsilon$ be called the point c.

|307| The distance between e and c is $\gamma\varepsilon$. But, in §17, it is proved that the distance between e and d_{-1} is $2\gamma\varepsilon$. Hence c bisects the intercept between e and d_{-1}, of which the length is $2\gamma\varepsilon$.

Similarly, when ρ is negative, and equal to $-\rho'$, equations (33) and (34) show that the remainder of the sheet is, in all respects, a reflexion in the plane ee_1e_2 of the part of the sheet for ρ positive. Thus there is an extreme parallel of latitude, of which the centre is the point $e\cos\varepsilon - e_3\sin\varepsilon (= c', \text{ say})$, and the point c' bisects the intercept between e and d_1 which is of length $2\gamma\varepsilon$.

The perpendicular distance, ξ, of any point x on the sheet from the axis ee_3 is given (*cf. Universal Algebra*, §226) by

$$\sin\frac{\zeta}{\gamma} = \sqrt{\frac{(xee_3|xee_3)}{(x|x)(ee_3|ee_3)}} = \sqrt{(xee_3|xee_3)} = \frac{\rho\cos\delta}{\gamma}. \qquad (38)[1]$$

Hence, as ρ increases from 0 to $\gamma\tan\delta$, the radii of the corresponding parallels of latitude increase from 0 to $\delta\gamma$.

Though the real part of each sheet is bounded by the two extreme parallels of latitude corresponding to $\rho^2 = \gamma^2\tan^2\delta$, the complete surface denoted by equation (23) is not discontinuous. For, since c_1 bisects the intercept between e and d_{-1}, the point c is also the centre of one of the extreme parallels of latitude of the sheet corresponding to the conical point d_{-1}, and the two parallels of latitude centre c coincide. Thus, at the termination of one sheet, another sheet commences, corresponding to the next conical point given by the series ... d_{-2}, d_{-1}, d_0, d_1, d_2, ... It will be proved that the junction of two sheets forms a cuspidal line on the surface.

20. The simpler properties of the meridian curves can now be easily stated. Consider the meridian curve in the plane ee_2e_3. Take any point p on the curve and draw pq perpendicular to ee_3.

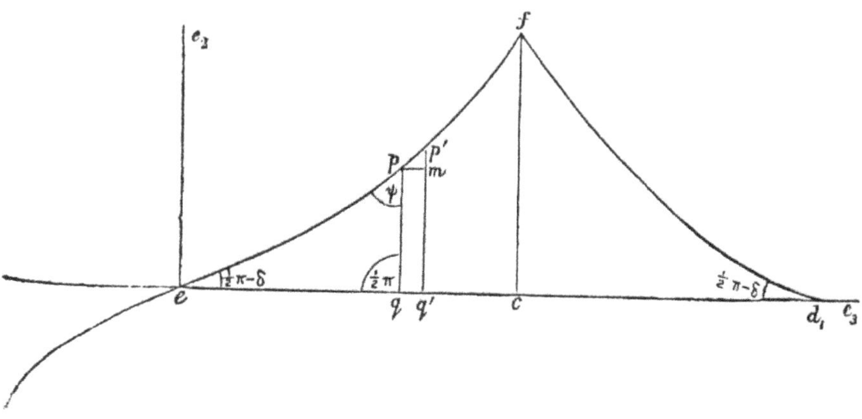

1. There is no equation (37).

|308| Let $\overline{ep}$ denote the distance between e and p, and ρ the length of the arc ep of the curve. At e the curve cuts the line ee_3 at an angle $\frac{1}{2}\pi - \delta$. From equation (38),

$$\sin\frac{\overline{pq}}{\gamma} = \frac{\rho\cos\delta}{\gamma} = \frac{\text{arc } ep\cdot\cos\delta}{\gamma}.$$

This may be looked on as the geometrical definition of the curve.

The angle ψ, which the tangent at p makes with pq, can be found from this equation. For, let p' be a neighbouring point to p on the curve; draw $p'q'$ perpendicular to ee_3. Let pm be an arc of the circle of equal distance from ee_3. Then the arc pm is ultimately a straight line perpendicular to pq. Also

$$p'm = p'q' - pq.$$

Hence, by differentiating equation (38),

$$\cos\frac{pq}{\gamma}\frac{p'm}{\gamma} = \frac{\delta\rho\cos\delta}{\gamma} = \frac{pp'\cos\delta}{\gamma}.$$

Hence $$\cos\psi = \text{Lt.}\frac{p'm}{pp'} = \frac{\cos\delta}{\cos\frac{\overline{pq}}{\gamma}} = \frac{\gamma\cos\delta}{\left\{\gamma^2 - \rho^2\cos^2\delta\right\}^{\frac{1}{2}}}. \qquad (39)$$

The angle ψ increases with ρ; hence the curve is convex to the axis ee_3.

The value of ψ is imaginary if

$$\frac{\gamma\cos\delta}{\left\{\gamma^4 - \rho^2\cos^2\delta\right\}^{\frac{1}{2}}} > 1,$$

that is, if $$\rho > \gamma\tan\delta.$$

This agrees with §19.

When $\rho = \gamma\tan\delta$, $\psi = 0$. Hence the tangent at f, the extreme point of the curve, is perpendicular to the axis ee_3. Hence as previously stated, the junction of two sheets is a cuspidal line.

The inverse measure of curvature (ρ_1) of the curve at p is most easily found from the formula

$$\frac{1}{\rho_1\rho_2} = -\frac{1}{\gamma^2},$$

|309| which expresses the condition that the curve is a meridian geodesic of a parabolic surface of constant curvature.

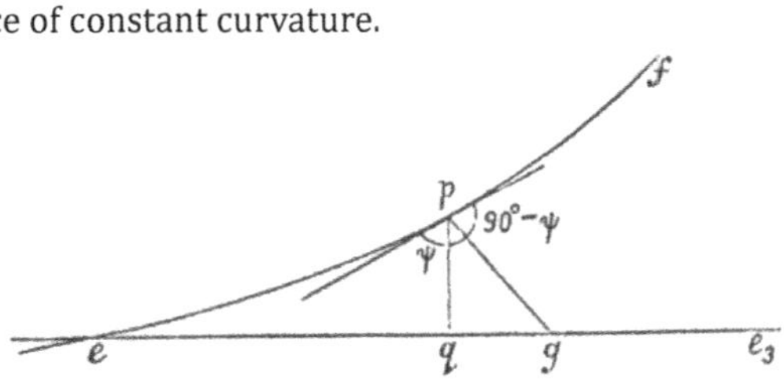

For draw the normal pg cutting ee_3 in g. Then, from Napier's formulæ for right-angled triangles,

$$\tan\frac{\overline{pg}}{\gamma} = \tan\frac{\overline{pq}}{\gamma}\sec(90° - \psi) = \tan\frac{\overline{pq}}{\gamma}\operatorname{cosec}\psi.$$

Hence, from equations (38) and (39),

$$\tan\frac{\overline{pg}}{\gamma} = \frac{\rho\cos\delta}{\{\gamma^2\sin^2\delta - \rho^2\cos^2\delta\}^{\frac{1}{2}}}.$$

But $\overline{pg}$ is the radius of circular curvature at *p* of the line of curvature perpendicular to the meridian; this must be distinguished [*cf. Universal Algebra*, §288 (4)] from the inverse measure of curvature o_2. The relation between the two is

$$\rho_2 = \gamma\tan\frac{\overline{pg}}{\gamma}.$$

Hence
$$\rho_2 = \frac{\gamma\rho\cos\delta}{\{\gamma^2\sin^2\delta - \rho^2\cos^2\delta\}^{\frac{1}{2}}}.$$

Hence
$$\rho_1 = -\frac{\gamma^2}{\rho_2} = -\frac{\gamma\{\gamma^2\sin^2\delta - \rho^2\cos^2\delta\}^{\frac{1}{2}}}{\rho\cos\delta}. \qquad (40)$$

Therefore ρ decreases numerically from ∞ at *e* to 0 at *f*.

21. The distance between successive conical points of the complete surface is (*cf.* §17) $2\gamma\varepsilon$, that is, $\pi\gamma(\sec\delta - 1)$. But the length of a complete straight line is $\pi\gamma$. Hence, if $\cos\delta = \frac{1}{2}$, that is, if $\delta = \frac{1}{3}\pi$, the surface has only one conical point. In this case the surface has only one sheet, and two extremities of the sheet, for ρ positive and ρ negative respectively, coincide along a parallel of latitude of radius |310| $\frac{1}{3}\pi\gamma$ [*cf.* equation (38)]. The total length of a meridian arc, measured from *e*, is $\gamma\tan\frac{1}{3}\pi$, that is, $\sqrt{3}\gamma$.

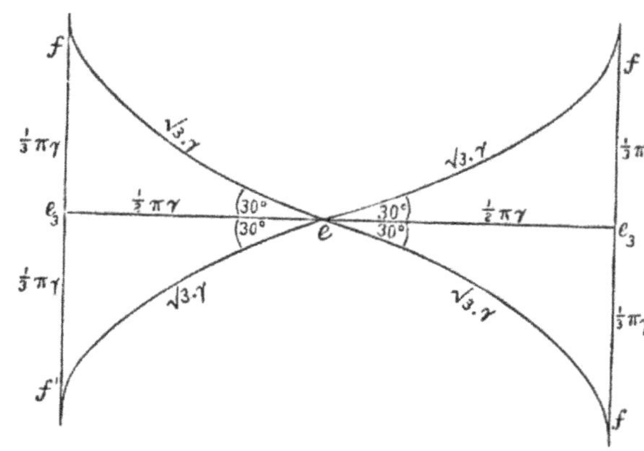

The accompanying figure, of course, cannot, in Euclidean space, represent the form of the meridian section of the surface in elliptic space; but it may help the imagination to realize the geometric relations of the parts. The two lines fe_3f', at the extremities of the figure, must be conceived as coinciding—thus the figure may be wrapped on a cylinder so that these lines become the same generator. The lines are tangents, so that the surface in elliptic space has a cuspidal line.

With the notation of the end of §18, $\theta=\phi\cos\delta=\frac{1}{2}\phi$. Now ϕ varies from 0 to 2π. Hence, each half of the sheet of a single-sheeted parabolic surface of revolution of constant curvature corresponds to a semi-circle of radius $\sqrt{3}\gamma$ on a Euclidean plane in Euclidean space.

The equation (23) can be rationalized in this case, since $\gamma=\frac{1}{3}\pi$, $\varepsilon=\frac{1}{2}\pi$. Thus, the equation of the one-sheeted parabolic surface of revolution of constant curvature is

$$3\left(\xi^2+\xi_3^2\right)^2\xi_3^2=3\left(3\xi^2+\xi_3^2\right)\left(\xi^2+\xi_3^2\right)\left(\xi_1^2+\xi_2^2\right)+\left(2\xi^2-\xi_3^2\right)\left(\xi_1^2+\xi_2^2\right)^2.$$

[*Added May*, 1898.—Professor Burnside, who has kindly read over this paper in manuscript, has pointed out that the detailed mention of the simplest case of all has been omitted. For in equation (16) put $\alpha=0$, so that $\pi=\beta$; this is equivalent to putting $\delta=\frac{1}{2}\pi$. Hence the equation for ϖ becomes

$$\frac{d\varpi}{d\rho}=\frac{1}{\gamma\left(1-\beta^2\right)^{\frac{1}{2}}},\quad \varpi=\frac{\rho+\varepsilon}{\gamma\left(1-\beta^2\right)^{\frac{1}{2}}}.$$

|311| Thus, instead of equations (21), we find

$$x=\left(1-\beta^2\right)^{\frac{1}{2}}\cos\frac{\rho+\varepsilon}{\gamma\left(1-\beta^2\right)^{\frac{1}{2}}}e+\beta\left(e_1\cos\phi+e_2\sin\phi\right)+\left(1-\beta^2\right)^{\frac{1}{2}}\cos\frac{\rho+\varepsilon}{\gamma\left(1-\beta^2\right)^{\frac{1}{2}}}e_3.$$

Thus the surface is a surface of equal distance from ee_3, and the constant distance from ee_3 is $\gamma\sin^{-1}\beta$. Professor Burnside makes this surface the basis of a synthetic treatment of the subject of this paper which is appended to this paper.][1]

Geodesics on Surfaces in Hyperbolic Space.

22. The preceding results can now easily be adapted to hyperbolic space; it will not be necessary to repeat the proofs in detail.

Let e be the origin of three mutually rectangular axes ee_1, ee_2, ee_3; and let e, e_1, e_2, e_3 be a system of mutually normal unit points. Then (*cf. Universal Algebra*, Book VI., chapter iv.) e_1, e_2, e_3 are points in antispace, and

$$(e|e)=1,\quad (e_1|e_1)=-1=(e_2|e_2)=(e_3|e_3).$$

Also, if x be any point, it can be expressed in the form $\xi e+\xi_1e_1+\xi_2e_2+\xi_3e_3$. If it be a unit point in (hyperbolic) space, as distinct from antispace,

$$\xi^2-\xi_1^2-\xi_2^2-\xi_3^2=1.$$

Also, if ρ be the distance[2] $\overline{ex}$, θ the angle xee_3, ϕ the angle between the planes xee_3 and e_1ee_3, then, x being a unit point,

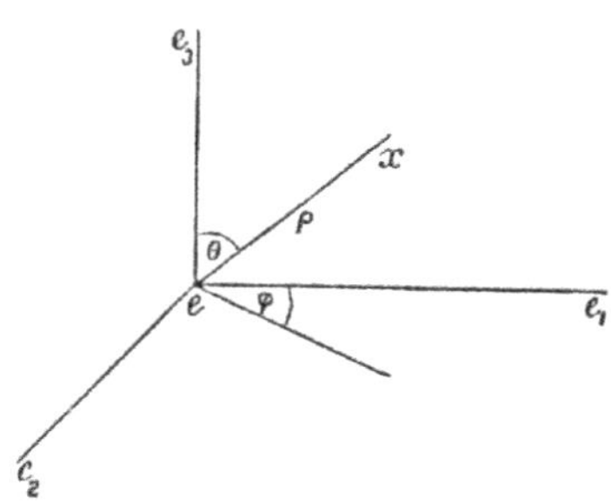

1. See pp. 84–86.
2. The accompanying diagram (right) erroneously labelled that distance P, not ρ.

$$\xi = \cosh\frac{\rho}{\gamma},\ \xi_1 = \sinh\frac{\rho}{\gamma}\sin\theta\cos\phi,\ \xi_2 = \sinh\frac{\rho}{\gamma}\sin\theta\sin\phi,\ \xi_3 = \sinh\frac{\rho}{\gamma}\cos\theta.$$

Hence γ is the "space constant" of the hyperbolic space.

Also let x, as before, be a unit point on a surface, defined by curvilinear coordinates θ and ϕ, distinct from the θ and ϕ of the previous explanation. Then $(x|x)=1$. The formulæ remain substantially unchanged from those of elliptic space; they can be |312| derived from the "elliptic" formulæ by writing $\gamma\sqrt{-1}$ instead of γ. Hence formulæ not involving γ will be absolutely unchanged, others involving γ^2 will have their signs changed, and some trigonometrical functions will be turned into hyperbolic functions. Then

$$\frac{\delta\sigma^2}{\gamma^2} = -(x_1|x_1)(\delta\theta)^2 - 2(x_1|x_2)\delta\theta\,\delta\phi - (x_2|x_2)(\delta\phi)^2.$$

Also
$$\frac{-\gamma^2\{(x_1|x_1)(x_2|x_2)-(x_3|x_2)\}^2}{\rho_1\rho_3}$$
$$= \Delta_1 - \Delta_2 + \{(x_1|x_1)(x_2|x_2)-(x_1|x_2)^2\}$$
$$\times\{(x_1|x_2)^2-(x_1|x_1)(x_2|x_2)+(x_1|x_2)_{12}-\tfrac{1}{2}(x_2|x_2)_{11}-\tfrac{1}{2}(x_1|x_1)_{22}\}, \quad (41)$$

where Δ_1 and Δ_2 have the same meanings as in the corresponding formula (equation 1) for elliptic geometry.

The condition that the curve $\theta = f(\tau),\ \phi = F(\tau)$ may be a geodesic is

$$(\dot{x}\ddot{x}|x_1x_2) = 0. \quad (42)$$

This can be expressed in the same form as equation (3).

23. The investigations of §§(4), (5), and (6) hold for hyperbolic space; that is to say, all the properties of semi-geodesic orthogonal coordinates. Thus, if the curves $\phi = \text{constant}$ be the geodesic family, and the curves $\rho = \text{constant}$ be the orthogonal family, we have

$$(x_1|x_1) = -\frac{1}{\gamma^2},\ (x_1|x_1) = 0. \quad (43)$$

Also
$$(\delta\sigma)^2 = (\delta\rho)^2 - \gamma^2(x_2|x_2)(\delta\phi)^2 = (\delta\rho)^2 + \gamma^2\Pi^2(\delta\phi)^2, \quad (44)$$

where
$$\Pi^2 = -(x_2|x_2).$$

The special properties of polar geodesic coordinates also hold, and with these coordinates the limit of Π, when ρ is small, is ρ/γ.

Equation (6) of §8 becomes, with semi-geodesic orthogonal coordinates,

$$\frac{\delta^2\Pi}{\delta\rho^2} + \left(\frac{1}{\rho_1\rho_2} - \frac{1}{\gamma^2}\right)\Pi = 0. \quad (45)$$

Also equations (7) and (8) of §9 for geodesics hold without any change of form. Hence Beltrami's theorem respecting the geodesic representation on planes of surfaces of constant curvature can be proved as in §10. Also, as in §11, putting

$$\frac{1}{\rho_1\rho_2} - \frac{1}{\gamma^2} = \frac{1}{\varepsilon^2}, \text{ or } -\frac{1}{\varepsilon^2}, \text{ or } 0,$$

|313| equations (10) hold without modification. Let these surfaces, as in elliptic space, be called respectively "elliptic," "hyperbolic,"[1] and "parabolic" surfaces of constant curvature.

The theorems of §§13 and 14 also hold, namely, the only surfaces in hyperbolic space which admit of two families of geodesics being drawn on them mutually orthogonal are parabolic surfaces of constant curvature, and, conversely, on any such surface two such families can be drawn, of which one family includes any assigned geodesic.

Let ξ = constant denote one such family of geodesics, and let η = constant denote the orthogonal family. Let the subscript 1 denote differentiation with respect to ξ and α with respect to η.

Then we may assume

$$(x_1|x_1) = -\frac{1}{\gamma^2} = (x_2|x_2), \quad (x_1|x_2) = 0;$$

and
$$(\delta\sigma)^2 = (\delta\xi)^2 + (\delta\eta)^2.$$

Thus, as in elliptic space, it may be seen that the geodesic geometry of the surface is Euclidean.

24. To find the equation of the general type of surfaces of revolution which are also parabolic surfaces of constant curvature. It may be noticed that Bolyai's limit-surfaces must be included as a special case of such surfaces.

Let ee_3 be the axis of revolution, and let ee_1, ee_2, ee_3 be three mutually rectangular axes of origin e. Let e, e_1, e_2, e_3 be a set of mutually normal unit points; thus e_1, e_2, e_3 are antispatial. Any unit spatial point

$$x = \xi e + \xi_1 e_1 + \xi_2 e_2 + \xi_3 e_3$$

satisfies the condition
$$(x|x) = 1,$$

that is,
$$\xi^2 - \xi_1^2 - \xi_2^2 - \xi_3^2 = 1,$$

and ξ, ξ_1, ξ_2, ξ_3 are the ordinary rectangular coordinates of hyperbolic geometry. Now, as in §15, take meridians and parallels of latitude as a system of semi-geodesic orthogonal coordinates. Let the length of the arc of the meridian through x, measured from a fixed parallel of latitude, be ρ, and let ϕ be the angle between the planes xee_3 and e_1ee_3. Then, as in §15, we have, since the last of equations (10) holds,

$$(\delta\sigma)^2 = (\delta\rho)^2 + \gamma^2\Pi^2(\delta\phi)^2;$$

and
$$\Pi = \alpha\rho + \beta, \tag{46}$$

1. The original erroneously had a single quote mark on the right rather than a double quote mark.

|314| where α and β are constants, since the surface is one of revolution. Also, since the surface is one of revolution, we may assume

$$x = \xi e + \mu(e_1 \cos\phi + e_2 \sin\phi) + \xi_3 e_3, \tag{47}$$

where ξ, μ, ξ_3 are functions of ρ only. Also, since

$$(x|x) = 1,$$

we have
$$\xi^2 - \mu^2 - \xi_3^2 = 1. \tag{48}$$

Since
$$(x_1|x_1) = -\frac{1}{\gamma^2},$$

we have
$$\left(\frac{d\mu}{d\rho}\right)^2 + \left(\frac{d\xi_3}{d\rho}\right)^2 - \left(\frac{d\xi}{d\rho}\right)^2 = \frac{1}{\gamma^2}. \tag{49}$$

Since
$$(x_2|x_2) = -\Pi^2 = -(\alpha\rho + \beta)^2,$$

we have
$$\mu = \alpha\rho + \beta. \tag{50}$$

The condition $(x_1|x_2) = 0$ is satisfied identically. From equations (48) and (50) we find

$$\xi^2 - \xi_3^2 = 1 + \mu^2 = 1 + (\alpha\rho + \beta)^2 = \sigma^2, \text{ say.} \tag{51}$$

Assume $\xi = \sigma \cosh\varpi$, $\xi_3 = \sigma \sinh\varpi$, where ϖ is some function of ρ to be determined. Substituting for μ, ξ, ξ_3 in equation (49),[1] we find

$$\sigma^2 \left(\frac{d\varpi}{d\rho}\right)^2 - \left(\frac{d\sigma}{d\rho}\right)^2 = \frac{1}{\gamma^2} - \alpha^2,$$

that is,
$$\frac{d\varpi}{d\rho} = \frac{\left\{\frac{1}{\gamma^2} - \alpha^2 + \frac{\mu^2}{\gamma^2}\right\}^{\frac{1}{2}}}{(1 + \mu^2)}. \tag{52}$$

Hence these cases arise for examination according as

$$(\text{Case I.})\ \alpha = \frac{1}{\gamma}, \quad (\text{Case II.})\ \alpha < \frac{1}{\gamma}, \quad (\text{Case III.})\ \alpha > \frac{1}{\gamma}.$$

25. *Case I.*—Let
$$\alpha = \frac{1}{\gamma}.$$

Equation (52) becomes

$$\frac{d\varpi}{d\rho} = \frac{\mu}{\gamma(1 + \mu^2)} = \frac{\rho + \kappa}{\gamma^2 + (\rho + \kappa)^2},$$

|315| where κ/γ is written for β. Hence

1. The original erroneously referenced equation (48).

$$\varpi = \log\{\gamma^2 + (\rho+\kappa)^2\}^{\frac{1}{2}} + \text{constant}$$
$$= \log\frac{\{\gamma^2 + (\rho+\kappa)^2\}^{\frac{1}{2}}}{\delta},$$

where δ is the constant introduced by integration. Hence

$$\left.\begin{aligned}
\xi &= \frac{\{\gamma^2 + (\rho+\kappa)^2\}^{\frac{1}{2}}}{\gamma}\cosh\varpi = \frac{\delta^2+\gamma^2+(\rho+\kappa)^2}{2\gamma\delta} \\
\xi_3 &= \frac{\{\gamma^2 + (\rho+\kappa)^2\}^{\frac{1}{2}}}{\gamma}\sinh\varpi = \frac{\gamma^2-\delta^2+(\rho+\kappa)^2}{2\gamma\delta} \\
\xi_1 &= \frac{\rho+\kappa}{\gamma}\cos\phi \\
\xi_2 &= \frac{\rho+\kappa}{\gamma}\sin\phi
\end{aligned}\right\}. \tag{53}$$

Now measure ρ from the evanescent parallel of latitude, which has hitherto been denoted by $\rho=-\kappa$, and choose *e* to be the point in which ee_3 cuts the axis ee_3, so that $\delta=\gamma$, and we find, without diminishing the generality of the type of surface,

$$\xi = 1+\tfrac{1}{2}\frac{\rho^2}{\gamma^2},\quad \xi_3 = \tfrac{1}{2}\frac{\rho^2}{\gamma^2},$$
$$\xi_1 = \frac{\rho}{\gamma}\cos\phi,\quad \xi_2 = \frac{\rho}{\gamma}\sin\phi.$$

Hence $$(x|x) = \{x|(e+e_3)\}^2.$$

But (*cf. Universal Algebra*, §299) this is the equation of Bolyai's limit-surface, which passes through the point *e*, and has its centre at the point $e+e_3$ on the absolute.

26. *Case II.*—Let $$\alpha = \frac{\cos\delta}{\gamma}.$$

Equation (52) becomes

$$\frac{d\varpi}{d\rho} = \frac{\{\sin^2\delta+\mu^2\}^{\frac{1}{2}}}{\gamma(1+\mu^2)} = \frac{\{\gamma^2\sin^2\delta+(\rho\cos\delta+\kappa)^2\}^{\frac{1}{2}}}{\gamma^2(\rho\cos\delta+\kappa)^2}.$$

Integrating by means of the transformation

$$\gamma\sin^2\delta+(\rho\cos\delta+\kappa)^2 = \zeta^2(\rho\cos\delta+\kappa)^2,$$

|316| we find

$$\varpi+\varepsilon = \sec\delta\coth^{-1}\frac{\{\gamma^2\sin\delta+(\rho\cos\delta+\kappa)^2\}^{\frac{1}{2}}}{\rho\cos\delta+\kappa}$$
$$-\coth^{-1}\frac{\{\gamma^2\sin^2\delta+(\rho\cos\delta+\kappa)^2\}^{\frac{1}{2}}}{(\rho\cos\delta+\kappa)\cos\delta}.$$

Hence, measuring ρ from the parallel of latitude hitherto denoted by

$$\rho=-\kappa\sec\delta,$$

we find the following set of equations to determine the surface,

$$\left.\begin{aligned} x&=\xi e+\frac{\rho\cos\delta}{\gamma}(e_1\cos\phi+e_2\sin\phi)+\xi_3 e_3\\ \xi&=\left\{1+\frac{\rho^2\cos^2\delta}{\gamma^2}\right\}^{\frac{1}{2}}\cosh\varpi\\ \xi_3&=\left\{1+\frac{\rho^2\cos^2\delta}{\gamma^2}\right\}^{\frac{1}{2}}\sinh\varpi\\ \varpi+\varepsilon&=\sec\delta\coth^{-1}\frac{\{\gamma^2\sin^2\delta+\rho^2\cos^2\delta\}^{\frac{1}{2}}}{\rho\cos\delta}\\ &\qquad-\coth^{-1}\frac{\{\gamma^2\sin^2\delta+\rho^2\cos^2\delta\}^{\frac{1}{2}}}{\rho\cos^2\delta} \end{aligned}\right\}. \qquad (54)$$

The parallel of latitude $\rho=0$ is the evanescent parallel surrounding the point in which the surface cuts the axis ee_3. Let this point be the point *e*; then this assumption determines ε by the equation

$$\varepsilon=\sec\delta\coth^{-1}\infty-\coth^{-1}\infty=0.$$

Also, in the neighbourhood of the point *e*, when ρ^2 is neglected,

$$\varpi=\rho\left(\frac{d\varpi}{d\rho}\right)_{\rho=0}=\frac{\rho\sin\delta}{\gamma};$$

and hence
$$\xi=1,\quad \xi_3=\frac{\rho\sin\delta}{\gamma}.$$

Thus, in the neighbourhood of *e*,

$$x=e+\frac{\rho\cos\delta}{\gamma}(e_1\cos\phi+e_2\sin\phi)+\frac{\rho\cos\delta}{\gamma}e_3.$$

Hence
$$x_1=\frac{\cos\delta}{\gamma}(e_1\cos\phi+e_2\sin\phi)+\frac{\sin\delta}{\gamma}e_3.$$

|317| Thus the tangent line at *e* to the meridian curve in the plane ϕ is denoted by the line element ex_1, that is, by

$$\frac{\cos\delta}{\gamma}(ee_1\cos\phi+ee_2\sin\phi)+\frac{\sin\delta}{\gamma}ee_3.$$

Hence (*cf. Universal Algebra*, §§244, 248) the angle between this line and the axis is given by

$$\cos\theta=\frac{(ee_3|ex_1)}{\sqrt{(ee_3|ee_3)(ex_1|ex_1)}}=\sin\delta.$$

Hence the tangents at e form a cone of semi-vertical angle $\frac{1}{2}\pi-\delta$, and the point e is a conical point on the surface. By putting $-\rho'$ instead of ρ in the equations it is easily seen that the surface is symmetrical with respect to the point e.

The plane of the parallel of latitude through the point x is perpendicular to ee_3, and is therefore represented by the planar element xe_1e_2, that is, by the planar element $(\xi ee_1e_2+\xi_3e_3e_1e_2)$. The centre of the parallel of latitude lies on ee_3, and is therefore the point $xe_1e_2\cdot ee_3$

But
$$xe_1e_2\cdot ee_3=\xi ee_1e_2\cdot ee_3+\xi_3e_3e_1e_2\cdot ee_3$$
$$=(\xi e+\xi_3e_3)(ee_1e_2e_3)\equiv\xi e+\xi_3e_3,$$

omitting the numerical factor.

The distance d of this point from e is given by

$$\cosh\frac{d}{\gamma}=\frac{\{e|(\xi e+\xi_3e_3)\}}{\sqrt{(e|e)\{(\xi e-\xi_3e_3)|(\xi e+\xi_3e_3)\}}}=\frac{\xi}{\sqrt{\xi^2-\xi_3^2}}=\cosh\varpi.$$

Hence
$$d=\varpi\gamma.$$

The radius (ζ) of the parallel of latitude is the perpendicular distance from x to the line ee_3. This is given by [*cf. Universal Algebra*, §254 (5)]

$$\sinh\frac{\zeta}{\gamma}=\sqrt{\frac{-(xee_3|xee_3)}{(x|x)(ee_3|ee_3)}}=\sqrt{(xee_3|xee_3)}=\frac{\rho\cos\delta}{\gamma}. \tag{55}$$

Now in the figure let x and x' be two neighbouring points on a

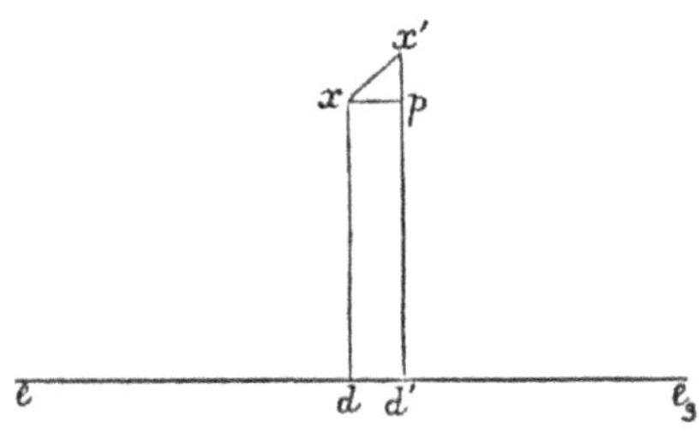

|318| meridian, let xd and $x'd'$ be the perpendiculars on to ee_3, and let xp be the arc of the curve of equal distance from ee_3, which is ultimately a straight line perpendicular to $x'd'$. Then ultimately

$$\overline{xx'}=\delta\rho,\quad \overline{x'p}=\delta\zeta.$$

Also, if ψ be the angle which the tangent to the meridian at x makes with xd, we have ultimately

$$\cos\psi=\frac{\overline{x'p}}{\overline{xx'}}=\frac{\delta\zeta}{\delta\rho}=\frac{\cos\delta}{\cosh\frac{\zeta}{\gamma}},$$

from equation (55). Thus

$$\cos\psi=\frac{\gamma\cos\delta}{\{\gamma^2+\rho^2\cos^2\delta\}^{\frac{1}{2}}}. \tag{56}$$

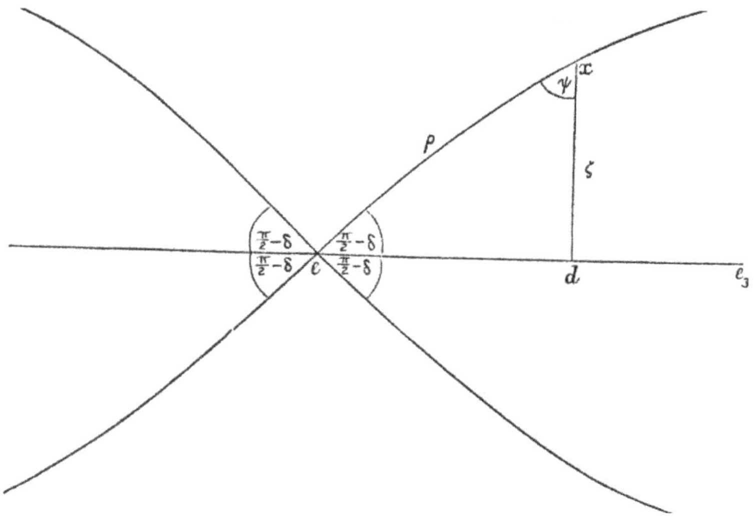

The form in hyperbolic space of the section of the surface by any meridian plane is suggested by the annexed figure. Both branches go to infinity at both ends. By putting $\delta = 0$, the surface becomes the limit surface investigated in Case I.

[*Added May*, 1898.—Here again, as in elliptic space, the simplest case has been omitted, namely, when $\delta = \frac{1}{2}\pi$. Then the equations become

$$\xi = \left(1+\beta^2\right)^{\frac{1}{2}} \cosh \frac{\rho+\varepsilon}{\gamma\left(1+\beta^2\right)^{\frac{1}{2}}}, \quad \xi_3 = \left(1+\beta^2\right)^{\frac{1}{2}} \sinh \frac{\rho+\varepsilon}{\gamma\left(1+\beta^2\right)^{\frac{1}{2}}},$$

$$\xi_1 = \beta\cos\phi, \quad \xi_2 = \beta\sin\phi.$$

|319| Thus the equation of the surface is

$$\beta^2\left(\xi^2 - \xi_3^2\right) - \left(1+\beta^2\right)\left(\xi_1^2 + \xi_2^2\right) = 0.$$

This is a surface of equal distance from the line ee_3.]

27. *Case III.*—Let
$$\alpha = \frac{\cosh\delta}{\gamma}.$$

Equation (52) becomes

$$\frac{d\varpi}{d\rho} = \frac{\left\{\mu^2 - \sinh^2\delta\right\}^{\frac{1}{2}}}{\gamma\left(1+\mu^2\right)} = \frac{\left\{\left(\rho\cosh\delta+\kappa\right)^2 - \gamma^2\sinh^2\delta\right\}^{\frac{1}{2}}}{\gamma^2 + \left(\rho\cosh\delta+\kappa\right)^2}.$$

Integrating by means of the transformation

$$\left(\rho\cosh\delta+\kappa\right)^2 - \gamma^2\sinh^2\delta = \xi^2\left(\rho\cosh\delta+\kappa\right)^2,$$

we find

$$\varpi + \varepsilon = \operatorname{sech}\delta \tanh^{-1} \frac{\left\{\left(\rho\cosh\delta+\kappa\right)^2 - \gamma^2\sinh^2\delta\right\}^{\frac{1}{2}}}{\rho\cosh\delta+\kappa}$$
$$- \tanh^{-1} \frac{\left\{\left(\rho\cosh\delta+\kappa\right)^2 - \gamma^2\sinh^2\delta\right\}^{\frac{1}{2}}}{\left(\rho\cosh\delta+\kappa\right)\cosh\delta}.$$

The generality of the surface will not be impaired by assuming $\kappa = 0 = \varepsilon$. Then the following set of equations determine the surface

$$\left.\begin{aligned} x &= \xi e + \frac{\rho\cosh\delta}{\gamma}(e_1\cos\phi + e_2\sin\phi) + \xi_3 e_3 \\ \xi &= \left\{1 + \frac{\rho^2\cosh^2\delta}{\gamma^2}\right\}^{\frac{1}{2}}\cosh\varpi \\ \xi_3 &= \left\{1 + \frac{\rho^2\cosh^2\delta}{\gamma^2}\right\}^{\frac{1}{2}}\sinh\varpi \\ \varpi &= \operatorname{sech}\delta\tanh^{-1}\frac{\{\rho^2\cosh\delta - \gamma^2\sinh^2\delta\}^{\frac{1}{2}}}{\rho\cosh\delta} \\ &\qquad - \tanh^{-1}\frac{\{\rho^2\cosh^2\delta - \gamma^2\sinh^2\delta\}^{\frac{1}{2}}}{\rho\cosh^2\delta} \end{aligned}\right\}. \tag{57}$$

The value of ϖ is imaginary if ρ lie between $\pm\gamma\tanh\delta$; hence the surface does not cut the axis ee_3 in real points.

Also $\varpi = 0$, when $\rho = \pm\gamma\tanh\delta$; and ϖ is real as ρ increases numerically from $\pm\gamma\tanh\delta$ to $\pm\infty$.

|320| Also, as ρ increases from $\gamma\tanh\delta$ to $+\infty$,

$$\frac{d\varpi}{d\rho} = \frac{\{\rho^2\cosh^2\delta - \gamma^2\sinh^2\delta\}^{\frac{1}{2}}}{\gamma^2 + \rho^2\cosh^2\delta},$$

and is therefore always positive. Hence ϖ continually increases, and

$$\varpi_{(\rho=\infty)} = \operatorname{sech}\delta\tanh^{-1}1 - \tanh^{-1}\{\operatorname{sech}\delta\} = \infty.$$

Hence when ρ is infinite, the limit of $\xi/\xi_3 = 1$.

Negative values of ρ give the same numerical values of ϖ, with the opposite sign, as the numerically equal positive values of ρ. Hence, for negative values of ρ, ξ/ξ_3 runs through the same numerical series of values only with the opposite sign, and the limit of ξ/ξ_3, when $\rho = -\infty$, is -1.

The perpendicular distance of x from ee_3 is given by

$$\sinh\frac{\zeta}{\gamma} = \sqrt{\frac{-(xee_3|xee_3)}{(x|x)(ee_3|ee_3)}} = \sqrt{(xee_3|xee_3)} = \frac{\rho\cos\delta}{\gamma}.$$

As ρ increases from $\gamma\tanh\delta$ to ∞, ζ increases from $\delta\gamma$ to ∞.

The centre of the parallel of latitude through x is the point $xe_1e_2 \cdot ee_3$, that is, the point $\xi e + \xi_3 e_3$. As ϖ varies from 0 to ∞, this point moves from e to $e + e_3$, a point on the absolute on ee_3; and, as ϖ varies from 0 to $-\infty$, this point moves from e to $e - e_3$, the other point on the absolute on ee_3.

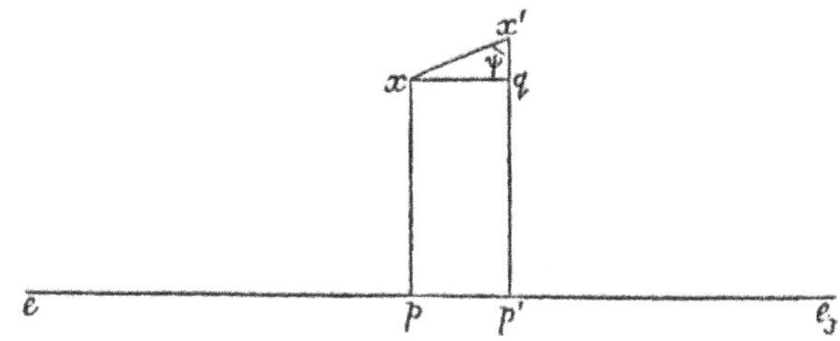

Let x and x' be neighbouring points on the same meridian curve, and draw xp and $x'p'$ perpendiculars to ee_3. Let ψ be the angle between the tangent at x and xp. Then, by the same construction as before,

$$\cos\psi = \frac{d\zeta}{d\rho} = \frac{\cosh\delta}{\cosh\frac{\zeta}{\gamma}} = \frac{\gamma\cosh\delta}{\{\gamma^2+\rho^2\cosh^2\delta\}^{\frac{1}{2}}}.$$

Hence, when $$\rho = \gamma\tanh\delta,$$
ψ is zero, and as ρ increases to ∞, ψ increases to $\frac{1}{2}\pi$.

|321| Thus the form in hyperbolic space of a meridian curve is suggested by the annexed figure.

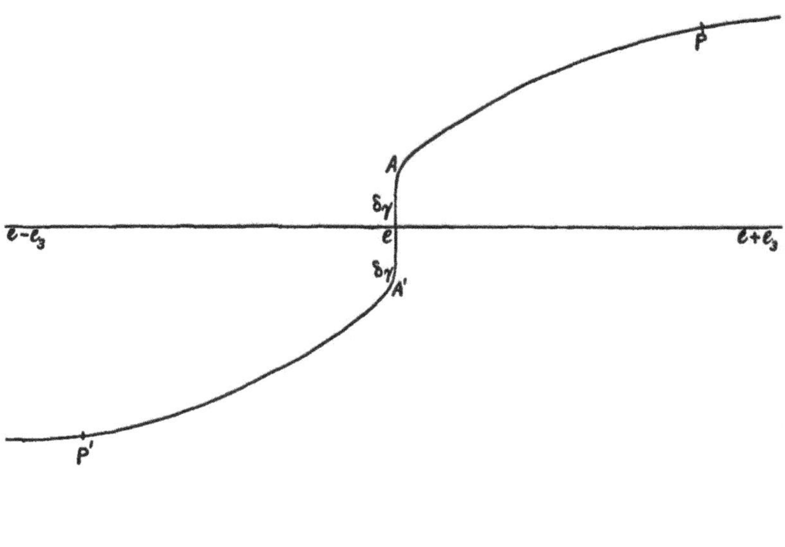

There are two detached branches, AP and $A'P'$, which both go to infinity; AA' is the tangent to the curve both at A and at A'. The arc $AP = \rho - \gamma\tanh\delta$, where the parameter ρ defines the point P; also, if ρ be negative and is written $-\rho'$, and if it defines a point P' on the other branch, then the arc $A'P' = \rho' - \gamma\tanh\delta$. The surface is formed by revolving this figure round ee_3. A meridian section is shown in the annexed figure.

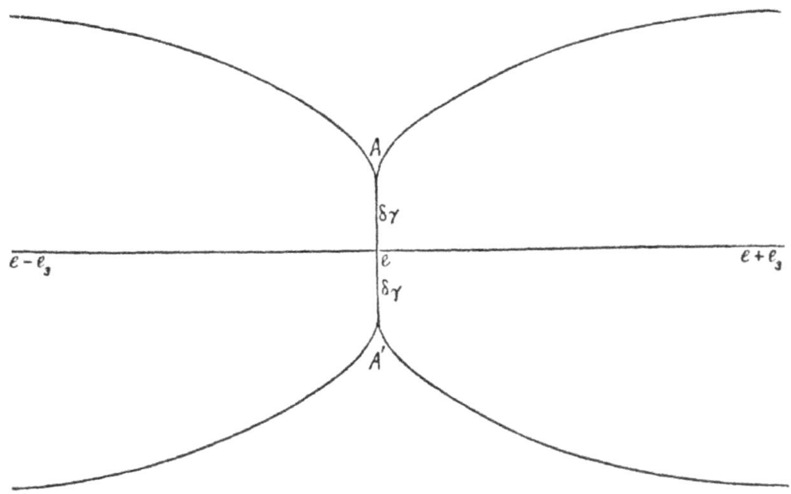

When $\delta = 0$, the surface becomes a Bolyai limit-surface.

|322| ***Appendix.***

Synthetic Proofs of some of the above results, communicated to me by Professor Burnside, F.R.S., April, 1898.

At Mr. Whitehead's request I have written out for publication at the end of his paper a note which the latter part of the paper suggested to me when I read it in manuscript. The note is the outcome of an attempt to approach the subject dealt with in the last half of Mr. Whitehead's paper from a synthetical point of view.

The locus of a point whose perpendicular distance from a given straight line is constant is called an equidistant surface; the given line is the axis of the surface, and the constant distance the radius.

An equidistance surface is clearly self-congruent for all translations along and rotations round the axis, and also for rotations through two right angles about any radius of the surface.

If the section of the surface by a plane through the axis be called a meridian, and that by a plane perpendicular to the axis a circular section, then, by every displacement for which the surface is self-congruent, a meridian is changed into a meridian and a circular section into a circular section. The distance between any two given circular sections (or meridians) measured along a meridian (or circular section) is, therefore, the same at all points.

Consider now a curve on the surface which meets all the meridians at a constant angle. Through a given point A on the surface, one, and only one, such curve can be drawn to meet the meridians at a given angle α $(0 \leq \alpha < \pi)$. If the surface be rotated through two right angles about the radius through A, the surface, and, therefore, also the curve, is changed into itself. Hence the osculating plane of the curve at A must contain the radius through A; in other words, the curve is a geodesic.

The geodesics on the surface are, therefore, the curves which meet the meridians (and circular sections) at a constant angle.

An infinite number of geodesics can clearly be drawn to join any two points on the surface. Among these, the shortest will be that one for which the (acute) angle α is least. Suppose, now, that any three points A, B, C on the surface are joined by shortest geodesics AB, BC, CA. Since each of these makes equal angles with each meridian it meets, it immediately follows that the sum of the angles of the geodesic triangle ABC is equal to two right angles. The infinitesimal geodesic geometry of the surface is, therefore, the same as that of the Euclidean plane.

|323| The radius of the equidistant surface considered may be any whatever; and therefore it follows that a portion of the surface of one equidistant may be cut out and applied without stretching or tearing on an equidistant of different radius.

The existence of surfaces, in elliptic or hyperbolic space, whose infinitesimal geometry is the same as that of the Euclidean plane being thus demonstrated, their form, when surfaces of revolution, may be determined by the consideration that, in a given space, a portion of any one such surface must be applicable on any other.

Let the figure represent a finite portion cut out from an equidistant surface; APB, CQD, ERF being a set of geodesics which meet in a point on the surface, while ACE, PQR, BDF are a family of curves on the surface which meet the set of geodesics everywhere at right angles. Since the geometry of a finite portion of the surface is the same as that of the Euclidean plane, such a set of orthogonal curves must exist, and the relations

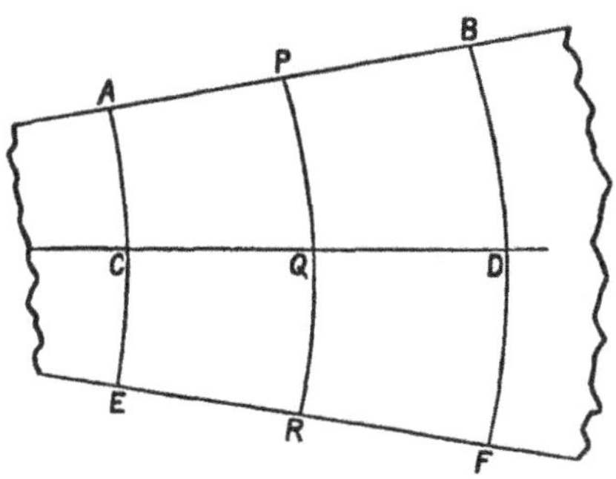

$$AP = CQ = ER,$$

$$\text{arc}\,PQR - \text{arc}\,ACE = \alpha \cdot AP,$$

must be satisfied for each curve PQR of the family, α being a constant.

Now bring opposite edges to coincidence so that A and E, P and R, B and F coincide, while ACE, PQR, BDF become circles with their centres on a straight line and their planes perpendicular to it. The surface will then form a portion of a surface of revolution whose infinitesimal geodesic geometry is that of the Euclidean plane.

Let AB be a meridian section of the surface so formed, and ab the |324| axis of the surface; Aa, Bb being the perpendiculars from A and B on the axis.

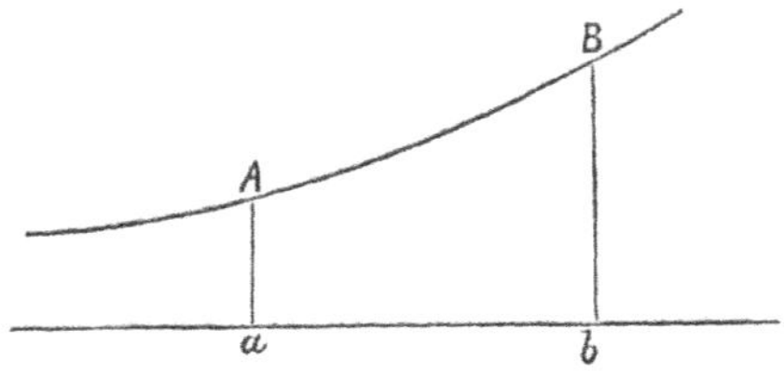

Then, taking suitable units of length, the meridian section is given (i.) in elliptic space by

$$\sin Bb - \sin Aa = \frac{\alpha}{2\pi} \operatorname{arc} AB;$$

and (ii.) in hyperbolic space by

$$\sinh Bb - \sinh Aa = \frac{\alpha}{2\pi} \operatorname{arc} AB.$$

It will be found that for elliptic space the relation

$$\frac{\alpha^2}{4\pi^2} < 1$$

is necessary in order that the surface may be real; and that, when $\alpha \neq 0$, the surface necessarily cuts the axis. Reckoning the arc s from a point where the curve cuts the axis and representing the ordinate by y, the equation is

$$\sin y = \frac{\alpha s}{2\pi}.$$

In hyperbolic space there is no limitation on α. If $\alpha^2 < 4\pi^2$, the curve cuts the axis; and, with the previous notation, the equation is

$$\sinh y = \frac{\alpha s}{2\pi}.$$

If $\alpha^2 > 4\pi^2$, the curve does not cut the axis. In this case, writing $\cosh \beta$ for $\frac{\alpha}{2\pi}$, the equation can be put in the form

$$\sinh y - \sinh \beta = s \cosh \beta.$$

The dividing case $\alpha = 2\pi$ gives the limit-surface, as it obviously should; for the limit-surface may be regarded as the equidistant of a straight line whose only real points are at infinity.

Sets of Operations in Relation to Groups of Finite Order

1899 *Proceedings of the Royal Society of London*, vol. 64, pp. 319–320 (abstract)

1899 Unpublished, Royal Society Archives, AP/74/4

This paper was submitted as a handwritten draft to the Royal Society in 1899 and marked as received on 19 January 1899, communicated by Andrew Forsyth,[1] *and read on 2 February 1899. However, it was ultimately rejected for publication, and so the full paper was forgotten for over a century, with only the paper's introduction appearing as an abstract in the Royal Society's* Proceedings. *Here, we describe its general mathematical content, the reasons for its rejection, and its connection with* Universal Algebra *and the actual historical course of mathematics.*

A group is a fundamental construct of abstract algebra. It consists, first, of a set together with an operation. One might exemplify this with the set of non-zero rational numbers – that is, non-zero fractions, which, incidentally, include the non-zero integers, with the denominator taken to be 1. An appropriate operation is then ordinary multiplication, which of course produces another non-zero fraction.

The set and operation are then subject to the following axioms. (a) Associativity. For example, (1·2)·4 = 1·(2·4) and similarly for any three rational numbers. (b) Existence of an identity element with respect to the given operation. For example, the rational number 1 has no multiplicative effect.) (c) The existence of inverses relative to the operation and its identity. For example, given the rational number 4, there exists another rational number, 1/4, such that 4·(1/4) = 1, and similarly for any other non-zero rational number; this is where excluding zero matters. Notice that these axioms do not require the commutative law (which the rational numbers, incidentally, do obey). The results achievable within the scope of these three axioms are astounding and pull in abstractions not only from the concept of number but also from geometry, particularly the notion of symmetry.

When two elements of a group are combined via the specified operation, we speak of their product in the group. This notion of product has no a priori *connection with ordinary multiplication, even though that is the case for the illustration chosen above.*

A group of finite order, *or, more succinctly, a* finite group, *is simply a group consisting of a finite number of elements. This turns out to be an amazingly strong rigidifying condition, but only one consequence need be mentioned here in connection with Whitehead's work. A* subgroup *of a group is no more than a subset of the given group that is a group in its own right under the operation specified for the ambient group. One should know that a non-empty subset of a finite group is a subgroup if*

1. Andrew Russell Forsyth (1858–1942) was a British mathematician and a close friend of Whitehead. For more on Forsyth, see p. 348n2.

and only if the subset is closed under the given operation. (This is not true for infinite groups.)

Whitehead's concern in this submission is to derive some fundamental properties of finite groups. His first move, however, is to extend the domain of discourse away from the individual group element to the set of all subsets of a fixed group; this original group generally remains in the background. The new domain of discourse is the set of all subsets of this background group. He then defines the product of two subsets to be the set of all products with the operands drawn from the respective sets in their given order. So, for instance, in the set of non-zero rational numbers, which is admittedly not finite, we have {1, 2}·{5, 7, 9} = {1·5, 1·7, 1·9, 2·5, 2·7, 2·9}, and so on. This extension of the group operation is, however, not sufficient to Whitehead's purposes: he also introduces a second operation on this collection of subsets, and this is no more than their set-theoretic union, which he denotes with a plus sign (+). Hence, {1, 2}+{5, 7, 9} = {1, 2, 5, 7, 9}. This move carries inherent problems, addressed below, but it does allow Whitehead to define positive exponents for any subset H *in play, and hence we have the power series* $H = H^1, H^2, H^3, \ldots$ *In the case that the background group is finite, this series has remarkable properties, the most important of which is this: after a transient segment with no repetitions, the series falls into a periodic segment that repeats ad infinitum (p. 30).*

The paper was rejected on three grounds stated by one of its referees, William Burnside (1852–1927), a mathematician famous for his early work in group theory. Here are Burnside's words, lightly edited for conciseness and to conform to the style of this volume:

> *It is quite clear that Mr. Whitehead is not acquainted with G. Frobenius's memoir 'Über Endliche Gruppen'* (Berliner Sitzungsberichte. *1895. pp. 163–94.) The definitions of a set of operations and of the addition and multiplication of such sets in Mr. Whitehead's sense are given explicitly on pp. 1, 2 of this memoir. From these definitions the whole of the results to the end of p. 39 in Mr. Whitehead's paper, are deduced, mainly without proof, as being self-evident consequences of the definition, between the middle of p. 2 and the middle of p. 3, in Frobenius's memoir. Mr. Whitehead's results between p. 40 and p. 56 are self-evident consequences of the fact that his group* H^{tm} *is a self-conjugate subgroup of* [H] *and that the corresponding factor-group is cyclical. This fact itself which follows, assuming the definitions of a self-conjugate subgroup and a cyclical group, from the very construction of a power sequence, is subsequently proved by Mr. Whitehead on pp. 65–70; the ordinary reproduction of the relation between a group and a self-conjugate subgroup, as given in the textbooks, being reproduced at length. The remainder of the paper consists in giving a few simple examples of power sequences and a paragraph pointing out the not very close analogy of the processes he has been using with the algebra of logic. Even in Herr Frobenius's practiced hands the method advocated does not, in my opinion, lend itself very readily to the discussion of really important properties of groups. Mr. Whitehead has not obtained from it any result which is not itself almost obvious. I do not think that the paper should be printed in the* Philosophical Transactions.[1]

1. William Burnside, 'Referee's Report by William Burnside on Paper "Sets of Operations in Relation to Groups of Finite Order" by Alfred North Whitehead', Royal Society Archives, RR/14/269.

This review was so persuasive that the second referee, E. B. Elliott (1851–1937), withdrew his original recommendation to publish the paper.[1] *(His report included a few corrections that have been verified and incorporated in the revision given in this volume.)*[2]

One sees in Burnside's comments three important objections. First and second, he attributes the priority of the results and the methods to Ferdinand Georg Frobenius (1849–1917); thus, there is nothing new in Whitehead's paper (and what is there is entirely elementary). The third objection is more subtle and more pervasive in the sense that it speaks to Whitehead's limited recognition as a mathematician (as opposed to his work in the foundations of mathematics or in philosophy). Burnside assails the very aptness of Whitehead's methods. In this, he would be proven right by the actual history of mathematics. By the middle of the twentieth century, group theory had become part of the undergraduate mathematics curriculum, and its methods took on characteristics that were far from the set-theoretic methods used by Whitehead. This is said here in hindsight, but in Burnside's case it was foresight, and this speaks to a mathematical talent of a kind independent of Whitehead's spectacular computational skills and encyclopaedic knowledge across several disciplines. Nonetheless, we can sensibly speculate on how Whitehead might have chosen the wrong path in this case.

In Universal Algebra, *Whitehead divides algebraic systems into two genera: the non-numerical genus and the numerical genus. The defining characteristic of the non-numerical genus is the equation* a + a = a, *which indeed holds if we are speaking of sets and the plus sign denotes union, or if we are speaking of logic, and the plus sign denotes the logical 'or' operator. What is patently non-numerical in this case is that we can dispense with numerical coefficients. For instance, if* 2a *is defined as* a + a, *and so on for larger natural numbers, then these coefficients are ignorable if in fact* a = 2a = 3a, *etc. Now this dichotomy, as he presents it, does not survive the present-day definition of an algebra, but, nonetheless, one sees at once that Whitehead's imposition of a second operation on the set of all subsets of a group moves group theory into the context of a non-numerical algebra. Given the emphasis that* Universal Algebra *places on these two types of structures and Whitehead's affinity for classes and relations, this is a natural move for him, and one for which he does not seem to be copying Frobenius, whose name is mentioned neither in Whitehead's submission nor in the index of* Universal Algebra *(which does include persons cited only in historical notes). Indeed, Elliott, the second-named referee above, seems to have been likewise unaware of Frobenius.*

Historically speaking, key algebraic objects with two or more operations, such as rings, fields, vector spaces, or algebras – an algebra in the present-day technical sense, not Whitehead's – are all built on top of additive groups. *This is a group with a commutative operation that is necessarily denoted with a plus sign. (The integers with respect to addition are the banner example.) Whitehead's non-numerical algebras immediately lose the possibility of being additive groups with respect to their addition with the defining equation* a + a = a, *for* a *non-zero. Hence his extension of group*

1. Edwin Bailey Elliott, 'Letter from Edwin Bailey to Andrew Forsyth, 11 February 1899', Royal Society Archives, RR/14/272.
2. Edwin Bailey Elliott, 'Corrections for Paper "Sets of Operations in Relation to Groups of Finite Order" by Alfred North Whitehead', Royal Society Archives, RR/14/271.

theory occurs in what is presently an exotic domain. By comparison, the study of groups is enormously enhanced by their embedding into group rings *(a natural type of algebra in the modern sense) and the associated theory of* group representations. *Perhaps ironically, one of the early key contributors to representation theory was, in fact, Ferdinand Georg Frobenius, while Whitehead seems never to have engaged group theory again.*

Readers should note that, as this paper by Whitehead has not previously been published, we have edited it differently from most of the other essays and articles in this volume. Whitehead made many changes by hand throughout the typescript – mostly deletions – and rather than footnote all of these, we have adopted corrections and deleted strikeouts silently. A back-of-book record tracks all silent corrections and deleted strikeouts.

Introduction.[1]

The present paper is concerned with the Theory of Groups of Finite Order.[2] The more general object of the paper is to place this theory in relation to a special algebra of the type of algebra considered in the general theory of Universal Algebra. This special algebra, which may be called the Algebra of Groups of Finite Order, has many affinities to the Algebra of Symbolic Logic; and a comparison of it with this algebra is given in the last section of this paper.

Mathematicians are accustomed in the study of quaternions to the idea of a vector symbol being considered from two points of view according to circumstances, namely, either as a geometrical entity, or as a symbol expressive of the operation of modifying some |2| geometrical entity into another geometrical entity. Now from the point of view of this paper, it is natural to abandon the idea of a group of N operations $S_0, S_1, \dots S_{N-1}$ on some unspecified object, as being an idea, which however vaguely, appertains to a special interpretation of the symbols. The N symbols $S_0, S_1, \dots S_{N-1}$ are to be considered, as in the similar case of quaternions, primarily as N distinct objects. When two of these objects are multiplied, as in $S_q S_r$, then a third object of the group, such as S_p, is produced: and in reference to this multiplication one of the symbols, say S_q, may be looked on as an operation on S_r modifying it into S_p. But this is not the sense in which the symbols are |3| usually called operations in the Theory of Groups. However in order not to disturb the well-understood nomenclature of the subject the N objects $S_0, S_1, \dots S_{N-1}$ will always be called the fundamental operations, or more shortly, the operations. But the word operation can simply be regarded as a name for the objects represented by these N symbols.

These N symbols are considered to be capable of addition according to the law

$$S + S = S.$$

This is the well-known law of addition in Symbolic Logic, and the introduction of numerical symbols as factors is thereby avoided.

---

1. The paper's published abstract (see reference at top) was simply the introduction, unaltered.
2. 'Orders' changed to 'Order' for consistency.

The sum of a selection of the N fundamental operations, such as, $S_p + S_q + S_r + S_t$, is called a set. If a set obeys certain special |4| conditions, it is called a group. The sum of the whole number (N) of fundamental operations, namely $S_0 + S_1 + ... + S_{N-1}$, obeys these conditions. This sum is called the complete group, and all other groups are its subgroups.

The first six sections of this paper are devoted to the detailed establishment of this purely algebraic view of the subject. At times the modification in treatment from that adopted in the standard Treatises on the subject, such as Burnside's *Theory of Groups of Finite Order*, is slight. Where the modification would be of no sufficient interest, it has been simply omitted and the theorems, when wanted, have been assumed as part of the general knowledge of the subject. Only so much reasoning has been given |5| as will establish the principles of the Algebra of Groups of Finite Order, viewed as an algebra independent of any interpretation, however vague.

The more special object of this paper follows directly from the changed point of view from which the Theory of Groups is here regarded. The idea of the group is no longer so absorbing: the set takes its place as the fundamental general entity which has to be investigated. A group is a special type of set. Accordingly in this paper some of the general properties of sets are investigated. A set of operations has numerous groups associated with it, and these groups have many relations with each other which this paper cannot pretend to have exhausted. The fundamental idea of this part of the paper (cf §7) is the |6| formation from a set H of an unending series of other sets, here called the successive powers of H, and in the notation of the algebra written $H^2, H^3, ...$ This series is called the power sequence of H. Any group which contains H also contains its power sequence. The power sequence is proved to have a periodic property (cf §8) which introduces a curious analogy to recurring decimals. This periodicity is the foundation of the rest of the paper. It governs the relations to each other of the various allied groups and sets. The periodicity is expressed by an equation of the form

$$H^{n+sm+q} = H^{n+q},$$

where m is called the period of H, and n the characteristic. The number of theorems relating to m is very large.

|7| *§ 1. The Fundamental Operations.*

Let $S_0, S_1, S_2, ... S_{N-1}$ be N distinct fundamental symbols. Let them be called, in accordance with the terminology of the *Theory of Groups*, the Operations. Let them obey the following algebraic Laws[1]†:

(i) No two of these symbols can be connected by the sign of equality.

(ii) A product is associative, but not necessarily commutative.

(iii) A product of any two, and therefore of any number, of these symbols is equal to some other symbol of the group, depending on the factors multiplied and in general also on their order of multiplication.

1. †Cf. Burnside's *Theory of Groups*, Chapter II.

|8| (iv) One of these symbols, say S_0, possesses the property,

$$S_0 S_p = S_p = S_p S_0,$$

where S_p is any symbol of the group. In particular

$$S_0^2 = S_0 .$$

Let S_0 be called the identical operation.

(v) To every symbol of the group, such as S_p, there corresponds one other symbol of the group, say S_p', such that

$$S_p S_p' = S_0 = S_p' S_p.$$

Hence we make the convention that

$$S_p^0 = S_0,$$

and employ the whole theory of positive and negative integral indices.

|9| Then remembering that multiplication is associative, we find that either of the equations

$$S_p S_q = S_p S_r, \; S_q S_p = S_r S_p$$

imply

$$S_q = S_r;$$

and that the equation

$$S_p^2 = S_p$$

implies

$$S_p = S_0.$$

Since there are only a finite number (N) of 'operations', the successive powers, $S, S^2, S^3, ...$ of any operation S cannot all be distinct. Assume that

$$S^{p+n} = S^n.$$

Multiplying by S^{-n}, we find

$$S^p = S^0 = S_0.$$

|10| Let p be the least integer, not zero, for which the equation

$$S^p = S_0$$

holds.

Then if q be an integer such that

$$S^q = S_0,$$

q is a multiple of p.

For write $q = sp + q'$, where $q' < p$.

Then

$$S^q = S^{sp} S^{q'} = S^{q'} = S_0.$$

But this is untrue by hypothesis, unless $q' = 0$. Hence q is a multiple of p. Let p be called the order of S.

|11| ***§ 2. Addition of Operations.***

(vi) Let the sum of r operations $S_1, S_2, ... S_r$, represent this set of operations conceived merely as a logical class of r things. Thus $S_1 + S_2 + ... + S_r$ represents in its commutative property the fact that the r operations in the set have no assigned order. It simply represents the statement, 'all the r operations $S_1, S_2, ... S_r$'.

Write

$$H = S_1 + S_2 + ... + S_r = S_2 + S_1 + ... + S_r$$

Then H will be called a set, or a sum, of operations.

Any of the operations, of which H is the sum, will be said to be contained in H.

Also the sum of any number of the operations contained in H will be called a set contained in H.

Thus H contains the operation S_1, and the set $S_1 + S_2$, and so on.

|12| Two sets H and H' are only equal if each set contains the same operations: in fact, only if the symbols H and H' represent the same set.

The repetition of an operation in addition does not alter the set of things, for $S_1 + S_1 + S_2 + ... + S_r$ according to definition merely represents all the assemblage of operations $S_1, S_2, ... S_r$, and this is also represented by $S_1 + S_2 + ... + S_r$.

(vii) Hence we arrive at the fundamental symbolic definition

$$S_1 + S_1 = S_1.$$

Thus numerical digits need never be used in this algebra as multipliers: their only use is as exponents. This of course does not take account of their use as suffixes, which is a mere question of nomenclature.

|13| Again there is no use in this algebra for a symbolism to represent subtraction, or a symbol to represent zero. A product or a sum has always one definite meaning which is not zero. The use of negative exponents is the appropriate symbolism for division.

(viii) If $H = S_1 + S_2 + ... + S_r$, and S be any operation, then by definition

$$SH = SS_1 + SS_2 + ... + SS_r,$$

and

$$HS = S_1S + S_2S + ... + S_rS.$$

Thus SH and HS represent two new sets each containing the same number of operations as H. It may avoid misconception to note that S is not necessarily contained in the set SH or the set HS. Thus the operations in SH are of the type SS_p.

|14| It follows that if

$$K = S'_1 + S'_2 + ... + S'_q,$$

Then

$$KH = \Sigma S'_m S_n,$$

And

$$HK = \Sigma S_n S'_m.$$

Thus KH is the set containing every operation of the type $S'_m S_n$, where S'_m is an operation from K and S_n is an operation from H.

It is evident that if H_1, H_2, H_3 be any three sets then

$$H_1H_2 \cdot H_3 = H_1 \cdot H_2H_3.$$

In other words, the multiplication of sets is associative.

Let *HH* be written H^2, *HHH* be written H^3, and so on. But a negative exponent has no meaning when *H* represents a set of operations as distinct from a single operation.

Since

$$H^m = H^{m-n+n} = H^{m-n}H^n, \text{ when } m > n,$$

it will be convenient to make the convention that $H^0 = S_0$, for then

$$H^m = H^{m-m+m} = H^{m-m}H^m = H^0H^m = S_0H^m = H^m.$$

|15| *§ 3. Multiplicity.*

Let the number of distinct operations $S_1, S_2, \ldots S_r$, which are contained in a set *H* be called the multiplicity of *H*.

Thus a single operation may be looked on as a set of multiplicity 1.

Again since the total number of operations is *N*, a set of multiplicity *N* can only be the set which contains all the operations. Call this set the 'complete group', and let it always be denoted by the letter *G*.

If *H* be any set of multiplicity *r*, and *S* any operation, then both *SH* and *HS* are of multiplicity *r*. For if S_1 and S_2 be any two distinct operations of *H*, it has been proved that SS_1 and SS_2 are distinct operations, and also that S_1S and S_2S are distinct operations.

|16| Now let *H′* be another set of multiplicity *r′*. Then it is evident that *HH′* may possibly be of multiplicity *rr′*, and it cannot be of greater multiplicity. But it may be of less multiplicity. For let S_1 and S_2 be any operation of *H* and S_1' and S_2' any operations of *H′*. Then it is possible that

$$S_1S_1' = S_2S_2'.$$

Hence

$$S_1S_1' + S_2S_2' = S_1S_1'.$$

Every such relation reduces the multiplicity of *HH′* by unity.

Also from the previous proposition the multiplicity of *HH′* cannot be less than the greatest of the two multiplicities *r* and *r′*. Thus if *p* be the multiplicity of *HH′*, and $r > r'$, we have

$$rr' \geq p \geq r.$$

Also if

$$p = rr',$$

then there is no |17| relation of the type

$$S_1S_1' = S_2S_2'.$$

Again if $p = r$, then putting

$$H' = S_1' + S_2' + \ldots + S_{r'}',$$

we must have satisfied the equations

$$HS_1' = HS_2' = ... = HS_{r'}' = HH'.$$

For each expression of the type HS_1' is of multiplicity r. Hence if they are not all equal, their sum must be of multiplicity greater than r.

Similar results hold when $r' > r$. In the special case when $r' = r$, we have the two systems of equations holding simultaneously, namely

$$HS_1' = HS_2' = ... = HS_r' = HH',$$

$$S_1H' = S_2H' = ... = S_rH' = HH'.$$

|18| *§4. Properties of G.*

Since G contains every operation and every set, it must satisfy the following equations where K is any set or any operation,

$$G + K = G,$$

$$GK = KG = G.$$

Thus there are two special symbols in this algebra, namely S_0 and G.

The characteristic property of S_0 is

$$S_0K = KS_0 = K,$$

and as a particular case

$$S_0^2 = S_0.$$

|19| *§5. Elementary Propositions.*

Let H be any set, and K any set or operation. The necessary and sufficient condition that K is contained in H is

$$H + K = H.$$

A particular case is

$$H + H = H.$$

It is to be noted that the equality of K and K' does not follow from

$$H + K = H + K';$$

nor does it follow from

$$KH = K'H;$$

nor does it follow from both these equations taken simultaneously.

If S be any operation, and H contain S_0, then

$$SH = S(H + S_0) = SH + SS_0 = SH + S.$$

Hence SH contains S. Similarly HS contains S.

Again if S be any operation and H be any set, and the equation

$$SH + S = SH$$

is satisfied, then H contains S_0.

For multiply this equation by S^{-1}, thus |20|

$$S^{-1}SH + S^{-1}S = S^{-1}SH.$$

Hence

$$H + S_0 = H.$$

Therefore H contains S_0.

Similarly if

$$HS + S = HS,$$

we can prove that H contains S_0.

Again if S, S' be operations, and H any set, and

$$H + S \neq H,$$

then

$$H + S = H + S',$$

implies

$$S = S'.$$

|21| If H and K be any two sets, or operations, then $(H+K)^n$ can evidently by mere multiplication be expanded into the sum of all possible terms of the type

$$H^{\alpha_1}K^{\beta_1}H^{\alpha_2}K^{\beta_2}H^{\alpha_3}K^{\beta_3} \ldots H^{\alpha_r}K^{\beta_r},$$

where $\alpha_1, \alpha_2, \ldots \alpha_r, \beta_1, \beta_2, \ldots \beta_r$ are any integers, including zero, satisfying

$$\alpha_1 + \alpha_2 + \ldots + \alpha_r + \beta_1 + \beta_2 + \ldots + \beta_r = n.$$

In particular, if α be any number less than n, there always exists in the expansion of $(H+K)^n$ a term of the form $H^{\alpha}K^{n-\alpha}$. If H and K are permutable, all the terms of the expansion are of this particular type.

|22| The total number of unequal sets and operations in this algebra is

$$2^N - 1.$$

For there are N distinct operations: there are

$$\tfrac{1}{2}N(N-1)$$

distinct sets of multiplicity 2, there are

$$\frac{N!}{r!(N-r)!}$$

distinct sets of multiplicity r, there is 1 set (the complete group) of multiplicity N. Hence the theorem follows immediately.

|23| §6 *Fundamental Properties of Groups.*

A set g of multiplicity r is called a group of order r when it has the property that if

then
$$\left.\begin{aligned} g+S=g, \\ gS=g=Sg. \end{aligned}\right\} \qquad (1)$$ [1]

Let $g=S_1+S_2+...+S_r$. Then equation (1) expresses that

$$gS_1=g, gS_2=g, ..., gS_r=g.$$

Hence by addition of these r equations

$$g^2=g \qquad (2)$$

This equation implies the condition (1) and is necessary and sufficient to secure that g is a group. For since g^2 has to be of the same multiplicity as g, the equation (2) implies that each of the r equations implied in (1) must be satisfied.

|24| Every group contains S_0. For multiply the equation

$$Sg=g$$

by S^{-1}, where S is any operation in g.

Then

$$g=S^{-1}g=S^{-1}(g+S)=S^{-1}g+S_0.$$

Hence g contains S_0.

Every group which contains S contains S^{-1}. For we have proved that

$$g=S^{-1}g+S_0.$$

Hence since $g=g+S_0$, we have

$$g=S^{-1}(g+S_0)+S_0=S^{-1}g+S^{-1}+S_0.$$

Thus g contains S^{-1}.

If g be a group and either

$$Sg=g, \text{ or } gS=g,$$

then

$$g+S=g.$$

For g contains S_0, hence

$$g=Sg=S(g+S_0)=Sg+S.$$

Hence g contains S.

Similarly from the equation

$$gS=g.$$

|25| If S and S' be two operations not necessarily contained in the group g, then

$$Sg=S'g,$$

implies that

$$g+S^{-1}S'=g.$$

1. We would now call such a non-empty subset g of G a *subgroup* of G, in which case Whitehead's equation (1) expresses the proposition that, for finite groups, closure suffices to ensure that a non-empty subset is a subgroup.

For multiply the given equation by S^{-1}. Hence

$$g = S^{-1}S'g.$$

Hence $S^{-1}S'$ is contained in g.

|26| *§7 The Power Sequence of a Set.*

Let H be any set of multiplicity r. Let any series of q consecutive powers, starting from any power, the pth, be called a power sequence of H consisting of q terms: thus this power sequence is

$$H^p,\ H^{p+1}, \ldots.\ H^{p+q-1}.$$

If no special power of H is mentioned as the initial term of the sequence, the first member will be always taken to be H. Thus the power sequence of H of q terms is

$$H, H^2, H^3, \ldots.\ H^q.$$

The unending power sequence

$$H, H^2, H^3, \ldots..$$

will be called the complete power sequence of H.

|27| The multiplicity of any term H^α in the power series cannot be less than the multiplicity of the preceding terms $H^{\alpha-1}$.

For if

$$H = S_1 + S_2 + \ldots + S_r,$$

then

$$H^\alpha = S_1 H^{\alpha-1} + S_2 H^{\alpha-1} + \ldots + S_r H^{\alpha-1}.$$

Now the multiplicity of $S_1 H^{\alpha-1}$ is equal to the multiplicity of $H^{\alpha-1}$; similarly for $S_2 H^{\alpha-1}$, and so on. Hence, unless $S_1 H^{\alpha-1}$, $S_2 H^{\alpha-1}$, ... $S_r H^{\alpha-1}$ are all equal, the multiplicity of H^α is greater than the multiplicity of $H^{\alpha-1}$. But if $S_1 H^{\alpha-1}$, $S_2 H^{\alpha-1}$, ... $S_r H^{\alpha-1}$ are all equal, then the multiplicities of H^α and $H^{\alpha-1}$ are also equal.

Hence as we ascend the complete power sequence of H, the multiplicities may increase, but can never diminish.

|28| Any group which contains H contains all members of the complete sequence of the form $H^{s\alpha}$, where $s = 1, 2, 3, \ldots$.

For if g be the group, then

$$g + H^\alpha = g,$$
$$H^\alpha g = g.$$

Hence

$$H^\alpha \left(g + H^\alpha\right) = H^\alpha g = g.$$

Therefore

$$g + H^{2\alpha} = g,$$

and so on.

Thus in particular any group which contains H contains the complete power sequence of H.

|29| *§8. The Period Equation.*

Since there are not more than $2^N - 1$ distinct symbols in this algebra, the terms of the complete power sequence of *H* cannot be all distinct.

Assume that H^{m+n} is the first term of the sequence which is equal to a former term: and let H^n be the term to which H^{m+n} is equal.

We have therefore

$$H^{m+n} = H^n.$$

|30| Now multiply this equation successively by $H, H^2 \ldots$. Then it is easily proved that for all integral values of *s*

$$H^{sm+n+r} = H^{n+r} \quad \ldots\ldots\ldots\ldots\ldots\ldots\ldots\ldots\ldots\ldots \quad (1)$$

Call equation (1) the period equation of the power sequence of *H*, or more shortly, the period equation of *H*.

Let *m* be called the period of *H*, and *n* the characteristic.

Then the complete power sequence of *H* consists only of $(m+n-1)$ distinct terms. The first $(n-1)$ terms $H, H^2, \ldots H^{n-1}$ are never repeated, the next in terms, $H^n, H^{n+1}, \ldots H^{m+n-1}$ are continually repeated in the same order. Call the sequence of *n* terms $H, H^2, \ldots H^{n-1}$ the non-recurring |31| part of the complete power sequence of *H*; and call the sequence of *m* terms $H^n, H^{n+1} \ldots H^{m+n-1}$, the recurring part of the complete power sequence of *H*.

The analogy to the theory of recurring decimals is obvious.

Let the $(m+n-1)$ distinct terms of the complete power sequence of *H* be called the digits of the sequence. Thus two powers of *H* which are equal to each other, such as H^{sm+n+r} and H^{n+r}, represent the same digit of the sequence. Thus the digits of the power sequence of *H* are divided into a recurring and a non-recurring part. |32| Now as we ascend the power series the successive multiplicities of the terms may increase, but can never diminish. But from the period equation

$$H^{n+sm+r} = H^{n+r},$$

for all values of *s* and *r*. For instance

$$H^{n+m} = H^n.$$

Thus the multiplicity of H^{n+m} is equal to the multiplicity of H^n. Hence the multiplicities of all the intervening digits are equal, for they cannot be less than that of H^n, or greater than that of H^{n+m}.

Accordingly the multiplicities of all the recurring digits are equal.

Let *p* be invariably used to denote this common multiplicity of the recurring digits. Then

$$p \geq r,$$

where *r* is, as usual, the multiplicity of *H*.

|33| *§9. The Group Generated by a Set.*

The sum of the recurring digits of *H* form a group.

For let this set be invariably written in the form [*H*]: so that [*H*] is one of the algebraic symbols. Now [*H*] contains every operation contained in any of the recurring digits. But $[H]^2$ is a sum of recurring digits, and is therefore a sum of operations contained in recurring digits of H. But the multiplicity of $[H]^2$ cannot be less than that of $[H]$. Hence

$$[H]^2 = [H].$$

Therefore $[H]$ is a group.

The group $[H]$ contains the complete power sequence of H. For

$$[H] = H^n + H^{n+1} + ... + H^{n+m-1}.$$

Hence
$$H[H] = H^{n+1} + H^{n+2} + ... + H^{n-m}$$
$$= H^n + H^{n+1} + ... + H^{n+m-1} = [H].$$

Therefore $[H]$ contains H: and therefore contains the complete power sequence of H.

Also it is evident that any group which contains the complete power sequence of H must contain $[H]$ as a subgroup. Let $[H]$ be called the group generated by H.

|34| Every operation in the group $[H]$ is commutative with each recurring digit, and the product is equal to another recurring digit.

For let H^α be any recurring digit, and let S be any operation in $[H]$. Let H^p be a digit of H in which S occurs: where H^p is not necessarily a recurring digit; for instance, it may be H itself. Then $H^{\alpha+p}$ is a recurring digit, and is therefore of the same multiplicity as H^α. But

$$H^{\alpha+p} = H^\alpha \cdot H^p = H^p \cdot H^\alpha.$$

Hence, since the multiplicity of the product is equal to that of H^α, the product of each operation of H^p into H^α is equal to $H^{\alpha+p}$. Thus

$$H^\alpha S = H^{\alpha+p} = SH^\alpha.$$

|35| *§10 The Power Group of a Set.*

To prove that one and only one digit of H is itself a group; and that this digit is represented by those powers of H of the form H^{tm} which lie within the recurring part of the digits.

For if H^p be a group, the sufficient and necessary condition is

$$H^{2p} = H^p$$

Hence by reference to the period equation

$$p \geq n,\ np \geq m+n.$$

Assume

$$2p = qm+n+q',\ p = sm+n+s',$$

where

$$q \neq 0, \text{ and } q' < m,\ s' < m.$$

Then from the period equation we have

$$H^{n+q'} = H^{n+s'}.$$

Hence, since both q' and s' are less than m,

$$q' = s'.$$

|36| Hence

$$(q-2s)m = n+q'. \quad \ldots\ldots\ldots\ldots \quad \text{(I)}$$

Firstly, assume that $m > n$.

Then since both n and q' are less than m, equation (I) gives

$$n+q' = m,\ q = (2s+1).$$

Hence the only digits which are groups are of the form $H^{(s+1)m}$, where s is any integer.

But since

$$m > n,\ H^{(s+1)m} = H^m.$$

Hence these powers of H all represent the same digit.

Secondly, assume that $m < n$.

Then from equation (I), $(q-2s)$ represents the least integer t for which $tm \geq n$. Let this integer be always represented by t: so that

$$(t-1)m < n \leq tm.$$

Then

$$q = 2s+t,\ p = (s+t)m.$$

|37| Hence digits of the form $H^{(s+t)m}$ are the only digits which represent groups. But since

$$tm \geq n,$$

$$H^{(s+t)m} = H^{tm}$$

Hence all these powers of H represent the same digit.

Thus the complete proposition is proved.

Let this group be called the power group of H. The power group of H can always be represented by H^{tm}, where m is the period of H and t in the first integer for which tm is equal to or greater than n, the characteristic of H. The group can also be represented in the form $H^{(s+t)m}$, where s is any integer.

It is evident that H^{tm} is a subgroup of [H]. |38| To prove that if H^p is contained in H^{tm}, then

$$p = sm,$$

where s is any integer.

For since H^{tm} is a group, the necessary and sufficient condition that H^p is contained in H^{tm} is

$$H^{p+tm} = H^{tm}.$$

Now we may always write

$$p = sm + s',$$

where s is any integer, including zero, and $s' < m$. Then, since $tm \geq n$,

$$H^{p+tm} = H^{sm+s'+tm} = H^{s'+tm}$$

Hence

$$H^{s'+tm} = H^{tm} \quad \ldots\ldots\ldots\ldots \quad \text{(A)}$$

But $tm = n + r$, where $r < m$. Hence from the period equation, equation (A) can only be satisfied by $s' = 0$.

|39| Thus the only powers of H included in H^{tm} are of the form H^{sm}.

If $s \geq t$, then $H^{sm} = H^{tm}$. But if s be any one of the series

$$1, 2, \ldots (t-1),$$

then H^{sm} represents a digit of the non-recurring digits of H which is contained in H^{tm}.

If $m > n$, then $t = 1$; and there is no such digit among the non-recurring digits.

Corollary. If $H^{tm} = [H]$, then the period of H is unity; and conversely if the period is unity the power group of H is equal to $[H]$.

For in this case H^{tm} contains H. Hence unity is contained among numbers of the form sm. Hence $m = 1$.

Conversely, if $m = 1$, then by the preceding proposition the power group of H contains H: and therefore it contains the complete power sequence of H.

|40| **§11. *Properties of the Period.***

The period (m) of H is a factor of the order of every operation contained in H.

For suppose that H contains the operation S of order p.

Then

$$H = H + S.$$

Now let s be any integer such that

$$(s+t)m > p.$$

Hence by §5

$$H^{tm} = H^{(t+s)m} = (H+S)^{(t+s)m}$$
$$= H^{(t+s)m} + \ldots + H^{(t+s)m-p} S^p + \ldots .$$

But

$$S^p = 1.$$

Hence $H^{(t+s)m-p}$ is contained in H^{tm}.

Hence

$$(t+s)m - p = s'm.$$

Hence m is a factor of p.

|41| *Corollary* I. If H contains S_0, its period is unity. For the order of S_0 is unity.

Corollary II. If H contains any group, its period is unity. For a group must contain S_0.

Corollary III. If H contains any two operations of which the orders are relatively prime, then its period is unity.

Corollary IV. The period of H is a factor of N. For the order of any operation is a factor of N.

|42| If p be the order of any operation contained in the digit H^{α}, considered as a sum of single operations, then m is a factor of $p\alpha$.

For let $(s+t)m > p\alpha$, then

$$H^{tm} = H^{(s+t)m-p\alpha} \cdot H^{p\alpha} = H^{(s+t)m-p\alpha} \cdot (H^{\alpha} + S)^p$$
$$= H^{tm} + H^{(s+t)m-p\alpha} \cdot S^p$$
$$= H^{tm} + H^{(s+t)m-p\alpha}.$$

Hence $(s+t)m - p\alpha = s'm$. Therefore m is a factor of $p\alpha$.
Corollary I. If α be prime to m, then m is a factor of p.
Corollary II. If p be prime to m, then m is a factor of α.

Hence the only digits in which operations of orders |43| relatively prime to m can be contained, are digits contained in H^{tm}. In other words all the operations in the group [H] of orders relatively prime to the period of H are contained in the subgroup H^{tm}.
Corollary III. The only digits in which S_0 can occur are digits contained in H^{tm}.

It is to be noticed that if m be unity, the power group is equal to [H]; and therefore it contains all the digits.

|44| If the two digits, H^α and H^β, when each is arranged as a sum of single operations, contain any operation in common, then m is a factor of $(\alpha - \beta)$.

For let H^α and H^β both contain the operation S, of the pth order. Assume that $\alpha > \beta$.

Let x and y be integers such that

$$x + y = p;$$

and let s be an integer such that

$$(s+t)m > x\alpha + y\beta.$$

Then

$$H^{tm} = H^{(s+t)m} = H^{(s+t)m - x\alpha - y\beta} \cdot H^{x\alpha} H^{y\beta}.$$

But

$$H^{x\alpha} = \left(H^\alpha\right)^x = \left(H^\alpha + S\right)^x = H^{x\alpha} + \ldots + S^x,$$

$$H^{y\beta} = \left(H^\beta\right)^y = \left(H^\beta + S\right)^y = H^{y\beta} + \ldots + S^y.$$

Hence

$$H^{tm} = H^{tm} + H^{(s+t)m - x\alpha - y\beta} \cdot S^p$$
$$= H^{tm} + H^{(s+t)m - x\alpha - y\beta}.$$

|45| Hence $H^{(s+t)m - x\alpha - y\beta}$ is contained in H^{tm}.

Therefore

$$(s+t)m - x\alpha - y\beta = s'm.$$

Accordingly by eliminating y,

$$x(\alpha - \beta) + p\beta = (s - s' + t)m.$$

But by the previous proposition m is a factor of $p\beta$. Also any integer less than p can be put for x. Put $x = 1$. Hence m is a factor of $(\alpha - \beta)$.

|46| *Corollary* I. No two digits in a power sequence of H of m terms, starting from any digit, can contain any operation in common.
Corollary II. In particular no two digits among the recurring digits of H can have any operation in common.
Corollary III. Since the multiplicity of each digit is not less than the multiplicity (r) of H, it follows that not less than mr distinct operations enter into a power sequence. Hence if r is greater than $\frac{1}{2}N$, the period must be unity.
Corollary IV. The order of [H] is not less than mr.

Corollary V. Since from Corollary II no two recurring digits contain any operation in common, and since the sum of the recurring digits contains every operation in the power sequence of H, it follows that any recurring digit H^{α} contains every digit of the form $H^{\alpha-qm}$.

|47| The quotient of the order of $[H]$ by the order of H^{tm} is the period (m) of H.

For the multiplicities of the recurring digits are all equal to p. Hence p is the order of H^{tm}, which is one of these digits. Again $[H]$ is the sum of these digits, and by Corollary II of the previous proposition no two of these digits have an operation in common. Hence the order of $[H]$ is mp.

Corollary. Since the order of any group is a factor of N, it follows that mp is a factor of N; and hence p is a factor of $\frac{N}{m}$.

|48| If the digit H^{α} contains an operation S, and the digit H^{β} contains the inverse operation S^{-1}, then m is a factor of $(\alpha+\beta)$.

For let x and y be integers such that

$$x-y=p$$

where p is the order of S. And let s be an integer such that

$$(s+t)m>x\alpha+y\beta.$$

Then

$$\begin{aligned} H^{tm} &= H^{(s+t)m} = H^{(s+t)m-x\alpha-y\beta}\cdot H^{x\alpha}\cdot H^{y\beta} \\ &= H^{tm}+H^{(s+t)m-x\alpha-y\beta}\cdot S^{x}\cdot S^{-y} \\ &= H^{tm}+H^{(s+t)m-x\alpha-y\beta}, \end{aligned}$$

since

$$S^{x-y}=S^{p}=1.$$

Thus, as before,

$$(s+t)m-x\alpha-y\beta=s'm.$$

Hence

$$(s+t)m-p\alpha-y(\alpha+\beta)=s'm.$$

But m is a factor of $p\alpha$, and y can be any integer. Hence m is a factor of $(\alpha+\beta)$.

|49| *Corollary* I. Hence if H^{α} contains the operation S, some digits of the form $H^{qm-\alpha}$ contain the operation S^{-1}.

Corollary II. If H^{α} be a recurring digit, then the recurring digit of the form $H^{(s+t)m-\alpha}$ is composed of all operations inverse to the operations of H^{α}; here s may be any integer, including zero, which is such that,

$$(s+t)m-\alpha\geq n.$$

Let such pairs of recurring digits be called inverse digits.

The product of two inverse digits is equal to the power group H^{tm}.

Corollary III. A recurring digit, such as H^{α}, necessarily contains all the operations of any digit of the form $H^{\alpha-qm}$; and necessarily contains all the inverse operations to the operations of any digit of the form $H^{qm-\alpha}$.

|50| *Corollary* IV. If a recurring digit be its own inverse, then

$$(s+t)m-\alpha=\alpha.$$

Therefore m is a factor of 2α.

Hence if m be odd, m is a factor of α. Therefore, when m is odd, the power group H^{tm} is the only recurring digit which is its own inverse.

If m be even, $\alpha = \frac{1}{2}qm$. Hence the power group H^{tm} and the digit $H^{\left(t+\frac{1}{2}\right)m}$ are the only digits which are their own inverses.

Corollary 5. If H contains both S and S^{-1}, then m is either 1 or 2. For m is a factor of 2.

Also if any operation in H be of odd order, m must be odd. Hence in this case, m is unity.

|51| *§12 Self-Conjugate Properties.*

Each recurring digit of H is a self-conjugate set in the group $[H]$.

For these digits are commutative with respect to any operation contained in $[H]$. Hence if S be any such operation, and H^{α} a recurring digit,

$$S^{-1}H^{\alpha}S = S^{-1}SH^{\alpha} = H^{\alpha}.$$

Corollary. In particular, the power group H^{tm} is a self-conjugate subgroup of $[H]$.

|52| Any set conjugate to H has the same period and characteristic as H. For let S be any operation, and let

$$K = S^{-1}HS.$$

Then

$$K^2 = S^{-1}HS.\,S^{-1}HS = S^{-1}H^2S.$$

Similarly

$$K^p = S^{-1}H^pS.$$

Thus

$$K^{sm+n+r} = S^{-1}H^{sm+n+r}S = S^{-1}H^{n+r}S$$
$$= K^{n+r},$$

where s and r are any integers including zero.

Thus if m' and n' be the period and characteristic of K,

$$n' \leq n,\ \ m = qm'.$$

But

$$H = SKS^{-1},\ H^p = SK^pS^{-1}.$$

Hence by a similar proof.

$$n \leq n',\ m' = q'm.$$

Thus

$$n = n',\ m = m'.$$

|53| *§13 Digital Groups.*

Let the power sequence of the digit H^{α} be called a digital power sequence of H, and let the group $[H^{\alpha}]$ be called a digital subgroup of $[H]$, or more shortly, a digital group.

The power group of every digital power sequence is H^{tm}. For

$$H^{\alpha tm} = H^{tm},$$

since

$$tm \geq n.$$

Thus H^{tm} is a common subgroup of every digital group. Let the period of H^α be m_α, and its characteristic be n_α. Then evidently

$$(n_\alpha - 1)\alpha < n \leq n_\alpha \alpha.$$

Thus if $\alpha \geq n$, that is, if H^α be one of the recurring digits of H, the characteristic of H^α is unity: and all the digits of H^α recur.

Again $m_\alpha \alpha$ must be the least common multiple of α and m. Hence m_α is a factor of m; and if α be prime to m, m_α is equal to m.

Then the order of $[H^\alpha]$ is $m_\alpha p$, where p is the order of H^{tm}.

|54| Since α can be any of the series 1, 2, 3, ... $(m+n-1)$, it follows that m_α will assume in turn the value of every divisor of m, and of m itself, more than once. Thus for different values of α, we obtain digital groups of the same order.

To prove that all digital groups of the same order are equal.

In other words, if $m_\alpha = m_\beta$, then

$$[H^\alpha] = [H^\beta].$$

Let $m_\alpha = m_\beta = \mu$, and let $m = \mu\mu'$.

Then by hypothesis $\alpha = \mu' p_1$, $\beta = \mu' p_2$, where p_1 and p_2 are prime to μ.

Now each digital group can be expressed entirely in terms of recurring digits of H. Hence the two digital groups are equal if we can always solve in integers the equation

$$x\alpha = y\beta + zm,$$

|55| where y is any given integer. For this secures that to every recurring digit of H^α there corresponds an equal recurring digit of H^β, and conversely.

This equation for x and z can be written

$$p_1 x - \mu z = p_2 y.$$

But p_1 and μ are prime to each other. Hence this equation always admits of solution in integers. Thus the two digital power sequences of H have the same recurring digits. Hence

$$[H^\alpha] = [H^\beta].$$

Corollary I. Hence the number of distinct digital groups of H is equal to the number of divisions of m, including unity and m. Here $[H]$ is considered as a digital group: also H^{tm} is another digital group which corresponds to the case when α is a multiple of, or is equal to, m.

Corollary II. Hence the series $[H^\alpha], [H^{\alpha+1}], [H^{\alpha+2}], \ldots$ contains every digital group, whatever α may be.

|56| Every digital group is a self-conjugate subgroup of $[H]$.

For every such group can be expressed as a sum of recurring digits of H; and every recurring digit of H is a self-conjugate set of H. Hence the proposition follows immediately.

In other words, every digital group is commutative with any operation contained in $[H]$.

|57| ***§14. The Absorbed Groups.***

When an operation S is such that either one of the two equations

$$SH^{\alpha} = H^{\alpha}, H^{\alpha}S = H^{\alpha},$$

holds, let it be called an operation absorbed by the digit H^{α}.

If only one of the two equations holds, let the absorption be called Unilateral. If only the first equation holds, let the absorption be called left-sided absorption: if only the second equation holds, let the absorption be called right-sided absorption.

If both equations hold, let the absorption be called Bilateral.

If S is absorbed by H^{α}, it is absorbed on the same side by all succeeding digits represented by higher powers of H.

For taking left-sided absorption as an example

$$SH^{\alpha+1} = SH^{\alpha} \cdot H = H^{\alpha}H = H^{\alpha+1};$$

and so on.

|58| Thus if S is absorbed by any digit, it is necessarily absorbed on the same side by all the recurring digits.

Hence if α is greater than n, and S is absorbed by H^{α}, it is also absorbed on the same side by $H^{\alpha-1}, H^{\alpha-2}, \dots H^{n}$.

But if H^{α} lies among the non-recurring digits, the fact that S is absorbed by H^{α} is no guarantee that it is absorbed by lower powers of H.

Corollary. Every operation absorbed by H^{α} is contained in the digital group $[H^{\alpha}]$.

For by the preceding proposition, if

$$SH^{\alpha} = H^{\alpha},$$

then

$$S[H^{\alpha}] = [H^{\alpha}].$$

But this implies that S is contained in the group $[H^{\alpha}]$.

|59| The set of operations absorbed by H^{α} on one side forms a group.

For if S_p and S_q be two such operations, then $S_pS_qH^{\alpha} = S_pH^{\alpha} = H^{\alpha}$.

Hence S_pS_q is another such operation.

Let the group of operations absorbed by H^{α} on its left side be written g_{α}, and the group absorbed by H^{α} on its right side be written g'_{α}.

Thus the groups absorbed by $H^1, H^2, \dots$ on their left sides are $g_1, g_2, \dots$; and the groups absorbed on their right sides are $g'_1, g'_2, \dots$.

Thus

$$g_{\alpha}H^{\alpha} = H^{\alpha}, H^{\alpha}g'_{\alpha} = H^{\alpha}.$$

Then it has been proved that g_1 is a subgroup of g_2, and g_2 of g_3, and so on: also g'_1 is a subgroup of g'_2, and g'_2 of g'_3, and so on.

|60| *Corollary* I. Also it follows from the corollary to the previous proposition that both g_{α} and g'_{α} are subgroups of $[H^{\alpha}]$. Hence also they are subgroups of $[H]$.

Corollary II. Since g_{α} is contained in $g_{\alpha+1}$, and $g_{\alpha+1}$ in $g_{\alpha+2}$, and so on, it follows that g_{α} is a common subgroup of $[H^{\alpha}], [H^{\alpha+1}], \dots$. But this series contains every digital group. Hence every absorbed group is contained in every digital group.

|61| Every operation absorbed by a recurring digit is absorbed bilaterally; and the absorbed group of each recurring digit is the power group.

Firstly since every absorbed group on either side is absorbed by all the recurring digits, the absorbed groups corresponding to the various recurring digits on their same side are the same.

But the power group is among the recurring digits. Now either of the equations

$$SH^{tm} = H^{tm},\ H^{tm}S = H^{tm}$$

can only be satisfied when S is contained in H^{tm}; and then S satisfies both equations. Hence the power group is the absorbed group of each recurring digit. Also since its operations are contained in [H], they are commutative with each recurring digit.

|62| *Corollary* I. Thus $g_1, g_2, \dots g_{n-1}$, and $g'_1, g'_2, \dots g'_{n-1}$, are all subgroups of H^{tm}.

This forms another proof of the proposition that every absorbed group is contained in every digital group.

|63| If H^{α} contains any operation of either of its absorbed groups, then α is a multiple of m.

For if S is contained in H^{α}, where

$$SH^{\alpha} = H^{\alpha},$$

then

$$H^{2\alpha} = \left(H^{\alpha} + S\right)H^{\alpha} = H^{2\alpha} + H^{\alpha};$$

also

$$H^{3\alpha} = \left(H^{\alpha} + S\right)H^{2\alpha} = H^{3\alpha} + H^{2\alpha},$$

and so on.

Thus

$$H^{(s+1)\alpha} = H^{(s+1)\alpha} + H^{s\alpha}.$$

But by making s large enough, $H^{s\alpha}$ and $H^{(s+1)\alpha}$ can be made to fall among the recurring digits.

Now one recurring digit cannot contain another recurring digit, since two recurring digits have no operation in common. Thus $(s+1)\alpha$ and $s\alpha$ must differ by a multiple of m. Hence α is a multiple of m.

Thus $H^m, H^{2m}, \dots H^{tm}$ are the only digits which can possibly contain any operation of either of their absorbed groups and of these digits H^{tm} is its own absorbed group.

|64| It is evident that the subgroup common to g_{α} and g'_{α} is absorbed bilaterally by H^{α}. Let this subgroup be written g''_{α}.

Thus $g''_1, g''_2, g''_3 \dots$ are the groups absorbed bilaterally by $H^1, H^2, H^3 \dots$ respectively. Also it is evident that g''_1 is a subgroup of g''_2, g''_2 of g''_3, and so on.

|65| *§15 Factor Groups.*

The theory of factor groups, or of multiple isomorphism can be expressed in this algebra as follows. Let g be any group and let k be any subgroup of g self-conjugate with respect to g. Let p be the order of k, and pq of g. Then it is easily proved that g can be expressed in the form

$$g = kS_0 + kS_1 + kS_2 + .. + kS_{q-1},$$

where $S_1, S_2, \dots S_{q-1}$ are $(q-1)$ operations of g, such that no two of the terms

$k, kS_1, \dots, kS_{q-1}$ have an operation in common. Also by hypothesis k is permutable with any operation S of g, so that

$$kS = Sk.$$

Also if

$$k = R_1 + R_2 + \dots + R_p,$$

it is evident that any operation of g can be written in the form RS, where R is contained in k and |66| S in the set $S_0, S_1, \dots S_{q-1}$.

Now

$$kS_1 \cdot kS_2 = k^2 S_1 S_2 = kS_1 S_2.$$

But

$$kS_1 S_2 \neq kS_1 \text{ or } kS_2.$$

For if

$$kS_1 S_2 = kS_1 = S_1 k,$$

then

$$S_1 S_2 = S_1 R,$$

where R is contained in k.

Hence $S_2 = R$, which is contrary to hypothesis, for then

$$kS_0 = k = kR = kS_2.$$

Similarly if

$$kS_1 S_2 = kS_2,$$

then

$$S_1 S_2 = RS_2,$$

where R is contained in k.

Hence

$$S_1 = R,$$

which is contrary to hypothesis. Thus

$$kS_1 \cdot kS_2 = kS_3,$$

where S_3 is some other member of the set $S_0, S_1, \dots S_{q-1}$.

Again corresponding to any term kS_1 in the equation

$$g = k + kS_1 + kS_2 + \dots + kS_{q-1},$$

there is a term kS_2 with the |67| property that

$$kS_1 \cdot kS_2 = k.$$

For multiplying the equation by S_2, we have

$$gS_2 = g = kS_2 + kS_1 S_2 + kS_2{}^2 + \dots + kS_{q-1} S_2.$$

Therefore one of the set of terms

$$kS_2, kS_1 S_2, \dots kS_{q-1} S_2$$

must be equal to k. Assume that

$$kS_1 S_2 = k.$$

But

$$kS_1S_2 = kS_1 \cdot kS_2.$$

In this case

$$kS_2 = kS^{-1}.$$

Now consider each of the set

$$k,\ kS_1,\ kS_2,\ \ldots.\ kS_{q-1}$$

as a single indecomposable symbol. When they are considered from this point of view write them in brackets, thus

$$(k), (kS_1), (kS_2), \ldots. (kS_{q-1}).$$

Then these symbols have all the properties of q fundamental operations of a complete |68| group of the qth order; where (k) is the identical operation. This complete group of the qth order is

$$(k)+(kS_1)+(kS_2)+\ldots+(kS_{q-1}).$$

Let the symbol g/k be written for it. Then

$$g/k=(k)+(kS_1)+(kS_2)+\ldots+(kS_{q-1}).$$

Now every transformation which is true of

$$(k), (kS_1), \ldots (kS_{q-1})$$

is true of

$$k,\ kS_1,\ \ldots\ kS_{q-1}.$$

Thus every theorem which is true of the factor group g/k implies an allied theorem concerning g.

|69| Now return to the set H of period m, characteristic n, and power group H^{tm}.

Let S be any operation in H. Then it has been proved that

$$SH^{tm} = H^{tm}S = H^{tm+1},$$
$$S^2H^{tm} = H^{tm}S^2 = H^{tm+2}$$
$$\vdots$$
$$S^{m-1}H^{tm} = H^{tm}S^{m-1} = H^{(t+1)m-1}$$

Thus

$$H^{tm}, SH^{tm}, S^2H^{tm}, \ldots. S^{m-1}H^{tm}$$

represent the m recurring digits. Hence

$$[H] = H^{tm} + SH^{tm} + S^2H^{tm} + \ldots + S^{m-1}H^{tm}$$

And H^{tm} is a self-conjugate subgroup of $[H]$.

Thus

$$[H]/H^{tm} = (H^{tm})+(SH^{tm})+(S^2H^{tm})+\ldots+(S^{m-1}H^{tm}).$$

|70| But

$$(S^{\alpha}H^{tm}) = (SH^{tm})^{\alpha}.$$

Hence the factor group $[H]/H^{tm}$ is a cyclical group of the mth order.

|71| *§16 Special Cases.*

(i) Let H be the single operation S, of order p. Then

$$S^p = S_0,\ S^{p+1} = S.$$

Hence

$$S^{sp+q} = S_0 S^q = S^q.$$

Thus S is of characteristic unity, and of period p. The group $[S]$ is the cyclical group of order p, generated by S, and the power group is S_0 of order 1. Hence the condition is fulfilled that the order of $[S]$ divided by the order of the power group is the period of S.

The total number of digits is p; and they are all recurring.

(ii) Let H be the group g. Then

$$g^2 = g.$$

Hence a group is a set of characteristic unity and of period unity. The power group and $[g]$ are both the given group g. There is only one digit, namely g.

|72| (iii) Let H be of period 1 and of characteristic n. Then the period equation is

$$H^{n+q} = H^n.$$

Hence all the terms from the nth onwards of the power sequence are equal, and represent but one digit.

Also H^n is the power group, and

$$H^n = [H].$$

Thus H^n contains all the previous $(n-1)$ digits.

The total number of digits is n, and only one is recurring.

If H contains S_0, then it is of period unity.

Also

$$H^2 = (H + S_0)^2 = H^2 + H + S_0 = H^2 + H.$$

Hence H^2 contains H. Similarly H^3 contains H^2, and so on. Thus each of the non-recurring digits contains all the preceding digits in the power sequence: and the single recurring digit contains the whole power sequence.

|73| Sets of this special type always exist in a given group. For let S be any operation of order p; and consider

$$H = S + S^{-1} + S_0$$

Then

$$H^2 = S^2 + S^{-2} + H + S_0 = S^2 + S^{-2} + H,$$
$$H^3 = S^3 + S^{-3} + H^2,$$
$$\vdots$$
$$H^p = S^p + S^{-p} + H^{p-1} = S_0 + H^{p-1} = H^{p-1}$$

Hence the characteristic of H is $p-1$ and its period is unity.

|74| (iv) The general type of set of characteristic unity and period m is given by

$$H^{m+1} = H,\ H^{m+n} = H^n.$$

In the special case in which $H^m = S_0$, H is a single operation.

All the m digits recur, therefore their multiplicities are all equal.

The power group is H^m, and its order is equal to r, the multiplicity of H. The order of the group $[H]$ is mr.

|75| (v) If H be any set of period m and characteristic n, and if H^{tm} be its power group, according to the usual rotation, so that

$$(t-1)m < n \leq tm.$$

Then the set

$$K = H^m + H^{2m} + \ldots + H^{(t-1)m},$$

is of characteristic 2 and of period 1. For

$$K^2 = \sum H^{jm} + H^{tm} = H^{tm}.$$

Also

$$K^3 = K^2.$$

Hence K^2 is the power group; and

$$K^2 = [K] = H^{tm}.$$

|76| (vi) Another simple example is found by taking

$$H = S + S^{-1},$$

where S is an operation of an even order (p).

Then

$$H^2 = S^2 + S^{-2} + S_0,$$
$$H^3 = S^3 + S^{-3} + H,$$
$$H^4 = S^4 + S^{-4} + H^2,$$
$$\vdots$$
$$H^p = S_0 + H^{p-2}.$$

Now it is evident that any even power of H contains all the preceding even powers of H, and any odd power of H contains all the preceding odd powers.

Therefore, since $p-2$ is even, H^{p-2} contains S_0. Hence

$$H^p = H^{p-2}.$$

Therefore the characteristic of H is $p-2$, the period is 2, the power group is H^{p-2}, and

$$[H] = H^{p-1} + H^{p-2} = [S].$$

|77| The order of $[H]$, that is, of $[S]$ is p. Hence the order of the power group H^{p-2} is p divided by the period, that is, $\frac{1}{2}p$.

Also no consecutive powers of H can contain the same operation. Therefore no even power of H contains the same operation as an odd power.

|78| *§17 Comparison with the Algebra of Symbolic Logic.*

This algebra is very analogous to the Algebra of Symbolic Logic.[1†] The law of addition, namely

$$x + x = x,$$

1. † cf. *A Treatise on Universal Algebra* by A. N. Whitehead, Book II.

is common to both of them. Thus this algebra as well as the algebra of logic belongs to the non-numerical genus[1†] of algebras. Also the multiplication in both obeys the associative law.

But here the identity of laws ends. For the numerical indices of this algebra have no place in the logical algebra. Again the multiplication in that algebra is commutative, here it is not necessarily so.

|79| Also in the Algebra of Symbolic Logic multiplication follows the law

$$xx = x$$

And hereby a reciprocity between multiplication and addition is introduced which has no place in the Algebra of Groups of Finite Order. The special symbol G is in some respects analogous to the special symbol i representing the Universe of Discourse in the Algebra of Symbolic Logic. For

$$i + x = i.$$

But in other respects S_0 is analogous to i. For

$$ix = x.$$

The null element 0 in Symbolic Logic has the properties

$$x + 0 = x, \quad 0x = 0.$$

This last equation is analogous to the equation

$$GH = G,$$

which is satisfied by GG whatever set or operation H may be.

|80| The inverse to S, namely S^{-1}, satisfies

$$SS^{-1} = S_0;$$

and the supplement of x, namely $\overline{x}$, satisfies

$$x\overline{x} = 0.$$

But S_0 and 0 have no other general properties analogous one to the other. Thus the analogy between S^{-1} and $\overline{x}$ is slight.

Both algebras are linear in the sense that the product of two things produces another thing of the same general type.

|81| The algebra of logic in the algebra of indefinite number: it deals with classes of things without consideration of the number of things in any class. In fact thought has not even discriminated as to whether it is dealing with regions of a continuum, or with classes containing numerable individuals.

Thus in symbolic logic number does not arise because counting has not begun.

But in the Algebra of Groups of Finite Order number hardly appears, except as regards exponents, because the counting has all been done. The algebra deals with a thoroughly finite Universe, with a definite number of individuals, here called the Operations in order to agree with the usual nomenclature of the theory of Groups. Also the only relations between individuals can be conceived as, or at least correlated with, other |82| individuals of the set. Thus the calculus which is required to explore such a Universe of thought is one which symbolizes the idea of 4 sets of these

1. † cf. *loc. cit.* p. 29, §22.

individuals, and also symbolizes the correlation between the relations between individuals and other individuals of the set.

The Algebra of Symbolic Logic in its entirety would be out of place here: for there is required no calculus to help thought to recognize the distinction between individuals. The appeal is to immediate intuition and not to general properties.

Memoir on the Algebra of Symbolic Logic (Part I)

1901 *American Journal of Mathematics*, vol. 23, no. 2 (April), pp. 139–165

Whitehead's 'Memoir on the Algebra of Symbolic Logic' was published in two parts in the American Journal of Mathematics *in 1901, while he was at Trinity College. The title may be misleading to some insofar as it does not in the least address the propositional calculus or quantification theory – the two principal components of logic as ordinarily understood – but rather the algebra of sets, or what Whitehead synonymously called classes. (These terms have subsequently diverged, with the latter becoming more general than the former. Hence in this introduction, we speak exclusively of sets.) Still, there is an obvious correspondence between the operators of logic (negation, conjunction, disjunction and material implication) and those of set theory (respectively, complementation, intersection, union and inclusion). Moreover, Whitehead is careful everywhere neither to invoke any set constructed by abstraction, or comprehension, which is to say the construction of a set by defining it as the collection of all elements that satisfy a given predicate, nor to argue by making mention of elements of any given set. (To explain this by contrast, the set of all even integers is defined by the property of divisibility by two; this is definition by abstraction. The assertion that this set is closed under addition proceeds by invoking two elements that satisfy this property and then deducing that this same property holds for their sum.) Thus, his work is not subject to the fatal toxin of Russell's paradox (also 1901) and depends only on the formalisms of the attendant operators. In light of these observations, his title remains appropriate to a present-day audience.*

In the 'Memoir', Whitehead seeks out mathematical properties of these formalisms that would generally be neglected by, say, a typical course in set theory and logic. His immediate concern is with set-theoretic functions of several variables and their derivative properties. Such a function has a list of sets as its domain and is built out of expressions using the standard set-theoretic operators. Hence $s(x, y) = x + y$ *is such a function of two variables, where the variables* x *and* y *denote sets and the given function represents their union. Whitehead speaks of the developed form of such functions (pp. 119–120), which amounts to a standard representation in terms of intersection, union and negation; he then goes on to develop their properties. Part I primarily introduces symmetric functions and the invariants of a multivariable set-theoretic function (p. 121) and subsequently introduces the notion of prime functions. These ideas are then used to examine the decomposition of set-theoretic functions both additively (that is, into sums) and multiplicatively (that is, into factors). Of particular importance in this context are Whitehead's evanescible functions (p. 122), which are multivariable set-theoretic functions whose output values admit the null set. (In more ordinary algebraic parlance, these are simply functions with roots.)*

That Whitehead is developing both additive and multiplicative decompositions provides us with a certain insight in light of the evolution of abstract algebra and number theory. Despite that he is working in a domain that naturally admits two commutative operations (union and intersection) with respect to which we have neutral elements (the null set and the universal class) as well as the distributive law (of intersection over union), he is not working in the context of what we would now call a commutative ring with unity. *Thus, the conditions for factorisation and associated arguments are necessarily highly technical and combinatoric insofar as they are not made in an ambient context that has had the benefit of historical evolution with its consequent grooming. His use of the term* prime *would indicate an intent to look for properties similar to those explored in number theory in the context of special classes of rings, but his methods must differ, and so his arguments may seem to the present-day mathematician far more technical and less subject to generalisation. This may partially explain the vanishingly small impact that this paper has had on mathematics. Nonetheless, we shall see that Part II of this 'Memoir' is much more forward-looking in its approach.*

Preface.

The present memoir is a purely mathematical investigation concerning the Algebra of Symbolic Logic. As a matter of history, this algebra has only been continuously studied since the publication of Boole's "Laws of Thought" (1854), and to C. S. Peirce[1†] and to Schröder,[2†] must be assigned the credit of perfecting its laws of operation. But, as a question of logical priority, this algebra must be considered as the first object of mathematical study. It holds this position by a double right.

First, its interpretation is concerned with the fundamental conceptions of classes, and of their mutual inclusions and exclusions; thus the terms of the algebra, such as $a, b, c, \ldots, x, y, z$, can be interpreted as each representing a class; a sum of terms, such as $a+b$, represents the class formed by the two classes a and b; the product of terms, such as ab, represents the class common to the two classes a and b; the symbol i represents all the classes which are the subject of discourse, so that not-a is conceived as a class composed of all i with the exception of a, and is denoted by $\bar{a}$; the symbol 0 means nonentity.

---

1. † Proc. of Amer. Acad. of Arts and Sciences, 1867, and Amer. Journ. of Math., vols. III and IV. [C. S. Peirce published three essays in March, April and May 1867 in *Proceedings of the American Academy of Arts and Sciences*, vol. 7 (May 1865–May 1868): C. S. Peirce, 'On an Improvement in Boole's Calculus of Logic', pp. 250–261; C. S. Peirce, 'On the Natural Classification of Arguments', pp. 261–287; C. S. Peirce, 'On a New List of Categories', pp. 287–298. The articles referenced from the *American Journal of Mathematics* are: C. S. Peirce, 'On the Algebra of Logic', vol. 3, no. 1 (March 1880), pp. 15–57; C. S. Peirce, 'On the Logic of Number', vol. 4, no. 1 (1881), pp. 85–95.]
2. † "Operationskreis des Logikkalküls," 1877, "Vorlesungen über die Algebra des Logik," vol. I, 1890, vol. II, 1891. [The full citations are Ernest Schröder, *Der Operationskreis des Logikkalkülus* (Leibzig: Teubner, 1877); *Vorlesungen über die Algebra der Logik, Volume I* (Leibzig: Teubner, 1890); *Volume II* (Leibzig: Teubner, 1891).]

Secondly, the symbolism of the algebra is the simplest of all algebraic systems. It is explained in §1, Part I, of this memoir. The laws $a+a=a$, $aa=a$, enable the use of numerals to be avoided, either as factors or indices. Also, the associative, commutative and distributive laws are preserved.

But the algebra has apparently had the defects natural to its simplicity. It is like argon in relation to the other chemical elements, inert and without |140| intrinsic activities. Accordingly, the purely mathematical study of the algebra has languished; the theory of Duality, the theory of Development by the process of Dichotomy, and the theory of Equations, have apparently exhausted its purely mathematical properties. The investigations concerning these properties were commenced by Boole, and have been brought to a high degree of perfection by Schröder. But the greater part of the literature relating to this algebra may be termed "applied mathematics;"[1†] it is primarily concerned with the relation of the symbolism to its interpretation in the field of Logic,[2†] and with its use as a practical means for the exact expression of deductive reasoning, especially in regard to the foundation of the various branches of mathematics.[3†]

In the present memoir, it is shown that the algebra has many more purely mathematical properties than those with which it has hitherto been credited. In the theories of ordinary algebraic symbols, the solution of equations is not the dominant subject of enquiry. Analogously, here the centre of interest has been shifted from the solution of equations to the study of the properties of functions of independent variables.

The keynote of this memoir is the prominence given to three ideas, namely, that of the "invariants" of a function of independent variables, that of "prime functions of independent variables," and that of the theory of "substitutions" of independent variables for independent variables. This last idea connects the algebra with the theory of Groups, and opens out a large field for investigation in that direction.

Invariants are defined in §2 of Part I: In the course of the memoir, the leading properties of functions of independent variables are shown to depend on them. Primes are defined in §3 of Part I. The expression here called a "primary prime" has been noticed by Jevons, and some of its elementary properties are discussed by Schröder[4†] and Peano.[5†] But here a different use is made |141| of them, and from

1. †Though the field of the application is itself "pure."
2. †Cf. Venn's "Symbolic Logic," 1st ed., 1881; 2d ed., 1894. [The full citations are John Venn, *Symbolic Logic*, 1st edition, (London: Macmillan, 1881); 2nd edition (London: Macmillan, 1894).]
3. †Cf. the admirable work in this direction of the Italian school, originated and inspired by Peano. The following are some of its principal works: "Arithmetices Principiae," by Peano, Turin, 1889; "Notations de Logique Mathématique," by Peano, Turin, 1894; "Formulaire de Mathématiques," Tome I, Turin, 1895; ibid., Tome II, No. 1; ibid., No. 2, 1898; ibid., No. 3, 1899; "Revue de Mathématiques," Turin, Tome VII, No. 1; ibid., passim. [The full citations are Giuseppe Peano, *Arithmetices Principia: Nova Methodo Exposita* (Rome: Fratres Bocca, 1889); *Notations de Logique Mathématique* (Turin: Guadagnini, 1894); *Formulaire de Mathématiques, Tome I* (Turin: Bocca frères, Ch. Clausen, 1895); *Tome II, No. 1: Logique mathématique* (Turin: Bocca frères, Ch. Clausen, 1897); *Tome II, No. 2* (Turin: Bocca frères, Ch. Clausen, 1898); *Tome II, No. 3* (Turin: Bocca frères, Ch. Clausen, 1899); *Revue de Mathématiques, Tome VII, No. 1* (Turin: Bocca frères, Ch. Clausen, 1900).]
4. †Cf. loc. cit., vol. I, "Neunte Vorlesung," pp. 370, 371; pp. 380, 381.
5. †Cf. "Formulaire de Math.," 1895, I, §3, propositions 24 to 30.

them are constructed other functions called "n-ary linear primes" and "n-ary separable primes."

In §5 of Part I, the properties of primes in relation to factorization and summation are developed; and the fundamental nature of primes is here made evident.

In §6, the factorization of any evanescible function of n variables into n-ary linear primes is discussed. An evanescible function is a function which can be made to vanish by an appropriate choice of values for its variables. This factorization of a given evanescible function into n-ary linear primes can be carried out in an indefinite number of ways. But the fundamental theorem is proved that, in general, the minimum number of such factors is 2^n. An exceptional evanescible function for which the minimum number is $2^n - r$ is said to be of deficiency r. The deficiency is shown to depend on the vanishing of r of the invariants. The term "deficiency" had previously been formally defined in §2 in relation to these invariants. The term has also a meaning, arising out of its present meaning, in regard to non-evanescible functions (cf. §8).

In §7, the theorems deduced from §6 by the method of duality are enunciated. These relate to the expression of functions of n variables as sums of separable primes. The property, analogous by the theory of duality to deficiency, is called "supplemental deficiency."

In §9, two special types of functions called "linear" and "separable" functions are discussed, and various theorems relating to them are proved.

Part II is devoted to the theory of "Substitutions." A function $\phi(x,y)$ is transformed into some function $\psi(u,v)$ by the transformation $x = f_1(u,v)$, $y = f_2(u,v)$. But, in general, it is not allowable to conceive x and y as independent variables, when this transformation is used. For the condition of the possibility of the two equations of transformation viewed as equations to find u and v, is an equation which x and y must satisfy. Thus x and y are restricted to be simultaneous solutions of this equation. But this equation reduces to the identity, $0 = 0$, for all values of x and y, if the coefficients of the functions $f_1(u,v)$ and $f_2(u,v)$ satisfy a certain condition. In this case, the transformation amounts to the substitution of one set of independent variables for another set of independent variables. The term "substitution" is exclusively used for this type of transformation. Each substitution is denoted by a single letter; also u and v are replaced by x and y. Thus we write $Tx = f_1(x,y)$, $Ty = f_2(x,y)$; also $T\phi(x,y)$ is written for $\phi(Tx,Ty)$. These explanations occupy §1 of Part II. |142| In §2, the relations between the coefficients of $f_1(x,y)$ and $f_2(x,y)$, that is, of Tx and Ty, are more fully discussed. It is proved that both Tx and Ty are of deficiency two and of supplemental deficiency two; and the two functions are also otherwise related.

In §3, it is proved that there is only one reverse substitution corresponding to a given substitution T, and its coefficients are determined. It follows (cf. §4) that all possible substitutions form a group. The group is not continuous, since the concepts of "real number" and of infinitesimal variations of real numbers have no place in this algebra. It is of finite order, if the number of distinct fundamental terms in the algebra, representing constants, is conceived as finite. It is of indefinite order in so far as these fundamental given constant terms are not brought into explicit definition; and also in so far as new fundamental constants may always be produced at discretion without violating any law of the algebra. The order of the subgroup $T, T^2, T^3, \ldots.$ is evidently finite and determinable in the general case.[1†]

In §6, the "congruence" of functions is defined. Two functions, $\phi(x,y)$ and $\Phi(x,y)$, are defined to be "congruent," if any substitution T exists such that $T\phi(x,y) = \Phi(x,y)$. The fundamental theorem is proved that all functions are congruent which have all their corresponding invariants equal. A complete set of congruent functions is called a congruent family.

The subgroup of substitutions (cf. §7) which leave a given function unchanged is called the identical group of the function. It is proved that the identical groups of all members of the same congruent family are simply isomorphic. It is also proved (cf. §8) that, except in assigned special cases, the subgroup common to the identical groups of any two functions, contains more than the single identical substitution T^0.

PART I.

The Theory of Primes

§1—Elementary Principles.

The following summary of the elements of the algebra may be useful:

|143| Let *a*, *b*, *c* be any terms subject to the laws of the algebra, then

$$a+b=b+a, \quad a+b+c=(a+b)+c=a+(b+c).$$
$$a+a=a, \quad a+0=a, \quad a+i=i.$$
$$ab=ba, \quad abc=ab\,.\,c=a\,.\,bc\,.$$
$$aa=a, \quad a0=0a=0, \quad ai=ia=a.$$

The supplement of any term a is written $\bar{a}$, and is defined by

$$a+\bar{a}=i, \quad a\bar{a}=0.$$

The supplement of a complex expression, such as $(a+b)$, is written $\overline{}(a+b)$
We have

$$\overline{}(a+b)=\bar{a}\bar{b}, \quad \overline{}(a+b)=\bar{a}+\bar{b}.$$

Also $$\bar{\bar{a}}=a, \quad \bar{0}=i, \quad \bar{i}=0.$$

The equation $$P+Q=0$$

implies $P=0$, $Q=0$, and conversely. Thus, any number of equations in which the right-hand sides are zero, can be combined into one equation by simple addition of their left-hand sides.

The equation $P=Q$ is equivalent to

$$P\bar{Q}+Q\bar{P}=0.$$

Thus any equation can be transformed into one with its right-hand side zero.

---

1. †I have since determined the order of this group, which is, in general, of the 12th order, and the orders of all other groups mentioned in this memoir. I hope shortly to publish these results, which depend on a new general method in connection with this subject. – *Note added February*, 1901.

The equation $P=Q$ implies $\overline{P}=\overline{Q}$, and conversely. The general "developed" form of a function of n independent variables $x, y, \ldots . t$ is

$$Axy \ldots . t+Bxy \ldots . \overline{t}+\ldots .+K\overline{x}\,\overline{y} \ldots . \overline{t},$$

where every possible product involving x or $\overline{x}, y$ or $\overline{y}, \ldots . t$ or $\overline{t}$, is represented, and the coefficients represent constants.

The supplement of this function is

$$\overline{A}xy \ldots . t+\overline{B}xy \ldots . \overline{t}+\ldots .+\overline{K}\,\overline{x}\,\overline{y} \ldots . \overline{t}.$$

For the sake of brevity, the argument will be conducted with respect to functions of two variables; but the leading theorems can easily be generalized for functions of any number of variables. The typical form of a function, $\Phi(x,y)$, of two variables will always be written[1]

$$Axy+B\overline{x}y+Cx\overline{y}+D\overline{x}\,\overline{y},$$

|144| and this notation for the coefficients will always be adhered to. Then[2]

$$\overline{\phi}(x,y)=\overline{A}xy+\overline{B}\,\overline{x}y+\overline{C}\,x\overline{y}+\overline{D}\,\overline{x}\,\overline{y}.$$

The condition that a proposed equation in n variables, $x, y, \ldots . t$, such as

$$\psi(x, y, \ldots . t)=0,$$

may be a possible equation, is found by substituting i or 0 in $\psi(x, y, \ldots . t)$ for each of $x, y, \ldots . t$ in every possible combination, and by equating to zero the product of the results of such substitutions. This equation of condition will be symbolically represented by

$$\prod \psi \begin{pmatrix} i, i, \ldots . i \\ 0,0, \ldots . 0 \end{pmatrix}=0.$$

Thus the condition for the possibility of

$$\phi(x,y)=0$$

is

$$ABCD=0,$$

and the condition for the possibility of

$$Hx+K\overline{x}=0$$

is

$$HK=0.$$

The general solution of this latter equation is

$$x=K+u\overline{H},$$

where u is an arbitrary unknown. Thus since an indefinite number of special values can be given to u, every equation has, in general, an indefinite number of particular roots which are all included in the general solution.

According to the duality,[3†] to every symbolic theorem involving $+$, $\times$ (the symbol for multiplication, usually omitted), i, 0, there corresponds a symbolic theorem in

1. In the following expression, the original erroneously had 'Cxy' instead of '$Cx\overline{y}$'.
2. In the following equation, the original erroneously omitted the overbar from the 'ϕ'.
3. † Cf. my "Universal Algebra," §24, for a full statement and proof of this theory. The theory is due to C. S. Peirce and to Schröder.

which + is replaced by ×, × by +, i by 0, 0 by i. An outcome of this theory in its application to functions of independent variables is as follows: From the theorem that $\phi(x,y)$ can be expressed in the form $\psi(x,y)$, when the condition $\chi(A,B,C,D)=0$ is fulfilled by the coefficients of $\phi(x,y)$, there can be derived the theorem that[1] $\bar{\phi}(x,y)$ can be expressed in the form $\bar{\psi}(x,y)$ when the condition $\chi(\bar{A},\bar{B},\bar{C},\bar{D})=0$ is fulfilled.

|145 |§2.—*Symmetric Functions.*

Consider n terms $a_1, a_2, \ldots . a_n$. Let their symmetric functions be defined as follows, by a notation which will be adhered to:

$$S_1=\sum_{p=1}^{n} a_p\ ,\ S_2=\sum_{p,q=1}^{n} a_p a_q\ ,\ S_3=\sum_{p,q,r=1}^{n} a_p a_q a_r, \ldots ., S_n=a_1 a_2 \ldots . a_n$$

where, in S_2, it is understood that the subscript p is not equal to q in the same product $a_p a_q$, and similarly for $S_3, S_4, \ldots . S_n$. This, or a similar supposition respecting the inequality of suffixes when typical products or sums are given, will be adhered to unless it is otherwise expressly stated. Then

$$\bar{S}_n=\sum_{p=1}^{n} \bar{a}_p,\ \bar{S}_{n-1}=\sum_{p,q=1}^{n} \bar{a}_p \bar{a}_q,\ \bar{S}_{n-2}=\sum_{p,q,r=1}^{n} \bar{a}_p \bar{a}_q \bar{a}_r, \ldots .\ \bar{S}_1=\bar{a}_1 \bar{a}_2 \ldots . \bar{a}_n.$$

Also, evidently,[2]

$$S_1 \mathrel{\dashv} S_2 \mathrel{\dashv} S_3 \ldots . \mathrel{\dashv} S_n,$$

$$\bar{S}_1 \mathrel{\vdash} \bar{S}_2 \mathrel{\vdash} \bar{S}_3 \ldots . \mathrel{\vdash} \bar{S}_n.$$

These subsumptions can all be expressed in the typical equation

$$S_{p+q}\bar{S}_p=0. \tag{1}$$

This equation is also equivalent to any one of the following forms:

$$S_p S_{p+q}=S_{p+q},\ S_p+S_{p+q}=S_p,\ \bar{S}_p \bar{S}_{p+q}=\bar{S}_p,\ \bar{S}_p+\bar{S}_{p+q}=\bar{S}_{p+q}. \tag{2}$$

All other symmetric functions of $a_1, a_2, \ldots . a_n$, and of their supplements, can be expressed in terms of these fundamental symmetric functions. Thus[3]

$$\left.\begin{aligned} &S_1\bar{S}_n=\Sigma a_p \bar{a}_q,\ S_1\bar{S}_{n-1}=\Sigma a_p \bar{a}_q \bar{a}_r,\ S_1\bar{S}_{n-2}=\Sigma a_p \bar{a}_q \bar{a}_r \bar{a}_s, \ldots . \\ &S_2\bar{S}_n=\Sigma a_p a_q \bar{a}_r,\ S_2 S_{n-1}=\Sigma a_p a_q \bar{a}_r \bar{a}_s, \ldots . \end{aligned}\right\} \tag{3}$$

It follows from equations (2) that if $S_p=0$, then $S_{p+1}, S_{p+2}, \ldots . S_n$ all vanish, and that if $\bar{S}_p=0$, then $\bar{S}_{p-1}, \bar{S}_{p-2}, \ldots ., \bar{S}_1$ all vanish.

The symmetric functions of the coefficients of the function $\phi(x,y, \ldots . t)$ will be called the invariants of the function. Thus, for the function $\phi(x,y)$ of two variables, the invariants are

1. In the following expression, the original erroneously omitted the overbar from the 'ϕ'.
2. The '$\dashv$' symbol represents the subset relation.
3. In the following equation, the original erroneously added an overbar to the first instance of 'a_q' on the second line.

$$\left.\begin{aligned} &S_1 = A+B+C+D,\ S_2 = AB+AC+AD+BC+BD+CD,\\ &S_3 = ABC+ABD+ACD+BCD,\ S_4 = ABCD,\\ &\overline{S}_1 = \overline{A}\overline{B}\overline{C}\overline{D},\ \overline{S}_2 = \overline{A}\overline{B}\overline{C}+\overline{A}\overline{B}\overline{D}+\overline{A}\overline{C}\overline{D}+\overline{B}\overline{C}\overline{D},\\ &\overline{S}_3 = \overline{A}\overline{B}+\overline{A}\overline{C}+\overline{A}\overline{D}+\overline{B}\overline{C}+\overline{B}\overline{D}+\overline{C}\overline{D},\ \overline{S}_4 = \overline{A}+\overline{B}+\overline{C}+\overline{D},\end{aligned}\right\} \quad (4)$$

|146| If the condition for any special property of a function can be expressed as a relation between its invariants, without the coefficients otherwise entering into the relation, the property will be called an invariant property. Such invariant properties will be proved in Part II to be analogous in many respects to invariant properties of rational integral algebraic forms in ordinary algebra.

The following definitions of technical terms may be stated at once, though their fitness will only be apparent when primes have been introduced. Let $\phi(x,y,\ldots.t)$ be a function of n variables, and let $S_1, S_2, \ldots. S_{2^n}$ be its 2^n invariants. Then if $\overline{S}_1 = 0$, $\phi(x,y,\ldots.t)$ will be said to be of "deficiency one" at least; if $\overline{S}_2 = 0$, the function will be said to be of "deficiency two" at least, and so on. Thus, if $\overline{S}_p = 0$, the function is of "deficiency p" at least.

Again, if $S_{2^n} = 0$, $\phi(x,y,\ldots.t)$ will be said to be of "supplemental deficiency one" at least; if $S_{2^n-1} = 0$, of "supplemental deficiency two" at least, and if $S_{2^n-p} = 0$, of "supplemental deficiency $p+1$" at least.

Now, if $S_{2^n} = 0$, the equation

$$\phi(x,y,\ldots.t) = 0$$

is a possible equation. In this case the function will also be called evanescible.

Thus an evanescible function is of supplemental deficiency one, and conversely. If $\overline{S}_1 = 0$, then equation

$$\overline{\phi}(x,y,\ldots.t) = 0, \text{ that is, } \phi(x,y,\ldots.t) = i$$

is a possible equation. In this case, the function will be said to be "capable of the value i." Thus a function capable of the value i is of deficiency one, and conversely.

§3.—Primes

A function of one variable x, which is of the special type

$$\overline{a}x + a\overline{x},$$

where a is any constant, will be called a "primary prime." It will be written for brevity in the form $p(a,x)$. It is obvious that $p(a,x) = p(x,a)$. Then

$$\overline{p}(a,x) = ax + \overline{a}\,\overline{x} = p(\overline{a},x).$$

Thus the supplement of a primary prime is itself a primary prime.

|147| The solution of the equation

$$p(a,x) = 0$$

is

$$x = a + ua = a.$$

Thus the solution is definite, involving no arbitrary unknown. In other words, the equation has only one root.

Conversely, if the equation

$$Hx + K\bar{x} = 0$$

has only one root, then the function $Hx + K\bar{x}$ is a primary prime. For the solution is

$$x = K + u\bar{H}.$$

Hence, since by hypothesis, the equation has only one root,

$$\bar{H} \nprec K.$$

But, since the equation is possible,

$$HK = 0.$$

that is, $$K \nprec \bar{H}.$$

Hence, $$\bar{H} = K.$$

Thus $$Hx + K\bar{x} = Hx + \bar{H}\,\bar{x} = p(\bar{H}, x).$$

A function of the n variables $x, y, z, \ldots.$ which can be expressed in the form

$$\bar{a}x + a\bar{x} + \bar{b}y + b\bar{y} + \bar{c}z + c\bar{z} + \ldots.$$

will be called an "n-ary linear prime." It can be written

$$p(a,x) + p(b,y) + p(c,z) + \ldots.;$$

and for brevity it will be written

$$p(a,x;b,y;c,z;\ \ldots.).$$

The supplement of an n-ary linear prime is

$$\begin{aligned}\bar{p}(a,x;b,y;c,z;\ \ldots.) &= \bar{p}(a,x)\bar{p}(b,y)\bar{p}(c,z)\ldots. \\ &= p(\bar{a},x)p(\bar{b},y)p(\bar{c},z)\ldots..\end{aligned}$$

This type of function of the n variables $x, y, z, \ldots.$ will be called an "n-ary separable prime." Thus for two variables x and y, we have secondary linear primes, symbolized thus:

$$p(a,x;b,y) = p(a,x) + p(b,y) = \bar{a}x + a\bar{x} + \bar{b}y + b\bar{y};$$

|148| and secondary separable primes, symbolized thus:

$$\bar{p}(a,x;b,y) = p(\bar{a},x)p(\bar{b},y) = (ax + \bar{a}\,\bar{x})(by + \bar{b}\,\bar{y}).$$

A primary prime can be looked on both as a linear prime of one variable, and as a separate prime of one variable.

It must be noted that an n-ary prime, either linear or separable, is not a degenerate form of a $(n+m)$-ary prime of the same type. Thus, $\bar{a}x + a\bar{x}$ is not a degenerate form of $\bar{a}x + a\bar{x} + \bar{b}y + b\bar{y}$; for no constant value of b can be found which will make $\bar{b}y + b\bar{y}$ vanish for all values of y. Similarly, $\bar{a}x + a\bar{x}$ is not a degenerate form of $(\bar{a}x + a\bar{x})(\bar{b}y + b\bar{y})$.

It is one of the leading objects of Part I of this memoir to prove that primes are to be considered as the fundamental types of function of this algebra, and that all functions of unknowns can be conveniently classified according as they are constructed of primes.

The condition that $\phi(x,y)$ may be a secondary linear prime is found by identifying it with $p(a,x;b,y)$. But we may write[1]

$$p(a,x;b,y)=(\bar{a}+\bar{b})xy+(\bar{a}+b)x\bar{y}+(a+\bar{b})\bar{x}y+(a+b)\bar{x}\,\bar{y}.$$

Hence, by comparison of coefficients,

$$A=\bar{a}+\bar{b},\quad B=\bar{a}+b,\quad C=a+\bar{b},\quad D=a+b.$$

These equations can be written as the single equation

$$(A+\bar{B}+\bar{C}+\bar{D})ab+(\bar{A}+B+\bar{C}+\bar{D})a\bar{b}+(\bar{A}+\bar{B}+C+\bar{D})\bar{a}b+(\bar{A}+\bar{B}+\bar{C}+D)\bar{a}\,\bar{b}=0$$

The resultant of this equation for *a* and *b* is

$$(A+\bar{B}+\bar{C}+\bar{D})(\bar{A}+B+\bar{C}+\bar{D})(\bar{A}+\bar{B}+C+\bar{D})(\bar{A}+\bar{B}+\bar{C}+D)=0.$$

Hence, after multiplying out and reducing, we find

$$S_4+\bar{S}_3=0;$$

which is the necessary and sufficient condition for a secondary linear prime.

Thus a secondary linear prime is evanescible and capable of the value *i*. Its invariants are $S_1=i,\ S_2=i,\ S_3=i,\ S_4=0.$ It is of deficiency three and of supplemental deficiency one.

|149| Reciprocally, we can state without further proof, that the necessary and sufficient condition that $\phi(x,y)$ may be a secondary separable prime is

$$\bar{S}_1+S_2=0. \tag{5}$$

Thus, a secondary separable prime is capable of the value *i* and is evanescible. Its invariant are $S_1=i,\ S_2=0,\ S_3=0,\ S_4=0.$ It is of deficiency one, and of supplemental deficiency three. It is evident that the same function cannot be both a secondary linear prime and a secondary separable prime.

It can be proved that a general form for a secondary linear prime is

$$(p+q+r)xy+(\bar{p}+q+r)x\bar{y}+(\bar{q}+r)\bar{x}y+\bar{r}\,\bar{x}\,\bar{y}; \tag{6}$$

and that a general form for a secondary separable prime is

$$pqrxy+\bar{p}qrx\bar{y}+\bar{q}rxy+\bar{r}\,\bar{x}\,\bar{y}. \tag{7}$$

Another form for the condition (4), that $\phi(x,y)$ may be a secondary linear prime, can be proved to be

$$\bar{A}=BCD,\quad \bar{B}=ACD,\quad \bar{C}=ABD,\quad \bar{D}=ABC. \tag{8}$$

Reciprocally, another form for the condition (5), that $\phi(x,y)$ may be a secondary separable prime is[2]

$$A=\bar{B}\bar{C}\bar{D},\quad B=\bar{A}\bar{C}\bar{D},\quad C=\bar{A}\bar{B}\bar{D},\quad D=\bar{A}\bar{B}\bar{C}. \tag{9}$$

The solution of the equation

$$p(a,x;b,y)=0$$

is $x=a,\ y=b.$ Thus the equation has only one set of roots.

1. In the following equation, the original erroneously omitted the comma in '*b*, *y*'.
2. In the following equation, the original erroneously omitted an overbar from the second '*C*': '$B=\bar{A}C\bar{D}$'.

Conversely, if the equation

$$\phi(x,y)=0$$

has only one set of roots, then $\phi(x,y)$ is a secondary linear prime.

For, by eliminating y, we find

$$ABx+CD\bar{x}=0.$$

This must have only one root, and accordingly

$$AB=\bar{C}+\bar{D}.$$

|150| Similarly, in order that the equation for x may only have one root, we must have

$$AC=\bar{B}+\bar{D}.$$

These equations are equivalent to the single equation

$$ABCD+(\bar{A}+\bar{B})(\bar{C}+\bar{D})+(\bar{A}+\bar{C})(\bar{B}+\bar{D})=0,$$

that is,

$$S_4+\bar{S}_3=0.$$

Hence, $\phi(x,y)$ is a secondary linear prime. A similar theorem holds for equations involving any number of unknowns.

The condition that $\phi(x,y)$ is a primary prime with x as sole variable, is evidently

$$\bar{A}=\bar{B}=C=D. \quad (10)$$

This is not an invariant condition, but it involves

$$S_1=i,\ S_2=i,\ S_3=0,\ S_4=0.$$

Thus a primary prime $p(a, x)$, if it is written in the form $(y+\bar{y})p(a, x)$ so as to appear as a function of two variables, x and y, is a function of deficiency two and of supplemental deficiency two.

§4.—Factorization and Expression as Sums.

Any function can be factorized in an indefinite number of ways. For assume

$$\phi(x)=Hx+K\bar{x}=(H_1x+K_1\bar{x})(H_2x+K_2\bar{x})=H_1H_2x+K_1K_2\bar{x}.$$

Then

$$H=H_1H_2,\quad K=K_1K_2.$$

Hence,

$$\bar{H}_1H=0,\quad H_1=H+p_1.$$

Also

$$H_2=H+p_2\bar{H}_1=H+\bar{p}_1p_2\bar{H}=H+\bar{p}_1p_2.$$

Similarly,

$$K_1=K+q_1,\ K_2=K+\bar{q}_1q_2.$$

|151| Thus,

$$\phi(x)=Hx+K\bar{x}=\{(H+p_1)x+(K+q_1)\bar{x}\}\{(H+\bar{p}_1p_2)x+(K+\bar{q}_1q_2)x\}. \quad (11)$$

Similarly,

$$\phi(x,y)=\{(A+p_1)xy+(B+q_1)x\bar{y}+(C+r_1)\bar{x}y+(D+s_1)\bar{x}\,\bar{y}\}$$
$$\times\{(A+\bar{p}_1p_2)xy+(B+\bar{q}_1q_2)x\bar{y}+(C+\bar{r}_1r_2)\bar{x}y+(D+\bar{s}_1s_2)\bar{x}\bar{y}\}. \qquad (12)$$

Thus the most general types of pairs of factors have been found. They can be expressed otherwise thus: Let $\psi_1(x,y)$ and $\psi_2(x,y)$ be any two functions of x and y, then the most general type of factor[1] of $\phi(x,y)$ is

$$\phi(x,y)+\psi_1(x,y);$$

and the most general type of a pair of factors is expressed by

$$\phi(x,y)=\{\phi(x,y)+\psi_1(x,y)\}\{\phi(x,y)+\bar{\psi}_1(x,y)\psi_2(x,y)\}. \qquad (13)$$

Equation (13) is evidently identical with equation (12) when we put

$$\psi_1(x,y)=p_1xy+q_1x\bar{y}+r_1\bar{x}y+s_1\bar{x}\bar{y},$$
$$\psi_2(x,y)=p_2xy+q_2x\bar{y}+r_2\bar{x}y+s_2\bar{x}\bar{y}.$$

Reciprocally, any function $\phi(x,y)$ can be expressed as the sum of two functions in an indefinite number of ways; thus, if we put

$$\phi(x,y)=\phi_1(x,y)+\phi_2(x,y),$$

the most general type for $\phi_1(x,y)$ is

$$\phi_1(x,y)=\phi(x,y)\psi_1(x,y); \qquad (14)$$

and if $\phi_1(x,y)$ be given by (14), the most general type for $\phi_2(x,y)$ is

$$\phi_2(x,y)=\phi(x,y)\{\bar{\psi}_1(x,y)+\psi_2(x,y)\}. \qquad (15)$$

Let $\phi_1(x,y)$ and $\phi_2(x,y)$ be called "summands" of $\phi(x,y)$, the word "summand" corresponding to the word "factor" in relation to products.

We have thus seen that any function $\phi(x,y)$ can be factorized, or expressed as a sum, in an indefinite number of ways. It remains to consider the special conditions which factors or summands can be made to obey.

It is evident from (12) that a factor of $\phi(x,y)$ cannot have supplemental deficiency of higher order than the supplemental deficiency of $\phi(x,y)$. For

$$(A+p_1)(B+q_1)=0,$$

|152| implies $$AB=0;$$

and $$(A+p_1)(B+q_1)(C+r_1)=0,$$

implies $$ABC=0;$$

and so on.

Again, from (12), it follows that a factor of $\phi(x,y)$ must have deficiency of at least equal order to that of $\phi(x,y)$, and may have deficiency of a higher order.

For if $$\bar{A}\bar{B}=0,$$

1. The original erroneously had 'fact or'.

it follows that

$$\overline{(A+p_1)}\,\overline{(B+q_1)}=\bar{A}\bar{B}\bar{p}_1\bar{q}_1=0;$$

and if

$$\bar{A}\bar{B}\bar{C}=0,$$

it follows that

$$\overline{(A+p_1)}\,\overline{(B+q_1)}\,\overline{(C+r_1)}=\bar{A}\bar{B}\bar{C}\,\bar{p}_1\bar{q}_1\bar{r}_1=0;$$

and so on.

Reciprocally, a summand of $\phi(x,y)$ cannot have deficiency of higher order than the deficiency of $\phi(x,y)$, and it must have supplemental deficiency of at least equal order to that of $\phi(x,y)$.

For example, no function can have an evanescible factor unless it be itself evanescible; and every factor of a function capable of the value i is itself capable of the value i. Reciprocally, no function can have a summand capable of the value i unless it be itself capable of the value i; and every summand of an evanescible function is itself evanescible.

If $\phi(x,y)$ be evanescible, then an evanescible companion factor to any possible factor can always be found. For any possible factor can be written in the form

$$(A+p_1)xy+(B+q_1)x\bar{y}+(C+r_1)\bar{x}y+(D+s_1)\bar{x}\bar{y}.$$

The most general type of its companion factor is

$$(A+\bar{p}_1p_2)xy+(B+\bar{q}_1q_2)x\bar{y}+(C+\bar{r}_1r_2)\bar{x}y+(D+\bar{s}_1s_2)\bar{x}\bar{y}.$$

This is evanescible if

$$(A+\bar{p}_1p_2)(B+\bar{q}_1q_2)(C+r_1\bar{r}_2)(D+\bar{s}_1s_2)=0.$$

|153| But this is always a possible equation for p_2,q_2,r_2,s_2, since, by hypothesis, one set of roots is $p_2=q_2=r_2=s_2=0$.

§5.—Properties of Primes.

A linear prime has no evanescible factor other than itself. And conversely, if an evanescible function has no evanescible factor other than itself, it is a linear prime. First consider the primary prime $p(a,x)$. Any factor must be of the form $(\bar{a}+p)x+(a+q)\bar{x}$. This factor is evanescible if

$$(\bar{a}+p)(a+q)=0,$$

that is, if

$$ap+\bar{a}q+pq=0.$$

But $ap=0$ implies $p=u\bar{a}$, and $\bar{a}q=0$ implies $q=va$. Thus

$$(\bar{a}+p)x+(a+q)\bar{x}=(\bar{a}+u\bar{a})x+(a+va)\bar{x}=\bar{a}x+a\bar{x}.$$

Hence, the evanescible factor is merely the original primary prime.

Again, consider the secondary linear prime $p(a,x;b,y)$. Any factor of it must be of the form[1]

1. The following expression contained multiple errors in the original. It originally read as follows: '$(\bar{a}+\bar{b}+p)xy+(\bar{a}+b+q)x\bar{y}+(\bar{a}+b+r)\bar{x}y+(a+b+s)xy$.'

$$(\bar{a}+\bar{b}+p)xy+(\bar{a}+b+q)x\bar{y}+(a+\bar{b}+r)\bar{x}y+(a+b+s)\bar{x}\,\bar{y}.$$

This factor is evanescible if

$$(\bar{a}+\bar{b}+p)(\bar{a}+b+q)(a+\bar{b}+r)(a+b+s)=0.$$

Hence, $$p(\bar{a}+b)(a+\bar{b})(a+b)=0,$$

that is, $$pab=0.$$

Hence,[1] $$p=u_1(\bar{a}+\bar{b}).$$

Similarly, $$q=u_2(\bar{a}+b),\ r=u_3(a+\bar{b}),\ s=u_4(a+b).$$

Thus the evanescible factor reduces to the original secondary linear prime. A similar proof holds for any n-ary linear prime.

Conversely, let $\phi(x,y)$ be an evanescible function which has no evanescible factor other than itself. Then, by hypothesis,

$$ABCD=0. \qquad \text{(a)}$$

Also any factor is of the form

$$(A+p)xy+(B+q)x\bar{y}+(C+r)\bar{x}y+(D+s)\bar{x}\,\bar{y}.$$

|154| And by hypothesis, if this factor is evanescible, it must reduce to the function $\phi(x,y)$. Hence, assuming evanescibility,

$$p=u_1A,\ q=u_2B,\ r=u_3C,\ s=u_4D;$$

that is, $$p\bar{A}=0,\ q\bar{B}=0,\ r\bar{C}=0,\ s\bar{D}=0. \qquad \text{(b)}$$

But the condition for evanescibility is

$$(A+p)(B+q)(C+r)(D+s)=0.$$

Eliminating q, r, s, we find

$$(A+p)BCD=0;$$

hence, $$p=ABCD+v(\bar{B}+\bar{C}+\bar{D})=v(\bar{B}+\bar{C}+\bar{D}),$$

by the use of equation (a).

But the first of equations (b) is to hold for every value of p which is consistent with the evanescibility of the factor. Hence,

$$v(\bar{B}+\bar{C}+\bar{D})\bar{A}=0,$$

for every value of v.

Hence, $$\bar{A}\bar{B}+\bar{A}\bar{C}+\bar{A}\bar{D}=0.$$

Similarly for q, r, s. Thus we find

$$\bar{S}_3=0.$$

By combining with equation (a), we deduce

1. The following equation originally had 'u' instead of 'u_1'. This is not strictly an error, but to be consistent with the next line and next page, 'u_1' is preferable, for reasons of clarity.

$$S_4+\overline{S}_3=0.$$

This is the condition that $\phi(x,y)$ may be a secondary linear prime. A similar proof of this converse part of the theorem holds for a function of any number of variables. Reciprocally, a separable prime has no summand capable of the value i other than itself. And conversely, if a function, capable of the value i has no summand capable of the value i, other than itself, it is a separable prime.

A linear prime cannot be expressed as the product of two factors, of which one is constant and other than i.

For assume

$$p(a,x;b,y)=d(A_1xy+B_1x\overline{y}+C_1\overline{x}y+D_1\overline{x}\overline{y}).$$

|155| Then $\quad dA_1=\overline{a}+\overline{b},\quad dB_1=\overline{a}+b,\quad dC_1=a+\overline{b},\quad dD_1=a+b.$

Thus
$$d(A_1+B_1+C_1+D_1)=i.$$

Hence,
$$d=i.$$

Reciprocally, a separable prime cannot be expressed as the sum of two summands, of which one is constant and other than 0.

The sum of two distinct linear primes, functions of the same variables, is not evanescible.

For if $p(a_1,x)+p(a_2,x)$ is evanescible, then

$$(\overline{a}_1+\overline{a}_2)(a_1+a_2)=0,$$

that is,
$$\overline{a}_1a_2+a_1\overline{a}_2=0.$$

Hence,
$$a_1=a_2.$$

Again, consider $p(a_1,x;b_1,y)+p(a_2,x;b_2,y)$. It is evident that this is only evanescible if

$$p(a_1,x)+p(a_2,x) \text{ and } p(b_1,y)+p(b_2,y)$$

are both evanescible. But this requires

$$a_1=a_2 \text{ and } b_1=b_2.$$

Reciprocally, the product of two distinct separable primes, functions of the same variables, is not capable of the value i.

The properties of primes proved in this section are the reason for the name "prime" here assigned to them: and they are also the foundation of the importance of primes, linear and separable, in the theory of factorization and summation respectively (cf. §§6, 7). For factorization into linear primes is ultimate in the sense that the factors cannot be further decomposed into evanescible factors; and linear primes are the only factors with this property. Similarly, for summation into separable primes, *mutatis mutandis*.

§6.—Factorization into Linear Primes.

Any evanescible function of n variables can, in general, be factorized into a product of a minimum number of 2^n n-ary linear primes; and the exceptional cases arise when $\bar{S}_r = 0$, $\bar{S}_{r+1} \neq 0$, and then the function can be factorized into a minimum number of $(2^n - r)$ n-ary linear primes. This proposition, which we |156| will proceed to prove, gives the reason of the term "of deficiency r" which has been applied to $\phi(x, y, \ldots. t)$ when $\bar{S}_r = 0$. For if $\phi(x, y, \ldots. t)$ be evanescible, the minimum number of n-ary linear prime factors has been reduced by r from that of the general case. Also, a very analogous meaning can be found (cf. §8) when the function is not evanescible.

It will be sufficient to write out the proof for a function of two variables: the method is evidently general.

First, to prove that any evanescible function of two variables can be expressed as a product of four factors, each of which is a secondary linear prime.

Assume

$$\phi(x,y) = \prod_{r=1}^{4} p(a_r, x; b_r, y).$$

Then, by comparison of coefficients, we have

$$A = \prod_{r=1}^{4}(\bar{a}_r + \bar{b}_r),\ B = \prod_{r=1}^{4}(\bar{a}_r + b_r),\ C = \prod_{r=1}^{4}(a_r + \bar{b}_r),\ D = \prod_{r=1}^{4}(a_r + b_r).$$

Hence,

$$\bar{A} = \sum_{r=1}^{4} a_r b_r,\ \bar{B} = \sum_{r=1}^{4} a_r \bar{b}_r,\ \bar{C} = \sum_{r=1}^{4} \bar{a}_r b_r,\ \bar{D} = \sum_{r=1}^{4} \bar{a}_r \bar{b}_r.$$

These equations for the coefficients are equivalent to the single equation

$$\bar{A}\prod_{r=1}^{4}(\bar{a}_r + \bar{b}_r) + A\sum_{r=1}^{4} a_r b_r + \bar{B}\prod_{r=1}^{4}(\bar{a}_r + b_r) + B\sum_{r=1}^{4} a_r \bar{b}_r$$

$$+\bar{C}\prod_{r=1}^{4}(a_r + \bar{b}_r) + C\sum_{r=1}^{4} \bar{a}_r b_r + \bar{D}\prod_{r=1}^{4}(a_r + b_r) + D\sum_{r=1}^{4} \bar{a}_r \bar{b}_r = 0. \qquad \text{(m)}$$

Consider this as an equation to find $a_1, b_1, a_2, b_2, a_3, b_3, a_4, b_4$; and put for its left-hand side

$$\lambda(a_1, b_1; a_2, b_2; a_3, b_3; a_4, b_4).$$

Now, since

$$\prod_{r=1}^{4}(\bar{a}_r + \bar{b}_r) = \overline{\sum_{r=1}^{4} a_r b_r},$$

with other analogous equations, we see that in each of the component factors of

$$\prod \lambda\begin{pmatrix} i, i & i, i & i, i & i, i \\ 0, 0 ; & 0, 0 ; & 0, 0 ; & 0, 0 \end{pmatrix},$$

either A or $\bar{A}$, and either B or $\bar{B}$, and either C or $\bar{C}$, and either D or $\bar{D}$, must appear; but not both of any pair.

|157| Again, any one of the four sets of values for a_1 and b_1 given in

$$a_1, b_1 = \left.\begin{matrix} i \\ 0 \end{matrix}\right\},$$

makes one of the four

$$\sum_{r=1}^{4} a_r b_r,\ \sum_{r=1}^{4} a_r \bar{b}_r,\ \sum_{r=1}^{4} \bar{a}_r b_r,\ \sum_{r=1}^{4} \bar{a}_r \bar{b}_r$$

take the value of *i*. Hence, in each of the before-mentioned component factors, at least one out of *A*, *B*, *C*, *D* must appear as a summand.

Also, it is easy to see that sets of values for $a_1, b_1, a_2, b_2, a_3, b_3, a_4, b_4$ can be chosen out of

$$a_1, b_1, a_2, b_2, a_3, b_3, a_4, b_4 = \left.\begin{matrix} i \\ 0 \end{matrix}\right\},$$

which respectively make three, two, one, none out of $\sum_{r=1}^{4} a_r b_r$, $\sum_{r=1}^{4} a_r \bar{b}_r$, $\sum_{r=1}^{4} \bar{a}_r b_r$, $\sum_{r=1}^{4} \bar{a}_r \bar{b}_r$, vanish.

Hence, the resultant of equation (m) is

$$\begin{aligned}&(A+B+C+D)(\bar{A}+B+C+D)(A+\bar{B}+C+D)(A+B+\bar{C}+D)\\ &\times(A+B+C+\bar{D})(\bar{A}+\bar{B}+C+D)(\bar{A}+B+\bar{C}+D)(\bar{A}+B+C+\bar{D})\\ &\times(A+\bar{B}+\bar{C}+D)(A+\bar{B}+C+\bar{D})(A+B+\bar{C}+\bar{D})(\bar{A}+\bar{B}+\bar{C}+D)\\ &\times(\bar{A}+\bar{B}+C+\bar{D})(\bar{A}+B+\bar{C}+\bar{D})(A+\bar{B}+\bar{C}+\bar{D})=0. \qquad \text{(n)}\end{aligned}$$

This reduces to $$ABCD=0.$$

But this is merely the condition for evanescibility. Hence, equation (m) can always be satisfied when $\phi(x,y)$ is evanescible.

Secondly, to prove that the evanescible function $\phi(x,y)$ can only be expressed as the product of three secondary linear primes when its coefficients satisfy

$$\bar{S}_1 + S_4 = 0; \qquad (16)$$

that is, when the function, in addition to being evanescible, is of deficiency one at least.

For, assume

$$\phi(x,y) = \prod_{r=1}^{3} p(a_r, x;\ b_r, y).$$

|158| Then, by comparison of coefficients, we have just as in the previous proposition[1]

$$\begin{aligned}&\bar{A}\prod_{r=1}^{3}(\bar{a}_r+\bar{b}_r)+A\sum_{r=1}^{3}a_r b_r+\bar{B}\prod_{r=1}^{3}(\bar{a}_r+b_r)+B\sum_{r=1}^{3}a_r\bar{b}_r\\ &+\bar{C}\prod_{r=1}^{3}(a_r+\bar{b}_r)+C\sum_{r=1}^{3}\bar{a}_r b_r+\bar{D}\prod_{r=1}^{3}(a_r+b_r)+D\sum_{r=1}^{3}\bar{a}_r\bar{b}_r=0. \qquad \text{(p)}\end{aligned}$$

The discussion of the form of the resultant of this equation is in all respects similar to that of the form of the resultant of equation (m), except that now it is not possible to choose a set of values out of

$$a_1, b_1, a_2, b_2, a_3, b_3 = \left.\begin{matrix} i \\ 0 \end{matrix}\right\},$$

1. In the following equation, the final '*D*' erroneously had an overbar.

which makes none out of $\sum_{r=1}^{3} a_r b_r$, $\sum_{r=1}^{3} a_r \bar{b}_r$, $\sum_{r=1}^{3} \bar{a}_r b_r$, $\sum_{r=1}^{3} \bar{a}_r \bar{b}_r$ to vanish. Thus the resultant of equation (p) is equation (n) with the first factor, namely, $(A+B+C+D)$, omitted. This resultant reduces to equation (16) given above.

Thus we have found the condition that the general minimum number of four secondary linear prime factors may be reduced to three. This condition may be stated: a function of which the field is not restricted,[1†] necessarily has both deficiency and supplemental deficiency.

It is easily proved, by solving for A in equation (16), that a general expression for an evanescible deficient function of two variables is

$$\{\bar{B}\bar{C}\bar{D}+q(\bar{B}+\bar{C}+\bar{D})\}xy+Bx\bar{y}+C\bar{x}y+D\bar{x}\bar{y}.$$

The evanescible function $\phi(x,y)$ can only be expressed as a product of two secondary linear primes when its coefficients satisfy

$$\bar{S}_2+S_4=0; \tag{17}$$

that is, when, in addition to being evanescible, it is of deficiency two at least.

For, assume

$$\phi(x,y)=\prod_{r=1}^{2} p(a_r, x;\, b_r, y).$$

|159| Then, by comparison of coefficients, we have

$$\bar{A}\prod_{r=1}^{2}(\bar{a}_r+\bar{b}_r)+A\sum_{r=1}^{2}a_r b_r+\bar{B}\prod_{r=1}^{2}(\bar{a}_r+b_r)+B\sum_{r=1}^{2}a_r\bar{b}_r$$
$$+\bar{C}\prod_{r=1}^{2}(a_r+\bar{b}_r)+C\sum_{r=1}^{2}\bar{a}_r b_r+\bar{D}\prod_{r=1}^{2}(a_r+b_r)+D\sum_{r=1}^{2}\bar{a}_r\bar{b}_r=0. \tag{q}$$

The discussion of the form of the resultant of this equation is in all respects similar to that of equations (m) and (p), except that now it is not possible to choose a set of values out of

$$a_1, b_1, a_2, b_2 = \left.\begin{matrix} i \\ 0 \end{matrix}\right\},$$

which makes none, or only one, of $\sum_{r=1}^{2} a_r b_r$, $\sum_{r=1}^{2} a_r \bar{b}_r$, $\sum_{r=1}^{2} \bar{a}_r b_r$, $\sum_{r=1}^{2} \bar{a}_r \bar{b}_r$, to vanish.

Hence, the resultant of equation (q) is

$$(\bar{A}+\bar{B}+C+D)(\bar{A}+B+\bar{C}+\bar{D})(A+B+C+\bar{D})(A+\bar{B}+\bar{C}+D)$$
$$\times(A+\bar{B}+C+\bar{D})(A+B+\bar{C}+\bar{D})(A+\bar{B}+\bar{C}+\bar{D})(\bar{A}+B+\bar{C}+\bar{D})$$
$$\times(\bar{A}+\bar{B}+C+\bar{D})(\bar{A}+\bar{B}+\bar{C}+D)=0.$$

This reduces to equation (17).

It has already been proved (cf. equation (4)) that the condition that the function $\phi(x,y)$ may be a secondary linear prime is

$$\bar{S}_3+S_4=0.$$

1. † Cf. my "Universal Algebra," §33.

It is easily proved from equation (17) that a general expression for an evanescible function of deficiency, two at least is

$$\{\bar{p}q+(p+\bar{q})(\bar{C}+\bar{D})+\overline{CD}\}xy+(\bar{p}+\overline{CD})\bar{x}\bar{y}+C\bar{x}y+D\bar{x}\bar{y}.$$

It must be carefully noticed that, to take the general case, $\phi(x,y)$ can be expressed as the product of four secondary linear primes in an indefinite number of ways: also, that can be expressed as the product of more than four such primes.

§7.—*Expression of Functions as Sums of Separable Primes.*

This article is devoted to the enunciation of the theorems reciprocal to those of the previous paragraph.

|160| Any function of n variables, capable of the value i, can, in general, be expressed as a sum of a minimum number of 2^n n-ary separable primes; and the exceptional cases arise when $S_{2^n-r+1}=0$ and $S_{2^n-r}\neq 0$, and then the function can be expressed as a sum of a minimum number of (2^n-r) n-ary separable primes.

This proposition gives the reason for the term "of supplemental deficiency r" which has been applied to $\phi(x,y\ldots.t)$ when $S_{2^n-r+1}=0$. For if $\phi(x,y\ldots.t)$ be capable of the value i, the general minimum number of n-ary separable prime summands has been reduced by r from that of the general case. The term has also an analogous meaning (cf. §8) when $\phi(x,y\ldots.t)$ is not capable of the value i.

The special enunciations for functions of two variables are as follows:

Any function capable of the value i can be expressed as a sum of four secondary separable primes.

The conditions that a function can be expressed as a sum of three, or two, secondary separable primes are respectively

$$\bar{S}_1+S_4=0, \tag{18}$$

or

$$\bar{S}_1+S_3=0. \tag{19}$$

The condition that a function may be a secondary separable prime has already (cf. equation (5)) been found to be

$$\bar{S}_1+S_2=0.$$

By a comparison of equations (16) and (18), we deduce the proposition that any function of n variables which can be expressed as a product of (2^n-1) n-ary linear primes, can also be expressed as a sum of (2^n-1) n-ary separable primes, and conversely.

§8.—*Non-Evanescible and Non-Deficient Functions.*

A non-evanescible function can be expressed as the sum of a constant summand and of an evanescible summand. The constant summand is definitely determined, but the general form of the evanescible summand has an ambiguity (that is, an arbitrary element) in its expression.

For, put $$\phi(x,y)=H+\phi_1(x,y),$$

where H is constant, and the coefficients (A_1, B_1, C_1, D_1) of $\phi_1(x, y)$ satisfy

$$A_1B_1C_1D_1 = 0. \qquad \text{(a)}$$

|161| Then, by comparison of coefficients,

$$A = A_1 + H, \ \ B = B_1 + H, \ \ C = C_1 + H, \ \ D = D_1 + H. \qquad \text{(b)}$$

Hence, by multiplication,

$$H + A_1B_1C_1D_1 = ABCD,$$

and thence, from (a), $$H = ABCD. \qquad \text{(c)}$$

Also, from (b),

$$\left.\begin{aligned} A_1 &= (\bar{H} + u_1)A = (\bar{S}_4 + u_1)A, \\ B_1 &= \qquad\qquad\quad (\bar{S}_4 + u_2)B, \\ C_1 &= \qquad\qquad\quad (\bar{S}_4 + u_3)C, \\ D_1 &= \qquad\qquad\quad (\bar{S}_4 + u_4)D, \end{aligned}\right\} \qquad \text{(d)}$$

where we find, by substitution in (a),

$$u_1u_2u_3u_4S_4 = 0.$$

Thus, finally, the general expression in the required form is[1]

$$\left.\begin{aligned} \phi(x,y) = S_4 + \{(\bar{S}_4 + u_1)Axy + (\bar{S}_4 + u_2)Bx\bar{y} + (\bar{S}_4 + u_3)C\bar{x}y + (\bar{S}_4 + u_4)D\bar{x}\bar{y}\} = 0, \\ u_1u_2u_3u_4S_4 = 0. \end{aligned}\right\} \qquad (20)$$

Let these two parts of $\phi(x, y)$ be called its constant summand and the general form of its evanescible summand.

It has already been proved in §4[2] that the evanescible summand of $\phi(x, y)$ cannot have a deficiency of higher order than that of $\phi(x, y)$. It is easily seen that the conditions that the evanescible summand may have the same deficiency as $\phi(x, y)$ are respectively as follows, where U_1, U_2, U_3, U_4 are the symmetric functions (cf. §2) of the four terms u_1, u_2, u_3, u_4:

If $$\bar{S}_1 = 0, \text{ then } S_4(U_4 + \bar{U}_1) = 0.$$

If $$\bar{S}_2 = 0, \text{ then } S_4(U_4 + \bar{U}_2) = 0.$$

If $$\bar{S}_3 = 0, \text{ then } S_4(U_4 + \bar{U}_3) = 0.$$

These conditions can always be satisfied by u_1, u_2, u_3, u_4 in an indefinite number of ways. It may be noted that if $\phi(x, y)$ is of deficiency three, its evanescible summand of the same deficiency is a secondary linear prime.

The corresponding theorem for n variables is as follows: It is always pos-|162|sible to express a function $\phi(x, y \ldots\ldots t)$ of n variables in the form

$$\phi(x, y, \ldots\ldots t) = S_{2^n} + \prod_{r=1}^{\mu} p(a_r, x; b_r, y; \ldots\ldots; k_r, t), \qquad (21)$$

1. The below equations contained multiple errors in the original. The closing, right curly brace on line 1 was omitted, and line 2 had a '+' instead of an '='.
2. The original erroneously referred to §5.

where μ has the minimum value of 2^n if $\overline{S}_1 \neq 0$, and of $(2^n - r)$, if $\overline{S}_r = 0$ and $\overline{S}_{r+1} \neq 0$. It is from this theorem that the term "deficiency" arises.

Reciprocally, it is always possible to express a function $\phi(x, y \ldots . t)$ of n variables in the form

$$\phi(x, y \ldots . t) = S_1 \sum_{r=1}^{\mu} \overline{p}(a_r, x; b_r, y; \ldots . ; k_r, t), \tag{22}$$

where μ has the minimum value of 2^n if $S_{2^n} \neq 0$; and of $(2^n - r)$, if $S_{2^n - r+1} = 0$ and $S_{2^n - r} \neq 0$. This theorem exhibits the meaning of the "supplemental deficiency."

Also, in the case of two variables, let v_1, v_2, v_3, v_4 be any arbitrary terms, and V_1, V_2, V_3, V_4 their symmetric functions. Then the reciprocal theorem to equation (20) is that $\phi(x, y)$ can be expressed in the form[1]

$$\left.\begin{aligned} \phi(x,y) = S_1\{(\overline{S}_1 v_1 + A)xy + (\overline{S}_1 v_2 + B)x\overline{y} + (\overline{S}_1 v_3 + C)\overline{x}y + (\overline{S}_1 v_4 + B)\overline{x}\overline{y}\} = 0, \\ \overline{S}_1 \overline{V}_1 = 0; \end{aligned}\right\} \tag{23}$$

so that the first factor is constant and the second is capable of the value i. Also the second factor has the same supplemental deficiency as $\phi(x, y)$, if v_1, v_2, v_3, v_4 are chosen to satisfy the following conditions, which are always possible:

If $\qquad S_4 = 0$, then $\overline{S}_1(\overline{V}_1 + V_4) = 0.$

If $\qquad S_3 = 0$, then $\overline{S}_1(\overline{V}_1 + V_3) = 0.$

If $\qquad S_2 = 0$, then $\overline{S}_1(\overline{V}_1 + V_2) = 0.$

By comparison of the first equations of (20) and (23), the following general theorem is deduced: Let u_1, u_2, u_3, u_4 be any set of four arbitraries, and let U_1, U_2, U_3, U_4 be their symmetric functions; and let v_1, v_2, v_3, v_4 be another set of four arbitraries, and let V_1, V_2, V_3, V_4 be their symmetric functions, then the following equation is an identity:[2]

$$\phi(x, y) = S_4 + S_1 (\overline{S}_4 + U_1)\phi'(x, y); \tag{24}$$

|163| where A', B', C', D' are the coefficients of $\phi'(x, y)$, S_1', S_2', S_3', S_4' its invariants, and

$$\left.\begin{aligned} A' &= (\overline{S}_1 + S_4\overline{U}_1)v_1 + (\overline{S}_4 + u_1)A, \\ B' &= (\overline{S}_1 + S_4\overline{U}_1)v_2 + (\overline{S}_4 + u_2)B, \\ C' &= (\overline{S}_1 + S_4\overline{U}_1)v_3 + (\overline{S}_4 + u_3)C, \\ D' &= (\overline{S}_1 + S_4\overline{U}_1)v_4 + (\overline{S}_4 + u_4)D, \end{aligned}\right\} \tag{25}$$

and, after some algebraic reduction,

$$\left.\begin{aligned} S_1' &= S_1\overline{S}_4 + S_1 U_1 + V_1, \\ S_2' &= S_2\overline{S}_4 + S_2 U_2 + \overline{S}_1 V_2 + S_4\overline{U}_1 V_2, \\ S_3' &= S_3\overline{S}_4 + S_3 U_3 + \overline{S}_1 V_3 + S_4\overline{U}_1 V_3, \\ S_4' &= S_4 U_4 + \overline{S}_4 V_4 + S_4\overline{U}_1 V_4. \end{aligned}\right\} \tag{26}$$

1. In the below equation, the original erroneously omitted the closing, right curly brace at the end of line 1.
2. In the following equation, the original erroneously had an overbar on the preceding term, S_1, and not on S_4.

Thus it is always possible to choose the arbitraries so that $S_1'=i$ and $S_4'=0$ simultaneously. But it is not possible to make S_2' have the value *i* unless $\overline{S}_1+S_2=i$; that is (cf. §2, equation (1)), unless $S_1=S_2$; and it is not possible to make S_3' vanish unless $S_3\overline{S}_4=0$; that is (cf. loc. cit.), unless $S_3=S_4$.

For instance, by choosing the arbitraries so that

$$U_1=U_2=i, \quad U_3=U_4=0, \quad V_1=V_2=i, \quad V_3=V_4=0,$$

we find

$$S_1'=i, \quad S_2'=S_2+\overline{S}_1, \quad S_3'=S_3\overline{S}_4, \quad S_4'=0. \tag{27}$$

§9.—Linear and Separable Functions.

A function of *n* variables *x*, *y*, *z*, ... , which can be expressed in the form

$$ax+b\overline{x}+cy+d\overline{y}+ez+f\overline{z}+\ldots\ldots, \tag{28}$$

will be called "linear," and this form will be called its "linear expression."

A function of *n* variables *x*, *y*, *z*, ... , which can be expressed in the form

$$(ax+b\overline{x})(cy+d\overline{y})(ez+f\overline{z})\ldots\ldots, \tag{29}$$

will be called "separable," and this form will be called its "separable expression."

The condition that $\phi(x,y)$ may be linear is found by comparing it with the form (28). Hence,

$$A=a+c, \quad B=a+d, \quad C=b+c, \quad D=b+d.$$

|164| Then, by eliminating *a*, *b*, *c*, *d*, we find

$$S_1(\overline{A}\overline{D}+\overline{B}\overline{C})=0. \tag{30}$$

It will be observed from the form of this condition that linearity is not an invariant property.

Reciprocally, the condition which the coefficients of $\phi(x,y)$ must satisfy in order that the function may be separable is

$$\overline{S}_4(AD+BC)=0. \tag{31}$$

Thus separability is not an invariant property.

By solving equation (30) for *A*, it can be proved that a general expression for a linear function is

$$(p+\overline{q})(B+C)xy+Bx\overline{y}+C\overline{x}y+D\overline{x}\,\overline{y}. \tag{32}$$

Also, from equation (31), a general expression for a separable function is

$$(p\overline{q}+BC)xy+Bx\overline{y}+C\overline{x}y+D\overline{x}\,\overline{y}. \tag{33}$$

It is easy to verify that a factor of a linear function is not necessarily itself linear, and that a factor of a separable function is not necessarily itself separable. Reciprocally, a summand of a linear or separable function is not necessarily itself linear or separable.

It can be proved that it is always possible to factorize any function $\phi(x,y)$ into a pair of linear factors; and, reciprocally, that it is always possible to express any function $\phi(x,y)$ as the sum of a pair of separable summands.

Also, the condition that every possible factor of $\phi(x,y)$ may be linear, is

$$\bar{B}\bar{C}+\bar{A}\bar{D}=0. \tag{34}$$

Reciprocally, the condition that every possible factor of $\phi(x,y)$ may be separable, is

$$BC+AD=0. \tag{35}$$

If a linear function is deficient, every possible factor is linear. For the condition are

$$\bar{S}_1=0, \quad S_1(\bar{A}\bar{D}+\bar{B}\bar{C})=0.$$

Hence, $$\bar{A}\bar{D}+\bar{B}\bar{C}=0.$$

Reciprocally, if a separable function has supplemental deficiency, every possible summand is separable.

|165| If the function $\phi(x, y)$ is linear, then $S_1=S_2$. For, by equation (30), we have

$$S_1=u\overline{}(\bar{A}\bar{D}+\bar{B}\bar{C})=u(A+D)(B+C).$$

But $$(A+D)(B+C) \mathrel{=\!\!\!\!\in} S_2.$$

Hence, $$S_1 \mathrel{=\!\!\!\!\in} S_2.$$

But by §2, $$S_2 \mathrel{=\!\!\!\!\in} S_1.$$

Hence, $$S_1=S_2.$$

It follows as a corollary that, if a linear function is deficient, it is of deficiency two at least.

Reciprocally, if the function $\phi(x,y)$ is separable, then $S_3=S_4$. Also, if a separable function has supplemental deficiency, it has supplemental deficiency two at least.

In general, there are an indefinite number of linear expressions of a linear function. For, let $\phi(x,y)$ be a linear function, and let $ax+b\bar{x}+cy+d\bar{y}$ be one linear expression of $\phi(x,y)$. Then it can be proved that the general form of linear expression of $\phi(x,y)$ is

$$\begin{aligned}&\{(\bar{a}+\bar{b}+u_1)a+p(ab+cd)\}x+\{(\bar{a}+\bar{b}+u_2)b+p(ab+cd)\}\bar{x}\\ &\quad+\{(\bar{c}+\bar{d}+v_1)c+(\bar{p}+q)(ab+cd)\}y+\{(\bar{c}+\bar{d}+v_2)d+(\bar{p}+q)(ab+cd)\}\bar{y}.\end{aligned} \tag{36}$$

But if $\phi(x,y)$ is evanescible, then evidently

$$ab=0, \quad cd=0.$$

Hence, each of the coefficients in (36) reduces to the corresponding coefficient of the given linear expression. Thus, when a linear function is evanescible, it has only one linear expression.

[To be continued.]

Memoir on the Algebra of Symbolic Logic (Part II)

1901 *American Journal of Mathematics*, vol. 23, no. 4 (October), pp. 297–316

These prefatory remarks inherit all of what we asserted in the first paragraph of the introduction to Part I as well as the first half of the second paragraph. As noted previously, Part II addresses the same set-theoretic functions, but its emphasis is entirely different. Here Whitehead introduces the idea of substitutions (pp. 139–140), which we might today describe as a coordinate transformation – an idea fundamental to both mathematics and physics. The key restriction is that such transformations should be invertible, or what Whitehead refers to as reversible. *The point of the discussion is to develop general conditions for an arbitrary transformation of variables to satisfy this condition.*

In comparison with Part I of the 'Memoir' and other elements of Whitehead's mathematical writing (in particular, UA), what is forward-looking in this work is the degree to which the transformations he discusses anticipate the related notions of covariance *in physics and* functoriality *in mathematics. Moreover, he explicitly makes use of what we now recognise as the general quotient construction that is inherent in any equivalence relation and, more importantly, the mathematical notion of an abstract* group, *a huge advance in informal axiomatic systems and relational mathematics. (Cayley introduced the idea of a group in the 1850s, but it caught fire only early in the following century.)*

In light of these modern attributes, one may wonder why at least the second part of the 'Memoir' in itself did not become more memorable. As with the first part, any remarks here must be sheerly speculative, but we might still make three plausible claims. First, the domain of this work was again set-theoretic, and sets, unlike groups and rings, know only one structural attribute: identity of elements. Hence the atmosphere here is perhaps too thin to support interesting life forms. Second, Whitehead was, with Russell, about to leave mathematics for the noble but ultimately insufficient exploration of its foundations, and so further development of relational mathematics[1] *was subordinated to the programme of logicism. Had PM succeeded – and Whitehead could not have known that this effort would inevitably fail until Gödel took centre stage in 1931 – it would, of course, have been more than worth the detour. Finally, the priority of relations over objects did subsequently burst forth from Whitehead, as if a conservation law were in place, but in the domain of process metaphysics, not mathematics.*

---

1. This is not to speak only of relationality in its purely set-theoretic sense, but beyond that, of relationality in the sense of morphisms (structure preserving maps within a class of objects) and, later, of functors across classes.

PART II.

The Theory of Substitutions.

§1.—Types of Transformation.

Any transformation of x and y into u and v can be represented by

$$x=f_1(u,v),\quad y=f_2(u,v). \tag{1}$$

Here $f_1(u,\,v)$ and $f_2(u,\,v)$ will be called the director functions of the transformation. They will always be represented by the following notation:

$$\left.\begin{aligned} f_1(u,v)&=a_1uv+a_2u\bar{v}+a_3\bar{u}v+a_4\bar{u}\,\bar{v},\\ f_2(u,v)&=b_1uv+b_2u\bar{v}+b_3\bar{u}v+b_4\bar{u}\,\bar{v}.\end{aligned}\right\} \tag{2}$$

This transformation will also be called the transformation $\left.\begin{matrix} a_1, a_2, a_3, a_4 \\ b_1, b_2, b_3, b_4 \end{matrix}\right\}$, and $a_1, \ldots . a_4;\ b_1, \ldots . b_4$ will be called the coefficients of the transformation.

In general, x and y cannot be considered as independent variables when they are submitted to this transformation. For, by eliminating u and v from (1) and (2), we deduce

$$p(a_1,x;\,b_1,y)\,p(a_2,x;\,b_2,y)\,p(a_3,x;\,b_3,y)\,p(a_4,x;\,b_4,y)=0. \tag{3}$$

Thus x and y are limited to be pairs of roots of equation (3). Conversely, equation (3) is the condition of the possibility of equations (1) and (2). Thus, if (3) is satisfied, (1) and (2) are satisfied. Hence, (1) and (2) form the general solution of (3).

Thus the theory of this type of transformation is simply the theory of equations. A few theorems connected with it will be given incidentally in this part.

|298| A distinct type of transformation exists when equation (3) is satisfied identically for all values of x and y. The condition for this is that each of the coefficients of xy, $x\bar{y}$, $\bar{x}y$, $\bar{x}\,\bar{y}$ in the development of the left-hand side of (3) should vanish. This condition can be exhibited in the single fundamental equation

$$\Pi(\bar{a}_r+\bar{b}_r)+\Pi(\bar{a}_r+b_r)+\Pi(a_r+\bar{b}_r)+\Pi(a_r+b_r)=0,\quad (r=1,2,3,4). \tag{4}$$

When this condition is satisfied, the transformation transforms the independent variables x and y into the independent variables u and v. Thus equations (1) and (2) can be solved for u and v in terms of x and y, and a reverse transformation is thus obtained of the form

$$u=F_1(x,y),\quad v=F_2(x,y), \tag{5}$$

where[1]

$$\left.\begin{aligned} F_1(x,y)&=\alpha_1xy+\alpha_2x\bar{y}+\alpha_3\bar{x}y+\alpha_4\bar{x}\,\bar{y},\\ F_2(x,y)&=\beta_1xy+\beta_2x\bar{y}+\beta_3\bar{x}y+\beta_4\bar{x}\,\bar{y}.\end{aligned}\right\} \tag{6}$$

Thus, this type of transformation is reversible. Let the term "substitution" be used exclusively for it.

1. In the following equation, the original erroneously had '$\beta_2\bar{x}y$' instead of '$\beta_2x\bar{y}$' on line 2.

In the case of substitutions, there is no advantage in changing the notations for the independent variables from x and y to u and v. Let a substitution T be defined to consist in the substitution of Tx for x and Ty for y, where

$$\left.\begin{aligned} Tx &= a_1xy + a_2x\bar{y} + a_3\bar{x}y + a_4\bar{x}\bar{y}, \\ Ty &= b_1xy + b_2x\bar{y} + b_3\bar{x}y + b_4\bar{x}\bar{y}; \end{aligned}\right\} \qquad (7)$$

and the coefficients of the substitution T satisfy equation (4). Also $T\phi(x,y)$ is defined to mean $\phi(Tx, Ty)$: for instance $T\bar{x}$ means $\overline{(Tx)}$.

§2.—*Relations between the Coefficients of a Substitution.*

We have now to consider more particularly the implications of equation (4). Let us first eliminate b_1, b_2, b_3, b_4 from the equation; put $\lambda(b_1, b_2, b_3, b_4)$ for the left-hand side. We find that the only factors composing $\Pi\lambda\begin{pmatrix} i, & i, & i, & i \\ 0, & 0, & 0, & 0 \end{pmatrix}$ which do not equal i are the six of the type $\lambda(0, 0, i, i)$, $\lambda(i, 0, 0, i)$, etc. Also,

$$\lambda(0, 0, i, i) = a_1a_2 + \bar{a}_1\bar{a}_2 + a_3a_4 + \bar{a}_3\bar{a}_4,$$

|299| with analogous values for the other factors of the same type. Hence,

$$\Pi\lambda\begin{pmatrix} i, & i, & i, & i \\ 0, & 0, & 0, & 0 \end{pmatrix} = \Sigma a_pa_qa_r + \Sigma\bar{a}_p\bar{a}_q\bar{a}_r, \quad (p, q, r = 1, 2, 3, 4),$$

where, as usual, p, q, r are unequal in the same product.

Hence, the resultant of (4), after eliminating b_1, b_2, b_3, b_4, is[1]

$$\Sigma a_pa_qa_r + \Sigma\bar{a}_p\bar{a}_q\bar{a}_r = 0, \quad (p, q, r = 1, 2, 3, 4). \qquad (8)$$

A similar equation holds of b_1, b_2, b_3, b_4. Hence, the director functions, Tx and Ty, of any substitution are each of deficiency two and of supplemental deficiency two. Thus, functions of this type play a fundamental rôle in the theory of substitutions.

The conditions (8), which hold between the coefficients of a function of deficiency two and supplemental deficiency two, can be put otherwise thus: We have from (8),

$$a_1a_2(a_3 + a_4) = 0,$$

hence,
$$a_1a_2 \Subset \bar{a}_3\bar{a}_4.$$

Also
$$\bar{a}_3\bar{a}_4(\bar{a}_1 + \bar{a}_2) = 0,$$

hence,
$$\bar{a}_3\bar{a}_4 \Subset a_1a_2.$$

Thus
$$a_1a_2 = \bar{a}_3\bar{a}_4.$$

Hence, equation (8) can be replaced by the set of equations

$$a_pa_q = \bar{a}_r\bar{a}_s, \quad (p, q, r, s = 1, 2, 3, 4). \qquad (9)$$

A similar set of equations hold for b_1, b_2, b_3, b_4.

---

1. In the following equation, the original erroneously had a closing parenthesis after 'p, q, r': '(p, q, r)'.

Secondly, these relations (8) or (9), between the coefficients of each of the director functions separately, do not exhaust the implications of equation (4). For, from equations (9), we deduce the two sets,

$$\left.\begin{aligned}\Sigma a_p a_q \bar{b}_r \bar{b}_s = \Sigma \bar{a}_p \bar{a}_q b_r b_s = \Sigma a_p a_q b_p b_q = \Sigma \bar{a}_p \bar{a}_q \bar{b}_p \bar{b}_q,\\ \Sigma a_p a_q b_r b_s = \Sigma a_p a_q \bar{b}_p \bar{b}_q = \Sigma \bar{a}_p \bar{a}_q \bar{b}_r \bar{b}_s = \Sigma \bar{a}_p \bar{a}_q b_p b_q;\end{aligned}\right\} \qquad (10)$$

where $(p, q, r, s = 1, 2, 3, 4)$.

Then, by the application to equation (4) of (8) and (10), we deduce

$$\Sigma a_p a_q b_p b_q + \Sigma \bar{a}_p \bar{a}_q b_p b_q = 0, \qquad (11)$$

|300| or equivalent forms deduced by the use of (10). A complete equivalent to equation (4) is given by

$$\Sigma\left(a_p a_q + \bar{a}_p \bar{a}_q\right)\left(b_p b_q + \bar{b}_p \bar{b}_q\right) = 0, \quad (p, q = 1, 2, 3, 4); \qquad (12)$$

that is,
$$\Sigma \bar{p}\left(a_p, a_q; b_p, b_q\right) = 0.$$

It is easy to see that this equation includes (11). It also includes (8), for we can deduce (8) from it by eliminating b_1, b_2, b_3, b_4.

Equation (8) may be solved for a_3 and a_4. The general solution is

$$\left.\begin{aligned} a_3 &= \bar{a}_1 \bar{a}_2 + p\left(\bar{a}_1 + \bar{a}_2\right),\\ a_4 &= \bar{a}_1 \bar{a}_2 + \bar{p}\left(\bar{a}_1 + \bar{a}_2\right).\end{aligned}\right.$$

Similarly,

$$\left.\begin{aligned} b_3 &= \bar{b}_1 \bar{b}_2 + q\left(\bar{b}_1 + \bar{b}_2\right),\\ b_4 &= \bar{b}_1 \bar{b}_2 + \bar{q}\left(\bar{b}_1 + \bar{b}_2\right).\end{aligned}\right\} \qquad (13)$$

But a_1, a_2, b_1, b_2, and p and q, cannot be assumed arbitrarily consistently with the equation (4). For from (12) we find

$$\left(a_1 a_2 + \bar{a}_1 \bar{a}_2\right)\left(b_1 b_2 + \bar{b}_1 \bar{b}_2\right) = 0 \qquad (14)$$

and

$$\left(a_1 \bar{a}_2 b_1 \bar{b}_2 + \bar{a}_1 a_2 \bar{b}_1 b_2\right)\left(pq + \bar{p}\bar{q}\right) + \left(a_1 \bar{a}_2 \bar{b}_1 b_2 + \bar{a}_1 a_2 b_1 \bar{b}_2\right)\left(p\bar{q} + \bar{p}q\right) = 0. \qquad (15)$$

Equation (15) does not require any relation to be fulfilled between a_1, a_2, b_1, b_2. Thus when[1] a_1, a_2, b_1, b_2 have been chosen to satisfy (14), p and q can be chosen to satisfy (15), and then a_3, a_4, b_3, b_4 are given by equation (13). By this process the coefficients of any substitution can be found.

It is to be noticed that any three of a_1, a_2, b_1, b_2 can be chosen arbitrarily, and equation (14) can then be satisfied by a proper choice of the fourth. Also, either p or q can be assumed arbitrarily and equation (15) satisfied by a proper choice of the other.

Again, if a_1, a_2, a_3 have been chosen so as to satisfy

$$a_1 a_2 a_3 + \bar{a}_1 \bar{a}_2 \bar{a}_3 = 0, \qquad \text{(m)}$$

then a_4 is definitely determined by the equation

1. The original erroneously had the following list as 'a_1, a_2, a_3, a_4'.

$$a_4 = \bar{a}_2\bar{a}_3 + \bar{a}_3\bar{a}_1 + \bar{a}_1\bar{a}_2. \quad (16)$$

For, solve the first of equations (13) to find the general value of p which is con-|301|sistent with the given values of a_1, a_2, a_3 and simplify by the use of (m). We find

$$p = a_1a_3 + a_2a_3 + u(a_3 + a_1a_2).$$

Thence, from the second of equations (13), we find

$$a_4 = \bar{a}_1\bar{a}_2 + \bar{a}_1\bar{a}_3 + \bar{a}_2\bar{a}_3 + u(\bar{a}_1\bar{a}_3 + \bar{a}_2\bar{a}_3 + \bar{a}_1\bar{a}_2),$$

which reduces to equation (16). Hence, if three coefficients of a function $\phi(x,y)$ of deficiency two and of supplemental deficiency two are given, the function is completely determined. Thus, if the six coefficients, $a_1, a_2, a_3, b_1, b_2, b_3$ of a substitution are given, the substitution is completely determined.

§3.—The Reverse Substitution.

There is only one reverse substitution corresponding to a given substitution T; that is, there is only one substitution T' such that[1]

$$T'Tx = x, \quad T'Ty = y.$$

This proposition requires proof, for if we solve the equations

$$x = f_1(u,v), \quad y = f_2(u,v),$$

for u and v we find

$$u = F_1(x, y, q_1, q_2), \quad v = F_2(x, y, q_1, q_2),$$

where q_1 and q_2 are arbitraries which have been introduced in the solution, and each particular choice of q_1 and q_2 would seem to determine a different substitution which is reverse to T. We have to show that there is only one set of roots for u and v in terms of x and y.

The equations for u and v can be written

$$p(a_1, x)uv + p(a_2, x)u\bar{v} + p(a_3, x)\bar{u}v + p(a_4, x)\bar{u}\bar{v} = 0,$$
$$p(b_1, y)uv + p(b_2, y)u\bar{v} + p(b_3, y)\bar{u}v + p(b_4, y)\bar{u}\bar{v} = 0.$$

These two equations can be combined into

$$p(a_1, x; b_1, y)uv + p(a_2, x; b_2, y)u\bar{v} + p(a_3, x; b_3, y)\bar{u}v + p(a_4, x; b_4, y)\bar{u}\bar{v} = 0.$$

The necessary and sufficient condition (cf. Part I, §3) that this equation for u and v should only have one set of roots is that the left-hand side should be a secondary linear prime. The condition for this is

$$\Sigma\bar{p}(a_q, x; b_q, y)\bar{p}(a_r, x; b_r, y) = 0, \quad (q, r = 1, 2, 3, 4).$$

|302| This equation can be written

1. In the following equation, the original erroneously had '$T'Ty = 0$.'

$$\Sigma\left(a_q a_r x + \bar{a}_q \bar{a}_r \bar{x}\right)\left(b_q b_r y + \bar{b}_q \bar{b}_r \bar{y}\right) = 0, \quad (q, r = 1, 2, 3, 4).$$

But from equation (12), this equation is satisfied identically for all values of x and y. Thus there is only one substitution reverse to T. Let it be written T^{-1}. It is evident that T is the reverse substitution to T^{-1}. Also, let the identical substitution be written T^0, so that

$$TT^{-1} = T^{-1}T = T^0.$$

Assume that[1]

$$T^{-1}x = \alpha_1 xy + \alpha_2 x\bar{y} + \alpha_3 \bar{x}y + \alpha_4 \bar{x}\bar{y},$$

$$T^{-1}y = \beta_1 xy + \beta_2 x\bar{y} + \beta_3 \bar{x}y + \beta_4 \bar{x}\bar{y}.$$

Then, by hypothesis,

$$T^{-1}f_1(x, y) = x, \quad T^{-1}f_2(x, y) = y. \; {}_0$$

Hence, for all values of x and y,

$$x = f_1(\alpha_1, \beta_1)xy + f_1(\alpha_2, \beta_2)x\bar{y} + f_1(\alpha_3, \beta_3)\bar{x}y + f_1(\alpha_4, \beta_4)\bar{x}\bar{y},$$
$$y = f_2(\alpha_1, \beta_1)xy + f_2(\alpha_2, \beta_2)x\bar{y} + f_2(\alpha_3, \beta_3)\bar{x}y + f_2(\alpha_4, \beta_4)\bar{x}\bar{y}.$$

Hence,

$$\bar{f}_1(\alpha_1, \beta_1) = 0, \quad \bar{f}_1(\alpha_2, \beta_2) = 0, \quad f_1(\alpha_3, \beta_3) = 0, \quad f_1(\alpha_4, \beta_4) = 0,$$
$$\bar{f}_2(\alpha_1, \beta_1) = 0, \quad f_2(\alpha_2, \beta_2) = 0, \quad \bar{f}_2(\alpha_3, \beta_3) = 0, \quad f_2(\alpha_4, \beta_4) = 0.$$

Hence, α_1 and β_1 satisfy the equation

$$\bar{f}_1(u, v) + \bar{f}_2(u, v) = 0;$$

and α_2 and β_2 satisfy the equation

$$\bar{f}_1(u, v) + f_2(u, v) = 0;$$

and α_3 and[2] β_3 satisfy the equation

$$f_1(u, v) + \bar{f}_2(u, v) = 0;$$

and α_4 and β_4 satisfy the equation

$$f_1(u, v) + f_2(u, v) = 0.$$

By the use of equations (8) and (10) and (11), it is easy to verify that the left-hand side of each of these equations is a secondary linear prime (cf. Part I, equation (4)) and, therefore, each equation has only one set of roots. Thus we obtain another proof of the previous proposition. Again, if $\phi(x, y)$ is a secondary linear prime, the solution of $\phi(x, y) = 0$ is

$$x = CD, \quad y = BD.$$

|303| Hence, the solution for the coefficients of the reverse substitution are

---

1. In the following equations, multiple overbars were missing. The original read as follows:

$$T^{-1}x = \alpha_1 xy + \alpha_2 xy + \alpha_3 xy + \alpha_4 \bar{x}\bar{y},$$

$$T^{-1}y = \beta_1 xy + \beta_2 x\bar{y} + \beta_3 xy + \beta_4 \bar{x}\bar{y}.$$

2. The original erroneously had 'β_2'.

$$\left.\begin{aligned}\alpha_1&=(\bar{a}_3+\bar{b}_3)(\bar{a}_4+\bar{b}_4), & \beta_1&=(\bar{a}_2+\bar{b}_2)(\bar{a}_4+\bar{b}_4),\\ \alpha_2&=(\bar{a}_3+b_3)(\bar{a}_4+b_4), & \beta_2&=(\bar{a}_2+b_2)(\bar{a}_4+b_4),\\ \alpha_3&=(a_3+\bar{b}_3)(a_4+\bar{b}_4), & \beta_3&=(a_2+\bar{b}_2)(a_4+\bar{b}_4),\\ \alpha_4&=(a_3+b_3)(a_4+b_4), & \beta_4&=(a_2+b_2)(a_4+b_4)\end{aligned}\right\} \tag{17}$$

An interesting example of a substitution is

$$\left.\begin{aligned}Tx&=p(a,x),\\ Ty&=p(b,y).\end{aligned}\right\} \tag{18}$$

It is easy to verify that in this case

$$T=T^{-1},\ \text{that is } T^2=T^0.$$

§4.—The Group of Substitutions.

If T_1 and T_2 are any two substitutions, then T_1T_2 and T_2T_1 are substitutions (as distinct from non-reversible transformations), also we have proved that one, and only one, reverse substitution corresponds to any given substitution. Hence, substitutions form a group.

The group of substitutions is not continuous since the concepts of quantity and of infinitesimal variations of quantities have no place among the concepts of this algebra. It is of finite order, if the number of distinct terms in the algebra representing given fundamental constants from which all reasoning starts, is conceived as finite. It is of indefinite order in so far as we may always suppose new constants to be produced without violating any of the laws of the algebra.

If T be any substitution the subgroup $T, T^2, T^3, \ldots.$ is necessarily of finite order, and is therefore, cyclical. For only a finite number of terms can be generated by algebraic combinations of the coefficients of T, and the coefficients of[1] $T^2, T^3, \ldots.$ must be selections from these coefficients. The order can be reduced by the assumption of additional relations among the coefficients of T. Thus, in the substitution of equation (18), the order is two.

§5.—Substitutions Satisfying Special Conditions

To find the condition which the functions $\psi_1(x,y)$, $\psi_2(x,y)$, $\psi_3(x,y)$, $\psi_4(x,y)$, must satisfy, if the coefficients of some substitution T are such that, with the |304| usual notation, a_1 and b_1 are a pair of roots of $\psi_1(x,y)$, a_2 and b_2 of $\psi_2(x,y)$, a_3 and b_3 of $\psi_3(x,y)$, a_4 and b_4 of $\psi_4(x,y)$. These equations can be expressed in the typical form

$$\psi_r(a_r,b_r)=0,\quad (r=1,2,3,4)\ldots. \tag{19}$$

Let $$\psi_r(x,y)=F_rxy+G_rx\bar{y}+H_r\bar{x}y+K_r\bar{x}\bar{y}.$$

1. The original erroneously had no comma separating 'T^2' and 'T^3'.

The complete condition satisfied by $a_r, b_r, (r=1, 2, 3, 4)$ is found by combining equations (4) and (19). It becomes.

$$\Pi(\bar{a}_r+\bar{b}_r)+\Pi(\bar{a}_r+b_r)+\Pi(a_r+\bar{b}_r)+\Pi(a_r+b_r)+\Sigma\psi_r(a_r, b_r)=0, \quad (r=1, 2, 3, 4). \tag{20}$$

We have to eliminate $a_r, b_r (r=1, 2, 3, 4)$ from this equation. Put $\lambda(a_1, a_2, a_3, a_4, b_1, b_2, b_3, b_4)$ for its left-hand side. Now,

$\Pi(\bar{a}_r+\bar{b}_r)=i$, unless at least one pair, $a_r=i,\ b_r=i$

$\Pi(\bar{a}_r+b_r)=i$, " " " " $a_r=i,\ b_r=0$,

$\Pi(a_r+\bar{b}_r)=i$, " " " " $a_r=0,\ b_r=i$,

$\Pi(a_r+b_r)=i$, " " " " $a_r=0,\ b_r=0$,

Hence, each of the factors composing $\Pi\lambda\begin{pmatrix} i, & i, & i, & i, & i, & i, & i, & i \\ 0, & 0, & 0, & 0, & 0, & 0, & 0, & 0 \end{pmatrix}$ is i, except those in which simultaneously $a_p=i,\ b_p=i;\ a_q=i,\ b_q=0;\ a_r=0,\ b_r=i,\ a_s=0,\ b_s=0$; where $(p, q, r, s=1, 2, 3, 4)$.

Hence the resultant of equation (20) is

$$\Pi(F_p+G_q+H_r+K_s)=0, \quad (p, q, r, s=1, 2, 3, 4). \tag{21}$$

This is the required condition which the coefficients of equations (19) must satisfy.

We can deduce as special cases of the above, the following theorems:

The condition which the coefficients of the first three of equations (19) must satisfy in order that (a_1, b_1) may be chosen to satisfy the first, (a_2, b_2) the second, and (a_3, b_3) the third, is deduced from equation (21) by putting zero for F_4, G_4, H_4, K_4.

The condition which the coefficients of the first two of equation (19) must satisfy in order that (a_1, b_1) may be chosen to satisfy the first and (a_2, b_2) the second, is found from equation (21) by putting zero for F_3, G_3, H_3, K_3 and also |305| for F_4, G_4, H_4, K_4. It can be written out in full in the form[1]

$$F_1G_1H_1K_1+F_2G_2H_2K_2+F_1G_1H_1\,.\,F_2G_2H_2+F_1G_1K_1\,.\,F_2G_2K_2 +F_1H_1K_1\,.\,F_2H_2K_2+G_1H_1K_1\,.\,G_2H_2K_2=0. \tag{22}$$

This condition in addition to securing that the two equations[2]

$$\psi_1(x, y)=0,\ \psi_2(x, y)=0,$$

are possible, also secures that the function $\psi_1(x, y)\psi_2(x, y)$ is of supplemental deficiency two. This latter part of the condition is satisfied if either of the functions is of supplemental deficiency two. For instance both (a_1, b_1) and (a_2, b_2) can be chosen to be pairs of roots of the same equation

$$\psi_1(x, y)=0,$$

---

1. The dots that appear in the following equation would now be written more naturally as the raised dot that frequently represents any formal multiplication operator.
2. In the following equation, the original erroneously had no comma between the second 'x' and 'y'.

if $\psi_1(x, y)$ is of supplemental deficiency two. Similarly from the previous theorem it follows that (a_1, b_1), (a_2, b_2), (a_3, b_3) can be chosen to be pairs of roots of the same equation, $\psi_1(x, y)=0$, if $\psi_1(x, y)$ is of supplemental deficiency three. But it follows from (21) that it is impossible that (a_1, b_1), (a_2, b_2), (a_3, b_3), (a_4, b_4) should be pairs of roots of the same equation.

This fact, that there are sets of four pairs of terms which cannot be severally pairs of roots of the same equation of two variables, is one of the fundamental facts of the algebra. The analogue for equations of one variable is that the two terms a and $\bar{a}$ cannot be both roots of the same equation of one variable. This fundamental fact is the correlative of the fact that there are functions which are not evanescible.

Again, if we assume that a_2 and b_2 are to satisfy the second of equations (19) and a_3 and b_3 the third, and a_4 and b_4 the fourth of equations (19), let us investigate the conditions which are thereby imposed on a_1 and b_1. We have to eliminate $a_2, b_2, a_3, b_3, a_4, b_4$, from these three equations and from equation (4); and the resulting equation for a_1, b_1 will be the required condition. It will be more convenient to substitute the equivalent equation (12) for equation (4). Thus the complete equation from which the elimination is to be performed can be written

$$\Sigma(a_1a_p+\bar{a}_1\bar{a}_p)(b_1b_p+\bar{b}_1\bar{b}_p)+\Sigma(a_pa_q+\bar{a}_p\bar{a}_q)(b_pb_q+\bar{b}_p\bar{b}_q)+\Sigma\psi_p(a_p, b_p)=0,$$
$$(p, q=2, 3, 4).$$

Now put $\lambda(a_2, a_3, a_4, b_2, b_3, b_4)$ for the left-hand side of this equation. Then any of the factors composing $\Pi\lambda\begin{pmatrix} i, & i, & i, & i, & i, & i \\ 0, & 0, & 0, & 0, & 0, & 0 \end{pmatrix}$, for which either $a_2=a_3$, and $b_2=b_3$ |306| simultaneously, or $a_2=a_4$ and $b_2=b_4$ simultaneously, or $a_3=a_4$ and $b_3=b_4$ simultaneously, it has the value i. Accordingly considering the factors which involve none of these three possibilities, we find that the required condition is

$$\Pi(G_p+H_q+K_r)a_1b_1+\Pi(F_p+H_q+K_r)a_1\bar{b}_1+\Pi(F_p+G_q+K_r)\bar{a}_1b_1$$
$$+\Pi(F_p+G_q+H_r)\bar{a}_1\bar{b}_1=0,\ (p, q, r=2, 3, 4). \qquad (23)$$

Thus if a_1b_1 is also to satisfy the first of equations (19), the complete condition which it satisfies is

$$\{F_1+\Pi(G_p+H_q+K_r)\}a_1b_1+\{G_1+\Pi(F_p+H_q+K_r)\}a_1\bar{b}_1$$
$$+\{H_1+\Pi(F_p+G_q+K_r)\}\bar{a}_1b_1+\{K_1+\Pi(F_p+G_q+H_r)\}\bar{a}_1\bar{b}_1=0. \qquad (24)$$

The condition for the possibility of this equation is simply equation (21), as should evidently be the case.

§6.—Congruence of Functions.

Two functions $\phi(x, y)$ and $\Phi(x, y)$ are said to be congruent, if one or more substitutions T exist, such that

$$T\phi(x, y)=\Phi(x, y).$$

The relation of congruence will be expressed by the symbolism

$$\phi(x, y) \leftrightarrow \Phi(x, y). \tag{25}$$

It is evident that if

$$\phi(x, y) \leftrightarrow \Phi(x, y),$$

and

$$\Phi(x, y) \leftrightarrow \psi(x, y),$$

then

$$\phi(x, y) \leftrightarrow \psi(x, y).$$

Thus the relation is transitive.

A set of functions such that any two are congruent, and which comprises all the functions congruent to members of the set, is call a congruent family.

We will now prove the fundamental theorem that a congruent family is composed of all the functions with a given set of invariants.

For let

$$\phi(x, y) = Axy + Bx\bar{y} + C\bar{x}y + D\bar{x}\bar{y},$$
$$\Phi(x, y) = Fxy + Gx\bar{y} + H\bar{x}y + K\bar{x}\bar{y}.$$

We require the necessary and sufficient condition that

$$\phi(x, y) \leftrightarrow \Phi(x, y).$$

|307| Now with the usual notation for the coefficients of the substitution T, we have

$$T\phi(x, y) = \phi(a_1, b_1)xy + \phi(a_2, b_2)x\bar{y} + \phi(a_3, b_3)\bar{x}y + \phi(a_4, b_4)\bar{x}\bar{y}.$$

Hence

$$\phi(a_1, b_1) = F,\ \phi(a_2, b_2) = G,\ \phi(a_3, b_3) = H,\ \phi(a_4, b_4) = K.$$

These equations can be written

$$\left.\begin{aligned} p(A, F)a_1b_1 + p(B, F)a_1\bar{b}_1 + p(C, F)\bar{a}_1b_1 + p(D, F)\bar{a}_1\bar{b}_1 &= 0,\\ p(A, G)a_2b_2 + p(B, G)a_2\bar{b}_2 + p(C, G)\bar{a}_2b_2 + p(D, G)\bar{a}_2\bar{b}_2 &= 0,\\ p(A, H)a_3b_3 + p(B, H)a_3\bar{b}_3 + p(C, H)\bar{a}_3b_3 + p(D, H)\bar{a}_3\bar{b}_3 &= 0,\\ p(A, K)a_4b_4 + p(B, K)a_4\bar{b}_4 + p(C, K)\bar{a}_4b_4 + p(D, K)\bar{a}_4\bar{b}_4 &= 0. \end{aligned}\right\} \tag{26}$$

Now by comparison with equations (19) and (21), we see that the requisite condition is

$$\Pi p(A, F; B, G; C, H; D, K) = 0, \tag{27}$$

where, in order to obtain the various factors, F, G, H, K are kept in the same positions, and A, B, C, D are permuted in every possible way, so that each appears in each factor.

In order to effect the solution of this equation, it will be convenient to alter the notation. Consider

$$\Pi p(x_1, c_p; x_2, c_q; x_3, c_r; x_4, c_s) = 0,\quad (p, q, r, s = 1, 2, 3, 4). \tag{28}$$

Let C_1, C_2, C_3, C_4 be the symmetric[1] functions of c_1, c_2, c_3, c_4; and let X_1, X_2, X_3, X_4 be

1. The original misspelled as 'symmmetric'.

the symmetric functions of x_1, x_2, x_3, x_4. Now, consider any one factor of the left-hand side of (28); for example, $p(x_1, c_p; x_2, c_q; x_3, c_r; x_4, c_s)$. Then[1]

$$p(x_1, c_p; x_2, c_q; x_3, c_r; x_4, c_s) = \bar{C}_4 x_1 x_2 x_3 x_4 + \Sigma\left[(\bar{c}_p + \bar{c}_q + \bar{c}_r + c_s) x_1 x_2 x_3 \bar{x}_4\right]$$
$$+ \Sigma\left[(\bar{c}_p + \bar{c}_q + c_r + c_s) x_1 x_2 \bar{x}_3 \bar{x}_4\right] + \Sigma\left[(\bar{c}_p + c_q + c_r + c_s) x_1 \bar{x}_2 \bar{x}_3 \bar{x}_4\right] + C_1 \bar{x}_1 \bar{x}_2 \bar{x}_3 \bar{x}_4.$$

Hence, equation (28) becomes

$$\bar{C}_4 X_4 + \overline{}(C_3\bar{C}_4) X_3\bar{X}_4 + \overline{}(C_2\bar{C}_3) X_2\bar{X}_3 + \overline{}(C_1\bar{C}_2) X_1\bar{X}_2 + C_1\bar{X}_1 = 0.$$

But, from the symmetry of (28), it is obvious that we may equally well write

$$\bar{X}_4 C_4 + \overline{}(X_3\bar{X}_4) C_3\bar{C}_4 + \overline{}(X_2\bar{X}_3) C_2\bar{C}_3 + \overline{}(X_1\bar{X}_2) C_1\bar{C}_2 + X_1\bar{C}_1 = 0.$$

By adding these two equations, and remembering that

$$P\bar{Q} + \bar{P}Q = 0,$$

|308| implies $P = Q$, we find

$$X_4 = C_4,\ X_3\bar{X}_4 = C_3\bar{C}_4,\ X_2\bar{X}_3 = C_2\bar{C}_3,\ X_1\bar{X}_2 = C_1\bar{C}_2,\ X_1 = C_1. \qquad \text{(m)}$$

The fourth and fifth of equations (m) give

$$C_1\bar{X}_4 = C_1\bar{C}_2.$$

Hence, $$X_2 = C_2 + q\bar{C}_1.$$

But $$X_2\bar{C}_1 = X_2\bar{X}_1 = 0,$$

and $$C_2\bar{C}_1 = 0.$$

Hence, $$q\bar{C}_1 = 0.$$

Thus $X_2 = C_2$. An exactly similar proof applied to the third of equations (m) will now prove that $X_3 = C_3$.

Hence, equation (27) is equivalent to the fact that the corresponding invariants of $\phi(x, y)$ and $\Phi(x, y)$ are equal.

As special examples of this theorem, we note that (1) secondary linear primes form a congruent family, (2) secondary separable primes form a congruent family, (3) functions both of deficiency two and of supplemental deficiency two form a congruent family. This last family includes as members all primary primes whether functions of *x* only or of *y* only.

A congruent family is entirely defined by its invariants. Accordingly, we shall name a family by its invariants, and speak of the family (S_1, S_2, S_3, S_4).

Any family (S_1, S_2, S_3, S_4) includes as a member the function

$$S_1 xy + S_2 x\bar{y} + S_3 \bar{x}y + S_4 \bar{x}\bar{y},$$

which can also be written

$$S_1 xy + S_2 x + S_3 y + S_4;$$

and also the family includes the twenty-three other functions of the same type found by permuting the invariants in their use as coefficients.

1. For the following equation, the '$\bar{c}_r$' originally had no overbar, but only a partial mark, likely a printing error.

Call such functions the "canonical functions" of the family.

There is no fundamental distinction in property between the various canonical functions of a family. Accordingly, we shall habitually use the one mentioned above, and will call it "the canonical function." Thus the canonical function of the family $(i, i, i, 0)$, which is the family of secondary linear primes, is $x+y$. The canonical function of the family $(i, 0, 0, 0)$ which is the family of secondary separable primes, is *xy*. The canonical function of the family |309| (*i*, *i*, 0, 0) is *x*. The only families which contain some linear functions as some members, have been proved (cf. Part I, §9) to be those for which $S_1 = S_2$. Thus the canonical function for such a family is $S_1x + S_3y + S_4$. Similarly, the only families which contain some separable functions as some members are those for[1] which $S_3 = S_4$. Thus the canonical function for such a family is

$$S_1xy + S_2x + S_4, \text{ that is, } (x+S_4)\ (S_1y+S_2).$$

Also, by reference to equations (23) and (24), we can find from equations (26) the complete conditions satisfied by a_1 and b_1 after the other coefficients have been eliminated. For let S_{11}, S_{12}, S_{13} be the symmetric functions of *B*, *C*, *D*; and let S_{21}, S_{22}, S_{23} be the symmetric functions of *A*, *C*, *D*, and S_{31}, S_{32}, S_{33} of *A*, *B*, *D* and S_{41}, S_{42}, S_{43} of *A*, *B*, *C*; and let R_{11}, R_{12}, R_{13} be the symmetric functions of *G*, *H*, *K*, and so on. Now, consider $\Pi p(B, G; C, H; D, K)$, where the various factors are formed by keeping *G*, *H*, *K* in their places and permuting *B*, *C*, *D*. By comparison with the evaluation of the left-hand side of (28), we see that

$$\Pi p(B, G; C, H; D, K) = p(S_{13}, R_{13}; S_{12}\bar{S}_{13}, R_{12}\bar{R}_{13}; S_{11}\bar{S}_{12}, R_{11}\bar{R}_{12}; S_{11}, R_{11}),$$

with similar expressions for other similar products. Thus we find the required equation to be

$$\begin{aligned} p(A, F; S_{13}, R_{13}; S_{12}\bar{S}_{13}, R_{12}\bar{R}_{13}; S_{11}\bar{S}_{12}, R_{11}\bar{R}_{12}; S_{11}, R_{11})a_1b_1 \\ +p(B, F; S_{23}, R_{13}; S_{22}\bar{S}_{23}, R_{12}\bar{R}_{13}; S_{21}\bar{S}_{22}, R_{11}\bar{R}_{12}; S_{21}, R_{11})a_1\bar{b}_1 \\ +p(C, F; S_{33}, R_{13}; S_{32}\bar{S}_{33}, R_{12}\bar{R}_{13}; S_{31}\bar{S}_{32}, R_{11}\bar{R}_{12}; S_{31}, R_{11})\bar{a}_1b_1 \\ +p(D, F; S_{43}, R_{13}; S_{42}\bar{S}_{43}, R_{12}\bar{R}_{13}; S_{41}\bar{S}_{42}, R_{11}\bar{R}_{12}; S_{41}, R_{11})\bar{a}_1\bar{b}_1 = 0. \end{aligned} \tag{29}$$

Also a similar equation can be found for a_2, b_2 by putting *G* for *F* and R_{21}, R_{22}, R_{23} for R_{11}, R_{12}, R_{13} respectively in equation (29). Also, similarly for a_3, b_3 and for a_4, b_4 by similar interchanges.

Thus a_1, b_1 can be chosen to be any pair of roots of equation (29), but when a_1, b_1 is once chosen, a_2, b_2 and a_3, b_3 and a_4, b_4 must be suitable pairs of roots of their corresponding equations.

It is easily proved by considering the invariants, that if

$$\phi_1(x, y) \leftrightarrow \Phi_1(x, y)$$

and

$$\phi_2(x, y) \leftrightarrow \Phi_2(x, y),$$

1. A duplicate 'for' has been deleted.

|310| then, for all values of λ,

$$\lambda\phi_1(x, y)+\lambda\bar{\phi}_2(x, y)\leftrightarrow\lambda\Phi_1(x, y)+\lambda\bar{\Phi}_2(x, y).$$

The conditions that $\phi(x, y)$ can be transformed into $\Phi(x, y)$ by some transformation which is not necessarily a substitution, are given by the resultants of equations (26) without the use of equation (4). These conditions are, if S_1, S_2, S_3, S_4 are the invariants of $\phi(x,y)$ and R_1, R_2, R_3, R_4 of $\Phi(x, y)$,[1]

$$\bar{S}_1F+S_4\bar{F}=0,\quad \bar{S}_1G+S_4\bar{G}=0,\quad \bar{S}_1H+S_4\bar{H}=0,$$
$$\bar{S}_1K+S_4\bar{K}=0.$$

Hence, by addition,

$$\bar{S}_1R_1+S_4\bar{R}_4=0.$$

Thus the required conditions can be written

$$R_1 \mathrel{+\!\!\!\in} S_1,\quad S_4 \mathrel{+\!\!\!\in} R_4. \tag{30}$$

In other words, the field of $\phi(x, y)$ must contain the field[2†] of $\Phi(x, y)$. Thus the conditions that $\phi(x, y)$ can be transformed into $\Phi(x, y)$, and that $\Phi(x, y)$ can be transformed into $\phi(x, y)$, are

$$R_1=S_1,\quad R_4=S_4. \tag{31}$$

Accordingly, the additional conditions required in order that these transformations may be substitutions, are[3]

$$R_2=S_2,\quad R_3=S_3.$$

§7.—The Identical Group of a Function.

Any function $\phi(x, y)$ can be conceived as congruent to itself. Also the substitutions such that

$$T\phi(x, y)=\phi(x, y)$$

evidently form a group. For, if *T* is such a substitution, T^{-1} also has the same property, and if T_1 is another such substitution, then T_1T has also the same property. Let this group be called the "identical group" of the function $\phi(x, y)$.

|311| By substituting *A*, *B*, *C*, *D* for *F*, *G*, *H*, *K* in equations (26), we see that the coefficients of a substitution of the identical group must satisfy the equations[4]

$$\left.\begin{aligned}
*+p(B, A)a_1\bar{b}_1+p(C, A)\bar{a}_1b_1+p(D, A)\bar{a}_1\bar{b}_1 &=0,\\
p(A, B)a_2b_2+*+p(C, B)\bar{a}_2b_2+p(D, B)\bar{a}_2\bar{b}_2 &=0,\\
p(A, C)a_3b_3+p(B, C)a_3\bar{b}_3+*+p(D, C)\bar{a}_3\bar{b}_3 &=0,\\
p(A, D)a_4b_4+p(B, D)a_4\bar{b}_4+p(C, D)\bar{a}_4b_4+* &=0,
\end{aligned}\right\} \tag{32}$$

1. For the equation below, we have corrected '$\bar{S}_kH+S_4\bar{H}=0$' to '$\bar{S}_1H+S_4\bar{H}=0$'.
2. † Cf. "Universal Algebra," §33.
3. For the equation below, we have corrected '$R_3=S_2$' to '$R_3=S_3$'.
4. In the following equation, the original had multiple errors on line 4. There was a period between '*A*' and '*D*' instead of a comma, and the first '+' had been replaced with a right parenthesis.

The coefficients must also, of course, satisfy equation (4).

Then equation (29), in the reduced form required for substitutions of the identical group of $\phi(x, y)$, becomes[1]

$$\begin{aligned} &p(B, A; S_{23}, S_{13}; S_{22}\bar{S}_{23}, S_{12}\bar{S}_{13}; S_{21}\bar{S}_{22}, S_{11}\bar{S}_{12}; S_{21}, S_{11})a_1\bar{b}_1 \\ &\quad + p(C, A; S_{33}, S_{13}; S_{32}\bar{S}_{33}, S_{12}\bar{S}_{13}; S_{31}\bar{S}_{32}, S_{11}\bar{S}_{12}; S_{31}, S_{11})\bar{a}_1 b_1 \qquad \text{(m)} \\ &\quad + p(D, A; S_{43}, S_{13}; S_{42}\bar{S}_{43}, S_{12}\bar{S}_{13}; S_{41}\bar{S}_{42}, S_{11}\bar{S}_{12}; S_{41}, S_{11})\bar{a}_1\bar{b}_1 = 0; \end{aligned}$$

with similar equations for a_2, b_2 and for a_3, b_3, and for a_4, b_4.

Now $$S_{11} = B + C + D,\ S_{12} = BC + BD + CD,\ S_{13} = BCD,$$

with similar meanings for analogous terms. Hence it is easily seen that

$$\begin{aligned} &p(S_{23}, S_{13}) = CDp(A, B),\ p(S_{22}\bar{S}_{23}, S_{12}\bar{S}_{13}) = p(A, B), \\ &p(S_{21}\bar{S}_{22}, S_{11}\bar{S}_{12}) = p(A, B),\ p(S_{21}, S_{11}) = \bar{C}\bar{D}p(A, B). \end{aligned}$$

Thus

$$p(B, A; S_{23}, S_{13}; S_{22}\bar{S}_{23}, S_{12}\bar{S}_{13}; S_{21}\bar{S}_{22}, S_{11}\bar{S}_{12}; S_{21}, S_{11}) = p(A, B);$$

with similar simplifications for the coefficients of[2] $\bar{a}_1 b_1$ and $\bar{a}_1\bar{b}_1$. Thus equation (m) reduces to the first of equation (32). Similarly for the equations for a_2, b_2 and for a_3, b_3, and for a_4, b_4. Thus in order to find substitutions of the identical group of $\phi(x, y)$, a_1 and b_1 can be chosen to be any pair of roots of the first of equations (32), and then a_2, b_2 and a_3, b_3, and a_4, b_4 must be chosen to be suitable roots of their corresponding equations in (32). Or we may start from a_2, b_2, or from a_3, b_3, or from a_4, b_4.

The identical group of every function is of order greater than one, that is contains substitutions other than the identical substitution.

For if the identical group of $\phi(x, y)$ is of order one the left-hand side of every equation of the set (32) must be a secondary linear prime, so as to give only one pair of roots for a_1, b_1, and one for a_2, b_2, and so on. Now, each of the left-hand sides of equation (32) lacks one term, thus the first of these equations lacks |312| the term $a_1 b_1$, and so on. Hence [cf. Part I, §3], the left-hand sides can only be secondary linear primes if each remaining coefficient is *i*.

Thus every equation of the type

$$p(A, B) = i,$$

must hold. But this gives

$$p(\bar{A}, B) = 0, \text{ that is, } A = \bar{B}.$$

Similarly $A = \bar{C}$, $B = \bar{C}$. But these equations are inconsistent. Hence the identical group of $\phi(x, y)$ cannot be of order one.

The identical groups of all members of a congruent family are simply isomorphic. For, let $\phi(x, y)$ and $\Phi(x, y)$ be two congruent functions; and let the members of

1. In the following equation, the original erroneously had '$\bar{a}_1 b_1$' at the end of line 1, and '$a_1\bar{b}_1$' at the end of line 2 (they had been reversed).
2. For the following expression, the original erroneously had '$a_1\bar{b}_1$'.

the identical group of $\phi(x, y)$ be written T_ϕ, T'_ϕ, etc., those of the identical group of $\Phi(x, y)$ be written T_Φ, T'_Φ, etc. Also let T be any substitution such that

$$T\phi(x, y) = \Phi(x, y).$$

Then $TT_\phi, TT'_\phi, \ldots.$ are evidently such substitutions.

Similarly $T_\Phi T, T'_\Phi T, \ldots.$ are such substitutions. Again Let T_1 be another substitution such that

$$T_1\phi(x, y) = \Phi(x, y).$$

Then
$$T^{-1}T_1\phi(x, y) = T^{-1}\Phi(x, y) = \phi(x, y).$$

Hence $T^{-1}T_1$ is a member of the identical group of $\phi(x, y)$. Accordingly we may write

$$T^{-1}T_1 = T_\phi, \text{ that is, } T_1 = TT_\phi.$$

Similarly $T_1 = T_\Phi T$.

Thus each of the sets $TT_\phi, TT'_\phi, \ldots.,$ and $T_\Phi T, T'_\Phi T, \ldots.$ includes every substitution which turns $\phi(x, y)$ into $\Phi(x, y)$. Accordingly, by a proper choice of T_ϕ, or of T_Φ, we can always write

$$TT_\phi = T_\Phi T.$$

Hence
$$T_\phi = T^{-1}T_\Phi T.$$

Hence the identical groups of $\phi(x, y)$ and of $\Phi(x, y)$ are simply isomorphic. Accordingly it is only necessary to study the structure of the identical group of one member of a congruence family; for instance, that of the canonical form.

|313| For instance, let us investigate the general expression for a substitution of the identical group of the canonical function of the family $(i, i, 0, 0)$. This canonical function is x. Thus if T is the required substitution, we have[1]

$$Tx = x,\ Ty = b_1xy + b_2x\bar{y} + b_3\bar{x}y + b_4\bar{x}\,\bar{y}.$$

Hence from equation (12),

$$b_1b_2 + \bar{b}_1\bar{b}_2 = 0,\ b_3b_4 + \bar{b}_3\bar{b}_4 = 0$$

Thus
$$Ty = xp(\bar{b}_1, y) + \bar{x}p(\bar{b}_3, y); \tag{33}$$

where b_1 and b_3 can be assumed arbitrarily. Thus the general form for a substitution of the group is found.

Now let T and T' be two substitutions of this group, so that

$$Tx = x,\ Ty = b_1xy + \bar{b}_1x\bar{y} + b_3\bar{x}y + \bar{b}_3\bar{x}\,\bar{y},$$
$$T'x = x,\ T'y = b'_1xy + \bar{b}'_1x\bar{y} + b'_3\bar{x}y + \bar{b}'_3\bar{x}\,\bar{y}.$$

Then

$$T'Ty = \bar{p}(b_1, b'_1)xy + p(b_1, b'_1)x\bar{y} + \bar{p}(b_3, b'_3)\bar{x}y + p(b_3, b'_3)\bar{x}\,\bar{y} = TT'y. \tag{34}$$

Hence the substitutions TT' and $T'T$ are the same. Thus this identical group is Abelian.

---

1. The following equation had multiple errors. It originally appeared as follows: $Tx = x,\ Ty = b_1xy + b_2\bar{x}y + \bar{b}_3xy + b_4\bar{x}\,\bar{y}$.

Also in equation (34), put $b_1' = b_1$, and $b_3' = b_3$; we find

$$T^2 y = y.$$

Thus $T^2 = T^0$. Hence every substitution of the group is of order two.
The equations satisfied by the coefficients of any substitution of the identical group of the canonical function of the family (S_1, S_2, S_3, S_4), are found from equations (32) to be[1]

$$\left.\begin{array}{l} * + S_1\bar{S}_2 a_1\bar{b}_1 + S_1\bar{S}_3\bar{a}_1 b_1 + S_1\bar{S}_4\bar{a}_1\bar{b}_1 = 0, \\ S_1\bar{S}_2 a_2 b_2 + * + S_2\bar{S}_3\bar{a}_2 b_2 + S_2\bar{S}_4\bar{a}_2\bar{b}_2 = 0, \\ S_1\bar{S}_3 a_3 b_3 + S_2\bar{S}_3 a_3\bar{b}_3 + * + S_3\bar{S}_4\bar{a}_3\bar{b}_3 = 0, \\ S_1\bar{S}_4 a_4 b_4 + S_2\bar{S}_4 a_4\bar{b}_4 + S_3\bar{S}_4\bar{a}_4 b_4 + * = 0, \end{array}\right\} \qquad (35)$$

together with equations (4); and it has been proved that any one pair, such as a_1 and b_1, can be assumed to be any pair of roots of their corresponding equation.

§8.—Common Subgroups of Identical Groups.

The identical groups of any two functions have a common subgroup which always includes other substitutions in addition to the identical substitution, |314| except in the case when the two functions are the two director functions of a substitution.

For the $\phi(x, y)$ and of $\Phi(x, y)$ be the two functions, where *A, B, C, D* are the coefficients of $\phi(x, y)$, and *F, G, H, K* are the coefficients of $\Phi(x, y)$. Then the coefficients of any substitution common to the two identical groups must satisfy two sets of equations of the type of (32). These two sets can be combined into the single set[2]

$$\left.\begin{array}{l} * + p(B, A; G, F)a_1\bar{b}_1 + p(C, A; H, F)\bar{a}_1 b_1 + p(D, A; K, F)\bar{a}_1\bar{b}_1 = 0, \\ p(A, B; F, G)a_2 b_2 + * + p(C, B; H, G)\bar{a}_2 b_2 + p(D, B; K, G)\bar{a}_2\bar{b}_2 = 0, \\ p(A, C; F, H)a_3 b_3 + p(B, C; G, H)a_3\bar{b}_3 + * + p(D, C; K, H)\bar{a}_3\bar{b}_3 = 0, \\ p(A, D; F, K)a_4 b_4 + p(B, C; G, K)a_4\bar{b}_4 + p(C, D; H, K)\bar{a}_4\bar{b}_4 + * = 0, \end{array}\right\} \qquad (36)$$

and, in addition, equation (4) must be satisfied.

Now, by reasoning in all respects the same as that in the previous article, if there is only one substitution satisfying these conditions, the left-hand side of each of equation (36) is a secondary linear prime. Hence, every equation of the type

$$\bar{p}(B, A; G, F) = 0,$$

must hold. But this typical equation is

1. In the following equation, the original erroneously had no overbar for the second 'b_4' on line 4.
2. There were multiple errors in the below equations. Line 1 contained no leading asterisk, and in lines 3 and 4, the second '$a_3\bar{b}_3$' and '$a_4\bar{b}_4$' erroneously placed an overbar on the 'a' and not on the 'b'. Finally, in line 4 omitted the overbar from the final 'b'.

$$\left(AB+\bar{A}\bar{B}\right)\left(FG+\bar{F}\bar{G}\right)=0.$$

Hence, by comparison with equation (12), we see that this condition requires that $\phi(x,\, y)$ and of $\Phi(x,\, y)$ should be a pair of director functions of a substitution. The above proposition can be stated thus: The Identical Group of any function $\phi(x,\, y)$ which does not belong to the family $(i, i, 0, 0)$ has a subgroup containing more than the one member T^0 in common with the identical group of any other function whatever. The same proposition is true of any function $\phi(x,\, y)$ which does belong to the family $(i, i, 0, 0)$, except in those cases when the second function also belongs to the same family $(i, i, 0, 0)$, and, in addition, is so related to $\phi(x,\, y)$ that the two functions are the director functions of a substitution.

It has been proved [cf. Part I, §8, equation[1]† (24)] that we can always write

$$\phi(x,\, y)=S_4+S_1\left(\bar{S}_4+U_1\right)\phi'(x,\, y),$$

|315| where the coefficients of $\phi'(x,\, y)$ are given by equations (25) of Part I, and the invariants by equations (26) of Part I.

Now $$T\phi(x,\, y)=S_4+S_1\left(\bar{S}_4+U_1\right)T\phi'(x,\, y).$$

Hence, if $$T\phi'(x,\, y)=\phi'(x,\, y),$$

it follows that $$T\phi(x,\, y)=\phi(x,\, y).$$

Hence, the identical group of $\phi'(x,\, y)$ is a subgroup of the identical group of $\phi(x,\, y)$. Now, the invariants of $\phi'(x,\, y)$ are given [cf. equation (26) of §8 Part I] by[2]

$$\left.\begin{aligned} S_1' &= S_1\bar{S}_4+S_1U_1+V_1, \\ S_2' &= S_2\bar{S}_4+S_2U_2+\bar{S}_1V_2+S_4\bar{U}_1V_2, \\ S_3' &= S_3\bar{S}_4+S_3U_3+\bar{S}_1V_3+S_4\bar{U}_1V_3, \\ S_4' &= S_4U_4+\bar{S}_1V_4+S_4\bar{U}_1V_4, \end{aligned}\right\} \qquad (37)$$

and U_1, U_2, U_3, U_4 are the symmetric functions of u_1, u_2, u_3, u_4 and V_1, V_2, V_3, V_4 of v_1, v_2, v_3, v_4.

Also, if $\phi(x,\, y)$ is the canonical function of the family (S_1, S_2, S_3, S_4), then equations (25) of §8, Part I, become[3]

$$\begin{aligned} A' &= \left(\bar{S}_1+S_4\bar{U}_1\right)v_1+\left(\bar{S}_4+u_1\right)+S_1, \\ B' &= \left(\bar{S}_1+S_4\bar{U}_1\right)v_2+\left(\bar{S}_4+u_2\right)+S_2, \\ C' &= \left(\bar{S}_1+S_4\bar{U}_1\right)v_3+\left(\bar{S}_4+u_3\right)+S_3, \\ D' &= \left(\bar{S}_1+S_4\bar{U}_1\right)v_4+\left(\bar{S}_4+u_4\right)+S_4. \end{aligned}$$

---

1. † A misprint in equation (24), Part I, is here corrected. [Whitehead's note here no longer applies, since this edition made the correction in Part I.]
2. In the following equation, the original erroneously had '$\bar{S}_4U_1V_2$' on line 2.
3. In the following equation, the original erroneously had 'S_3' as the final term on line 4.

Hence, in general, $\phi'(x, y)$ does not become the canonical function of (S_1', S_2', S_3', S_4'). But if we make[1]

$$u_1 = U_1,\quad u_2 = U_2,\quad u_3 = U_3,\quad u_4 = U_4,$$
$$v_1 = V_1,\quad v_2 = V_2,\quad v_3 = V_3,\quad v_4 = V_4,$$

which can be done without altering U_1, U_2, U_3, U_4 or V_1, V_2, V_3, V_4, then $\phi'(x, y)$ becomes the canonical function of the family (S_1', S_2', S_3', S_4'). Hence, the identical group of the canonical function of the family (S_1', S_2', S_3', S_4') is a subgroup of the identical group of the canonical function of the family (S_1, S_2, S_3, S_4).

For instance, put

$$U_1 = U_2 = i,\quad U_3 = U_4 = 0,$$
$$V_1 = V_2 = i,\quad V_3 = V_4 = 0.$$

|316| We deduce that the identical group of the canonical function of the family $(i, S_2 + \bar{S}_1, S_3\bar{S}_4, 0)$ is a subgroup of the identical group of the canonical function of the family[2] (S_1, S_2, S_3, S_4).

Now, it is proved in §9 of Part I, that if $S_1 = S_2$ and $S_3 = S_4$, the family contains both linear members and separable members. Also, in this case,

$$S_2 + \bar{S}_1 = i,\quad S_3\bar{S}_4 = 0,$$

and the canonical function of the family $(i, i, 0, 0)$ is x. Hence, the identical group of x is a subgroup of the identical groups of the canonical functions of all families which contain both linear and separable members. Thus, from the discussion at the end of the previous paragraph, all these identical groups have a common Abelian subgroup.

1. In the following equation, the original erroneously had a period after 'V_4' instead of a comma.
2. The original erroneously omitted the following left parenthesis.

On Cardinal Numbers

1902 *American Journal of Mathematics*, vol. 24, no. 4 (October), pp. 367–394

This work is expository, rather than research oriented, and Whitehead directly credits Cantor, Peano and Russell for most of its results (p. 162). The style is notable for its lack of prose: after introducing his system of notation in Section I, Whitehead proceeds for page after page of statements and proofs reduced to logical or relational formulas. Nonetheless, this is an impressive treatment of cardinal arithmetic and goes beyond what one might find in a standard undergraduate course on set theory and logic. For instance, it takes the ideas of combinations and permutations into the realm of the infinite. The work does suffer from a flaw in connection with Russell's paradox and exhibits a fascinating implicit entanglement with the Axiom of Choice; both are discussed below.

The non-technical reader will require the definition of cardinality to make sense of the topic at hand. Figure 1 will help.

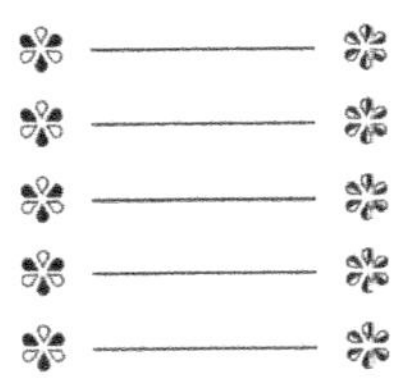

Figure 1.

Figure 1 consists of two columns of flowers with alternating petal colours on the left and variegated petals on the right. The lines connecting serve to pair the members of the respective columns in such a way that every alternating flower is associated with one variegated flower and, likewise, every variegated flower is associated, using the same connecting line, with one alternating flower. This special kind of pairing is called a bijective correspondence *or a* bijection, *and it has consequences. Notice that if any line is removed, the corresponding flowers are orphaned: they no longer take part in any pairing. If we adjust just one of the lines to connect a left-hand object with a different right-hand object, we profoundly disturb the symmetry: one of variegated flowers is orphaned and another acquires the distinction of occurring in two different pairings. (Imagine, for instance, redrawing the top line so that it connects the first flower on the left with the second flower on the right.) Now, with additional adjustments we could in fact create many other bijective pairings using non-parallel lines, but the diagram above suffices to make the intended point.*

First notice that if Figure 1 went on for thousands of lines, without the tedium of counting, one could say with certainty that the number of flowers on the left is equal to the number of flowers on the right. Thus, the existence of such a pairing implies that the left-hand set is the same size as the right-hand set. Moreover, if one wanted to check that inference in a more familiar way then, say, for Figure 1, one could simply count the flowers of each type: one, two, three, four, five. But a little reflection shows that in making such a count, what one is really doing is making a bijective correspondence with a column of flowers and the numbers in the set {1, 2, 3, 4, 5}. A little more reflection then shows that if two sets of any objects have the same count, one could make a diagram like Figure 1 using the numbers implicit in the counting procedure. (For example, alternating flower number 4 is paired with variegated flower number 4.) The upshot is that both in the sense of asserting (a) that two finite sets have the same number of elements, and (b) counting a finite set to get the precise size of its membership, the paramount underlying construction is a bijective correspondence. (Indeed, someone who could not count at all could still assert equality of size by making an explicit pairing, and this is exactly how we can compare apples to oranges.)

One of Georg Cantor's huge contributions to mathematics was to extend counting to infinite sets by a direct extension of the considerations above. Simply put, one says that the sets S *and* T *have the same* cardinality *if there is a bijective correspondence between them. This corresponds precisely with the use of bijections to assert the equality of size above, and yet this extended notion of size has some distinctly non-intuitive aspects. For example, the set of natural numbers {0, 1, 2,...} has the same cardinality as the full set of integers {0, ±1, ±2,...}. The set of integers in turn has the same cardinality as the set of rational numbers (fractions), even though between any two numbers there are an infinite number of rationals. More astounding still is that, according to Cantor's* diagonalisation argument, *the set of real numbers in a very definite sense has cardinality greater than the set of natural numbers. This means that there is not just one kind of infinity and, in fact, one can show that there are infinitely many kinds of infinity.*

The preceding paragraph describes what it means for two sets to have the same cardinality, but many take a kind of ontological comfort in going a step further. Just as one might want to say that the two flower collections shown in Figure 1 both have five elements – that is, each set has cardinality five – one might want to attach a cardinal number to an arbitrary set, thus making cardinality a matter of objects as well as relations, and this is possible with some care. Thus, for instance, the cardinal number 2 becomes the set of all sets containing two members. The cardinal number α_0 *(to use Whitehead's notation) becomes the set of all sets that can be put into a bijective correspondence with the natural numbers. With that in mind, it becomes natural to do cardinal arithmetic, for instance, to add or multiply cardinal numbers, but, again, this arithmetic has some surprising features. (For example,* $\alpha_0 + \alpha_0 = \alpha_0$.*) Indeed, this eerie arithmetic is precisely what Whitehead's paper addresses.*

While most of the statements and proofs that Whitehead develops translate into present-day mathematical theorems, the world has moved beyond his notation, terminology and virtually stenographic style of proof. These cosmetic elements do not essentially infirm the paper, but they do make for difficult reading. The real worth of the paper from a historical perspective is its intersection with the two foundational issues, as noted in the leading paragraph above. The first involves Russell's paradox, which one can easily describe.

A common occurrence in mathematics is the definition of a set by abstraction or comprehension. (The two terms are synonymous in this context.) Thus, for example, one might begin with, 'Let S *be the set of all even integers'. The set is defined by the property of being an even integer, and accordingly the number 4 is in* S, *while 5 is not. Russell's paradox shows that definition by comprehension has its limits and proceeds as follows: Let* S *be the set of all sets, a construction that Whitehead uses freely (p. 372). Note that* S, *being a set, belongs to itself, and this suggests another definition by comprehension: let* M *(for monster) be the set of all sets that do not belong to themselves. Now, according to the law of the excluded middle, either* M *belongs to itself, or it does not. But if* M *does belong to itself, then, by definition,* M *does not belong to itself. And if* M *does not belong to itself, then, again by definition,* M *does belong to itself. In either case, we have a contradiction, and this sounds the death knell for unencumbered definition by comprehension and creates a crisis in the foundations of mathematics.*

In the long term, the most compelling response to this crisis was to formalise set theory in a way that prohibits unrestricted comprehension and does not allow a set to belong to itself. In formal set theory, we may speak informally of classes, which are essentially sets that do not belong to other sets. The construction of sets is thus rigidified in a way that prevents Russell's paradox, but at the cost of engaging Gödel's limitative theorems. Regarding Whitehead's paper on cardinality, the calendar and the contents tell the tale: Russell is said not to have discovered his paradox until late spring of 1901, he wrote his famous letter communicating this to Frege about a year later, and Russell and Frege first mention the paradox in print in 1903 (Russell in The Principles of Mathematics *and Frege in his second volume of* Grundgesetze der Arithmetik*). It is likely that Whitehead would have submitted his manuscript to an American journal either in late 1901 or early 1902. In either case, the contents strongly suggest that he was unaware of Russell's paradox – or at least of its seriousness – and its consequences for his article. While this invalidates many elements of his exposition, in the main his results persist in the context of formal set theory.*

The second foundational problem in Whitehead's development of cardinality is its entanglement with the Axiom of Choice (AC), a principle that allows the construction of new sets from a prior collection. For example, consider the sets S = *{0, 1, 2, 3} and* T = *{4, 5, 6, 7}. (These sets have no elements in common, but this is inessential.) We can then construct a set which has exactly one element in common with* S *and one element in common with* T: *{0, 4} and {2, 7} are two examples of, in this case, sixteen possibilities. Both sets serve as a kind of 'parliament' (used here as an informal, metaphorical and unoriginal term) for the two given sets, with one member from each. What is less clear is that we can construct such a parliament for an infinite collection of sets (that is, choose one element from each set to lie in our newly constructed parliamentary set) and, indeed, as formal set theory developed, it emerged that the other axioms of set theory were insufficient to guarantee this construction. Hence these axioms are generally supplemented by the AC.*

Whitehead engages the AC twice in this article, but both times the engagement is implicit. When made explicit, one finds one of the most remarkable features latent in the work. Thus, on the one hand, on page 172 he laments that no one has yet proven that every infinite cardinal number is in a sense made precise by cardinal arithmetic, a multiple of the smallest infinite cardinal α_0*. In hindsight, one knows that this statement is true and a consequence of the AC; moreover, one, again, cannot reach it with the prior axioms of formal set theory. Yet, on the other hand, given a possibly infinite* *collection of non-intersecting sets, Whitehead blithely constructs the set of all sets that intersect each of the component sets exactly once (p. 174); that is, he asserts the full complement of parliamentary sets in the sense explained above. Here he is implicitly using the AC, exactly the assertion one needs for his conjecture that every infinite cardinal is a multiple of* α_0*. So, in one and the same work,* Whitehead is unable to confirm an assertion because he is missing the AC, while asserting the existence of entities that require the AC.

In light of its idiom and historical placement at the end-of-life of naïve set theory, one is unlikely to turn to Whitehead's introduction to cardinality and cardinal arithmetic for either its methods or its results. Yet for those very reasons this paper is of enormous historical value. Its treatment of the fundamentals of set theory, like Whitehead's treatment of the foundations of geometry, is the product of a profoundly adept mathematician working in a time of transition, in this case, a transition away from an approach to mathematics that had become invisibly and fatally suffused by intuitive constructs unable to survive the rigours of consistency. Hence the stage was set for the departure into logicism and formalism. In turn, these approaches would ultimately have to give way – as announced three decades later by Gödel – to the ineluctable limitations of epistemically pure, but semantically devoid systems.

Preface.

The following memoir is concerned solely with the theory of Cardinal numbers, and does not touch upon the theory of ordered aggregates or of ordinal types which form the other parts of the general theory of aggregates (mengenlehre).[1†]

Sections I and II are preliminary, to explain the notation and methods of reasoning. Incidentally they may be of independent service as explaining the elements of Peano's developments of mathematical logic[2†] (Section I) and of Russell's symbolism for the Logic of Relations[3†] (Section II). I believe that these two methods are almost

1. † Due almost entirely to G. Cantor. Almost all that is known respecting Cardinal Numbers, apart from their relations to the other parts of the theory, is given in his memoir in Math. Annal., Bd. XLVI, 1895: this and its sequel on Ordinal Types, loc. cit. Bd. XLIX, can be obtained as one pamphlet in a French translation, translated by F. Marotte, Paris, A. Hermann, 1899. For a general sketch of the whole theory, with bibliography, cf. "Die Entwickelung der Lehre von den Punktmannigfaltigkeiten": Bericht con A. Schönflies, "Yahresbericht der Deutschen Mathematiker Vereinigung," 1900. For a development of the theory of Ordinal Types in respect to condensed series, by the methods of the present memoir, cf. Bertrand Russell, *Revue de Mathématiques*, vol. VII, Turin, 1901, where the methods of reasoning used here were first elaborated. [Full citations: Georg Cantor, 'Beiträge zur Begründung der transfiniten Mengenlehre (1)', *Mathematische Annalen*, vol. 46, no. 4 (November 1895), pp. 481–512; Georg Cantor, 'Beiträge zur Begründung der transfiniten Mengenlehre (2)', *Mathematische Annalen*, vol. 49, no. 2 (June 1897), pp. 207–246; Georg Cantor, *Sur les fondements de la théorie des ensembles transfinis*, traduction de F. Marotte (Paris: A. Hermann, 1899); Arthur Schoenflies, 'Die Entwickelung der Lehre von den Punktmannigfaltigkeiten', *Jahresbericht der Deutschen Mathematiker-Vereinigung*, vol. 8 (1900), pp. 1–250; Bertrand Russell, 'Sur la logique des relations avec des applications à la théorie des series', *Revue de Mathématiques*, vol. 7, nos 2–3 (1901), pp. 115–136, 137–148.]
2. † For references to Peano's articles, cf. my memoir on the "Algebra of Symbolic Logic," in this Journal, vol. XXIII [see p. 117n2].
3. † Cf. Russell, loc. cit.

indispensable for the development of the theory of Cardinal numbers. The abstract nature of the subject makes ordinary language totally ineffective, only gaining precision by verbosity, and imagination is very misleading, since it presents to us special aggregates which are denumerable or of the power of the continuum. Thus we are thrown back onto a strict logical deduction by a symbolical method. I believe that the invention of the Peano and Russell symbolism, used here, forms an epoch in mathematical reasoning.

|368| Section III is entirely due to Russell, and is written by him throughout. It deals with the principles of the subject, and effects a considerable simplification in them. Two classes are defined [cf. ★1·1 (all references are to the series of propositions commencing with Section III and thence numbered consecutively)] as similar when a one–one relation can be established between their numbers. The cardinal number (or power) of a class u is defined simply as the class of those classes similar to u (cf. ★1·3); thus, for instance, the cardinal number 2 is the class of those classes which (in ordinary language) contain two members, and we shall say that a class containing two members is a 2. The finite cardinals are defined [cf. ★1·4 to ★1·61] as those cardinal numbers which can be obtained by induction.[1†] The class of infinite cardinals is defined [cf. ★1·8] as the class of cardinals which are not finite. The whole class of cardinal numbers is denoted by *Nc*. The properties of finite and infinite cardinals are deduced by strict logical reasoning from these definitions. In particular, if the cardinal number of the class of finite cardinals is denoted by α_0, it is proved that every infinite class contains a class whose cardinal number is α_0, [cf. ★2·6], and that no finite class is similar to any part of itself [cf. ★2·75], and that every infinite class is similar to some part of itself [cf. ★2·76]. The cardinal and ordinal theories of finite number are connected in ★3·3.

The known properties of α_0 are proved [★4·1 to ★4·24]. The theorems that if α and β are *any* two cardinals, and α is infinite and $\alpha \geqq \beta$, then[2]

$$\alpha+\beta=\alpha \quad \text{and} \quad \alpha\times\beta=\alpha$$

cannot be deduced from these premises, and, as far as I am aware, have never been proved. Russell points out that the first of them follows from another unproved theorem [★4·3]. The class of numbers for which the first theorem holds is considered (cf. ★30). The second of the theorems cannot be thus deduced; but in Section V, ★20·1, it is proved that if α belong to the class for which the first theorem holds and $\beta \geqq \delta$, then

$$\beta^{\alpha}\times\delta^{\alpha}=\beta^{\alpha}.$$

The similar cases for $\alpha+\beta=\alpha$ are examined in ★30·0 to ★31·3.

Russell also gives exact proofs of such theorems as, $\beta<\gamma$ implies |369| $\alpha+\beta\leqq\alpha+\gamma$, and $\alpha\times\beta\leqq\alpha\times\gamma$, and $\alpha^{\beta}\leqq\alpha^{\gamma}$, and $\beta^{\alpha}\leqq\gamma^{\alpha}$ [cf. ★4·37, ★4·39, ★4·4, ★4·41].

Also ★4·5 contains another proof (assuming ★4·3) of Bernstein's and Schröder's theorem, that $\alpha\geqq\beta$ and $\beta\geqq\alpha$ imply $\alpha=\beta$.

1. †The recognition of the importance of Induction is due to Dedekind, cf. "Was sind and was sollen die Zahlen?" 1887. [Richard Dedekind, *Was sind und was sollen die Zahlen?* (Braunschweig: Friedrich Vieweg und Sohn, 1887).]
2. In the second equation below, a meaningless end-of-line colon has been deleted.

Throughout Sections IV and V, I am indebted to Russell for great improvements in the symbolism of the proofs and for the correction of many oversights. Section IV is devoted to the general consideration of addition and multiplication when the number of summands or of factors is not necessarily finite or ever denumerable. Cantor's definitions only apply to these cases [cf. ★7·0 to ★7·21]. The associative laws are proved [cf. ★8·1 to ★8·31, and ★10·0 to ★10·22] both for addition and multiplication when the number of brackets, as well as the number of numbers in each bracket, is not necessarily finite or denumerable. With these definitions the commutative laws become entirely irrelevant. The ordinary connection between addition and multiplication, exemplified in $\alpha+\alpha=2\times\alpha$, is proved [cf. ★7·3] to hold generally; also the distributive law [cf. ★11·0]. These definitions and theorems require several preliminary definitions; thus for instance [★5·1], that of a "class of classes exclusive," and [★6·0] of the "multiplicative class" of a class of classes exclusive; this last conception assumes great importance in the sequel; also, the general definitions of sums and products depend on the definition [★8·22] of "a class of classes of classes arithmetical." The remainder of Section IV is devoted to theorems following from these definitions.

Section V opens with the definition [★12·1] of a power (β^α); this definition follows from that of a product; and [★14·1] Cantor's definition is deduced from it. Thus, by this procedure, the ordinary relations between addition, multiplication, and powers of numbers are preserved in their full generality. The theory of combinations is then extended to infinite cardinals; thus if α, β are numbers, and $\alpha \geqq \beta$, C^α_β is defined [cf. ★15·1 and ★16·0]: the extension of Vandermonde's theorem is proved [cf. ★16·1]; if α is infinite and such that $\alpha+\alpha=\alpha$, $C^\alpha_\beta=2^\alpha$ [cf. ★16·23]: extensions of the binomial theorem are given in ★17·0 to ★19·7; it is proved [cf. ★17·4] that if α and β are any cardinals,

$$(1+\beta)^\alpha=\sum^{\nu\leqq\alpha} C^\alpha_\nu \times \beta^\nu.$$

If α is infinite and ν finite [cf. ★21·4][1], then $\nu^\alpha=2^\alpha$, and when α is infinite and such that $\alpha^2=\alpha$, then [cf. ★32·4], if $\beta \leqq 2^\alpha$, we have $\beta^\alpha=2^\alpha$.

|370| The idea of the number of permutations (P^α_β) of α things, β at a time, is extended [cf. ★22·0 and ★22·01] to the case when α and β are infinite: then [cf. ★22·1] ${}^\beta P_\beta \times C^\alpha_\beta = P^\alpha_\beta$; also, if α is infinite, and $\alpha \geqq \beta$ and with certain other limitations, then [cf. ★22·2]

$$C^{\alpha\times\beta}_\beta = P^{\alpha\times\beta}_\beta = \alpha^\beta.$$

Hence [cf. ★22·21], if $\alpha\times\beta=\alpha$, we have

$$C^\alpha_\beta = P^\alpha_\beta = \alpha^\beta.$$

In ★30·0 to ★32·5[2] certain classes of numbers which, without assuming ★4·3, satisfy $\beta+\gamma=\beta$ and $\beta\times\gamma=\beta$ when $\beta\geqq\gamma$ are considered.

The Italian system of numeration of propositions, on the left-hand margin, by integral and decimal numbers, is adopted. It is unnecessary for me to emphasise

1. The original erroneously had '★20·4'.
2. The original erroneously had '★32·6'.

further, what is evident from inspection, how largely this article is based upon the ideas of G. Cantor, and of Peano, and of Russell.

Section I.
Account of Peano's Logical Symbolism.

Logical addition and multiplication are denoted by ⌣ and by ⌢ instead[1†] of by + and by × ; when these signs connect propositions, they indicate, as usual, that the propositions are to be taken disjunctively or conjunctively. In this memoir simple juxtaposition of terms will never be used to indicate the logical multiplication of classes; by the multiplication of propositions (which is symbolised by ⌢) will also be indicated by separating them by one dot, thus $a.b$, or two dots, thus $a:b$, or three dots, thus $a\therefore b$, or four dots, thus $a::b$; it is to be noted that ∴ never stands for therefore.

A conjunctive or disjunctive complex of many propositions is itself one proposition. The different arrangements of dots in multiplication are therefore useful in the place of brackets.

The symbol ⊃ between classes means "is contained in," and between propositions means "implies"; it is the same in signification as Schröder's ⋲. Thus if the letters denote propositions,

$$a \supset b . x . y : \supset . u \smile v \therefore \supset . w,$$

|371| is analysable into the hypothesis

$$a \supset b . x . y : \supset . u \smile v,$$

and into the protasis[2] w. That symbol ⊃, in a given chain of symbols, which is preceded or succeeded by a greater number of dots than any other symbol ⊃ in the chain, is the main logical copula of the proposition connecting the hypothesis and the protasis: and the proposition is that the hypothesis implies the protasis.

In the above example, the hypothesis is as follows:

The proposition that the proposition a implies the proposition b, together with the proposition x and y, implies the proposition u or v or both. Thus the hypothesis is itself separable into a hypothesis and protasis. In other examples the same may be true of the protasis. In the above example, if a and b represent classes, the hypothesis becomes,

The proposition that the class a is contained in the class b, together with the propositions x and y, implies the proposition u or v or both.

The classes contained in a given class A are never to be confused with the entities which form the extension of the class A. The fact that x is a member of the class A is denoted by

$$x \in A.$$

1. † For example, cf. my previous memoir "On Symbolic Logic" in this Journal, vol. XXIII, No. 2. [See p. 115.]
2. Whitehead uses the word 'protasis' here in an almost vanishingly rare and idiosyncratic sense. Most would assimilate this word to 'antecedent' or 'hypothesis', but he clearly intends 'conclusion'.

Here $\in$ means "is a member of", or simply "is a"; $\in$ is chosen for this signification as being the initial letter of $\varepsilon\sigma\tau\iota$.[1]

Thus $y \rangle A . x \in A$ may coexist with $x \in y$, but not in general with $x \rangle y$ or with $x = y$, though exceptional cases arise in which $x \in y$ and $x \rangle y$ coexist. As an example, let N be the class "Finite Integer", and Np the class "Prime Number", we have

$$2 \in N . Np \rangle N . 2 \in Np$$

But 2 is a particular number and not a class of numbers; hence the symbolism, $2 \rangle N$, is wrong.

Accordingly, $x \rangle y$ implies either that x and y are both classes or that they are both propositions; and in this memoir single letters are not required to denote propositions.

If x and y are both members of the class A, then $x, y \in A$ is an abbreviation for $x \in A . y \in A$.

There are classes containing only one single member as their extension. If A is such a class, then $\imath A$ is used to denote the single member of the class A. |372| Also, conversely, if x is any entity, ιx is the class of which x is the single member. Hence, in symbols,

$$y \in \iota x . \rangle . y = x \quad \text{and} \quad y = x . \rangle . y \in \iota x,$$

and these two propositions are conveniently combined into

$$y \in \iota x . = . y = x;$$

thus the combination $\in \iota$ between classes is equivalent in signification to =.

For instance, the class of even primes contains the single member 2. If A denote this class, then

$$\iota 2 = A . 2 \in A . 2 = \imath A : x \in A . \rangle . x = 2.$$

We shall have occasion [★11·0] to use the symbol $\iota\iota x$. This signifies the class whose single member is the class whose single member is x.

Again, if u is a class, $u \smile \iota x$ is the class whose members are the members of u together with the entity x. The symbolism $u \smile x$ for this meaning would have been wrong, for $\smile$ can only connect classes to form a new class whose extension is the joint extension of the two classes. Thus if u and x are both classes, the two symbols $u \smile \iota x$ and $u \smile x$ have different meanings; we have

$$x \in u \smile \iota x . x \rangle u \smile x;$$

also $$y \in x . \rangle . y \in u \smile x,$$

but $y \in x$ does not imply $y \in u \smile \iota x$.

The symbol elm, short for "element," stands for that class whose extension is formed by all classes whose extension is some single member. Thus, elm is a class whose members are themselves classes, but each containing only one member. Thus,

$$A \in \text{elm} . \rangle : x \in A . y \in A . \rangle . x = y : \rangle . \imath A \in A.$$

1. Throughout, we use the now standard symbol for set membership as given in the previous line; this is a stylised version of the Greek epsilon. The traditional epsilon was used in Whitehead's original publication. Reflected traditional epsilons have likewise been replaced with the stylised form.

The symbol cls,[1] short for class, stands for that class whose extension is formed by all classes. Thus $x \in \mathrm{cls}$ simply means x is a class. We get here an exception to the rule that $x \in A$ is inconsistent with $x \supset A$, for cls is itself a class, hence,[2]

$$\mathrm{cls} \in \mathrm{cls} . \mathrm{cls} = \mathrm{cls}.$$

If p_x denotes a proposition containing a letter x which may be given various significations consistent with the truth of the proposition, then

$$x \ni (p_x)$$

|373| means the class whose extension is all the entities x consistent with the truth of p_x. Thus $\ni$ may be read "such that." Thus $x \ni (x^2 = 1)$ is the class whose extension is formed by ± 1, and $x \ni (x = 2)$ is the class $\iota 2$. The symbol $x \ni (p_x)$ is always to be treated as one whole: thus, if A is a class, $A \frown x \ni (p_x)$ means the class common to A and to the class $x \ni (p_x)$. Thus,

$$A \frown x \ni (p_x) . = . x \ni (x \in A . p_x).$$

This symbol is especially useful in definitions, thus, for ιx, we have

$$\iota x = y \ni (y = x),$$

and for elm we have

$$\mathrm{elm} = \mathrm{cls} \frown x \ni (y, z \in x . \supset . y = z).$$

If p and q are propositions containing a variable x, then

$$p . \supset_x . q$$

means that p implies q for every value of x satisfying p: this subscript to $\supset$ is not required for the main implication of a proposition, and can sometimes be omitted when it is strictly required. Similarly, if p and q are propositions containing the variables x and y, then

$$p . \supset_{x,y} . q$$

means that p implies q for every value of x and y satisfying p.

Λ is the symbol for the null class: thus,

$$A \in \mathrm{cls} . \supset_A . \Lambda \supset A.$$

If p is a proposition, $p = \Lambda$ is the proposition which is the denial of the proposition p.

If A is a class, $\exists A$ is[3] the proposition that "A is not Λ": thus,

$$A \in \mathrm{cls} . \supset : \exists A . = . (A = \Lambda) = \Lambda.$$

The legitimacy of $x \in A$ implies $\exists A$; thus, from the definition $u = x \ni (p_x)$, we cannot derive the legitimacy of some proposition $x \in u$ without the premise $\exists u$.

If A is a class, $-A$ is the class of all entities which are not A; if A is a proposition, $-A$ is the proposition which is the denial of A; also $-\in$ means "is not a member of", and $-\supset$ means "is not contained in"; thus,

$$A \in \mathrm{cls} . \supset . -A = x \ni (x -\in A) = x \ni [(x \in A) = \Lambda] = x \ni [-(x \in A)],$$

and

$$A, B \in \mathrm{cls} . A - \supset B . \supset . \exists A - B.$$

1. Uppercase instances of 'Cls' in this paragraph and the line below have been changed to lowercase 'cls' to give consistent usage in this article.
2. Regarding the assertion below, see the discussion of Russell's paradox in the introduction to this article.
3. '∃' was italicised in original throughout.

|374| The symbol $\frown$ between classes is replaced in this memoir by simple juxtaposition in the sole case when $\frown$ immediately precedes the symbol $\bar{}$; thus,

$$A,\bar{}B \in \text{cls} \,.\supset\!\bar{:}\, A \;.\; B = A \frown \;\; B$$

Also thus, if u is a class and $x \in u$, then $u - \iota x$ is the class u with the exception of the entity x. Note that $u - x$ for this meaning is wrong.

If A is a class, cls'A means the class whose members are the classes contained in A; thus,

$$A \in \text{cls}.\supset.\text{cls'}A = \text{cls} \frown x \ni (x \supset A);$$

we might also have written

$$A \in \text{cls}.\supset.\text{cls'}A = x \ni (x \supset A);$$

leaving it to be inferred from the use of $\supset$ in $x \supset A$, that $x \in \text{cls}$; thus,

$$\text{cls'cls} = \text{cls} \frown x \ni (x \supset \text{cls});$$

accordingly, cls'cls means the complete class of those classes whose members are themselves classes; thus,

$$A \in \text{cls'cls}.x \in A.\supset.x \in \text{cls}.$$

Similarly, $$\text{cls'cls'cls} = \text{cls} \frown x \ni (x \supset \text{cls'cls}),$$

and thus $$A \in \text{cls'cls'cls}.x \in A.\supset.x \in \text{cls'cls}.$$

These conceptions are essential in the general consideration of cardinal numbers. We shall abbreviate cls'cls into cls^2 and cls'cls'cls into cls^3.

We may note that

$$\text{cls}^2 \in \text{cls} \quad \text{and} \quad \text{cls}^2 \supset \text{cls}$$

also that[1]

$$\text{cls}^3 \in \text{cls}^2, \text{ and } \text{cls}^3 \in \text{cls}, \text{ and } \text{cls}^3 \supset \text{cls}^2, \text{ and } \text{cls}^3 \supset \text{cls},$$

$$A \in \text{cls'cls}.\supset.\smile\text{‘}A = x \ni \left[\exists A \frown P \ni (x \in P)\right],$$

$$A \in \text{cls'cls}.\supset.\smile\text{‘}A = x \ni \left[P \ni A.\supset_P.\, x \in P\right].$$

The class $\smile$‘A is thus the logical sum of the classes which form the extension of A; thus, if the extension of A consists of only two classes B and C, then $\smile\text{‘}A = B \smile C$. The class $\frown$‘A is the logical product of the classes which form the extension of A; thus, if the extension of A consists only of the two classes, B and C, then $\frown\text{‘}A = B \frown C$.

We shall, following Russell, employ frequently the following notation: $x \in^2 A$ means that A is a cls'cls, and $x \in B$ where $B \in A$: thus in symbols,

$$A \in \text{cls'cls}.\supset: x \in^2 A. = .\exists A \frown B \ni (x \in B).$$

|375| Note that $\exists A \frown B \ni (x \in B)$ affirms the existence of a class of B's such that $B \in A . x \in B$; it is obvious that to affirm the existence of a class of entities is the same as affirming the existence of at least one of them.

But this use of $\in^2$ brings us to Russell's "Logic of Relations", which we will now proceed to explain, shortening the explanations by the use of Peano's symbolism and numbering the definitions and propositions for reference. Df. (short for

1. In the third line below, the original erroneously had a lowercase subscript 'p'. It has been changed to uppercase.

Definition) will be written after a definition, and Pp.[1] (short for Primitive proposition) after a proposition assumed without proof (cf. ★4·3 of Section III).

Section II.
Russell's "Logic of Relations."

Russell's[2†] "Logic of Relations" may conveniently be considered as a generalization of the ideas suggested by Peano's proposition $x \in A$. This proposition asserts a special relation between x and A, namely, that A is a class and x is a member of its extension. Thus, according to Russell, if R is any relation, xRy signifies that x has the relation R to y. Let rel[3] denote the complete class whose members are relations, then

1·00 $R \in \text{rel.} \supset : xRy. = . x$ has the relation R to y. Df.

Relations will always be denoted by capital letters to which there is a corresponding small Greek letter, and this Greek letter will have a fixed meaning, explained in 1·01, in relation to the corresponding capital italic letter. Small Greek letters will also be used in this memoir to represent cardinal numbers, but not those letters whose corresponding capital italics have been used for relations. The letters R, S will be confined to relations.

1·01 $R \in \text{rel.} \supset . \rho = x \ni \left[\exists y \ni (xRy) \right],$ Df.

1·02 $R \in \text{rel.} \supset . \breve{\rho} = y \ni \left[\exists x \ni (xRy) \right].$ Df.

Here $\breve{\rho}$ is read as "ρ converse". Then ρ is the class of entities such that $x \in \rho$ implies that a class of entities (and thus at least one entity) such as y exists, such that x has the relation R to y. Similarly for $\breve{\rho}$. Here ρ is called the domain of R and $\breve{\rho}$ the converse domain of R.

|376|

1·03 $R \in \text{rel.} x \in \rho . \supset . \breve{\rho}x = y \ni (xRy),$ Df.

1·04 $R \in \text{rel.} y \in \breve{\rho} . \supset . \rho y = x \ni (xRy).$ Df.

Here $\breve{\rho}x$ and ρy are each to be considered as one symbol, and x and y are of the nature of indices.

1·05 $u \in \text{cls.} R \in \text{rel.} u \supset \rho . \supset . \breve{\rho}u = y \ni \left[\exists u \frown x \ni (xRy) \right],$ Df.

1·06 $u \in \text{cls.} R \in \text{rel.} u \supset \breve{\rho} . \supset . \rho u = x \ni \left[\exists u \frown y \ni (xRy) \right],$ Df.

1·10 $R, R' \in \text{rel.} \supset \therefore R \supset R' \quad = : xRy . \supset_{x.y} . xR'y.$ Df.

This defines the meaning of "the relation R is contained in the relation R'"; thus, relations have some properties analogous to those of classes.

1·11 $R, R' \in \text{rel.} \supset \therefore R = R' . = : R \supset R' . R' \supset R.$ Df.

---

1. The original erroneously omitted a period after 'Pp'.
2. † Cf. *Revue de Mathématiques* (Turin), loc. cit.
3. 'Rel.' here and in the line below has been changed to lowercase for consistency with the rest of the article. The period has been removed in this instance because the period is not serving as a delimiter.

This defines the equivalence of relations.

1·12 $R \in \mathrm{rel.} \supset . \breve{R} = \iota\ \mathrm{rel} \frown R' \ni (xR'y . = . yRx)$, Df.

$\breve{R}$ is the converse relation to R.

1·13 $R \in \mathrm{rel.} \supset . \breve{\breve{R}} = R$,

1·20 $R_1, R_2 \in \mathrm{rel.} \supset : x(R_1 \smile R_2)y . = . xR_1y . \smile . xR_2y$. Df.

This defines the "logical sum" of two relations; the logical sum $R_1 \smile R_2$ is itself a relation.

1·21 $R_1, R_2 \in \mathrm{rel.} \supset : x(R_1 \frown R_2)y . = . xR_1y . xR_2y$. Df.

This defines the "logical product" of two relations; the logical product $R_1 \frown R_2$ is itself a relation.

1·22 $R_1 . R_2 \in \mathrm{rel.} \supset : xR_1R_2z . = . \exists y \ni (xR_1y . yR_2z)$. Df.

This defines the "relational product" of two relations; the relational product. R_1R_2 is itself a relation.

1·23 $K \in \mathrm{cls'rel.} \supset : x \smile 'Ky . = . \exists K \frown R \ni (xRy)$. Df.

$\smile 'K$ is a relation; it is the "logical sum" of the relations forming the extension of the class K; this definition agrees with 1·20 when the extension of K is formed by two relations only.

1·30 $R \in \mathrm{rel.} \supset . R^2 = RR$, Df.

1·31 $R, S \in \mathrm{rel.} \supset . {}^{\cup}(RS) = \breve{S}\breve{R}$, Df.

1·40 $R \in \mathrm{rel.} \supset : x - Ry . = . -(xRy)$. Df.

|377| Thus $-R$ is the relation of not having the relation R.

2·01 $\in \in \mathrm{rel}$, ($\in$ is a relation).

2·02 $x \in \breve{\in}\, y . = . \exists z \ni (x \in z . y \in z)$,

2·03 $x\, \breve{\in} \in y . = . \exists z \ni (z \in x . z \in y) . = . \exists xy$.

2·04 $y \in \mathrm{cls'cls.} \supset . \smile 'y = x \ni (x \in^2 y)$,

2·11 $1' = \mathrm{identity}$ Df.

1' is the relation of identity; thus, $x1'x$; and $a \in \mathrm{elm}. x, y \in A. \supset . x1'y$.

2·12 $0' = -1'$. Df.

Thus 0' is the relation of diversity, that is, of "being different from".

2·13[1] $R, P \in \mathrm{rel.} \supset : R\breve{P} \supset 0' . = . P\breve{R} \supset 0'$.

3·01 $Nc \to 1 . = . \mathrm{rel} \frown R \ni \{xRy . xRz . \supset_x . y1'z\}$. Df.

$Nc \to 1$ is the symbol for a certain class of relations, namely, the class of "many-one" relations; any x in the domain of R has the relation R to only one member of the converse domain.

3·02[2] $1 \to Nc . = . \mathrm{rel} \frown R \ni \{yRx . zRx . \supset_x . y1'z\}$. Df.

---

1. The original erroneously had $\breve{R}P$ instead of $P\breve{R}$.
2. The original erroneously did not subscript the third 'x'.

Thus $1 \to Nc$ is the class of one-many relations

3·03 $$R \in Nc \to 1 . = . \breve{R} \in 1 \to Nc.$$

3·04 $$R \in 1 \to Nc . \supset . R\breve{R} \supset 1',$$

that is, $$xRy . y\breve{R}z . \supset . x1'z.$$

3·11 $$1 \to 1 . = . (Nc \to 1) \frown (1 \to Nc).$$ Df.[1]

$1 \to 1$ is the class of one-one relations.

3·12 $$R \in 1 \to 1 . \supset . \breve{R} \in 1 \to 1.$$

3·13 $$R, S \in 1 \to 1 . \supset . RS \in 1 \to 1,$$

4·11 $$S \in Nc \to 1 . R = S\breve{S} . \supset . R^2 \supset R . R = \breve{R}.$$

4·12 $$R \in \mathrm{rel} . R^2 \supset R . R = \breve{R} . \exists R . \supset . \exists Nc \to 1 \frown S \ni (R = S\breve{S}).$$

Note that $$R^2 \supset R . = : xRy . yRz . \supset . xRz,$$

and that $$R = \breve{R} . = : xRy . = . yRx.$$

5·11 $$R \in \mathrm{rel} . u \supset \rho . \supset : xR_u y . = . x \in u . xRy.$$ Df.

5·12 $$R \in \mathrm{rel} . u \supset \rho . \supset . \rho_u = u . \breve{\rho}_u = \breve{\rho} u.$$

|378| R_u is the relation R with its field limited, namely, the domain is limited to u, and thus the converse domain is limited to $\breve{\rho} u$.

For the rest of this memoir the propositions are numbered progressively onwards, and all references to propositions without indication of section refer to the series of numbers. In the future it will occur that a general hypothesis applies to a number of propositions: such a hypothesis will be stated only once, after mention of the integral number belonging to that set of propositions, and will be understood to hold till the integral number is changed. Demonstrations are placed in square brackets after the enunciations; subsidiary propositions used only in the demonstrations are numbered on the right instead of on the left, and only integral numbers in brackets are used for such numeration; thus, (2) would refer to a subsidiary proposition used in that demonstration and ★3·11 would refer to a proposition in the main body of the article; thus, (1) . (2). ★3·11 . $\supset$. means that "the subsidiary propositions (1) and (2), together with the main proposition 3·11 imply" the protasis. The hypothesis of the proposition to be proved is cited by "Hp"; thus, Hp . (1) means "the hypothesis together with the subsidiary proposition (1)"; but Hp(1) means the hypothesis of the subsidiary proposition (1) and Hp(3·11) the hypothesis of the previous proposition ★3·11.

1. The original erroneously omitted the 'Df.'.

Section III.
On Finite and Infinite Cardinal Numbers.

This section III is entirely due to Russell, with the exception of some of the notes.

★1

·1 $$u \operatorname{sim} v . = . \ u, v \in \text{cls} . \exists 1 \rightarrow 1 \frown R \ni \left(u \supset \rho . v = \breve{\rho} u\right).$$ Df.

Note: sim is contracted from similar, and $u \operatorname{sim} v$ can be read: u is similar to v. Thus sim is a symbol for a relation, and has all the properties ascribed to such symbols in Section II.

·2 $$u \in \text{cls} . \supset . \mu u = \text{cls} \frown v \ni \left(u \operatorname{sim} v\right).$$ Df.

·3 $$Nc = \text{cls}^2 \frown z \ni \left\{\exists \text{cls} \frown u \ni \left(z = \mu u\right)\right\}$$ Df.

·4 $$0 \ = \iota \Lambda.$$ Df.

·41 $$1 \ = \text{cls} \frown u \ni \left(x \in u . \supset . \ u - \iota x \in 0\right).$$ Df.

|379| Note: 1 has the same signification as elm (cf. Section I) and will now replace that symbol.

★1·42 $2 = \text{cls} \frown u \ni \left(x \in u . \supset . u - \iota x \in 1\right).$ Df.

·43 $3 = \text{cls} \frown u \ni \left(x \in u . \supset . u - \iota x \in 2\right).$ Df.

·5[1] $n \in Nc . \supset . n + 1 = \text{cls} \frown u \in \left(x \in u . \supset . u - \iota x \in n\right).$ Df.

·6 $Nc \, \text{fin} = Nc \frown n \ni \left\{s \in \text{cls} . 0 \in s : m \in Nc \frown s . \supset_m . m + 1 \in s : \supset . n \in s\right\}.$ Df.

·61 $\text{Induct} . = .\therefore s \in \text{cls} . 0 \in s : m \in Nc \frown s . \supset_m . m + 1 \in s : \supset_s . Nc \, \text{fin} \supset s$ Df.

Note: Induct is contracted from Induction.

·7 $0 \in Nc$ fin.

·71 $1 \in Nc$ fin, $\left[1 = 0 + 1 . 0 \in Nc \text{ fin} . \supset . \text{Prop}\right].$

·72 $2 \in Nc$ fin.

·73 $n \in Nc \text{ fin} . \supset . n + 1 \in Nc$ fin, [Induct].

·8 Nc infin $= Nc - Nc$ fin. Df.

1.81 cls fin $= \smile ‘Nc$ fin. Df.

·82 cls infin $= \smile ‘Nc$ infin. Df.

·83 cls $=$ cls fin $\smile$ cls infin.

·84 cls fin $\frown$ cls infin $= \Lambda$.

·91 $u \operatorname{sim} u' . \ v \operatorname{sim} v' . \ u \frown v = \Lambda . \ u' \frown v' = \Lambda . \supset . \ u \smile v \operatorname{sim} u' \smile v',$

$[S, S' \in 1 \rightarrow 1 . u = \sigma . u' = \breve{\sigma} . v = \sigma' . v' = \breve{\sigma}' . S \smile S' = S'' . \supset .$

$S'' \in 1 \rightarrow 1 . \sigma'' \in u \smile v . \breve{\sigma}'' = u' \smile v' . \supset . \text{Prop}].$

·92 $\alpha, \beta \in Nc . \supset . \alpha + \beta = \imath Nc \frown z \ni \left\{u \in \alpha . \ v \in \beta . \ u \frown v = \Lambda . \supset . u \smile v \in z\right\}.$ Df.

·93 $m \in Nc . \ n \in Nc \text{ fin} . \supset . \ m + \left(n + 1\right) = \imath Nc \frown z \ni \{u \in m . \ v \in n .$

$u \frown v = \Lambda . \ x - \in u \smile v . \ v \smile \iota x = v' . \supset . \ u \smile v' \in z\}.$

1. The original erroneously had a minus sign instead of an equality sign.

·94 $(m+n)+1 = ℩Nc \frown z \ni \{w \in m+n \,.\, x - \in w \,.\supset.\, w \smile \iota x \in z\}$.

·95 $m+(n+1) = (m+n)+1$.

★2

·1 $n \in Nc\ \mathrm{fin} .\supset. n- = n+1$,

$[0- = 1 \ldots (1)$ $\quad n- = n+1 .=: u \in n \,.\, x - \in u \,.\supset.\, u \smile \iota x - \mathrm{sim}\ u :$

$=: u \smile \iota x \in n+1 \,.\, y - \in u \smile \iota x .\supset.$

$u \smile \iota x \smile \iota y - \mathrm{sim}\ u \smile \iota x :$

$=: n+1- = (n+1)+1$, (2)

(1) . (2) . Induct $.\supset.$ Prop].

·2 $R = ℩1 \rightarrow 1 \frown R' \ni (\rho' = Nc\ \mathrm{fin} : xR'y .=.\, x+1 = y)$. Df.

|380| Note: This special signification for R, and thus for ρ and $\breve{\rho}$, is only retained throughout Section III.

★2·3 $R \supset 0$', $\quad$ [★ 2·1$.\supset.$ Prop].

·31 $\breve{\rho} \supset \rho \,.\, \exists \rho - \breve{\rho}$, $[\rho = \rho - \iota 0]$.

·32 Induct $.=: s \in \mathrm{cls} \,.\, 0 \in s \,.\, \breve{\rho}(s \frown \rho) \supset s .\supset_s. \rho \supset s$.

·4 $\alpha_0 = \mathrm{cls} \frown s \ni \{\exists 1 \rightarrow 1 \frown S \ni (\breve{\sigma} \supset \sigma .\, \exists \sigma - \breve{\sigma} : s \supset \sigma .\, \breve{\sigma} s \supset s . \exists s - \breve{\sigma} s .\supset.\, s = \sigma)\}$. Df.

·41 $\alpha_0 \in Nc$, $\quad$ [cf. Russell, Révue de Math., loc. cit., §3, Prop. 1·9].

·42 $Nc\ \mathrm{fin} \in \alpha_0$, $\quad$ [★2·31.★2·32 $.\supset.$Prop].

·43 $\alpha_0 \in Nc$ infin, $\quad$ $[\alpha_0 = \mu Nc\ \mathrm{fin} \,.\, Nc\ \mathrm{fin} = \rho \,.\, \mu\rho = \mu\breve{\rho} = \mu\breve{\rho}+1$. ★2·1$.\supset.$ Prop].

·5[1] $n \in Nc - \iota 0 .\supset. n-1 = ℩Nc \frown x \ni (x+1 = n)$. Df.

·51 $n \in Nc$ infin$.\supset.n-1 \in Nc$ infin, $\quad$ [★1·8. ★1·73$.\supset.$Prop].

·6 $n \in Nc$ infin $.\, u \in n .\supset. \exists \alpha_0 \frown \mathrm{cls}$'$u$,

$[u \in n \,.\, x \in u$. ★2·51$.\supset.$ $u - \iota x \in$ cls infin, (1)

$m \in Nc$ fin . $u_m \in m \,.\, u_m \supset u \,.\, u - u_m \in$ cls infin . $x \in u - u_m .\supset.\, u - u_m - \iota x \in$ cls infin, (2)

(1) . (2) . Induct$.\supset: m \in Nc$ fin . $u_m \in m \,.\, u_m \supset u .\supset.\, u - u_m \in$ cls infin, (3)

$K =$ cls'cls $\frown v \ni \{v \supset$ cls'$u \frown$ cls fin $:: m \in Nc$ fin $.\supset_m \therefore v \frown m \in 1 : p, p' \in v$.

$p \in m \,.\, p' \in m+1 .\supset.\, p \supset p'\} .\, v \in K .\supset. \smile$'$v \in \alpha_0 \,.\smile$'$v \supset u$. (4)

(3)$\supset \exists K \ldots (5)$, $\quad$ (4) . (5) $.\supset.$ Prop].

·7 $u \in$ cls infin . $v \in$ cls fin . $v \supset u .\supset.\, u$ sim $u - v$,

[★2·6 $.\supset.\, \exists \alpha_0 \frown$ cls'u, (1)

$w \in \alpha_0 \frown$ cls'u . $v' \in$ cls fin . $v' \supset w .\supset.\, w$ sim $w - v' .\supset.\, u$ sim $u - v'$, (2)

v' sim $v .\supset. u - v'$ sim $u - v \ldots (3)$, $\quad$ (2) . (3) $.\supset.$ Prop].

---

1. The original erroneously had 'ιo' instead of '$\iota 0$'.

·71	$\alpha \in Nc$ infin . $n \in Nc$ fin .›. $\alpha - n = \alpha = \alpha + n$,	[★2·7.›.Prop].
·72	$m, n \in Nc$ fin.›. $m + n - = m$,	[★2·4.›.Prop].
·73	$m + n \in Nc$ fin .›. $m, n \in Nc$ fin.	
·74	$m, n \in Nc$.›: $m + n = m$.=. $m \in Nc$ infin.	
·75	Nc fin $= Nc \frown n \ni (n - = n + 1)$,	[★2·1 . ★2·71 . Transp .›. Prop].
·76	Nc infin $= Nc \frown \alpha \ni (\alpha = \alpha - 1)$,	[★2·1 . Transp . ★2·71 .›. Prop].

The Props ★2·75 and ★2·76 show that from the Df of finite numbers by induction we can deduce the properties of infinity with regard to similarity of whole and part.

★3

·1 $\alpha, \beta \in Nc$.›: $\alpha < \beta$.=. $\alpha \frown \beta = \Lambda$. $\exists \alpha \frown u \ni \{\exists \beta \frown v \ni (u \rangle v)\}$. Df.

·12 $\alpha, \beta \in Nc$.›: $\alpha > \beta$.=. $\beta < \alpha$, Df.

·13[1] $\alpha, \beta \in Nc$.›: $\alpha \leqq \beta$.=. $\exists \alpha \frown u \ni \{\exists \beta \frown v \ni (u \rangle v)\}$. Df.

|381|

★3·2 $\alpha, \beta \in Nc$. $\alpha < \beta$. $v \in \beta$.›. $\exists \alpha \frown$ cls'v

[$u_0 \in \alpha$. $v_0 \in \beta$. $u_0 \rangle v_0$. $v \in \beta$.›. $\exists 1 \to 1 \frown S \ni (\sigma = v_0$. $\breve{\sigma} = v$. $\breve{\sigma} u_0 \in \alpha \frown$ cls'$v)$].

·3 $m, n \in Nc$ fin .›: $m \leqq n$.=. $mR^{n-m}n$, [cf. ★2·2],

[$a, b \in Nc$ fin . aRb .›. $\exists a \frown u \ni (x - \in u$.›. $u \smile \iota x \in b)$.›: $v \in b$.›$_v$. $\exists a \frown$ cls'v, (1)

$a, b, c \in Nc$ fin . $aR^c b$. $v \in b$. $\exists a \frown$ cls'v .›: $x - \in v$.›. $\exists a \frown$ cls'$(v \smile \iota x)$. (2)

(1) . (2) . Induct . ›: $a, b, c \in Nc$ fin . $aR^c b$. $v \in b$.›. $\exists a \frown$ cls'v, (3)

aRb .›. $a = b - 1$. $1 = b - a$. (4)

$aR^c b$.›. $c = b - a$: ›: $aR^{c'}(b+1)$.›. $c' = b + 1 - a$, (5)

(4) . (5) . Induct .›: $aR^c b$.›.$c = b - a$, (6)

(3) . (6) . ›: $a, b \in Nc$ fin . $aR^{b-a}b$.›. $a \leqq b$, (7)[2]

$a = b$.›. $aR^0 b$.›. $aR^{b-a}b$, (8)

$a < b$. $u \in a$. $v \in b$. $u \rangle v$. $v - u \in c$.›. $aR^c b$. $c = b - a$, (9)

(7) . (8) . (9) .›. Prop].

Note: This prop. connects the cardinal and ordinal theories of finite number.

★4

·1 $\alpha \in Nc$ infin . ›. $\alpha \geqq \alpha_0$. [★2·6 .›. Prop].

· 2 $\alpha_0 = \alpha_0 + \alpha_0$,

[$R' = \imath$Rel $\frown P \ni \{xPy$.=. $\exists Nc$ fin $\frown n \ni (xR^{2n}y)\}$.

$\rho_0 = \breve{\rho}'0$. $\rho_1 = \breve{\rho}'1$.›. $\rho_0 \frown \rho_1 = \Lambda$. $\rho_0 \smile \rho_1 = \rho$. $\rho_0, \rho_1 \in \alpha_0$.›. Prop].

1. The original erroneously repeated '.12'.
2. The original erroneously had 'a, b, c' instead of 'a, b'.

·21 $n \in Nc$ fin .›. $\alpha_0 = \alpha_0 \times n$, [★4·2.Induct. ›.Prop].

·22 $\alpha_0 = \alpha_0 \times \alpha_0$,

$[M = x \ni \{\exists Nc \text{ fin} \frown (m,n) \ni (x = \iota m \smile \iota n)\}$.›. $\mu M = \alpha_0 \times \alpha_0$, (1)

$R' \in 1 \to 1$. $\rho' = M$ ∴ $\iota m \smile \iota n R' \iota m' \smile \iota n'$. =: $m > n$. $m - n > 1$.

$m' + n' = m + n . m' = m - 1 . \smile . m \geqq n . m - n \leqq 1$.

$m' + n' = m + n + 1$. $n' = 0$ ∴›. $\rho' \in \alpha_0$.›. Prop].

·23 $n \in Nc$ fin .›. $\alpha_0 = \alpha_0^n$, [Induct].

·24[1] $u, v \in$ cls'Nc fin .›. $\sum^{x \in u . y \in u} x \times \alpha_0^y = \alpha_0$.

·3[2] $u \in$ cls infin.›.∃ cls'α_0 excl $\frown v \ni (\smile$'$v = u)$. Pp.

(For definition of cls'α_0 excl, cf ★5·1 later.)

Note: It has been shown that every infinite class contains classes whose number is α_0. Thus if a class v of α_0's has been found which contains all but a finite number of u's, these may be added on to the v's; but if an infinite number of u's remain, new α_0's may be found out of them, and to this process there can |382| be no end as long as any u's remain. It appears not impossible that a theorem may be discovered proving that this process can always be completed. In any case, it is much to be desired that the possibility in all cases of this completion of the above process should be either proved or disproved. In order to show the importance of this investigation, the theorem is here hypothetically assumed stated as a primitive proposition (★4·3), and some immediate deductions (★4·31 to ★4·51) are made from it. The subject is resumed in the sets of propositions ★30, ★31, and ★32.

$\alpha \in Nc$ infin . ★4·3 .›:

★4·31 $\exists Nc \frown \beta \ni (\alpha = \alpha_0 \beta)$ [★4·3.›. Prop].

·32 $\alpha + \alpha = \alpha$, $[\alpha = \alpha_0 \beta = (\alpha_0 + \alpha_0)\beta = \alpha + \alpha]$.

·33 $\alpha\alpha_0 = \alpha$, $[\alpha = \alpha_0 \beta = \alpha_0^2 \beta = \alpha\alpha_0]$.

·34 $\alpha \in Nc \frown \beta \ni (\alpha = \alpha_0 \beta)$, [★4·33.›. Prop].

·35 $\alpha > \alpha_0 . \beta < \alpha$.›. $\alpha_0 \beta < \alpha$,

$[\beta \in Nc$ fin .›. $\alpha_0 \beta = \alpha_0$.›. $\alpha_0 \beta < \alpha$, (1)

$\beta \in Nc$ infin .›. $\alpha_0 \beta = \beta$.›. $\alpha_0 \beta < \alpha$, (2)

(1) . (2) .›. Prop].

·36 $\alpha > \alpha_0$.›. $Nc \frown \beta \ni (\alpha = \alpha_0 \beta) = \iota\alpha$, [★4·35 .›. Prop].

·37 $\beta < \gamma$.›. $\alpha + \beta \leqq \alpha + \gamma$,

$[u \in \alpha . v \in \beta . w \in \gamma . v › w . u \frown w = \Lambda$.›. $u \smile v › u \smile w$.›. Prop].

·38 $\beta < \alpha$.›. $\alpha + \beta = \alpha$, $[\alpha \leqq \alpha + \beta \leqq \alpha + \alpha . \alpha = \alpha + \alpha$.›. Prop].

1. The original erroneously had 'α^y' instead of 'α_0^y'.
2. An errant left parenthesis at the start of this line has been moved to its intended place in the line below.

·39 $\beta<\gamma .\supset. \alpha\beta \leqq \alpha\gamma,$

$[v \in \beta . w \in \gamma . v \supset w . w \supset \alpha .\supset. \smile`v \supset \smile`w .\supset. \text{Prop}].$

·4 $\beta<\gamma .\supset. \alpha^{\beta} \leqq \alpha^{\gamma},$

$[u \in \alpha . v \in \beta . w \in \gamma . v \supset w .\supset. \alpha^{\beta} = \mu Nc \rightarrow 1 \frown S \ni (\sigma = v . \breve{\sigma} \supset u) .\supset.$

$\alpha^{\gamma} = \mu Nc \rightarrow 1 \frown S \ni (\sigma = w . \breve{\sigma} \supset u) .\supset. \text{Prop}].$

Note: Here Cantor's definition of the power of a cardinal number (e. g. α^{β}) is used, cf. ★14·1 of Section V.

·41 $\beta<\gamma .\supset. \beta^{\alpha} \leqq \gamma^{\alpha},$

$[u \in \alpha . v \in \beta . w \in \gamma . v \supset w .\supset. Nc \rightarrow 1 \frown S \ni (\sigma = u . \breve{\sigma} \supset v) \in \beta^{\alpha}.$

$Nc \rightarrow 1 \frown S \ni (\sigma = u . \breve{\sigma} \supset w) \in \gamma^{\alpha},$ (1)

$(1) . Nc \rightarrow 1 \frown S \ni (\sigma = u . \breve{\sigma} \supset v) \supset Nc \rightarrow 1 \frown S \ni (\sigma = u . \breve{\sigma} \supset w) .\supset. \text{Prop}].$

·42 $n \in Nc \text{ fin} .\supset. n\alpha = \alpha . \alpha\alpha_0^n = \alpha.$

·43 $\beta, \gamma \in Nc . \beta \leqq \alpha .\supset. \alpha^{\gamma} = (\alpha+\beta)^{\gamma}.$

·44 $\beta \in Nc . \gamma \in Nc \text{ infin} . \beta \leqq \alpha .\supset. \alpha^{\gamma}\beta^{\gamma} = \alpha^{\gamma}.$

|383| [This theorem is enunciated and proved in Section V, ★20·1; it is mentioned here for the sake of completeness.]

★4·5 $u, v \in \text{cls} . \exists \mu u \frown \text{cls}`v . \exists \mu v \frown \text{cls}`u .\supset. u \text{ sim } v.$

Note: This is Schröder's and Bernstein's theorem; cf. Borel, "Théorie des Fonctions," Paris, 1898, p. 108 ff.[1]

$[S \in 1 \rightarrow 1 . \sigma = u . \breve{\sigma} \supset v . S' \in 1 \rightarrow 1 . \sigma' = v . \breve{\sigma}' \supset u .\supset. \sigma(\sigma') = u . \breve{\sigma}'(\breve{\sigma}) \supset \breve{\sigma}' \supset u,$ (1)

$(1) . \mu u = \alpha . \mu v = \beta . \mu u - \breve{\sigma}' = \gamma . \mu\breve{\sigma}' - \breve{\sigma}'(\breve{\sigma}) = \delta .\supset.$

$\alpha \geqq \beta . \beta \geqq \alpha . \alpha = \beta+\gamma . \beta = \alpha+\delta .\supset. \alpha = \alpha+\gamma+\delta,$ (2)

$(2) . ★4{\cdot}38 .\supset. \gamma+\delta \leqq \alpha .\supset. \delta \leqq \alpha .\supset. \alpha+\delta = \alpha .\supset. \beta = \alpha .\supset. \text{Prop}].$

Note: Bernstein's proof of Schröder's and Bernstein's theorem given by Borel (loc. cit.) does not depend upon ★4·3.

·51 $\alpha, \beta \in Nc . \alpha > \beta .\supset. -(\beta \geqq \alpha),$ $[★4{\cdot}5 .\supset. \text{Prop}].$

Section IV.

★5

·1 $\text{cls}^2 \text{excl} .=. \text{cls}^2 \frown d \ni \{q, q' \varepsilon d .\supset. q = q' . \smile . q \frown q' = \Lambda\}.$ Df.

Note: excl is contracted from "exclusive".

·2 $a \in \text{cls}^2 .\supset. \text{cls}`a \text{ excl} = \text{cls}`a \frown \text{cls}^2 \text{ excl}.$ Df.

---

1. Whitehead misspelled the original reference as 'Théorie des Functions'. See Emile Borel, *Leçons sur la Théorie des Fonctions* (Paris: Gauthier-Villars et fils, 1898).

★6 $d \in \text{cls}^2\, \text{excl} .\supset\therefore$

·0 $d^\times = \text{cls} \cap m \ni \{p \in d .\supset_p. p \cap m \in 1 : m \supset \cup\text{'}d\}$. Df.

Note: $d^\times$ is the "multiplicative" class of d, and can be read as "d multiplicative": it has no existence, according to this definition, unless d is a cls^2 excl.

·11 $p \in d . x \in p . d_1 = d - \iota p . m_1 \in d_1^\times . m = m_1 \cup \iota x .\supset. m \in d^\times$,

$[\text{Hp} . p' \in d . p' -= p .\supset. p' \in d_1 . x -\in p' .\supset. m \cap p' = m_1 \cap p' .\supset. m \cap p' \in 1.$ (1)

$\text{Hp} .\supset. m \cap p = \iota x .\supset. m \cap p \in 1.$ (2)

$(1) . (2) .\supset: p'' \in d .\supset. p'' \cap m \in 1 \ldots . (3)$ $m \supset \cup\text{'}d,$ (4)

$(3) . (4) .\supset. \text{Prop}].$

·12 $u \supset \cup\text{'}d : x, y \in u . x0\text{'}y .\supset_{x,y}. -\exists d \cap p \ni (x, y \in p) :\supset. \exists d^\times \cap m \ni (u \supset m),$

$[\text{Hp} . p \in d . \exists p \cap u .\supset. p \cap u \in 1,$ (1)

$d_u = d \cap p \ni (p \cap u = \Lambda) .\supset. d_u \in \text{cls}^2\, \text{excl} .\supset. \exists d_u^\times,$ (2)

$(1) . m_1 \in d_u^\times . m = m_1 \cup u .\supset: p \in d .\supset_p. p \cap m \in 1,$ (3)

$m \supset \cup\text{'}d \ldots . (4),$ $(3) . (4) .\supset. m \in d^\times .\supset. \text{Prop}].$

·13[1] $m \in d^\times .\supset. -\exists d^\times \cap m' \ni (m'0\text{'}m . m' \supset m),$

$[m'0\text{'}m . m' \supset m .\supset. \exists d \cap p \ni (m' \cap p = \Lambda) .\supset. m' -\in d^\times .\supset. \text{Prop}].$

|384|

★6·14 $l \in \text{cls} .\supset\therefore d^\times \cap \text{cls'}l \in 1 .=: p \in d .\supset_p. p \cap l \in 1,$

$[★6{\cdot}0 . \text{Hp} .\supset: p \in d .\supset_p. \exists p \cap l,$ (1)

$m = \text{℩} d^\times \cap \text{cls'}\, l . p \in d . x = \text{℩} p \cap m .\supset. x \in l,$ (2)

$\text{Hp}\,(2) . y \in p \cap l .\supset. m - \iota x \cup \iota y \in d^\times \cap \text{cls'}l .\supset. m - \iota x \cup \iota y = m .\supset. x1\text{'}y,$ (3)

$(1) . (2) . (3) .\supset\therefore d^\times \cap \text{cls'}\, l \in 1 .\supset: p \in d .\supset_p. p \cap l \in 1$ (4)[2]

$★6{\cdot}0 .\supset\therefore p \in d .\supset_p. p \cap l \in 1 :\supset. d^\times \cap \text{cls'}\, l \in 1,$ (5)

$(4) . (5) .\supset. \text{Prop}].$

·21 $\cup\text{'}d^\times = \cup\text{'}d.$

·22 $f \in \text{cls}^2\, \text{excl} .\supset: d^\times = f^\times .=. d = f,$

$[★6{\cdot}21. \text{Hp} .\supset. \cup\text{'}d = \cup\text{'}f,$ (1)

$p \in d . x, y \in p .\supset. -\exists d^\times \cap m \ni (x, y \in m) .\supset. -\exists f^\times \cap m \ni (x, y \in m),$ (2)

$x, y \in^2 f . -\exists f^\times \cap m \ni (x, y \in m) .\supset. \exists f \cap q \ni (x, y \in q).$ (3)

$(1) . (2) . (3) .\supset: d^\times = f^\times .\supset. d = f,$ (4)

$(4) : d = f .\supset. d^\times = f^\times :\supset. \text{Prop}].$

1. The original erroneously had 'o' instead of '0' (twice).
2. The original erroneously did not italicise the '*l*' at end of line.

★7

·0 $R \in 1 \rightarrow 1 \,.\, R \rangle \mathrm{sim} \,.\, \rho, \breve{\rho} \in \mathrm{cls}^2 \,\mathrm{excl} \,.\rangle.\, \smile`\rho \;\mathrm{sim} \smile`\breve{\rho}$,

$$\left[u \in \rho \,.\rangle.\, \exists 1 \rightarrow 1 \frown R' \ni \left(\rho' = u \,.\, \breve{\rho}' = \imath\breve{\rho}u\right),\right. \qquad (1)$$

$$K = x \ni \left[\exists \rho \frown u \ni \left\{x = 1 \rightarrow 1 \frown R' \ni \left(\rho' = u \,.\, \breve{\rho}' = \imath\breve{\rho}u\right)\right\}\right].$$

$$\left. L \in K^{\times} \,.\, S = \smile `L \,.\rangle.\, S \in 1 \rightarrow 1 \,.\, \sigma = \smile `\rho \,.\, \breve{\sigma} = \smile `\breve{\rho} \,.\rangle.\, \mathrm{Prop}\right].$$

This prop proves that if d is a cls^2 excl, $\mu \smile `d$ is defined entirely by the numbers involved in d and its classes; thus we get the conception of a sum of numbers.

·1 $$d \in \mathrm{cls}^2 \,\mathrm{excl} \,.\rangle.\, \sum^{p \in d} \mu p = \mu \smile `d. \qquad \text{Df.}$$

Thus $\sum^{p \in d} \mu p$ is the sum of all the numbers of the classes forming the extension of the class d. The usual notation $\alpha + \beta + \gamma + \ldots.$ is only suitable for a finite, or at least for a well-ordered set of numbers; also, it introduces unnecessarily the idea of order among the numbers added, and thus renders necessary the totally irrelevant commutative law. But we shall use it, when convenient (e. g. Section III above), for a finite number of numbers; thus,

$$\alpha, \beta \in Nc \,.\, p \in \alpha \,.\, q \in \beta \,.\, d = \iota p \smile \iota q \,.\rangle.\, \sum^{u \in d} \mu u = \alpha + \beta = \beta + \alpha = \mu \smile `d.$$

·2 $R \in 1 \rightarrow 1 \,.\, R \rangle \mathrm{sim} \,.\, \rho, \breve{\rho} \in \mathrm{cls}^2 \,\mathrm{excl} \,.\rangle.\, \rho^{\times} \mathrm{sim}\, \breve{\rho}^{\times}$,

$$\left[u \in \rho \,.\rangle.\, \exists 1 \rightarrow 1 \frown R' \ni \left(\rho' = u \,.\, \breve{\rho}' = \imath\breve{\rho}u\right).\right. \qquad (1)$$

$$K = x \ni \left[\exists \rho \frown u \ni \left\{x = 1 \rightarrow 1 \frown R' \ni \left(\rho' = u \,.\, \breve{\rho}' = \imath\breve{\rho}u\right)\right\}\right] . L \in K^{\times}.$$

$$S = \smile `L \,.\rangle \therefore m \in \rho^{\times} \,.\rangle_m.\, \breve{\rho}^{\times} \frown n \ni \left(n = \breve{\sigma}m\right) \in 1 : n \in \breve{\rho}^{\times} \,.\rangle_n.$$

$$\left.\rho^{\times} \frown m \ni \left(m = \sigma n\right) \in 1 \therefore \rangle.\, \mathrm{Prop}\right].$$

|385| Analogous remarks to those on ★7·0 apply here; we derive from this prop the conception of a product of numbers.

★7·21 $$d \in \mathrm{cls}^2 \,\mathrm{excl} \,.\rangle.\, \prod^{p \in d} \mu p = \mu d^{\times}. \qquad \text{Df.}$$

Thus, $\prod^{p \in d} \mu p$ is the product of all the numbers of the classes forming the extension of the class d. Analogous remarks to those on ★7·1 apply to the notation $\alpha \times \beta \times \gamma \ldots.$ for a product of cardinal numbers.

·3 $\alpha, \beta \in Nc \,.\, d \in \alpha \frown \mathrm{cls}`\beta \;\mathrm{excl} .\rangle.\, \smile `d \in \alpha \times \beta$,

$$\left[d \rangle \beta \,.\rangle : p, p' \in d \,.\rangle.\, \exists 1 \rightarrow 1 \frown R \ni \left(p' = \rho \,.\, p = \breve{\rho}\right),\right. \qquad (1)$$

$$p \in d \,.\rangle.\, R(p) = x \ni \left\{\exists d \frown p' \ni \left[x = 1 \rightarrow 1 \frown R \ni \left(p' = \rho \,.\, p = \breve{\rho}\right)\right]\right\} :$$

$$P \in R(p)^{\times} \,.\, S = \smile `P \,.\, k = x \ni \left\{\exists p \frown y \ni \left(x = \sigma y\right)\right\} : \rangle \therefore \sigma = \smile `d.$$

$$\breve{\sigma} = p \,.\, k \in \beta \frown \mathrm{cls}`\, \alpha \;\mathrm{excl} \,.\, k \rangle d^{\times}, \qquad (2)^{1}$$

1. The original erroneously skipped the number '2' here. Hence we have renumbered equations 3–6 as 2–5.

$$(2)\,.\, h = \iota d \smile \iota k \,.\supset.\, h^{\times} \in \alpha \times \beta, \qquad (3)$$

$$k \supset d^{\times} \,.\, x \in h^{\times} \,.\supset.\, \exists d \frown p' \in \{\exists k \frown q \ni (x = \iota p' \smile \iota q \,.\, p' \frown q \in 1 \,.\, \iota p' \frown q \in \smile {}^{\backprime} d)\}, \qquad (4)$$

$$y \in \smile {}^{\backprime} d \,.\supset.\, h^{\times} \frown x \ni (\frown {}^{\backprime} x = \iota y) \in 1, \qquad (5)$$

$$(3)\,.\,(4)\,.\,(5) \,.\supset.\, \text{Prop}].$$

Note: The proposition connects addition with the multiplication of two numbers. It also gives a signification to the commutative law $\alpha \times \beta = \beta \times \alpha$, namely, it can be construed to signify

$$d \in \alpha \frown \text{cls}{}^{\backprime}\beta \text{ excl} \,.\, d' \in \beta \frown \text{cls}{}^{\backprime}\, \alpha \text{ excl} \,.\supset.\, \smile {}^{\backprime} d \text{ sim} \smile {}^{\backprime} d'.$$

Also incidentally the demonstration shows how to construct a cls^2 excl contained in $d^{\times}$; for $k \in \text{cls}^2\text{excl}$ and $k \supset d^{\times}$.

★8 $((a)) \in \text{cls}^3 \,.\supset \therefore$

·1 $(\smile {}^{\backprime}(a)) = q \ni \{\exists((a)) \frown p \ni (q = \smile {}^{\backprime} p)\}$. Df.

Note: In $((a))$ the symbol a is devoid of independent signification.

·11 $\smile {}^{\backprime}(\smile {}^{\backprime}(a)) = \smile {}^{\backprime} \smile {}^{\backprime}((a))$.

·12 $p, q \in ((a)) \,.\, p -= q \,.\supset_{p,q}.\, \smile {}^{\backprime} p -= \smile {}^{\backprime} q :\supset.\, \mu(\smile {}^{\backprime}\, (a)) = \mu((a))$.

$$[\text{Hp} \,.\, x \in (\smile {}^{\backprime}(a)) \,.\supset.\, ((a)) \frown y \ni (x = \smile {}^{\backprime} y) \in 1, \qquad (1)$$

$$\text{Hp} \,.\, y \in ((a)) \,.\supset.\, (\smile {}^{\backprime}(a)) \frown x \ni (x = \smile {}^{\backprime} y) \in 1, \qquad (2)$$

$$(1)\,.\,(2) \,.\supset.\, \text{Prop}].$$

·2 $\text{cls}^3 \text{ excl} = \text{cls}^3 \frown \text{cls}^2 \text{ excl}$. Df.

·21 $\text{cls}^3 \text{ excl excl} = \text{cls}^3 \text{ excl} \frown \text{cls}{}^{\backprime}\, (\text{cls}^2 \text{ excl})$. Df.

·22 $((a)) \in \text{cls}^3 \text{ arithm} \,.=.\, (\smile {}^{\backprime}(a)) \in \text{cls}^2 \text{ excl} \,.\, ((a)) \supset \text{cls}^2 \text{ excl}$. Df.

|386| Note: "Arithm" is contracted from "arithmetical"; this definition is essential to the statement and proof of the associative laws in their full generality.

★8·23 $((a)) \in \text{cls}^3 \text{arithm} \,.\supset.\, \smile {}^{\backprime}((a)) \in \text{cls}^2 \text{excl}$,

$$[\star 8{\cdot}22 \,.\, \text{Hp} \,.\supset: p \in ((a)) \,.\, q, q' \in p \,.\, q -= q' \,.\supset.\, q \frown q' = \Lambda, \qquad (1)$$

$$\star 8{\cdot}22 \,.\, \text{Hp} \,.\, p, p' \in ((a)) \,.\, p -= p' \,.\supset.\, \smile {}^{\backprime} p \frown \smile {}^{\backprime} p' = \Lambda \,.\supset:$$

$$q \in p \,.\, q' \in p' \,.\supset_{q,q'}.\, q \frown q' = \Lambda, \qquad (2)$$

$$(1)\,.\,(2) \,.\supset.\, \text{Prop}].$$

·3 $((a)) \in \text{cls}^3 \text{arthm} \,.\supset.\, \sum^{p \in ((a))} \sum^{q \in p} \mu q = \mu \smile {}^{\backprime}(\smile {}^{\backprime}(a))$. Df.

·31 $((a)) \in \text{cls}^3 \text{arthm} \,.\supset.\, \sum^{p \in ((a))} \sum^{q \in p} \mu q = \sum^{q \in \smile {}^{\backprime}((a))} \mu q$, [★8·11 .›. Prop].

Note: This is the associative law for addition when the number of brackets and the number of numbers in each bracket is not necessarily finite or denumerable.

★9 $((a)) \in \mathrm{cls}^3 . ((a)) \rightarrowtail \mathrm{cls}^2 \, \mathrm{excl} \; .\!\supset\!.\therefore$

·0 $\left((a)^{\times}\right) = m \in \left\{\exists((a)) \frown p \ni \left(m = p^{\times}\right)\right\}.$ Df.

·1 $\left((a)^{\times}\right) = ((c)) \; .\!\supset\!. \; (\smile `(c)) = (\smile `(a)).$ [★6·21 .⊃ Prop].

★10 $((a)) \in \mathrm{cls}^3 \, \mathrm{arithm} \; .\!\supset\!.\therefore$

·0 $\mu\left((a)^{\times}\right) = \mu((a)) = \mu(\smile `(a)),$

$\Big[\mathrm{Hp}\, .★8\cdot12 \; .\!\supset\!. \; \mu((a)) = \mu(\smile `(a)),$ (1)

$\mathrm{Hp} \, . \, p, q \in \left((a)^{\times}\right) . \, p -= q \; .\!\supset\!. \smile `p -= \smile `q,$ (2)

$(2) \, . ★8\cdot12 \, . ★9\cdot1 \, . \left((a)^{\times}\right) = ((c)) \; .\!\supset\!. \; \mu\left((a)^{\times}\right) = \mu(\smile `(c)) = \mu(\smile `(a));$ (3)

$(1) \, . \, (3) \; .\!\supset\!. \; \mathrm{Prop}\Big].$

·1 $\left((a)^{\times}\right) \in \mathrm{cls}^3 \, \mathrm{excl}.$

·11 $p, q \in \left((a)^{\times}\right)^{\times} . \, p -= q \; .\!\supset\!. \smile `p -= \smile `q,$

$\Big[m, m' \in \left((a)^{\times}\right) .\!\supset\!. \smile `m \frown \smile `m' = \Lambda,$ (1)[1]

$\mathrm{Hp} \; .\!\supset\!. \; \exists \left((a)^{\times}\right) \frown m \ni (p \frown m -= q \frown m),$ (2)

$\mathrm{Hp} \, . \, m \in \left((a)^{\times}\right) . \, p \frown m -= q \frown m \, . \, u = \iota p \frown m . v = \iota q \frown m \; .\!\supset\!: \exists u - v \, .\!\smile\!. \, \exists v - u,$ (3)

$(1) \, . \, \mathrm{Hp}(1) \, . \, \mathrm{Hp}(3) \; .\!\supset\!. \; u \frown \iota \, (q \frown m') = \Lambda \, . \, v \frown \iota (p \frown m') = \Lambda,$ (4)

$(2) \, . \, (3) \, . \, (4) \; .\!\supset\!: \exists \smile `p - \smile `q \, .\!\smile\!. \, \exists \smile `q - \smile `p :\supset\!. \; \mathrm{Prop}\Big].$

·12 $\left((a)^{\times}\right)^{\times} = ((b)) \; .\!\supset\!. \; \mu(\smile `(b)) = \mu\left((a)^{\times}\right)^{\times},$

$\big[★8\cdot12 \, . ★10\cdot11 \; .\!\supset\!. \; \mathrm{Prop}\big].$

|387|

★10·13[2] $\smile `\smile `\left((a)^{\times}\right) = \smile `\smile `((a)),$

$\Big[y \in \smile `\left((a)^{\times}\right) . = . \exists \, ((a)) \frown p \ni \left(y \in p^{\times}\right),$ (1)

$z \in \smile `\smile `\left((a)^{\times}\right) . = . \exists \, ((a)) \frown p \ni \left(z \in \smile `p^{\times}\right),$ (2)

$(2) \, . ★6\cdot21 \; .\!\supset\!: z \in \smile `\smile `\left((a)^{\times}\right) . = . \exists \, ((a)) \frown p \ni (z \in \smile `p) :\supset\!. \; \mathrm{Prop}\Big].$

1. The original erroneously omitted the '∈'.
2. The original erroneously omitted a period at the end.

$\cdot 131 \quad q,\, q' \in \cup`\left((a)^{\times}\right) .\supset: q - = q' .\supset. \smallsmile`q - = \smallsmile`q',$

$\left[p_1,\, p_2 \in ((a)) \,.\, p_1 - = p_2 \,.\, q \in p_1^{\times} \,.\, q' \in p_2^{\times} .\supset. \smallsmile`q - = \smallsmile`q'\right.$ (1)[1]

$p \in ((a)) \,.\, q,\, q' \in p^{\times} \,.\, p \in \mathrm{cls}^2 \text{ excl} .\supset. \smallsmile`q - = \smallsmile`q',$ (2)

$\left.(1) \,.\, (2) .\supset. \text{Prop}\right].$

$\cdot 14 \quad \left((a)^{\times}\right)^{\times} = ((b)) .\supset. \left(\smallsmile`(b)\right) = \left\{\smallsmile`((a))\right\}^{\times},$

$\left[\star 6\cdot 0 \,.\, \star 10\cdot 11 .\supset: x \in \left(\smallsmile`(b)\right) .=. \; ((b)) \frown y \ni (x = \smallsmile`y) \in 1.\right.$

$= . y \ni \left\{y \supset \smallsmile`\left((a)^{\times}\right) : p \in \left((a)^{\times}\right) .\supset_p. y \frown p \in 1 : x = \smallsmile`y\right\} \in 1.$ (1)

$\star 6\cdot 0 \,.\, \star 10\cdot 13 \,.\, \star 10\cdot 131 .\supset: x \in \left\{\smallsmile`((a))\right\}^{\times} .=.$

$x \supset \smallsmile`\smallsmile`((a)) : q \in \smallsmile`((a)) .\supset_q. x \frown q \in 1 :=:$

$x \supset \smallsmile`\smallsmile`\left((a)^{\times}\right) : q \in \smallsmile`((a)) .\supset_q. x \frown q \in 1 :=:$

$y \ni \left\{y \supset \smallsmile`\left((a)^{\times}\right) : p \in \left((a)^{\times}\right) .\supset_p. y \frown p \in 1 : x = \smallsmile`y\right\} \in 1,$ (2)

$\left.(1) \,.\, (2) .\supset. \text{Prop}\right].$

$\cdot 2 \quad \mu \left((a)^{\times}\right)^{\times} = \mu \left\{\smallsmile`((a))\right\}^{\times},$ $\quad [\star 10\cdot 12 \,.\, \star 10\cdot 14 .\supset. \text{Prop}].$

$\cdot 21 \quad \prod^{p \in ((a))} \prod^{q \in p} \mu q = \mu \left((a)^{\times}\right)^{\times}.$ $\quad$ Df.

$\cdot 22$[2] $\quad \prod^{p \in ((a))} \prod^{q \in p} \mu q = \prod^{q \in \smallsmile`((a))} \mu q,$ $\quad [\star 10\cdot 2 .=. \text{Prop}].$

Note: This proposition, or its equivalent ★10·2, is the associative law of multiplication when the number of brackets and the number of factors in each bracket is not necessarily finite or denumerable.

★11

·0[3]

$d \in \mathrm{cls}^2 \text{ excl} \,.\, f \in \mathrm{cls} \,.\, f \frown \smallsmile`d = \Lambda \,.\, h = \iota\left(\smallsmile`d\right) \smallsmile \iota f \,.\, ((g)) = \left(\iota\iota f \smallsmile \iota d\right)^{\times} .\supset. \smallsmile`\left((g)^{\times}\right) = h$

$\left[\smallsmile`\left((g)^{\times}\right) = m \ni \left\{f \frown m \in 1 \,.\, d \frown p \ni (p \frown m \in 1) \in 1 \,.\, m \supset f \smallsmile \left(\smallsmile`d\right)\right\}\right.$

$\left. = m \ni \left\{f \frown m \in 1 \,.\, \smallsmile`d \frown m \in 1 \,.\, m \supset f \smallsmile \left(\smallsmile`d\right)\right\} = h^{\times}\right].$

Note: This is the distributive law for addition and multiplication; it can be symbolized thus:

$\cdot 1 \quad \alpha \in Nc \,.\, d \in \mathrm{cls}^2 \text{ excl} .\supset. \; \alpha \times \mu\left(\smallsmile`d\right) = \sum^{p \in d} \alpha \times \mu p.$

---

1. The original erroneously failed to italicise the variable in the expression '$p_1^{\times}$'.

2. The original erroneously began with a 'v': $\prod^{v \in ((a))}$.

3. In the second line below, the original erroneously had '$f \,\underline{\frown}\, m$' instead of '$f \frown m$'.

|388| ***Section V.***
On Powers, Combinations and Permutations.

★12 $\alpha, \beta \in Nc \,.\supset\therefore$

·0 $a, b \in \alpha \frown \text{cls}'\beta \text{ excl} \,.\supset.\, \mu a^{\times} = \mu b^{\times},$ $[★7\cdot2 \,.\supset.\, \text{Prop}]$.

·1 $\beta^{\alpha} = \imath Nc \frown x \ni \{a \in \alpha \frown \text{cls}'\beta \text{ excl} \,.\supset.\, x = \mu a^{\times}\}.$ Df.

★13 $((a)) \in \text{cls}^3 \text{ arithm} \,.\, \varpi \in Nc : q \in^2 ((a)) .\supset.\, q \in \varpi : \supset ::$

·0 $\mu\left(\left(a\right)^{\times}\right)^{\times} = \prod^{p \in ((\alpha))} \varpi^{\mu p},$ $[★12\cdot1 \,.\supset.\, \text{Prop.}]$

·1 $\gamma \in Nc \,.\smile`((a)) \supset \gamma .\supset.\, \prod^{p \in ((a))} \gamma^{\mu p} = \gamma^{\sum^{p \in ((a))} \mu p},$ $[★10\cdot22 \,.\supset.\, \text{Prop}]$.

·2 $\alpha, \beta, \gamma \in Nc \,.\supset.\, (\gamma^{\beta})^{\alpha} = \gamma^{\alpha \times \beta},$

$[\text{Hp}\,(★13\cdot1) \,.\, ((a)) \in \alpha \frown \text{cls}'\beta \,.\, ★13\cdot1 \,.\, ★7\cdot3 \,.\supset.\, \text{Prop}]$.

★14 $u, v \in \text{cls} \,.\, u \frown v = \Lambda \,.\supset:$

·0 $v^{u} = Nc \to 1 \frown R \ni (\rho = u \,.\, \breve{\rho} \supset v).$ Df.

·1 $\mu(v^{u}) = \mu v^{\mu u},$

$[x \in u \,.\supset_x.\, w_x = (\iota x \smile \iota v)^{\times} : w = k \ni \{\exists u \frown x \ni (k = w_x)\} : \supset \therefore$

$p \in w^{\times} .=: x \in u \,.\supset_x.\, w_x \frown p \in 1 : p \supset \smile `w \therefore \supset.\, \mu w^{\times} = \mu(v^{u})$ (1)

$w \in \mu u \frown \text{cls}'\mu v \,.\supset.\, \mu w^{\times} = \mu v^{\mu u}, \ldots .$ (2) $(1) \,.\, (2) \,.\supset.\, \text{Prop}]$.

Note: This proposition deduces Cantor's definition of a power [cf. Math. Ann., Bd. XLVI] from the definition of a power given in ★12·1.

★15 $a \in \text{cls} \,.\supset\therefore$

·0 $\mu\ \text{cls}'a = 2^{\mu a},$

$[x \in a \,.\supset_x.\, a_x = \text{cls}'a \frown u \ni (x \in u) \,.\, a'_x = \text{cls}'a - a_x :$

$c = k \ni \{\exists a \frown x \in (k = \iota a_x \smile \iota a'_x)\} : \supset. \mu c^{\times} = 2^{\mu a},$ (1)

$m \in c^{\times} .\supset.\, d_m = \imath \text{cls}'a \frown u \ni \{x \in a \,.\supset\therefore a_x \in m \,.\supset.\, x \in u : a'_x \in m \,.\supset.\, x - \in u\} :$

$\supset.\, \text{cls}'a = w \ni \{\exists c^{\times} \frown m \ni (w = d_m)\} .\supset.\, \mu\ \text{cls}'a = \mu c^{\times},$ (2)

$(1) \,.\, (2) \,.\supset.\, \text{Prop}]$.

·1 $\beta \in Nc \,.\supset.\, C^{a}_{\beta} = \text{cls}'a \frown \beta.$ Df.

·2 $\text{cls}'a = x \ni \{\exists Nc \frown \beta \ni (x \in C^{a}_{\beta})\}.$

·3 $\sum^{\beta \leq \mu a} \mu C^{a}_{\beta} = \mu\ \text{cls}'a = 2^{\mu a}.$

·4 $a' \text{sim}\, a \,.\, \beta \in Nc \,.\supset.\, \mu C^{a}_{\beta} = \mu C^{a'}_{\beta}.$

★16 $\alpha, \beta, \gamma \in Nc . \supset ::$

|389|

★16·0 $C^\alpha_\beta = \iota Nc \frown x \ni \{a \in \alpha . \supset . C^a_\beta \in x\}$. Df.

·1 $\sum^{\rho+\sigma=\gamma} C^\alpha_\rho \times C^\beta_\sigma = C^{\alpha+\beta}_\gamma$,

$[a \in \alpha . b \in \beta . a \frown b = \Lambda . u = \gamma \frown z \ni \{\exists(x, y) \ni (x \supset a . y \supset b . z = x \smile y)\}.$

$\supset . u = \gamma \frown z \ni (z \supset a \smile b) . \supset . \text{Prop.}]$

·2 $\alpha \in Nc$ fin $. \supset . C^\alpha_\alpha = 1$.

·21[1] $\alpha \in Nc$ infin $. \rho \in Nc . \rho \leqq \alpha . a \in \alpha . \supset . C^a_{\alpha_\rho} = C^a_\alpha \frown p \ni (a - p \in \rho)$. Df.

Note: The class of infinite numbers denoted in this memoir by $N\alpha_0$ is defined below (cf. ★30·0); it is the class of numbers such that if β be any one of its members, $\beta + \beta = \beta$, and hence if γ be any smaller number $\beta + \gamma = \beta$. In Section III these propositions are deduced by the use of the hypothetical primitive proposition ★4·3; and in the sets of propositions ★30, ★31 and ★32, properties of the class are discussed. It will be noticed that in the proof of ★19·7 there is a forward reference to ★31·3.

·22 $\alpha \in N\alpha_0 . \rho \in Nc . \rho < \alpha . a \in \alpha . \supset . \mu C^a_{\alpha_\rho} = C^\alpha_\rho$,

$[$★$4{\cdot}38 . \supset . \alpha - \rho = \alpha . \supset . \text{Prop}]$.

·23 $\alpha \in N\alpha_0 . a \in \alpha . \supset . \mu C^a_{\alpha_\alpha} = C^\alpha_\alpha$.

$[\alpha + \alpha = \alpha . \supset . \exists \text{ cls' } a \frown \alpha \frown u \ni (a - u \in \alpha) . \supset . C^u_\alpha \supset C^a_{\alpha_\alpha} . \supset . C^\alpha_\alpha \leqq \mu C^a_{\alpha_\alpha}$, (1)

$C^\alpha_{\alpha_\alpha} \supset C^a_\alpha . \supset . \mu C^a_{\alpha_\alpha} \leqq C^\alpha_\alpha$, ... (2) (1) . (2) $. \supset .$ Prop].

·24 $\alpha \in N\alpha_0 . \supset . C^\alpha_\alpha = 2^\alpha$, $\left[a \in \alpha . \supset . C^a_\alpha = \sum^{\rho \leqq \alpha} \mu C^a_{\alpha_\rho} = \sum^{\rho \leqq \alpha} C^\alpha_\rho = 2^\alpha\right]$.

·3 $\alpha, \beta \in Nc . \supset . \alpha^\beta \leqq C^{\alpha\beta}_\beta$,

$[a \in \beta \frown \text{cls'}\alpha \text{ excl} . \supset . \mu \smile ‘a = \alpha\beta . \mu a^\times = \alpha^\beta$, (1)

$a^\times \supset \beta . \supset . a^\times \supset C^{\smile ‘a}_\beta$, ... (2) (1) . (2) $\supset$ Prop.]

·31 $\alpha \times 2^\beta \leqq C^{\alpha\beta}_\beta$,

$[a \in \alpha \frown \text{cls'}\beta \text{ excl} . \supset . \mu \smile ‘a = \alpha\beta$, (1)[2]

$b = q \ni \{\exists a \frown p \ni (q = C^p_\beta)\} . \supset . b \in \alpha \frown \text{cls' } 2^\beta . \supset . \mu \smile ‘b = \alpha \times 2^\beta$, (2)

$\smile ‘b \supset C^{\smile ‘a}_\beta$, ... (3) (1) . (2) . (3) $. \supset .$ Prop].

Binomial Theorem.—The next three sets of props (under integral numbers 17 to 19) are devoted to proving forms of the Binomial Theorem. The main results are

1. The original erroneously had ‘$p \in (a - p \in \rho)$’ instead of ‘$p \ni (a - p \in \rho)$’.
2. The original erroneously did not separate this line from the one above.

given in ★17·32 and ★17·4 and ★18·3[1], ★19·4 and ★19·7.

★17 $\alpha, \beta, \delta, \nu \in Nc \,.\, R \in 1 \rightarrow 1 \,.\, \rho \in \alpha \frown \text{cls}'\, \beta \text{ excl} \,.\, \breve{\rho} \in \alpha \frown \text{cls}'\, \delta \text{ excl} \,.$
$\smile{}' \rho \frown \smile{}' \breve{\rho} = \Lambda \,.\, a = p \ni \{\exists \rho \frown q \ni (p = q \smile \imath \breve{\rho} q)\} \,.\!\supset\! \therefore$

·0 $a \in \alpha \frown \text{cls}'(\beta + \delta)$.

·1 $a^{\times} \in (\beta + \delta)^{\alpha} \frown \text{cls}'\, \alpha$.

|390|

★17·11 $m \in a^{\times} \,.\, q \in \rho \,.\!\supset\!.\, m \frown q \in 1 \smile 0$.

·2[2] $a_{\rho_\nu} = a^{\times} \frown m \ni (\smile{}' \rho \frown m \in \nu)$ Df.

·21 $l_\nu \in C_\nu^\rho \,.\!\supset\!.\, \breve{l}_\nu = \breve{\rho} - \breve{\rho}\ (l_\nu)$. Df.

·211 $l_\nu \in C_\nu^\rho \,.\!\supset\!.\, \mu \breve{l}_\nu \in \sigma \ni (\sigma + \nu = \alpha)$.

·212 $l_\nu \in C_\nu^\rho \,.\!\supset\!.\, \mu \breve{l}_\nu^{\times} = \delta^{\mu \breve{l}_\nu}$

·22 $a_{\rho_\nu} = m \ni \left[\exists\ (p, q) \ni \{\exists C_\nu^\rho \frown l_\nu \ni (p \in l_\nu^{\times} \,.\, q \in \breve{l}_\nu^{\times} \,.\, m = p \smile q)\}\right]$.

·23[3] $\mu a_{\rho_\nu} = \sum^{l_\nu \in C_\nu^\rho} \mu l_\nu^{\times} \times \mu \breve{l}_\nu^{\times} = \sum^{l_\nu \in C_\nu^\rho} \beta^\nu \times \mu \breve{l}_\nu^{\times} = \beta^\nu \times \sum^{l_\nu \in C_\nu^\rho} \delta^{\mu \breve{l}_\nu}$.

·30 $a_\rho = k \ni \{\exists\, Nc \frown \nu \ni (k = a_{\rho_\nu})\}$. Df.

·31 $a_\rho \in \text{cls}^2 \text{ excl} \,.\, \smile{}' a_\rho = a^{\times}$.

·32 $(\beta + \delta)^\alpha = \mu a^{\times} = \mu \smile{}' a_\rho = \sum^{\nu \leqq \alpha} \mu a_{\rho_\nu} = \sum^{\nu \leqq \alpha} \beta^\nu \times \sum^{l_\nu \in C_\nu^\rho} \delta^{\mu \breve{l}_\nu}$

·4 $\delta = 1 \,.\!\supset\!.\, (1 + \beta)^\alpha = \sum^{\nu \leqq \alpha} \beta^\nu \times \mu C_\nu^\rho = \sum^{\nu \leqq \alpha} C_\nu^\alpha \times \beta^\nu$.

★18 Hp (★17) . $\alpha \in Nc$ fin $.\!\supset\!:$

·1 $\nu \leqq \alpha \,.\!\supset\!.\, \sigma \ni (\sigma + \nu = \alpha) \in 1 \,.\, \mu \breve{l}_\nu = \alpha - \nu$.

[The second part of the prop follows from ★17·211].

·2 $\sum^{l_\nu \in C_\nu^\rho} \delta^{\mu \breve{l}_\nu} = C_\nu^\alpha \times \delta^{\alpha - \nu}$.

·3 $(\beta + \delta)^\alpha = \sum^{\nu \leqq \alpha} C_\nu^\alpha \times \beta^\nu \times \delta^{\alpha - \nu}$.

★19 Hp (★17) . $\alpha \in N\alpha_0 \,.\!\supset\! \therefore$

·0 $\nu < \alpha \,.\, \sigma + \nu = \alpha \,.\!\supset\!.\, \sigma = \alpha$.

·1 $\nu < \alpha \,.\!\supset\!. \sum^{l_\nu \in C_\nu^\rho} \delta^{\mu \breve{l}_\nu} = C_\nu^\alpha \times \delta^\alpha$,

[★17·211 . ★19·0 .⊃. Prop].

·2 $\nu \leqq \alpha \,.\!\supset\!: l_\alpha \in C_{\alpha_\nu}^\rho \,.=.\, \mu \breve{l}_\alpha = \nu$.

·3 $\sum^{l_\alpha \in C_\alpha^\rho} \delta^{\mu \breve{l}_\alpha} = \sum^{\nu \leqq \alpha} \mu C_{\alpha_\nu}^\rho \times \delta^\nu = \sum^{\nu \leqq \alpha} C_\nu^\alpha \times \delta^\nu = (1 + \delta)^\alpha$. [★17·4 .⊃. Prop]

1. The original erroneously had '★18·53'.
2. The original erroneously used curly brackets instead of parentheses.
3. The original erroneously subscripted the multiplication sign below the second beta ('$\beta_{\times}^{\nu}$').

·4 $\quad (\beta+\delta)^{\alpha} = \sum^{\nu<\alpha} C^{\alpha}_{\nu} \times \beta^{\nu} \times \delta^{\alpha-\nu} + \beta^{\alpha}(1+\delta)^{\alpha}$,

$\quad\quad [\bigstar 17{\cdot}32 \,.\, \bigstar 19{\cdot}3 \,.\!\supset\!.\, \text{Prop}]$.

|391|

★19·7 $\quad (\beta+\delta)^{\alpha} = \delta^{\alpha} \times (1+\beta)^{\alpha} + \beta^{\alpha} \times (1+\delta)^{\alpha}$,

$$\Big[(\beta+\delta)^{\alpha} = \sum^{\nu<\alpha} C^{\alpha}_{\nu} \times \beta^{\nu} \times \delta^{\alpha-\nu} + \beta^{\alpha} \times (1+\delta)^{\alpha}$$

$$= \sum^{\nu<\alpha} C^{\alpha}_{\nu} \times \delta^{\nu} \times \beta^{\alpha-\nu} + \delta^{\alpha} \times (1+\beta)^{\alpha}, \qquad (1)$$

$$\sum^{\nu<\alpha} C^{\alpha}_{\nu} \times \beta^{\nu} \times \delta^{\alpha-\nu} = \delta^{\alpha} \sum^{\nu<\alpha} C^{\alpha}_{\nu} \times \beta^{\nu} \leqq \delta^{\alpha} \sum^{\nu\leqq\alpha} C^{\alpha}_{\nu} \times \beta^{\nu} \leqq \delta^{\alpha} \times (1+\beta)^{\alpha}. \qquad (2)^{1}$$

$$\sum^{\nu<\alpha} C^{\alpha}_{\nu} \times \delta^{\nu} \times \beta^{a-\nu} \leqq \beta^{\alpha}(1+\delta)^{\alpha}, \qquad (3)$$

$\bigstar 31{\cdot}3 \,.\!\supset\!:\, \beta^{\alpha} \times (1+\delta)^{\alpha} + \delta^{\alpha} \times (1+\beta)^{\alpha} = \beta^{\alpha} \times (1+\delta)^{\alpha} \,.\!\vee\!.$

$$\beta^{\alpha} \times (1+\delta)^{\alpha} + \delta^{\alpha} \times (1+\beta)^{\alpha} = \delta^{\alpha} \times (1+\beta)^{\alpha}, \qquad (4)$$

$(1) \,.\, (2) \,.\, (3) \,.\, (4) \,.\!\supset\!.\, \text{Prop}\Big]$.

★20·1 $\quad \alpha \in N\alpha_0 \,.\, \beta, \delta \in Nc \,.\, \beta \geqq \delta \,.\!\supset\!.\, (\beta\times\delta)^{\alpha} = \beta^{\alpha}$,

$$\Big[\alpha \in N\alpha_0 \,.\, \beta \geqq \delta \,.\!\supset\!.\, \beta^{\alpha} = \beta^{\alpha+\alpha} = (\beta\times\beta)^{\alpha} \geqq (\beta\times\delta)^{\alpha}, \qquad (1)$$

$$(\beta\times\delta)^{\alpha} \geqq \beta^{\alpha}, \qquad (2)$$

$(1) \,.\, (2) \,.\!\supset\!.\, \text{Prop}\Big]$.

★21 $\quad \alpha \in N\alpha_0 \,.\, \beta \in Nc \,.\!\supset\!\therefore$

·1 $\quad (2\beta)^{\alpha} = \beta^{\alpha}(1+\beta)^{\alpha}$, $\quad [\bigstar 19{\cdot}7 \,.\, \beta = \delta \,.\, \text{Prop}]$.

·2 $\quad (1+\beta)^{\alpha} = (2\beta)^{\alpha} + (1+\beta)^{\alpha}$, $\quad [\bigstar 19{\cdot}7 \,.\, \delta = 1 \,.\!\supset\!.\, \text{Prop}]$.

·3 $\quad (2\beta)^{\alpha} = (1+\beta)^{\alpha}$, $\quad [\bigstar 21{\cdot}1 \,.\, \bigstar 21{\cdot}2 \,.\!\supset\!.\, \text{Prop}]$.

·4 $\quad \nu \in Nc\ \text{fin} \,.\!\supset\! \nu^{\alpha} = 2^{\alpha}$,

$\quad\quad \big[\sigma \in Nc\ \text{fin} - \iota 0 \,.\, \beta = 2^{\sigma}+1 \,.\, \bigstar 21{\cdot}3 \,.\, \text{Induct} \,.\!\supset\! .$

$$\left(2^{\sigma+1}+2\right)^{\alpha} = \left(2^{\sigma}+2\right)^{\alpha} = (2+2)^{\alpha} = 3^{\alpha}, \qquad (1)$$

$$(2+2)^{\alpha} = 2^{2\alpha} = 2^{\alpha}, \qquad (2)$$

$$\nu \in Nc\ \text{fin} - \iota 0 \,.\, 2^{\sigma+1}+2 \geqq \nu \geqq 2^{\alpha}+2 \,.\!\supset\!.\, \left(2^{\sigma+1}+2\right)^{\alpha} \geqq \nu^{\alpha} \geqq \left(2^{\sigma}+2\right)^{\alpha}, \qquad (3)$$

$(1) \,.\, (2) \,.\, (3) \,.\!\supset\!.\, \text{Prop}\big]$.

★22 $\quad \alpha, \beta \in Nc \,.\, \alpha \geqq \beta \,.\, a \in \alpha \,.\, b \in \beta \,.\!\supset\!\therefore$

·0 $\quad P^{a}_{b} = 1 \to 1 \frown R \ni (\rho = b \,.\, \breve{\rho} \supset a)$. $\quad$ Df.

---

1. The original erroneously had $\sum^{\nu-\alpha} C^{\alpha}_{\nu}$ for the first term instead of $\sum^{\nu<\alpha} C^{\alpha}_{\nu}$.

·01[1] $P_\beta^\alpha = \text{℩}Nc \frown \gamma \ni \left(a \in \alpha \,.\, b \in \beta \,.\!\supset\!.\, P_b^a \in \gamma\right).$ Df.

·02 $P_b^a \supset a^b,$ $\left[\bigstar 14{\cdot}0 \,.\, \bigstar 22{\cdot}0 \,.\!\supset\!.\, \text{Prop}\right].$

·03 $P_\beta^\alpha \leqq \alpha^\beta,$ $\left[\bigstar 14{\cdot}1 \,.\, \bigstar 22{\cdot}02 \,.\!\supset\!.\, \text{Prop}\right].$

·04 ${}^\beta P_\beta = \text{℩}Nc \frown \gamma \ni \left\{1 \to 1 \frown R \ni \left(\rho = b \,.\, \breve{\rho} = b\right) \in \gamma\right\}.$ Df.

·05 ${}^\beta P_\beta \leqq P_\beta^\beta.$

·06 $\beta \in Nc \text{ fin} \,.\!\supset\!.\, {}^\beta P_\beta = P_\beta^\beta.$

|392|

★22·1[2] ${}^\beta P_\beta \times C_\beta^\alpha = P_\beta^\alpha,$

$\left[a \in \alpha \,.\, b \in \beta : l \in C_\beta^\alpha \,.\!\supset_l\!.\, k_l = 1 \to 1 \frown R \ni \left(\rho = b \,.\, \breve{\rho} = l\right):\right.$

$k = p \ni \left\{\exists C_\beta^\alpha \frown l \ni \left(p = k_l\right)\right\} :\!\supset\!.\, \mu k = \mu C_\beta^\alpha = C_\beta^\alpha.$

$\left. k \supset {}^\beta P_\beta \,.\smile{}^{\backprime}k = P_\beta^\alpha :\supset P_\beta^\alpha = \mu P_b^\alpha = \mu \smile{}^{\backprime}k = {}^\beta P_\beta \times C_\beta^\alpha\right].$

·2 $\alpha \in N\alpha_0 \,.\, \beta \in Nc \text{ fin} \,.\smile.\, \beta \in N\alpha_0 \,.\, \alpha \leqq \beta :\!\supset\!.\, C_\beta^{\alpha\times\beta} = P_\beta^{\alpha\times\beta} = \alpha^\beta,$

$\left[\bigstar 16{\cdot}3 \,.\, \bigstar 22{\cdot}03 \,.\, \bigstar 22{\cdot}1 \,.\!\supset\!.\, \alpha^\beta \leqq C_\beta^{\alpha\times\beta} \leqq P_\beta^{\alpha\times\beta} = \left(\alpha \times \beta\right)^\beta,\right.$ (1)[3]

$\text{Hp} \,.\, \bigstar 4{\cdot}42 \,.\, \bigstar 20{\cdot}1 \,.\!\supset\!.\, \alpha^\beta = \left(\alpha \times \beta\right)^\beta, \ldots.\ (2)$ $\left.(1) \,.\, (2) \,.\!\supset\!.\, \text{Prop}\right],$

·21 $\alpha \in N\alpha_0 \,.\, \beta \in Nc \frown \beta' \ni \left(\alpha \times \beta' = \alpha\right) .\!\supset\!. C_\beta^\alpha = P_\beta^\alpha = \alpha^\beta,$

$\left[\bigstar 22{\cdot}2 \,.\!\supset\!.\, \text{Prop}\right].$

·22[4] $\alpha \in N\alpha_0 \,.\, \alpha^2 = \alpha \,.\!\supset\!.\, \alpha^\alpha = 2^\alpha,$ $\left[C_\alpha^\alpha = 2^\alpha . \bigstar 22{\cdot}21 .\!\supset\!. \text{Prop}\right]$

★23·1 $\beta \in Nc \,.\!\supset\!.\, \beta < 2^\beta,$

$\left[a \in \beta \frown \text{cls' 2 excl} \,.\, q,\, q' \in a^\times \,.\, q \frown q' = \Lambda : p \in a \,.\!\supset_p\!.\right.$

$p' = \text{℩}a^\times \frown m \ni \left(m = q - p \smile q' \frown p\right) : b = x \ni \left\{\exists a \frown p \ni \left(x = p'\right)\right\} :\!\supset\!.$

$b \supset a^\times . \mu b = \mu a :\!\supset\!. \mu a^\times \geqq \mu a,$ (1)[5]

$a \in \beta \frown \text{cls' 2 excl} \,.\, S \in 1 \to 1 . \sigma = a . \breve{\sigma} \supset a^\times : p \in a .\!\supset\!. p_1 = \text{℩}\left(p \frown \iota \breve{\sigma} p\right).$

$p_2 = \text{℩}\left(p - \iota p_1\right) : q = x \in \left\{\exists a \frown p \ni \left(x 1' p_2\right)\right\} :\!\supset\!.\!\cdot\!.$

$q \in a^\times : p \in a \,.\!\supset_p\!.\, p_2 \in q - \breve{\sigma} p \,.\!\cdot\!.\!\supset\!.\, q \in a^\times - \breve{\sigma} a,$ (2)[6]

$\left.(1) \,.\, (2) \,.\!\supset\!.\, \text{Prop}\right].$

Note: This proposition is the extension for any cardinal number of Cantor's proof [cf. Jahresber. d. Deutsch. Math. Ver. I, S. 77] that the power of the continuum $\left(2^{\alpha_0}\right)$

1. The original erroneously had '$P_b^a \ni \gamma$' instead of '$P_b^a \in \gamma$'.
2. On the last line of ★22·1, the original erroneously had $\beta_b^\alpha :\supset P_\beta$ instead of $P_\beta^\alpha :\supset P_\beta^\alpha$.
3. The original had '★23.03' instead of '★23·03'.
4. The original erroneously had '$a^\alpha = 2^{\alpha}$' instead of '$\alpha^\alpha = 2^\alpha$'.
5. The original erroneously had '$ma^\times \geqq \mu a$' instead of '$\mu a^\times \geqq \mu a$'.
6. The original erroneously began with 'α' instead of 'a'.

is greater than the first infinite cardinal (α_0). The adaptation of the proof and its reduction to symbols are due to Russell.

In Section IV and the present Section V, all the proofs involving any characteristic properties of the class *Nc* infin have been limited to the class $N\alpha_0$ and based on the theorem that

$$\alpha \in N\alpha_0 \,.\, \beta \leqq \alpha \,.\!\supset\!.\, \alpha+\beta=\alpha.$$

It becomes important therefore to investigate for what class contained within *Nc*, the theorem

$$\beta \leqq \alpha \,.\!\supset\!.\, \alpha+\beta=\alpha$$

can be asserted, since the universal validity of ★4·3 is not proved. Furthermore, we have not succeeded in deducing, even from ★4·3, the allied proposition

$$\alpha \in Nc \text{ infin} \,.\, \beta \leqq \alpha \,.\!\supset\!.\, \alpha\times\beta=\alpha.$$

The nearest approach to it as yet is ★20·1.

|393| The sets of propositions with integral numbers 30 to 32 will be devoted to these investigations; the proofs of some propositions practically proved by Russell in Section III will be omitted. In these sets of propositions (30 to 32) it will be found that the only properties of an *Nc* infin used are that [cf. ★4·1]

$$\alpha \in Nc \text{ infin} \,.\!\supset\!.\, \alpha \geqq \alpha_0 \,.\, \alpha+\alpha_0=\alpha,$$

where α_0 is the first infinite[1] cardinal number.

★30

·0 $N\alpha_0 = Nc \frown \beta \ni \left[\exists Nc \frown \delta \ni (\delta>0 . \beta=\alpha_0\times\delta)\right].$ Df.

Note: As long as all the numbers considered belong to $N\alpha_0$, the propositions in Section III, ★4·31 to 4·51, hold without modification.

·1 $\alpha \in N\alpha_0 \,.\, \beta \in Nc \,.\!\supset\!\therefore\, \alpha \geqq \beta \,.\!\supset\!.\, \alpha+\beta=\alpha : \alpha \leqq \beta \,.\!\supset\!.\, \alpha+\beta=\beta,$

$[\alpha \geqq \beta \,.\!\supset\!.\, \alpha \leqq \alpha+\beta \leqq \alpha+\alpha \leqq \alpha \,.\!\supset\!.\, \alpha+\beta=\alpha,$ (1)

$\alpha \leqq \beta \,.\!\supset\!.\, \exists Nc \frown \gamma \ni (\alpha+\gamma=\beta) \,.\!\supset\!.\, \alpha+\beta=\alpha+\alpha+\gamma=\alpha+\gamma=\beta,$ (2)

$(1) . (2) \,.\!\supset\!.\, \text{Prop}].$

·2 $\beta, \gamma \in Nc \,.\, \exists N\alpha_0 \frown \alpha \ni [\beta \leqq \alpha \leqq \gamma] \,.\!\supset\!.\, \beta+\gamma=\gamma,$

$[\gamma \leqq \beta+\gamma \leqq \alpha+\gamma \leqq \gamma \,.\!\supset\!.\, \text{Prop}].$

Note: This theorem is important, and enables the enunciations of some theorems of ★21 and ★22 to be extended.

·3 $\beta \in Nc \,.\, \alpha \in N\alpha_0 \frown z \ni (z \geqq \beta) \,.\!\supset\!.\, \alpha \geqq \alpha_0\times\beta,$

$[\alpha \geqq \beta \,.\!\supset\!.\, \alpha_0\times\alpha \geqq \alpha_0\times\beta \,.\!\supset\!.\, \alpha \geqq \alpha_0\times\beta].$

·4 $\beta \in Nc \,.\, \gamma \in Nc \frown z \ni (\beta \leqq \gamma \leqq \alpha_0\times\beta) \,.\!\supset\!.\, \alpha_0\times\beta=\alpha_0\times\gamma,$

$[\bigstar 30\cdot 3 \,.\!\supset\!.\, \text{Prop}].$

1. The original erroneously omitted the word 'infinite'.

·5 $\alpha,\ \alpha' \in N\alpha_0\ .\ \alpha < \alpha'\ .\ \beta \in Nc \frown \beta \ni (\alpha = \alpha_0 \times \beta).$

$\beta' \in Nc \frown \beta' \ni (\alpha' = \alpha_0 \times \beta')\ .\supset.\ \beta \leqq \alpha_0 \times \beta < \beta' \leqq \alpha_0 \times \beta',$

$[\bigstar 30 \cdot 4\ .\supset.\ \text{Prop}].$

★31

·1 $Nc\exp = Nc \frown \alpha \ni [\exists Nc \text{ infin} \frown \beta \ni \{\exists Nc \frown \gamma \ni (\gamma > 1\ .\ \alpha = \gamma^\beta)\}].$ Df.

·11 $u \in \text{cls infin}\ .\supset.\ \mu\ \text{cls}'\ u \in Nc \exp,$

$[\bigstar 15 \cdot 0\ .\supset.\ \text{Prop}].$

·2 $\alpha \in Nc\exp .\supset.\ \alpha > \alpha_0.$

·3 $Nc\exp .\supset.\ N\alpha_0,$

$[\gamma^\beta \in Nc\exp .\supset.\ \gamma > 1\ .\supset.\ \gamma^{\alpha_0} \in Nc\exp .\supset.\ \gamma^{\alpha_0} > \alpha_0,$ (1)

$(1)\ .\supset.\ \gamma^\beta \leqq \alpha_0 \times \gamma^\beta \leqq \gamma^{\alpha_0} \times \gamma^\beta \leqq \gamma^{\alpha_0+\beta},$ (2)

$\gamma^\beta \in Nc\exp .\supset.\ \beta \in Nc \text{ infin}\ .\supset.\ \alpha_0 + \beta = \beta,$ (3)

$(2)\ .\ (3)\ .\supset.\ \gamma^\beta = \alpha_0 \times \gamma^\beta .\supset.\ \text{Prop}].$

|394|

★32

·1 $Nc\exp^{\alpha_0} = N\alpha_0 \frown \alpha \ni [\exists N\alpha_0 \frown \beta \ni \{\exists Nc \frown \gamma \ni (\gamma > 1\ .\ \alpha = \gamma^\beta)\}]$ Df.

·2 $\beta \in Nc\exp^{\alpha_0}\ .\supset.\ \beta^2 = \beta,$

$[\beta = \gamma^\delta .\supset.\ \beta^2 = \gamma^{2\delta} = \gamma^\delta = \beta].$

·3[1] $\beta \in Nc\ .\ \gamma \in Nc\exp^{\alpha_0}\ .\ \beta \leqq \gamma .\supset.\ \beta\gamma = \gamma.$

·4 $\alpha \in Nc\exp^{\alpha_0}\ .\ \beta \in Nc - \iota 0 - \iota 1\ .\ \beta \leqq 2^\alpha .\supset.\ \beta^\alpha = 2^\alpha,$

$[2^\alpha \leqq \beta^\alpha\ \ldots.\ (1)\ \text{Hp}\ .\ \bigstar 32 \cdot 2\ .\supset.\ \beta^\alpha \leqq (2^\alpha)^\alpha \leqq 2^{\alpha^2} \leqq 2^\alpha,$ (2)

$(1)\ .\ (2)\ .\supset.\ \text{Prop}].$

·5 $\delta \in Nc\exp^{\alpha_0}\ .\ \gamma \in Nc - \iota 0 - \iota 1\ .\ \alpha \in Nc\exp^{\alpha_0}\ .\ \delta \geqq \alpha\ .\ \beta = \gamma^\delta .\supset.\ \beta^\alpha = \beta,$

$[\beta^\alpha = \gamma^{\alpha \times \delta} = \gamma^\delta = \beta].$

1. The original erroneously had $\gamma \in Nc\exp_{\alpha^0}$.

The Logic of Relations, Logical Substitution Groups, and Cardinal Numbers

1903 *American Journal of Mathematics*, vol. 25, no. 2 (April), pp. 157–178

Published in April 1903, a note at the bottom specified that this paper was submitted on 4 July 1901.

This paper relies heavily on Whitehead's two prior papers 'Memoir on the Algebra of Symbolic Logic, Part I and Part II' (1901)[1] *and 'On Cardinal Numbers' (1902).*[2] *Insofar as its main objective is a kind of counting, one might well consider the paper under consideration here to be an extension of the latter work. To understand Whitehead's objectives, the non-technical reader of the present introduction will require the explanation of cardinal numbers (and of a few other notions) given in this volume in the material prefatory to the above-mentioned article on cardinality. In fact, before discussing 'The Logic of Relations' (LR) specifically, a brief review of cardinality will be helpful.*[3]

Recall that 'cardinality' is doubly infirmed by foundational flaws related to the definition of sets by comprehension (that is, the set of all elements that have some given attribute) and the Axiom of Choice (the assertion that given any collection of non-overlapping sets, we can declare a set consisting of exactly one element from each). Nonetheless, beyond its historical significance in appearing at the end of naïve set theory, Whitehead's paper on cardinality does have some interesting mathematics. For example, the well-known idea of counting the number of subsets of size k *that can be drawn from a fixed set of size* n *is extended to infinite sets. Whether this is as useful as the finite case (where high-school algebra students encounter it in raising two-termed expressions to integer powers), the question of the extension of this idea to transfinite numbers is natural and both asked and answered by Whitehead.*

LR inherits all the foundational infirmities of 'cardinality' but again asks and answers interesting questions. The principal matter is that of counting solutions to what Whitehead calls logical equations (as opposed to our usual numerical equations) in transfinite domains (see p. 188). Since Whitehead develops his methods with respect to a universal set i*, which, if not too big, circumvents the infelicities of unrestrained definition by comprehension, it is possible, albeit unlikely, that some of the combinatorial methods used here have not been superseded by a subsequent, more comprehensive approach and are therefore still of some mathematical interest. Nonetheless, Whitehead's technical prerequisites lie in algebra and set theory, as he understood them.*

The connection of logical equations with set theory (hence cardinality) may not be surprising because of the well-known correspondence between the propositional calculus

1. See pp. 115 and 138.
2. See p. 156.
3. See also editorial introduction to 'On Cardinal Numbers' (p. 156).

in logic and the ideas of subset, negation, union and intersection. Yet Whitehead's approach is ingenious insofar as he refers his problem of solution-counting back to his theorems on cardinality by introducing an idea that equates logical equations with those of classes and relations. He does so by extension of a device called divisionals, introduced in a more limited context by W. E. Johnson (see pp. 189 and 195).

Preface.

In Section I, the theory of logical equations is generalized; any definite logical equation is proved to correspond to a definite class of relations [cf. ★1·1 and ★1·2], and to each relation of the class corresponds a solution of the equation. But from the relational point of view the theory is equally simple, whether the number of variables in the corresponding logical equation is finite or infinite. Accordingly, we obtain a theory of logical equations when the cardinal number of the variables has any infinite value. The solution of this general type of equation is found [cf. ★3·32 and ★4·04]. This is effected by the help of some important definitions [cf. ★2·0, and ★2·05, and ★2·22, and ★3·10, and ★3·20]. In ★5, the application to equations with a finite number of variables is considered.

In Section II, Cantor's theory of cardinal as developed in my paper on "Cardinal Numbers" in Vol. XXIV, p. 367[1] of this Journal, is applied; and after determining the cardinal numbers of various classes of relations in ★10, in ★11 the number of solutions of any logical equation is determined [cf. ★11·13 and ★11·25]. In ★12, these results are considered for the special case of a finite number of variables [cf. ★12·01, and ★12·02, and ★12·2], and some examples for one and two variables are appended. In ★13, the following problem is considered: i is a given class, a and b are given classes contained in i, required the number of classes x contained in i such that the cardinal number of the class $(a \frown x) \smile (b \frown \bar{x})$, where $\bar{x}$ is the part of i not x, is some given number α. This number is determined [cf. ★13·20 to ★13·24]. Thence |158| [cf. ★13·30], with the same suppositions, the sum of the following series is determined:

$$\sum^{x \ni i} 2^{\mu\{(a \frown x) \smile (b \frown \bar{x})\}}.$$

These two sections are written out in the notations of Peano and Russell, explained in the memoir on "Cardinal Numbers" (loc. cit.).

Section III considers the orders of the Logical Substitution Groups, considered in my memoir on "Symbolic Logic," in Vol. XXIII, p. 297,[2] of this Journal; the order of the complete group is $24^{\mu i}$ (cf. ★20·1); the order of the identical group of a function with invariants s_1, s_2, s_3, s_4 is

$$24^{\mu(\bar{s}_1 \smile s_4)} \times 6^{\mu\{(s_1 \frown \bar{s}_2) \smile (s_3 \frown \bar{s}_4)\}} \times 4^{\mu(s_2 \frown \bar{s}_3)}.$$

Also the orders of other groups are determined.

Section IV deals with some properties of a certain simple type of substitutions.

---

1. See p. 156.
2. See p. 138.

My memoir on "Symbolic Logic" in this Journal, Part I in Vol. XXIII, p. 139,[1] and Part II, Vol. XXIII, p. 297, is always cited as Symb. Log., Part I or Symb. Log., Part II; the memoir "On Cardinal Numbers" in Vol. XXIV of this[2] Journal is cited as Card. Numb.[3]

Section I.

★1 $\quad i, h \in \text{cls} \,.\, P \in \text{rel} \,.\, \pi › i \,.\, \breve{\pi} › h \,.› ∴$

·1 $\quad \text{equ}\,(i, h, P) = \text{rel} \frown R \ni [\rho = i \,.\, \breve{\rho} › h \,.\, \breve{R}P › 0'].$ $\qquad$ Df.

Note: $\breve{R}P › 0' . = . R\breve{P} › 0'$ [cf. Card. Numb., Section II, 2·13]. Here "equ" is contracted from "equation." The connection between this definition and the ordinary theory of logical equations is most easily seen from the next proposition,

·2 $\quad \text{equ}\,(i, h, P) = \text{rel} \frown R \ni [\rho = i . \breve{\rho} › h : k \in h .›_k . \pi k \frown \rho k = \Lambda],$

[★1·1 . = . Prop.]

Note: To establish the connection between these propositions and the theory of logical equations, consider *h* as the class of indices not necessary finite or denumerable in number: *i* is the class called the universe, and all the classes appearing in the equation as factors or as summands are contained in *i*; *P* is the relation determining the known |159| coefficients of the various terms, thus πk may be written a_k, where a_k is a known class contained in *i* and corresponding to the index *k*; since $\breve{\pi} › h$ and is not necessarily equal to *h*, it may happen that $a_k = \Lambda$; R is the relation determining the unknowns of the equation, thus: let ρk be written x_k, where x_k is a class contained in *i* and corresponding to the index *k*: then ★1·1 and the general hypothesis assert that for every product of the type $a_k \frown x_k$ we have

$$a_k \frown x_k = \Lambda,$$

and that the logical sum of all classes of the type x_k is equal to *i*. For example, if the number of indices is two, so that $\mu h = 2$ and these indices are 1 and 2, then

$$a_1 \frown x_1 \smile a_2 \frown x_2 = \Lambda, \quad x_1 \smile x_2 = i.$$

In logical equations, as ordinarily considered, we should also have $x_1 \frown x_2 = \Lambda$, so that $x_2 = \overline{x}_1$ (putting $\overline{x}_1$ for $i \sim x_1$), and the equation becomes

$$a_1 \frown x_1 \smile a_2 \frown \overline{x}_1 = \Lambda.$$

This further specialization of the general idea will be considered later; but meanwhile we shall prove a series of propositions which belong equally to the more general conception here defined.

---

1. The original erroneously referenced p. 140, when the article in fact began on p. 139.
2. The original erroneously had 'his'.
3. Epsilon and reversed epsilon have been changed to '∈' and '∋' throughout.

★2 $i,\ h \in \text{cls} \,.\, P \in \text{rel} \,.\, \pi \rangle i \,.\, \breve{\pi} \rangle h \,.\rangle.\!\!:\,.$

·0 $b \rangle h \,.\rangle.\, (b,\ \text{div}\,P) = i \frown x \ni (\breve{\pi}x = h \sim b).$ Df.

·01 $b \rangle h \,.\, \exists h \sim b \sim \breve{\pi} \,.\rangle.\, (b,\ \text{div}\,P) = \Lambda,$

$[\text{Hp} \,.\rangle.\!\sim \exists i \frown x \ni (h \sim b = \breve{\pi}x) \,.\rangle.\, \text{Prop}].$

·02 $\breve{\pi} = h \,.\rangle.\, (\Lambda,\ \text{div}\,P) = x \ni (\breve{\pi}x = h).$

·03 $\breve{\pi} \sim= h \,.\rangle.\, (\Lambda,\ \text{div}\,P) = \Lambda,$

$[\text{Hp} \,.\rangle.\!\sim \exists i \frown x \ni (\breve{\pi}x = h) \,.\rangle.\, \text{Prop}].$

·04 $(h,\ \text{div}\,P) = i \sim \pi,$

$[(h,\ \text{div}\,P) = i \frown x \ni (\breve{\pi}x = \Lambda) \,.\rangle.\, \text{Prop}].$

·05 $(Nc,\ \text{div}\,P) = y \ni [\exists\ \text{cls}'\ h \frown b \ni \{y = (b,\ \text{div}\,P)\}].$ Df.

Note: "div" is contracted from "divisional": the importance of a similar conception in relation to logical equations containing a finite number of variables was exemplified by W. E. Johnson in a paper read |160| before the International Congress of Philosophy, Paris, 1900, in the section dealing with "Logique et Histoire des Sciences" (published by Armand Colin, Paris).[1] The definitions, ★2·0 and ★2·05, have essential reference to i, h which are given in the general hypothesis; in other connections it might be necessary to express these classes and to write $(b,\ \text{div}^i_h P)$ for $(b,\ \text{div}\,P)$ and $(Nc,\ \text{div}^i_h P)$ for $(Nc,\ \text{div}\,P)$.

·10 $P_{\breve{\pi}} = \iota\ \text{rel} \frown P_1 \ni (\pi_1 = i : x \in i \,.\, xP_1y \,.\rangle.\, y = \breve{\pi}x).$ Df.

·11 $\breve{\pi}_{\breve{\pi}} \rangle \text{cls}'\ h \,.\smile{}' \breve{\pi}_{\breve{\pi}} = \breve{\pi}.$

·12 $P_{\breve{\pi}} \in Nc \rightarrow 1.$

·13 $b \rangle h \,.\rangle.\, (b,\ \text{div}\,P) = \pi_{\breve{\pi}}(h \sim b).$

·20[2] $(Nc,\ \text{div}\,P) \in \text{cls}^2\ \text{excl},$

$[P_{\breve{\pi}} \varepsilon\ Nc \rightarrow 1 \,.\rangle:\, xP_{\breve{\pi}}y \,.\, x'P_{\breve{\pi}}y' \,.\, y0'y' \,.\rangle.\, x0'x' :\rangle.\, \text{Prop}].$

·21 $\smile{}'(Nc,\ \text{div}\,P) = i,$

$[x \in \pi .\langle: b = h \sim \breve{\pi}x \,.\rangle.\, x \in (b,\ \text{div}\,P) :\rangle.\, x \in^2 (Nc,\ \text{div}\,P),$ (1)

★2·04 $.\rangle: x \in i \sim \pi .\rangle.\, x \in (h,\ \text{div}\,P) \,.\rangle.\, x \in^2 (Nc,\ \text{div}\,P),$ (2)

(1) . (2) $.\rangle.$ Prop].

Note: The importance of the class $(Nc,\ \text{div}\,P)$ depends upon ★2·20 and ★2·21.

·22 $\beta \in Nc \,.\rangle.\, (\beta,\ \text{div}\,P) = x \ni [\exists\ \text{cls}'\ h \frown b \ni \{b \in \beta \,.\, x = (b,\ \text{div}\,P)\}].$ Df.

★3 $i,\ h \in \text{cls} \,.\, P \in \text{rel} \,.\, \pi \rangle i \,.\, \breve{\pi} \rangle h \,.\, \exists\ \text{equ}\,(i,\ h,\ P) \,.\, R \in \text{equ}\,(i,\ h,\ P) \,.\rangle.\!\!:\,.$

·01 $b \rangle h \,.\, a \in (b,\ \text{div}\,P) \,.\, aRk \,.\rangle.\, k \in b,$

$[\text{Hp} \,.\rangle: a \in i \,.\, h \sim b = \breve{\pi}a \,.\rangle: x \in h \sim b \,.\rangle_x.\, aPx,$ (1)

1. William Ernest Johnson, 'Sur la théorie des equations logiques', in *Bibliothèque du Congrès International de Philosophie, Tome III, Logique et Histoire des Sciences* (Paris: Armand Colin, 1901), pp. 185–199.

2. In the following expression, the original erroneously had 'o' instead of '0' in two places.

$(1) . x \in h \sim b . aRk . \breve{R}P \rangle 0' . \rangle . x0'k,$ (2)[1]
$\text{Hp} . (2) . \rangle . \text{Prop}].$

·02 $(\Lambda, \text{div}\, P) = \Lambda,$
$[\bigstar 3\cdot 01 . \rho = i . \rangle : b \rangle h . \exists\, (b, \text{div}\, P) . \rangle . \exists b : \rangle . \text{Prop}].$

·10 $b \rangle h . \exists\, (b, \text{div}\, P) . \rangle . (b, \text{rel}\, P) = \text{rel} \frown S \ni [\sigma = (b, \text{div}\, P) . \breve{\sigma} \rangle b].$ Df.

Note: $(b, \text{rel}\, P)$, like $(b, \text{div}\, P)$, refers essentially to i and h which are given in the general hypothesis. If it were necessary to render these classes explicit in the notation, we could write $(b, \text{rel}_h^i\, P)$ for $(b, \text{rel}\, P)$.

·11 $\beta \in Nc . \rangle . (\beta, \text{rel}\, P) = x \ni [\exists\, \text{cls}' h \frown b \ni \{b \in \beta . x = (b, \text{rel}\, P)\}].$ Df.

·12 $(Nc, \text{rel}\, P) = x \ni [\exists\, \text{cls}' h \frown b \ni \{x = (b, \text{rel}\, P)\}].$ Df.

|161|

·13 $S \in^2 (Nc, \text{rel}\, P) . = . S \in \text{rel} . \exists\, \text{cls}' h \frown b \ni [\sigma = (b, \text{div}\, P) . \breve{\sigma} \rangle b].$

·14[2] $S, S' \in^2 (Nc, \text{rel}\, P) . S0' S' . \rangle : \sigma \frown \sigma' = \Lambda . \smile . \sigma = \sigma',$
$[\bigstar 2\cdot 20 . \bigstar 3\cdot 13 . \rangle . \text{Prop}].$

·20 $\{\text{rel}\, (Nc, \text{rel}\, P)^{\times}\} = \text{rel} \frown S \ni [\exists\, (Nc, \text{rel}\, P)^{\times} \frown M \ni (S = \smile ' M)].$ Df.

Note: For an explanation of this use of the symbol $\times$, cf. Card. Numb. ★6·0.

·21 $S \in \{\text{rel}\, (Nc, \text{rel}\, P)^{\times}\} . \rangle . \sigma = \smile '(Nc, \text{div}\, P) = i.$

·22 $S \in \{\text{rel}\, (Nc, \text{rel}\, P)^{\times}\} . b \rangle h . \exists\, (b, \text{div}\, P) . \rangle .$
$S_b = \imath (b, \text{rel}\, P) \frown T \ni [x\ \varepsilon\ b . \rangle : xTy . = . xSy].$ Df.

·23 $\text{Hp} \bigstar 3\cdot 22 . \rangle . \sigma_b = (b, \text{div}\, P) . \breve{\sigma}_b \rangle b.$

·24[3] $S \in \{\text{rel}\, (Nc, \text{rel}\, P)^{\times}\} . b \rangle h . b' \rangle h . b0'b' . \rangle . \sigma_b \frown \sigma_{b'} = \Lambda.$

·25 $S \in \{\text{rel}\, (Nc, \text{rel}\, P)^{\times}\} . \rangle . \breve{\sigma} \rangle h.$

·26[4] $S \in \{\text{rel}\, (Nc, \text{rel}\, P)^{\times}\} . \rangle . \breve{S}P \rangle 0',$
$[x \in \breve{\sigma} . \rangle . \exists\, \text{cls}' h \frown b \ni \{x \in b . \exists\, (b, \text{div}\, P)\},$ (1)
$(1) . x\breve{S}z . \rangle . z \in (b, \text{div}\, P),$ (2)
$z \in (b, \text{div}\, P) . zPy . \rangle . y \in h \sim b,$ (3)
$(1) . (2) . (3) . \rangle : x\breve{S}Py . \rangle . \exists\, \text{cls}' h \frown b \ni \{x \in b . y \in h \sim b\} : \rangle . \text{Prop}].$

·30 $\{\text{rel}\, (Nc, \text{rel}\, P)^{\times}\} \rangle \text{equ}(i, h, P),$
$[\bigstar 3\cdot 21 . \bigstar 3\cdot 25 . \bigstar 3\cdot 26 . \rangle . \text{Prop}].$

1. In this expression, the original erroneously had 'o' instead of '0' in two places.
2. In the following equation, the original erroneously had 'o' instead of '0'.
3. In the following equation, the original erroneously had 'o' instead of '0'.
4. In the following expression, the original erroneously had 'o' instead of '0'.

·31[1] $\text{equ}\,(i,\,h,\,P) > \left\{\text{rel}\,(Nc,\text{rel}\,P)^{\times}\right\}$,

$[R \in \text{equ}\,(i,\,h,\,P)\ .>.\!:\!.$

$\bigstar 2{\cdot}20\ .\bigstar 2{\cdot}21\ .\ a \in i\ .>.\ \text{cls'}\ h \frown b \ni \left\{a \in (b,\ \text{div}\,P)\right\} \in 1,$ (1)

$(1)\ .\bigstar 3{\cdot}01\ .\ a \in i\ .\ b \in \iota\ \text{cls'}\ h \frown b \ni \left\{a \in (b,\ \text{div}\,P)\right\}\ .\ aRk\ .>.\ k \in b,$ (2)

$(2)\ .\ b > h\ .\ \exists\ (b,\ \text{div}\,P)\ .>.\ R_b \ni [\rho_b = (b,\ \text{div}\,P)\ .\ \breve{\rho}_b > b\ .\!:\!.$

$a \in (b,\ \text{div}\,P)\ .>_a : aRk\ .=.\ aR_b k] \in 1 \frown \text{cls'}\ (b,\ \text{rel}\,P),$ (3)

$\rho = i\ .\ \bigstar 2{\cdot}21\ .\ (4)\ .>.\ R \in \left\{\text{rel}\ (Nc,\ \text{rel}\,P)^{\times}\right\}\ .>.\ \text{Prop}].$

·032 $\text{equ}\,(i,\,h,\,P) = \left\{\text{rel}\ (Nc,\ \text{rel}\,P)^{\times}\right\}$,

$[\bigstar 3{\cdot}30\ .\ \bigstar 3{\cdot}31\ .>.\ \text{Prop}].$

Note: ★3·32 gives the general solution for the class of relations indicated by $\text{equ}(i,\,h,\,P)$, in the sense that $\left\{\text{rel}\ (Nc,\ \text{rel}\,P)^{\times}\right\}$, which has been proved to be the same class, is defined by indicating a method for the construction of any member of the class, whereas the definition of $\text{equ}(i,\,h,\,P)$ simply indicates the general property of any member |162| of the class: we have here an example of two different class-concepts with the same extension.

We now proceed to specialize these ideas in the direction of ordinary logical equations.

★4 $i,\,h \in \text{cls}\ .\ P \in \text{rel}\ .\ \pi > i\ .\ \breve{\pi} > h\ .\ \exists\ \text{equ}\ (i,\,h,\,P)\ .>.\!:\!.$

·0[2] $Nc \to 1 \frown \text{equ}\ (i,\,h,\,P) = Nc \to 1 \frown R \ni \left\{\rho = i\ .\ \breve{\rho} > h\ .\ \breve{R}P > 0'\right\}.$

·01 $b > h\ .\ \exists\,(b,\ \text{div}\,P)\ .>.\ (b,\ Nc \to 1,\ P) = Nc \to 1 \frown (b,\ \text{rel}\,P).$ Df.

·02 $(Nc,\ Nc \to 1,\ P) = x \ni [\exists\ \text{cls'}\ h \frown b \ni \left\{x = (b,\ Nc \to 1,\ P)\right\}].$ Df.

·03 $\left\{\text{rel}\ (Nc,\ Nc \to 1,\ P)^{\times}\right\}$

$= \text{rel} \frown S \ni [\exists (\ Nc,\ Nc \to 1,\ P)^{\times} \frown M \ni \{S = \frown\text{'}M\}].$ Df.

·04 $Nc \to 1 \frown \text{equ}\ (i,\,h,\,P) = \left\{\text{rel}\ (Nc,\ Nc \to 1,\ P)^{\times}\right\},$

$[\bigstar 3{\cdot}32\ .>.\ \text{Prop}].$

Note: With the notation of the note on ★1·2, we have, if $R \in Nc \to 1 \frown \text{equ}\ (\iota,\,h,\,P)$,

$$a_k \frown x_k = \Lambda,$$

and the logical sum of all the classes of the type x_k is equal to *i*, and the logical product of any two different classes of the type x_k, say x_k and $x_{k'}$, where *k* is different from k', is null,[3] that is,[4]

$$k,\ k' \varepsilon\ h\ .\ k\ 0'\ k'\ .>.\ x_k \frown x_{k'} = \Lambda.$$

1. In the following expression, the original erroneously had '$(Nc, \text{r}1P)$' instead of '$(Nc, \text{rel}\,P)$'.
2. In the following equation, the original erroneously had 'o' instead of '0'.
3. The original erroneously had 'nu l'.
4. In the following equation, the original erroneously had 'o' instead of '0'.

Thus the class of classes of the type x_k is exhaustive of i and the classes are mutually exclusive. For instance, if the number of indices is 2, so that $h \in 2$, and if $\bar{x}_1$ is put for $i \sim x_1$, then $x_1 \smile x_2 = i$ and $x_1 \frown x_2 = \Lambda$; hence $x_2 = \bar{x}_1$, and the equation becomes

$$a_1 \frown x_1 \smile a_2 \frown \bar{x}_1 = \Lambda.$$

The general relation of the above theorems to logical equations with a finite number of unknowns is considered in the next set of propositions ★5·0 to ★5·11; and equations with a finite number of variables are again considered in set ★12.

We shall use the following notation wherever the symbol i represents a class

$$x \rangle i \;.\rangle_x.\; \bar{x} = i \sim x.$$

★5[1] $i \in \text{cls} : x \rangle i \;.\rangle_x.\; \bar{x} = i \sim x : \nu \in Nc \text{ fin} \,.\, h = Nc \frown \beta \ni (0 < \beta \leqq \nu) :$
$P \in \text{rel} \,.\, \pi \rangle i \,.\, \breve{\pi} \rangle h \,.\, \exists\, \text{equ}\,(i, h, P) : \beta \in h \;.\rangle.\; a_\beta = \pi\beta :$
$R \in \text{equ}\,(i, h, P) \,.\, \beta \in h \;.\rangle.\; x_\beta = \rho\beta :\rangle \therefore$

|163|

·0 $\lambda \in Nc \,.\, 0 < \lambda \leqq \nu \,.\, b = \iota 1 \smile \iota 2 \cup \iota 3 \cup \;\ldots\; \iota\lambda \;.\rangle.$
$(b, \text{div}\, P) = \bar{a}_1 \cap \bar{a}_2 \frown \ldots \frown \bar{a}_\lambda \frown a_{\lambda+1} \frown a_{\lambda+2} \frown \ldots \frown a_\nu,$
[★2·0 .›. Prop].

·01 $(\Lambda, \text{div}\, P) = a_1 \frown a_2 \frown \ldots \frown a_\nu = \Lambda,$
[★2·0 . ★3·02 . Hp .›. Prop].

·02 $(h, \text{div}\, P) = \bar{a}_1 \frown \bar{a}_2 \frown \ldots \frown \bar{a}_\nu.$

·03 $\beta \in h \,.\, A = Z \ni [\exists\, h \frown \lambda \ni (Z = a_\lambda)].$

$D^A_\beta = y \ni [\exists\, C^A_\beta \frown u \ni (y = \frown`u)] \;.\rangle.\; S_\beta = \smile`D^A_\beta.$ Df.

Note: $S_1, S_2, \ldots S_\nu$ are the symmetric functions of[2] $a_1, a_2, \ldots, a_\nu$. as defined in "Symb. Logic," Part I, §2: thus, $S_1 = a_1 \smile a_2 \smile \ldots \smile a_\nu$ and $S_\nu = a_1 \frown a_2 \frown \ldots \frown a_\nu$.

·04 $S_0 = i.$ Df.

Note: This definition is convenient to preserve the generality of certain formulæ.

·05 $\beta \in Nc \,.\, \beta \leqq \nu \;.\rangle.\smile`(\beta, \text{div}\, P) = S_{\nu-\beta} \frown \bar{S}_{\nu-\beta+1},$
[★2·22 . ★5·0 . ★5·03 .›. Prop].

·06 $S_\nu = \Lambda,$ [★5·01 .=. Prop].

·10 $R \in \text{equ}\,(i, h, P) \;.\rangle: (a_1 \frown x_1) \smile (a_2 \frown x_2) \smile \ldots \smile (a_\nu \frown x_\nu) = \Lambda.$
$x_1 \smile x_2 \smile \ldots \smile x_\nu = i,$
[Hp (★5) .›. Prop].

1. In the following equation, the original erroneously had 'o' instead of '0'.
2. In the following expression, the original erroneously omitted the subscript 'ν'.

·11[1] $R \in Nc \rightarrow 1 \frown \text{equ}\,(i,\,h,\,P) \,.\!\supset\!.\!\because\! \star 5\cdot 10 : \lambda,\,\lambda' \in h.$

$\lambda 0\text{'}\lambda' .\!\supset\!.\; x_{\lambda} \frown x_{\lambda'} = \Lambda.$

Note: Comparing this with the ordinary type of logical equation, for instance, in two variables,

$$(a \frown x \frown y) \smile (b \frown x \frown \bar{y}) \smile (c \frown \bar{x} \frown y) \smile (d \frown \bar{x} \frown \bar{y}) = \Lambda,$$

we see that $x_1 = x \frown y,\ x_2 = x \frown \bar{y},\ x_3 = \bar{x} \frown y,\ x_4 = \bar{x} \frown \bar{y}$. Thus $x = x_1 \smile x_2,\ y = x_1 \smile x_3$. Also for the comparison to hold, ν must be a number of the type 2^{δ}, and then δ is the number of unknowns in the ordinary logical equation. But whatever ν may be, the equation

$$(a_1 \frown x_1) \smile (a_2 \frown x_2) \smile \ldots\ldots \smile (a_{\nu} \frown x_{\nu}) = \Lambda,$$

where $x_1 \smile x_2 \smile \ldots\ldots \smile x_{\nu} = i$ and[2] $x_{\lambda} \frown x_{\lambda'} = \Lambda,\ (\lambda 0\text{'}\lambda')$ can always be modified into an equation of the required type.

|164| For let $\delta \leqq \nu \leqq 2^{\delta}$, and let $a_{\nu+1},\ a_{\nu+2},\ \ldots\ldots a_{2^{\delta}}$ be each equal to i, so that

$$\nu < \lambda \leqq 2^{\delta} .\!\supset\!.\; a_{\lambda} = i,$$

then
$$\nu < \lambda \leqq 2^{\delta} .\, x_{\lambda} \supset i \,.\, a_{\lambda} \frown x_{\lambda} = \Lambda \,.\!\supset\!.\, x_{\lambda} = \Lambda \,.$$

Hence by adding on to $a_1,\ \ldots\ldots a_{\nu}$ the $(2^{\delta} - \nu)$ terms $a_{\nu+1},\ \ldots\ldots a_{2^{\delta}}$ (all equal to i), and to $x_1,\ \ldots\ldots x_{\nu}$ the $(2^{\delta} - \nu)$ terms $x_{\nu+1},\ \ldots\ldots x_{2^{\delta}}$ (all equal to Λ), we obtain

$$(a_1 \frown x_1) \smile (a_2 \frown x_2) \smile \ldots\ldots \smile (a_{2^{\delta}} \frown x_{2^{\delta}}) = \Lambda,$$

where $x_1 \smile x_2 \smile \ldots\ldots \smile x_{2^{\delta}} = i$ and[3] $x_{\lambda} \frown x_{\lambda'} = \Lambda,\ (\lambda 0\text{'}\lambda')$, and $x_1,\, x_2,\ \ldots\ldots x_{\nu}$ can be any set of terms satisfying the unmodified equation and can be no other set. Hence there is no loss of generality in supposing that ν is always of the form 2^{δ}. The next set of propositions (★6) will deal with the generalization of this reasoning for the case when ν may be infinite.

★6 $\quad i,\, h \in \text{cls} \,.\, \nu \in Nc \,.\, h \in \nu \,.\, R \in Nc \rightarrow 1 \,.\, \rho = i \,.\, \breve{\rho} \supset h \,.\!\supset\!.\!\therefore$

·10 $\quad \exists\, Nc \frown \delta \ni (\delta \leqq \nu \leqq 2^{\delta}).$

·2 $\quad P \in \text{rel} \,.\, \pi \supset i \,.\, \breve{\pi} \supset h \,.\, \exists\, \text{equ}\,(i,\,h,\,P) \,.\, \delta \in Nc \,.\, \delta \leqq \nu \leqq 2^{\delta}.$

$h' \in \text{cls} \,.\, h \frown h' = \Lambda \,.\, h \smile h' = 2^{\delta}.$

$P' = \iota\, \text{rel} \frown P'' \ni [\pi'' \supset i \,.\, \breve{\pi}'' \supset h \smile h' \therefore z \in h \,.\!\supset\!: xP'z \,.\!=\!.\, xPz \therefore$

$z \in h' .\, x \in i \,.\!\supset_{x,z}\!.\, xP'z] \,.\!\supset\!:\ \text{equ}\,(i,\, h \smile h',\, P') = \text{equ}\,(i,\,h,\,P).$

Note: This proposition, of which the proof is easy, shows that there is no loss of generality in always assuming, when convenient, ν to be of the form 2^{δ}.

---

1. In the following expression, the original erroneously had 'o' instead of '0'.
2. In the following equation, the original erroneously had 'o' instead of '0'.
3. In the following equation, the original erroneously had 'o' instead of '0'.

Section II.
The Cardinal Numbers of Various Classes.

★10 $u,\ v \in \mathrm{cls}\ .\ u \frown v = \Lambda\ .\rangle \therefore$

·01 $(\rangle u,\ \mathrm{rel},\ \rangle v) = \mathrm{rel} \frown R \ni (\rho \rangle u\ .\ \breve{\rho} \rangle v)$. Df.

·11 $(u,\ \mathrm{rel},\ \rangle v) = \mathrm{rel} \frown R \ni (\rho = u\ .\ \breve{\rho} \rangle v)$. Df.

·12 $(\rangle u,\ \mathrm{rel},\ v) = \mathrm{rel} \frown R \ni (\rho \rangle u\ .\ \breve{\rho} = v)$. Df.

·13 $(u,\ \mathrm{rel},\ v) = \mathrm{rel} \frown R \ni (\rho = u\ .\ \breve{\rho} = v)$. Df.

·2 $(u\ ;\ v)(x,\ y) \ni (x \in u\ .\ y \in v)$. Df.

|165|

·30 $\mu(u\ ;\ v) = \mu u \times \mu v$, [cf. Card. Numb. ★7·21].

·31 $\mu(\rangle u,\ \mathrm{rel},\ \rangle v) = \mu\ \mathrm{cls'}\ (u\ ;\ v) = 2^{\mu u \times \mu v}$, [cf. Card. Numb. ★15·0].

·32 $\mu(u,\ \mathrm{rel},\ \rangle v) = (2^{\mu v} - 1)^{\mu u}$,

$[x \in u\ .\rangle.\ k_x = l \ni \{l \rangle \iota x \smile \mathrm{cls'}\ v \sim \iota\Lambda\ .\ x \in l\ .\ l \frown \mathrm{cls'}\ v \in 1\} \therefore$

$k = p \ni \{\exists u \frown x \ni (p = k_x)\} :\rangle.\ k \in \mu\ u\ .\ k \rangle \mu\ \mathrm{cls'}\ v \sim \iota\Lambda,$ (1)

$m \in k^{\times} .\rangle \therefore x \in u\ .\rangle_x.\ m \frown k_x \in 1 : m \rangle u \smile \mathrm{cls'}\ v \sim \iota\Lambda\ .\ u \rangle m :: \rangle.$

$\mu k^{\times} = \mu(u,\ \mathrm{rel},\ \rangle v),$ (2)

(Card. Numb. ★12·1) . (1) . (2) .›. Prop].

·33 $\mu u + \mu v \in Nc\ \mathrm{infin}\ .\ \mu u > 1\ .\ \mu v > 1\ .\rangle.\ \mu(u,\ \mathrm{rel},\ \rangle v) = 2^{\mu u \times \mu v} = \mu(\rangle u,\ \mathrm{rel},\ v)$.

·40 $Nc \to 1 \frown (u,\ \mathrm{rel},\ \rangle v) = v^u$, [cf. Card. Numb. ★14·0].

·41 $\mu\{Nc \to 1 \frown (u,\ \mathrm{rel},\ \rangle v)\} = \mu v^{\mu u}$, [cf. Card. Numb. ★14·1].

·51[1] $\mu\{Nc \to 1 \frown (\rangle u,\ \mathrm{rel},\ \rangle v)\} = (1 + \mu v)^{\mu u}$,

$[w \in C^u_\beta .\rangle.\ \mu\{Nc \to 1 \frown (w,\ \mathrm{rel},\ \rangle v)\} = (\mu v)^\beta,$ (1)

$(1)\ .\rangle.\ \mu\{Nc \to 1 \frown (\rangle u,\ \mathrm{rel},\ \rangle v)\} = \sum^{\beta \leqq \mu u} C^{\mu u}_\beta \times (\mu v)^\beta,$ (2)

(Card. Numb. ★17·4) . (2) .›. Prop].

·52 $\mu\{1 \to Nc \frown (\rangle u,\ \mathrm{rel},\ \rangle v)\} = (1 + \mu u)^{\mu v}$.

·61 $\mu u > 1\ .\ \mu v > 1 .\rangle: (2^{\mu v - 1} - 1)^{\mu u - 1} \leqq \mu(u,\ \mathrm{rel},\ v) \leqq (2^{\mu v} - 1)^{\mu u}$.

$(2^{\mu u - 1} - 1)^{\mu v - 1} \leqq \mu(u,\ \mathrm{rel},\ v) \leqq (2^{\mu u} - 1)^{\mu v}$,

$[x \in u\ .\ y \in v\ .\ R \in (u \sim \iota x,\ \mathrm{rel},\ \rangle v \sim \iota y)\ .\ R' \in \mathrm{rel}\ .\ \rho' = \iota x.$

$\breve{\rho}' = v \sim \breve{\rho}(u \sim \iota x)\ .\rangle.\ \exists \breve{\rho}' .\rangle.\ R \smile R' \in (u,\ \mathrm{rel},\ v),$ (1)

$(1)\ .\ x \in u\ .\ y \in v\ .\rangle.\ \mu(u,\ \mathrm{rel},\ v) \geqq \mu(u \sim \iota x,\ \mathrm{rel},\ \rangle v \sim \iota y),$ (2)

$(2)\ .\ \bigstar 10 \cdot 32\ .\ (u,\ \mathrm{rel},\ v) \rangle (u,\ \mathrm{rel},\ \rangle v)\ .\rangle.$ Prop].

1. In the following expression, the original erroneously omitted the first right parenthesis.

·62 $\mu u > 1 \,.\, \mu v > 1 \,.\, \mu u + \mu v \in Nc$ infin $.\!\supset\!.\ \mu(u, \text{rel}, v) = 2^{\mu u \times \mu v}$,

[★10·61 $.\!\supset\!.$ Prop].

★11 $i, h \in \text{cls} \,.\, P \in \text{rel} \,.\, \pi \supset i \,.\, \bar{\pi} \supset h \,.\, \exists\ \text{equ}\ (i, h, P)\ .\!\supset\therefore$

·0 $\mu\ \text{equ}\ (i, h, P) = \mu\left\{\text{rel}\ (Nc, \text{rel}\ P)^{\times}\right\} = \mu(Nc, \text{rel}\ P)^{\times}$.

[★3·32 . ★3·20 $.\!\supset\!.$ Prop].

·01[1] $\mu\ \text{equ}\ (i, h, P) = \prod^{\beta \leqq \mu h} \mu(\beta, \text{rel}\ P)^{\times}$,

[(Card. Numb. ★10·22) . ★11·0 . ★3·11 . ★3·12 $.\!\supset\!.$ Prop].

·11 $b \supset h\ .\!\supset\!.\ \mu(b, \text{rel}\ P) = \left(2^{\mu b} - 1\right)^{\mu(b,\, \text{div}\ P)}$,

[★3·10 . ★10·32 $.\!\supset\!.$ Prop].

|166|

·12 $\beta \in Nc\ .\!\supset\!.\ \mu(\beta, \text{rel}\ P)^{\times} = \left(2^{\beta} - 1\right)^{\mu\smile`(\beta,\, \text{div}\ P)}$,

[★3·11 . ★2·22 . ★11·11 . (Card. Numb. ★10·22 . ★13·1) $.\!\supset\!.$ Prop].

·13[2] $\mu\ \text{equ}\ (i, h, P) = \prod^{\beta \leqq \mu h} \left(2^{\beta} - 1\right)^{\mu\smile`(\beta,\, \text{div}\ P)}$,

[★11·01 . ★11·12 $.\!\supset\!.$ Prop].

Note: This is the general formula for the number of relations belonging to the class $\text{equ}(i, h, P)$, and thus also for the number of solutions of the corresponding logical equation.

·21 $\mu\{Nc \to 1 \frown \text{equ}\ (i, h, P)\} = \mu\left\{\text{rel}\ (Nc, Nc \to 1, P)^{\times}\right\} = \mu(Nc, Nc \to 1, P)^{\times}$,

[★4·03 . ★4·04 $.\!\supset\!.$ Prop].

·22 $\mu\{Nc \to 1 \frown \text{equ}\ (i, h, P)\} = \prod^{\beta \leqq \mu h} \mu(\beta, Nc \to 1, P)^{\times}$,

[(Card. Numb. ★10·22) $.\!\supset\!.$ Prop].

·23 $b \supset h\ .\!\supset\!.\ \mu(b, Nc \to 1, P) = (\mu b)^{\mu(b,\, \text{div}\ P)}$,

[★4·01 . ★3·10 . ★10·41 $.\!\supset\!.$ Prop].

·24[3] $\beta \in Nc\ .\!\supset\!.\ \mu(\beta, Nc \to 1, P)^{\times} = \beta^{\mu\smile`(\beta,\, \text{div}\ P)}$,

[★3·11 . ★2·22 . (Card. Numb. ★10·22 ★13·1) $.\!\supset\!.$ Prop].

·25 $\mu\{Nc \to 1 \frown \text{equ}\ (i, h, P)\} = \prod^{\beta \leqq \mu h} \beta^{\mu\smile`(\beta,\, \text{div}\ P)}$,

[★11·22 . ★11·24 $.\!\supset\!.$ Prop].

Note: This is the general formula for the number of relations belonging to the class $Nc \to 1 \frown \text{equ}\ (i, h, P)$, and thus also for the number of solutions of the corresponding logical equation. M. Poretsky[4] has given the number of solutions of a

1. In the following equation, the original erroneously omitted the comma after the 'β'.
2. In the following equation, the original erroneously had '$\mu\smile(\beta, \text{div}\ P)$' instead of '$\mu\smile`(\beta, \text{div}\ P)$'.
3. In the following equation, the original erroneously had '$\smile\cdot\mu(\beta, \text{div}\ P)$' instead of '$\mu\smile`(\beta, \text{div}\ P)$'.
4. Platon Sergeevich Poretsky (1846–1907) was a Russian astronomer and mathematician.

logical equation in one variable $\left(\text{viz., } a \frown x \smile b \frown \bar{x} = \Lambda\right)$ in the *Revue de Mathématiques*, Turin, Tome VI, 1896, in his paper, "La Loi des racines en Logique."[1] The solution given now holds for any finite or infinite number of variables. We proceed to state the propositions ★11·13 and ★11·25 in forms convenient for the case where the number of variables in the logical equation is finite; this case has already been partially considered in ★5.

★12 Hp(★5) .›∴

·01 $$\mu \text{ equ } (i,\, h,\, P) = \prod_{\beta > \nu}^{\beta \leqq \nu} \left(2^{\beta} - 1\right)^{\mu\left(S_{\nu-\beta} \frown \bar{S}_{\nu-\beta+1}\right)},$$

[★5·05 . ★11·13 .›. Prop].

|167|

·02[2] $$\mu\left\{Nc \to 1 \frown \text{equ}(i,\, h,\, P)\right\} = \prod_{\beta > 1}^{\beta \leqq \nu} \beta^{\mu\left(S_{\nu-\beta} \frown \bar{S}_{\nu-\beta+1}\right)},$$

[★5·05 . ★11·25 .›. Prop].

Note: ★12·01 and ★12·02 give the number of solutions of the two types of logical equation when the number of variables is finite; ★12·02 is of fundamental importance, especially in the theory of Logical Substitution Groups, developed in Section III. It can be verified (the number of variables being finite) by another method.

·03 $$\mu \bar{S}_{\nu-1} \in Nc \text{ infin } .›. \ \mu \text{ equ } (i,\, h,\, P) = \mu\left\{Nc \to 1 \frown \text{equ } (i,\, h,\, P)\right\} = 2^{\mu \bar{S}_{\nu-1}}$$

$[\alpha_2, \ldots ., \alpha_\nu \in Nc \ . \ \alpha_2 + \ \ldots . + \alpha_\nu \in Nc \text{ infin } .›.$

$$\prod_{\beta > 1}^{\beta \leqq \nu} (2^{\beta} - 1)^{\alpha_\beta} = \prod_{\beta > 1}^{\beta \leqq \nu} \beta^{\alpha_\beta} = 2^{\Sigma \alpha_\beta} \qquad (1)$$

$$1 < \beta \leqq \nu \ .›. \ \bar{S}_{\nu-\beta} \smile \left(S_{\nu-\beta} \frown \bar{S}_{\nu-\beta+1}\right)$$

$$= \bar{S}_{\nu-\beta} \smile \left(\bar{S}_{\nu-\beta} \frown \bar{S}_{\nu-\beta+1}\right) \smile \left(S_{\nu-\beta} \frown \bar{S}_{\nu-\beta+1}\right)$$

$$= \bar{S}_{\nu-\beta} \smile \bar{S}_{\nu-\beta+1} = \bar{S}_{\nu-\beta+1}. \qquad (2)$$

$$(2) \ . \ S_0 = i \ .›. \ \mu\left(S_{\nu-2} \frown \bar{S}_{\nu-1}\right) + \mu\left(S_{\nu-3} \frown \bar{S}_{\nu-2}\right) + \ \ldots .$$

$$+ \mu\left(S_0 \frown \bar{S}_1\right) = \mu \bar{S}_{\nu-1}, \qquad (3)$$ [3]

Hp . (1) . (3) . ★12·01 . ★12·02 .›. Prop].

Note: This proposition is a great simplification of ★12·01 and of ★12·02 in the most important case.

·1 $$\mu \text{ equ } (i,\, h,\, P) \in Nc \text{ fin } . \ \mu\left\{Nc \to 1 \frown \text{equ } (i,\, h,\, P)\right\} \in Nc \text{ fin } :\smile:$$

$$\mu \bar{S}_{\nu-1} \in Nc \text{ infin},$$

[Demonst (★12·03) .›. Prop].

1. Platon Poretsky, 'La loi des racines en Logique', in *Revue de Mathématiques, Tome VI* (Turin: Bocca Frères Libraires, 1899), pp. 5–8. (*Tome VI* spanned 1896–1899.)
2. In the following equation, the original erroneously omitted the '⌢'.
3. The original erroneously misnumbered this as '(4)' (and in the line below).

Note: It follows from ★12·03 and ★12·1 that the number of solutions of a logical equation is either finite or is a number not less than that of the continuum.

·2 $$2^{\mu\bar{S}_{\nu-1}} \leqq \mu\{Nc \rightarrow 1 \frown \text{equ}\,(i, h, P)\} \leqq \nu^{\mu\bar{S}_{\nu-1}},$$

[★12·02 . (demonstration of ★12·03) .›. Prop].

Examples. (A) of ★12·02,

$$(a \frown x) \smile (b \frown \bar{x}) = \Lambda.$$

Here $\nu = 2$, $S_1 = a \smile b$, $S_2 = a \frown b = \Lambda$; hence the number of solu-|168|tions is $2^{\mu\bar{S}_1} = 2^{\mu(\bar{a} \frown \bar{b})}$. This example is the case given by Poretsky.

(B) of ★12·02,

$$(a \frown x \frown y) \smile (b \frown x \frown \bar{y}) \smile (c \frown \bar{x} \frown y) \smile (d \frown \bar{x} \frown \bar{y}) = \Lambda.$$

Here $\nu = 4$, $S_1 = a \smile b \smile c \smile d, \ldots\ldots, S_\nu = a \frown b \frown c \frown d = \Lambda$; hence the number of its solutions is [1]

$$2^{\mu(S_2 \frown \bar{S}_3)} \times 3^{\mu(S_1 \frown \bar{S}_2)} \times 4^{\mu\bar{S}_4} = 2^{2 \times \mu S_1 + \mu(S_2 \frown \bar{S}_3)} \times 3^{\mu(S_1 \frown \bar{S}_2)}$$

and if $\mu\bar{S}_3$ is infinite, it follows from ★12·03 that the number of solutions can be written in the simplified form $2^{\mu\bar{S}_3}$, where

$$\bar{S}_3 = (\bar{a} \frown \bar{b}) \smile (\bar{a} \frown \bar{c}) \smile (\bar{a} \frown \bar{d}) \smile (\bar{b} \frown \bar{c}) \smile (\bar{b} \frown \bar{d}) \smile (\bar{c} \frown \bar{d}).$$

(C) of ★12·01,

$$(a \frown x_1) \smile (b \frown x_2) = \Lambda, \quad x_1 \smile x_2 = i.$$

Here $\nu = 2$, $S_1 = a \smile b$, $S_2 = a \frown b = \Lambda$; and the number of solutions is $(2^2 - 1)^{\mu\bar{S}_1} = 3^{\mu(\bar{a} \frown \bar{b})}$.

(D) of ★12·01,[2]

$$(a \frown x_1) \smile (b \frown x_2) \smile (c \frown x_3) \smile (d \frown x_4) = \Lambda, \quad x_1 \smile x_2 \smile x_3 \smile x_4 = i.$$

Here $\nu = 4$, $S_1 = a \smile b \smile c \smile d, \ldots\ldots, S_\nu = a \frown b \frown c \frown d = \Lambda$; the number of solutions is[3]

$$(2^2 - 1)^{\mu(S_2 \frown \bar{S}_3)} \times (2^3 - 1)^{\mu(S_1 \frown \bar{S}_2)} \times (2^4 - 1)^{\mu\bar{S}_1}.$$

Since, in ★12·02, the coefficients of the equation only enter into the answer through the invariants $S_1, \ldots. S_\nu$, it follows that all equations whose left-hand sides are members of the same congruent family (cf. Symb. Log., Part II, §6), have the same number of solutions; for instance, considering an equation with two unknowns, such as that in example (B) above, for the family of secondary linear primes (cf. Symb. Log., Part I, §3), $S_1 = i$, $S_2 = i$, $S_3 = i$, $S_4 = \Lambda$; hence the number of solutions is 1, as is otherwise known (cf. Symb. Log., Part I, §3). For the family of secondary separable primes, $S_1 = i$, $S_2 = \Lambda$, $S_3 = \Lambda$, $S_4 = \Lambda$; hence the number of

1. In the following equation, the original erroneously had '$4^{\mu\bar{S}}$' instead of '$4^{\mu\bar{S}_4}$'.
2. In the line below, the original erroneously had '$x_1 \smile x_2 \smile x_3 \smile x = \Lambda.^4$' instead of '$x_1 \smile x_2 \smile x_3 \smile x_4 = i.$'
3. In the following expression, the original erroneously omitted the second '$\frown$'.

solutions is $3^{\mu i}$. This can be verified by considering the equation $x \frown y = \Lambda$. For the family of deficiency two and of supplemental deficiency two, $S_1 = i$, $S_2 = i$, $S_3 = \Lambda$, $S_4 = \Lambda$; and hence the number of solutions is $2^{\mu i}$. This is |169| immediately obvious from considering the equation (in two variables), $x \frown (y \smile \bar{y}) = \Lambda$, that is, $x = \Lambda$, and y can be any class subordinate to i.

★13 $i \in \text{cls} : x \rangle i \;.\rangle_x .\; \bar{x} = i \sim x :\rangle\therefore$

·01 $a,\, b \in \text{cls' } i \;.\; u = z \ni [\exists \text{ cls' } i \frown x \ni \{z = (a \frown x) \smile (b \frown \bar{x})\}] \;.\rangle:\; \mu u = 2^{\mu(\bar{a} - b)}$,

$[\mu u = \mu \text{ cls' } (\bar{a} \frown b) \;.\;$ (Card. Numb. ★15·0) $.\rangle.$ Prop].

·11 $a \in \text{cls' } i \;.\; \beta \in Nc \;.\; \beta \leqq \mu a \;.\; u = \text{cls' } i \frown x \ni \left(x \frown a \in C^i_\beta\right) \;.\rangle.\; \mu u = C^{\mu a}_\beta \times 2^{\mu \bar{a}}$,

$[p = \iota C^\alpha_\beta \smile \iota \text{ cls' } \bar{a} \;.\rangle.\; u = x \ni [\exists p^\times \frown m \ni (x = \smile\text{‘}m)] \;.\rangle.$ Prop].

·20 $a,\, b \in \text{cls' } i \;.\; a \frown b = \Lambda \;.\; \alpha \in Nc \;.\; \alpha \leqq \mu(a \smile b)$.

$u = \text{cls' } i \frown x \ni [(a \frown x) \smile (b \frown \bar{x}) \in C^i_\alpha] \;.\rangle.\; \mu u = C^{\mu(a \smile b)}_\alpha \times 2^{\mu p(a,\, \bar{b})}$.

Note:[1] $p(a,\, b) = (a \frown \bar{b}) \smile (\bar{a} \smile b) \;.\; p(a,\, \bar{b}) = (a \frown b) \smile (\bar{a} \frown \bar{b})$ (cf. Symb. Log., Part I, §3). The proof is as follows:

$[\text{Hyp} \;.\rangle.\; u = y \ni [\exists\, Nc \frown (\xi,\, \eta) \ni \{\xi + \eta = \alpha \;.\; \exists(x_1,\, x_2,\, u) \ni (x_1 \in C^a_\xi \;.$

$x_2 \in C^b_\eta \;.\; u \rangle i \;.\; y = x_1 \smile \bar{x}_2 \smile u \frown p(a,\, \bar{b}))\}]$. (1)

(1) . (Card. Numb. ★7·21) $.\rangle.\; \mu u = \sum^{\xi + \eta = \alpha} C^{\mu a}_\xi \times C^{\mu b}_\eta \times 2^{\mu p(a,\, b)}$, (2)

(2) . (Card. Numb. ★16·1) $.\rangle.$ Prop].

·21 $a,\, b \in \text{cls' } i \;.\; \alpha \in Nc \text{ fin} \smile N\alpha_0 \;.\; \mu(a \frown b) < \alpha \leqq \mu(a \smile b)$.

$u = \text{cls' } i \frown x \ni [(a \frown x) \smile (b \frown \bar{x}) \in C^i_\alpha] \;.\rangle.\; \mu u = C^{\mu p(a,\, b)}_{\alpha - \mu(a \frown b)} \times 2^{\mu p(\bar{a},\, b)}$,

$[(a \frown x) \smile (b \frown \bar{x}) \in C^i_\alpha .=. (a \frown \bar{b} \frown x) \smile (\bar{a} \frown b \frown \bar{x}) \in C^i_{\alpha - \mu(a \frown b)}$, (1)

(1) . ★13·20 $.\rangle.$ Prop].

For the definition of $N\alpha_0$, cf. Card. Numb. ★30·0. The point of the limitation to Nc fin or to $N\alpha_0$ is that then $\alpha - \mu(a \frown b)$ is a definite number.

·22 $a,\, b \in \text{cls' } i \;.\; \alpha \in N\alpha_0 \;.\; \mu(a \frown b) < \alpha \leqq \mu(a \smile b)$.

$u = \text{cls' } i \frown x \ni [(a \frown x) \smile (b \frown \bar{x}) \in C^i_\alpha] \;.\rangle.$

$\mu u = C^{\mu p(a,\, b)}_\alpha \times 2^{\mu p(a,\, \bar{b})}$,

$[\alpha - \mu(a \frown b) = \alpha \;.\;$ ★13·21 $.\rangle.$ Prop].

·23 $a,\, b \in \text{cls' } \iota \;.\; \alpha \in Nc \text{ fin} \;.\; \mu(a \frown b) = \alpha \;.$

$u = \text{cls' } i \frown x \ni [(a \frown x) \smile (b \frown \bar{x}) \in C^i_\alpha] \;.\rangle.\; \mu u = 2^{\mu p(a,\, \bar{b})}$,

[★12·02 . $u = \text{cls' } i \frown x \ni [(a \frown b \frown x) \smile (\bar{a} \frown b \frown \bar{x}) = \Lambda] \;.\rangle.$ Prop].

---

1. The final part of this equation erroneously had ‘$(\bar{a} \frown b)$’ instead of ‘$(\bar{a} \frown \bar{b})$’.

|170|

·24[1] $a, b \in \mathrm{cls}'\, i \,.\, \alpha \in N\alpha_0 \,.\, \mu(a \frown b) = \alpha \,.$

$$u = \mathrm{cls}'\, i \frown x \ni [(a \frown x) \frown (b \smile \bar{x}) \in C^{i}_{\alpha}] \,.\supset.$$

$$\mu u = \sum^{\nu \leqq a} C^{\mu p(a,\, b)}_{\nu} \times 2^{\mu p(a,\, \bar{b})},$$

$[\nu \leqq \alpha \,.\supset.\, \alpha + \nu = \alpha \,.\supset.\, u = \mathrm{cls}'\, i \frown x \ni [\exists\, Nc \frown \nu \ni \{\nu \leqq \alpha \,.$

$(a \frown \bar{b} \frown x) \smile (\bar{a} \frown b \frown \bar{x}) \in C^{i}_{\nu}\}],$ (1)

$\bigstar 13{\cdot}20 \,.\, (1) \,.\supset.$ Prop].

·30[2] $a, b \in \mathrm{cls}'\, i \,.\, a \frown b = \Lambda \,.\supset. \sum^{x \supset i} 2^{\mu\{(a \frown x) \smile (b \frown \bar{x})\}} = 2^{\mu p(a,\, \bar{b})} \times 3^{\mu p(a,\, b)},$

$$[\bigstar 13{\cdot}20 \,.\supset. \sum^{x \supset i} 2^{\mu\{(a \frown x) \smile (b \frown \bar{x})\}} = \sum^{\xi \leqq \mu(a \smile b)} C^{\mu(a \smile b)}_{\xi} \times 2^{\mu p(a,\, \bar{b})} \times 2^{\xi}$$

$$= 2^{\mu p(a,\, \bar{b})} \times \sum^{\xi \leqq \mu(a \smile b)} C^{\mu(a \smile b)}_{\xi} \times 2^{\xi}, \quad (1)$$

$(1) \,.\,$ (Card. Numb. $\bigstar 17{\cdot}4$) $.\, a \smile b = p(a,\, b) \,.\supset.$ Prop].

·31[3] $a, b \in \mathrm{cls}'\, i \,.\supset. \sum^{x \supset i} 2^{\mu\{(a \frown x) \smile (b \frown \bar{x})\}} = 2^{\mu(a \frown b) + \mu p(\bar{a},\, b)} \times 3^{\mu p(a,\, b)},$

$[2^{\mu\{(a \frown x) \smile (b \frown \bar{x})\}} = 2^{\mu(a \frown b)} \times 2^{\mu\{(a \frown \bar{b} \frown \bar{x}) \smile (a \frown b \frown \bar{x})\}},$ (1)[4]

$p(a \frown \bar{b},\, \bar{a} \frown b) = p(a,\, b) \,.\, p\{a \frown \bar{b},\, (\bar{a} \frown b)\} = p(a,\, \bar{b}),$ (2)

$\bigstar 13{\cdot}30 \,.\, (1) \,.\, (2) \,.\supset.$ Prop].

Section III.

Orders of Various Logical Substitution Groups.

The properties of these groups have been investigated (cf. Symb. Log., Part II) for the case of functions of two variables. In the present section the orders of the various groups, discussed in the memoir referred to, will be determined. In the reference the theory of substitution groups is not investigated by symbolic methods, accordingly, these methods will be largely abandoned in the present section. A group is a class of operations and the order of the group is the cardinal number of the class. The class *i* will be assumed to contain the classes denoted by the two variables *x* and *y*, and also the classes denoted by the coefficients of any function of one or both of these variables. Also as before,[5]

$$z \supset i \,.\supset_z.\, \bar{z} = i \sim z.$$

1. In the following equation, the original erroneously omitted the '$\frown$'.
2. In the following equations, the original erroneously had '$\breve{x}$' instead of '$\bar{x}$' (twice).
3. In the following equation, the original erroneously had '$\breve{x}$' instead of '$\bar{x}$'.
4. The original erroneously had '$[2^{\mu\{(a \quad x) \smile (b \frown \breve{x})\}} = 2^{\mu(a \frown b)} \times 2^{\mu\{(a \frown \bar{b} \frown \bar{x}) \quad (a \frown b \frown \breve{x})\}}$,' instead of '$[2^{\mu\{(a \frown x) \smile (b \frown \bar{x})\}} = 2^{\mu(a \frown b)} \times 2^{\mu\{(a \frown \bar{b} \frown \bar{x}) \smile (a \frown b \frown \bar{x})\}}$,'.
5. In the following equation, the original erroneously omitted the overbar above the '*z*'.

|171| The statement that $\left.\begin{matrix}\breve{\xi}_1, \breve{\xi}_2, \breve{\xi}_3, \breve{\xi}_4 \\ \eta_1, \eta_2, \eta_3, \eta_4\end{matrix}\right\}$ are the coefficients of a substitution T, means that

$$Tx = \left(\breve{\xi}_1 \frown x \frown y\right) \smile \left(\breve{\xi}_2 \frown x \frown \bar{y}\right) \smile \left(\breve{\xi}_3 \frown \bar{x} \frown y\right) \smile \left(\breve{\xi}_4 \frown \bar{x} \frown \bar{y}\right),$$
$$Ty = \left(\eta_1 \frown x \frown y\right) \smile \left(\eta_2 \frown x \frown \bar{y}\right) \smile \left(\eta_3 \frown \bar{x} \frown y\right) \smile \left(\eta_4 \frown \bar{x} \frown \bar{y}\right),$$

where $\left.\begin{matrix}\breve{\xi}_1, \ldots . \breve{\xi}_4 \\ \eta_1, \ldots . \eta_4\end{matrix}\right\}$ satisfy the equation (cf. Symb. Log., Part II, §2, equation (12)).

★20·01 $\sum\left\{\left(\breve{\xi}_p \frown \breve{\xi}_q\right) \smile \left(\bar{\breve{\xi}}_p \frown \bar{\breve{\xi}}_q\right)\right\} \frown \left\{\left(\eta_p \frown \eta_q\right) \smile \left(\bar{\eta}_p \frown \bar{\eta}_q\right)\right\} = \Lambda, \ (p, q = 1, 2, 3, 4),$

which can also be written in the form

·02 $\sum \bar{p}\left(\breve{\xi}_p, \breve{\xi}_q\right) \frown \bar{p}\left(\eta_p, \eta_q\right) = \Lambda, \ (p, q = 1, 2, 3, 4),$

and also in the form

·03 $\Pi\left(\bar{\breve{\xi}}_r \smile \bar{\eta}_r\right) \smile \Pi\left(\bar{\breve{\xi}}_r \smile \eta_r\right) \smile \Pi\left(\breve{\xi}_r \smile \bar{\eta}_r\right) \smile \Pi\left(\breve{\xi}_r \smile \eta_r\right) = \Lambda, \ (r = 1, 2, 3, 4),$

and also in the form

·04 $$\Pi\left\{\left(\bar{\breve{\xi}}_r \frown \eta_r\right) \smile \left(\bar{\breve{\xi}}_r \frown \bar{\eta}_r\right) \smile \left(\breve{\xi}_r \frown \bar{\eta}_r\right)\right\} \smile \Pi\left\{\left(\bar{\breve{\xi}}_r \frown \eta_r\right) \smile \left(\bar{\breve{\xi}}_r \frown \bar{\eta}_r\right) \smile \left(\breve{\xi}_r \frown \eta_r\right)\right\}$$
$$\smile \Pi\left\{\left(\breve{\xi}_r \frown \eta_r\right) \smile \left(\breve{\xi}_r \frown \bar{\eta}_r\right) \smile \left(\bar{\breve{\xi}}_r \frown \bar{\eta}_r\right)\right\}$$
$$\smile \Pi\left\{\left(\breve{\xi}_r \frown \eta_r\right) \smile \left(\breve{\xi}_r \frown \bar{\eta}_r\right) \smile \left(\bar{\breve{\xi}}_r \frown \eta_r\right)\right\} = \Lambda.$$

Any one of these forms will be called the equation of condition for the coefficients of a substitution. When this equation of condition is fully developed in terms of its eight unknowns $\breve{\xi}_1 \ldots . \breve{\xi}_4, \eta_1 \ldots . \eta_4$, it has 2^8 terms; the coefficients of these various terms are either i or Λ. Considering the form ★20·04, it is easily seen that the first product (i. e., $\Pi\left\{\left(\bar{\breve{\xi}}_r \frown \eta_r\right) \smile \left(\bar{\breve{\xi}}_r \frown \bar{\eta}_r\right) \smile \left(\breve{\xi}_r \frown \bar{\eta}_r\right)\right\}$) gives 3^4 terms with coefficient i; the second product gives $\left(3^4 - 2^4\right)$ other terms with coefficient i, the third product gives $\left(3^4 - 2^4 - 2^4 + 1\right)$ other terms with coefficient i, the fourth product gives $\left(3^4 - 2^4 - 2^4 - 2^4 + 3\right)$ other terms with coefficient i. Hence, there are 232 terms with coefficient i and 24 with with coefficient Λ.

Thus, calculating the invariants $S_0, \ldots . S_{2^8}$ of this equation from S_0 to S_{2^8-24}, they are each equal to i, and from S_{2^8-23} to S_{2^8} they are each equal to Λ. Hence from ★12·02 we deduce

·1 The order of the complete logical substitution group for functions of two variables is $24^{\mu i}$. Thus, if the order is infinite, it is equal to the power of the continuum at least.

|172| The identical group of $\phi(x, y)$ is simply isomorphic with that of the canonical function of the congruent family (cf. Symb. Log., Part II, §7) to which $\phi(x, y)$ belongs. Hence, in order to determine the order of the identical group of $\phi(x, y)$, we have only to determine it for the congruent family. Let s_1, s_2, s_3, s_4 be the invariants of this family, so that $s_4 \succ s_3 \succ s_2 \succ s_1$. Then the coefficients of a substitution of the identical group of the canonical function,

in addition to satisfying the equation of condition (★20·04) must satisfy (cf. Symb. Log., Part II, §7, equ. (37)), the four equations

$$\left.\begin{aligned}
&(s_1 \frown \bar{s}_2 \frown \breve{\xi}_1 \frown \bar{\eta}_1) \smile (s_1 \frown \bar{s}_3 \frown \bar{\breve{\xi}}_1 \frown \eta_1) \smile (s_1 \frown \bar{s}_4 \frown \bar{\breve{\xi}}_1 \frown \bar{\eta}_1) = \Lambda,\\
&(s_1 \frown \bar{s}_2 \frown \breve{\xi}_2 \frown \eta_2) \smile (s_2 \frown \bar{s}_3 \frown \bar{\breve{\xi}}_2 \frown \eta_2) \smile (s_2 \frown \bar{s}_4 \frown \bar{\breve{\xi}}_2 \frown \bar{\eta}_2) = \Lambda,\\
&(s_1 \frown \bar{s}_3 \frown \breve{\xi}_3 \frown \eta_3) \smile (s_2 \frown \bar{s}_3 \frown \breve{\xi}_3 \frown \bar{\eta}_3) \smile (s_3 \frown \bar{s}_4 \frown \bar{\breve{\xi}}_3 \frown \bar{\eta}_3) = \Lambda,\\
&(s_1 \frown \bar{s}_4 \frown \breve{\xi}_4 \frown \eta_4) \smile (s_2 \frown \bar{s}_4 \frown \breve{\xi}_4 \frown \bar{\eta}_4) \smile (s_3 \frown \bar{s}_4 \frown \bar{\breve{\xi}}_4 \frown \eta_4) = \Lambda.
\end{aligned}\right\} \qquad (1)$$

The equations can be combined with the equation of condition into one single equation of condition with 2^8 terms when fully developed. By noticing the symmetry of equations (1) and that of ★20·04, and reducing by the relations between s_1, s_2, s_3, s_4, we find that the values of the 2^8 coefficients are given by the following table, the values of the coefficients being on the upper line and the corresponding number on the lower line being the number of coefficients with that value:

i	Λ	$s_1 \frown \bar{s}_4$	$s_1 \frown \bar{s}_3$	$s_2 \frown \bar{s}_4$	$s_2 \frown \bar{s}_3$	$s_3 \frown \bar{s}_4$	$s_1 \frown \bar{s}_2$	$(s_1 \frown \bar{s}_2) \smile (s_3 \smile \bar{s}_4)$
232'	1'	13 '	3 '	3 '	1 '	1 '	1 '	1

Hence the values of the invariants of the equation $S_0, S_1, \ldots . S_{2^8}$ can be calculated. After some reduction we find that from $\beta = 0$ to $\beta = 2^8 - 24$ inclusive, $S_\beta = i$; and that from $\beta = 2^8 - 23$ to $\beta = 2^8 - 6$, inclusive, $S_\beta = s_1 \frown \bar{s}_4$; and that for $\beta = 2^8 - 5$ and $\beta = 2^8 - 4$, $S_\beta = s_2 \frown \bar{s}_3$; and that from $\beta = 2^8 - 3$ to $\beta = 2^8$, inclusive, $S_\beta = \Lambda$. Hence, from ★12·02, we deduce:

·2 The order of the identical group of any member of the congruent family (s_1, s_2, s_3, s_4), where $s_4 \succ s_3 \succ s_2 \succ s_1$ is[1]

$$24^{\mu(\bar{s}_1 \smile s_4)} \times 6^{\mu\{(s_1 \frown \bar{s}_2) \smile (s_3 \frown \bar{s}_4)\}} \times 4^{\mu(s_2 \frown \bar{s}_3)},$$

which can also be written

$$12^{\mu(\bar{s}_1 \smile s_4)} \times 3^{\mu\{(s_1 \frown \bar{s}_2) \smile (s_3 \frown \bar{s}_4)\}} \times 2^{\mu(s_2 \frown \bar{s}_3)} \times 2^{\mu i}.$$

|173| And from ★12·03, or from ★20·2, we deduce

★20·21 If i is an infinite class, the order of the identical group of any function is $2^{\mu i}$.

Examples. The order of the identical group of a linear secondary prime [of congruent family (i, i, i, Λ)] is $6^{\mu i}$; this is also the order for a separable secondary prime [of congruent family $(i, \Lambda, \Lambda, \Lambda)$].

The order of the identical group of a function of deficiency two and of supplemental deficiency two [of congruent family (i, i, Λ, Λ)] is $4^{\mu i}$.

If $\phi(x, y)$ and $\psi(x, y)$ be any two functions of x and y, we shall always denote by $\Theta(\phi, \psi)$ a certain important function of their coefficients, defined as follows:

Let

$$\phi(x, y) = (a_1 \frown x \frown y) \smile (a_2 \frown x \frown \bar{y}) \smile (a_3 \frown \bar{x} \frown y) \smile (a_4 \frown \bar{x} \frown \bar{y}),$$

---

1. In the below expression, the original erroneously had '$(\bar{s}_1 \frown s_4)$' instead of '$(\bar{s}_1 \smile s_4)$'.

and

$$\psi(x, y) = (b_1 \frown x \frown y) \smile (b_2 \frown x \frown \bar{y}) \smile (b_3 \frown \bar{x} \frown y) \smile (b_4 \frown \bar{x} \frown \bar{y}),$$

then $\quad \Theta(\phi, \psi) = \sum \bar{p}(a_r, a_s; b_r, b_s), \qquad (r, s = 1, 2, 3, 4).$

Thus $\quad \Theta(\phi, \psi) = \Theta(\psi, \phi).$

Also, $\Theta(\phi, \psi)$ is the same as the left-hand side of ★20·02, after substituting a for ξ and b for η; hence, after the same substitution, $\Theta(\phi, \psi)$ can be written in the form of the left-hand side of ★20·01, or of ★20·03, or of ★20·04. Also $\Theta(\phi, \psi) = \Lambda$ is the condition that $\phi(x, y)$ and $\psi(x, y)$ should be a pair of director-functions (cf. Symb. Log., Part II, §1) of some substitution.

We shall now prove the following theorem:

·22 If $\phi(x, y)$ and $\psi(x, y)$ are two functions of x and y, such that $\mu\Theta(\phi, \psi)$ is infinite, the order of the common subgroup of the identical groups of $\phi(x, y)$ and of $\psi(x, y)$ is $2^{\mu\Theta(\phi, \psi)}$; and otherwise the order of the subgroup is finite.

For (cf. Symb. Log., Part II, §8, equ. (36))[1] the coefficients $\left.\begin{matrix}\xi_1, \xi_2, \xi_3, \xi_4\\ \eta_1, \eta_2, \eta_3, \eta_4\end{matrix}\right\}$, of any substitution of the common subgroup, satisfy in |174| addition to ★20·04, the four equations[2]

$$\{p(a_2, a_1; b_2, b_1) \frown \xi_1 \frown \bar{\eta}_1\} \smile \{p(a_3, a_1; b_3, b_1) \frown \bar{\xi}_1 \frown \eta_1\}$$
$$\smile \{p(a_4, a_1; b_4, b_1) \frown \bar{\xi}_1 \frown \bar{\eta}_1\} = \Lambda,$$

$$\{p(a_1, a_2; b_1, b_2) \frown \xi_2 \frown \eta_2\} \smile \{p(a_3, a_2; b_3, b_2) \frown \bar{\xi}_2 \frown \eta_2\}$$
$$\smile \{p(a_4, a_2; b_4, b_2) \frown \bar{\xi}_2 \frown \bar{\eta}_2\} = \Lambda,$$

$$\{p(a_1, a_3; b_2, b_3) \frown \xi_3 \frown \eta_3\} \smile \{p(a_2, a_3; b_2, b_3) \frown \xi_3 \frown \bar{\eta}_3\}$$
$$\smile \{p(a_4, a_3; b_4, b_3) \frown \bar{\xi}_3 \frown \bar{\eta}_3\} = \Lambda,$$

$$\{p(a_1, a_4; b_1, b_4) \frown \xi_4 \frown \eta_4\} \smile \{p(a_2, a_4; b_2, b_4) \frown \xi_4 \frown \bar{\eta}_4\}$$
$$\smile \{p(a_3, a_4; b_3, b_4) \frown \bar{\xi}_4 \frown \bar{\eta}_4\} = \Lambda.$$

Thus the complete condition, satisfied by the coefficients, is an equation in eight variables $\xi_1 \ldots \xi_4, \eta_1 \ldots \eta_4$ with 2^8 terms, of which $2^8 - 24$ are i, one is Λ, and the remaining 23 are equal either to single coefficients of the four equations above, or to sums of these coefficients. Now if $d_1, d_2 \ldots d_{23}$ are these remaining coefficients and $S_0, S_1, \ldots S_{2^8}$ are the invariants of the equation, it is easy to see that

$$\bar{S}_{2^8-1} = \bar{d}_1 \smile \bar{d}_2 \smile \ldots \smile \bar{d}_{23},$$

and hence

1. The original erroneously referenced equation (37).
2. In the final equation below, the original erroneously omitted the overbar from the final '$\bar{\eta}_4$'.

$$\bar{S}_{2^8-1} = \bar{p}(a_1, a_2; b_1, b_2) \smile \bar{p}(a_1, a_3; b_1, b_3) \smile \bar{p}(a_1, a_4; b_1, b_4)$$
$$\smile \bar{p}(a_2, a_3; b_2, b_3) \smile \bar{p}(a_2, a_4; b_2, b_4) \smile \bar{p}(a_3, a_4; b_3, b_4) = \Theta(\theta, \psi).$$

Accordingly, the required proposition follows from ★12·03.

Note: I have not succeeded in shortening the labor of calculating the 256 invariants $S_0, S_1, \ldots . S_{2^8}$ of the complete equation satisfied by the coefficients of any substitution of the common subgroup of the identical groups of two functions. Accordingly, I have not deduced the order, when finite, of this common subgroup. But from ★12·2, we deduce:

★20·3 The order of the common subgroup of the identical groups of and two functions $\phi(x, y)$ and $\psi(x, y)$ is not less than $2^{\mu\Theta(\phi, \psi)}$ and not greater than $2^{8\times\mu\Theta(\phi, \psi)}$.

From this we deduce, as a corollary, that if the order of this common subgroup is unity, the functions $\phi(x, y)$ and $\psi(x, y)$ are a pair of director-functions of some substitution, a proposition already known (cf. Symb. Log., Part II, §8).

|175|

★21·1 The cyclical group generated by any substitution T is, in general, of the 12th order.

For, let

$$Tx = (a_1 \frown x \frown y) \smile (a_2 \frown x \frown \bar{y}) \smile (a_3 \frown \bar{x} \frown y) \smile (a_4 \frown \bar{x} \frown \bar{y}),$$
$$Ty = (b_1 \frown x \frown y) \smile (b_2 \frown x \frown \bar{y}) \smile (b_3 \frown \bar{x} \frown y) \smile (b_4 \frown \bar{x} \frown \bar{y}).$$

Since (cf. Symb. Log., Part II, §2),

$$a_p \frown a_q \frown a_r = \Lambda = \bar{a}_p \frown \bar{a}_q \frown \bar{a}_r, \quad (p, q, r \text{ unequal}),$$

and $$a_p \frown a_q = a_r \frown a_s, \quad (p, q, r, s \text{ unequal}),$$

it follows that in the complete development of i in terms of a_1, a_2, a_3, a_4 (2^4 terms), the only terms not vanishing can be written in the form $a_p \frown a_q$. Similarly for b_1, b_2, b_3, b_4.

Let $$A_{pq} = a_p \frown a_q, \qquad B_{pq} = b_p \frown b_q, \qquad (p, q = 1, 2, 3, 4).$$

Then, from the condition for a substitution,

$$A_{pq} \frown B_{pq} = \Lambda, \quad A_{pq} \frown B_{rs} = \Lambda,$$

where, as in the sequel, different subscripts are unequal. Also

$$a_1 = A_{12} \smile A_{13} \smile A_{14}, \quad b_1 = B_{12} \smile B_{13} \smile B_{14},$$
$$\bar{a}_1 = A_{23} \smile A_{34} \smile A_{42}, \quad \bar{b}_1 = B_{23} \smile B_{34} \smile B_{42},$$

with similar equations for other subscripts.

Also put X_1 for $x \frown y$, X_2 for $x \frown \bar{y}$, X_3 for $\bar{x} \frown y$, X_4 for $\bar{x} \frown \bar{y}$.

Then

$$X_p \frown X_q = \Lambda.$$

Hence $$Tx = \sum A_{pq} \frown (X_p \smile X_q), \qquad (p, q = 1, 2, 3, 4),$$
$$Ty = \sum B_{pq} \frown (X_p \smile X_q), \qquad \text{" "}$$
$$T\bar{x} = \sum A_{pq} \frown (X_r \smile X_s), \qquad (p, q, r, s = 1, 2, 3, 4),$$
$$T\bar{y} = \sum B_{pq} \frown (X_r \smile X_s), \qquad \text{" "}$$

and $TX_1 = Tx \frown Ty = \sum A_{pq} \frown B_{pr} \frown X_p,$ " "

$TX_2 = Tx \frown T\overline{y} = \sum A_{pq} \frown B_{pr} \frown X_q,$ " "

$TX_3 = T\overline{x} \frown Ty = \sum A_{pq} \frown B_{pr} \frown X_r,$ " "

$TX_4 = T\overline{x} \frown T\overline{y} = \sum A_{pq} \frown B_{pr} \frown X_s,$ " "

|176| Thus[1] $T\{A_{pq} \frown B_{qr} \frown X_1\} = A_{pq} \frown B_{qr} \frown X_p,$ $(p, q, r, s = 1, 2, 3, 4),$

$T\{A_{pq} \frown B_{qr} \frown X_2\} = A_{pq} \frown B_{qr} \frown X_q,$ " "

$T\{A_{pq} \frown B_{qr} \frown X_3\} = A_{pq} \frown B_{qr} \frown X_r,$ " "

$T\{A_{pq} \frown B_{qr} \frown X_4\} = A_{pq} \frown B_{qr} \frown X_s,$ " "

Hence if P is one of the terms

$A_{pq} \frown B_{qr} \frown X_p$, or $A_{pq} \frown B_{qr} \frown X_q$, or $A_{pq} \frown B_{qr} \frown X_r$, or $A_{pq} \frown B_{qr} \frown X_s$,

we can easily verify that, either $TP = P$, or $T^2P = P$, or $T^3P = P$, or $T^4P = P$; for instance,

$$T^3\{A_{23} \frown A_{24} \frown X_1\} = T^2\{A_{23} \frown A_{24} \frown X_3\} = T\{A_{23} \frown A_{24} \frown X_4\} = A_{23} \frown A_{24} \frown X_1,$$

$$T^4\{A_{23} \frown A_{24} \frown X_3\} = T^3\{A_{23} \frown A_{24} \frown X_4\}$$

$$= T^2\{A_{23} \frown A_{24} \frown X_1\} = T\{A_{23} \frown A_{24} \frown X_2\} = A_{23} \frown A_{24} \frown X_3.$$

Hence the smallest number n for which the equation $T^nP = P$ holds for every term P of the type defined above is 12.

But remembering the conditions satisfied by a_1, a_2, a_3, a_4 b_1, b_2, b_3, b_4, we see that we can write

$$\phi(x, y) = \sum g \frown A_{pq} \frown B_{qr} \frown X,$$

where $p, q, r = 1, 2, 3, 4$, g is any coefficient and X is any one of X_1, X_2, X_3, X_4. Hence the proposition follows.

Section IV.
The Group of Primary Prime Substitutions.

Consider a substitution T such that Tx and Ty are each functions of one variable only, not the same for both; for instance, we will suppose that Tx is a function of x only, and Ty is a function of y only. Then, if $\left.\begin{matrix}\breve{\xi}_1, \breve{\xi}_2, \breve{\xi}_3, \breve{\xi}_4 \\ \eta_1, \eta_2, \eta_3, \eta_4\end{matrix}\right\}$ are the coefficients of T, we must have

$$\breve{\xi}_1 = \breve{\xi}_2,\ \breve{\xi}_3 = \breve{\xi}_4;\quad \eta_1 = \eta_3,\ \eta_2 = \eta_4,$$

and hence from ★20·01, $\breve{\xi}_3 = \overline{\breve{\xi}}_1$, $\eta_2 = \overline{\eta}_1$.

Thus, $Tx = \left(\overline{\breve{\xi}} \frown x\right) \smile \left(\breve{\xi} \frown \overline{x}\right) = p\left(\breve{\xi}, x\right),\ Ty = p(\eta, y),$

1. In the second line below, the original erroneously had 'X_p' instead of 'X_q'.

is the general form for such a substitution; both Tx and Ty are pri-|177|mary primes. Let such a substitution be called a primary prime substitution.

★22·0 Substitutions of the type $Tx = p(\breve{\xi}, x)$, $Ty = p(\eta, y)$ form an abelian group, in which every substitution is of the second order.

For if T' be the substitution, $T'x = p(\breve{\xi}', x)$, $T'y = p(\eta', y)$, then

$$T'Tx = \{\bar{p}(\breve{\xi}, \breve{\xi}') \frown x\} \smile \{p(\breve{\xi}, \breve{\xi}') \frown \bar{x}\},$$

$$T'Ty = \{\bar{p}(\eta, \eta') \frown y\} \smile \{p(\eta, \eta') \frown \bar{y}\}.$$

Hence $T'T$ is a substitution of the same form.

Further, these equations show that

$$TT' = T'T \text{ and } T^2 = T_0,$$

where T_0 is the identical substitution. Hence the group is abelian, and every primary prime substitution is of order two.

★22·1 The order of the complete group of primary prime substitutions is $4^{\mu i}$.

For whatever classes contained in i, $\breve{\xi}$ and η may be $Tx = p(\breve{\xi}, x)$

$Ty = p(\eta, y)$ belongs to the group.

·2 The class of congruent families (s_1, s_2, s_3, s_4) such that if $\phi(x, y)$ and $\psi(x, y)$ are members of the same family of the class, a primary prime substitution T can be found such that $T\phi(x, y) = \psi(x, y)$, is the class of congruent families for which $s_2 = s_3$.

For if a_1, a_2, a_3, a_4 are the coefficients of $\phi(x, y)$ and b_1, b_2, b_3, b_4 of $\psi(x, y)$, and $\breve{\xi}, \eta$ are the parameters of the required substitution, then (cf. Symb. Log., Part II, §6, equ (31))[1] the condition for these two functions is[2]

$$[\{p(a_1, b_1) \smile p(a_2, b_2) \smile p(a_3, b_3) \smile p(a_4, b_4)\} \frown \breve{\xi} \frown \eta] \smile$$

$$[\{p(a_2, b_1) \smile p(a_1, b_2) \smile p(a_4, b_3) \smile p(a_3, b_4)\} \frown \breve{\xi} \frown \bar{\eta}] \smile$$

$$[\{p(a_3, b_1) \smile p(a_4, b_2) \smile p(a_1, b_3) \smile p(a_2, b_4)\} \frown \bar{\breve{\xi}} \frown \eta] \smile$$

$$[\{p(a_4, b_1) \smile p(a_3, b_2) \smile p(a_2, b_3) \smile p(a_1, b_4)\} \frown \bar{\breve{\xi}} \frown \bar{\eta}] = \Lambda.$$

Now we know that the functions must be congruent, hence all we have to do is to seek the condition that any function $\phi(x, y)$ can be so transformed into the canonical function of its family; hence we may put

$$b_1 = s_1, \quad b_2 = s_2, \quad b_3 = s_3, \quad b_4 = s_4,$$

where s_1, s_2, s_3, s_4 are the invariants of the family and $s_4 \succ s_3 \succ s_2 \succ s_1$.

|178| Then the resultant of the above equation, i.e., the condition for its possibility reduces to

$$(a_1 \frown a_3 \frown \bar{a}_2 \frown \bar{a}_4) \smile (a_2 \frown a_3 \frown \bar{a}_1 \frown \bar{a}_4) \smile (a_1 \frown a_4 \frown \bar{a}_2 \frown \bar{a}_3) \smile (a_2 \frown a_4 \frown \bar{a}_1 \frown \bar{a}_3) = \Lambda.$$

Hence, remembering that functions of the same family exist with a_1, a_2, a_3, a_4 interchanged, we find $s_2 \frown \bar{s}_3 = \Lambda$, that is, $s_2 \succ s_3$. But $s_3 \succ s_2$, hence $s_2 = s_3$.

1. See also equation (29) from the same section.
2. In the third line below, the original erroneously had '$\breve{\xi}$' instead of '$\bar{\breve{\xi}}$'.

We notice that the families (i, i, i, Λ) and $(i, \Lambda, \Lambda, \Lambda)$ both belong to this class of families.

★22·3 The class of congruent families (s_1, s_2, s_3, s_4), such that $s_2 = s_3$, is such that if $\phi(x, y)$ and $\psi(x, y)$ be any two members of the same family, a substitution T can be found such that

$$T\phi(x, y) = \psi(x, y) \quad \text{and} \quad T\psi(x, y) = \phi(x, y).$$

This follows from ★22·0 and ★22·2.

·4 The identical group of any function of the family (s_1, s_2, s_3, s_4) contains a primary prime subgroup of order

$$2^{\mu(s_2 \frown \bar{s}_3)} \times 4^{\mu(\bar{s}_1 \smile s_4)}.$$

For in the demonstration of ★22·2 make $\phi(x, y)$ and $\psi(x, y)$ identical by putting a_1, a_2, a_3, a_4 for b_1, b_2, b_3, b_4, then the parameters, $\breve{\xi}$ and η, of the required primary prime substitution must satisfy[1]

$$[\{p(a_1, a_2) \smile p(a_3, a_4)\} \frown \breve{\xi} \frown \bar{\eta}] \smile [\{p(a_1, a_3) \smile p(a_2, a_4)\} \frown \bar{\breve{\xi}} \frown \eta] \smile$$
$$[\{p(a_1, a_4) \smile p(a_2, a_3)\} \frown \bar{\breve{\xi}} \frown \bar{\eta}] = \Lambda.$$

This equation is always possible, and if S_1, S_2, S_3, S_4 are its invariants, we find

$$S_2 \frown \breve{S}_3 = s_2 \frown \bar{s}_3, \quad S_2 \frown \bar{S}_2 = \Lambda, \quad \bar{S}_1 = \bar{s}_1 \smile s_4.$$

Hence from ★12·02 the proposition follows.

1. In the second line below, the original erroneously had 'ξ' instead of '$\bar{\breve{\xi}}$'.

Theorems on Cardinal Numbers

1904 *American Journal of Mathematics*, vol. 26, no. 1 (January), pp. 31–32

Published in January 1904, a note at the bottom specified that this article was composed in January 1903. The significance of the submission date will be seen below. This is essentially an addendum to 'On Cardinal Numbers' (1902)[1] *and adds a few new results to the longer paper. The brief discussion here assumes the ideas summarised in the editors' introduction to that earlier work.*

The principal result of this paper is its second proposition (★2), which is preceded by one and followed by four less salient, but by no means unimportant, results. The focus here is this second result, which is easily explained in non-technical language. The point is that the notion of summing transfinite numbers respects strict inequalities. To illustrate in a less forbidding context, consider the strict inequalities 19 < 28 and 68 < 79, where we think of all four participating operands as counting the number of elements in a given set. We may then make the inference that 19 + 68 < 28 + 79 without completing the indicated sums. Whitehead's second proposition extends this idea to infinite cardinals; in his previous paper he was limited to asserting that in summing a pair of strict inequalities of transfinite numbers, one could infer only the corresponding non-strict inequality. In terms of the example just given, this would be like limiting our conclusion to the weaker statement that 19 + 68 ≤ 28 + 79: true, but the possibility of equality is superfluous; the stronger statement should prevail.

Whitehead notes in the third paragraph that his results here are conditional on dealing with classes of cardinal numbers that satisfy a natural trichotomy. Given cardinals α *and* β *drawn from such classes, then exactly one of the following three statements must hold:* $\alpha < \beta$, $\alpha = \beta$, *or* $\alpha > \beta$. *In other words, cardinal numbers are always comparable in the same sense that ordinary numbers are. (In contrast, the subset relationship has many of the key properties of an ordering, but it is simply not true that, given two sets, one is necessarily a subset of the other; accordingly, this trichotomy does not hold.)*

Of more interest to the historian or philosopher than the results themselves is the crystallisation of this key property of cardinal numbers shortly after the appearance of this paper. In 1904, Ernst Zermelo introduced a form of the Axiom of Choice (AC) precisely to prove that even infinite cardinals are always comparable. What is remarkable is that AC was lurking implicitly in Whitehead's prior paper on cardinality in his 'multiplicative class' construction,[2] *so that all that he was lacking was the correct formulation of this axiom, and that would indeed soon be forthcoming.*

---

1. See p. 156.
2. See p. 174.

In this paper it is proved (★2) that for cardinal numbers (finite or infinite) if $\alpha < \alpha'$ and $\beta < \beta'$, then $\alpha + \beta < \alpha' + \beta'$. It is already known (cf. ★4·37 in my article on *Cardinal Numbers*, in this Journal, Vol. XXIV, 1902)[1] that in this case $\alpha + \beta \leqq \alpha' + \beta'$. Another form of ★2 is ★1, which is the most useful form for deductions. By the help of ★1, it is proved (★3 and ★4) that if *n* be a finite number and α_0 (as usual) the first infinite number, and if $n \times \beta = \alpha_0 \times \beta$, then $\beta = \alpha_0 \times \beta$. It is already known (cf. Zermelo, Gott. Nachricht., 1901)[2] that if $\beta = n \times \beta$, then $\beta = \alpha_0 \times \beta$; and also it is obvious that if $\beta = \alpha_0 \times \beta$, then $\beta = n \times \beta$: then the present theorem completes this set of results.

Furthermore it is proved (★6) that if α is of the form $\alpha_0 \times \gamma$ [i. e. an $N\alpha_0$, cf. *Cardinal Numbers*, ★30] and $\beta < \alpha$, then $\beta + \sigma = \alpha$ implies $\sigma = \alpha$. This proposition was assumed without proof in my article on *Cardinal Numbers* (cf. ★16·22, for example). It is already known that $\beta + \alpha = \alpha$.

It will be noticed that a condition attaches itself to the hypothesis of ★1, and thence to the hypotheses of the succeeding theorems, namely (stating it generally), that any two numbers less than the numbers considered, are such that one of them is either greater than, or equal to, or less than the other. I am not aware of any proof that one of these relations must hold for any two cardinal numbers; but classes of infinite cardinal numbers are known for which this condition is true.

The proofs are written in Peano's notation explained in the article on *Cardinal Numbers*, already cited; except that here ‹ is used for "is contained in," as applied to classes, and › is used for "implies," as applied to propositions, instead of › for both these ideas. Also, in the proofs where a hypothesis is made to hold for the remainder of the proof "constr" for "constructive" is written after it.

|32|

★1 $\alpha, \beta, \alpha', \beta' \in Nc \,.\, \alpha + \beta = \alpha' + \beta' : \mu \leqq \alpha \,.\, \nu \leqq \beta \,.\text{›}_{\mu,\nu}.$

$$\mu \gtreqless \nu :\text{›}: \alpha \geqq \alpha' . \vee . \beta \geqq \beta',$$

$[R \in 1 \to 1 \,.\, u \cap v = \Lambda \,.\, u \cup v = \rho \,.\, u' \cap v' = \Lambda \,.\, u' \cup v' = \breve{\rho},$ constr (1)

$S \in 1 \to 1 \,.\, \sigma \text{‹} u \cap \rho v' \,.\, \breve{\sigma} = v \cap \rho u',$ constr (2)

$p = (u \cup v) - (\sigma \cup \breve{\sigma}).$ constr (3)

$R' = R_p \cup SR \cup \breve{S}R,$ constr (4)

$(4) .\text{›}. R' \in 1 \to 1 \,.\, \breve{\rho}'u = \breve{\rho}(u - \sigma) \cup \breve{\rho}(v \cap \rho u'),$ (5)

$(5) .\text{›}. u' \text{‹} \breve{\rho}'u .\text{›}. Nc\text{'}u' \leqq Nc\text{'}u,$ (6)

$(2) . (6) .\text{›}: Nc\text{'}(u \cap \rho v') \geqq Nc\text{'}(v \cap \rho u') .\text{›}. Nc\text{'}u' \leqq Nc\text{'}u.$ (7)

Similarly $Nc\text{'}(v \cap \rho u') \geqq Nc\text{'}(u \cap \rho v') .\text{›}. Nc\text{'}v' \leqq Nc\text{'}v,$ (8)

$(7) . (8) : \mu \leqq \alpha \,.\, \nu \leqq \beta \,.\text{›}_{\mu,\nu}. \mu \gtreqless \nu : u \in \alpha \,.\, v \in \beta \,.\, u' \in \alpha' . v' \in \beta' :\text{›}:$ Prop],

1. See p. 156.
2. Ernst Zermelo, 'Über die Addition transfiniter Cardinalzahlen', *Nachrichten von der Königl. Gesellschaft der Wissenschaften zu Göttingen. Mathematisch-physikalische Klasse aus dem Jahre 1901* (1902), pp. 34–38. The original misspelled as 'Nashricht'.

★2 $\alpha, \beta, \alpha', \beta' \in Nc \,.\, \alpha < \alpha' . \beta < \beta' : \mu \leqq \alpha \,.v \leqq \beta \,.\!\supset_{\mu,v} \mu \gtreqless v :\supset.\ \alpha + \beta < \alpha' + \beta'$,
[★1 .⊃. Prop.]

★3 $m, n \in Nc$ fin $- \iota 0 \,.\, m \geqq n \,.\, \beta \in Nc : \mu < \beta \,.\, v < \beta \,.\!\supset_{u,v} \mu \gtreqless v :$

$(m+n) \times \beta = \alpha_0 \times \beta :\supset. m \times \beta = \alpha_0 \times \beta,$

[★1 .⊃. Hyp .⊃: $m \times \beta + n \times \beta = \alpha_0 \times \beta + \alpha_0 \times \beta \,.\!\supset:$

$m \times \beta \geqq \alpha_0 \times \beta . \vee . n \times \beta \geqq \alpha_0 \times \beta,$ (1)

$l \in Nc$ fin . $l \times \beta \geqq \alpha_0 \times \beta .\!\supset\!. l \times \beta = \alpha_0 \times \beta,$ (2)

$m \geqq n \,.\!\supset\!: n \times \beta = \alpha_0 \times \beta .\!\supset\!. m \times \beta = \alpha_0 \times \beta,$ (3)

(1) . (2) . (3) .⊃. Prop].

★4 $n \in Nc$ fin $- \iota 0 \,.\, \beta \in Nc : \mu < \beta \,.\, v < \beta \,.\!\supset_{u,v} . \mu \gtreqless v :\ n \times \beta = \alpha_0 \times \beta :\supset. \beta = \alpha_0 \times \beta.$

[★3 .⊃. Prop.]

★5[1] $m, n \in Nc$ fin $- \iota 0 \,.\, m0'n \,.\, \beta \in Nc : \mu < \beta \,.\, v < \beta \,.\!\supset_{u,v} \mu \gtreqless v :$
$m \times \beta = n \times \beta :\supset. \beta = \alpha_0 \times \beta,$

[$m \times \beta = n \times \beta .\!\supset\!. m \times \beta = \alpha_0 \times \beta.$ (1)

Note: This theorem (1) is given by Zermelo, Gott. Nachrichten, 1901.

(1) . ★4 .⊃. Prop.]

★6 $\alpha \in N\alpha_0 : \mu \leqq \alpha \,.\, v \leqq \alpha \,.\!\supset_{\mu,v} . \mu \gtreqless v : \beta < \alpha \,.\, \beta + \sigma = \alpha :\supset: \sigma = \alpha,$

[Hyp . $\beta + \sigma = \alpha \,.\!\supset\!. \beta + \sigma = \alpha + \alpha,$ (1)

(1) . ★1 . $\beta < \alpha \,.\!\supset\!. \sigma \geqq \alpha,$ (2)

$\beta + \sigma = \alpha \,.\!\supset\!. \sigma \leqq \alpha,$ (3)

(2) . (3) .⊃. Prop.]

1. The original erroneously had '*mo'n*' instead of '*m*0'*n*'.

Note in *Revue de Métaphysique et de Morale*

1905 *Revue de Métaphysique et de Morale*, vol. 13, no. 6 (November), pp. 916–917

In this short note in a French philosophy journal, Whitehead clarifies some points of his mathematical philosophy on which he felt he had been misrepresented. The offending author was Pierre Léon Boutroux (1880–1922) – son of the more famous French philosopher Émile Boutroux – who had written an article in the same journal in July (two issues earlier) entitled 'Correspondance Mathématique: Et Relation Logique' ('Mathematical Correspondence: And Logical Relation').[1] *Boutroux cited an informal conversation with Whitehead, a practice of which Whitehead did not much approve; nearly forty years later he would chide Victor Lowe for attempting to do the same thing.*[2]

The exact occasion upon which the cited conversation took place is unclear. But it is worth noting that Boutroux's uncle was the famed mathematician and physicist Henri Poincaré, and that the latter stayed at Whitehead's home in Grantchester at least once.[3] *Also, in 1900 Russell and Whitehead had attended the First International Congress of Philosophy in Paris, which was attended by Émile Boutroux (who gave the inaugural address), Poincaré and Xavier Léon (who edited the* Revue de Métaphysique et de Morale*).*

According to Whitehead, mathematics is the study of patterns. The basis of mathematics is the practice of pattern recognition by mathematicians; its resulting theory of patterns can be expressed in terms of the logic of relations. Hence in his comments to Boutroux, Whitehead emphasised that the symbolism of the logic of relations was not of practical use in the development of analysis – it did not yet exist and did not guide Newton and Leibniz when they discovered its basic concepts – but is certainly of theoretical use in formulating analysis, and in preventing analytical reasoning from some of the fallacies of the mathematician's intuitive recognition. It also follows that in an informal conversation, the logically exact formulation of the notion of a continuous function is not at play and, hence, that such a conversation is more liable to misunderstandings.

1. Pierre Léon Boutroux, 'Correspondance Mathématique: Et Relation Logique', *Revue de Métaphysique et de Morale*, vol. 13, no. 4 (July 1905), pp. 620–637.
2. 'You must not take my random statements as evidence of my original meaning at the time of writing. And I am *not* prepared to sit down and comment on my past work. Please do *not* quote my letters' (Alfred North Whitehead, 'Letter from Alfred North Whitehead to Victor Lowe, 21 August 1942', Alfred North Whitehead Collection, MS-0282, Box 2, Special Collections, Sheridan Libraries, Johns Hopkins University). By implication, this would also seem to put Lucien Price's *Dialogues of Alfred North Whitehead* (Boston: Little, Brown, 1954) in a negative light. See Brian G. Henning, 'Revisiting Lucien Price's Dialogues', Whitehead Research Project, 3 January 2020, <https://whiteheadresearch.org/2020/01/03/revisiting-lucien-prices-dialogues/> (accessed 4 February 2025).
3. 'I rely on North's memory in reporting that Henri Poincaré was once a guest, and that as he had neglected to pack shirts, after a few days Evelyn offered to lend him one of Alfred's'. Victor Lowe, *Alfred North Whitehead: The Man and His Work, Volume I: 1861–1910* (Baltimore: Johns Hopkins University Press, 1985), p. 205.

Whitehead's comments are important for understanding his stance on the logic of relations. Unlike Russell, he did not aspire to reduce mathematics to logic, though this did not stop him from engaging in the Principia Mathematica *project.*

Whitehead did not have much facility with French, and so it seems somewhat unlikely that he himself composed the note in French. However, as with Whitehead's 1916 article in the same journal,[1] *if Whitehead did indeed originally compose it in English, that original is now lost. This being the case, we are presenting the French and an English translation on facing pages. The translation is by Michael Halewood.*

---

1. 'La Théorie Relationniste de L'Espace' (The Relational Theory of Space), see p. 401.

M. P. Boutroux m'a fait l'honneur de citer quelques remarques de moi dans son article. Je remercie l'éditeur de la *Revue* de me fournir l'occasion de m'expliquer sur elles. P. 627 (note 1), M. P. |917| Boutroux dit: « M. Whitehead, qui a eu l'obligeance de me fournir quelques indications à ce sujet, ne semble pas penser que la théorie des fonctions appartienne à l'heure actuelle au domaine de la Logistique ». Je crains que quelque maladresse d'expression de ma part, dans notre conversation, n'ait induit M. P. Boutroux en erreur. J'ai voulu exprimer l'opinion que le *symbolisme* de la Logique n'était pas *pratiquement* utile dans le développement technique de l'Analyse; mais je maintiens avec assurance que les idées de l'Analyse sont capables d'être exprimées en termes logiques, et qu'on peut exprimer tous les raisonnements dans le symbolisme logique.

P. 633, sont citées des indications, que j'ai notées au cours de la même conversation, sur la méthode générale par laquelle le problème de la « continuation » d'une fonction pourrait s'exprimer en termes logiques. Le problème de la construction de définitions précises est un des plus délicats qui se posent à un logicien-mathématicien, et je serais bien fâché qu'une assertion émise au cours d'une conversation fût présentée comme une définition mûrement réfléchie. Elle ne peut être prise que comme un aperçu grossier qui n'est nullement à l'abri des possibilités de malentendu.

Mr P. Boutroux has done me the honour of citing several remarks of mine in his article. I thank the editor of the *Revue* for providing me with the opportunity to clarify my position. P. 627 (note 1), Mr P. |917| Boutroux wrote: 'Mr. Whitehead, who was kind enough to furnish me with several observations on the matter, does not seem to think that the theory of functions belongs presently to the domain of symbolic logic'. I think that some awkwardness of expression on my part, during our conversation, may have misled Mr P. Boutroux. I had wanted to express the opinion that the *symbolism* of Logic was not *practically* useful in the technical development of the Analysis; but I am confident in maintaining that the ideas of Analysis are capable of being expressed in logical terms, and that all reasoning can be expressed within logical symbolism.

P. 633 cites points that I had made during the course of the same conversation regarding the general method by which the problem of the 'continuation' of a function could be expressed in logical terms. The construction of precise definitions is one of the most delicate problems for a logician-mathematician and I would be rather angry if an assertion made during the course of a conversation were taken as a carefully considered definition. It should only be regarded as a general overview which is liable to the possibility of misunderstanding.

On Mathematical Concepts of the Material World

1906 *Proceedings of the Royal Society of London, Series A, Containing Papers of a Mathematical and Physical Character*, vol. 77, no. 517 (29 March), pp. 290–291 (Abstract)

1906 ***Philosophical Transactions of the Royal Society of London*, vol. 205 (May), pp. 465–525**

A note just below the paper's title specified that it was 'Received September 22, – Read December 7, 1905'. In it, Whitehead literally attempts to make space for the development of physics. He uses the word concepts here in a technical but ultimately informal sense that we shall summarise shortly and moves through a hierarchy of such concepts. He reaches, in the fifth and final instance, one which he speculates might be a serious candidate for the exposition of physics as it was then known, while at the same time he retains the axiomatic aspects of geometry. Regarding the latter, Whitehead works with geometry as an extension of the theory of classes and relations (that is, naïve set theory) and this may strike the present-day reader as suspect. Nonetheless, as described later, his programme was likely not adopted by physicists, for reasons that had almost nothing to do with any latent foundational problems.

This introduction addresses: (1) the Whiteheadian notion of a concept of nature*; (2) the first two such concepts; (3) remarks on the nature of time in physical theories; (4) Whitehead's key moves and corresponding rationale in ascending from what is virtually an axiomatisation of Euclidean geometry to his subsequent, far more sophisticated later formulations; finally, (5) the reception of his framework in the development of twentieth-century physics.*

For Whitehead, a concept has three primary and one supplementary component. He is aware that their logical ordering is not intuitive. Logically, a concept, in this technical sense, begins with a bare relation among objects, called the essential relation, *generally involving three or more relata. The objects occurring in the arguments of the relation then constitute the ontological deployment relative to said relation. This is not to say that Whitehead is doing metaphysics here, but he does objectify the relata relative to his fundamental relation at least in his choice of language (speaking, for instance, of the ultimate existents and objective reals). The relation is then given semantic content in the next step, namely by declaring what axioms or hypotheses it satisfies; inevitably, some definitions must be introduced at this step that will appear in the axioms and again subsequently. To give a most modest example, R(*abc*) might, via axiomatisation, be interpreted as the relation that* a, b *and* c *are collinear points, and that* b *is between* a *and* c*. Many axioms will be required to converge on the plausibility of this meaning, and points become the de facto objective reals in the space defined by this relation.*

Once the axiomatisation is complete, it is then possible to pass on to the third primary component, namely the propositions that follow from the axioms subject to ordinary set theory and logic. In this step it is key not to be led by the semantics latent in the axioms, but rather to confirm the intended semantics by deductions that are semantically neutral; one also uncovers other features of the concept that may surpass any intuition. Finally, the supplementary component of a concept is a set of extraneous relations. *One might think of these as the kind of contingent laws that inevitably separate science from mathematics. In this connection, Whitehead hews rigorously to Ockham's razor, or the principle of parsimony, to develop concepts which require as few as possible extraneous relations.*

Whitehead calls Concept I (the classical concept; p. 227) and Concept II (p. 231) the punctual concepts because, apart from time, their objective reals consist of points and particles. Concept I is basically an axiomatic development of Euclidean geometry (without the explicit notion of distance) within the theory of sets and relations. Its essential relation is that given in the example above, but its principal defect is its dualism: the strictly geometric properties make no mention of time, but to make use of Euclidean space in physics, the particles must be linked to space via an extravagant set of extraneous relations. Concept II, suggested by Russell, is similar, but particles are omitted in favour of explicit relations between spatial and temporal points. Whitehead rejects this on the grounds of the imperceptibility of such relations.

Before proceeding, a brief note on the appearance of time in physical theories is in order; this is a necessity, as Whitehead acknowledges, if we are to be capable of describing change. Note, first, that in Euclidean geometry, time plays no role. The Pythagorean theorem carries no parameter t *to denote the instances in time in which it holds. Geometry, in this sense, is eternal. We might contrast this with the law that describes how far a body falls in an idealised environment, and in this case we will explicitly need t as a parameter in the description: we say an object is here, at this position, at time* t. *Still, the way that* t *appears in the equation is simply parametric: the form of the equation is undisturbed by the passage of any sort of absolute time or the type of clock or temporal units that we use. Thus, paradoxically, this element of dynamics is also essentially timeless in physics. In contrast, consider that if at some point in the history of the universe the form of the position dependency abruptly changed to what it is now (*t^2*) from another distinct form (say,* t^3*). One might say that* t *intrudes in an entirely different way, describing this dependence as epochal rather than merely parametric.*

In Concept III (the Leibnizian concept; p. 231) and both subsequent concepts, time appears explicitly in the essential relation and, in this sense, all of geometry is relativised to time. Thus, the incidence example above now becomes $R(abct)$*, with the altered meaning that the points* a, b *and* c *are collinear at time instant* t. *This explicit inclusion of time begs the question of whether its use is parametric or epochal in the sense described above, and the clear answer is that it is merely parametric. This is noteworthy to the extent that the* t *becomes virtually invisible in geometric deductions, and Whitehead still, for the most part, is essentially doing geometry both here and in his subsequent concepts. Nonetheless, the* t *reminds us that this is a space in which physics is to be done with its attendant contingencies and, indeed, Whitehead characterises this move in Concept III as 'a monistic variant of the classical concept, obtained by abandoning the prejudice against points moving' (p. 231).*

Time continues as a key relatum in Concepts IV (p. 238) and V (p. 253), but now geometric points are deprecated in favour of lines, which become the objective reals of both systems, with points then defined as classes of lines. In Concept IV, this move gives a clever alternative formulation of geometry in which ordinary points are defined via interpoints, a construction of some complexity. There are some technical advantages to this approach, but one might well wonder what has been gained for physicists. That is answered in Concept V, for which Concept IV is only a staging point.

Whitehead prepares for his final concept with a general theory of dimension (p. 241) relativised to a ϕ-class, *where* ϕ *is a property that may or may not hold with respect to some given set. (His banner example is the Euclidean notion of flatness, later to be extended to the more general notion of* homaloty *(p. 254) as the characterisation of Concept V takes shape.) He mentions in his prefatory remarks (p. 218) that this theory of dimension is new and, indeed, it is a remarkable achievement to develop such a fundamental geometric idea in the context of sets and relations. Nonetheless, the more usual development of dimension in linear algebra is lurking nearby in the work's clear analogues of the notions of maximally linearly independent sets and minimal generating sets. Whitehead then uses dimension to give a new development of points from lines, and these bear a direct relationship to the earlier notion of points. The big advance is that when the property* ϕ *is specified as homaloty,* points may have parts. *This in turn allows Whitehead to speculate on different species of points corresponding to different species of particles in nature (p. 270). The possibility is astounding, and Whitehead concludes that, should his programme ultimately succeed, '[T]hese laws [of physics] should not presuppose geometry, but create it' (p. 271).*

Whitehead's paper, for all its ingenuity, did not create a revolution in the foundations of physics, and it is worthwhile to ask why. Whitehead, in fact, anticipates this (p. 217), recognising that his language is not the language of physicists. To explain the work's lack of impact in broadest terms, one might look at two features of the evolution of twentieth-century physics. The first merely acknowledges that geometry in the style of Euclid – not necessarily Euclidean geometry – was subsumed by analytic geometry, and the diaspora of this movement certainly included physicists.[1] *The second is a matter of the development of physics as the systematic investigation of the natural world. While, in principle, Whitehead's species of points may have been extended to the entire deployment of the Standard Model, the very notion of a particle has been subverted by quantum theory and that, together with the dynamic geometry of general relativity, would seem to invalidate the programme that Whitehead envisioned. Still, this must be regarded as a remarkable achievement in the mathematical and scientific world of 1905.*

---

1. See the introduction to 'Axioms of Geometry' in this volume, p. 299.

PREFACE.[1]

THE object of this memoir is to initiate the mathematical investigation of various possible ways of conceiving the nature of the material world. In so far as its results are worked out in precise mathematical detail, the memoir is concerned with the possible relations to space of the ultimate entities which (in ordinary language) constitute the "stuff" in space. An abstract logical statement of this limited problem, in the form in which it is here conceived, is as follows: Given a set of entities which form the field of a certain polyadic (*i.e.*, many-termed) relation R, what "axioms" satisfied by R have as their consequence, that the theorems of Euclidean geometry are the expression of certain properties of the field of R? If the set of entities are themselves to be the set of points of the Euclidean space, the problem, thus set, narrows itself down to the problem of the axioms of Euclidean geometry. The solution of this narrower problem of the axioms of geometry is assumed (*cf.* Part II., Concept I.) without proof in the form most convenient for this wider investigation. But in Concepts III., IV., and V., the entities forming the field of R are the "stuff," or part of the "stuff," constituting the moving material world. POINCARÉ[2†] has used language which might imply the belief that, with the proper definitions, Euclidean geometry can be applied to express properties of the field of any polyadic relation whatever. His context, however, suggests that his thesis is, that in a certain sense (obvious to mathematicians) the Euclidean and certain other geometries are interchangeable, so that, if one can be applied, then each of the others can also be applied. Be that as it may, the problem, here discussed, is to find various formulations of axioms concerning R, from which, with appropriate definitions, the Euclidean geometry issues as expressing properties of the field of R. In view of the existence of change in the material world, the investigation has to be so conducted as to introduce, in its abstract form, the idea of time, and to provide for the definition of velocity and acceleration.

The general problem is here discussed purely for the sake of its logical (*i.e.*, mathematical) interest. It has an indirect bearing on philosophy by disentangling the essentials of the idea of a material world from the accidents of one particular concept. The problem might, in the future, have a direct bearing upon physical science if a concept widely different from the prevailing concept could be elaborated, which |466| allowed of a simpler enunciation of physical laws. But in physical research so much depends upon a trained imaginative intuition, that it seems most unlikely that existing physicists would, in general, gain any advantage from deserting familiar habits of thought.

Part I. (i) consists of general considerations upon the nature of the problem and the method of procedure. Part I. (ii) contains a short explanation of the symbols used. Part II. is devoted to the consideration of three concepts, which embody the ordinary prevailing ideas upon the subject and slight variants from them. The present investigation has, as a matter of fact, grown out of the *Theory of Interpoints*, which is presented in Part III. (ii), and of the *Theory of Dimensions* of Part IV. (i).

1. The paper's published abstract (see reference at top) was simply the preface, unaltered.
2. † *Cf.* 'La Science et l'Hypothèse,' chap. III., at the end [Henri Poincaré, *La Science et l'Hypothèse* (Paris: Flammarion, 1902)].

These contain two separate answers to the question: How can a point be defined in terms of lines? The well-known definition[1†] of the projective point, as a bundle of lines, assumes the descriptive point. The problem is to define it without any such assumption. By the aid of these answers two concepts, IV. and V., differing very widely from the current concepts, have been elaborated. Concept V., in particular, appears to have great physical possibilities. Indeed, its chief difficulty is the bewildering variety of material which it yields for use in shaping explanations of physical laws. It requires, however, the discovery of some appropriate laws of motion before it can be applied to the ordinary service of physical science.

The Geometry throughout is taken to be three-dimensional and Euclidean. In Concept V. the definition of parallel lines and the "Euclidean" axiom receive new forms; also the "points at infinity" are found to have an intimate connection with the theory of the order of points on any straight line. The *Theory of Dimensions* is based on a new definition of the dimensions of a space.

The main object of the memoir is the development of the *Theory of Interpoints*, of the *Theory of Dimensions*, and of *Concept V.* The other parts are explanatory and preparatory to these, though it is hoped that they will be found to have some independent value.

PART I.—(i) GENERAL CONSIDERATIONS.

Definition.—The *Material World* is conceived as a set of relations and of entities which occur as forming the "fields" of these relations.

Definition.—The *Fundamental Relations* of the material world are those relations in it, which are not defined in terms of other entities, but are merely particularized by hypotheses that they satisfy certain propositions.

Definition.—The hypotheses, as to the propositions which the fundamental relations satisfy, are called the *Axioms* of that concept of the material world.

Definition.—Each complete set of axioms, together with the appropriate definitions and the resulting propositions, will be called a *Concept of the Material World.*

|467| *Definition.*—The complete class of those entities, which are members of the fields of fundamental relations, is called the class of *Ultimate Existents.* This technical name is adopted without prejudice to any philosophic solution of the question of the true relation to existence of the material world as thus conceived.

Every concept of the material world must include the idea of time. Time must be composed of *Instants* (*cf.* BERTRAND RUSSELL, 'Mind,' N.S., vol. 10, No. 39).[2] Thus *Instants of Time* will be found to be included among the ultimate existents of every concept.

Definition.—The class of ultimate existents, exclusive of the instants of time, will be called the class of *Objective Reals.*

1. †Here in "Descriptive Geometry" straight lines are open, and three collinear points have a non-projective relation of order; in "Projective Geometry" straight lines are closed, and four collinear points have a projective relation of separation.
2. 'Is Position in Time and Space Absolute or Relative?', July 1901, pp. 293–317.

The relation of a concept of the material world to some perceiving mind is not to be part of the concept. Also we have no concern with the philosophic problem of the relation of any, or all, of these concepts to existence.

In Geometry, as derived from the Greeks, the simple elements of space are *points*, and the science is the study of the relations between points. Points occur as members of the fields of these relations. Then matter (the ultimate "stuff" which occupies space) in its final analysis, even if it is continuous, consists of entities, here called *particles*, associated with the points by relations which are expressed by saying that a particle *occupies* (or *is at*) a point. Thus matter merely occurs as one portion of the field of this relation of occupation; the other portion consists of points of space and of instants of time. Thus "occupation" is a triadic relation holding in each specific instance between a particle of matter, a point of space, and an instant of time. According to this concept of a material world, which we will call the *Classical Concept*, the class of ultimate existents is composed of three mutually exclusive classes of entities, namely, *points of space*, *particles of matter*, and *instants of time*. Corresponding to these classes of entities there exist the sciences of *Geometry*, of *Chronology*, which may be defined as the theory of time considered as a one-dimensional series ordinally similar to the series of real numbers, and of *Dynamics*. There appears to be no science of matter apart from its relations to time and space.

Opposed to the classical concept stands LEIBNIZ's theory of the Relativity of Space. This is not itself a concept of the material world, according to the narrow definition here given. It is merely an indication of a possible type of concepts alternative to the classical concept. It is not very obvious how to state this theory in the precise nomenclature here adopted. The theory at least means that the points of space, as conceived in the classical concept, are not to be taken among the objective reals. But a wider view suggests that it is a protest against dividing the class of objective reals into two parts, one part (the space of the classical concept) being the field of fundamental relations which do not include instants of time in their fields, and the other part (the particles) only occurring in the fields of fundamental relations which do include instants of time. In this sense it is a protest against exempting any part of the universe from change. But it is not probable that this is the light in which |468| LEIBNIZ himself regarded the theory. This theory, though at present it is nominally the prevailing one, has never been worked out in the form of a precise mathematical concept. It is on this account criticized severely by BERTRAND RUSSELL (*cf. loc. cit.* and 'Philosophy of LEIBNIZ,' Cambridge, 1900, p. 120), who, however, has gone further than any of its upholders to give it mathematical precision. Of course, from the point of view of this paper, we are not concerned with upholding or combatting any theory of the material world. Our sole purpose is to exhibit concepts not inconsistent with some, if not all, of the limited number of propositions at present believed to be true concerning our sense-perceptions.

Definition.—Any concept of the material world which demands two classes of objective reals will be called a *Dualistic* concept; whereas a concept which demands only one such class will be called a *Monistic* concept.

The classical concept is dualistic; Leibnizian concepts will be, in general, monistic (cf. however Concept IVA.). OCCAM's razor—Entia non multiplicanda præter necessitatem[1]—formulates an instinctive preference for a monistic as against a dualistic concept. Concept III. below is an example of a Leibnizian monistic concept. The objective reals in it may be considered to represent either the particles or the points of the classical concept. But they change their spatial relations. Perhaps Leibniz was restrained from assimilating his ideas more closely to Concept III. by a prejudice against anything, so analogous to a point of space, moving—a prejudice which arises from confusing the classical dualistic concept with the monistic concepts. It is of course essential that at least some members of the class of objective reals should have different relations to each other at different instants. Otherwise we are confronted with an unchanging world. Concept V. is another Leibnizian monistic concept.

The Time-Relation.—In every concept a dyadic serial relation, having for its field the instants of time and these only, is necessary. The properties of this *Time-Relation* form the pure science of chronology. The time-relation is, in all concepts, a serial relation ordinally similar to the serial relation which generates the series of negative and positive real numbers.[2]† This fact need not be further specified during the successive consideration of the various concepts, nor need any of the propositions of pure chronology be enunciated.

Definition.—The class of instants of time is always denoted by T in every concept.

The Essential Relation.—In every concept at least all the propositions of geometry will be exhibited as properties of a single polyadic relation, here called the *essential relation.* The field of the essential relation will consist, *either* of the whole class of ultimate existents (*e.g.*, in Concepts III., IVB. and V.), *or* of part of the class of objective reals together with the instants of time (*e.g.*, in Concept IVA.), *or* of the whole class of objective reals (*e.g.*, in Concept II.), *or* of part of the class of objective |469| reals (*e.g.*, in Concept I.). The essential relation of any one concept will be a relation between a definite finite number of terms, for example, between three terms in Concepts I. and II., between four terms in Concept III., and between five terms in Concepts IV. and V.[3]†

Definition.—In the exposition of every concept, the essential relation of that concept is denoted by R.

1. 'Entities must not be multiplied beyond necessity'.
2. † For interesting reflections on this subject, influenced by the Kantian Philosophy and previous to the modern "Logicization of Mathematics," cf. Hamilton, 'Lectures on Quaternions,' preface [William Rowan Hamilton, *Lectures on Quaternions* (Dublin: Hodges and Smith, 1853)].
3. † The idea of deriving geometry (at least projective geometry without reference to order) from a single triadic relation was (I believe) first enunciated and investigated by Mr. A. B. Kempe, F.R.S., in 1889, *cf.* "On the Relation between the Logical Theory of Classes and the Geometrical Theory of Points," 'Proc. Lond. Math. Soc.,' vol. XXI [No. 1, (November), pp. 147–182; date corrected from 1890]. It has since been worked out in detail for Euclidean geometry by Dr. O. Veblen, *cf.* "A System of Axioms for Geometry," Trans. Amer. Math. Soc.,' vol. 5, 1904 [no. 3 (July), pp. 343–384]. Also *cf.* Professor J. Royce on "The Relation of the Principles of Logic to the Foundations of Geometry," 'Trans. Amer. Math. Soc.,' 1905 [vol. 6, no. 3 (July), pp. 353–415]. Professor Royce emphasises the importance of Kempe's work and considerably extends it. This memoir (which unfortunately only came into my hands after the completion of the present investigation) anticipates a general line of thought of the present paper in the emphasis laid on the derivation of geometry from a single polyadic relation; otherwise our papers are concerned with different problems.

The Extraneous Relations.—In all the concepts here considered, other relations, here called the *extraneous relations*, will be required in addition to the time-relation and the essential relation. In Concepts I. and II. and IV. an indefinite (if not infinite) number of extraneous relations are required, determining the positions of particles. In Concepts III, IV. and V. one tetradic extraneous relation is required to determine the "kinetic axes" of reference for the measurement of velocity.

The time-relation, the essential relation and the extraneous relations form the fundamental relations of any concept in which they occur.

It will now be necessary to define geometry anew, since the previous definition has essential reference to the dualism of the classical concept. A proposition of geometry is any proposition (1) concerning the essential relation; (2) involving one, and only one, instant of time; (3) true for any instant of time.

In the classical concept everything is sacrificed to simplicity in reference to geometry, probably because it arose when geometry was the only developed science. The result is that, when the properties of matter are dealt with, an appalling number of extraneous relations are necessary.

Judged by Occam's principle, this class of extraneous relations forms a defect in Concepts I. and II. and IV. Also, in both forms of the classical concept (viz., in Concepts I. and II.) geometry is segregated from the other physical sciences to a greater degree than in the other concepts.

In the study of any concept there are four logical stages of progress. The first stage consists of the *definition* of those entities which are capable of definition in terms of the fundamental relations. These definitions are logically independent of any axioms concerning the fundamental relations, though their convenience and importance are chiefly dependent upon such axioms. The second stage consists of the |470| *deduction* of those properties of the defined entities which do not depend upon the axioms. The third stage is the *selection* of the group of *axioms* which determines that concept of the material world. The fourth stage is the *deduction* of propositions which involve among their hypotheses some or all of the axioms of the third stage. Psychologically the order of study is apt to be inverted, by first choosing propositions of the second and fourth stages because of their parallelism with the propositions of sense-perception and then by considering the first and third stages. The essential part of our task in passing concepts in review is the exhibition of the first and third stages. The second and fourth stages will only be so far touched upon as seems desirable for the purposes of elucidation.

Thus in respect to each concept considered the investigation will proceed as follows:—A certain relation R (the essential relation of the concept in question), which holds between a certain definite number of entities, is considered. The class of entities, between sets of which this relation holds, is called the "field" of R. Definitions of entities allied to R and to entities of the field of R are then given. These definitions involve no hypotheses as to the properties of R, but are of no importance unless R has as a matter of fact certain properties. For example, it may happen that the classes, thus defined, are all the null class (*i.e.*, the class with no members) unless R has the requisite properties. Again deductions (in the second stage), made without any hypothesis as to the properties of R, may be entirely trivial unless R has certain properties. If R has not the requisite properties the deductions often sink into the assertion that a certain proposition which is false implies some other proposition. This is true[1†] but trifling. The "axioms" respecting R are then given. These are the hypotheses as to the properties of R which are required in the concept under consideration. Finally such deductions are given as are necessary to elucidate the concept.

None of the reasoning of this memoir depends on any special logical doctrine which may appear to be assumed in the form in which it is set out. Furthermore certain contradictions recently discovered have thrown grave doubt upon the current doctrine of classes as entities. Any recasting of our logical ideas upon the subject of classes must of course simply issue in a change of our ideas as to the true logical analysis of propositions in which classes appear. The propositions themselves, except a few extreme instances which lead to contradictions, must be left intact. Accordingly the present memoir in no way depends upon any theory of classes.

The above considerations as to method must essentially hold for any investigation respecting axioms of geometry or of physics, viewed purely as deductive sciences, and apart from the question of experimental verification.

In Concepts I., II., and III. the members of the "field" of R are to be considered as points, except those members of the field which are instants of time. In these concepts the lines and planes are classes of points. In Concepts IV. and V. the |471| members of the "field" of R, other than the instants of time, are to be considered as lines taken as simple entities. Points are classes of these simple lines. But the ordinary line of geometry which has parts and segments is a class of points, and so is the ordinary plane of geometry. In Concept III., which is Leibnizian and monistic, the points (perhaps "particles" is here a better word) move, and the straight lines and planes disintegrate from instant to instant. In Concepts IV. and V. the points similarly disintegrate.

(ii) Explanation of Symbolism.

This explanation is only concerned with the general logical symbolism. The special symbols which arise out of the ideas of the paper are defined in their proper places. Peano's[2†] chief symbols are used. The changes and developments from Peano, which will be found here, are due to Russell and myself working in collaboration for another purpose. It would be impossible to disentangle our various contributions.[3†]

None of the reasoning of the paper is based upon any peculiarity of the symbolism. It is used here only as an alternative form for enunciations, for the sake of its conciseness and (above all) its precision. In the verbal enunciations precision has been to some extent sacrificed to lucidity; and the exact statement of what is meant

1. † *Cf.* Russell, 'The Principles of Mathematics,' § 16 [Bertrand Russell, *The Principles of Mathematics* (Cambridge: Cambridge University Press, 1903), vol. I].
2. † *Cf.* 'Notations de Logique Mathématique,' Turin, 1894; and 'Formulaire Mathématique,' Turin, 1903.
3. † See, however, Russell's articles, "Sur la Logique des Relations," 'Revue de Mathématiques,' vol. VII., 1900–1901, Turin [nos 2–3, pp. 115–136, 137–148].

is always to be sought in the symbolic alternative form. The proofs have been translated into words out of the symbolic form in which they were mostly elaborated.

On $\supset$, $\equiv$, $\subset$, $\in$, =, = Df

There are five copulas, namely, $\supset$, $\equiv$, $\subset$, $\in$, =. Here $x \supset y$ means x *implies* y; and $x \equiv y$ means x *implies* y *and* y *implies* x; and $x \subset y$ means *all* x*'s are* y*'s*; and $x \in y$ means x *is a member of* y; and $x, y \in u$ means x *and* y *are members of* u. Note that $x \subset y$ implies that x is a subclass of y; a class will be said to *contain* a subclass and to *possess* a member. Lastly, $x = y$ means x *is identical with* y. Note that, if Df, short for Definition, is placed at the end of the line, thus,

$$x = y \qquad \text{Df}$$

the symbols mean that x *is defined to stand for* y. In such a case y is some complex of symbols, and x will be an abbreviated symbol standing for y.

On $\phi!x$, $(x).\phi!x$, $(\exists x).\phi!x$, (x, y), $(\exists x, y)$, $\supset_x$

Propositional Functions.— $\phi!x$ means x *has the property* ϕ, where ϕ is given different forms corresponding to different properties; $\exists!x$ means x *has the property* |472| *of being a class possessing at least one member*; and $(x) . \phi!x$ means $\phi!x$ *is true for every value of* x; and $(\exists x).\phi!x$ means *there exists a value of* x *for which* $\phi!x$ *is true.* Note that (x) and $(\exists x)$, written before any proposition involving x, give the above meanings, even if the proposition is not in the symbolic form $\phi!x$. If the proposition involves[1] both x and y, then (x, y) prefixed means that the proposition is true for all values of x and y; and similarly for (x, y, z), and so on. Also $(\exists x, y)$ prefixed means that there exist values of x and y, such that the proposition is true; and similarly for $(\exists x, y, z)$, and so on. Furthermore $\phi!x \supset_x \psi!x$ stands for $(x).\{\phi!x \supset \psi!x\}$, and similarly for two and three variables.

On the Use of Dots, viz., ., :, .:., : :

$p . q$ or $p : q$ or $p .\!:\!. q$ or $p :: q$ all mean p *and* q *are both true propositions.* As an example, $x, y \in u$, which has been defined above, is really the proposition $x \in u . y \in u$; and $x, y, z \in u$ is the proposition $x \in u . y \in u . z \in u$.

Dots as Brackets.—The different symbolic forms for the joint assertion of propositions arise from the fact that dots are also used as bracket forms for propositions according to the following rules:—

(i) The larger aggregation of dots represents the exterior bracket. (ii) The dots at the end of a complete sequence of symbols are omitted. (iii) The dots immediately preceding or succeeding the implication sign, viz., $\supset$, are exterior brackets to any equal number of dots occurring in other capacities (*e.g.*, as above in the joint assertion of propositions). (iv) The dots which also serve to indicate the joint assertion of propositions are interior brackets to any equal number of dots occurring in other

1. The original erroneously had 'involve'.

capacities. (v) The dots after (x) and $(\exists x)$ are increased in number according to the necessity for their use as brackets.

In reading a symbolic proposition it is best to begin by searching for that implication sign, viz., $\supset$, which is preceded or succeeded by the greatest number of dots. This splits up the proposition into hypothesis and consequent; and so on with these subsidiary propositions, if necessary.

On $\vee,\ \sim,\ \sim\in,\ \neq,\ \sim(x).\phi!x,\ \sim(\exists x).\phi!x$

Again $p \vee q$ means *one or other or each of p and q is a true proposition*; and $\sim p$ means *p is not true*. Thus $\sim\phi!x$ means *x has not the property* ϕ; also $x \sim\in u$ stands for $\sim(x \in u)$; and $x \neq y$ stands for $\sim(x = y)$; and $\sim(x).\phi!x$ stands for $\sim\{(x).\phi!x\}$; and $\sim(\exists x).\phi!x$ stands for $\sim\{(\exists x).\phi!x\}$.

On $\hat{x}(\phi!x),\ (\imath x)(\phi!x),\ \iota`,\ \overset{\frown}{\iota}`,\ \cup,\ \cap,\ \cup`,\ \cap`$

Non-Propositional Functions.— $\hat{x}(\phi!x)$ denotes *the class of terms which have the property* ϕ, and $(\imath x)(\phi!x)$ denotes *the single entity (if there is such) which, when* |473| *substituted for x, makes* $\phi!x$ *to be a true proposition*. It is not necessary for the above symbolism that the proposition involving x should be in the symbolic form $\phi!x$. Again, $\iota`x$ denotes *the class possessing x as its sole member*, and $\overset{\frown}{\iota}`x$ denotes *the sole member of the class x*, and $u \cup v$ denotes *the logical sum of u and v*, that is, the class possessing all members of u and all members of v and no other members. Thus, $\iota`a \cup \iota`b$ denotes the class whose sole members are a and b. Again, $u \cap v$ denotes *the logical product of u and v*, that is, the complete common subclass of u and v; and $\cup`u$ denotes *the class which is the logical sum of all members of u*, that is, the class which has as members all members of members of $u$; and $\cap`u$ denotes *the class which is the logical product of all members of u*. The exact symbolic definition of $\cap`u$ is

$$\cap`u = \hat{x}\{v \in u .\supset_v . x \in v\} \qquad \text{Df}$$

It follows from this definition that, if u possess no members, $\cap`u$ is the class of all entities.

On Λ, *cls*`, $-$, Nc`

Again, Λ denotes the *null class*, that is, the class with no members; cls`$u$ denotes *the class whose members are the subclasses contained in u, including u itself and the null class*. It follows that the propositions, $v \in$ cls`u and $v \subset u$, have practically identical meanings. Again, $u - v$ denotes *the class u with the exception of those members which it possesses in common with v*.

The cardinal numbers[1†] are themselves classes. Thus, 1 is the class whose members are the unit classes, 2 is the class whose members are couples. Accordingly, $x \in 2$ means *x is a class with two members*; Nc`u denotes *the cardinal number of the class u*.

1. † *Cf.* Russell, 'Principles of Mathematics,' chap. XI, and Frege, 'Grundlagen der Arithmetik,' Breslau, 1884, pp. 79, 85.

On $\phi‘$, $\phi‘‘$, $\iota‘‘$, ?‘‘, $\cup‘‘$, *and so on.*

The general form for a non-propositional function whose value depends on x is $\phi‘x$, where ϕ receives different forms for different functions, as has been illustrated by the particular cases considered above. The apostrophe may be read as "of"; it is the general symbol for the connection of the preceding functional sign with the succeeding argument. According to this rule we should write $\sin‘x$ for $\sin x$ and $\log‘x$ for $\log x$. Again, $\phi‘‘u$ denotes *the class of values of* $\phi‘x$ *, when the various members of u are substituted for x*; it may be read "the class of ϕ's of u's." Thus, if $\phi‘x$ is "the head of x," and u is "the class of horses," then $\phi‘‘u$ is "the class of heads of horses." The exact symbolic definition of $\phi‘‘u$ is as follows:

$$\phi‘‘u = \dot{z}\{(\exists x) \,.\, x \in u \,.\, z = \phi‘x\} \qquad \text{Df}$$

It follows from the definition, by substituting for ϕ, that $\iota‘‘u$, $\imath‘‘u$, $\cup‘‘u$, $\cap‘‘u$, $\mathrm{cls}‘‘u$, $\mathrm{N}c‘‘u$ are now defined.

|474| *On* $(\mathrm{E}x)\phi‘y)$

Again, $(\mathrm{E}x)\phi‘y)$ means *there exists an entity which is denoted by the non-propositional function* $\phi‘z$, *when z has the particular value y*. For example, if u is a class, there is such an entity as its cardinal number, denoted by $\mathrm{N}c‘u$; but if u is not a class, there is no such entity as its cardinal number.[1†]

On Symbols of the Type $R^{;}(\quad)$

Relations.— $R^{;}(xyz)$ means *x, y, z form an instance in which the triadic relation* R *holds, the special "positions" of x, y, z in this instance of that relationship being indicated by their order of occurrence in the symbol* $R^{;}(xyz)$. Again, $R^{;}(\cdot\, yz)$ means *there exists an entity, x say, such that* $R^{;}(xyz)$. The symbolic definitions of $R^{;}(\cdot\, yz)$. and of analogous symbols, are

$$R^{;}(\cdot\, yz) \,.=.\, (\exists x) \,.\, R^{;}(xyz) \qquad \text{Df}$$

$$R^{;}(x \cdot z) \,.=.\, (\exists y) \,.\, R^{;}(xyz) \qquad \text{Df}$$

$$R^{;}(xy\, \cdot) \,.=.\, (\exists z) \,.\, R^{;}(xyz) \qquad \text{Df}$$

$$R^{;}(x \cdot\cdot) \,.=.\, (\exists y,z) \,.\, R^{;}(xyz) \qquad \text{Df}$$

$$R^{;}(\cdot\, \mathrm{y} \,\cdot) \,.=.\, (\exists x,z) \,.\, R^{;}(xyz) \qquad \text{Df}$$

and so on. Again, $R^{;}(;yz)$ denotes *the class of terms, such as x (say), which satisfy* $R^{;}(xyz)$, and $R^{;}(\cdot;z)$ denotes *the class of terms, such as y (say), such that there exists a term, x say, such that* $R^{;}(xyz)$ *holds*. The symbolic definitions of $R^{;}(;yz)$ and of analogous entities, and of $R^{;}(\cdot;z)$ and of analogous entities, are

1. †The difficult question of the import of a proposition, which contains a non-propositional function (with some particular entity as argument) to which no entity corresponds, has recently been elucidated by Russell, cf. 'Mind,' October, 1905 [Bertrand Russell, 'On Denoting', *Mind*, vol. 14, no. 56, pp. 479–493]. All propositions containing such a function are untrue, unless the function is merely a constituent of a subsidiary proposition whose truth is not implied by the proposition in question.

$$R^{;}(;yz) = \dot{x}\{R^{;}(xyz)\} \qquad \text{Df}$$
$$R^{;}(x;z) = \dot{y}\{R^{;}(xyz)\} \qquad \text{Df}$$
$$R^{;}(xy;) = \dot{z}\{R^{;}(xyz)\} \qquad \text{Df}$$
$$R^{;}(;\cdot z) = \dot{x}\{(\exists y).R^{;}(xyz)\} \qquad \text{Df}$$
$$R^{;}(\cdot;z) = \dot{y}\{(\exists x).R^{;}(xyz)\} \qquad \text{Df}$$
$$R^{;}(;y\cdot) = \dot{x}\{(\exists z).R^{;}(xyz)\} \qquad \text{Df}$$

and so on:

$$R^{;}(;\cdot\cdot) = \dot{x}\{(\exists y,z)\,.\,R^{;}(xyz)\} \qquad \text{Df}$$
$$R^{;}(\cdot;\cdot) = \dot{y}\{(\exists x,z)\,.\,R^{;}(xyz)\} \qquad \text{Df}$$
$$R^{;}(\cdot\cdot;) = \dot{z}\{(\exists x,y)\,.\,R^{;}(xyz)\} \qquad \text{Df}$$

|475| Thus, $R^{;}(;\cdot\cdot)$ denotes *the class of terms, such as x (say), such that there exist terms, y and z say, such that* $R^{;}(xyz)$ *holds*.

Again, $R^{;}(;;z)$ denotes *the class which is the logical sum of* $R^{;}(;\cdot z)$ *and* $R^{;}(\cdot;z)$; and $R^{;}(;;\cdot)$ denotes *the class which is the logical sum of* $R^{;}(;\cdot\cdot)$ *and* $R^{;}(\cdot;\cdot)$; and $R^{;}(;;;)$ denotes *the class which is the logical sum of* $R^{;}(;\cdot\cdot)$ *and* $R^{;}(\cdot;\cdot)$ *and* $R^{;}(\cdot\cdot;)$, that is, the "field" of the relation R. The symbolic definitions of the above, and of similar entities, are:

$$R^{;}(;;z) = R^{;}(;\cdot z) \cup R^{;}(\cdot;z) \qquad \text{Df}$$
$$R^{;}(;y;) = R^{;}(;y\cdot) \cup R^{;}(\cdot y;) \qquad \text{Df}$$
$$R^{;}(x;;) = R^{;}(x;\cdot) \cup R^{;}(x\cdot;) \qquad \text{Df}$$
$$R^{;}(;;\cdot) = R^{;}(;\cdot\cdot) \cup R^{;}(\cdot;\cdot) \qquad \text{Df}$$

and so on, and

$$R^{;}(;;;) = R^{;}(;\cdot\cdot) \cup R^{;}(\cdot;\cdot) \cup R^{;}(\cdot\cdot;) \qquad \text{Df}$$

This notation, which has been explained for triadic relations, can obviously be extended to any polyadic relations. Thus, $R^{;}(abcd)$ and $R^{;}(abcdt)$ are defined in a similar manner, and so are the symbols for the allied propositions and classes.

On 1→1.

A dyadic relation, S say, is called *one-one*, when each referent has only one relatum, and each relatum has only one referent. The class of one-one relations is denoted by 1→1. The symbolic definition is

$$1 \to 1 \,.\!=.\, \dot{S}\{S \in \text{relation} : x \in S^{;}(;\cdot)\,.\supset_x.\, S^{;}(x;) \in 1 : y \in S^{;}(\cdot;)\,.\supset_y.\, S^{;}(;y) \in 1\} \qquad \text{Df}$$

On $\vdash$

The Assertion Sign.—A proposition, which is stated in symbols as being true, *i.e.*, which is *asserted* as distinct from being *considered*, has the symbol $\vdash$ prefixed to it, with as many dots following as will serve to bracket off the proposition. This symbol ⊦ is called the *assertion sign*.[1†]

Part II. — The Punctual Concepts.

Those concepts of the material world in which the class of objective reals is composed of *points*, or *particles*, or of *both*, will be called the *punctual* concepts. The classical concept is a punctual concept, and will be considered first. The other punctual concepts can be explained briefly by reference to the classical concept.

Concept I.—(*The Classical Concept*).—This is dualistic, the class of objective reals |476| being subdivided into *points of space* and *particles of matter*. The essential relation has for its field the points of space only. Slight variants (not considered here) can be given to the concept by varying the properties of the essential relation, so as to make the geometry non-Euclidean, or, retaining Euclidean geometry, so as to give various forms to the essential relation and the resulting axioms. In the exposition of a system of geometrical axioms for Concept I., Veblen's memoir (*cf. loc. cit.*), to which I am largely indebted, will be followed. The changes which are made from Veblen's treatment are (i) in the addition of the symbolism which emphasizes the idea of the essential relation, and (ii) in the fact that the question of the independence of axioms is here ignored, through a desire not to overload this memoir with difficulties (both for the author and reader) belonging to another part of the subject. As the result of (ii), some of Veblen's definitions and axioms have been simplified (and, in a sense, spoiled). The axioms thus obtained for Concept I. will shorten our investigations of other concepts by serving as a standard of comparison to determine whether the axioms of the other concepts are sufficient to yield three-dimensional Euclidean geometry.[2†]

The essential relation (called R) is triadic. R‘(*abc*) means *the points a, b, c are in the linear order (or the* R*-order) abc*. The relation R, when R‘(*abc*) holds, is not symmetrical as between the three points *a*, *b*, and *c*; namely, it will be found that *a* and *b* (or *b* and *c*) cannot be interchanged.

---

1. †This symbol is due to Frege, who first drew attention to the necessity of the idea which it symbolizes; *cf.* his 'Begriffsschrift,' Halle, 1879 [the original erroneously had 'Begriffschrift'; Gottlob Frege, *Begriffsschrift: eine der arithmetischen nachgebildete Formelsprache des reinen Denkens* (Halle an der Saale: Verlag von Louis Nebert, 1879)], and Russell, 'Principles of Mathematics,' p. 35.
2. †On the philosophical questions connected with the mathematical analysis of geometry *cf.* 'A Critical Exposition of the Philosophy of Leibniz,' Cambridge, 1900, and 'The Principles of Mathematics,' Cambridge, 1903, both by Bertrand Russell; and also two articles by L. Couturat in the 'Revue de Métaphysique et de Morale' (Paris) for May and September, 1904, one entitled "La philosophie des Mathématiques de Kant" [no. 3, pp. 321–383] and the other "Les principes des Mathématiques — VI. La géometrie" [no. 5, pp. 810–844]; also Poincare's 'Science and Hypothesis,' Part II., English translation, London, 1905.

   For expositions of exact systems of axioms *cf.* 'Vorlesungen über neuere Geometrie,' Leipzig, 1882, by Pasch; also 'I Principii di Geometria,' Turin, 1889, by Peano; also "I Principii della Geometria di Posizione," 'Trans. Acad, of Turin,' 1899 [vol. 48, no. 2, pp. 1–62; date corrected from 1898], by Pieri; also Hilbert's 'Foundations of Geometry,' Engl. Transl., Chicago, 1902; also Professor E. H. Moore, "On the Projective Axioms of Geometry," 'Transact. of the Amer. Math. Soc.,' 1902 [vol. 3, no. 1 (January), pp. 142–158; date corrected from 1903]; also Dr. O. Veblen (*loc. cit.*); also Professor J. Royce (*loc. cit.*).

Definitions of Concept I.

Definition.—The class $R'(a;b)$ is the *segment between a and b*; and the class $R'(ab;)$ is the *segmental prolongation of ab beyond b*; and the class $R'(;ab)$ is the *segmental prolongation of ab beyond a*. It follows from the subsequent axioms that $R'(ab;)$ is identical with $R'(;ba)$.

Definition.—The *straight line ab* is the logical sum of $R'(a;b)$ and $R'(;ab)$ and $R'(ab;)$ together with a and b themselves. Its symbol is $R'\overline{ab}$. The definition in symbols is

$$R'\overline{ab} = R'(;ab) \cup R'(a;b) \cup R'(ab;) \cup \iota`a \cup \iota`b \qquad \text{Df}$$

|477| *Definition.*—The class whose members are the various straight lines is denoted by lin_R. The definition in symbols is[1]

$$\text{lin}_R = \dot{v}\{(\exists a,\, b)\,.\, a,\, b \in R'(;;;)\,.\, a \neq b\,.\, v = R'\overline{ab}\} \qquad \text{Df}$$

Definition.—Any class of points [*i.e.*, members of $R'(;;;)$] is called a *figure*.

Definition.—The *class of lines defined by a figure u* is the class of lines defined by any two distinct points of u. Its symbol is $\text{ln}_R`u$. The definition in symbols is

$$\text{ln}_R`u = \dot{v}\{(\exists x,\, y)\,.\, x,\, y \in u \cap R'(;;;)\,.\, x \neq y\,.\, v = R'\overline{xy}\} \qquad \text{Df}$$

Definition.—The *linear figure defined by a figure u* is the logical sum of all the lines defined by u (*i.e.*, is all the points on such lines). Thus its symbol is $\cup`\text{ln}_R`u$.

Definition.—Three points form a triangle, if there is no line which possesses them all. The symbol expressing that a, b, c are points forming a triangle is $\Delta_R'(abc)$. The definition in symbols is

$$\Delta_R'(abc)\,.\!=.\, a,\, b,\, c \in R'(;;;)\,.\, {\sim}(\exists v)\,.\, v \in \text{lin}_R\,.\, a,\, b,\, c \in v \qquad \text{Df}$$

Definition.—The *space defined by the triangle abc* is the linear figure defined by the linear figure defined by the three points a, b, c. Its symbol is $\Pi_R(abc)$. The definition in symbols is

$$\Pi_R(abc) = \cup`\text{ln}_R`\cup`\text{ln}_R`(\iota`a \cup \iota`b \cup \iota`c) \qquad \text{Df}$$

Definition.—The *class of planes* is the class of spaces defined by any three points a, b, c when they form a triangle. Its symbol is ple_R. The definition in symbols is

$$\text{ple}_R = \dot{v}\{(\exists a,\, b,\, c)\,.\, \Delta_R'(abc)\,.\, v = \Pi_R(abc)\} \qquad \text{Df}$$

Definition.—The space defined by the four points a, b, c, d is the linear figure defined by the figure which is the logical sum of $\Pi_R(bcd)$ and $\Pi_R(acd)$ and $\Pi_R(abd)$ and $\Pi_R(abc)$. Its symbol is $\Pi_R(abcd)$. The symbolic definition is

$$\Pi_R(abcd) = \cup`\text{ln}_R`\{\Pi_R(bcd) \cup \Pi_R(acd) \cup \Pi_R(abd) \cup \Pi_R(abc)\} \qquad \text{Df}$$

The above definitions are sufficient to exhibit the dependence of the various geometrical entities on the essential relation, and also to enable us, as far as geometry is concerned, to pass on to the third stage. Owing to the simplicity of the definitions, the second stage for this concept is of very small importance.

It will be noticed that none of the definitions contain any reference to length, distance, area, or volume. This is because none of these ideas appear in the axioms, and only such definitions are given here as are necessary for the enunciation of the

1. In the line below, the original erroneously had 'v' instead of '$\dot{v}$'.

axioms. According to the well-known[1†] method of projective metrics, the ideas are introduced by definition and require no special axiom.

|478| *Axioms of Concept I.*

The axioms, it must be remembered, are merely an enumeration of various propositions concerning the properties of the fundamental relations, which will occur as hypotheses in the propositions of the fourth stage. In this instance we are merely considering the axioms of geometry, and these concern the essential relation (R) only. The axioms will be named systematically thus, I H*p* R, II H*p* R, III H*p* R, and so on. Their enumeration will take the form of defining these *names* as abbreviations standing for the various statements, which will be used subsequently as hypotheses.

I H*p* R is the statement that there is at least one set of entities, *a*, *b*, *c*, such that $\mathrm{R}^{;}(abc)$ is true. The definition in symbols is

$$\mathrm{I\ H}p\ \mathrm{R} . = . \exists ! \mathrm{R}^{;}(;;;) \qquad \mathrm{Df}$$

II H*p* R is the statement that $\mathrm{R}^{;}(abc)$ implies $\mathrm{R}^{;}(cba)$. The definition in symbols is

$$\mathrm{II\ H}p\ \mathrm{R} . = : (a, b, c) : \mathrm{R}^{;}(abc) . \supset . \mathrm{R}^{;}(cba) \qquad \mathrm{Df}$$

III H*p* R is the statement that $\mathrm{R}^{;}(abc)$ is inconsistent with $\mathrm{R}^{;}(bca)$. The definition in symbols is

$$\mathrm{III\ H}p\ \mathrm{R} . = : (a, b, c) : \mathrm{R}^{;}(abc) . \supset . \sim \mathrm{R}^{;}(bca) \qquad \mathrm{Df}$$

IV H*p* R is the statement that $\mathrm{R}^{;}(abc)$ implies that *a* is distinct from *c*. The definition in symbols is

$$\mathrm{IV\ H}p\ \mathrm{R} . = : (a, b, c) : \mathrm{R}^{;}(abc) . \supset . a \neq c \qquad \mathrm{Df}$$

V H*p* R is the statement that, if *a* and *b* are distinct points, the segmental prolongation of *ab* beyond *b* possesses at least one member. The definition in symbols is

$$\mathrm{V\ H}p\ \mathrm{R} . = : (a, b) : a, b \in \mathrm{R}^{;}(;;;) . a \neq b . \supset . \exists ! \{\mathrm{R}^{;}(ab;)\} \qquad \mathrm{Df}$$

VI H*p* R is the statement that, if *c* and *d* are distinct points, possessed by the line defined by the points *a* and *b*, then *a* is possessed by the line defined by *c* and *d*. The definition in symbols is

$$\mathrm{VI\ H}p\ \mathrm{R} . = : (a, b, c, d) : c, d \in \mathrm{R}^{;}\overline{ab} . c \neq d . \supset . a \in \mathrm{R}^{;}\overline{cd} \qquad \mathrm{Df}$$

VII H*p* R is the statement that there exist at least three points forming a triangle. The definition in symbols is

$$\mathrm{VII\ H}p\ \mathrm{R} . = . (\exists a, b, c) . \Delta_R{}^{;}(abc) \qquad \mathrm{Df}$$

VIII H*p* R is the statement that, if *a*, *b*, *c* be three points forming a triangle, and $\mathrm{R}^{;}(bcd)$ and $\mathrm{R}^{;}(cea)$ hold, then there exists a point possessed both by the segment *ab*, and by the line defined by *d* and *e*. The definition in symbols is

$$\mathrm{VIII\ H}p\ \mathrm{R} . = : (a, b, c, d, e) : \Delta_{\mathrm{R}}{}^{;}(abc) . \mathrm{R}^{;}(bcd) . \mathrm{R}^{;}(cea) . \supset . \exists ! \{\mathrm{R}^{;}\overline{de} \cap \mathrm{R}^{;}(a;b)\} \qquad \mathrm{Df}$$

|479| IX H*p* R is the statement that there exists a point and a plane, such that the plane does not possess the point. The definition in symbols is

1. † *Cf.* Veblen, *loc. cit.*; also 'Vorlesungen über Geometrie,' by Clebsch, third part [edited by Ferdinand Lindemann (Leipzig: Teubner, 1891)]; also 'The Principles of Mathematics,' by Russell, chap. XLVIII.

$$\text{IX H}p\,\text{R} \,.\!=\,: (\exists p, d) \,.\, p \in \text{ple}_{\text{R}} \,.\, d \in \text{R}^{\text{;}}(;;;) - p \qquad \text{Df}$$

X H*p* R is the statement that there exist four points *a*, *b*, *c*, *d*, such that the three-dimensional space $\Pi_{\text{R}}(abcd)$ contains the whole class of points. The definition in symbols is,

$$\text{X H}p\,\text{R} \,.\!=\!.\, (\exists a, b, c, d) \,.\, \text{R}^{\text{;}}(;;;) \subset \Pi_{\text{R}}(abcd) \qquad \text{Df}$$

XI H*p* R is some statement which secures the continuity (in CANTOR'S sense) of the points on a line. The axiom need not be given here, since there will be no reasoning in this memoir connected with it.

XII H*p* R is the statement that, if α be any plane and *a* a line contained in it, then there exists a point *c* in α, such that there is not more than one line, possessing *c* and contained in the plane α and not intersecting *a*. The definition in symbols is

$$\text{XII H}p\,\text{R} \,.\!=\therefore \alpha \in \text{ple}_{\text{R}} \,.\, a \in \text{lin}_{\text{R}} \cap \text{cls}`\alpha \,.\supset_{a,\alpha}: (\exists c): c \in \alpha :$$
$$l, l' \in \text{lin}_{\text{R}} \cap \text{cls}`\alpha \,.\, c \in l \cap l' \,.\, l \cap a = \Lambda \,.\, l' \cap a = \Lambda \,.\supset_{l,l'} .\, l = l' \qquad \text{Df}$$

Of these axioms, IX H*p* R secures that space is of three dimensions at least, and X H*p* R secures that it is of three dimensions at most, and XII H*p* R is the "Euclidean" axiom. From these twelve axioms the whole of geometry[1†] can be deduced. The well-known parabolic (*i.e.*, Euclidean) *definition* of distance (not given here) assumes an important significance, and all the usual metrical properties follow.

The Extraneous Relations.—Nothing could be more beautiful than the above issue of the classical concept, if only we limit ourselves to the consideration of an unchanging world of space. Unfortunately, it is a changing world to which the complete concept must apply, and the intrusion at this stage into the classical concept of the necessity of providing for change can only spoil a harmonious and complete whole. Owing to the fact that the instants of time are not members of the field of the essential relation, the time relation and the essential relation have (so to speak) no point of contact. To remedy this, another subdivision of the class of objective reals is conceived, namely, the class of *particles* (where the particles are the ultimate entities composing the fundamental "stuff" which moves in space). These particles must form part of the fields of a class of extraneous relations. Each such extraneous relation is conceived as a triadic relation, which in any particular instance holds between a particle, a point of space, and an instant of time. Also the field of each such extraneous relation only possesses one particle, and no particle belongs to the field of two such relations. Thus each extraneous relation is peculiar to one particle. Also, as has been pointed out by RUSSELL,[2†] to whom the above analysis of these extraneous relations of the classical concept is in substance due, the impenetrability |480| of matter is secured by the axiom that two different extraneous relations cannot both relate the same instant of time to the same point. The general laws of dynamics, and all the special physical laws, are axioms concerning the properties of this class of extraneous relations.

Thus the classical concept is not only dualistic, but has to admit a class of as many extraneous relations as there are members of the class of particles.

---

1. † *Cf.* VEBLEN, *loc. cit.*
2. † *Cf.* 'Principles of Mathematics,' vol. I, § 440.

Instead of the specific relations of occupation for the various particles, one general triadic relation of occupation can be considered. Thus, $\mathrm{O}^{;}(p\mathrm{A}t)$ may be considered as the statement the particle p occupies the point A at the instant t. Then for any given A and t there is either one only or no particle p for which $\mathrm{O}^{;}(p\mathrm{A}t)$ is true. Then the laws of physics are the properties of this single extraneous relation O. But the use of this single relation apparently introduces no real simplification, differing in this respect from the use of the essential relation which so simplifies the statement of the axioms of geometry. The general relation O remains a mere alternative statement of the facts respecting the various specific relations of occupation.

Concept II.—This concept is a monistic variant of the classical concept suggested by RUSSELL.[1†] In the classical concept the particles only occur as terms in the triadic extraneous relations. If we abolish the particles (in the "classical" sense), and transform the extraneous relations into dyadic relations between points of space and instants of time, everything will proceed exactly as in the classical concept. The reason for the original introduction of "matter" was, without doubt, to give the senses something to perceive. If a relation can be perceived, this Concept II. has every advantage over the classical concept. Otherwise the material world, as thus conceived, would appear to labour under the defect that it can never be perceived. But this is a philosophic question with which we have no concern.

Concept III.—This is a Leibnizian concept, and also a monistic variant of the classical concept, obtained by abandoning the prejudice against points moving.

This concept can be otherwise considered, as obtained from the modern (and Cartesian) point of view of the ether, as filling all space. The particles of ether (or moving points) compose the whole class of objective reals. The essential relation (R) is a tetradic relation, and, in each specific instance of the relation holding, three of the terms are objective reals and the remaining term is an instant of time. $\mathrm{R}^{;}(abct)$ may be read as stating *the objective reals a, b, c are in the* R*-order abc at the instant t.* Instead of $\mathrm{R}^{;}(abct)$, it will be convenient to write $\mathrm{R}_t{}^{;}(abc)$. Then the geometrical definitions are exactly those of Concept I., replacing R everywhere by R_t. Also the geometrical axioms are those of Concept I.; except (i) that R is replaced by R_t, (ii) that I H*p* R, and VII H*p* R, and IX H*p* R, are further modified by the introduction of the hypothesis $t \in \mathrm{T}$ — thus I H*p* R of Concept I. becomes

$$\mathrm{I\ H}p\ \mathrm{R} .=: t \in \mathrm{T} . \supset_t . \exists! \mathrm{R}_t{}^{;}(;;:) \qquad \mathrm{Df}$$

|481| and similarly for the other two axioms, and (iii) that one additional axiom (the axiom of persistence) must be introduced, namely,

$$\mathrm{XIII\ H}p\ \mathrm{R} .=: t \in \mathrm{T} . \supset_t . \mathrm{R}^{;}(;;;\cdot) \subset \mathrm{R}^{;}(;;;t) \qquad \mathrm{Df}$$

This axiom of persistence is unnecessary for the geometrical reasoning, but is an integral part of the "physical" side of the concept. Also the hypothesis $t \in \mathrm{T}$, which is introduced in the three axioms (I, VII, IX) H*p* R, is unnecessary in the other axioms, since it is implied by the hypotheses already existing. The same explanation holds of the absence of the hypothesis, $t \in \mathrm{T}$, from many axioms and propositions of subsequent concepts.

1. † *Cf.* 'Principles of Mathematics,' vol. I., § 441.

Thus at each instant the objective reals may be considered as the points of the classical concept, and the whole of Euclidean geometry holds concerning them. But at another instant the points will not have preserved the same geometrical relations as held between them at the previous instant. Thus, in the comparison of the states of the objective reals at different instants, the objective reals assume the character of particles.

The Extraneous Relation.—A single extraneous relation is necessary to obviate the difficulty of comparing straight lines and planes at one instant with similar entities at another instant. In what sense can a point at one instant be said to have the *same position* as a point at another instant? This definition can be effected by introducing into the concept a single tetradic extraneous relation S, so that, when $S^{;}(uvwt)$ holds, t is an instant of time, and u, v, w are intersecting straight lines mutually at right angles. Also corresponding to any instant t in the fourth term, there is one and only one line for each of the other terms respectively. This last condition, expressed in symbols, is

$$t \in \mathrm{T} \,.\supset_t .\, S^{;}(;\cdot\cdot t) \in 1 \,.\, S^{;}(\cdot;\cdot t) \in 1 \,.\, S^{;}(\cdot\cdot;t) \in 1$$

The straight lines indicated at each instant by this relation are to be taken as the "kinetic axes."[1†] Velocity and acceleration can now be defined, and a general continuity of motion (in some sense) must be included among the axioms.

This concept has the advantage over Concepts I. and II. that it has reduced the class of extraneous relations to one member only, in the place of the innumerable and perhaps infinite number of extraneous relations in the other two concepts. The concept pledges itself to explain the physical world by the aid of motion only. It was indeed a dictum with some eminent physicists of the nineteenth century that "motion is of the essence of matter." But this concept takes them rather sharply at their word. There is absolutely nothing to distinguish one part of the objective reals from another part except differences of motion. The "corpuscle" will be a volume in which some peculiarity of the motion of the objective reals exists and persists. Two |482| different developments, viz., Concept IIIA. and Concept IIIB., are now possible, according as the persistence is taken to be of one or other of two possible types.

Concept IIIA.—Here the persistence is that of the *same objective reals* in the same special type of motion. KELVIN'S vortex ring theory of matter can be adapted to such a concept.

Concept IIIB. — Here the persistence is that of the *type of motion* in some volume, but not necessarily of the identity of the objective reals in the volume. The *continuity of motion* of a corpuscle as a whole becomes then the *definition of the identity* of a corpuscle at one instant with a corpuscle at another instant.

PART III.—(i) GENERAL EXPLANATIONS OF LINEAR CONCEPTS.

These concepts depart widely from the classical concept. The objective reals (at least those which, with the instants of time, form the field of the essential relation)

1. † *Cf.* W. H. MACAULAY, 'Bulletin of the Amer. Math. Soc.,' 1897 ['Newton's theory of kinetics', vol. 3, no. 10, pp. 363–371].

have properties which we associate with straight lines, considered throughout their whole extent as single indivisible entities. These objective reals, which in Concept V. are all the objective reals, will be called *linear objective reals*. Perhaps, however, a closer specification of the linear objective reals of these concepts is to say that they are the lines of force of the modern physicist, here taken to be ultimate unanalysable entities which compose the material universe, and that geometry is the study of a certain limited set of their properties. But this mode of realizing the nature of the linear objective reals has also its pitfalls, for a line of force suggests *ends*, while these linear objective reals have no properties analogous to the properties of the ends of lines of force. The whole of a straight line, viewed as a point-locus, will be found to be associated with a linear objective real. The "linear" concepts here considered are all Leibnizian.

Concept IVA. is dualistic, and requires among the objective reals a class of "particles" in addition to the linear objective reals. Concept IVB. is the monistic variant of Concept IVA., obtained exactly as Concept II. is derived from Concept I. Both of the Concepts IVA. and IVB. labour under the same defect as Concepts I. and II. in requiring an indefinitely large class of extraneous relations. Concept V. is monistic, and is by far the most interesting of the set of linear concepts. It requires only one extraneous relation to perform a similar office to that of the extraneous relation in Concept III.

Points are now defined complex entities, being certain classes of linear objective reals. Geometers are already used to the idea of the point as complex. In projective geometry, as derived from descriptive geometry, the projective point is nothing but a class of straight lines.[1]† This idea will now be extended to all |483| points; and the descriptive point, from which in the current theory the projective point is ultimately derived, is here abolished. The "Theory of Interpoints" [*cf.* Part III. (ii)] and the "Theory of Dimensions" [*cf.* Part IV. (i)] represent two distinct methods of overcoming the following initial and obvious difficulty of these "linear" concepts:—A point is to be defined as the class of objective reals "concurrent" at a point. But this definition is circular. How can this circularity be removed? The Theory of Interpoints and the Theory of Dimensions give two separate answers to this question. The points in the linear concepts, being only classes of objective reals, are capable of disintegration. In fact, when motion is considered, it will be found that the points of one instant are, in general, different from the points of another instant, not in the sense of Concept III. that they are the same entities with different relations, but in the sense that they are different entities. More difficulty will probably be felt in conceiving anything analogous to a line as a simple unity. Here it is to be observed that a linear objective real does not replace a line of points of ordinary geometry. On the contrary, the class of those points (here called a *punctual line*), which have a given linear objective real as a common member, is this ordinary geometrical line. A punctual line has parts and segments in the ordinary way. The idea of a single unity

1. † *Cf.* Pasch, *loc. cit.*, and Schur, "Ueber die Einführung der sogennanten idealen Elemente in die projective Geometrie," 'Math. Annal.,' vol. XXXVIII [1891, pp. 113–124; volume number corrected from XXXIX], and Bonola, "Sulla Introduzione degli Enti improprii in Geometria projettiva," 'Giorn. di Mat.,' vol. XXXVIII [1900, pp. 105–116].

underlying a straight line is not wholly alien to ordinary language. The idea of a direction, as it could also be used in non-Euclidian geometries where each line will have its own peculiar direction, may be conceived as being that of a line taken as a unit. But it is unnecessary to elaborate these considerations, as they have no relation to the logic of the subject.

In the dualistic Concept IVA. the particles form another class of objective reals in addition to the linear objective reals. Each particle is associated at each instant with some one point, that is, with some class of linear objective reals. Thus the two points, respectively associated at any instant with two particles, have in common one linear objective real. Thus, when mutually determined motions are considered, these linear objective reals assume the aspect of lines of force. In the monistic Concept V. the analogy of objective reals to lines of force arises in a similar way. In this case particles, in the sense used above, do not exist. Corpuscles, to use another term, are defined entities, analogous to the corpuscles of Concept III.; any general consideration of them is best deferred till the definitions can be understood.

In Concepts IV. and V. the conception of an ether is (in a sense) rendered unnecessary, or (in another sense) is largely modified. The collection of linear objective reals (*i.e.*, in Concepts IVB. and V., of *all* objective reals) now forms the entity (the ether) which "lies between" the corpuscles of gross matter. These corpuscles must be conceived as volumes with some peculiarity either of motion or of structure. Of course it might be found useful, for the explanation of physical phenomena, to assume that corpuscles of some sort are generally distributed between bodies of gross matter, thus forming an ether in a secondary sense. The ancient controversy concerning action at a distance becomes irrelevant in these concepts. In |484| one sense there is something, not mere space, between two distant corpuscles, namely, the objective reals possessed in common; in another sense there is a direct action between two distant corpuscles not depending on intervening corpuscles. In fact, the premises common to both bands of disputants are swept away.

The Essential Relation.—In both of the Concepts IV. and V. the essential relation (R) is a pentadic relation, and has for its field both the class of instants of time and that of linear objective reals, that is, in Concept V. the field is the complete class of ultimate existents. The proposition R'(*abcdt*) can be read as the statement that *the objective real a intersects the objective reals b, c, d in the order bcd at the instant t.* This conception of "the intersection in order of three linear objective reals by a fourth at an instant of time" must be taken as a fundamental relation between the five entities. But the properties of the relation are not to be limited by the suggestion of the technical name "intersection." The axioms will be so assumed that R'(*abcdt*) implies that *a*, *b*, *c*, and *d* are distinct. Also, when points are defined, it will be found that the axioms secure that *a* intersects *b*, *c*, and *d* in distinct points. Furthermore, in general, *b*, *c*, and *d* are *not* co-punctual; so that the case when *a* is a transversal of the pencil *b*, *c*, *d* of *co-punctual* lines is only a particular case of the satisfaction of R'(*abcdt*).

Definitions.—The notation of the general symbolism provides us with the symbol R'(;;;;·) for the class of linear objective reals, and with R'(····;) for the class of instants. But these symbols are long. Accordingly O will be defined to stand for the class of linear objective reals, and T for the class of instants. Thus, in symbols,

$$\mathrm{O} = \mathrm{R}^{\prime}(;;;;\cdot) \qquad \text{Df}$$

$$\mathrm{T} = \mathrm{R}^{\prime}(\cdots\,;) \qquad \text{Df}$$

When "particles" (in Concept IV.) are not being directly considered, the term "objective real" will be used instead of "linear objective real," or "member of O."

(ii) The Theory of Interpoints.[1†]

*1. The theory of *intersection-points* (shortened into *interpoints*) is required in both of the Concepts IV. and V. Accordingly, it is convenient to investigate it before the special consideration of either concept. In Concept IV. the interpoints are the points, and there are no other points. In Concept V. the interpoints are, in general, only portions of points, and a point may contain no interpoint or many interpoints. Thus |485| the axioms of Concept IV. (*cf.* *2) and those of Concept V. (*cf.* *22) are two alternative sets of hypotheses as to the properties of R in connection with which the theory of interpoints, as given in the present *1, assumes importance. Some axioms, involving interpoints in their statements, are identical in Concept IV. and Concept V. These axioms are stated now in *1, and their simple consequences are deduced. The theory of interpoints depends on that of "similarity of position" in a relation. This general idea will only be explained in the special form in which it is here required in respect to the essential relation R.

*1·11. *Definition.*—An entity, *y*, will be said to have a *position* in the pentadic relation R, *similar* to that of the entity *x*, with *a* as first term and *t* as last (fifth) term, if, whenever the relation holds between five terms, *a* being the first term and *t* the last term, and either *x* or *y* or both occurring among the other terms, the relation also holds when *x* is substituted for *y* (whenever *y* occurs), and also holds when *y* is substituted for *x* (whenever *x* occurs). The symbol $\mathrm{R}^{\prime}\left(\frac{a???t}{x}\right)$ denotes the class of entities with positions similar to that of *x* in the relation R, *a* being first term and *t* last term. The definition in symbols is

$$\mathrm{R}^{\prime}\left(\frac{a???t}{x}\right) . =: \dot{y}\{(\xi,\ \eta) : \mathrm{R}^{\prime}(ax\xi\eta t) . \equiv . \mathrm{R}^{\prime}(ay\xi\eta t) :$$

$$\mathrm{R}^{\prime}(a\xi x\eta t) . \equiv . \mathrm{R}^{\prime}(a\xi y\eta t) : \mathrm{R}^{\prime}(a\xi\eta xt) . \equiv . \mathrm{R}^{\prime}(a\xi\eta yt)\} \qquad \text{Df}$$

*1·12. *Proposition.*—If *y* is a member of $\mathrm{R}^{\prime}\left(\frac{a???t}{x}\right)$, then $\mathrm{R}^{\prime}\left(\frac{a???t}{y}\right)$ is identical with $\mathrm{R}^{\prime}\left(\frac{a???t}{x}\right)$. In symbols,

$$\vdash : y \in \mathrm{R}^{\prime}\left(\frac{a???t}{x}\right) . \supset . \mathrm{R}^{\prime}\left(\frac{a???t}{y}\right) = \mathrm{R}^{\prime}\left(\frac{a???t}{x}\right)$$

*1·13. *Proposition.*—*x* is a member of $\mathrm{R}^{\prime}\left(\frac{a???t}{x}\right)$

*1·21. *Definition.*—A class P of objective reals is called an *intersection-point on a* (shortened into *interpoint on a*), when there exists an objective real *x*, which is a

1. †From this point a continuous argument commences, and the sections and included propositions are numbered by a combined integral and decimal system, the whole number for the section and the decimal part for the proposition, also the symbol (*) is placed before an integral number marking a section. All the easier proofs of propositions are omitted, those proofs remaining being retained either as specimens, or as containing some point of difficulty. The omitted proofs are often replaced by references to the preceding propositions used in them, as a guide to their reconstruction. Note that "*cf.* *2·31·41·5" is used as a shortened form of "cf. *2·31 and *2·41 and *2·5."

member of $\mathrm{R}^{;}(a;;;t)$, and P is the class whose members are a together with all the members of the class $\mathrm{R}^{;}\left(\frac{a???t}{x}\right)$. The symbol $\mathrm{R}^{;}(a???t)$ stands for *the class of interpoints on a at the instant t.* The definition in symbols is

$$\mathrm{R}^{;}(a???t) = \dot{\mathrm{P}}\left\{(\exists x) \,.\, x \in \mathrm{R}^{;}(a;;;t) \,.\, \mathrm{P} = \iota^{\backprime}a \cup \mathrm{R}^{;}\left(\tfrac{a???t}{x}\right)\right\} \qquad \mathrm{Df}$$

*1·22. *Definition.*—P is called an *interpoint* of the relation R at the instant t, if there exists an objective real a, such that P is a member of $\mathrm{R}^{;}(a???t)$. The symbol $\mathrm{intpnt}_{\mathrm{R}t}$ stands for *the class of interpoints of* R *at the instant t.* In symbols,

$$\mathrm{intpnt}_{\mathrm{R}t} = \dot{\mathrm{P}}\left\{(\exists a).\ \mathrm{P} \in \mathrm{R}^{;}(a???t)\right\} \qquad \mathrm{Df}$$

|486| *1·23. *Proposition.*— If P and Q are distinct members of $\mathrm{R}^{;}(a???t)$, then a is the sole member common to P and Q. In symbols,

$$\vdash\!: \mathrm{P}, \mathrm{Q} \in \mathrm{R}^{;}(a???t).\ \mathrm{P} \neq \mathrm{Q} \,.\!\supset.\, \iota^{\backprime}a = \mathrm{P} \cap \mathrm{Q}$$

Proof.—*Cf.* *1·11·12·21.

*1·31. *Definition.*—The interpoints B, C, D will be said to be in the *interpoint-order* BCD at the instant t with respect to the relation R, when there exist objective reals a, x, y, z, such that (1) B, C, D are members of $\mathrm{R}^{;}(a???t)$; (2) x is a member of B, y of C, z of D; (3) $\mathrm{R}^{;}(axyzt)$ holds. The symbol $\mathrm{R}_{\mathrm{in}}{}^{;}(\mathrm{BCD}t)$ stands for the statement that *the interpoints* B, C, D *are in the interpoint-order* BCD *at the instant t with respect to the relation* R. In symbols,

$$\mathrm{R}_{\mathrm{in}}{}^{;}(\mathrm{BCD}t) .=. (\exists a, x, y, z) \,.\, \mathrm{B}, \mathrm{C}, \mathrm{D} \in \mathrm{R}^{;}(a???t).\ x \in \mathrm{B} \,.\, y \in \mathrm{C} \,.\, z \in \mathrm{D} \,.\, \mathrm{R}^{;}(axyzt) \qquad \mathrm{Df}$$

*1·32. $\vdash.\, \mathrm{intpnt}_{\mathrm{R}t} = \mathrm{R}_{\mathrm{in}}{}^{;}(;;;t)$

Proof.—The class $\mathrm{R}_{\mathrm{in}}{}^{;}(;;;t)$ is part (or all) of the class $\mathrm{intpnt}_{\mathrm{R}t}$ (*cf.* *1·31). Again (*cf.* *1·22), if B is a member of $\mathrm{intpnt}_{\mathrm{R}t}$, objective reals a and x exist, such that x is a member of $\mathrm{R}^{;}(a;;;t)$, and B is the interpoint possessing a and x. Hence there are objective reals y and z, such that *either* $\mathrm{R}^{;}(axyzt)$ or $\mathrm{R}^{;}(ayxzt)$ or $\mathrm{R}^{;}(ayzxt)$. Hence (*cf.* *1·31), there are interpoints C and D, such that *either* $\mathrm{R}_{\mathrm{in}}{}^{;}(\mathrm{BCD}t)$ *or* $\mathrm{R}^{;}(\mathrm{CBD}t)$ *or* $\mathrm{R}^{;}(\mathrm{CDB}t)$. Hence B is a member of $\mathrm{R}_{\mathrm{in}}{}^{;}(;;;t)$.

The theory of interpoints has its chief interest when the following axiom is satisfied. It is named intpnt H*p* R.

*1·41. Intpnt H*p* R is the statement that *if* A *be an interpoint at the instant t, and a be any member of* A, *then* A *is a member of* $\mathrm{R}^{;}(a???t)$. In symbols,

$$\mathrm{Intpnt\ H}p\ \mathrm{R} .=: \mathrm{A} \in \mathrm{intpnt}_{\mathrm{R}t} .\, a \in \mathrm{A} \,.\!\supset_{a,\mathrm{A},t} .\, \mathrm{A} \in \mathrm{R}^{;}(a???t) \qquad \mathrm{Df}$$

*1·42. *Proposition.*—Assuming intpnt H*p* R, then if A and B are distinct members of $\mathrm{intpnt}_{\mathrm{R}t}$, A and B have *either* no members in common *or* one only. In symbols,

$$\vdash .\because \mathrm{intpnt\ H}p\ \mathrm{R} \,.\!\supset\!: \mathrm{A}, \mathrm{B} \in \mathrm{intpnt}_{\mathrm{R}t} .\, \mathrm{A} \neq \mathrm{B} \,.\!\supset.\, \mathrm{A} \cap \mathrm{B} \in 0 \cup 1.$$

Proof.— *Cf.**1·23·41.

The interest of the relation of interpoint-order $(\mathrm{R}_{\mathrm{in}})$ arises when the relation R satisfies four axioms specifying the idea that $\mathrm{R}^{;}(abcdt)$ expresses that *a intersects b, c, d in the order bcd.* These axioms (together with intpnt H*p* R) will be employed both in Concept IV. and in Concept V. They will be named α H*p* R, β H*p* R, γ H*p* R, δ H*p* R.

*1·51. α H*p* R is the statement that *a is not a member of* $\mathrm{R}^{;}(a;;;t)$. In symbols,

$$\alpha\ \mathrm{H}p\ \mathrm{R} .=. (a,t) \,.\, a \sim\in \mathrm{R}^{;}(a;;;t) \qquad \mathrm{Df}$$

*1·52. β *Hp* R is the statement that $R'(abcdt)$ *implies* $R'(adcbt)$. In symbols,

$$\beta \text{ Hp R} . = : (a, b, c, d, t) : R'(abcdt) . \supset . R'(adcbt) \qquad \text{Df}$$

|487| *1·53. γ *Hp* R is the statement that $R'(abcdt)$ *and* $R'(acdbt)$ *are inconsistent.* In symbols,

$$\gamma \text{ Hp R} . = : (a, b, c, d, t) : R'(abcdt) . \supset . \sim R'(acdbt) \qquad \text{Df}$$

*1·54. δ *Hp* R is the statement that $R'(abcdt)$ *implies that b and d are distinct.* In symbols,

$$\delta \text{ Hp R} . = : (a, b, c, d, t) : R'(abcdt) . \supset . b \neq d \qquad \text{Df}$$

*1·61. *Proposition.*—Assuming $(\alpha, \beta, \gamma, \delta)$ *Hp* R, then $R'(abcdt)$ implies that a, b, c, d are all distinct. In symbols,

$$\vdash \therefore (\alpha, \beta, \gamma, \delta) \text{Hp R} . \supset : R'(abcdt) . \supset . a \neq b . a \neq c . a \neq d . b \neq c . b \neq d . c \neq d$$

*1·62. *Proposition.*—Assuming $(\alpha, \beta, \gamma, \delta)$ *Hp* R, then $R_{in}'(BCDt)$ implies that B, C, D are all distinct. In symbols,

$$\vdash \therefore (\alpha, \beta, \gamma, \delta) \text{Hp R} . \supset : R_{in}'(BCDt) . \supset . B \neq C . B \neq D . C \neq D$$

Proof.— By definition (*cf.* *1·31) $R_{in}'(BCDt)$ implies that a, x, y, z exist such that B, C, D are members of $R'(a???t)$, x is a member of B, y of C, z of D, and $R'(axyzt)$. Hence (*cf.* *1·61) a, x, y, z are distinct. Now if any two of B, C, D are identical, *e.g.*, B and C, then x and y are both members of B. Hence (*cf.* *1·12·21), since a is distinct from x and y, x can be substituted for y in $R'(axyzt)$. Hence $R'(axxzt)$, which contradicts *1·61.

*1·63. *Proposition.*—Assuming β *Hp* R, then $R_{in}'(BCDt)$ implies $R_{in}'(DCBt)$. In symbols,

$$\vdash \therefore \beta \text{ Hp R} . \supset : R_{in}'(BCDt) . \supset . R_{in}'(DCBt)$$

Proof.—*Cf.* *1·31·52.

*1·64. *Proposition.*—Assuming $(\alpha, \beta, \gamma, \delta)$*Hp*R, then $R_{in}'(BCDt)$ and $R_{in}'(CDBt)$ are inconsistent. In symbols,

$$\vdash \therefore (\alpha, \beta, \gamma, \delta) \text{Hp R} . \supset : R_{in}'(BCDt) . \supset . \sim R_{in}'(CDBt)$$

Proof.— $R_{in}'(BCDt)$ implies (*cf.* *1·31) that a, x, y, z exist such that a is a common member of B, C, D, x is a member of B, y of C, z of D, and $R'(axyzt)$, and B, C, D are members of $R'(a???t)$. Hence (*cf.* *1·61) a, x, y, z are all distinct. Similarly also if $R_{in}'(CDBt)$, then a', x', y', z' exist with similar properties, viz., x' a member of B, &c., except that $R'(a'y'z'x't)$. Hence (*cf.* *1·23·62) a and a' are identical. Thus $R'(axyzt)$ and $R'(ay'z'x't)$. But (*cf.* *1·21) x can be substituted for x', y for y', and z for z'. Hence $R'(axyzt)$ and $R'(ayzxt)$. But this contradicts γ *Hp* R.

*1·65. *Proposition.*—Assuming (intpnt, $\alpha, \beta, \gamma, \delta$) *Hp* R, the classes $R'(;\cdots t)$ and $R'(;;;;t)$ are identical. In symbols,

$$\vdash : (\text{intpnt}, \alpha, \beta, \gamma, \delta) \text{Hp R} . \supset . R'(;\cdots t) = R'(;;;;t)$$

|488| *Proof.*—If x is a member of $R'(\cdot;;;t)$, y exists such that x is a member of $R'(y;;;t)$; also (*cf.* *1·61) x and y are distinct. Hence (*cf.* *1·21) P exists such that it is a member of $R'(y???t)$, and x is a member of it. Hence (*cf.* *1·41) P is a member of

R'$(x???t)$, and hence (*cf.* *1·21) y is a member of R'$(x;;;t)$. Hence x is a member of R'$(;\cdots t)$.

*1·71. *Proposition.*—Assuming αHpR, every interpoint possesses at least two members. In symbols,

$$\vdash .\therefore \alpha \text{ Hp R} . \supset : \text{A} \in \text{intpnt}_{\text{R}t} . \supset . \text{Nc}`\text{A} \geqq 2$$

Proof.—*Cf.* *1·13·21·22·51.

*1·72. *Proposition.*—Assuming (intpnt, α, β, γ, δ) Hp R, then on every objective real there exist at least three interpoints. In symbols,[1]

$$\vdash .\therefore (\text{intpnt}, \alpha, \beta, \gamma, \delta)\text{H}p \text{ R} . \supset : a \in \text{R}'(;;;;t) . \supset_a . \text{Nc}`\text{R}'(a???t) \geqq 3$$

Proof.—*Cf.* *1·21·31·62·65.

*1·73. *Proposition.*—Assuming (intpnt, α, β, γ, δ) Hp R, then, if there are any objective reals, the interpoints are not all on any one objective real. In symbols,

$$\vdash .\therefore (\text{intpnt}, \alpha, \beta, \gamma, \delta)\text{H}p \text{ R} . \supset : \exists ! \text{R}'(;;;;t) . \supset . \exists ! \{\text{intpnt}_{\text{R}t} - \text{R}'(a???t)\}$$

Proof.—*Cf.* *1·42·71·72.

(iii) Concept IV.

*2. This concept bifurcates into two alternate forms, namely IVA. and IVB. Concept IVB. is related to IVA. just as Concept II. is related to the classical concept. Thus Concept IVA. is dualistic, and Concept IVB. is the monistic variant of it. Both concepts can initially be considered together as Concept IV. In Concept IV. the essential relation (R) is pentadic, one of the terms being an instant of time. R'$(abcdt)$ can be read as a intersects b, c and d, in the order bcd at the instant t. The class of those entities, appearing among the first four terms in any instance of the relation holding, is called the class (O) of "linear objective reals." The remaining class of objective reals, required for Concept IVA., is called the class of "particles."

The geometrical points of this concept are simply the interpoints of R, as defined above (*cf.* *1). During the consideration of this concept they will be called *points*. The further definitions, beyond those of *1, required for a concise statement of the geometrical axioms are almost exactly those of Concept I., with the R_{in} of this Concept IV. written for the R of Concept I., and modified by the mention of t, as in Concept III. This mention of t can be managed in a similar (though not identical) way to that in Concept III. by writing

*2·01 $$\text{R}_t{}'(\text{ABC}) = \text{R}_{\text{in}}{}'(\text{ABC}t) \qquad \text{Df}$$

|489| Then the definitions of Concept I. will be assumed to apply to R_t. For example, the *punctual line joining the points* A *and* B is the class of points which is the logical sum of $\text{R}_t{}'(;\text{AB})$ and $\text{R}_t{}'(\text{A};\text{B})$ and $\text{R}_t{}'(\text{AB};)$ together with A and B themselves. Its symbol is $\text{R}_t{}'\overline{\text{AB}}$. The definition in symbols is

1. In the below equation, the original erroneously had '$\supset$' instead of '$\supset_a$'.

$$R_t{}^{\prime}\overline{AB} = R_t{}^{\prime}(;AB) \cup R_t{}^{\prime}(A;B) \cup R_t{}^{\prime}(AB;) \cup \iota^{\prime}A \cup \iota^{\prime}B \qquad \text{Df}$$

It will follow (*cf.* *1·23·31) from the axioms that a punctual line is the class of those points with some member of O as sole common member. The other definitions can be managed in like manner, only in the symbolism a suffix to a suffix will be avoided by writing $\Delta_{Rt}{}^{\prime}(ABC)$, and so on, instead of $\Delta_{R_t}{}^{\prime}(ABC)$, and so on.

The Axioms.—The earlier axioms have to be modified from those of Concept I., but the later axioms are simply those of Concept I. with the R of that concept replaced by the R_t of Concept IV.

I $Hp\ R .=: t \in T . \supset_t . O \subset R^{\prime}(;;;;t)$ Df

II $Hp\ R .= . \exists ! R^{\prime}(\cdots;)[i.e., \exists ! T]$ Df

III $Hp\ R .= . \alpha\ Hp\ R$ Df (*cf.* *1·51)

IV $Hp\ R .= . \beta\ Hp\ R$ Df (*cf.* *1·52)

V $Hp\ R .= . \gamma\ Hp\ R$ Df (*cf.* *1·53)

VI $Hp\ R .= . \delta\ Hp\ R$ Df (*cf.* *1·54)

VII $Hp\ R .= .$ intpnt $Hp\ R$ Df (*cf.* *1·41)

VIII $Hp\ R .=\therefore (A, B, C) \therefore A, B, C \in R_t{}^{\prime}(;;;) . A \neq B . A \neq C . B \neq C .$

$$\exists !(A \cap B \cap C) . \supset : R_t{}^{\prime}(ABC) . \vee . R_t{}^{\prime}(BCA) . \vee . R_t{}^{\prime}(CAB) \qquad \text{Df}$$

IX $Hp\ R .=: (A, B) : A, B \in R_t{}^{\prime}(;;;) . A \neq B . \supset . \exists ! R_t{}^{\prime}(AB;)$ Df

X $Hp\ R. =: (A, B, C, D, E) : \Delta_{Rt}{}^{\prime}(ABC) . R_t{}^{\prime}(BCD) . R_t{}^{\prime}(CEA) . \supset .$

$$\exists ! \{R_t{}^{\prime}\overline{DE} \cap R_t{}^{\prime}(A;B)\} \qquad \text{Df}$$

XI $Hp\ R .=: t \in T . \supset_t . (\exists p, D) . p \in ple_{Rt} . D \in R_t{}^{\prime}(;;;) - p$ Df

XII $Hp\ R .= . (\exists A, B, C, D) . R_t{}^{\prime}(;;;) \subset \Pi_{Rt}(ABCD)$ Df

XIII $Hp\ R .= .$ the axiom of continuity, *cf.* XII Hp R of Concept I. Df

XIV $Hp\ R .=\therefore \alpha \in ple_{Rt} . a \in lin_{Rt} \cap cls^{\prime}\alpha . \supset_{a,\alpha} : (\exists C) : C \in \alpha :$

$$l, l^{\prime} \in lin_{Rt} \cap cls^{\prime}\alpha . C \in l \cap l^{\prime} . l \cap a = \Lambda . l^{\prime} \cap a = \Lambda . \supset_{l,l^{\prime}} . l = l^{\prime} \qquad \text{Df}$$

Note that only I Hp R and XI Hp R require the hypothesis $t \in T$; in all the other axioms there is a hypothesis which can only be true when $t \in T$. For the purpose of comparison with the axioms of Concept I., the following propositions are required:—

|490|

*2·11. $\vdash \therefore$ (III, IV, V, VI, VII) $Hp\ R . \supset : t \in T . \supset . Nc^{\prime}R_t{}^{\prime}(;;;) \geqq 2$

Proof.—*Cf.* *1·72.

*2·21. $\vdash \therefore$ IV $Hp\ R . \supset : R_t{}^{\prime}(ABC) . \supset . R_t{}^{\prime}(CBA)$

Proof.—*Cf.* *1·63.

*2·22. $\vdash \therefore$ (III, IV, V, VI) $Hp\ R . \supset : R_t{}^{\prime}(ABC) . \supset . \sim R_t{}^{\prime}(BCA)$

Proof.—*Cf.* *1·64.

*2·23. $\vdash \therefore$ (III, IV, V, VI) $Hp\ R . \supset : R_t{}^{\prime}(ABC) . \supset . A \neq C$

Proof.—*Cf.* *1·62.

*2·31. *Proposition.*—Assuming (VII, IX) H*p* R, if A and B are two distinct points at the time *t*, they possess one, and only one, common member. In symbols,

$$\vdash \therefore (\text{VII, IX})\ \text{H}p\ \text{R} . \supset : \text{A, B} \in \text{R}_t{}^{\backprime}(;;;) . \text{A} \neq \text{B} . \supset . \text{A} \cap \text{B} \in 1$$

Proof. —*Cf.* *1·31·42 and (VII, IX) H*p* R.

*2·32. *Proposition.*— Assuming (VII, VIII, IX) H*p* R, a line at any instant *t* (*i.e.*, a member of lin_{Rt}) is the complete class of points (interpoints) possessing some linear objective real. In symbols,[1]

$$\vdash : (\text{VII, VIII, IX})\text{H}p\ \text{R} . \supset . \text{lin}_{Rt} = \hat{p}[(\exists a) . a \in \text{R}^{\backprime}(;;;;t) . p = \hat{\text{A}}\{\text{A} \in \text{R}_t{}^{\backprime}(;;;) . a \in \text{A}\}]$$

Proof.—*Cf.* VIII H*p* R and *2·31.

Propositions *2·31·32 effect the identification of the punctual line, as defined above, and the class of points on some linear objective real. Thus a straight line considered as an entity with parts is a punctual line, and considered as a simple unit is a linear objective real.

*2·33. *Proposition.*—Assuming (VII, VIII, IX) H*p* R, if C and D are two points in the punctual line $\text{R}_t{}^{\backprime}\overline{\text{AB}}$, then A is a point in the punctual line $\text{R}_t{}^{\backprime}\overline{\text{CD}}$. In symbols,

$$\vdash \therefore (\text{VII, VIII, IX})\ \text{H}p\ \text{R} . \supset : \text{C, D} \in \text{R}_t{}^{\backprime}\overline{\text{AB}} . \text{C} \neq \text{D} . \supset . \text{A} \in \text{R}_t{}^{\backprime}\overline{\text{CD}}$$

Proof.—*Cf.* *2·32.

*2·41. $\vdash \therefore (\text{III-IX})\ \text{H}p\ \text{R} . \supset : t \in \text{T} . \supset . (\exists \text{A, B, C}) . \Delta_{Rt}{}^{\backprime}(\text{ABC})$

Proof.—*Cf.* *1·72·73. *2·32.

*2·5. *Proposition.*—Assuming (III-XIV) H*p* R of Concept IV., then all the axioms of Concept I. hold when the R_t of Concept IV. is substituted for the R of Concept I., and *t* is a member of T.

Proof.—*Cf.* *2·11·21·22·23·33·41 and (IX-XIV) H*p* R (of Concept IV.) and (I-XII) H*p* R of Concept I.

It will be noticed that I H*p* R (of Concept IV.) is not required in the above |491| comparison. It does not belong to the purely geometrical side of the concept, but is a necessary part of the "physical" ideas. II H*p* R (of Concept IV.), though it does not occur explicitly in the above comparison, is required to give the geometry "existence." Thus the geometry of Concept IV. requires thirteen axioms.

For the purpose of the transition to projective geometry (*cf.* VEBLEN, *loc. cit.*), it is now unnecessary to conceive a new class of "projective points." The points already on hand are exactly the entities required. All that is necessary is to define the class of those linear objective reals (*cf.* XIV H*p* R), coplanar with any given linear objective real and not intersecting it, as *the point at infinity* on that objective real. Then with these new points at infinity, and the old points, the complete set of "projective points" is obtained.

The Extraneous Relation.—For the purpose of the definition of motion, one extraneous tetradic relation is required, exactly as in Concept III. Also the same hypotheses must hold respecting it. The three mutually rectangular and intersecting punctual lines, thus indicated at each instant, are to be taken as the "kinetic axes," and all motion measured by reference to them. A given set of kinetic axes does not,

1. In the line below, the original erroneously had '$\llcorner$' instead of '$\vdash$'.

in general, correspond to the same three linear objective reals at different instants of time.

Matter.—It is necessary to assume that the points in this concept disintegrate, and do not, in general, persist from instant to instant. For otherwise the only continuous motion possible would be representable by linear transformations of coordinates; and it seems unlikely that sense-perceptions could be explained by such a restricted type of motions. We have therefore to consider what, in this concept, can represent the permanence of matter. A "corpuscle," as we may call it, may be conceived to be a volume with some special property in respect to the linear objective reals "passing through" it. This is the procedure adopted in Concept V.; and the methods of overcoming the obvious difficulties which suggest themselves will be considered in detail there. It is sufficient here to notice that, in this Concept IV., the special property of the volume must relate merely to the motion of the objective reals. For the only alternative is to make the property consist of the permanence of the points within the volume. But then the difficulty of permanent collineations, mentioned above, recurs. To find a special property of motion, we require a kinematical science for linear objective reals in this concept analogous to the kinematical parts of hydrodynamics. In the absence at the present time of such a science, we proceed to other alternatives.

Concept IVa.—Conceive a class of *particles*, each particle being associated at each instant with some point, but not necessarily each point with some particle. Then the particles represent the "matter" which "occupies" space. Laws of motion must then be stated (i) for the particles and (ii) for the linear objective reals. Also the motion of the particles may be conceived to be influenced by that of the linear objective reals, and *vice versâ*. The endeavour to state such laws appears to reduce |492| itself to rewriting with appropriate changes a chapter of any modern treatise of electricity and magnetism. It would seem necessary to subdivide the class of particles into "positive" and "negative" particles, a charged volume containing an excess of one type. The conception of an ether conveying lines of force is replaced by the class of the linear objective reals. The details can be managed much as in the analogous case of Concept V., considered later. An indefinite number of extraneous relations are required to "locate" the particles, just as in Concept I. This concept (as thus developed with "particles") is not completely a "linear" concept. It is a hybrid between the "linear" and "punctual" concepts. In its dualism it is not superior to the classical concept. But, in possessing moving linear objective reals as well as moving particles, it is richer in physical ideas.

Concept IVb. — In this concept, just as in Concept II., each triadic extraneous relation of Concept IVa. between an instant of time, a particle, and a point is replaced by a dyadic extraneous relation between a point and an instant of time.

Part IV.— (i) The Theory Of Dimensions.

*3. Concept V. depends upon a treatment of the theory of dimensions different from that which at present obtains. The theory here developed is relevant to any definite property which (1) is a property of classes only, and (2) is only a property of some classes. It will be clearer, and no longer, to explain the theory in its full generality, and in Concept V. to make the special application required.

This general theory of dimensions may, perhaps, have a range of importance greater than that which is assigned to it in the sequel. In *10 a set of hypotheses are given respecting the property ϕ; and when these are true of ϕ, the propositions and definitions of *3 to *8 acquire importance and emerge from triviality, also in this case further deductions of propositions can be made. The Concept V. to which this theory is applied is explained in the definitions of *20 and the axioms of *22. In this Concept V. a special property ϕ is taken, which is termed "Homaloty" (*cf.* *20·11·12), and (*cf.* *22) in the axioms a relation R is considered such that "homaloty," defined in respect to R, has the properties of the axioms in *10.

*3·01. *Definition.*—If $\phi!x$ is some proposition involving the entity x, which may be varied, so that $\phi!x$ and $\phi!y$ make the same statement (ϕ) about x and y respectively, then any entity z, for which $\phi!z$ is true, is said to *possess the property* ϕ.

*3·02. *Definition.*—A *ϕ-class* is a class with the property ϕ, that is to say, if u is a ϕ-class, then $\phi!u$ is true.

*3·11. *Definition.*—The *ϕ-region* is the logical sum of all classes which possess the property ϕ. The symbol O_ϕ will denote the ϕ-region. The symbolic definition is

$$O_\phi = \cup\text{‘}\hat{u}\{\phi!u\} \qquad \text{Df}$$

*3·12. The *common ϕ-subregion for u* is that class which is the common subclass |493| of all ϕ-classes with the class u as a subclass. The symbol $\text{cm}_\phi\text{‘}u$ will denote the common ϕ-subregion for u. The symbolic definition is

$$\text{cm}_\phi\text{‘}u = \cap\text{‘}\hat{v}\{\phi!v \,.\, u \in \text{cls‘}v\} \qquad \text{Df}$$

Note.—If no class v, with the property ϕ and containing u as a subclass, exists, then $\text{cm}_\phi\text{‘}u$ will be the class of all entities. But if a class v exists which has the property ϕ and contains u, then $\text{cm}_\phi\text{‘}u$ is a subclass of O_ϕ. In the sequel it will be found that this latter is the only relevant case for our purposes.

Elucidatory Note.—Assuming our ordinary geometrical ideas, let the property of the "flatness" of a class of straight lines be defined thus: A class of straight lines is flat, *either*, when it is a necessary and sufficient condition for membership that a straight line meets two members of the class, not at their point of meeting, *or*, when the class is a unit class with one line as its sole member. Thus a plane (as a line-locus) is flat, a three-dimensional space (as a line-locus) is flat, and so on. Now let the property ϕ in the above definition be the property of flatness. If then u is a class consisting only of two straight lines, the common ϕ-subregion for u is either a three-dimensional space or a plane, according as the two lines are not, or are, coplanar. Also in a space of higher dimensions than three, if u be a class consisting of three straight lines, the common ϕ-subregion for u may be either (1) a plane, or (2) a three-dimensional space, or (3) a four-dimensional space, or (4) a five-dimensional space, according to the circumstances of the lines. It will be noticed that, in the application of this theory of the common ϕ-subregion to the particular case of geometrical flatness, the common ϕ-subregion of any class of lines is itself flat. But this is not, in general, the case when any property not flatness is considered. It is this peculiar property of flatness which has masked the importance in geometry of the theory of common ϕ-subregions.

*3·121. *Definition.*—Two classes u and v have *ϕ-equivalence* if $\text{cm}_\phi\text{‘}u=\text{cm}_\phi\text{‘}v$. The class of those classes (including u itself as a member), which have ϕ-equivalence with u is denoted by $\text{equiv}_\phi\text{‘}u$. The symbolic definition is[1]

$$\text{equiv}_\phi\text{‘}u=\hat{v}\{\text{cm}_\phi\text{‘}v=\text{cm}_\phi\text{‘}u\} \qquad \text{Df}$$

*3·13. *Definition.*—A class u (not the null class) is *ϕ-prime*, when, if v be any proper part (part, not the whole) of u, v is *not* ϕ-equivalent to u. The class of those classes which are ϕ-prime will be denoted by the symbol prm_ϕ. The symbolic definition is

$$\text{prm}_\phi=\hat{u}\{\exists!u:v\subset u\,.\,\exists!(u-v)\,.\supset_v.\,\text{cm}_\phi\text{‘}v\neq\text{cm}_\phi\text{‘}u\} \qquad \text{Df}$$

Elucidatory Note.—With the assumptions of the elucidatory note on $\text{cm}_\phi\text{‘}u$, it is at once obvious that two straight lines form a ϕ-prime (where ϕ is flatness) class, whether they are or are not coplanar. But if u consist of three straight lines, (1) u is not ϕ-prime if $\text{cm}_\phi\text{‘}u$ is a plane, (2) u is not, in general, ϕ-prime if $\text{cm}_\phi\text{‘}u$ is a space of three dimensions, but u is (in this case) ϕ-prime if the three lines are concurrent, |494| (3) u is ϕ-prime if $\text{cm}_\phi\text{‘}u$ is of four dimensions, (4) u is ϕ-prime if $\text{cm}_\phi\text{‘}u$ is of five dimensions.

*3·21. The *ϕ-dimension number* (or the ϕ-dimensions) of a class u is the greatest of the cardinal numbers of all classes (including possibly u itself) which are both ϕ-equivalent to u and ϕ-prime. The ϕ-dimension number of u will be denoted by $\text{dim}_\phi\text{‘}u$. The symbolic definition is

$$\text{dim}_\phi\text{‘}u\,.=:(\iota\alpha):\alpha\in\text{Nc‘‘}(\text{prm}_\phi\cap\text{equiv}_\phi\text{‘}u):\rho\in\text{Nc‘‘}(\text{prm}_\phi\cap\text{equiv}_\phi\text{‘}u)\,.\supset_\rho.\,\rho\leqq\alpha \qquad \text{Df}$$

Elucidatory Note.—With the assumptions of the previous elucidatory notes (where ϕ is flatness), we see that those ϕ-prime classes, the common ϕ-subregions for which are spaces of three dimensions (as ordinarily understood), are all pairs of non-intersecting lines and all trios of concurrent non-coplanar lines; also no class of four lines in such a space can be prime. Thus three is the greatest cardinal number of any ϕ-prime class of lines for which the common ϕ-subregion is such a space. Hence, according to the above definition, three is the ϕ-dimension number of the space.

*3·22. *Definition.*— A class u is ϕ-axial when (1) it is ϕ-prime and (2) its cardinal number is equal to its ϕ-dimensions. The class of ϕ-axial classes is denoted by the symbol ax_ϕ. The symbolic definition is

$$\text{ax}_\phi=\hat{u}\{u\in\text{prm}_\phi\cap\text{dim}_\phi\text{‘}u\} \qquad \text{Df}$$

Elucidatory Note.—With the assumptions of the previous elucidatory notes (where ϕ is flatness), we see that two coplanar lines form a ϕ-axial class, and so also do three concurrent non-coplanar lines.

*3·23. *Definition.*— A class u is *ϕ-maximal* when (1) all those of its subclasses (possibly including u itself), which are both ϕ-prime and ϕ-equivalent to u, are ϕ-axial, and (2) there are such subclasses. The class of ϕ-maximal classes will be denoted by mx_ϕ. The symbolic definition is

$$\text{mx}_\phi=\hat{u}\{\exists!\in(\text{prm}_\phi\cap\text{equiv}_\phi\text{‘}u\cap\text{cls‘}u)\,.\,\text{prm}_\phi\cap\text{equiv}_\phi\text{‘}u\cap\text{cls‘}u\subset\text{ax}_\phi\} \qquad \text{Df}$$

1. In the below equation, the original erroneously used parentheses instead of curly brackets.

Elucidatory Note.—Referring to the previous elucidatory notes (where ϕ is flatness), we see that any set of coplanar lines form a ϕ-maximal class; similarly any set of concurrent lines form a ϕ-maximal class.

*3·31. *Definition.*—The *ϕ-concurrence* of u and v, where u and v are classes, is that subclass of u (possibly u itself), such that any couple, formed by any member of it and any member of v, is ϕ-axial. The ϕ-concurrence of u with v is denoted by the symbol $\tilde{u}_\phi$'v. The definition in symbols is

$$\tilde{u}_\phi\text{‘}v = \acute{x}\{x \in u : y \in v . \supset_y . \iota\text{‘}x \cup \iota\text{‘}y \in \mathrm{ax}_\phi\} \qquad \text{Df}$$

The ϕ-concurrence of the ϕ-region (0_ϕ) with any class v will be written $\tilde{0}_\phi$'v instead of $\tilde{0}_{\phi_\phi}$'v.

Elucidatory Note.—Referring to the previous elucidatory notes (when ϕ is flatness), |495| we see that, when u and v are classes of straight lines, $\tilde{u}_\phi$'v (*i.e.*, the ϕ-concurrence of u with v) is that complete set of lines of u which is such that any member of it is coplanar with every member of v.

*3·32. *Definition.*—A class u is a *self-ϕ-concurrence* if the ϕ-concurrence of u with itself is the whole class u. The class of those classes which are self-ϕ-concurrences will be denoted by conc_ϕ. The symbolic definition is

$$\mathrm{conc}_\phi = \acute{u}\{u = \tilde{u}_\phi\text{‘}u\} \qquad \text{Df}$$

*3·33. *Definition.*—x will be said to be *ϕ-concurrent with* y, if the class composed of x and y only (*i.e.*, the class ι'$x \cup \iota$'y) is ϕ-axial.

*3·41. *Definition.*—A *ϕ-plane* is a class u such that there exists a class v, which (1) is ϕ-axial, (2) is composed of two members only, and (3) is such that u is the class cm_ϕ'v. The class of those classes which are ϕ-planes is denoted by ple_ϕ. The symbolic definition is

$$\mathrm{ple}_\phi = \acute{u}\{(\exists v) . v \in 2 \cap \mathrm{ax}_\phi . u = \mathrm{cm}_\phi\text{‘}v\} \qquad \text{Df}$$

Note.—It requires an axiom to establish that a ϕ-plane is a self-ϕ-concurrence (cf. *16·11).

*3·42. *Definition.*—A class u is a *ϕ-point*, if there exists a class v, which (1) is ϕ-axial, (2) is composed of three members only, and (3) is such that u is the ϕ-concurrence of the ϕ-region with v. The class of those classes which are ϕ-points is denoted by the symbol pnt_ϕ. The symbolic definition is

$$\mathrm{pnt}_\phi = \acute{u}\{(\exists v) . v \in 3 \cap \mathrm{ax}_\phi . u = \tilde{0}_\phi\text{‘}v\} \qquad \text{Df}$$

Note.—It requires axioms to establish that a ϕ-point is ϕ-maximal and is a self-ϕ-concurrence (*cf.* *14·11·12). Also note that this definition does not apply unless the number of dimensions of 0_ϕ is at least three, but then applies *unchanged* however great this number may be.

Elucidatory Note.—Referring to the previous elucidatory notes (where ϕ is flatness), we see that a ϕ-point now becomes simply that class of straight lines concurrent at a point. The analogy with KLEIN'S "ideal," or "projective," points is obvious. Only when the present theory is applied, it will be found that the original "descriptive" point has entirely vanished.

*3·43. *Definition.*—A class is *ϕ-coplanar* if there exists a ϕ-plane of which it is a subclass. The symbol $\mathrm{cople}_\phi!u$ denotes that the class u is ϕ-coplanar. The definition in symbols is

$$\text{cople}_\phi!u. = .(\exists p).p \in \text{ple}_\phi.u \in \text{cls}`p \qquad \text{Df}$$

*3·44. *Definition.*—A class is *ϕ-copunctual* if there exists a ϕ-point of which it is a subclass. The symbol $\text{copnt}_\phi!u$ denotes that the class u is ϕ-copunctual. The definition in symbols is

$$\text{copnt}!u. = .(\exists P).P \in \text{pnt}_\phi.u \in \text{cls}`P \qquad \text{Df}$$

|496| ***General Deductions Concerning Dimensions.***

A large chapter of interesting propositions concerning the entities defined above in *3 can be compiled. The following are chosen as being directly wanted in the subsequent investigations:—

* 4. *On Common ϕ-subregions.*

*4·21. *Proposition.*— If v is a subclass of u, then $\text{cm}_\phi`v$ is a subclass of $\text{cm}_\phi`u$. In symbols,

$$\vdash : v \subset u. \supset .\text{cm}_\phi`v \subset \text{cm}_\phi`u$$

Proof.—*Cf.* *3·12.

*4·25. *Proposition.*—A class u is itself a subclass of $\text{cm}_\phi`u$. In symbols

$$\vdash .u \subset \text{cm}_\phi`u$$

Proof.—*Cf.* *3·12.

*4·27. *Proposition.*—If u is a class with the property ϕ, then u is identical with $\text{cm}_\phi`u$. In symbols,

$$\vdash : \phi!u. \supset .u = \text{cm}_\phi`u$$

Proof.—*Cf.* *3·12.

*4·28. *Proposition.*—If there exist two classes, both with the property ϕ, which possess no common member, then $\text{cm}_\phi`\Lambda$ is itself the null class (Λ). In symbols,

$$\vdash : (\exists u, v).u \cap v = \Lambda.\phi!u.\phi!v. \supset .\text{cm}_\phi`\Lambda = \Lambda$$

Proof.—Note that $\text{cm}_\phi`\Lambda$ is the common part of all ϕ-classes.

Corollary.—If x and y exist such that they are distinct, and the two unit classes with them as members respectively each have the property ϕ, then $\text{cm}_\phi`\Lambda$ is Λ.

Note that when *4·28 is appealed to, it will be this corollary which is directly used.

*4·31. *Proposition.*—The common ϕ-subregion for the common ϕ-subregion for u is the common ϕ-subregion for u. In symbols,

$$\vdash .\text{cm}_\phi`\text{cm}_\phi`u = \text{cm}_\phi`u$$

Proof.—For $\text{cm}_\phi`u$ is contained in every $\phi$-class containing $u$. Hence (*cf.* *3·12) $\text{cm}_\phi`\text{cm}_\phi`u$ is contained in $\text{cm}_\phi`u$. Also (*cf.* *4·25·21) $\text{cm}_\phi`u$ is contained in $\text{cm}_\phi`\text{cm}_\phi`u$.

*4·32. *Proposition.*—If u and v are ϕ-equivalent, and w is any class, then the common ϕ-subregion for the logical sum of u and w is identical with the common ϕ-subregion for the logical sum of v and w. In symbols,

$$\vdash : \text{cm}_\phi`u = \text{cm}_\phi`v. \supset .\text{cm}_\phi`(u \cup w) = \text{cm}_\phi`(v \cup w)$$

Proof.—For (cf. *4·21) $\mathrm{cm}_\phi\text{‘}v$ is contained in $\mathrm{cm}_\phi\text{‘}(v \cup w)$, and hence (*hypothesis*) $\mathrm{cm}_\phi\text{‘}u$ is contained in $\mathrm{cm}_\phi\text{‘}(v \cup w)$, and hence (*cf.* *4·25) $u \cup w$ is contained in $\mathrm{cm}_\phi\text{‘}(v \cup w)$, and hence (*cf.* *4·21) $\mathrm{cm}_\phi\text{‘}(u \cup w)$ is contained in $\mathrm{cm}_\phi\text{‘}\mathrm{cm}_\phi\text{‘}(v \cup w)$, and |497| hence (cf. *4·31) $\mathrm{cm}_\phi\text{‘}(u \cup w)$ is contained in $\mathrm{cm}_\phi\text{‘}(v \cup w)$. Then interchanging u and v, and combining the two results, the proposition follows.

The following propositions are not cited subsequently, so their verbal enunciations are omitted:—

*4·41. $\vdash . \mathrm{cm}_\phi\text{‘}(\mathrm{cm}_\phi\text{‘}u \cap \mathrm{cm}_\phi\text{‘}v) = \mathrm{cm}_\phi\text{‘}u \cap \mathrm{cm}_\phi\text{‘}v$

*4·42. $\vdash . \cap\text{‘}\mathrm{cm}_\phi\text{‘‘}p = \mathrm{cm}_\phi\text{‘}\cap\text{‘}\mathrm{cm}_\phi\text{‘‘}p$

*4·43. $\vdash . \mathrm{cm}_\phi\text{‘}(\mathrm{cm}_\phi\text{‘}u \cup \mathrm{cm}_\phi\text{‘}v) = \mathrm{cm}_\phi\text{‘}(u \cup v)$

*4·44. $\vdash . \mathrm{cm}_\phi\text{‘}\cup\text{‘}\mathrm{cm}_\phi\text{‘‘}p = \mathrm{cm}_\phi\text{‘}\cup\text{‘}p$

*5. *On ϕ-Primes.*

*5·23. *Proposition.*—If u is a ϕ-prime, and v is a subclass of u, and is not the null class, then v is a ϕ-prime. In symbols,

$$\vdash : u \in \mathrm{prm}_\phi . v \in \mathrm{cls}\text{‘}u . \exists ! v . \supset . v \in \mathrm{prm}_\phi$$

Proof.—For if w be any subclass (not the null class) of v, then (*cf.* *3·13) $\mathrm{cm}_\phi\text{‘}(u - w)$ is not $\mathrm{cm}_\phi\text{‘}u$. But $(u - w)$ can be written $\{(v-w) \cup (u-v)\}$, and u can be written $\{v \cup (u-v)\}$. Hence $\mathrm{cm}_\phi\text{‘}\{(v-w)\cup(u-v)\}$ is not $\mathrm{cm}_\phi\text{‘}\{v \cup (u-v)\}$. Hence (*cf.* *4·32) $\mathrm{cm}_\phi\text{‘}(v-w)$ is not $\mathrm{cm}_\phi\text{‘}v$. Hence (*cf.* *3·13) v is a ϕ-prime.

Note.—This theorem, together with *4·31·32, is the foundation of the whole theory. It is remarkable that it requires no axiom concerning ϕ. The companion theorem (*cf.* *12·42), with ax_ϕ substituted for prm_ϕ, requires axioms respecting ϕ.

*5·231. *Proposition.*—Necessary and sufficient conditions, that a class u may be ϕ-prime, are: (1) u is not the null class, and (2) if x be any member of u, then $\mathrm{cm}_\phi\text{‘}(u - \iota\text{‘}x)$ is not $\mathrm{cm}_\phi\text{‘}u$. In symbols,

$$\vdash .\because \exists ! u : x \in u . \supset_x . \mathrm{cm}_\phi\text{‘}(u - \iota\text{‘}x) \neq \mathrm{cm}_\phi\text{‘}u : \equiv . u \in \mathrm{prm}_\phi$$

Proof.—*Cf.* *3·13 and *4·21.

*5·233. *Proposition.*—If $\mathrm{cm}_\phi\text{‘}\Lambda = \Lambda$, then every unit class is a ϕ-prime. In symbols,

$$\vdash : \mathrm{cm}_\phi\text{‘}\Lambda = \Lambda . \supset . 1 \subset \mathrm{prm}_\phi$$

Proof.—*Cf.* *3·13 and *4·25.

*5·235. *Proposition.*— If x and y are distinct, and the unit classes $\iota\text{‘}x$ and $\iota\text{‘}y$ have the property ϕ, then the class, which is the couple composed of x and y, is a ϕ-prime. In symbols,

$$\vdash : x \neq y . \phi ! \iota\text{‘}x . \phi ! \iota\text{‘}y . \supset . \iota\text{‘}x \cup \iota\text{‘}y \in \mathrm{prm}_\phi$$

Proof.—*Cf.* *3·13 and *4·25·27.

*6. *On ϕ-Dimensions and ϕ-Axial Classes.*

*6·23. *Proposition.*—The ϕ-dimension of u, if there is such an entity, is a cardinal number not zero. In symbols,

$$\vdash : (\mathrm{Ex})\mathrm{dim}_\phi\text{‘}u) . \supset . \mathrm{dim}_\phi\text{‘}u \in Nc - \iota\text{‘}0$$

Proof.— *Cf.* *3·21.

|498|*6·25. *Proposition.*—If v is a ϕ-prime and has a ϕ-dimension number, then the cardinal number of v is less than, or equal to, $\dim_\phi$‘v. In symbols,

$$\vdash : (\text{Ex})\dim_\phi\text{‘}v).v \in \text{prm}_\phi . \supset . \text{Nc‘}v \leqq \dim_\phi\text{‘}v$$

Proof.— *Cf.* *3·21.

*6·26. *Proposition.*— If v is ϕ-axial and is ϕ-equivalent to u, then the cardinal number of v is equal to $\dim_\phi$‘u. In symbols,

$$\vdash : v \in \text{ax}_\phi \cap \text{equiv}_\phi\text{‘}u . \supset . \text{Nc‘}v = \dim_\phi\text{‘}u \cdot$$

Proof.—*Cf.* *3·21·22.

*8. *On ϕ-Concurrences.*

*8·21. *Proposition.*—If u is contained in w, then the ϕ-concurrence of u with v is contained in the ϕ-concurrence of w with v. In symbols,

$$\vdash : u \subset w . \supset . \tilde{u}_\phi\text{‘}v \subset \tilde{w}_\phi\text{‘}v$$

Proof.—*Cf.* *3·31.

*8·22. *Proposition.*—If v is contained in w, then the ϕ-concurrence of u with w is contained in the ϕ-concurrence of u with v. In symbols,

$$\vdash : v \subset w . \supset . \tilde{u}_\phi\text{‘}w \subset \tilde{u}_\phi\text{‘}v$$

Proof.—*Cf.* *3·31.

*10. *Geometrical Properties.*—A property ϕ is called *geometrical* if it satisfies the five axioms $(\lambda, \mu, \nu, \pi, \rho)$ Hp ϕ stated below. The axiom ν Hp ϕ takes the special form for three dimensions. It is to be noticed that three dimensions is the lowest number for which a ϕ-point (*cf.* *3·42) can be defined. The reasoning can be applied to higher dimensions, only more elaborate inductions and an extra axiom are required. Other axioms and definitions are wanted to enable all the propositions of projective geometry to be proved. These will not be considered here as such an investigation would involve some repetition when we come to Concept V. The class O_ϕ is the class of straight lines of the geometry, conceived as simple unities. The class pnt_ϕ is the class of points, each point being a class of lines. The class ple_ϕ is the class of planes, each plane being a class of lines.

*10·1. λ Hp ϕ is the statement that O_ϕ *has the property* ϕ. In symbols,

$$\lambda \text{ Hp } \phi . = . \phi ! \text{O}_\phi \qquad \text{Df}$$

*10·2. μ Hp ϕ is the statement that, *if x is any member of* O_ϕ, *the unit class* ι‘x *has the property* ϕ. In symbols,

$$\mu \text{ Hp } \phi . = : x \in \text{O}_\phi . \supset_x . \phi ! \iota\text{‘}x \qquad \text{Df}$$

*10·3. ν Hp ϕ is the statement that *the ϕ-dimension number of* O_ϕ *is three.* In symbols,

$$\nu \text{ Hp } \phi . = . \dim_\phi\text{‘O}_\phi = 3 \qquad \text{Df}$$

|499| *10·4. π Hp ϕ is the statement that, *if u is a subclass of* O_ϕ, *and v is ϕ-axial and contained in* cm_ϕ‘u, *then there exists a class w (possibly the null class) such that the logical sum of v and w is ϕ-axial and ϕ-equivalent to u.* In symbols,

$$\pi \text{ Hp } \phi . = : u \in \text{cls‘O}_\phi . v \in \text{ax}_\phi \cap \text{cls‘cm}_\phi\text{‘}u . \supset_{u,v} . (\exists w) . v \cup w \in \text{ax}_\phi \cap \text{equiv}_\phi\text{‘}u \qquad \text{Df}$$

*10*5. ρ Hp ϕ is the statement that *if u and v are both ϕ-axial, and if they possess at least two members in common, then their logical sum is ϕ-maximal.* In symbols,

$$\rho \text{ Hp } \phi . = : u,v \in \text{ax}_\phi . \text{Nc‘}(u \cap v) \geqq 2 . \supset_{u,v} . u \cup v \in \text{mx}_\phi \qquad \text{Df}$$

Elucidatory Note.—Referring to the previous elucidatory notes (where ϕ is flatness), we see that *10·4 in effect assumes that a line can always be added (1) to two concurrent lines to form a set of three concurrent non-coplanar lines, and (2) to one line in a plane to form a set of two concurrent lines in that plane. Also *10·5 assumes that, if two sets of three concurrent lines have two members in common, the four lines are concurrent.

Deductions from the Axioms.

*11. *Preliminary Propositions.*

*11·11. *Proposition.*— Assuming (λ,ν)Hp ϕ, O_ϕ has at least three members. In symbols,

$$\vdash : (\lambda,\nu)\text{Hp } \phi . \supset . \text{Nc‘O}_\phi \geqq 3$$

Proof.—*Cf.* *4·27 and *10·1·3.

*11·12. *Proposition.*—Assuming (λ,μ,ν)Hp ϕ, cm_ϕ‘Λ is the null class (Λ). In symbols,

$$\vdash : (\lambda,\mu,\nu)\text{Hp } \phi . \supset . \text{cm‘}\Lambda = \Lambda$$

Proof.—*Cf.* *4·28 and *10·2 and *11·11.

*11*21. *Proposition.*—Assuming (λ,μ)Hp ϕ, all ϕ-prime classes with more than one member are contained in O_ϕ. In symbols,

$$\vdash \therefore (\lambda,\mu)\text{Hp } \phi . \supset : v \in \text{prm}_\phi . \text{Nc‘}v > 1 . \supset . v \in \text{cls‘O}_\phi$$

Proof.—For if x is not a member of O_ϕ, then cm_ϕ‘ι‘x is the class of all entities. Hence (*cf.* *4·21·27 and *10·1·2) the conclusion follows.

*12. *On ϕ-Axial Classes and ϕ-Dimensions.*

*12·11. *Proposition.*— Assuming (λ,μ,ν)Hp ϕ, every unit class whose single member belongs to O_ϕ is ϕ-axial. In symbols,

$$\vdash \therefore (\lambda,\mu,\nu)\text{Hp } \phi . \supset : x \in \text{O}_\phi . \supset_x . \iota\text{‘}x \in \text{ax}_\phi$$

Proof.—*Cf.* *3·21·22 and *4·27 and *5·233 and *10·2.

*12·12. *Proposition.*—Assuming $(\lambda-\pi)$Hp ϕ, every subclass of O_ϕ, not the null |500| class, has a set of ϕ-axes. (*Note.*—A class which is ϕ-axial and ϕ-equivalent to a class u is said to be a set of ϕ-axes of u.) In symbols,

$$\vdash \therefore (\lambda-\pi)\text{Hp } \phi . \supset : u \in \text{cls‘O}_\phi . \exists ! u . \supset . \exists !(\text{ax}_\phi \cap \text{equiv}_\phi\text{‘}u)$$

Proof.—Since there is at least one member of u, there is (*cf.* *12·11) a ϕ-axial class contained in u. Hence (*cf.* *10·4) this class can be augmented so as to become a set of ϕ-axes of u.

*12·13. *Proposition.*—Assuming $(\lambda-\pi)$Hp ϕ, if u and v are subclasses of O_ϕ, and v is not the null class, and cm_ϕ‘v is contained in cm_ϕ‘u, then there exist two subclasses of O_ϕ, w and w', say, such that w is a set of ϕ-axes of v, and $w \cup w'$ is a set of ϕ-axes of u. In symbols,

$$\vdash \therefore (\lambda - \pi)\text{Hp}\ \phi . \supset : u,v \in \text{cls}`O_\phi . \exists ! v . \text{cm}_\phi`v \subset \text{cm}_\phi`u . \supset .$$
$$(\exists w,w') . w \in \text{ax}_\phi \cap \text{equiv}_\phi`v . w \cup w' \in \text{ax}_\phi \cap \text{equiv}_\phi`u.$$

Proof.—*Cf.* *10·4 and *12·12.

*12·21. *Proposition.*—Assuming $(\lambda - \pi)\text{Hp}\ \phi$, if u and v are subclasses of O_ϕ, and v is not the null class, and $\text{cm}_\phi`v$ is contained in $\text{cm}_\phi`u$, then the ϕ-dimension number of v is less than, or equal to, the ϕ-dimension number of u. In symbols,

$$\vdash \therefore (\lambda - \pi)\text{Hp}\ \phi . \supset : u,\ v \in \text{cls}`O_\phi . \exists ! v . \text{cm}_\phi`v \subset \text{cm}_\phi`u . \supset . \dim_\phi`v \leqq \dim_\phi`u.$$

Proof.—From *6·26 and *12·13, w and w' exist (assuming $w \cap w' = \Lambda$) such that $\text{Nc}`w = \dim_\phi`v$ and $\text{Nc}`w + \text{Nc}`w' = \dim_\phi`u$. Hence $\dim_\phi`v < \dim_\phi`u$, unless $w'$ is the null class, or unless the numbers are not finite, in which cases $\dim_\phi`v = \dim_\phi`u$ is possible.

*12·23. *Proposition.*—Assuming $(\lambda - \pi)\text{Hp}\ \phi$, if u and v are subclasses of O_ϕ, and v is not the null class, and $\text{cm}_\phi`v$ is contained in $\text{cm}_\phi`u$, then, if $\dim_\phi`v = \dim_\phi`u$, we have $\text{cm}_\phi`v = \text{cm}_\phi`u$, and conversely. In symbols,

$$\vdash :: (\lambda - \pi)\text{Hp}\ \phi . \supset \therefore v,u \in \text{cls}`O_\phi . \exists ! v . \text{cm}_\phi`v \subset \text{cm}_\phi`u . \supset :$$
$$\text{cm}_\phi`v = \text{cm}_\phi`u . \equiv . \dim_\phi`v = \dim_\phi`u$$

Proof.—Assuming $\dim_\phi`v = \dim_\phi`u$, and also assuming the notation of the proof of *12·21, then w and w' are such that (1) w is ϕ-equivalent to v and $w \cup w'$ to u, (2) $\text{Nc}`w = \dim_\phi`v$ and $\text{Nc}`w + \text{Nc}`w' = \dim_\phi`u$. Hence, by hypothesis and (2), $\text{Nc}`w + \text{Nc}`w' = \text{Nc}`w$. Also (*cf.* *10·3 and *12·21) $\text{Nc}`w + \text{Nc}`w' \leqq 3$. Hence $\text{Nc}`w' = 0$, that is, $w' = \Lambda$. Hence from (1), $\text{cm}_\phi`u = \text{cm}_\phi`v$. The converse is obvious.

*12·33. *Proposition.*—Assuming $(\lambda - \pi)\text{Hp}\ \phi$, if u and v are subclasses of O_ϕ, and v is not the null class, and $\text{cm}_\phi`v$ is contained in, but is not identical with, $\text{cm}_\phi`u$, then $\dim_\phi`v$ is less than $\dim_\phi`u$. In symbols,

$$\vdash \therefore (\lambda - \pi)\text{Hp}\ \phi . \supset : u,v \in \text{cls}`O_\phi . \exists ! v . \text{cm}_\phi`v \subset \text{cm}_\phi`u .$$
$$\text{cm}_\phi`v \neq \text{cm}_\phi`u . \supset . \dim_\phi`v < \dim_\phi`u$$

Proof.—*Cf.* *12·21·23.

|501| The following proposition should be compared with *12·12:—

*12·37. *Proposition.*—Assuming $(\lambda - \pi)\text{Hp}\ \phi$, if u is a subclass of O_ϕ, and is not the null class, there exists a subclass of u which is ϕ-prime and ϕ-equivalent to u. In symbols,

$$\vdash \therefore (\lambda - \pi)\text{Hp}\ \phi . \supset : u \in \text{cls}`O_\phi . \exists ! u . \supset . \exists ! (\text{prm}_\phi \cap \text{cls}`u \cap \text{equiv}_\phi`u)$$

Proof.—From *12·12·21, u is either of one, or of two, or of three ϕ-dimensions. If u is of one ϕ-dimension, the conclusion follows from *5·233 and *10·2 and *11·12. If u is of two ϕ-dimensions, then (*cf.* *5·235 and *10·2) any two members of it form a ϕ-prime class, and (*cf.* *12·21) the ϕ-dimension number of this class is not greater than two, and hence (*cf.* *6·25) it is two, and hence (*cf.* *12·23) this subclass is ϕ-equivalent to u. If u is of three ϕ-dimensions, it must contain at least one subclass v consisting of two members, and, as before, v must be ϕ-prime. If v is of three ϕ-dimensions, then (*cf.* *12·23) u and v are ϕ-equivalent. If v is of two

ϕ-dimensions, then there is a member of u, x say, which is not a member of $\text{cm}_\phi\text{‘}v$. Then, either $v \cup \iota\text{‘}x$ is ϕ-prime and (*cf.* *12·23) ϕ-equivalent to u, or the class composed of x and some one (not necessarily any one) of the members of v is ϕ-prime and ϕ-equivalent to u.

The following proposition should be compared to *3·13 and *12·33:—

*12·41. *Proposition.*—Assuming $(\lambda - \pi)\text{Hp}\ \phi$, if u is ϕ-axial, and v is a subclass of u, and both v and (u-v) are not the null class, then $\text{dim}_\phi\text{‘}v$ is less than $\text{dim}_\phi\text{‘}u$. In symbols,

$$\vdash :. (\lambda - \pi)\text{Hp}\ \phi . \supset : u \in \text{ax}_\phi . v \in \text{cls‘}u . \exists ! v . \exists !(u - v) . \supset . \text{dim}_\phi\text{‘}v < \text{dim}_\phi\text{‘}u$$

Proof.—*Cf.* *3·13 and *4·21 and *11·21 and *12·21·23.

The following proposition should be compared to *5·23:—

*12·42. *Proposition.*—Assuming $(\lambda - \pi)\text{Hp}\ \phi$, then, if u is a subclass of O_ϕ and is ϕ-axial, any subclass of u, not the null class, is ϕ-axial. In symbols,

$$\vdash :. (\lambda - \pi)\text{Hp}\ \phi . \supset : u \in \text{ax}_\phi \cap \text{cls‘O}_\phi . v \in \text{cls‘}u . \exists ! v . \supset . v \in \text{ax}_\phi$$

Proof.—From *6·26 we have $\text{Nc‘}u = \text{dim}_\phi\text{‘}u$; from *5·23 and *6·25 we have $\text{Nc‘}v \leqq \text{dim}_\phi\text{‘}v$. Hence (*cf.* *12·41), if v is not identical with u, we have

$$\text{Nc‘}v \leq \text{dim}_\phi\text{‘}v < \text{Nc‘}u \qquad (1)$$

Firstly, assume that v omits one member of u only. Then $\text{Nc‘}v + 1 = \text{Nc‘}u$. Hence, from (1), $\text{Nc‘}v = \text{dim}_\phi\text{‘}v$, and hence (*cf.* *3·21) v is ϕ-axial.

Secondly, if v omits two members of u, then it is a unit class, and (*cf.* *12·11) is ϕ-axial.

It is convenient to conclude this section (*12) with three theorems which are fundamental to the theory of ϕ-points and of ϕ-planes.

*12·51. *Proposition.*—Assuming $(\lambda - \pi)\text{Hp}\ \phi$, if u is of two ϕ-dimensions, and x and y are members of $\text{cm}_\phi\text{‘}u$, then the class composed of x and y is ϕ-axial and is a subclass of O_ϕ. In symbols,

$$\vdash :. (\lambda - \pi)\text{Hp}\ \phi . \supset : \text{dim‘}u = 2 . x, y \in \text{cm}_\phi\text{‘}u . \supset . \iota\text{‘}x \cup \iota\text{‘}y \in \text{ax}_\phi \cap \text{cls‘O}_\phi$$

|502| *Proof.*— $\text{cm}_\phi\text{‘}u$ is contained in O_ϕ (*cf.* *4·21·27 and *10·1 and *11·21). If x and y are identical, *cf.* *12·11. If x is distinct from y, then *cf.* *3*21 and *12·11·21·23.

*12·52. *Proposition.*— Assuming $(\lambda - \pi)\text{Hp}\ \phi$, if u has three members only, and is a self-ϕ-concurrence, and its ϕ-dimension number is three, then u is ϕ-axial and a subclass of O_ϕ. In symbols,

$$\vdash :. (\lambda - \pi)\text{Hp}\ \phi . \supset : u \in 3 \cap \text{conc}_\phi . \text{dim}_\phi = 3 . \supset . u \in \text{ax}_\phi \cap \text{cls‘O}_\phi$$

Proof.—If v is contained in u and possesses two members, then (*cf.* *3·31·32) v is ϕ-axial, and (*cf.* *12·23) is not ϕ-equivalent to u. Hence (*cf.* *5·231) u is ϕ-prime, and hence (*cf.* *3·22) is ϕ-axial, and also (*cf.* *11·21) is a subclass of O_ϕ.

*12·53. *Proposition.*—Assuming $(\lambda - \pi)\text{Hp}\ \phi$, if x, y, and z are three distinct entities forming a ϕ-axial class, then the common subclass of $\text{cm}_\phi\text{‘}(\iota\text{‘}x \cup \iota\text{‘}y)$ and $\text{cm}_\phi\text{‘}(\iota\text{‘}x \cup \iota\text{‘}z)$ is the unit class $\iota\text{‘}x$. In symbols,

$$\vdash :. (\lambda - \pi)\text{Hp}\ \phi . \supset : \iota\text{‘}x \cup \iota\text{‘}y \cup \iota\text{‘}z \in 3 \cap \text{ax}_\phi . \supset . \text{cm}_\phi\text{‘}(\iota\text{‘}x \cup \iota\text{‘}y) \cap \text{cm}_\phi\text{‘}(\iota\text{‘}x \cup \iota\text{‘}z) = \iota\text{‘}x$$

Proof.—*Cf.* *4·21·27 and *10·2 and *12·23·42.

*13. *On ϕ-Maximal Classes and Self-ϕ-Concurrences.*

13·11. *Proposition.*—Assuming $(\lambda-\pi)\mathrm{Hp}\,\phi$, if p is a ϕ-maximal class and a subclass of O_ϕ, and q is a subclass of p, not the null class, then q is a ϕ-maximal class. In symbols,[1]

$$\vdash\therefore(\lambda-\pi)\mathrm{Hp}\,\phi\,.\supset:p\in\mathrm{mx}_\phi\cap\mathrm{cls}\text{‘}\mathrm{O}_\phi\,.\,q\in\mathrm{cls}\text{‘}p\,.\,\exists!q\,.\supset.\,q\in\mathrm{mx}_\phi$$

Proof.—The class q must (*cf.* *10·3 and *12·21) be of one, or two, or three ϕ-dimensions. If the ϕ-dimension number of q is one or two, then (*cf.* *10·2 and *12·11·51) q is a ϕ-maximal class. If the ϕ-dimension number of q is three, then (*cf.* *12·23) q is ϕ-equivalent to p. Hence if v be a subclass of q, which is ϕ-prime and ϕ-equivalent to q, it is ϕ-equivalent to p, and hence (*cf.* *3·23) it is ϕ-axial. Hence (*cf.* *3·23 and *12·37) q is ϕ-maximal.

*13·31. *Proposition.*—Assuming $(\lambda-\pi)\mathrm{Hp}\,\phi$, if u is a self-ϕ-concurrence and a subclass of O_ϕ, then u is a ϕ-maximal class. In symbols,

$$\vdash\therefore(\lambda-\pi)\mathrm{Hp}\,\phi.\supset:u\in\mathrm{conc}_\phi\cap\mathrm{cls}\text{‘}\mathrm{O}_\phi.\supset.\,u\in\mathrm{mx}_\phi$$

Proof.—There exists (*cf.* *12·37) a subclass (v) of u, which is ϕ-prime and ϕ-equivalent to u. If v is a unit class, then (*cf.* *10·2 and *12·11) v and u are identical, and u is of one ϕ-dimension and ϕ-maximal. If v is a couple, then (*cf.* *3·31·32) v is ϕ-axial, and (*cf.* *3·22·23) u is of two ϕ-dimensions and is ϕ-maximal. If v is composed of three members, then u is of three ϕ-dimensions, and neither of the previous cases can hold. Hence again u is ϕ-maximal.

|503| *13·32. *Proposition.*—Assuming $(\lambda-\pi)\mathrm{Hp}\,\phi$, all subclasses of O_ϕ which are ϕ-maximal are self-ϕ-concurrent, and conversely. In symbols,

$$\vdash\therefore(\lambda-\pi)\mathrm{Hp}\,\phi.\supset.\,\mathrm{mx}_\phi\cap\mathrm{cls}\text{‘}\mathrm{O}_\phi=\mathrm{conc}_\phi\cap\mathrm{cls}\text{‘}\mathrm{O}_\phi$$

Proof.—*Cf.* *13·11·31.

*14. *On Points.*

*14·11. Assuming $(\lambda-\pi)\mathrm{Hp}\,\phi$, every ϕ-point is a self-ϕ-concurrence and a subclass of O_ϕ. In symbols,

$$\vdash\therefore(\lambda-\rho)\mathrm{Hp}\,\phi.\supset.\,\mathrm{pnt}_\phi\subset\mathrm{conc}_\phi\cap\mathrm{cls}\text{‘}\mathrm{O}_\phi$$

Proof.—Every ϕ-point (*cf.* *11·21) is a subclass of O_ϕ. Again let x and y be two distinct members of a ϕ-point P. Then (*cf.* *3·42) a, b, c exist such that $\iota\text{‘}a\cup\iota\text{‘}b\cup\iota\text{‘}c$ is ϕ-axial and of three dimensions, and x and y are each ϕ-concurrent with each of a, b, and c. Hence (*cf.* *12·53) at least one pair of a, b, and c exist (say a and b) such that $\iota\text{‘}x\cup\iota\text{‘}a\cup\iota\text{‘}b$ and $\iota\text{‘}y\cup\iota\text{‘}a\cup\iota\text{‘}b$ are both three ϕ-dimensional and self-ϕ-concurrences. Hence (*cf.* *12·52) $\iota\text{‘}x\cup\iota\text{‘}a\cup\iota\text{‘}b$ and[2] $\iota\text{‘}y\cup\iota\text{‘}a\cup\iota\text{‘}b$ are both ϕ-axial. Hence (*cf.* *10·5 and *13·32) $\iota\text{‘}x\cup\iota\text{‘}y$ is ϕ-axial. Hence P is a self-ϕ-concurrence.

*14·12. *Proposition.*—Assuming $(\lambda-\rho)\mathrm{Hp}\,\phi$, every ϕ-point is ϕ-maximal. In symbols,

$$\vdash:(\lambda-\rho)\mathrm{Hp}\,\phi.\supset.\,\mathrm{pnt}_\phi\subset\mathrm{mx}_\phi$$

Proof.—*Cf.* *13·32 and *14·11.

---

1. In the line below, the second-to-last 'q' was erroneously a 'p' in the original.
2. In the following expression, the original erroneously had '$\iota\text{‘}x$' instead of '$\iota\text{‘}y$'.

*14·13. *Proposition.*—Assuming $(\lambda-\rho)\mathrm{Hp}\,\phi$, if P is a ϕ-point, then P is the ϕ-concurrence of O_ϕ with P. In symbols,

$$\vdash\therefore(\lambda-\rho)\mathrm{Hp}\,\phi.\supset:\mathrm{P}\in\mathrm{pnt}_\phi.\supset.\mathrm{P}=\tilde{\mathrm{O}}_\phi\text{‘}\mathrm{P}$$

Proof.—*Cf.* *3·42 and *8·21·22 and *14·11.

*14·14. *Proposition.*—Assuming $(\lambda-\pi)\mathrm{Hp}\,\phi$, if P is a ϕ-point, it possesses at least three members. In symbols,

$$\vdash\therefore(\lambda-\pi)\mathrm{Hp}\,\phi.\supset:\mathrm{P}\in\mathrm{pnt}_\phi.\supset.Nc\text{‘}\mathrm{P}\geqq 3$$

Proof.—P possesses (*cf.* *3·42) every member of O_ϕ which is ϕ-concurrent with each of a certain ϕ-axial set of three members. Hence (*cf.* *12·42) P possesses this set of three members.

*14·21. *Proposition.*—Assuming $(\lambda-\rho)\mathrm{Hp}\,\phi$, ϕ-points with more than one member in common are identical. In symbols,

$$\vdash\therefore(\lambda-\rho)\mathrm{Hp}\,\phi.\supset:\mathrm{P},\mathrm{Q}\in\mathrm{pnt}_\phi.Nc\text{‘}(\mathrm{P}\cap\mathrm{Q})>1.\supset.\mathrm{P}=\mathrm{Q}$$

Proof.—Let a and b be two distinct members of $\mathrm{P}\cap\mathrm{Q}$. Then (*cf.* *3·42 and *12·52 and *14·11) c and d exist, such that c is a member of P and d of Q, and |504| $\iota\text{‘}a\cup\iota\text{‘}b\cup\iota\text{‘}c$ and $\iota\text{‘}a\cup\iota\text{‘}b\cup\iota\text{‘}d$ are both of them three ϕ-dimensional and ϕ-axial. Hence (cf. *10·5 and *13·32) d is a member of $\tilde{\mathrm{O}}_\phi\text{‘}(\iota\text{‘}a\cup\iota\text{‘}b\cup\iota\text{‘}c)$, and hence (*cf.* *3·42 and *14·11·13) d is a member of P. Thus P and Q are identical.

*16. *On ϕ-Planes.*

*16·11. *Proposition.*—Assuming $(\lambda-\pi)\mathrm{Hp}\,\phi$, every ϕ-plane is ϕ-maximal, self-ϕ-concurrent, and a subclass of O_ϕ. In symbols,

$$\vdash:(\lambda-\pi)\mathrm{Hp}\,\phi.\supset.\mathrm{ple}_\phi\subset\mathrm{mx}_\phi\cap\mathrm{conc}_\phi\cap\mathrm{cls}\text{‘}\mathrm{O}_\phi$$

Proof.—*Cf.* *12·21·23·51.

*16*21. *Proposition.*—Assuming $(\lambda-\pi)\mathrm{Hp}\,\phi$, ϕ-planes with more than one member in common are identical. In symbols,

$$\vdash\therefore(\lambda-\pi)\mathrm{Hp}\,\phi.\supset:p,\,q\in\mathrm{ple}_\phi.Nc\text{‘}(p\cap q)>1.\supset.p=q$$

Proof.—*Cf.* *12·23 and *16·11.

*16·31. *Proposition.*—Assuming $(\lambda-\pi)\mathrm{Hp}\,\phi$, every self-$\phi$-concurrence is either ϕ-copunctual or ϕ-coplanar. In symbols,

$$\vdash\therefore(\lambda-\pi)\mathrm{Hp}\,\phi.\supset:u\in\mathrm{conc}_\phi.\supset.\mathrm{copnt}_\phi!u.\vee.\mathrm{cople}_\phi!u$$

Proof.—A proof is only required when u is of three ϕ-dimensions. Then a, b, c exist, such that they are three distinct members of u and are not a ϕ-coplanar class. Hence (*cf.* *12·52) they form a ϕ-axial class of three members. Hence (*cf.* *3·42) u is, in this case, ϕ-copunctual.

*16·32. *Proposition.*—Assuming $(\lambda-\rho)\mathrm{Hp}\,\phi$, if p is a ϕ-plane, and P and Q are distinct ϕ-points, and p and P have common members, and also p and Q, then the member (if any) common to P and Q is a member of p. In symbols,

$$\vdash\therefore(\lambda-\rho)\mathrm{Hp}\,\phi.\supset:p\in\mathrm{ple}_\phi.\mathrm{P},\mathrm{Q}\in\mathrm{pnt}_\phi.\mathrm{P}\neq\mathrm{Q}.\exists!(p\cap\mathrm{P}).\exists!(p\cap\mathrm{Q}).\supset.\mathrm{P}\cap\mathrm{Q}\subset p$$

Proof.—If $\mathrm{P}\cap\mathrm{Q}$ is the null class, then $\mathrm{P}\cap\mathrm{Q}$ is contained in p. If $\mathrm{P}\cap\mathrm{Q}$ is not null, let c be a member; also let a and b be, respectively, members of $p\cap\mathrm{P}$ and of

$p \cap Q$, which exist by hypothesis. (i) If c is identical with a or b, then (*cf.* *14·21) $P \cap Q$ is contained in p. Again (ii) if c is not identical with a or b, then (*cf.* *14·11 and *16·11) a, b and c form a self-ϕ-concurrence. Hence (*cf.* *16·31) this class is either ϕ-copunctual or ϕ-coplanar. If the class is ϕ-copunctual, then (*cf.* *14·21) P and Q are identical. Hence it is ϕ-coplanar, and hence (*cf.* *16·21) c is a member of p.

*16·33. *Proposition.*—Assuming $(\lambda - \rho)\text{H}p\ \phi$, if P is a ϕ-point and p and q are distinct ϕ-planes, and P and p have common members, and so, also, have P and q, then the member (if any) common to p and q is a member of P. In symbols,

$$\vdash \therefore (\lambda - \rho)\text{H}p\ \phi . \supset : \text{P} \in \text{pnt}_\phi . p, q \in \text{ple}_\phi . p \neq q . \exists!(\text{P} \cap p) . \exists!(\text{P} \cap q) . \supset . p \cap q \subset \text{P}$$

Proof.—The proof is in all respects similar to that of *16·32.

|505| *16·42. *Proposition.*—Assuming $(\lambda - \rho)\text{H}p\ \phi$, if p is a ϕ-plane and is not ϕ-co-punctual, then p is the ϕ-concurrence of O_ϕ with p. In symbols,

$$\vdash \therefore (\lambda - \rho)\text{H}p\ \phi . \supset : p \in \text{ple}_\phi . \sim \text{copnt}_\phi ! p . \supset . p = \tilde{\text{O}}_\phi \text{`} p$$

Proof.—From *8·21 and *16·11 we have

$$p \subset \tilde{\text{O}}_\phi \text{`} p \qquad (1)$$

Let x be any member of the ϕ-concurrence of O_ϕ with p. Hence (*cf.* *16·11) $p \cup \iota \text{`} x$ is a self-$\phi$-concurrence. Hence (*cf.* *16·31) $p \cup \iota \text{`} x$ is either ϕ-copunctual or ϕ-coplanar. But on the first alternative p is ϕ-copunctual. Hence $p \cup \iota \text{`} x$ is ϕ-coplanar. Hence x is a member of p. Hence from (1) the proposition follows.

Note.—In Concept V. the hypothesis of *16·42, that a ϕ-plane is not copunctual, is verified (*cf.* *28·11), where ϕ represents "homaloty," and the axioms of that concept are assumed.

Summary of the Complete Development of this Subject.—By the use of further axioms the whole theory of projective geometry, apart from "order" and apart from FANO's axiom[1†] respecting the distinction of harmonic conjugates, can be proved for ϕ-points and the associated geometrical entities. Then FANO's axiom can be added, and the theory of order and continuity can be introduced, as in PIERI's memoir (*loc. cit.*). In the sequel a somewhat different line of development is adopted, suitable for the special ideas of Concept V.

(ii) Concept V.

This concept is linear and monistic. It makes use both of the theory of inter-points and of the theory of dimensions. The points are classes of objective reals, and disintegrate from instant to instant. The corpuscles are capable of various and complicated structures, and are thus well fitted to bear the weight of modern physical ideas. The concept is Leibnizian, and only requires one extraneous relation for the same purposes as that of Concept III.

The essential relation is the pentadic relation $\text{R}'(abcdt)$, as explained at the commencement of Part III. The four first terms, namely, a, b, c, d, are objective reals and are mutually distinct, the fifth term is an instant of time.

1. † *Cf.* PIERI, *loc. cit.*

The relation $R'(abcdt)$ can be read, *a intersects b, c, d in the order bcd at the instant t.* In this concept *copunctual* objective reals do *not* necessarily *intersect*, though two intersecting objective reals are necessarily copunctual. The relation of *intersection* is not to be limited in properties by the mere geometrical suggestion of its technical name.

Since points are defined by the aid of the theory of dimensions, it follows (*cf.* note |506| to *3·42) that the geometry cannot be of less than three dimensions. Hence in this concept geometry of three dimensions occupies a position of unique simplicity. The points at infinity, here called *cogredient points*, are points in exactly the same sense as the other points. They are defined by a property not hitherto taken as fundamental. The properties of cogredient points play an essential part in the construction of a relation which assigns an *order* to the points on any straight line.

*20. *Definitions.*

*20·11. *Definition.*—An objective real p is *doubly secant* with a class u at an instant t if there exist two objective reals, members of u (x and y, say), which are both intersected by p at the instant t, and are such that there exists no interpoint on p of which x and y are both members. The symbol $(\widetilde{uu})_{Rt}!p$ will denote that p is doubly secant with u at the instant t. The symbolic definition is

$$(\widetilde{uu})_{Rt}!p. = :(\exists x,y).x \neq y.x,y \in u \cap R'(p;;;t).\sim(\exists v).v \in R'(p???t).x,y \in v \qquad \text{Df}$$

*20·12. *Definition.*—A class u is *homalous* at an instant t, *either* when a necessary and sufficient condition, that x should be a member of u, is that x should be doubly secant with u, *or* when u is a unit class contained in $R'(;;;;t)$. The symbol $\mu_{Rt}!u$ will denote that u has the property of homaloty at the instant t. The symbolic definition is

$$\mu_{Rt}!u. = \therefore x \in u. \equiv_x .(\widetilde{uu})_{Rt}!x : \vee : u \in 1 \cap \text{cls}\text{‘}R'(;;;;t) \qquad \text{Df}$$

This property (μ_{Rt}) of homaloty will now be taken as the special value of ϕ, to which the theory of dimensions will be applied. The common μ_{Rt}-subregion for u is denoted, according to the definition of *3·12, by $\text{cm}_{\mu_{Rt}}$‘u. But a suffix to a suffix will be avoided by using the simpler symbol cm_{Rt}‘u and similarly for the other entities defined in *3. Thus the following symbols are also defined, namely,

$$O_{Rt},\ \text{equiv}_{Rt}\text{‘}u,\ \text{prm}_{Rt},\ \text{dim}_{Rt},\ \text{ax}_{Rt},\ \text{mx}_{Rt},\ \tilde{u}_{Rt}\text{‘}v,\ \text{conc}_{Rt},\ \text{ple}_{Rt},$$
$$\text{pnt}_{Rt},\ \text{cople}_{Rt}!u,\ \text{copnt}_{Rt}!u.$$

With regard to the nomenclature, the term "ϕ-equivalence" should be particularized into "homaloty-equivalence," and "ϕ-prime" into "homaloty-prime," and so on. But, except where confusion is likely to occur, the term "homaloty" will be dropped; and the terms "equivalence," "prime," "dimensions," "axial," "maximal," "concurrence of u with v," "self-concurrence," "plane," "point," "coplanar," "copunctual" will be used in the senses defined in *3, with ϕ particularized into homality.

Elucidatory Note.—This definition of *homaloty* should be compared with the definition of the *flatness* of a class of punctual lines which has been used in the elucidatory notes of *3. Thus a class of punctual lines is flat, *either* when it is a unit class whose single member is a straight line, *or* when it is a necessary and |507| sufficient condition of a straight line x being a member of it, that x should meet two members of the class in points which are not their point of meeting (if they have a point of meeting). Owing to the fact that "intersection" (as used here) is wider in intension and narrower in extension than the idea of the "meeting" of two punctual lines, two punctual lines may "meet" without the corresponding objective reals "intersecting." The result is that homaloty and flatness have some different properties, for example, *cf.* *21·21.

*20·21. *Definition.*—The *punctual associate* of a class *u* is the class of those points which have a member in common with *u*. The punctual associate of *u* is denoted by ass_{Rt}'u. The definition in symbols is

$$\text{ass}_{Rt}\text{'}u = \dot{P}\{P \in \text{pnt}_{Rt} \,.\, \exists!(P \cap u)\} \qquad \text{Df}$$

Note.—The punctual associate of the class ι'a, *where a is an objective real*, will be called the punctual associate of *a*. Its symbol is $\text{ass}_{Rt}\text{'}\iota\text{'}a$.

*20·22. *Definition.*—A *punctual line* is a class of points such that there exist two planes, *p* and *q*, which are distinct and are such that the class of points is the common subclass of the punctual associates of *p* and *q*. The class of punctual lines at any instant *t* is denoted by lin_{Rt}. The symbolic definition is

$$\text{lin}_{Rt} = \dot{m}\{(\exists p, q) \,.\, p, q \in \text{ple}_{Rt} \,.\, p \neq q \,.\, m = \text{ass}_{Rt}\text{'}p \cap \text{ass}_{Rt}\text{'}q\} \qquad \text{Df}$$

Note.—Those punctual lines which are not "lines at infinity" (to be explained later) will be proved as the result of the axioms to be the punctual associates of the various objective reals.

*20·23. *Definition.*—The point, if there is one and one only, which contains a class *u* is called the *dominant point of u*. The dominant point of *u* is denoted by u_{Rt}. The symbolic definition is

$$u_{Rt} = (\imath p)\{p \in \text{pnt}_{Rt} \,.\, u \in \text{cls}\text{'}p\} \qquad \text{Df}$$

Note.—The idea of a dominant point obtains its importance from the fact that, according to the axioms given below, each interpoint is contained in one and only one point.

*20·231. *Definition.*—The *nonsecant part of u* is that subclass of *u* of which no member is a member of any interpoint which is a subclass of *u*. The nonsecant part of *u* is denoted by nsc_{Rt}'u. The symbolic definition is

$$\text{nsc}_{Rt}\text{'}u = \hat{x}\{x \in u \,.\, \sim(\exists v) \,.\, v \in \text{intpnt}_{Rt} \cap \text{cls}\text{'}u \,.\, x \in v\} \qquad \text{Df}$$

Note.—This definition takes its importance from the fact that (assuming the subsequent axioms) a point in general consists of a nonsecant part and of a part made up of interpoints contained in it. Either the interpoints or the nonsecant part may be wholly absent.

*20·232. *Definition.*—A class of points is called a *Figure.*

|508| *20·233. *Definition.*—A point, which is a member of a figure, will be said to *lie in* that figure.

*20·234. *Definition.*—A point, which lies in the punctual associate of a class of objective reals, will be said to *be on*, or *upon*, that class.

*20·235. *Definition.*—A punctual line is said to *join* two points if both the points lie in it.

*20·236. *Definition.*—Two punctual lines, which possess a common point, will be said to *meet at* that point. Similarly, any two classes of points will be said to *meet in* their common subclass, and this subclass will be called their *meeting*.

*20·24. *Definition.*—A class of points is called *collinear* if there exists[1] a punctual line in which they all lie. The symbol $\mathrm{coll}_{\mathrm{R}t}!u$ will denote that u is a class of collinear points at the instant t. The symbolic definition is

$$\mathrm{coll}_{\mathrm{R}t}!u \,.=.\, (\exists m) \,.\, m \in \mathrm{lin}_{\mathrm{R}t} \,.\, u \in \mathrm{cls}‘m \qquad \mathrm{Df}$$

*20·31. *Definition.*—Two figures are *in perspective* if (i) they have a one-one correspondence to each other, (ii) the joint figure formed by the two figures combined is not collinear, and (iii) there exists a point (the centre of perspective) which lies in every punctual line joining two *distinct* corresponding points. The statement that u and v are in perspective with each other at the instant t, and that S is the requisite one-one correspondence, will be denoted by $u(\mathrm{S\ persp})_{\mathrm{R}t}v$. The symbolic definition is

$$u(\mathrm{S\ persp})_{\mathrm{R}t}v \,.=:\, u, v \in \mathrm{cls}‘\mathrm{pnt}_{\mathrm{R}t} \,.\, \sim \mathrm{coll}_{\mathrm{R}t}!(u \cup v) \,.\, \mathrm{S} \in 1 \rightarrow 1 \,.\, u = \mathrm{S}^{;}(;\cdot) \,.\, v = \mathrm{S}^{;}(\cdot;):$$
$$(\exists \mathrm{V}) : m \in \mathrm{lin}_{\mathrm{R}t} \,.\, \mathrm{S}^{;}(\mathrm{AA}') \,.\, \mathrm{A} \neq \mathrm{A}' \,.\, \mathrm{A}, \mathrm{A}' \in m \,.\supset_{m, \mathrm{A}, \mathrm{A}'} .\, \mathrm{V} \in m \qquad \mathrm{Df}$$

*20·32. *Definition.*—The symbol $[\mathrm{AB}]\mathrm{persp}_{\mathrm{R}t}[\mathrm{A'B'}]$ denotes that A, B, A′, B′ are points, and that the figure formed by A and B is in perspective with the figure formed by A′ and B′, and that the one-one correspondence of the perspective is of A to A′ and of B to B′. Also $[\mathrm{ABC}]\mathrm{persp}_{\mathrm{R}t}[\mathrm{A'B'C'}]$ has a similar meaning, and so on. In symbols,

$$[\mathrm{AB}]\mathrm{persp}_{\mathrm{R}t}[\mathrm{A'B'}] \,.=.\, (\exists \mathrm{S}) \,.\, (\iota‘\mathrm{A} \cup \iota‘\mathrm{B})(\mathrm{S\ persp})_{\mathrm{R}t}(\iota‘\mathrm{A}' \cup \iota‘\mathrm{B}') \,.\, \mathrm{S}^{;}(\mathrm{AA}') \,.\, \mathrm{S}^{;}(\mathrm{BB}') \qquad \mathrm{Df}$$

$$[\mathrm{ABC}]\mathrm{persp}_{\mathrm{R}t}[\mathrm{A'B'C'}] \,.=.\, (\exists \mathrm{S}) \,.\, (\iota‘\mathrm{A} \cup \iota‘\mathrm{B} \cup \iota‘\mathrm{C}](\mathrm{S\ persp})_{\mathrm{R}t}(\iota‘\mathrm{A}' \cup \iota‘\mathrm{B}' \cup \iota‘\mathrm{C}') \,.$$
$$\mathrm{S}^{;}(\mathrm{AA}') \,.\, \mathrm{S}^{;}(\mathrm{BB}') \,.\, \mathrm{S}^{;}(\mathrm{CC}') \qquad \mathrm{Df}$$

*20·33. *Definition.*—The symbol $u\ \mathrm{persp}_{\mathrm{R}t}\ v$ denotes that there exists a one-one relation S such that, at the instant t, u is in perspective with v and S is the requisite one-one correspondence. In symbols,

$$u\ \mathrm{persp}_{\mathrm{R}t}\ v \,.=.\, (\exists \mathrm{S}) \,.\, u\,(\mathrm{S\ persp})_{\mathrm{R}t}v \qquad \mathrm{Df}$$

*20·41. *Definition.*—Two objective reals, a and c, are called *cogredient* at an instant t when (1) if u, v, w are three interpoints on a, and u', v', w' are three interpoints on c, and the dominant points $u_{\mathrm{R}t}$, $v_{\mathrm{R}t}$, $w_{\mathrm{R}t}$ are a trio of points in |509| perspective with the trio of dominant points $u'_{\mathrm{R}t}$, $v'_{\mathrm{R}t}$, $w'_{\mathrm{R}t}$, then the interpoint relation (R_{in}), if it arranges either trio of interpoints in an interpoint order, arranges both trios of interpoints in the same interpoint order (*i.e.*, either *uvw* and *u′v′w′*, or *vwu* and *v′w′u′*, or so on), and (2) there exist three interpoints u, v, w on a in the interpoint order *uvw*, and three interpoints u', v', w' on c such that $u_{\mathrm{R}t}$, $v_{\mathrm{R}t}$, $w_{\mathrm{R}t}$ are in perspective with $u'_{\mathrm{R}t}$, $v'_{\mathrm{R}t}$, $w'_{\mathrm{R}t}$. The symbol $\mathrm{cogrd}_{\mathrm{R}t}‘a$ denotes the class of objective reals cogredient with a.[2] The symbolic definition is

1. The original erroneously had ‘exist’.
2. The original erroneously had ‘α’.

$$\text{cogrd}_{\text{R}t}`a = \hat{x}\{(u, v, w, u', v', w'): u, v, w \in \text{R}^{;}(a???t).\, u', v', w' \in \text{R}^{;}(x???t).$$
$$[u_{\text{R}t}v_{\text{R}t}w_{\text{R}t}]\text{persp}_{\text{R}t}[u'_{\text{R}t}v'_{\text{R}t}w'_{\text{R}t}] . \supset . \text{R}_{\text{in}}{}^{;}(uvwt) \equiv \text{R}^{;}(u'v'w't):$$
$$(\exists\, u, v, w, t, u', v', w') . u, v, w \in \text{R}^{;}(a???t) . u', v', w' \in \text{R}^{;}(x???t).$$
$$\text{R}_{\text{in}}{}^{;}(uvwt) . [u_{\text{R}t}v_{\text{R}t}w_{\text{R}t}]\text{persp}_{\text{R}t}[u'_{\text{R}t}v'_{\text{R}t}w'_{\text{R}t}]\} \qquad \text{Df}$$

Note.—The class $\text{cogrd}_{\text{R}t}`a$ does not include a itself (*cf.* *27·43). It will be noticed that universal preservation of order by ranges in perspective on a pair of lines is a characteristic of a pair of parallel lines in Euclidean space, and of nonsecant lines in hyperbolic space. The choice of this property for the definition of parallelism (or nonsecancy) arises from the facts that (1) any two coplanar objective reals are copunctual (according to the subsequent axioms), so that the property of nonsecancy (in its ordinary acceptation) is not available, (2) we do not wish to make "cogredience" synonymous with "nonintersection" (using "intersection" in the special sense here defined), as this would impose an unnecessary limitation on the concept. The idea of cogredience is an essential element in the definition of a relation which, with the aid of axioms, distributes the points in any punctual line into an order.

*20·42. *Definition.—A Cogredient Point* is the class of objective reals cogredient with some objective real a, together with a itself. The symbol $\infty_{\text{R}t}$ denotes the class of cogredient points at the instant t. The definition in symbols is

$$\infty_{\text{R}t} = \hat{u}\{(\exists a) . a \in \text{O}_{\text{R}t} . u = \iota`a \cup \text{cogrd}_{\text{R}t}`a\} \qquad \text{Df}$$

Note.—In the case of Euclidean geometry, which is the only case considered here, each cogredient point is a point according to the definition of *3·42. The present definition would be very inconvenient, unless this were the case. The symbol $\infty_{\text{R}t}$ is reminiscent of the fact that the cogredient points are the points at infinity.

*20·51. *Definition.*—The *Point-Ordering Relation* is a tetradic relation holding between three points and an instant of time. Its symbol is R_{pn}, and $\text{R}_{\text{pn}}{}^{;}(\text{ABC}t)$ is defined to mean that, at the instant t, (1) A, B, C are non-cogredient points upon the same objective real, a say, and (2) there exist an objective real x and three interpoints u, v, w on x such that (i) x is cogredient with a, and (ii) u, v, w are in the interpoint order uvw, and (iii) A, B, C are in perspective with the dominant points $u_{\text{R}t}$, $v_{\text{R}t}$, $w_{\text{R}t}$. The definition in symbols is
|510|

$$\text{R}_{\text{pn}}{}^{;}(\text{ABC}t) . = . \text{A, B, C} \in \text{pnt}_{\text{R}t} - \infty_{\text{R}t} . (\exists a, x, u, v, w) . a \in \text{A} \cap \text{B} \cap \text{C} . x \in \text{cogrd}_{\text{R}t}`a.$$
$$u, v, w \in \text{R}^{;}(x???t) . \text{R}_{\text{in}}{}^{;}(uvwt) . [\text{ABC}]\text{persp}_{\text{R}t}[u_{\text{R}t}v_{\text{R}t}w_{\text{R}t}] \qquad \text{Df}$$

Note.—Since (*cf.* *27·43) x in the above definition is distinct from a, three collinear (*i.e.*, on a) points, A, B, C, cannot directly take their point-order from three interpoints which they themselves may severally contain (*cf.* however *21·51). The point-order of A, B, and C must arise from the order communicated (in a sense) to a copunctual pencil of three punctual lines by three interpoints contained respectively in points in these lines, and all three interpoints possessing an objective real (x) in common. The punctual lines of this pencil must possess A, B, and C respectively. This intervention of a pencil for the communication of point-order is necessary for the comparison of the orders of different ranges. If the apparently simpler plan is adopted, inextricable difficulties seem to arise. Also it will be remembered that not every point will necessarily contain an interpoint.

*20·61. *Definition.—A Punctual Plane* is a figure which is *either* the punctual associate of some plane, *or* is the class ∞_{Rt}. The class of punctual planes is denoted by pple_{Rt}. The definition in symbols is

$$\text{pple}_{Rt} = \text{ass}_{Rt}``\text{ple}_{Rt} \cup \iota`\infty_{Rt} \qquad \text{Df}$$

Note.—This definition is only convenient for Euclidean geometry.

*20·72. *Definition.*—A figure is called *Punctually Coplanar* if there is a punctual plane containing it. The symbol $\text{copple}_{Rt}!u$ will denote that u is a punctually coplanar figure at the instant t. In symbols,

$$\text{copple}_{Rt}!u \,.=.\, (\exists p) \,.\, p \in \text{pple}_{Rt} \,.\, u \in \text{cls}`p \qquad \text{Df}$$

Note.—This definition should be compared with that of $\text{cople}_{Rt}!u$ in *3·43.

*21. *General Deductions.*

*21·01. *Proposition.*—All the general deductions in the theory of dimensions, namely, *4 to *8, hold.

The following propositions, dependent on the special definition of homaloty, also hold:—

*21·11. *Proposition.*— O_{Rt} is the class $R`(;;;;t)$. In symbols,

$$\vdash . O_{Rt} R`(;;;;t)$$

Proof.—Cf. *20·12.

Note.—If t is not an instant of time, the classes O_{Rt} and $R`(;;;;t)$ are both the null class, and are thus identical. Accordingly the hypothesis, $t \in T$, is not required in this proposition. A similar explanation of the absence of the hypothesis, $t \in T$, holds for many other propositions.

|511| *21·21. *Proposition.*—An objective real, which is doubly secant with the common subregion for u, is a member of the common subregion for u. In symbols,

$$\vdash : w = \text{cm}_{Rt}`u \,.\, \widetilde{(ww)}_{Rt}!x \,.\supset.\, x \in \text{cm}_{Rt}`u$$

Proof.—Cf. *3·12 and *20·11.

Note.—The converse is not in general true, namely, that, if u is a class of objective reals, and x is a member of $\text{cm}_{Rt}`u$, then $x$ is doubly secant with $\text{cm}_{Rt}`u$. Nor does this converse follow from subsequent axioms. In the absence of this converse proposition the properties of homaloty differ from those of "flatness" for classes of punctual lines. For if u is a class of punctual lines, and ϕ stands for the property of flatness, then $\text{cm}_{\phi}`u$ is flat.

*21·31. *Proposition.*—The proposition μ Hp ϕ is true when μ_{Rt} is substituted for ϕ. In symbols,

$$\vdash . \mu \text{ Hp } \mu_{Rt}$$

Proof.—Cf. *10·2 and *20·12.

*21·41. *Proposition.*—If a is an objective real cogredient with c, then c is cogredient with a. In symbols,[1]

1. In the line below, the original erroneously had "a' instead of "c'.

$$\vdash : a \in \text{cogrd}_{Rt}`c . \equiv . c \in \text{cogrd}_{Rt}`a$$

Proof.—Cf. *20·41.

*21·51. *Proposition.*—If *u, v, w* be three interpoints, possessing the same objective real, and with dominant points u_{Rt}, v_{Rt}, w_{Rt} then $\text{R}_{pn}{}^{;}(u_{Rt}v_{Rt}w_{Rt}t)$ implies $\text{R}_{in}{}^{;}(uvwt)$. In symbols,

$$\vdash \therefore u, v, w \in \text{R}^{;}(a???t) . \supset : \text{R}_{pn}{}^{;}(u_{Rt}v_{Rt}w_{Rt}t) . \supset . \text{R}_{in}{}^{;}(uvwt)$$

Proof.—By definition (*cf.* *20·51) $\text{R}_{pn}{}^{;}(u_{Rt}v_{Rt}w_{Rt}t)$ implies (1) the existence of an objective real *x*, cogredient with *a*, and also of three interpoints, *u′*, *v′*, *w′*, all possessing *x*, and (2) that $\text{R}_{in}{}^{;}(u'v'w't)$, and (3) that *u′*, *v′*, *w′* are contained in dominant points $u'_{Rt}, v'_{Rt}, w'_{Rt}$ in perspective with u_{Rt}, v_{Rt}, w_{Rt}. Hence by the definition of cogredience (*cf.* *20·41) also $\text{R}_{in}{}^{;}(uvwt)$ holds.

*22. *The Axioms.*—Just as in Concept III., the axiom of persistence (*cf.* *22·1) does not enter into the geometrical reasoning, but it is essential to the physical side of the concept.

*22·1. I H*p* R is the statement that, *if t be an instant of time,* O *is contained in* O_{Rt}. In symbols,

$$\text{I } \textit{Hp} \text{ R} . = : t \in \text{T} . \supset_t . \text{O} \subset \text{O}_{Rt} \qquad \text{Df}$$

The next four axioms, viz. (II.-V.) H*p* R, are the axioms of order. They have already been explained in *1·51·52·53·54.

*22·21. II H*p* R $= \alpha$ H*p* R Df

*22·22. III H*p* R $= \beta$ H*p* R Df

*22·23. IV H*p* R $= \gamma$ H*p* R Df

*22·24. V H*p* R $= \delta$ H*p* R Df

|512| The next three axioms, viz. (VI. -VIII.) H*p* R, are the axioms establishing the relation of interpoints to points. Intpnt H*p* R has been defined in *1·41.

*22·31. VI H*p* R = intpnt H*p* R Df

*22·32. VII H*p* R is the statement that, *if u is an interpoint, there exists a point containing u.* In symbols,

$$\text{VII } \textit{Hp} \text{ R} . = : (t, u) : u \in \text{intpnt}_{Rt} . \supset . (\exists p) . p \in \text{pnt}_{Rt} . u \subset p \qquad \text{Df}$$

*22·33. VIII H*p* R is the statement that, *if p be a point, and u and v be two distinct interpoints contained in p, then u and v possess no common member.* In symbols,

$$\text{VIII } \textit{Hp}\text{R} . = : (p, u, v, t) : p \in \text{pnt}_{Rt} . u, v \in \text{intpnt}_{Rt} \cap \text{cls}`p . u \neq v . \supset . u \cap v = \Lambda \qquad \text{Df}$$

The next set of three axioms, viz. (IX.–XI.) H*p* R, supplies the missing hypotheses requisite to make homaloty a "geometrical property," as defined in *10.

*22·41. IX H*p* R is the statement that, *if t is an instant of time,* $\nu \text{Hp} \mu_{Rt}$ *is true.* In symbols,

$$\text{IX } \textit{Hp} \text{ R} . = : t \in \text{T} . \supset_t . \nu \text{ Hp } \mu_{Rt} \qquad \text{Df } (\textit{cf.} \text{ *10·3})$$

*22·42. X H*p* R is the statement that, *if t is an instant of time,* π Hp μ_{Rt} *is true.* In symbols,

$$\text{X } \textit{Hp} \text{ R} . = : t \in \text{T} . \supset_t . \pi \text{ Hp } \mu_{Rt} \qquad \text{Df } (\textit{cf.} \text{ *10·4})$$

*22·43. XI H*p* R is the statement that, *if t is an instant of time,* ρ Hp μ_{Rt} *is true.* In symbols,

$$\text{XI Hp R} \,.\!=:\, t \in \mathrm{T} \,.\supset_t .\, \rho \text{ Hp } \mu_{\mathrm{R}t} \qquad \text{Df } (cf. \ *10{\cdot}5)$$

*22·51. XII Hp R is the statement that, *if p and q are distinct planes, and there exists a point, not a cogredient point, which is a member of the punctual associates of both planes, then p and q possess a common member.* In symbols,

$$\text{XII Hp R} \,.\!=:\, p,q \in \mathrm{ple}_{\mathrm{R}t} \,.\, p \neq q \,.\, \exists!\{(\mathrm{ass}_{\mathrm{R}t}{}^{\backprime}p \cap \mathrm{ass}_{\mathrm{R}t}{}^{\backprime}q) - \infty_{\mathrm{R}t}\}.\supset_{p,q,t} .\, \exists!(p \cap q) \qquad \text{Df}$$

The next axiom, XIII Hp R, is the "Euclidean" axiom.

*22·61. XIII Hp R is the statement that *the cogredient points are points.* In symbols,

$$\text{XIII Hp R} \,.\!=.\, \infty_{\mathrm{R}t} \subset \mathrm{pnt}_{\mathrm{R}t} \qquad \text{Df}$$

The next three axioms, namely (XIV.–XVI.) Hp R, establish the theory of the order of points as determined by the point-ordering relation (*cf.* *20·51). Incidentally some existence theorems can be deduced from them, which would else have to be provided for elsewhere.

*22·71. XIV Hp R is the statement that, *if* A *and* B *are two distinct non-*|513|*cogredient points, then there exists at least one point* C *such that* A, B, C *are in the point-order* ABC *at the instant considered.* In symbols,

$$\text{XIV Hp R} \,.\!=:\, (\mathrm{A}, \mathrm{B}, t) : \mathrm{A}, \mathrm{B} \in \mathrm{pnt}_{\mathrm{R}t} - \infty_{\mathrm{R}t} \,.\supset.\, \exists!\mathrm{R}_{\mathrm{pn}}{}^{\backprime}(\mathrm{AB};t) \qquad \text{Df}$$

*22·72. XV Hp R is the statement that, *if* A, B, C *are three distinct non-cogredient points, on the same objective real, then at the instant considered one of the*[1] *t point-orders* ABC, *or* BCA, *or* CAB *holds.* In symbols,

$$\text{XIV Hp R} \,.\!=\therefore (\mathrm{A}, \mathrm{B}, \mathrm{C}, t) \therefore \mathrm{A}, \mathrm{B}, \mathrm{C} \in \mathrm{pnt}_{\mathrm{R}t} - \infty_{\mathrm{R}t} \,.\, \exists!(\mathrm{A} \cap \mathrm{B} \cap \mathrm{C}).$$
$$\mathrm{A} \neq \mathrm{B} \,.\, \mathrm{B} \neq \mathrm{C} \,.\, \mathrm{C} \neq \mathrm{A} \,.\supset: \mathrm{R}_{\mathrm{pn}}{}^{\backprime}(\mathrm{ABC}t) \,.\vee.\, \mathrm{R}_{\mathrm{pn}}{}^{\backprime}(\mathrm{BCA}t) \,.\vee.\, \mathrm{R}_{\mathrm{pn}}{}^{\backprime}(\mathrm{CAB}t) \qquad \text{Df}$$

The next axiom, XVI Hp R, is the well-known "transversal" axiom.

*22·73. XVI Hp R is the statement that, *if at the instant t the points* B, C, D *are in the point-order* BCD, *and the points* C, E, A *are in the point-order* CEA, *and the points* A, B, C *are not collinear, and* F *lies in the punctual associates both of* A∩B *and of* D∩E, *then the points* A, F, B *are in the point-order* AFB. In symbols,

$$\text{XVI Hp R} \,.\!=: (\mathrm{A}, \mathrm{B}, \mathrm{C}, \mathrm{D}, \mathrm{E}, \mathrm{F}, t) : \mathrm{R}_{\mathrm{pn}}{}^{\backprime}(\mathrm{BCD}t) \,.\, \mathrm{R}_{\mathrm{pn}}{}^{\backprime}(\mathrm{CEA}t).$$
$$\mathrm{A} \cap \mathrm{B} \cap \mathrm{C} = \Lambda \,.\, \mathrm{F} \in \mathrm{ass}_{\mathrm{R}t}{}^{\backprime}(\mathrm{A} \cap \mathrm{B}) \cap \mathrm{ass}_{\mathrm{R}t}{}^{\backprime}(\mathrm{D} \cap \mathrm{E}) \,.\supset.\, \mathrm{R}_{\mathrm{pn}}{}^{\backprime}(\mathrm{AFB}t) \qquad \text{Df}$$

*22·74. As XVII Hp R, an axiom of continuity will be wanted.

Note.—The above axioms are all axioms of *geometry*, in the sense of "geometry" as defined in the sense definition of it given in Part I. (i.). But geometry in this Concept V. includes more than does geometry in Concept I. For in Concept I. geometry has only to do with points, punctual lines, and punctual planes; but in Concept V. geometry has, in addition, to consider the relation of the objective reals (which are all "linear") and of interpoints to the above entities. In this respect, geometry in Concept V. merges into physics more than does geometry in Concept I. Thus the excess of the number of axioms in Concept V. over the number in Concept I. arises from the fact that there is a larger field to be covered. Also, I Hp R is not required in the geometrical reasoning.

1. The original omitted 'the'.

*25. *Preliminary Propositions.*

*25·11. *Proposition.*—Assuming (II.–VI.)[1] Hp R, all the propositions of the theory of interpoints (*cf.* *1) hold of the interpoints of this Concept.

*25·12. *Proposition.*—Assuming (II.–VI.) Hp R, if t be an instant of time, then O_{Rt} possesses at least *four* members. In symbols,

$$\vdash .\because \text{IX Hp R} . = : t \in \text{T} . \supset . \text{Nc}`\text{O}_{\text{R}t} \geqq 4$$

Proof.—*Cf.* *1·61·71·72·73 and *21·11 and *25·11.

|514| *25·13. *Proposition.*—Assuming (II.–VI.) Hp R, if t be an instant of time, then O_{Rt} is homalous. In symbols,

$$\vdash .\because (\text{II.-VI.}) \text{ Hp R} . \supset : t \in \text{T} . \supset . \mu_{\text{R}t} ! \text{O}_{\text{R}t}$$

Proof.—From *1·65 and *21·11, O_{Rt} is identical with $\text{R}`(;\cdots t)$. Hence (*cf.* *1·31·61·62) every member of O_{Rt} is doubly secant with O_{Rt}. Again (*cf.* *21·11), every objective real which is doubly secant with O_{Rt} is a member of it.

*25·14. *Proposition.*—Assuming (II.–VI., IX–XL) Hp R, if t be an instant of time, then all the special deductions of the theory of dimensions, namely *11 to *16, hold respecting homaloty, that is, with μ_{Rt} substituted for ϕ.

Proof.—*Cf.* *21·31 and *22·41·42·43 and *25·13.

*25·21. *Proposition.*—Assuming I Hp R, if t is an instant of time, then $0 = O_{Rt}$. In symbols,

$$\vdash .\because \text{I Hp R} . \supset : t \in \text{T} . \supset . 0 = \text{O}_{\text{R}t} .$$

Proof.—*Cf.* *21·11 and *22·1.

Note.—The above theorem is not used in any geometrical reasoning.

*25·31. *Proposition.*—Assuming (II.–VIII.) Hp R, if a be a member of O_{Rt}, then the number of points on a is at least three. In symbols,

$$\vdash .\because (\text{II.-VIII.}) \text{ Hp R} . \supset : a \in \text{O}_{\text{R}t} . \supset . \text{Nc}`\text{ass}_{\text{R}t}`\iota`a \geqq 3$$

Proof.—*Cf.* *1·72 and *21·11 and *22·32·33 and *25·1.

*25·32. *Proposition.*—Assuming (II.–VII., IX.–XI.) Hp R, if u be an interpoint, there is one and only one point containing it. In symbols,

$$\vdash .\because (\text{II.-VII., IX.-XI.}) \text{ Hp R} . \supset . u \in \text{intpnt}_{\text{R}t} . \supset . \text{P}\{\text{P} \in \text{pnt}_{\text{R}t} . u \subset \text{P}\} \in 1$$

Proof.—*Cf.* *1·71 and *14·21 and *22·32.

*26. *On Cogredient Points.*

*26·11. *Proposition.*—Assuming (II.–VI., IX–XI., XIII.) Hp R, if a point possesses two members which are cogredient to each other, it is a cogredient point. In symbols,

$$\vdash .\because (\text{II.-VI., IX.-XI., XIII.}) \text{ Hp R} . \supset :$$

$$\text{A} \in \text{pnt}_{\text{R}t} . a, b \in \text{A} . b \in \text{cogrd}_{\text{R}t}`a . a \neq b . \supset . \text{A} \in \infty_{\text{R}t}$$

Proof.—*Cf.* *14·21 and *20·42 and *22·61.

*26·22. *Proposition.*—Assuming (II.–VI., IX.–XI., XIII.) Hp R, if A is a cogredient point and a is a member of A, then A is identical with $\iota`a \cup \text{cogrd}_{Rt}`a$. In symbols,

1. The original omitted a period after 'VI'.

$$\vdash .\therefore (\text{II.-VI., IX.-XI., XIII.})\ Hp\ R.\ \ : A \in \infty_{Rt} . a \in A . \supset . A = \iota`a \cup \text{cogrd}_{Rt}`a$$

Proof.—*Cf.* *14·21 and *20·42 and *21·41 and *22·61.

|515| *26·23. *Proposition.*—Assuming (II.-VI., IX.-XI., XIII.) Hp R, there is one and only one cogredient point lying in the punctual associate of an objective real. In symbols,

$$\vdash .\therefore (\text{II.-VI., IX.-XI., XIII.})\ Hp\ R . \supset : a \in 0_{Rt} . \supset . \infty_{Rt} \cap \text{ass}_{Rt}`\iota`a \in 1$$

Proof.—*Cf.* *26·22.

*26·24. *Proposition.*—Assuming (II.-XI., XIII.) Hp R, there are at least two points, not cogredient points, lying in the punctual associate of an objective real. In symbols,

$$\vdash .\therefore (\text{II.-XI., XIII.})\ Hp\ R . \supset :\ a \in 0_{Rt} . \supset . Nc`\{\text{ass}_{Rt}`\iota`a - \infty_{Rt}\} \geqq 2$$

Proof.—*Cf.* *25·31 and *26·23.

*27. *On Punctual Lines.*

*27·11. *Proposition.*—Assuming (II.-VI., IX.-XI.) Hp R, if p and q are distinct planes, and $p \cap q$ possesses a member, then $\text{ass}_{Rt}`p \cap \text{ass}_{Rt}`q$ is identical with $\text{ass}_{Rt}`(p \cap q)$. In symbols,

$$\vdash .\therefore (\text{II.-VI., IX.-XI.})\ Hp\ R . \supset :$$
$$p,\ q \in \text{ple}_{Rt} . p \neq q . \exists!(p \cap q) . \supset . \text{ass}_{Rt}`p \cap \text{ass}_{Rt}`q = \text{ass}_{Rt}`(p \cap q)$$

Proof.— *Cf.* *16·33 and *20·21.

*27·12. *Proposition.*—Assuming (II.-VI., IX.-XII.) Hp R, if p and q are distinct planes, and a point, not a cogredient point, lies in the punctual associates both of p and also of q, then $p \cap q$ possesses one and only one member. In symbols,

$$\vdash .\therefore (\text{II.-VI., IX.-XII.})\ Hp\ R . \supset :$$
$$p,\ q \in \text{ple}_{Rt} . p \neq q . \exists!\{(\text{ass}_{Rt}`p \cap \text{ass}_{Rt}`q) - \infty_{Rt}\} . \supset . p \cap q \in 1$$

Proof.—*Cf.* *16·21 and *22·51.

*27·13. *Proposition.*—Assuming (II.-XIII.) Hp R, if p and q are distinct planes, then if $p \cap q$ possesses one member, there are non-cogredient points lying in the punctual associates both of p and of q; and also conversely. In symbols,

$$\vdash :: (\text{II.-XIII.})\ Hp\ R . \supset .\therefore p,\ q \in \text{ple}_{Rt} . p \neq q . \supset :\ \exists!\{(\text{ass}_{Rt}`p \cap \text{ass}_{Rt}`q) - \infty_{Rt}\} . \equiv . p \cap q \in 1$$

Proof.—*Cf.* *26·24 and *27·12.

*27·21. *Proposition.*—Assuming (II.-VI., IX.-XII.) Hp R, if m is a punctual line possessing a non-cogredient point, then there exists an objective real such that m is its punctual associate. In symbols,

$$\vdash .\therefore (\text{II.-VI., IX.-XII.})\ Hp\ R . \supset :\ m \in \text{lin}_{Rt} . \exists!(m - \infty_{Rt}) . \supset . (\exists a) . a \in 0_{Rt} . m = \text{ass}_{Rt}`\iota`a$$

Proof.—*Cf.* *27·11·12.

|516| *27·22. *Proposition.*—Assuming (II.-XIII.) Hp R, a punctual line possesses either more than one non-cogredient point or no such point. In symbols,

$$\vdash .\therefore (\text{II.-XIII.})\ Hp\ R . \supset : m \in \text{lin}_{Rt} . \exists!\{m - \infty_{Rt}\} . \supset . Nc`(m - \infty_{Rt}) > 1$$

Proof.—*Cf.* *20·22 and *26·24 and *27·21.

*27·23. *Proposition.*—Assuming (II.-VI., IX.-XIII.) Hp R, a punctual line, possessing a non-cogredient point, possesses one and only one cogredient point. In symbols,

$$\vdash\therefore \text{(II.-VI., IX.-XIII.) Hp R} . \supset : m \in \text{lin}_{Rt} . \exists!(m - \infty_{Rt}) . \supset . m \cap \infty_{Rt} \in 1$$

Proof.—*Cf.* *26·23 and *27·21.

*27·31. *Proposition.*—Assuming (II.-VI., IX.-XII.) Hp R, if m is a punctual line possessing a non-cogredient point, and A and B are two distinct points lying in it, then m is the punctual associate of $A \cap B$. In symbols,

$$\vdash\therefore \text{(II.-VI., IX.-XII.) Hp R} . \supset :$$
$$m \in \text{lin}_{Rt} . \exists!(m - \infty_{Rt}) . A, B \in m . A \neq B . \supset . m \in \text{ass}_{Rt}\text{‘}(A \cap B)$$

Proof.—*Cf.* *14·21 and *27·11·12.

*27·41. *Proposition.*—Assuming (II.-XI.) Hp R, if a is any objective real, then there exist two planes p and q such that a is the sole member of $p \cap q$. In symbols,

$$\vdash\therefore \text{(II.-XI.) Hp R} . \supset : a \in O_{Rt} . \supset . (\exists p, q) . p, q \in \text{ple}_{Rt} . p \cap q = \iota\text{‘}a$$

Proof.—*Cf.* *12·21 and *14·12·14 and *22·41 and *25·31.

*27·42. *Proposition.*—Assuming (II.-XIII.) Hp R, if a be an objective real, then $\text{ass}_{Rt}\text{‘}\iota\text{‘}a$ is a punctual line with a non-cogredient point, and conversely, if m is a punctual line with a non-cogredient point, there exists an objective real a such that $m = \text{ass}_{Rt}\text{‘}\iota\text{‘}a$. In symbols,

$$\vdash\therefore \text{(II.-XIII.) Hp R} . \supset : m \in \text{lin}_{Rt} . \exists!(m - \infty_{Rt}) . \equiv . (\exists a) . a \in O_{Rt} . m \in \text{ass}_{Rt}\text{‘}\iota\text{‘}a$$

Proof.—*Cf.* *27·11·13·21·41.

Note.—This proposition, *27·42, establishes the connection between the objective reals and the punctual lines.

*27·43. *Proposition.*—Assuming (II.-XIII.) Hp R, if a and c are cogredient, they are distinct. In symbols,

$$\vdash\therefore \text{(II.-XII.) Hp R} . \supset : c \in \text{cogrd}_{Rt}\text{‘}a . \supset . a \neq c$$

Proof.—*Cf.* *20·31·41 and *27·42.

*27·51. *Proposition.*—Assuming (II.-XIV.) Hp R, if A is any non-cogredient point |517| and B is any other point, then $A \cap B$ possesses one and only one member. In symbols,

$$\vdash\therefore \text{(II.-XIV.) Hp R} . \supset : A \in \text{pnt}_{Rt} \text{-} \infty_{Rt} . B \in \text{pnt}_{Rt} . A \neq B . \supset . A \cap B \in 1$$

Proof.—(i) If B is non-cogredient, *cf.* *14·21 and *20·51 and *22·71. (ii) If B is cogredient, then (*cf.* *14·14) let b be a member of B. Then (*cf.* *26·24) there is on b a non-cogredient point D. Hence by (i) $A \cap D$ possesses a single objective real, d say. Hence (*cf.* *14·12) b and d are coplanar and $\text{cm}_{Rt}\text{‘}(\iota\text{‘}b \cup \iota\text{‘}d)$ is a plane whose punctual associate possesses both A and B. Also, since a point is three-dimensional, B possesses another objective real, c say, not coplanar with b and d. Hence by similar reasoning $\text{cm}_{Rt}\text{‘}(\iota\text{‘}b \cup \iota\text{‘}c)$ is a plane, not identical with $\text{cm}_{Rt}\text{‘}(\iota\text{‘}b \cup \iota\text{‘}d)$, whose punctual associate also possesses A and B. Hence (cf. *27·13) these two planes have one objective real in common, and hence (*cf.* *16·33) this objective real is a member of $A \cap B$, and hence (*cf.* *14·21) $A \cap B$ possesses one and only one member.

*27·52. *Proposition.*—Assuming (II.-XIV.) H*p* R, if A be a non-cogredient point and B be any other point, then ass_{Rt}‘(A ∩ B) is a punctual line with a non-cogredient point, and conversely, if *m* be a punctual line with a non-cogredient point, then there exist two points A and B, such that A is not cogredient and *m* is identical with ass_{Rt}‘(A ∩ B). In symbols,

$$\vdash \therefore \text{(II.-XIV.) H}p\ \text{R} . \supset : m \in \mathrm{lin}_{Rt} . \exists!(m - \infty_{Rt}) . \equiv .$$
$$(\exists \text{A, B}) . \text{A} \in \mathrm{pnt}_{Rt} - \infty_{Rt} . \text{B} \in \mathrm{pnt}_{Rt} . \text{A} \neq \text{B} . m = \mathrm{ass}_{Rt}‘(\text{A} \cap \text{B})$$

Proof.—*Cf.* *27·22·31·42·51.

*28. *On Figures.*

*28·01. *Proposition.*—Assuming (II.-VI., IX.-XI.) H*p* R, if *t* be an instant of time, there exists at least one punctual plane, not the plane ∞_{Rt}. In symbols,

$$\vdash \therefore \text{(II.-VI., IX.-XI.) H}p\ \text{R} . \supset : t \in \text{T} . \supset . \exists!(\mathrm{pple}_{Rt} - \iota‘\infty_{Rt})$$

Proof.—*Cf.* *12·42 and *22·41.

*28·11. *Proposition.*—Assuming (II.-XIII.) H*p* R, if *p* be any punctual plane, not the plane ∞_{Rt}, it possesses at least three non-cogredient points, which are not collinear. In symbols,

$$| \therefore \text{(II.-XIII.) H}p\ \text{R} . \supset : p \in \mathrm{pple}_{Rt} - \iota‘\infty_{Rt} . \supset .$$
$$(\exists u) . u \in (3 \cap \mathrm{cls}‘p) . u \cap \infty_{Rt} = \Lambda . \sim \mathrm{coll}_{Rt}!u$$

Proof.—*Cf.* *14·21 and *26·24 and *27·21·31.

Note.—*Cf.* *16·42 and the note on it.

*28·12. *Proposition.*—Assuming (II.-XIII.) H*p* R, if *p* be any punctual plane, not |518| the plane ∞_{Rt}, there exists at least one non-cogredient point not lying in it. In symbols,

$$\vdash \therefore \text{(II.-XIII.) H}p\ \text{R} . \supset : p \in \mathrm{pple}_{Rt} - \iota‘\infty_{Rt} . \supset . (\exists \text{A}) . \text{A} \in (\mathrm{pnt}_{Rt} - \infty_{Rt}) - p$$

Proof.—*Cf.* *12·21 and *16·32 and *22·41 and *26·24 and *28·11.

*28·21. *Proposition.*—Assuming (II.-XIV.) H*p* R, if *x* be any objective real and *p* be any plane, then either the punctual associates of *x* and *p* have one and only one common member, or *x* is a member of *p*. In symbols,

$$\vdash :: \text{(II.-XIV.) H}p\ \text{R} . \supset \therefore x \in \text{O}_{Rt} . p \in \mathrm{ple}_{Rt} . \supset : \mathrm{ass}_{Rt}‘\iota‘x \cap \mathrm{ass}_{Rt}‘p \in 1 . \vee . x \in p$$

Proof.—Take (*cf.* *26·24 and *28·11) two non-cogredient points A and B upon *x*, and a non-cogredient point C in $\mathrm{ass}_{Rt}‘p$ but not on *x*. If either A or B lie in $\mathrm{ass}_{Rt}‘p$, then *cf.* *16·32. If neither A nor B lie in $\mathrm{ass}_{Rt}‘p$, then (*cf.* *16·32 and *27·51) cm_{Rt}‘{(B ∩ C) ∪ (A ∩ C)} is a plane possessing *x*, and its punctual associate possesses C. Hence (*cf.* *22·51) there is a common member of *p* and this plane, *y* say, and *x* and *y* are coplanar, hence (*cf.* *10·4 and *16·11) the punctual associates of *x* and *y* possess a common point. Hence $\mathrm{ass}_{Rt}‘\iota‘x$ and $\mathrm{ass}_{Rt}‘p$ possess a point in common, and then *cf.* *16·32.

*28·22. *Proposition.*—Assuming (II.-XIV.) H*p* R, if *p* and *q* are punctual planes, and *p* is not identical with ∞_{Rt}, and $p \cap q$ is contained in ∞_{Rt}, then $p \cap q$ and $p \cap \infty_{Rt}$ are identical. In symbols,

$$\vdash .\therefore (\text{II.-XIV.})\ \text{H}p\ \text{R} . \supset : p, q \in \text{pple}_{\text{R}t} . p \neq \infty_{\text{R}t} . p \cap q \subset \infty_{\text{R}t} . \supset . p \cap q = p \cap \infty_{\text{R}t}$$

Proof.—If p and q are identical, then $p \subset \infty_{\text{R}t}$, but (cf. *28·11) this is impossible. Hence $p \neq q$. If $q = \infty_{\text{R}t}$, then $p \cap q = p \cap \infty_{\text{R}t}$. Assume $q \neq \infty_{\text{R}t}$. Then (*cf.* *20·61) there exist planes, p' and q' say, such that p is $\text{ass}_{\text{R}t}$'p' and $\text{ass}_{\text{R}t}$'q'. Since $p \cap q \subset \infty_{\text{R}t}$, there is (*cf.* *26·24) no objective real common to p' and q'. Hence (*cf.* *28·21) upon every objective real possessed by p' there is one and only one point lying in q. Hence (*cf.* *26·23) $p \cap q = p \cap \infty_{\text{R}t}$.

*28·31. *Proposition.*—Assuming (II.-XIV.) Hp R, in every punctual line there lie at least three points. In symbols,

$$\vdash .\therefore (\text{II.-XIV.})\ \text{H}p\ \text{R} . \supset : m \in \text{lin}_{\text{R}t} . \supset . \text{Nc}\text{'}m \geqq 3$$

Proof.—If m possesses non-cogredient points, then *cf.* *27·22·23. If m is contained in $\infty_{\text{R}t}$, then (cf. *28·22) m is identical with $p \cap \infty_{\text{R}t}$, where p is a punctual plane. But (cf. *27·51·52 and *28·11) there are three distinct punctual lines contained in p, meeting two by two in three non-cogredient points; then cf. *27·23.

*28·32. *Proposition.*—Assuming (II.-XIV.) Hp R, a punctual line is the common meeting of two punctual planes, and conversely. In symbols,

$$\vdash .\therefore (\text{II.-XIV.})\ \text{H}p\ \text{R} . \supset : \text{lin}_{\text{R}t} = \dot{m}\{(\exists p, q) . p, q \in \text{pple}_{\text{R}t} . p \neq q . m = p \cap q\}$$

|519| *Proof.*—The direct proposition follows from *20·22·61. For the converse, let p and q be a pair of punctual planes. If neither p nor q be $\infty_{\text{R}t}$, then *cf.* *20·22·61. Consider now $p \cap \infty_{\text{R}t}$, where p is distinct from $\infty_{\text{R}t}$. Take (*cf.* *28·12) a non-cogredient point A, not in p. Also (*cf. proof* of *28·31) there are two distinct points B and C in $p \cap \infty_{\text{R}t}$. Hence (cf. *27·51) $\text{cm}_{\text{R}t}$'$\{(\text{A} \cap \text{B}) \cup (\text{A} \cap \text{C})\}$ is a plane, and its punctual associate, q say, possesses A and B and is distinct from p. Hence (cf. *27·23 and *28·22) $p \cap q$ is identical with $p \cap \infty_{\text{R}t}$. But (*cf.* *28·22) $p \cap q$ is a punctual line, and hence $p \cap \infty_{\text{R}t}$ is a punctual line.

*28·33. *Proposition.*—Assuming (II.-XIV.) Hp R, two distinct points lie in one and only one punctual line. In symbols,[1]

$$\vdash .\therefore (\text{II.-XIV.})\ \text{H}p\ \text{R} . \supset : \text{A}, \text{B} \in \text{pnt}_{\text{R}t} . \text{A} \neq \text{B} . \supset . \dot{m}\{m \in \text{lin}_{\text{R}t} . \text{A}, \text{B} \in m\} \in 1$$

Proof.—Firstly, only one punctual line (if any) possesses both A and B (*cf.* *27·31 and *28·22·32). Secondly, to prove that a punctual line exists possessing both A and B. If either point is non-cogredient, cf. *27·52. If both points are cogredient, then (cf. *28·11·12) two non-cogredient points C and D exist such that the four points A, B, C, D are not punctually coplanar. Hence (cf. *27·51) the meeting of the punctual associates of $\text{cm}_{\text{R}t}$'$\{(\text{A} \cap \text{C}) \cup (\text{B} \cap \text{C})\}$ and of $\text{cm}_{\text{R}t}$'$\{(\text{A} \cap \text{D}) \cup (\text{B} \cap \text{D})\}$ is a punctual line possessing A and B.

*28·41. *Proposition.*—Assuming (II.-XIV.) Hp R, three points, which are not collinear, lie in one and only one plane. In symbols,

$$\vdash .\therefore (\text{II.-XIV.})\ \text{H}p\ \text{R} . \supset : u \in 3 \cap \text{pnt}_{\text{R}t} . \sim \text{coll}_{\text{R}t} !u . \supset . \dot{p}\{p \in \text{pple}_{\text{R}t} . u \subset p\} \in 1$$

Proof.—If the three points are all cogredient, then (*cf.* *28·22·32) $\infty_{\text{R}t}$ is the only punctual plane which possesses them all. If the three points are A, B, C,

1. In the below expression, the original erroneously had 'm' instead of '$\dot{m}$'.

and A be non-cogredient, then (*cf.* *27·51 and *28·32) the punctual associate of cm_{Rt}‘{(A∩B)∪(A∩C)} is a punctual plane, possessing A, B, and C, and is the only one.

*28·42. *Proposition.*—Assuming (II.-XIV.) H*p* R, three punctual planes, which do not meet in a punctual line, meet in one point. In symbols,

$$\vdash .\because (\text{II.-XIV.})\ \text{H}p\ \text{R} . \supset : u \in 3 \cap \text{cls‘pple}_{Rt} . \supset . \cap‘u \in \text{lin}_{Rt} \cup 1$$

Proof.—Let *p*, *q*, *r* be the three punctual planes. Assume that *p* and *q* are neither the punctual plane ∞_{Rt}. If $q \cap r$ is contained in ∞_{Rt}, then *cf.* *20·22·61 and *27·23 and *28·22·32. If $q \cap r$ is not contained in ∞_{Rt}, then *cf.* *27·12 and *28·21.

*30. *Perspective.*

A few propositions on perspective (*cf.* *20·31·32·33) are required as a preliminary to the discussion of the point-ordering relation (*cf.* *20·51).

|520| *30·1. *Proposition.*—Assuming (II.-XIV.) H*p* R, if two figures are in perspective, their cardinal numbers are equal and each greater than one. In symbols,

$$\vdash .\because (\text{II.-XIV.})\ \text{H}p\ \text{R} . \supset : u\ \text{persp}_{Rt} v . \supset . \text{Nc‘}u = \text{Nc‘}v . \text{Nc‘}u > 1$$

Proof.—The equality of the cardinal numbers follows from the definition; also if both figures were unit classes, then (*cf.* *28·33) they would be collinear.

*30·3. *Proposition.*—Assuming (II.-XIV.) H*p* R, if the figure *u* is in perspective with the figure *v*, and also with the figure *w*, and if *u*, *v*, *w* are respectively collinear, and the punctual lines respectively containing *u*, *v*, *w* possess a common meeting, then either *v* is in perspective with *w*, or the joint class of *v* and *w* (*i.e.*, $v \cup w$) is collinear. In symbols,

$$\vdash :: (\text{II.-XIV.})\ \text{H}p\ \text{R} . \supset \therefore u\ \text{persp}_{Rt} v . u\ \text{persp}_{Rt} v . m, m', m'' \in \text{lin}_{Rt} .$$
$$u \subset m . v \subset m' . w \subset m'' . \exists!(m \cap m' \cap m'') . \supset : v\ \text{persp}_{Rt} w . \vee . m' = m''$$

Proof.—DESARGUES'[1] well-known propositions respecting triangles in perspective being coaxial, and its converse, can now (*cf.* *28·11·12·31·32·33·41·42) be proved. Then by drawing a figure for the present proposition the conclusion easily follows from some pure geometrical reasoning.

*31. *The Point-Ordering Relation.*

It will be proved in this section that the point-ordering relation (R_{pn}) has at any instant the same properties as the essential relation of Concept I. (*cf.* *31·3). It follows that the ordinary Euclidean geometry holds of the figures of Concept V., the points at infinity being the points of the punctual plane ∞_{Rt}, and the metrical ideas being introduced by appropriate definitions.

*31·11. *Proposition.*—Assuming (II.-XI., XIII., XIV.) H*p* R, the class R_{pn}‘(;;;*t*) is identical with the class of non-cogredient points. In symbols,

$$\vdash .\because (\text{II.-XI., XIII., XIV.})\ \text{H}p\ \text{R} . \supset . \text{R}_{pn}‘(;;;t) = \text{pnt}_{Rt} - \infty_{Rt}$$

Proof—Cf. *3·42 and *20·51 and *22·71 and *26·24.

*31·12. *Proposition.*—Assuming (II.-XV.) H*p* R, if *a* is a punctual line, and A and B are two non-cogredient points on it, then *a*, without its cogredient point, is identical

1. Girard Desargues (1591–1661) was a French mathematician who was one of the founders of projective geometry.

with the whole class formed by $R_{pn}'(;ABt)$ and $R_{pn}'(A;Bt)$ and $R'(AB;t)$ together with A and B added as members. In symbols,

$$\vdash\therefore (\text{II.-XV.})\ \text{Hp}\ R\,.\supset : a \in \text{lin}_{Rt}\,.\,A, B \in a - \infty_{Rt}\,.\,A \neq B\,.\supset .$$
$$a - \infty_{Rt} = R_{pn}'(;ABt) \cup R_{pn}'(A;Bt) \cup R_{pn}'(AB;t) \cup \iota`A \cup \iota`B$$

Proof.—The identity is to be proved by showing that each class contains the other. For one half of the proof, *cf.* *22·72 and *27·21. For the other half, *cf.* *20·51 and *27·21·31.

|521| *31·21. *Proposition.*—Assuming (II.-VI.) Hp R, the point-order ABC implies the point-order CBA. In symbols,

$$\vdash\therefore (\text{II.-VI.})\ \text{Hp}\ R\,.\supset : R_{pn}'(ABCt)\,.\supset .\,R_{pn}'(CBAt)$$

Proof.—*Cf.* *1·63 and *20·51.

*31·22. *Proposition.*—Assuming (II.-VII., IX.-XI.) Hp R, if A, B, C are in the point-order ABC, then A, B, and C are distinct. In symbols,

$$\vdash\therefore (\text{II.-VII., IX.-XI.})\ \text{Hp}\ R\,.\supset : R_{pn}'(ABCt)\,.\supset .\,A \neq B\,.\,A \neq C\,.\,B \neq C$$

Proof.—There are (*cf.* *20·51) interpoints *u*, *v*, *w* on a common objective real *a*, such that $R_{in}'(uvwt)$ and $[ABC]\text{persp}_{Rt}[u_{Rt}v_{Rt}w_{Rt}]$, where (*cf.* *20·23 and *25·32) u_{Rt}, v_{Rt}, w_{Rt} are the dominant points of *u*, *v*, *w*. But (*cf.* *1·62 and *22·33) u_{Rt}, v_{Rt}, w_{Rt} are distinct points. Hence (*cf.* *20·32) A, B, C are distinct points.

*31·23. *Proposition.*—Assuming (II-XV.) Hp R, the point-order ABC is inconsistent with the point-order BCA. In symbols,

$$\vdash\therefore (\text{II.-XV.})\ \text{Hp}\ R\,.\supset : R_{pn}'(ABCt)\,.\supset .\sim R'(BCAt)$$

Proof.—Since this proof is long, the paragraphs will be numbered for reference by (i), (ii), &c., prefixed.

(i) If *u*, *v*, *w* are interpoints on the objective real *a*, and *u'*, *v'*, *w'* interpoints on the objective real *a'*, and *a* and *a'* are cogredient, and $[u_{Rt}v_{Rt}w_{Rt}]\text{persp}_{Rt}[u'_{Rt}v'_{Rt}w'_{Rt}]$, then (*cf.* *20·23·41 and *25·32) *u*, *v*, *w* and *u'*, *v'*, *w'* must agree in interpoint order, and (*cf.* *1·64) each set of interpoints has only one (if any) interpoint order (counting *uvw* and *wvu* as the same order), and (*cf.* *27·43) *a* and *a'* are distinct.

(ii) From *20·51, $R_{pn}'(ABCt)$ implies (α) that A, B, C are on an objective[1] real *d*; (β) that there are interpoints *u*, *v*, *w* on an objective real *a*, cogredient with *d*; (γ) that $R_{in}'(uvwt)$; (δ) that $[ABC]\text{persp}_{Rt}[u_{Rt}v_{Rt}w_{Rt}]$; and ($\varepsilon$) (*cf.* *27·43) that *a* and *d* are distinct.

(iii) Assume that $R_{pn}'(BCAt)$ also holds. Then, in addition to the entities of (ii), there exist interpoints *u'*, *v'*, *w'* on an objective real *a'*, satisfying all the conditions of (ii) without changes, except that (ii, γ) becomes $R_{in}'(v'w'u't)$. This assumption (iii) will now be proved to be absurd.

(iv) From XIII Hp R (*cf.* *22·61) and *26·11·22, *d*, *a*, *a'* are copunctual and *a* and *a'* are either cogredient or identical.

1. The original erroneously had 'object'.

(v) Hence (*cf.* *30·3) *either* (Case I.) $[u_{Rt}v_{Rt}w_{Rt}]\text{persp}_{Rt}[u'_{Rt}v'_{Rt}w'_{Rt}]$ *or* (Case II.), a and a' are identical.

(vi). *Case I.*—We have $[u_{Rt}v_{Rt}w_{Rt}]\text{persp}_{Rt}[u'_{Rt}v'_{Rt}w'_{Rt}]$. Hence (*cf.* *20·41) the interpoint orders of u, v, w and u', v', w' must agree. Hence, from (ii, γ), $\mathrm{R}_{in}{}'(u'v'w't)$ holds. But (*cf.* *1·64) this is inconsistent with $\mathrm{R}_{in}{}'(v'w'u't)$. Hence Case I. cannot hold.

|522| (vii). *Case II.*—We have a and a' identical. Now (*cf.* *31·22) A, B, C are distinct points. Hence [*cf.* (ii, δ) and *30·1] u_{Rt}, v_{Rt}, w_{Rt} are distinct points, and [*cf.* (ii, β)] they are collinear. Hence, by XV Hp R (*cf.* *22·72), they are in some point-order. Hence [*cf.* (ii)] three interpoints u'', v'', w'' exist on a common objective real x, which is distinct from a and cogredient with it, and also $[u''_{Rt}v''_{Rt}w''_{Rt}]\text{persp}_{Rt}[u_{Rt}v_{Rt}w_{Rt}]$.

(viii). Hence Case II. divides into two subclasses, *either* (*Case II.*, α) x is not identical with d, *or* (Case II., β) x is identical with d.

(ix). *Case II,* α .—a and a' are identical and distinct from both d and x, and d and x are distinct; also a, d, x are cogredient and therefore (*cf.* *26·11·22) copunctual. Hence [*cf.* (ii, δ) and (vii) and *30·3] we have $[\mathrm{ABC}]\text{persp}_{Rt}[u''_{Rt}v''_{Rt}w''_{Rt}]$. Hence [*cf.* (iii, δ) and *30·3] $[u'_{Rt}v'_{Rt}w'_{Rt}]\text{persp}_{Rt}[u''_{Rt}v''_{Rt}w''_{Rt}]$. Hence [*cf.* *20·41 and (ii, γ) and (vii)] $\mathrm{R}_{in}{}'(u'v'w't)$. Hence (*cf.* *1·64) Case II., α, cannot hold.

(x). *Case II.,* β .—a and a' are identical, x and d are identical, and a and d are cogredient and distinct. Since (*cf.* *20·51) A, B, C are not cogredient points, they are distinct from[1] u'_{Rt}, v'_{Rt}, w'_{Rt} since the only point common to the punctual associates of a and d is a cogredient point. Thus none of u'_{Rt}, v'_{Rt}, w'_{Rt} can be cogredient points. Hence (*cf.* *28·33) there is a punctual line joining B and w'_{Rt} which does not possess A or u'_{Rt} or V (*cf.* figure annexed).

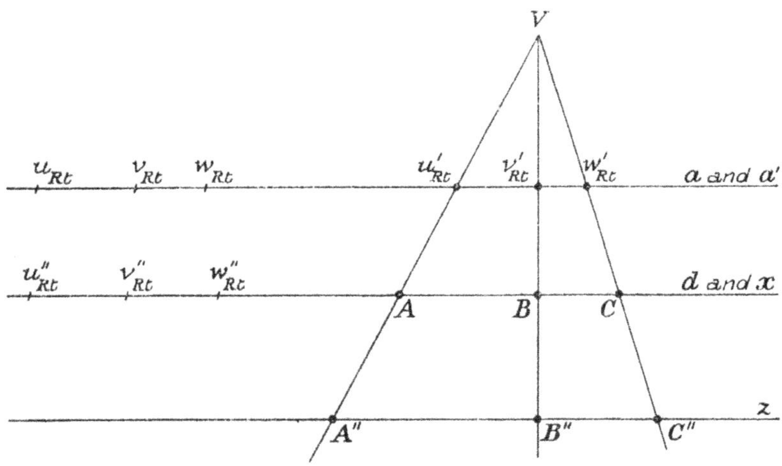

Hence (*cf.* *28·42) there is at least one other point, A″ say, lying in the punctual line joining A and u'_{Rt} in addition to V and u'_{Rt} and A. Hence (*cf.* *28·33) there is a punctual line, z say, joining A″ to the cogredient point common to the punctual associates of a and d. This punctual line must meet (cf. *28·42) the punctual lines VB and VC (*cf.* figure annexed) in the points B″ and C″. Hence

1. For the following expression, the original erroneously had 'u'_R, v'_R, w'_R'.

$$[A''B''C'']persp_{Rt}[ABC],$$

and hence (*cf.* *30·3) we have

$$[A''B''C'']persp_{Rt}[u_{Rt}v_{Rt}w_{Rt}],$$

and hence (*cf.* *30·3) we have

$$[A''B''C'']persp_{Rt}[u''_{Rt}v''_{Rt}w''_{Rt}].$$

|523| But also

$$[A''B''C'']persp_{Rt}[u'_{Rt}v'_{Rt}w'_{Rt}],$$

and hence (*cf.* *30·3) we have

$$[u'_{Rt}v'_{Rt}w'_{Rt}]persp_{Rt}[u''_{Rt}v''_{Rt}w''_{Rt}].$$

Thence, by the same reasoning as that in (ix) for Case II., α, it follows that Case II., β, cannot hold.

(xi). Hence neither Case I. nor Case II. can hold; and therefore the proposition follows.

*31·3. *Proposition.*—The point-ordering relation (R_{pn}) satisfies all the axioms satisfied by the essential relation of Concept III., except the axiom of persistence for that concept.

Proof.—In order to prove this, we make a comparison, as in Concept III., with the axioms of Concept I.

For	I	H*p* R of Concept I., *cf.* *25·12 and *26·24 and *31·11.
For	II	H*p* R of Concept I., *cf.* *31·21.
For	III	H*p* R of Concept I., *cf.* *31·23
For	IV	H*p* R of Concept I., *cf.* *31·22
For	V	H*p* R of Concept I., *cf.* XIV H*p* R (*22·71).
For	VI	H*p* R of Concept I., *cf.* *28·33 and *31·12.
For	VII	H*p* R of Concept I., *cf.* *28·01·11.
For	VIII	H*p* R of Concept I., *cf.* XVI H*p* R (*22·73).
For	IX	H*p* R of Concept I., *cf.* *28·12.
For	X	H*p* R of Concept I., cf. *28·42.
For	XI	H*p* R of Concept I., cf. *XVII H*p* R (*22·74).
For	XII	H*p* R of Concept I., cf. *26·23 and *28·33.

Note.—In order to complete this comparison, it must be noticed that it follows from *31·12 that the punctual line, with its cogredient point excepted, is the line as defined on the analogy of Concept I. Also, it follows, from *28·32 and *28·42 and the propositions of *31, that the punctual plane, with its cogredient points excepted, is the plane as defined on the analogy of Concept I. Then the transition to projective geometry is made, not by constructing a fresh type of points (the projective points), but simply by putting back the class (∞_{Rt}) of cogredient points. Metrical geometry is then constructed in the well-known way,[1†] making the plane (∞_{Rt}) of cogredient points to be the plane at infinity.

1. † *Cf.* Veblen, *loc. cit.*, for a sketch of this method; also Clebsch and Lindemann, *loc. cit.*

The Extraneous Relation.— For the purpose of enabling velocity and acceleration to be measured, an extraneous relation is required, in all respects similar to those required in Concepts III. and IV., and the description already given need not be repeated.

|524| *The Corpuscles.*—We may distinguish five types of points. A point of *Type* (1) contains no interpoints, and consists only of its nonsecant part (*cf.* *20·231). A point of *Type* (2) contains a single interpoint and no nonsecant part. Such a point is a single interpoint. A point of *Type* (3) contains a single interpoint together with a nonsecant part. A point of *Type* (4) contains many interpoints with no nonsecant part. A point of *Type* (5) contains many interpoints together with a nonsecant point.

We seem to be precluded from considering the "particles" to be stable points by the same difficulty as to the resulting permanence of collinearity, which was explained in considering the corpuscles of Concept IV. It is evident that at this stage many subdivisions of Concept V. are possible, in respect to the ideas which may be formed of the nature of the corpuscle. The following sketch of a possible development is given because of its superior simplicity, and also because of a certain consonance which it possesses with some modern physical ideas.

It is evident that volumes, in which, in some sense, there is an excess or a defect of interpoints, can be conceived as being charged with one or other of the two sorts of electricity. This idea is taken as the basis of the following brief outline of a possible development of the concept. Let the interpoints be identified with negative electricity and the nonsecant parts of points with positive electricity. A point of type (1) is a negative electron; a point of type (2) is a positive electron. The persistence of existence of an isolated electron of either type is to be *defined* by persistence of type and continuity of motion. If the electron is not isolated, consider, for example, a volume in which electrons of type (2) either compose all the points, or, at least, are everywhere dense. Then the persistence of such a collection of electrons must be considered as a whole, and is *defined*, as in the simpler case, by persistence of type and continuity of motion.

Three methods of procedure now suggest themselves, *either* (*Case I.*) to assume that the electrons consist of single points, so that a corpuscle is a volume containing a large finite number of points of type (2), and a small finite number of points of type (1), *or* (*Case II.*) to assume that a corpuscle is a volume in which points of type (2) are (at least) everywhere dense, and which contains a finite number of points of type (1), *or* (*Case III*) to assume that an electron of either type is essentially a volume (possibly with internal boundaries) in which points of the appropriate type are at least everywhere dense. In Case III. a corpuscle will be a relatively large electron of type (2) containing within it a finite number of relatively small electrons of type (1). Case III. has the merit, such as it is, of making the "inverse square" law of electricity appear somewhat natural. The field of force "at a point" produced by an electron may be conceived as proportional to the number of objective reals shared in common by the point and the "electric points" in the electron, and also to the number of these electric points. The number of electric points would be measured by the mass of the electron, the number of objective reals by the solid angle subtended at the point by the electron.

|525| What is wanted at this stage is some simple hypothesis concerning the motion of objective reals and correlating it with the motion of electric points and

electrons. From such a hypothesis the whole electromagnetic and gravitational laws might follow with the utmost simplicity. The complete concept involves the assumption of only one class of entities as forming the universe. Properties of "space" and of the physical phenomena "in space" become simply the properties of this single class of entities. In regard to the simplification of the preceding axioms, viz., of (I.-XVI.) H*p* R, the ideal to be aimed at would be to deduce some or all of them from more general axioms which would also embrace the laws of physics. Thus these laws should not presuppose geometry, but create it.

Liberty and the Enfranchisement of Women

1906 Pamphlet, Cambridge Women's Suffrage Association (Cambridge: W. Seaton & Son, Mill Rd., 5 November)

On 5 November 1906, Whitehead delivered a speech in favour of the enfranchisement of women at the annual meeting of the Cambridge Women's Suffrage Association. The Association subsequently published an extract as a pamphlet that could be obtained for the price of a schilling for fifty copies, post free.

A copy of the original document was discovered in 1976 by Professor John G. Slater of the University of Toronto.[1] *Knowing that Whitehead's biographer, Victor Lowe, would be interested, Slater immediately sent him a photocopy. Lowe in turn sent a copy to Barry Woodbridge at the Center for Process Studies in Claremont, who was busy assembling a Whitehead bibliography.*[2] *The text of the pamphlet was subsequently published in the Center's journal,* Process Studies *(albeit with a few errors).*[3] *Slater's original pamphlet now appears to be missing, as it is not in his papers at the University of Toronto, but Lowe's papers at Johns Hopkins University still contain the photocopy that was sent to him.*

Lowe aptly summarises this piece: 'The pamphlet is a vigorous, but careful, statement of the basic argument for women's suffrage and a conclusive answer to the various arguments . . . then heard against it'.[4]

The year after this speech was given (1907), Whitehead became Chairman of the Cambridge branch of the Men's League for Women's Suffrage, which had started in London the same year,[5] *and gave another speech as the Chairman at a meeting of the Liberal Women's Social Council, with an audience of 800–900 and at least two Liberal MPs in attendance.*[6]

Women were finally given limited voting rights in 1918 with the Representation of the People Act. Universal suffrage was not achieved until the Equal Franchise Act of 1928.

---

1. John G. Slater, 'Letter from John G. Slater to Victor Lowe, 5 December 1976', MS-0284, Victor Lowe Papers, Box 2.3, 'Women's Suffrage, W on' folder, Special Collections, Sheridan Libraries, Johns Hopkins University.
2. Barry A. Woodbridge, *Alfred North Whitehead: A Primary–Secondary Bibliography* (Bowling Green: Philosophy Documentation Center, 1977).
3. Alfred North Whitehead, 'Liberty and the Enfranchisement of Women', *Process Studies*, vol. 7, no. 1 (spring 1977), pp. 37–39.
4. Victor Lowe, *Alfred North Whitehead: The Man and His Work, Volume I: 1861–1910* (Baltimore: Johns Hopkins University Press, 1985), p. 313.
5. Lowe, *Alfred North Whitehead, Volume I*, p. 314. As a point of interest, in that same year, Whitehead's friend and collaborator, Bertrand Russell, ran for Parliament in the Wimbledon by-election (due to the resignation of Conservative MP Eric Hambro) as the first candidate of the National Union of Women's Suffrage Societies. Russell's candidacy was symbolic, as he had no real chance of winning, and he lost to the Conservative candidate, Henry Chaplin, by more than fifty percentage points.
6. The MPs were John Simon (1873–1954) and Stanley Buckmaster (1861–1934). Woman Citizen Publishing Society, 'Notes and Comments', *Women's Franchise*, 20 February 1908, p. 394. *Women's Franchise* was a weekly suffragette newspaper that ran from 27 June 1907 to at least 27 July 1911.

"A few years ago the political emancipation of women was merely the dream of a few political idealists, to-day it is a living political issue, popular among large masses of voters. This is a fact which should give us confident hope and renewed energy. I will not waste time speculating upon the causes which have produced this progress, but will pass to a consideration of the reasons which animate the Women's Suffrage agitation.

"Speaking for myself, I base my adherence to the cause upon the old-fashioned formula of Liberty: that is, upon the belief that in the life of a rational being it is an evil when the circumstances affecting him are beyond his control, and are not amenable to his intelligent direction and comprehension. External constraint upon the rational self-direction of conduct is, indeed, inextricably interwoven in the nature of things. But wherever it exists, and is removable without some corresponding loss of liberty, it is the evil, it is the enemy. Progress in science, progress in thought, progress |2|[1] in civilization depend for their good ultimately upon this, that they deliver the life of man into his own hands.

"But the chief and noblest of the external activities of human beings is concerned with the life of the State, that great organism on whose well-being depends the nature of the opportunities which life can offer to each individual.

"It follows, therefore, that except for plain, overmastering reasons connected with the necessary efficiency of Government, it is a crime against Liberty deliberately to deprive any portion of the population of possibilities of political action. That such overmastering reasons for limitation of political functions do exist in many states, perhaps in all states, I am not concerned to deny. They may arise when there is a cleavage in the population produced by inferiority of race, inferiority of civilization, or by deficiency in goodwill.

"But what we are here to-day to assert is that in the case of women in England at the present time, there is no reason for any exceptional treatment which does not also exist for the corresponding class of men.

"For what reasons have been alleged against the enfranchisement of women? It can no longer be maintained that women are prevented by social usage from interfering in Politics. Both the great parties in the state now make use of large political organizations of women.

|3| "Again, Mr. Asquith[2] tells us that women have now a vast political influence, and that it will be destroyed if they possess the suffrage. I grant that it is true that women have great indirect influence, and for this very reason they should be accorded a definite and determinate place in the political organization.

"Their influence as a class, at present, is necessarily irresponsible and often uninformed. When they have the vote they will take themselves seriously, and form their opinions with a sense of responsibility.

1. The original pamphlet was a single page folded in half to create four pages of text. The pages were not numbered, so we have added numbers.
2. At the time, H. H. Asquith was Chancellor of the Exchequer, but would become Prime Minister in April 1908, about a year and half from the time of Whitehead's speech. Opposed to women's suffrage, women would not receive the vote until 1918, during the premiership of David Lloyd George. George would prove to be the last Liberal Prime Minister before the Labour Party overtook the Liberals as the main opposition to the Conservatives in the mid-1920s under Ramsay MacDonald.

"During the last General Election nothing has struck me more than the educative effect of the vote upon the agricultural labourers in the village in which I live. They listened to both sides with care and attention, and their random political ideas were transformed into an instructed body of opinion. So we may agree that women now have influence, and from this we deduce that they should be admitted within that class to which direct political teaching is addressed.

"Again we are told that women are virtually represented by the votes of the men. This theory of virtual representation is ever the last bad argument of prejudice. In another connection it was long ago examined and torn to shreds by the historian Macaulay.[1] *Either* the virtual representative will act as his so-called virtual constituents |4| wish, and then, why not give them the direct vote? *Or* the virtual representative will not act as his virtual constituents desire, and then in what sense is he their representative.

"Lastly, there is the physical force argument; namely, that the ultimate direction of affairs must be vested in those who in their own persons possess the fighting strength of the nation. This argument goes far. In fact it goes much too far, and it proves that the vast majority of Governments, that have ever existed, were impossible. It proves that the government of Louis the Great of France was impossible, because in his own person he was unable to coerce the French nation. It proves that any despotism is impossible, and that any oligarchy is impossible. And when I contemplate these results of the argument, I wish it were true. But unfortunately it is manifestly false.

"In the name of Liberty, then, we demand the Suffrage for Women."

1. This likely refers to the historian, politician and poet Thomas Babington Macaulay (1800–1859). He gave a speech in the House of Commons on 2 March 1831 which criticised the idea of virtual representation.

Non-Euclidean Geometry (with Bertrand Russell)

1910 Hugh Chisholm (ed.), *Encyclopaedia Britannica*, 11th edition, vol. 11 (Cambridge: Cambridge University Press), pp. 724–730

1947 *Essays in Science and Philosophy* (New York: Philosophical Library), pp. 289–312

The first thing that must be understood about this article is its authorship. It was attributed at the end of the article to both Bertrand Russell and Whitehead,[1] *but this does not tell the whole story.*

Russell had written the entry for 'Geometry, Non-Euclidean' for the previous (10th) edition of the Encyclopaedia Britannica *in 1902.*[2] *It is important to note that Russell actually composed the article in late 1899 and early 1900, after he had abandoned neo-Hegelianism but before he discovered Peano.*

When in late 1905 the editor of the Encyclopaedia Britannica, *Hugh Chisholm,*[3] *asked for a revision of the article for the 11th edition, Russell suggested that Whitehead take over in his stead. This is entirely unsurprising given that the two men were right in the middle of collaborating on the* Principia Mathematica, *which would be published in three volumes in 1910, 1912 and 1913. Chisholm responded: 'I would naturally prefer to have you carry out the work proposed. But I know Mr. Whitehead's ability, and if you would prefer to hand it over to him, I should acquiesce.'*[4]

In essence, Whitehead used Russell's 1902 article as a basis for the new entry, but had a free hand to make what changes he wanted. He maintained some sections from Russell's entry entirely unchanged, deleted or changed others, and added new material that was entirely his own. But the finished article was also shorter by about a third than Russell's (about 7,500 versus 12,000 words); possibly Chisholm had some influence here, as he wrote to Russell in 1905 that 'I do not want the article "Geometry, Non-Euclidean" lengthened if I can help it'.[5]

To give a very broad overview, the introductory portion of the article which does not appear under any heading (about 2,500 words) is all Whitehead's, while the remainder (about 5,000 words) was carried over and/or adapted from the first half of Russell's 1902 entry. Aside from Russell's penultimate section on 'Scientific evidence as to the nature

1. Specifically by their initials: '(B. A. W. R.; A. N. W.)'.
2. Vol. 28, pp. 664–674.
3. Hugh Chisholm (1866–1924) was a British journalist who edited the 10th, 11th, and 12th editions of the *Encyclopaedia Britannica*.
4. Hugh Chisholm, 'Letter from Hugh Chisholm to Bertrand Russell, 20 November 1905', Victor Lowe Papers, MS-0284, Series 2, Box 2.1, Special Collections, Sheridan Libraries, Johns Hopkins University.
5. Hugh Chisholm, 'Letter from Hugh Chisholm to Bertrand Russell, 30 October 1905', Victor Lowe Papers, MS-0284, Series 2, Box 2.1, Special Collections, Sheridan Libraries, Johns Hopkins University.

of actual space', Whitehead also deleted the entire second half of Russell's article (about 7,000 words).

In a certain sense, the changes are almost as interesting as the finished product itself, since they reveal possible points of divergence between Whitehead and Russell (or at least differences of emphasis). This being the case, we have taken pains to exhaustively collate Russell's 1902 entry and Whitehead's 1910 one.[1]

Material that Whitehead deleted from the first half of Russell's 1902 entry will appear in footnotes.[2] *A summary of how much of the text can be attributed to Russell and how much to Whitehead can be found as a note to each major heading, with more granular changes noted in the appropriate places as needed. Purely stylistic changes (to capitalisation, punctuation, etc.) have been relegated to the back-of-book record.*

This is the sixth part of the 'Geometry' entry in the 11th edition of the Encyclopaedia Britannica. *As with the other* Encyclopaedia *articles to which Whitehead contributed, this one is more technical than one might expect, but not relentlessly so. The work is a properly integrated essay that might be described as bipolar to the extent that extremely technical content, likely not of great interest to the general reader, is blended with an account of the historical evolution of the topic. This introduction will informally recount some of the mathematical content and summarise some of the historical elements.*[3]

Whitehead begins the article with the announcement of a most sensible strategy: he acknowledges that the roots of non-Euclidean geometry lie in the analyses of the independence of its axioms, particularly the Parallel Postulate (stated below), but that these lead to interpretations that become part of an abstractly construed form of analytic geometry. Thus, he moves forward immediately to illustrate hyperbolic and elliptic geometry via two spaces that are described by four coordinates and a modification of the distance between two points defined by equations that admit only the squares of these coordinates. The equations also specify when a point is at infinity, or represents an 'improper ideal point'.[4] *These spaces demonstrate both ways that the Parallel Postulate might fail: there are, respectively, more than one or no lines parallel to a given line through a point not on that line.*[5] *Whitehead gives roughly two pages to the technical aspects of these constructions, including the metrical matters of distance and angle not involving ideal points, the key aspects of parallels, triangles and orientability. He also makes two other key distinctions between the spaces: (1) that the hyperbolic case admits lines of infinite extension, whereas in the elliptic case every line is a finite closed curve; (2) that the sum of the interior angles of a triangle is always less than a straight*

1. In doing the collation, we have used the version of Russell's article that appears in the *Collected Papers of Bertrand Russell*: Bertrand Russell, 'Geometry, Non-Euclidean', in Bertrand Russell, *The Collected Papers of Bertrand Russell, Volume III: Toward the 'Principles of Mathematics': 1900–1902*, edited by Gregory H. Moore (London: Routledge, 1993), pp. 470–504.
2. The text deleted from the second half, totalling about 7,000 words, is too voluminous to be put into footnotes; the reader may refer to Russell's original article to read it. For a listing of the section headings for the cut material see p. 296n3.
3. Whitehead's 1898 paper 'The Geodesic Geometry of Surfaces in non-Euclidean Space' (p. 42) is also germane to this joint article.
4. See also *Universal Algebra*, §247.
5. Regarding hyperbolic geometry, the reader may find the well-known half-plane and disk models far easier to visualize. These are attributed to Eugenio Beltrami.

angle in the first case but greater than a straight angle in the second. Nonetheless, both spaces contradict the notion of similarity; namely, that two triangles can have the same proportions but different sizes.

The technical content of the entry does not end at this point, but from p. 283 onward the emphasis passes to history. The history of non-Euclidean geometry is, of course, a tale oft told that would seem invariably to begin with the Parallel Postulate: 'And that, if a straight line falling on two straight lines make the angles, internal on the same side, less than two right angles, the two straight lines, being produced indefinitely, meet on the side on which are the angles less than two right angles'.[1] *The article, after a few prefatory remarks, then accepts the division of this history as having 'been aptly divided by Felix Klein into three very distinct periods'.*[2] *These three phases are summarised in a single paragraph and then treated in greater detail in the balance of the article. Our introduction will simply characterise each period and describe some of the highlights from their extended exposition.*

The first period identified is characterised by a kind of convergence via synthetic methods to the conclusion that hyperbolic geometry is consistent with Euclid in the absence of the Parallel Postulate; more specifically the possibility that multiple lines, none of which intersect some given line, may pass through a point not lying on that given line.[3] *Indeed, Gauss asserted in 1831 that this species of geometry is consistent with Euclid minus the Parallel Postulate. Almost simultaneously but independently of Gauss and each other, Nicolai Lobachevsky and John Bolyai produced extended analyses of such a system.*

The article describes the second period as analytical, with two main exemplars. Bernhard Riemann assimilates the idea of ordinary space to that of a manifold, which resembles Euclidean space only locally. It correctly points out that Riemann's great idea is not yet complete, but nonetheless describes the idea as 'profound and epoch-making'.[4] *Yet it gives pride of place to this period to Eugenio Beltrami, who indeed worked out a model for hyperbolic geometry from which one concludes its consistence with and independence from the other axioms of Euclid. Given the importance of Beltrami's interpretation, once again the article engages in many technical details.*[5]

Finally, the third period begins with the introduction of projective methods, to which the metrics of the second period are subordinate. Here the works of Arthur Cayley, Felix Klein and Sophus Lie are called out, but further explication is largely left for Whitehead's following article ('Axioms of Geometry'), the seventh part of the Encyclopaedia *entry, the next article in this present volume.*[6]

To conclude, a word on historical perspective. The innocent categorisation of the history of non-Euclidean geometry as 'a tale oft told' is literally true, but not quite fair in this case to the extent that it might suggest that Russell and Whitehead had little

1. See p. 284; the translation from the original Greek appears to be by Whitehead.
2. p. 285.
3. p. 289.
4. p. 290.
5. pp. 291–294.
6. 'Non-Euclidean Geometry' was the sixth article under 'Geometry', with 'Axioms of Geometry' following. Whitehead wrote the latter article without Russell, and he references it numerous times throughout the text. See p. 299.

to add. The point is that over a century ago some of the key elements of this story were recent and some had yet to occur. That alone gives this article a special relevance and explains some of its arcana. Moreover, a noted problem with many expositions of the history and philosophy of mathematics is that they are often not written by working mathematicians. Now, perhaps as a matter of timing or culture – and certainly in part due to the failure of Principia Mathematica *– neither Russell nor Whitehead had much impact on mathematics apart from its foundations and the extension of their work into mathematical logic. Yet their historical grasp of mathematics, as one can see so clearly in this article, is energised and vivified by their extraordinary technical capabilities.*[1]

The various metrical geometries are concerned with the properties of the various types of congruence-groups, which are defined in the study of the axioms of geometry and of their immediate consequences. But this point of view of the subject is the outcome of recent research, and historically the subject has a different origin. Non-Euclidean geometry arose from the discussion, extending from the Greek period to the present day, of the various assumptions which are implicit in the traditional Euclidean system of geometry. In the course of these investigations it became evident that metrical geometries, each internally consistent but inconsistent in many respects with each other and with the Euclidean system, could be developed. A short historical sketch will explain this origin of the subject, and describe the famous and interesting progress of thought on the subject. But previously a description of the chief characteristic properties of elliptic and of hyperbolic geometries will be given, assuming the standpoint arrived at below under VII. Axioms of Geometry.[2]

First assume the equation to the absolute (cf. *loc. cit.*) to be $w^2 - x^2 - y^2 - z^2 = 0$. The absolute is then real, and the geometry is hyperbolic.

The distance (d_{12}) between the two points (x_1, y_1, z_1, w_1) and (x_2, y_2, z_2, w_2) is given by

1. We are indebted to Nicholas Griffin, Canada Research Chair in Philosophy emeritus at McMaster University and former general editor of the *Collected Papers of Bertrand Russell*, for pointing us to Russell's 1902 entry, of which we were unaware when we first began editing this article.
2. The first part of this article, about 2,500 words, is entirely Whitehead's. Starting with the first heading within the article – 'Theory of Parallels before Gauss' – Whitehead begins reusing portions of Russell's 1902 article. By contrast, Russell had only a single lead paragraph of about 150 words before a heading, as follows:

 > Non-Euclidean Geometry, in the narrower sense, may be defined as the study of the consequences which follow from denying one or more of the (explicit or tacit) assumptions upon which the Euclidean system is founded. In a broader sense, for which the name Pangeometry is more appropriate, the word is used to describe systems which merely dispense with one or more of Euclid's assumptions without making any decision as to the truth or falsity of these assumptions. The motives and scope of non-Euclidean Geometry can hardly be grasped without some knowledge of its growth and development. The present article, therefore, will begin with a historical sketch; it will then attempt to dispel some common confusions as to the meaning or the assumptions of non-Euclidean systems; and finally, it will briefly indicate the bearing of these systems upon Epistemology and the philosophy of space.

$$\cosh(d_{12}/\gamma)=(w_1w_2-x_1x_2-y_1y_1-z_1z_2)/\{(w_1^2-x_1^2-y_1^2-z_1^2)(w_2^2-x_2^2-y_2^2-z_2^2)\}^{\frac{1}{2}} \quad (1)$$

The only points to which the metrical geometry applies are those within the region enclosed by the quadric; the other points are "improper ideal points." The angle (θ_{12}) between two planes, $l_1x+m_1y+n_1z+r_1w=0$ and $l_2x+m_2y+n_2z+r_2w=0$, is given by

$$\cos\theta_{12}=(l_1l_2+m_1m_2+n_1n_2-r_1r_2)/\{(l_1^2+m_1^2+n_1^2-r_1^2)\,(l_2^2+m_2^2+n_2^2-r_2^2)\}^{\frac{1}{2}} \quad (2)$$

These planes only have a real angle of inclination if they possess a line of intersection within the actual space, *i.e.* if they intersect. Planes which do not intersect possess a shortest distance along a line which is perpendicular to both of them. If this shortest distance is δ_{12}, we have[1]

$$\cosh(\delta_{12}/\gamma)=(l_1l_2+m_1m_2+n_1n_2-r_1r_2)/\{(l_1^2+m_1^2+n_1^2-r_1^2)(l_2^2+m_2^2+n_2^2-r_2^2)\}^{\frac{1}{2}} \quad (3)$$

Thus in the case of the two planes one and only one of the two, θ_{12} and δ_{12}, is real. The same considerations hold for coplanar straight lines (see VII. *Axioms of Geometry*). Let O (fig. 67) be the point $(0, 0, 0, 1)$, OX the line y = 0, z = 0, OY the line z = 0, x = 0, and OZ the line x = 0, y = 0. These are the coordinate axes and are at right angles to each other. Let P be any point, and let ρ be the distance OP, θ the angle POZ, and ϕ the angle between the planes ZOX and ZOP. Then the coordinates of P can be taken to be $\sinh(\rho/\gamma)\sin\theta\cos\varphi, \sinh(\rho/\gamma)\sin\theta\sin\varphi, \sinh(\rho/\gamma)\cos\theta, \cosh(\rho/\gamma)$.

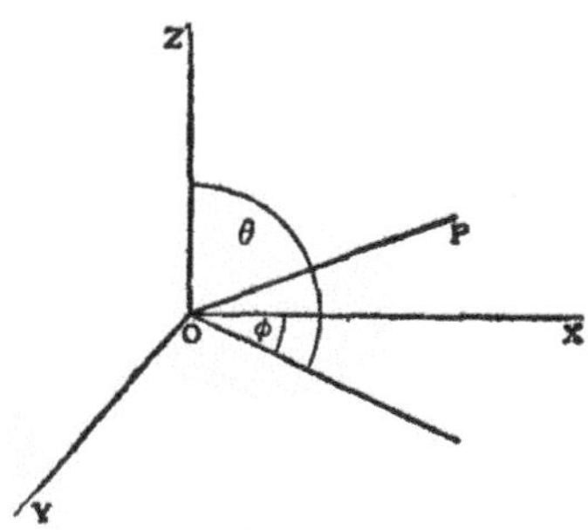

FIG. 67.

If ABC is a triangle, and the sides and angles are named according to the usual convention, we have

$$\sinh(a/\gamma)/\sin A=\sinh(b/\gamma)/\sin B=\sinh(c/\gamma)/\sin C, \quad (4)$$

and also

$$\cosh(a/\gamma)=\cosh(b/\gamma)\cosh(c/\gamma)-\sinh(b/\gamma)\sinh(c/\gamma)\cos A, \quad (5)$$

with two similar equations. The sum of the three angles of a triangle is always less than two right angles. The area of the triangle ABC is $\lambda^2(\pi-A-B-C)$. If the base BC of a triangle is kept fixed and the vertex A moves in the fixed plane ABC so that the area ABC is constant, then the locus of A is a line of equal distance from BC. This locus is not a straight line. The whole theory of similarity is inapplicable; two triangles are either congruent, or their angles are not equal two by two. Thus the elements of a triangle are determined when its three angles are given. By keeping

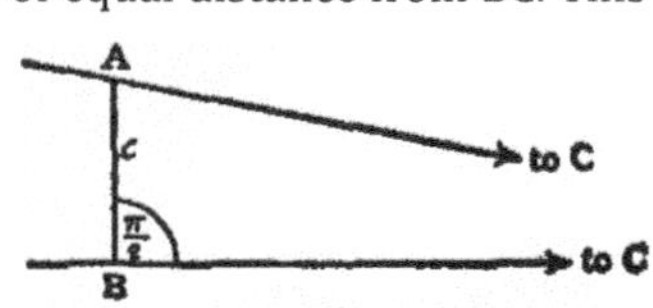

FIG. 68.

1. In the below expression we have supplied a missing closing curly bracket.

A and B and the line BC fixed, but by making C move off to infinity along BC, the lines BC and AC become parallel, and the sides *a* and *b* become infinite. Hence from equation (5) above, it follows that two parallel lines (cf. Section VII. *Axioms of Geometry*) must be considered as making a zero angle with each other. Also if B be a right angle, from the equation (5), remembering that, in the limit,

$$\cosh(a/\gamma)/\cosh(b/\gamma)=\cosh(a/\gamma)/\sinh(b/\gamma)=1,$$

|725a| we have $$\cos \mathrm{A}=\tanh(c/2\gamma) \qquad (6)^1$$

The angle A is called by N. I. Lobatchewsky[2] the "angle of parallelism."

The whole theory of lines and planes at right angles to each other is simply the theory of conjugate elements with respect to the absolute, where ideal lines and planes are introduced.

Thus if *l* and *l′* be any two conjugate lines with respect to the absolute (of which one of the two must be improper, say *l′*), then any plane through *l′* and containing proper points is perpendicular to *l*. Also if *p* is any plane containing proper points, and P is its pole, which is necessarily improper, then the lines through P are the normals to P. The equation of the sphere, centre (x_1, y_1, z_1, w_1) and radius *ρ*, is

$$\left(w_1^{\,2}-x_1^{\,2}-y_1^{\,2}-z_1^{\,2}\right)\left(w^2-x^2-y^2-z^2\right)\cosh^2(\rho/\gamma)=\left(w_1w-x_1x-y_1y-z_1z\right)^2 \qquad (7)$$

The equation of the surface of equal distance (*σ*) from the plane $lx+my+nz+rw=0$ is

$$\left(l^2+m^2+n^2-r^2\right)\left(w^2-x^2-y^2-z^2\right)\sinh^2(\sigma/\gamma)=\left(rw+lx+my+nz\right)^2 \qquad (8)$$

A surface of equal distance is a sphere whose centre is improper; and both types of surface are included in the family

$$k^2\left(w^2-x^2-y^2-z^2\right)=\left(ax+by+cz+dw\right)^2 \qquad (9)$$

But this family also includes a third type of surfaces, which can be looked on either as the limits of spheres whose centres have approached the absolute, or as the limits of surfaces of equal distance whose central planes have approached a position tangential to the absolute. These surfaces are called limit-surfaces. Thus (9) denotes a limit-surface, if $d^2-a^2-b^2-c^2=0$. Two limit-surfaces only differ in position. Thus the two limit-surfaces which touch the plane YOZ at O, but have their concavities turned in opposite directions, have as their equations

$$w^2-x^2-y^2-z^2=\left(w\pm x\right)^2.$$

The geodesic geometry of a sphere is elliptic, that of a surface of equal distance is hyperbolic, and that of a limit-surface is parabolic (*i.e. Euclidean*). The equation of the surface (cylinder) of equal distance (*δ*) from the line OX is

$$\left(w^2-x^2\right)\tanh^2(\delta/\gamma)-y^2-z^2=0.$$

1. We have deleted periods after the numbers (6)–(11) for consistency.
2. Nikolai Ivanovich Lobachevsky (1792–1856) was a Russian mathematician best known for his work on hyperbolic geometry.

This is not a ruled surface. Hence in this geometry it is not possible for two straight lines to be at a constant distance from each other.

Secondly, let the equation of the absolute be $x^2+y^2+z^2+w^2=0$. The absolute is now imaginary and the geometry is elliptic.

The distance (d_{12}) between the two points (x_1, y_1, z_1, w_1) and (x_2, y_2, z_2, w_2) is given by

$$\cos(d_{12}/\gamma)=\pm(x_l x_2+y_l y_2+z_1 z_2+w_1 w_2)/\left\{(x_1^{\,2}+y_1^{\,2}+z_1^{\,2}+w_1^{\,2})\right.$$
$$\left.(x_2^{\,2}+y_2^{\,2}+z_2^{\,2}+w_2^{\,2})\right\}^{\frac{1}{2}} \quad (10)$$

Thus there are two distances between the points, and if one is d_{12}, the other is $\pi\gamma-d_{12}$. Every straight line returns into itself, forming a closed series. Thus there are two segments between any two points, together forming the whole line which contains them; one distance is associated with one segment, and the other distance with the other segment. The complete length of every straight line is $\pi\gamma$.

The angle between the two planes $l_1 x+m_1 y+n_1 z+r_1 w=0$ and $l_2 x+m_2 y+n_2 z+r_2 w=0$ is

$$\cos\theta_{12}=(l_1 l_2+m_1 m_2+n_1 n_2+r_1 r_2)/\left\{(l_1^{\,2}+m_1^{\,2}+n_1^{\,2}+r_1^{\,2})\,(l_2^{\,2}+m_2^{\,2}+n_2^{\,2}+r_2^{\,2})\right\}^{\frac{1}{2}} \quad (11)$$

The polar plane with respect to the absolute of the point (x_1, y_1, z_1, w_1) is the real plane $x_1 x+y_1 y+z_1 z+w_1 w=0$, and the pole of the plane $l_1 x+m_1 y+n_1 z+r_1 w=0$ is the point (l_1, m_1, n_1, r_1). Thus (from equations 10 and 11) it follows that the angle between the polar planes of the points $(x_1, \ldots)$ and $(x_2, \ldots)$ is d_{12}/γ, and that the distance between the poles of the planes $(l_1, \ldots)$ and $(l_2, \ldots)$ is $\gamma\theta_{12}$. Thus there is complete reciprocity between points and planes in respect to all properties. This complete reign of the principle of duality is one of the great beauties of this geometry. The theory of lines and planes at right angles is simply the theory of conjugate elements with respect to the absolute. A tetrahedron self-conjugate with respect to the absolute has all its intersecting elements (edges and planes) at right angles. If *l* and *l′* are two conjugate lines, the planes through one are the planes perpendicular to the other. If P is the pole of the plane *p*, the lines through P are the normals to the plane *p*. The distance from P to *p* is $\frac{1}{2}\pi\gamma$. Thus every sphere is also a surface of equal distance from the polar of its centre, and conversely. A plane does not divide space; for the line joining any two points P and Q only cuts the plane once, in L say, then it is always possible to go from P to Q by the segment of the line PQ which does not contain L. But P and Q may be said to be separated by a plane *p*, if the point in which PQ cuts *p* lies on the shortest segment between P and Q. With this sense of "separation," it is possible[1†] to find three points P, Q, R such that P and Q are separated |725b| by the plane *p*, but P and R are not separated by *p*, nor are Q and R.

Let A, B, C be any three non-collinear points, then four triangles are defined by these points. Thus if *a*, *b*, *c* and A, B, C are the elements of any one triangle, then the four triangles have as their elements:

1. †Cf. A. N. Whitehead, *Universal Algebra*, Bk. vi. (Cambridge, 1898).

(1)	a,	b,	c,	A,	B,	C.
(2)	a,	$\pi\gamma-b$,	$\pi\gamma-c$,	A,	$\pi-$B,	$\pi-$C.
(3)	$\pi\gamma-a$,	b,	$\pi\gamma-c$,	$\pi-$A,	B,	$\pi-$C.
(4)	$\pi\gamma-a$,	$\pi\gamma-b$,	c,	$\pi-$A,	$\pi-$B,	C.

The formulae connecting the elements are

$$\sin A/\sin(a/\gamma)=\sin B/\sin(b/\gamma)=\sin C/\sin(c/\gamma), \tag{12}$$

and

$$\cos(a/\gamma)=\cos(b/\gamma)\cos(c/\gamma)+\sin(b/\gamma)\sin(c/\gamma)\cos A, \tag{13}$$

with two similar equations.

Two cases arise, namely (I.) according as one of the four triangles has as its sides the shortest segments between the angular points, or (II.) according as this is not the case. When case I. holds there is said to be a "principal triangle."[1†] If all the figures considered lie within a sphere of radius $\frac{1}{4}\pi\gamma$ only case I. can hold, and the principal triangle is the triangle wholly within this sphere, also the peculiarities in respect to the separation of points by a plane cannot then arise. The sum of the three angles of a triangle ABC is always greater than two right angles, and the area of the triangle is $\gamma^2(A+B+C-\pi)$. Thus as in hyperbolic geometry the theory of similarity does not hold, and the elements of a triangle are determined when its three angles are given. The coordinates of a point can be written in the form

$$\sin(\rho/\gamma)\sin\theta\cos\phi,\ \sin(\rho/\gamma)\sin\theta\sin\phi,\ \sin(\rho/\gamma)\cos\theta,\ \cos(\rho/\gamma),$$

where ρ, θ and ϕ have the same meanings as in the corresponding formulae in hyperbolic geometry. Again, suppose a watch is laid on the plane OXY, face upwards with its centre at O, and the line 12 to 6 (as marked on dial) along the line YOY. Let the watch be continually pushed along the plane along the line OX, that is, in the direction 9 to 3. Then the line XOX being of finite length, the watch will return to O, but at its first return it will be found to be face downwards on the other side of the plane, with the line 12 to 6 reversed in direction along the line YOY. This peculiarity was first pointed out by Felix Klein.[2] The theory of parallels as it exists in hyperbolic space has no application in elliptic geometry. But another property of Euclidean parallel lines holds in elliptic geometry, and by the use of it parallel lines are defined. For the equation of the surface (cylinder) of equal distance (δ) from the line XOX is

$$(x^2+w^2)\tan^2(\delta/\gamma)-(y^2+z^2)=0.$$

This is also the surface of equal distance, $\frac{1}{2}\pi\gamma-\delta$ from the line conjugate to XOX. Now from the form of the above equation this is a ruled surface, and through every point of it two generators pass. But these generators are lines of equal distance from XOX. Thus throughout every point of space two lines can be drawn which are lines of equal distance from a given line l. This property was discovered by W. K. Clifford.[3] The two lines are called Clifford's right and left parallels to l through the point. This property of parallelism is reciprocal, so that if m is a left parallel to l, then l is

1. † Cf. A. N. Whitehead, *loc. cit.*
2. Felix Christian Klein (1849–1925) was a German mathematician.
3. William Kingdon Clifford (1845–1879) was a British mathematician and philosopher.

a left parallel to *m*. Note also that two parallel lines *l* and *m* are not coplanar. Many of those properties of Euclidean parallels, which do not hold for Lobatchewsky's parallels in hyperbolic geometry, do hold for Clifford's parallels in elliptic geometry. The geodesic geometry of spheres is elliptic, the geodesic geometry of surfaces of equal distance from lines (cylinders) is Euclidean, and surfaces of revolution can be found[1†] of which the geodesic geometry is hyperbolic. But it is to be noticed that the connectivity of these surfaces is different to that of a Euclidean plane. For instance there are only ∞^2 congruence transformations of the cylindrical surfaces of equal distance into themselves, instead of the ∞^3 for the ordinary plane. It would obviously be possible to state "axioms" which these geodesics satisfy, and thus to define independently, and not as loci, quasi-spaces of these peculiar types. The existence of such Euclidean quasi-geometries was first pointed out by Clifford.[2†]

In both elliptic and hyperbolic geometry the spherical geometry, *i.e.* the relations between the angles formed by lines and planes passing through the same point, is the same as the "spherical trigonometry" in Euclidean geometry. The constant γ, which appears in the formulae both of hyperbolic and elliptic geometry, does not by its variation produce different types of geometry. There is only one type of elliptic geometry and one type of hyperbolic geometry; and the magnitude of the constant γ in each case simply depends upon the magnitude of the arbitrary unit of length in comparison with the natural unit of length |726a| which each particular instance of either geometry presents. The existence of a natural unit of length is a peculiarity common both to hyperbolic and elliptic geometries, and differentiates them from Euclidean geometry. It is the reason for the failure of the theory of similarity in them. If γ is very large, that is, if the natural unit is very large compared to the arbitrary unit, and if the lengths involved in the figures considered are not large compared to the arbitrary unit, then both the elliptic and hyperbolic geometries approximate to the Euclidean. For from formulae (4) and (5) and also from (12) and (13) we find, after retaining only the lowest powers of small quantities, as the formulae for any triangle ABC,

$$a / \sin \mathrm{A} = b / \sin \mathrm{B} = c / \sin \mathrm{C},$$

and

$$a^2 = b^2 + c^2 - 2bc \cos \mathrm{A},$$

with two similar equations. Thus the geometries of small figures are in both types Euclidean.

Theory of Parallels before Gauss[3]

History.—"In pulcherrimo Geometriae corpore," wrote Sir Henry Savile in 1621,[4] "duo sunt naevi, duae labes nec quod sciam plures, in quibus eluendis et

1. † Cf. A. N. Whitehead, *loc. cit.*
2. † Cf. Klein, "Zur nicht-Euklidischen Geometrie," *Math. Annal.* vol. xxxvii [no. 4 (December 1890), pp. 544–572].
3. Whitehead kept most of the material from Russell for this section while expanding and editing it slightly. He also moved the latter half to the next two sections below, 'Saccheri' and 'Lambert'.
4. Henry Savile, *Praelectiones tresdecim in principium Elemetorum Euclidis: Oxonii habitae M.DC.XX.* (Oxford: Iohannes Lichfield & Iacobus Short, 1621).

emaculendis cum veterum turn recentiorum . . . vigilavit industria."[1] These two blemishes are the theory of parallels and the theory of proportion. The "industry of the moderns," in both respects, has given rise to important branches of mathematics, while at the same time showing that Euclid is in these respects[2] more free from blemish than had been previously credible. It was from endeavours to improve the theory of parallels that non-Euclidean geometry arose; and though it has now acquired a far wider scope, its historical origin remains instructive and interesting.[3] Euclid's "axiom of parallels" appears as Postulate V. to the first book of his *Elements*, and is stated thus, "And that, if a straight line falling on two straight lines make the angles, internal and on the same side, less than two right angles, the two straight lines, being produced indefinitely, meet on the side on which are the angles less than two right angles." The original Greek is *καὶ ἐὰν εἰς δύο εὐθείας εὐθεῖα ἐμπίπτουσα τὰς ἐντὸς καὶ ἐπὶ τὰ αὐτὰ μέρη γωνίας δύο ὀρθῶν ἐλάσσονας ποιῇ, ἐκβαλλομένας τὰς δύο εὐθείας ἐπ' ἄπειρον συμπίπτειν, ἐφ' ἃ μέρη εἰσὶν αἱ τῶν δύο ὀρθῶν ἐλάσσονες.*

To Euclid's successors this axiom[4] had signally failed to appear self-evident, and had failed equally to appear indemonstrable. Without the use of the postulate its converse is proved in Euclid's 28th proposition, and it was hoped that by further efforts the postulate itself could be also proved. The first step consisted in the discovery of equivalent axioms. Christoph Clavius[5] in 1574 deduced the axiom from the assumption that a line whose points are all equidistant from a straight line is itself straight. John Wallis[6] in 1663 showed that the postulate follows from the possibility of similar triangles on different scales. Girolamo Saccheri (1733)[7] showed that it is sufficient to have a single triangle, the sum of whose angles is two right angles. Other equivalent forms may be obtained, but none shows any essential superiority to Euclid's. Indeed plausibility, which is chiefly aimed at, becomes a positive demerit where it conceals a real assumption.

SACCHERI[8]

A new method, which, though it failed to lead to the desired goal, proved in the end immensely fruitful, was invented by Saccheri, in a work entitled *Euclides ab omni naevo vindicatus* (Milan, 1733). If the postulate of parallels is involved in Euclid's other assumptions, contradictions must emerge when it is denied while the others are maintained. This led Saccheri to attempt a *reductio ad absurdum*, in which

1. 'In the most beautiful body of Geometry, there are two blemishes, two decays, and I do not know of more, in which the washings and besmirchings of the ancients turn to the moderns. . . He watched the industry.'
2. Whitehead added 'in these respects' to Russell's version.
3. From this point to the end of the current paragraph is a Whitehead addition to Russell's version.
4. Whitehead replaced ', the axiom, or more properly the postulate, of parallels' with 'this axiom'.
5. Christoph Clavius (1538–1612) was a German mathematician and astronomer, and also a Jesuit priest.
6. John Wallis (1616–1703) was an English mathematician who helped develop infinitesimal calculus.
7. Giovanni Girolamo Saccheri (1667–1733) was an Italian philosopher, mathematician and Jesuit priest. His work helped lead to the development of non-Euclidean geometry.
8. Aside from stylistic changes, Whitehead carried this section over from Russell's version unchanged, though Russell had this material under the former heading.

he mistakenly believed himself to have succeeded. What is interesting, however, is not his fallacious conclusion, but the non-Euclidean results which he obtains in the process. Saccheri distinguishes three hypotheses (corresponding to what are now known as Euclidean or parabolic, elliptic and hyperbolic geometry), and proves that some one of the three must be universally true. His three hypotheses are thus obtained: equal perpendiculars AC, BD are drawn from a straight line AB, and CD are joined. It is shown that the angles ACD, BDC are |726b| equal. The first hypothesis is that these are both right angles; the second, that they are both obtuse; and the third, that they are both acute. Many of the results afterwards obtained by Lobatchewsky and Bolyai[1] are here developed. Saccheri fails to be the founder of non-Euclidean geometry only because he does not perceive the possible truth of his non-Euclidean hypotheses.

Lambert[2]

Some advance is made by Johann Heinrich Lambert[3] in his *Theorie der Parallellinien* (written in 1766; posthumously published 1786). Though he still believed in the necessary truth of Euclidean geometry, he confessed that, in all his attempted proofs, something remained undemonstrated. He deals with the same three hypotheses as Saccheri, showing that the second holds on a sphere, while the third would hold on a sphere of purely imaginary radius. The second hypothesis he succeeds in condemning, since, like all who preceded Bernhard Riemann,[4] he is unable to conceive of the straight line as finite and closed. But the third hypothesis, which is the same as Lobatchewsky's, is not even professedly refuted.[5†]

Three periods of non-Euclidean geometry[6]

Non-Euclidean geometry proper begins with Karl Friedrich Gauss.[7] The advance which he made was rather philosophical than mathematical: it was he (probably) who first recognized that the postulate of parallels is possibly false, and should be empirically tested by measuring the angles of large triangles. The history of non-Euclidean geometry has been aptly divided by Felix Klein into three very distinct periods. The first—which contains only Gauss, Lobatchewsky and Bolyai—is

1. János Bolyai (1802–1860) was a Hungarian mathematician who developed 'absolute geometry'.
2. Whitehead carried this section over from Russell's version mostly unchanged, though Russell had this material under the 'Theory of parallels before Gauss' heading.
3. Johann Heinrich Lambert (1728–1777) was a French-Swiss philosopher, mathematician, physicist and astronomer. Whitehead inserted his full name in place of Russell's 'Lambert (the demonstrator of the irrationality of "TT")'.
4. Georg Friedrich Bernhard Riemann (1826–1866) was a German mathematician who made foundational contributions to number theory and differential geometry.
5. † On the theory of parallels before Lobatchewsky, see Stäckel and Engel, *Theorie der Parallellinien von Euklid bis auf Gauss* (Leipzig, 1895). The foregoing remarks are based upon the materials collected in this work.
6. Whitehead carried this section over from Russell's version with only minor edits.
7. Johann Carl Friedrich Gauss (1777–1855) was a German mathematician, astronomer and physicist.

characterized by its synthetic method and by its close relation to Euclid. The attempt at indirect proof of the disputed postulate would seem to have been the source of these three men's discoveries; but when the postulate had been denied, they found that the results, so far from showing contradictions, were just as self-consistent as Euclid. They inferred that the postulate, if true at all, can only be proved by observations and measurements. Only one kind of non-Euclidean space is known to them, namely, that which is now called hyperbolic. The second period is analytical, and is characterized by a close relation to the theory of surfaces. It begins with Riemann's inaugural dissertation, which regards space as a particular case of a *manifold*;[1] but the characteristic standpoint of the period is chiefly emphasized by Eugenio Beltrami.[2] The conception of measure of curvature is extended by Riemann[3] from surfaces to spaces, and a new kind of space, finite but unbounded (corresponding to the second hypothesis of Saccheri and Lambert), is shown to be possible. As opposed to the second period, which is purely metrical, the third period is essentially projective in its method. It begins with Arthur Cayley,[4] who showed that metrical properties are projective properties relative to a certain fundamental quadric, and that different geometries arise according as this quadric is real, imaginary or degenerate. Klein, to whom the development of Cayley's work is due, showed further that there are two forms of Riemann's space, called by him the elliptic and the spherical. Finally, it has been shown by Sophus Lie,[5] that if figures are to be freely movable throughout all space in ∞^6 ways, no other three-dimensional spaces than the above four are possible.

Gauss[6]

Gauss published nothing on the theory of parallels, and it was not generally known until after his death that he had interested himself in that theory from a very early date. In 1799 he announces that Euclidean geometry would follow from the assumption that a triangle can be drawn greater than any given triangle. Though unwilling to assume this, we find him in 1804 still hoping to prove the postulate of parallels. In 1830 he announces his conviction that geometry is not an a priori science; in the following year he explains that non-Euclidean geometry is free from contradictions, and that, in this system, the angles of a triangle diminish without limit when all the sides are increased. He also gives for the |727a| circumference of a circle of radius *r* the formula[7] $\pi k\left(e^{r/k} - e^{-r/k}\right)$, where *k* is a constant depending upon the nature of the space. In 1832, in reply to the receipt of Bolyai's *Appendix*, he gives

1. The remainder of this sentence after the semicolon is a Whitehead addition to Russell's version.
2. Eugenio Beltrami (1835–1900) was an Italian mathematician best known for his work on differential geometry.
3. 'by Riemann' is a Whitehead addition to Russell's version.
4. Arthur Cayley (1821–1895) was a British mathematician. A professor at Trinity College, many of his papers were edited by Whitehead's friend Andrew Forsyth.
5. Marius Sophus Lie (1842–1899) was a Norwegian mathematician credited with the theory of continuous symmetry.
6. Aside from stylistic changes, Whitehead carried this section over from Russell's version unchanged.
7. The original erroneously had 'r–' instead of '–r'.

an elegant proof that the amount by which the sum of the angles of a triangle falls short of two right angles is proportional to the area of the triangle. From these and a few other remarks it appears that Gauss possessed the foundations of hyperbolic geometry, which he was probably the first to regard as perhaps true. It is not known with certainty whether he influenced Lobatchewsky and Bolyai, but the evidence we possess is against such a view.[1†]

Lobatchewsky[2]

The first to publish a non-Euclidean geometry was Nicholas Lobatchewsky, professor of mathematics in the new university of Kazañ.[3†] In the place of the disputed postulate he puts the following: "All straight lines which, in a plane, radiate from a given point, can, with respect to any other straight line in the same plane, be divided into two classes, the *intersecting* and the *non-intersecting*. The *boundary line* of the one and the other class is called *parallel to the given line*." It follows that there are two parallels to the given line through any point, each meeting the line at infinity, like a Euclidean parallel. (Hence a line has two distinct points at infinity, and not one only as in ordinary geometry.) The two parallels to a line through a point make equal acute angles with the perpendicular to the line through the point. If p be the length of the perpendicular, either of these angles is denoted by $\Pi(p)$. The determination of $\Pi(p)$ is the chief problem (cf. equation (6) above);[4] it appears finally that, with a suitable choice of the unit of length,

$$\tan\tfrac{1}{2}\Pi(p)=e^{-p}.$$

Before obtaining this result it is shown that spherical trigonometry is unchanged, and that the normals to a circle or a sphere still pass through its centre. When the radius of the circle or sphere becomes infinite all these normals become parallel, but the circle or sphere does not become a straight line or plane. It becomes what Lobatchewsky calls a limit-line or limit-surface. The geometry on such a surface is shown to be Euclidean, limit-lines replacing Euclidean straight lines. (It is, in fact, a surface of zero measure of curvature.) By the help of these propositions Lobatchewsky obtains the above value of $\Pi(p)$, and thence the solution of triangles. He

1. †See Stäckel and Engel, *op. cit.*, and "Gauss, die beiden Bolyai, und die nicht-Euklidische Geometrie," *Math. Annalen*, Bd. xlix [vol. 49, no. 2 (June 1897), pp. 149–167]; also Engel's translation of Lobatchewsky (Leipzig, 1898), pp. 378 ff. [Paul Stäckel and Friedrich Engel, *Urkunden zur Geschichte der nichteuklidischen Geometrie, Vol. I, Nikolaj Iwanowitsch Lobatschefskij* (Leipzig: Teubner, 1898)].
2. Aside from stylistic changes and one reference to an equation from his new opening section, Whitehead carried this section over from Russell's version unchanged.
3. †Lobatchewsky's works on the subject are the following:—"On the Foundations of Geometry," *Kazañ Messenger*, 1829–1830; "New Foundations of Geometry, with a complete Theory of Parallels," *Proceedings of the University of Kazañ*, 1835 (both in Russian, but translated into German by Engel, Leipzig, 1898 [as *Urkunden zur Geschichte der nichteuklidischen Geometrie, Erster Theil, Nikolaj Iwanowitsch Lobatschefskij*]); "Géométrie imaginaire," *Crelle's Journal*, 1837 [vol. 17, pp. 295–320]; *Theorie der Parallellinien* (Berlin, 1840; 2nd ed., 1887; translated by Halsted, Austin, Texas, 1891 [as *Geometrical Researches on the Theory of Parallels*]). His results appear to have been set forth in a paper (now lost) which he read at Kazañ in 1826. [See also p. 301n2.]
4. The parenthetical is a Whitehead addition to Russell's version.

points out that his formulae result from those of spherical trigonometry by substituting *ia*, *ib*, *ic*, for the sides *a*, *b*, *c*.

Bolyai[1]

John Bolyai, a Hungarian, obtained results closely corresponding to those of Lobatchewsky. These he published in an appendix to a work by his father, entitled *Appendix Scientiam spatii absolute veram exhibens: a veritate aut falsitate Axiomatis XI. Euclidei (a priori haud unquam decidenda) independentem: adjecta ad casum falsitatis, quadratura circuli geometrica.*[2†] This work was published in 1832,[3] but its conception dates from 1823. It reveals a profounder appreciation of the importance of the new ideas, but otherwise differs little from Lobatchewsky's. Both men point out that Euclidean geometry is a limiting case of their own more general system, that the geometry of very small spaces is always approximately Euclidean, that no a priori grounds exist for a decision, and that observation can only give an approximate answer. Bolyai gives also, as his title indicates, a geometrical construction, in hyperbolic space, for the quadrature of the circle, and shows that the area of the greatest possible triangle, which has all its sides parallel and all its angles zero, is $\pi\iota^2$, where ι is what we should now call the space-constant.

|727b| *Riemann*[4]

The works of Lobatchewsky and Bolyai, though known and valued by Gauss, remained obscure and ineffective until, in 1866, they were translated into French by J. Hoüel.[5] But at this time Riemann's dissertation, Über *die Hypothesen, welche der Geometrie zu Grunde liegen*,[6†] was already about to be published. In this work Riemann, without any knowledge of his predecessors in the same field, inaugurated a far more profound discussion, based on a far more general standpoint; and by its publication in 1867 the attention of mathematicians and philosophers was at last secured. (The dissertation dates from 1854, but owing to changes which Riemann wished to make in it, it remained unpublished until after his death.)

---

1. Aside from one case change, Whitehead carried this section over from Russell's version unchanged.
2. †Translated by Halsted (Austin, Texas, 4th ed., 1896) [*The Science Absolute of Space: Independent of the Truth or Falsity of Euclid's Axiom XI (Which Can Never Be Decided* A Priori*)*].
3. Corrected from 1831.
4. Whitehead carried this section over from Russell's version unchanged.
5. Jules Hoüel, 'Etudes géométriques sur la théorie des parallèles', *Mémoires de la Société des Sciences Physiques et Naturelles de Bordeaux, Tome IV* (Paris: Gauthier-Villars, 1867), pp. 83–128; Jules Hoüel, 'La Science absolue de l'espace', *Mémoires de la Société des Sciences Physiques et Naturelles de Bordeaux, Tome V* (Paris: Gauthier-Villars, 1867), pp. 189–248.
6. †*Abhandlungen d. Königl. Ges. d. Wiss. zu Göttingen*, Bd. xiii. [vol. 13 (1868), pp. 133–150]; *Ges. math. Werke*, pp. 254–269 [edited by Heinrich Weber (Leipzig: Teubner, 1876)]; translated by Clifford, *Collected Mathematical Papers* [originally translated in Bernhard Riemann, 'On the Hypotheses which lie at the Bases of Geometry', translated by William Kingdon Clifford, *Nature*, vol. 8, nos 183–184, pp. 14–17, 36–37)].

Definition of a Manifold[1]

Riemann's work contains two fundamental conceptions, that of a manifold and that of the *measure of curvature* of a continuous manifold possessed of what he calls flatness in the smallest parts. By means of these conceptions space is made to appear at the end of a gradual series of more and more specialized conceptions. Conceptions of magnitude, he explains, are only possible where we have a general conception capable of determination in various ways. The manifold consists of all these various determinations, each of which is an element of the manifold. The passage from one element to another may be discrete or continuous; the manifold is called discrete or continuous accordingly. Where it is discrete two portions of it can be compared, as to magnitude, by counting; where continuous, by measurement. But measurement demands superposition, and consequently some magnitude independent of its place in the manifold. In passing, in a continuous manifold, from one element to another in a determinate way, we pass through a series of intermediate terms, which form a one-dimensional manifold. If this whole manifold be similarly caused to pass over into another, each of its elements passes through a one-dimensional manifold, and thus on the whole a two-dimensional manifold is generated. In this way we can proceed to n dimensions. Conversely, a manifold of n dimensions can be analysed into one of one dimension and one of $(n-1)$ dimensions. By repetitions of this process the position of an element may be at last determined by n magnitudes.[2] We may here stop to observe that the above conception of a manifold is akin to that due to Hermann Grassmann in the first edition (1847) of his *Ausdehnungslehre*.[3†]

Measure of Curvature[4]

Both concepts have been elaborated and superseded by the modern procedure in respect to the axioms of geometry, and by the conception of abstract geometry involved therein.[5] Riemann proceeds to specialize the manifold by considerations as to measurement. If measurement is to be possible, some magnitude, we saw, must be independent of position; let us consider manifolds in which lengths of lines are such magnitudes, so that every line is measurable by every other. The coordinates of a point being $x_1, x_2, \ldots x_n$, let us confine ourselves to lines along which the ratios $dx_1 : dx_2 : \ldots : dx_n$ alter continuously. Let us also assume that the element of length, *ds*, is unchanged (to the first order) when all its points undergo the same infinitesimal motion. Then if all the increments *dx* be altered in the same ratio, *ds* is also altered in this ratio. Hence *ds* is a homogeneous function of the first degree of the increments *dx*. Moreover, *ds* must be unchanged when all the *dx* change sign. The

1. Russell's heading was 'Definition of a manifold. Measure of a curvature.' Whitehead kept all of Russell's text from under this heading, but split it into two discrete sections, while also adding some material.
2. The next sentence (the last of the section) is a Whitehead addition to Russell's version.
3. † Cf. *Gesamm. math. and phys. Werke*, vol. i. (Leipzig, 1894).
4. The text of this section is mostly Russell's, but Whitehead added a new first sentence and a new concluding paragraph.
5. This first sentence is a Whitehead addition to Russell's version.

simplest possible case is, therefore, that in which *ds* is the square root of a quadratic function of the *dx*. This case includes space, and is alone considered in what follows. It is called the case of flatness in the smallest parts. Its further discussion depends upon the measure of curvature, the second of Riemann's fundamental conceptions. This conception, derived from the theory of surfaces, is applied as follows. Any one of the shortest lines which issue from a given point (say the origin) is completely determined by the initial ratios of the *dx*. Two such lines, defined by *dx* and δx say, determine a pencil, or one-dimensional series, of shortest lines, any one of which is defined |728a| by $\lambda dx + \mu\delta x$, where the parameter $\lambda{:}\mu$ may have any value. This pencil generates a two-dimensional series of points, which may be regarded as a surface, and for which we may apply Gauss's formula for the measure of curvature at any point. Thus at every point of our manifold there is a measure of curvature corresponding to every such pencil; but all these can be found when $n(n-1)/2$ of them are known. If figures are to be freely movable, it is necessary and sufficient that the measure of curvature should be the same for all points and all directions at each point. Where this is the case, if[1] α be the measure of curvature, the linear element can be put into the form

$$ds = \sqrt{\left(\Sigma dx^2\right)} / \left(1 + \tfrac{1}{4}\alpha\Sigma x^2\right).$$

If α be positive, space is finite, though still unbounded, and every straight line is closed—a possibility first recognized by Riemann. It is pointed out that, since the possible values of α form a continuous series, observations cannot prove that our space is strictly Euclidean. It is also regarded as possible that, in the infinitesimal, the measure of curvature of our space should be variable.[2]

There are four points in which this profound and epoch-making work is open to criticism or development—(1) the idea of a manifold requires more precise determination; (2) the introduction of coordinates is entirely unexplained and the requisite presuppositions are unanalysed; (3) the assumption that *ds* is the square root of a quadratic function of $dx_1, dx_2, \ldots$ is arbitrary; (4) the idea of superposition, or congruence, is not adequately analysed. The modern solution of these difficulties is properly considered in connexion with the general subject of the axioms of geometry.[3]

---

1. The original erroneously had 'it'.
2. The paragraph below is a Whitehead addition to Russell's version.
3. Here Whitehead deleted a section from Russell's version with the heading 'Criticism of Riemann'. This section reads as follows (footnotes have been moved to parentheticals):

> There is one point, at least, upon which this profound work seems open to criticism, namely, as regards the introduction of coordinates. If we are to be able, without a vicious circle, to presuppose coordinates in discussing distance, it is evident that our coordinates must not presuppose distance. It then remains a question how they are to be defined—a question satisfactorily answered (as we shall see) by the projective method, but unanswerable from Riemann's metrical standpoint. Moreover, if our coordinates are to represent any kind of spatial magnitudes, we must assume the possibility of equal quantities in different places, and hence, it will be found, we shall be compelled to regard the measure of curvature as constant. Let us examine the consequences of supposing it variable. In the first place, the method of superposition would have become impossible, so that measurements could no longer be effected. As regards areas and volumes this is obvious, since congruent areas or volumes could not, generally speaking, exist in two different assigned regions. As regards lengths, it seems equally true that superposition would be impossible. For straight lines corresponding to different values of the measure of curvature differ, as we shall see later, not only in quantity, but

Helmholtz[1]

The publication of Riemann's dissertation was closely followed by two works of Hermann von Helmholtz,[2†] again undertaken in ignorance of the work of predecessors. In these a proof is attempted that *ds* must be a rational integral quadratic function of the increments of the coordinates. This proof has since been shown by Lie to stand in need of correction (see VII. *Axioms of Geometry*). Helmholtz's remaining works on the subject[3†] are of almost exclusively philosophical interest. We shall return to them later.

Beltrami[4]

The only other writer of importance in the second period is Eugenio Beltrami, by whom Riemann's work was brought into connexion with that of Lobatchewsky and Bolyai. As he gave, by an elegant method, a convenient Euclidean interpretation of hyperbolic plane geometry, his results will be stated at some length.[5†] The *Saggio* shows that Lobatchewsky's plane geometry holds in Euclidean geometry[6]

also in quality. Hence two such lines cannot be fitted together, and are, strictly, neither equal nor unequal. Hence it would be impossible, in general, to divide a given length into equal portions, or to compare it numerically with other lengths. Thus a metrical coordinate system would become impossible. Moreover, Geometry would become akin to Geography; it would not consist of general theorems, but of descriptions of various localities. The variation of the space-constant would be not quantitative merely, but qualitative, and quantities in different places would be of different kinds. Thus the constancy of the measure of curvature is a precondition of any metrical coordinate system, and cannot be held doubtful while such a system is retained (See B. Russell, *Foundations of Geometry*, Cambridge, 1897). Another respect in which Riemann's language, though not fallacious, has proved unfortunate, is the use of the word curvature. This has led philosophers, and even mathematicians (*e.g.* Newcomb), to suppose non-Euclidean spaces curved, and therefore necessarily contained in Euclidean spaces of more dimensions. So far is this from being the case that there is perfect reciprocity. Parabolic spaces can be contained both in elliptic and in hyperbolic spaces of one more dimension (See Whitehead, "Geodesic Geometry of Surfaces in non-Euclidean Space", *Proc. Lond. Math. Soc.*, Vol. xxix.) [see p. 42]. It has, consequently, become customary to speak of the reciprocal of the square root of Riemann's measure of curvature as the *space-constant*, in order to avoid all appearance of implying a curvature of non-Euclidean spaces.

1. Aside from stylistic changes, Whitehead carried this section over from Russell's version unchanged.
2. † *Wiss. Abh.* vol. ii. pp. 610, 618 (1866, 1868). [The page numbers and dates refer to two articles – 'Ueber die thatsächlichen Grundlagen der Geometrie' and 'Ueber die Thatsachen, die der Geometrie zum Grunde liegen' – later published in Herman von Helmholtz, *Wissenschaftliche Abhandlungen, Volume II* (Leipzig: Barth, 1883).]
3. † *Mind*, O.S., vols. i. and iii [Herman von Helmholtz, 'The Origin and Meaning of Geometrical Axioms', *Mind*, vol. 1, no. 3 (July 1876), pp. 301–321; Herman von Helmholtz, 'The Origin and Meaning of Geometrical Axioms (II)', *Mind*, vol. 3, no. 10 (April 1878), pp. 212–225]; *Vorträge und Reden*, vol. ii. pp. 1, 256. (Braunschweig: Friedrich Vieweg und Sohn, 1903).
4. Aside from stylistic changes, Whitehead carried this section over from Russell's version mostly unchanged. He did insert a pair of very minor phrases to further clarify the text and corrected two errors in Russell's equations.
5. † His papers are "Saggio di interpretazione della geometria non-Euclidea," *Giornale di matematiche*, vol. vi. (1868) [pp. 284–312]; "Teoria fondamentale degli spazii di curvatura costante," *Annali di matematica*, vol. ii (1868) [pp. 232–255]. Both were translated into French by J. Hoüel, *Annales scientifiques de l'École Normale supérieure*, vol. vi. (1869) ['Essai d'interprétation de la géométrie noneuclidéenne', pp. 251–288; 'Théorie fondamentale des espaces de courbure constante', p. 347–375].
6. 'in Euclidean Geometry' is a Whitehead addition to Russell's version.

on surfaces of constant negative curvature, straight lines being replaced by geodesics. Such surfaces are capable of a conformal representation on a plane, by which geodesics are represented by straight lines. Hence if we take, as coordinates on the surface, the Cartesian coordinates of corresponding points on the plane, the geodesics must have linear equations.

Hence it follows that

$$ds^2 = \mathrm{R}^2 w^{-4}\left\{\left(a^2 - v^2\right)du^2 + 2uvdudv + \left(a^2 - u^2\right)dv^2\right\}$$

where $w^2 = a^2 - u^2 - v^2$, and $-1/\mathrm{R}^2$ is the measure of curvature of our surface (note that $k = \gamma$ as used above).[1] The angle between two geodesics u = const., v = const. is θ, where

$$\cos\theta = uv / \sqrt{\left(a^2 - u^2\right)\left(a^2 - v^2\right)}, \sin\theta = aw / \sqrt{\left(a^2 - u^2\right)\left(a^2 - v^2\right)}.$$

Thus $u = 0$ is orthogonal to all geodesics v = const., and vice versa. In order that sin θ may be real, w^2 must be positive; thus geodesics have no real intersection when the corresponding straight lines intersect outside the circle $u^2 + v^2 = a^2$. When they intersect on this circle, $\theta = 0$. Thus Lobatchewsky's parallels are represented by straight lines intersecting on the circle. Again, transforming to polar coordinates $u = r\cos\mu$, $v = r\sin\mu$, and calling ρ the geodesic |728b| distance of u, v from the origin, we have, for a geodesic through the origin,

$$d\rho = \mathrm{R}adr / \left(a^2 - r^2\right), \rho = \tfrac{1}{2}\mathrm{R}\log\frac{a+r}{a-r}, r = a\tanh\frac{\rho}{\mathrm{R}}.$$

Thus points on the surface corresponding to points in the plane on the limiting circle $r = a$, are all at an infinite distance from the origin. Again, considering r constant, the arc of a geodesic circle subtending an angle μ at the origin is

$$\sigma = \mathrm{R}r\mu / \sqrt{a^2 - r^2} = \mu\mathrm{R}\sinh\left(\rho / R\right),$$

whence the circumference of a circle of radius ρ is $2\pi\mathrm{R}\sinh\left(\rho / \mathrm{R}\right)$. Again, if α be the angle between any two geodesics

$$\mathrm{V} - v = m\left(\mathrm{U} - u\right), \mathrm{V} - v = n\left(\mathrm{U} - u\right),$$

then
$$\tan\alpha = a\left(n - m\right)w / \left\{\left(1 + mn\right)a^2 - \left(v - mu\right)\left(v - nu\right)\right\}.$$

Thus α is imaginary when u, v is outside the limiting circle, and is zero when, and only when, u, v is on the limiting circle. All these results agree with those of Lobatchewsky and Bolyai. The maximum triangle, whose angles are all zero, is represented in the auxiliary plane by a triangle inscribed in the limiting circle. The angle of parallelism is also easily obtained. The perpendicular to $v = 0$ at a distance δ from the origin is $u = a\tanh\left(\delta / \mathrm{R}\right)$, and the parallel to this through the origin is $u = v\sinh\left(\delta / \mathrm{R}\right)$. Hence $\Pi\left(\delta\right)$, the angle which this parallel makes with $v = 0$, is given by

$$\tan\Pi\left(\delta\right)\cdot\sinh\left(\delta / \mathrm{R}\right) = 1, \text{ or } \tan\tfrac{1}{2}\Pi\left(\delta\right) = e^{-\delta/\mathrm{R}}$$

which is Lobatchewsky's formula. We also obtain easily for the area of a triangle the formula $R^2\left(\pi - A - B - C\right)$.

1. The parenthetical is a Whitehead addition to Russell's version.

Beltrami's treatment connects two curves which, in the earlier treatment, had no connexion. These are limit-lines and curves of constant distance from a straight line. Both may be regarded as circles, the first having an infinite, the second an imaginary radius. The equation to a circle of radius ρ and centre u_0v_0 is

$$\left(a^2-uu_0-vv_0\right)^2=\cosh^2\left(\rho/\mathrm{R}\right)w_0{}^2w^2=\mathrm{C}^2w^2 \qquad \text{(say).}$$

This equation remains real when ρ is a pure imaginary, and remains finite when $w_0=0$, provided ρ becomes infinite in such a way that $w_0\cosh\left(\rho/\mathrm{R}\right)$ remains finite. In the latter case the equation represents a limit-line. In the former case, by giving different values to C, we obtain concentric circles with the imaginary centre u_0v_0. One of these, obtained by putting C = 0, is the straight line $a^2-uu_0-vv_0=0$. Hence the others are each throughout at a constant distance from this line. (It may be shown that all motions in a hyperbolic plane consist, in a general sense, of rotations; but three types must be distinguished according as the centre is real, imaginary or at infinity. All points describe, accordingly, one of the three types of circles.)

The above Euclidean interpretation fails for three or more dimensions. In the *Teoria fondamentale*, accordingly, where n dimensions are considered, Beltrami treats hyperbolic space in a purely analytical spirit. The paper shows that Lobatchewsky's space of any number of dimensions has, in Riemann's sense, a constant negative measure of curvature. Beltrami starts with the formula (analogous to that of the *Saggio*)

$$ds^2=\mathrm{R}^2x^{-2}\left(dx^2+dx_1{}^2+dx_2{}^2+\ \dots\ +dx_n{}^2\right)$$

where

$$x^2+x_1{}^2+x_2{}^2+\ \dots\ +x_n{}^2=a^2.$$

He shows that geodesics are represented by linear equations between $x_1, x_2, \dots, x_n$, and that the geodesic distance ρ between two points x and x' is given by

$$\cosh\frac{\rho}{\mathrm{R}}=\frac{a^2-x_1x_1'-x_2x_2'-\dots-x_nx_n'}{\left\{\left(a^2-x_1{}^2-x_2{}^2-\dots-x_n{}^2\right)\left(a^2-x_1'^2-x_2'^2-\dots-x_n'^2\right)\right\}^{\frac{1}{2}}}$$

(a formula practically identical with Cayley's, though obtained by a very different method). In order to show that the measure of curvature is constant, we make the substitutions

$$x_1=r\lambda_1,\ x_2=r\lambda_2\ \dots\ x_n=r\lambda_n,\ \text{where } \Sigma\lambda^2=1.$$

Hence

$$ds^2=\left(\mathrm{R}a\,dr/\left(a^2-r^2\right)\right)^2+\mathrm{R}^2r^2d\Lambda^2/\left(a^2-r^2\right).$$

where

$$d\Lambda^2=\Sigma d\lambda^2.$$

Also calling ρ the geodesic distance from the origin, we have

$$\cosh\left(\rho/\mathrm{R}\right)=\frac{a}{\sqrt{\left(a^2-r^2\right)}},\ \sinh\left(\rho/\mathrm{R}\right)=\frac{r}{\sqrt{\left(a^2-r^2\right)}}.$$

Hence

$$ds^2=d\rho^2+\left(\mathrm{R}\sinh\left(\rho/\mathrm{R}\right)\right)^2d\Lambda^2.$$

Putting

$$z_1=\rho\lambda_1,\ z_2=\rho\lambda_2,\ \dots\ z_n=\rho\lambda_n,$$

we obtain

$$ds^2 = \Sigma dz^2 + \frac{1}{\rho^2}\left\{\left(\frac{R}{\rho}\sinh\frac{\rho}{R}\right)^2 - 1\right\}\Sigma(z_i dz_k - z_k dz_i)^2.$$

Hence when ρ is small, we have approximately

$$ds^2 = \Sigma dz^2 + \frac{1}{3R^2}\Sigma\left(z_i dz_k - z_k dz_i\right)^2 \qquad (1).$$

Considering a surface element through the origin, we may choose our axes so that, for this element,

$$z_3 = z_4 = \ldots = z_n = 0.$$

Thus $$ds^2 = dz_1{}^2 + dz_2{}^2 + \frac{1}{3R^2}\left(z_1 dz_2 - z_2 dz_1\right)^2 \qquad (2).$$

Now the area of the triangle whose vertices are $(0, 0), (z_1, z_2), (dz_1, dz_2)$ is $\frac{1}{2}(z_1 dz_2 - z_2 dz_1)$. Hence the quotient when the terms of the fourth order in (2) are divided by the square of this triangle is |729a| $4/3R^2$; hence, returning to general axes, the same is the quotient when the terms of the fourth order in (1) are divided by the square of the triangle whose vertices are $(0, 0, \ldots 0), (z_1, z_2, z_3, \ldots z_n)$, $(dz_1, dz_2, dz_3 \ldots dz_n)$. But $-\frac{3}{4}$ of this quotient is defined by Riemann as the measure of curvature.[1]† Hence the measure of curvature is $-1/R^2$, *i.e.* is constant and negative. The properties of parallels, triangles, &c., are as in the *Saggio*. It is also shown that the analogues of limit surfaces have zero curvature; and that spheres of radius ρ have constant positive curvature $1/R^2 \sinh^2(\rho/R)$, so that spherical geometry may be regarded as contained in the pseudo-spherical (as Beltrami calls Lobatchewsky's system).

Transition to the Projective Method[2]

The *Saggio*, as we saw, gives a Euclidean interpretation confined to two dimensions. But a consideration of the auxiliary plane suggests a different interpretation, which may be extended to any number of dimensions. If, instead of referring to the pseudosphere, we merely *define* distance and angle, in the Euclidean plane, as those functions of the coordinates which gave us distance and angle on the pseudosphere, we find that the geometry of our plane has become Lobatchewsky's. All the points of the limiting circle are now at infinity, and points beyond it are imaginary. If we give our circle an imaginary radius the geometry on the plane becomes elliptic. Replacing the circle by a sphere, we obtain an analogous representation for three dimensions. Instead of a circle or sphere we may take any conic or quadric. With this definition, if the fundamental quadric be $\Sigma_{xx} = 0$, and if $\Sigma_{xx'}$ be the polar form of Σ_{xx}, the distance ρ between x and x' is given by the projective formula

$$\cos(\rho/k) = \Sigma_{xx'}/\{\Sigma_{xx} \cdot \Sigma_{x'x'}\}^{\frac{1}{2}}.$$

1. † Beltrami shows also that this definition agrees with that of Gauss.
2. Whitehead carried over much of the text in this section from Russell's version, but did make significant cuts at the end and added a new final sentence.

That this formula is projective is rendered evident by observing that $e^{-2i\rho/k}$ is the anharmonic ratio of the range consisting of the two points and the intersections of the line joining them with the fundamental quadric. With this we are brought to the third or projective period. The method of this period is due to Cayley; its application to previous non-Euclidean geometry is due to Klein. The projective method contains a generalization of discoveries already made by Laguerre[1†] in 1853 as regards Euclidean geometry.[2] The arbitrariness of this procedure of deriving metrical geometry from the properties of conics is removed by Lie's theory of congruence. We then arrive at the stage of thought which finds its expression in the modern treatment of the axioms of geometry.

The Two Kinds of Elliptic Space[3]

The projective method leads to a discrimination, first made by Klein,[4†] of the two varieties of Riemann's space; Klein calls these elliptic and spherical. They are also called the polar and antipodal forms of elliptic space. The latter names will here be used. The difference is strictly analogous to that between the diameters and the points of a sphere. In the polar form two straight lines in a plane always intersect in one and only one point; in the antipodal form they intersect always in two points, which are antipodes.[5] According to the definition of geometry adopted in section VII. (*Axioms of Geometry*), the antipodal form is not to be termed "geometry," since any pair of coplanar straight lines intersect each other in two points. It may be called a "quasi-geometry." Similarly in the antipodal form[6] two diameters always determine a plane, but two points on a sphere do not determine a great circle when they are antipodes, and two great circles always intersect in two points. Again, a plane does

1. † "Sur la théorie des foyers," *Nouv. Ann.* vol. xii [Edmon Laguerre-Verly, 'Sur la théorie des foyers', *Nouvelles annales de mathématiques*, Series 1, vol. 12 (1853), pp. 57–66].
2. The following sentence is a Whitehead addition to Russell's version after deleting Russell's final paragraph, which reads as follows:

   Metrical properties, in Euclidean space, are projective properties relative, in two dimensions, to the circular points and the line at infinity considered as a degenerate conic, or, in three dimensions, to the circle and the plane at infinity considered as a degenerate quadric. Substituting a proper conic, or quadric, real or imaginary, we obtain the recognized non-Euclidean results. (In three dimensions some restriction as to the kind of quadric is necessary, if the measurement of angles is to be unchanged, and if a finite rotation is to cause a return to the initial position. Tangent lines from any point to the Absolute or fundamental quadric make an infinite angle with all other lines through the point; it is therefore necessary that they should be imaginary. Hence the absolute, when expressed as the sum or difference of squares, must have all its signs positive, or three positive and one negative. This excludes the hyperboloid of one sheet.)
3. In this final section, Whitehead carried much of Russell's text over for the first paragraph, but also made significant additions and deletions. He then deleted most of the rest of Russell's version, totalling about 7,000 words, though he did pull in material from a Russell section with the heading 'Scientific evidence as to the nature of actual space' for the remainder. For more on the deleted Russell material, see p. 296n3.
4. † *Math. Annalen*, iv. vi., 1871–1873 [corrected from 1871–1872. 'Ueber die sogenannte Nicht-Euklidische Geometrie', vol. 4, no. 4 (December 1871), pp. 573–625, and vol. 6, no. 2 (June 1873), pp. 112–145].
5. The next two sentences (ending in 'quasi-geometry') are a Whitehead addition to Russell's version.
6. 'in the antipodal form' is a Whitehead addition to Russell's version.

not form a boundary among lines through a point: we can pass from any one such line to any other without passing through the plane. But a great circle does divide the surface of a sphere. So, in the polar form, a complete straight line does not divide a plane, and a plane does not divide space, and does not, like a Euclidean plane, have two sides.[1†] But, in the antipodal form, a plane is, in these respects, like a Euclidean plane.[2]

It is explained in section VII. in what sense the metrical geometry of the material world can be considered to be determinate and not a matter of arbitrary choice.[3] The scientific |729b| question as to the best available evidence concerning the nature of this geometry is one beset with difficulties of a peculiar kind.[4] We are obstructed by the fact that all existing physical science assumes the Euclidean hypothesis. This hypothesis has been involved in all actual measurements of large distances, and in all the laws of astronomy and physics. The principle of simplicity would therefore lead us, in general, where an observation conflicted with one or more of those laws, to ascribe this anomaly, not to the falsity of Euclidean geometry, but to the falsity of the laws in question. This applies especially to astronomy. On the earth our means of measurement are many and direct, and so long as no great accuracy is sought they involve few scientific laws. Thus we acquire from such direct measurements, a very high degree of probability that the space-constant, if not infinite, is yet large as compared with terrestrial distances. But astronomical distances and triangles can only be measured by means of the received laws of

1. † For an investigation of these and similar properties, see Whitehead, *Universal Algebra* (Cambridge, 1898), bk. vi., ch. ii. The polar form was independently discovered by Simon Newcomb in 1877.
2. Whitehead deleted the rest of section from Russell's version, which reads as follows:

> The polar form alone, among the various spaces, gives absolute duality between points and planes, or, in two dimensions, between points and lines. The reason of this is simple. The measurement of angles is always elliptic; among the lines or planes through a point, those that make an infinite angle with the rest are imaginary, and for real values angles are periodic. Thus it is only where the measurement of length is also elliptic that complete duality prevails.

3. This sentence is a Whitehead addition from Russell's version. At this point, after the conclusion of Russell's shorter section on 'The two kinds of elliptic space', Whitehead deleted most of the rest of Russell's version, totalling about 7,000 words. The remainder of the article is taken from Russell's penultimate section on 'Scientific evidence as to the nature of actual space' (with some sentences deleted near the beginning). At 7,000 words, the deleted material is too voluminous to appear in a footnote; readers may refer to Russell's original entry to read it (referenced in the introduction this article). However, to give a general idea of the content, the following is a listing of all the headings for the deleted material: 'Projective coordinates', 'Sophus Lie', 'Spaces not allowing free mobility in Lie's sense', 'Clifford's surface of zero curvature and finite extent', 'Three prevalent misconceptions', 'Measure of curvature', 'Motion irrelevant to Geometry', 'Relation between different space-constants', 'Projective theory of distance', 'Rigid bodies unnecessary in Geometry', 'Four stages in specifying a space', 'Projective axioms', 'Axioms of distance', 'Free mobility', 'Differences among the four principal spaces', 'Two general questions', 'Supposed à priori grounds for Euclidean space' and, finally, following the section 'Scientific evidence as to the nature of actual space', which Whitehead mostly kept, 'Philosophical value of non-Euclidean Geometry'.
4. Following this sentence, which matched the first sentence of Russell's version for the section 'Scientific evidence as to the nature of actual space', Whitehead deleted the following from Russell's version:

> Admitting non-Euclidean spaces to be possible, we must also admit, since Euclidean space is merely a limiting case to which other spaces may indefinitely approximate, that the space we live in, though approximately Euclidean, may be found, on closer examination, to be of one or other of the non-Euclidean types. But in making this closer examination, we...

Whitehead's version capitalised the 'we' and picked up from there.

astronomy and optics, all of which have been established by assuming the truth of the Euclidean hypothesis. It therefore remains possible (until a detailed proof of the contrary is forthcoming) that a large but finite space-constant, with different laws of astronomy and optics, would have equally explained the phenomena. We cannot, therefore, accept the measurements of stellar parallaxes, &c., as conclusive evidence that the space-constant is large as compared with stellar distances. For the present, on grounds of simplicity, we may rightly adopt this view; but it must remain possible that, in view of some hitherto undiscovered discrepancy, a slight correction of the sort suggested might prove the simplest alternative. But conversely, a finite parallax for very distant stars, or a negative parallax for any star, could not be accepted as conclusive evidence that our geometry is non-Euclidean, unless it were shown—and this seems scarcely possible—that no modification of astronomy or optics could account for the phenomenon. Thus although we may admit a probability that the space-constant is large in comparison with stellar distances, a conclusive proof or disproof seems scarcely possible.

Finally, it is of interest to note that, though it is theoretically possible to prove, by scientific methods, that our geometry is non-Euclidean, it is wholly impossible to prove by such methods that it is accurately Euclidean. For the unavoidable errors of observation must always leave a slight margin in our measurements. A triangle might be found whose angles were certainly greater, or certainly less, than two right angles; but to prove them *exactly* equal to two right angles must always be beyond our powers. If, therefore, any man cherishes a hope of proving the exact truth of Euclid, such a hope must be based, not upon scientific, but upon philosophical considerations.

Bibliography.—The bibliography appended to section VII,[1] should be consulted in this connexion. Also, in addition to the citations already made, the following works may be mentioned:

For Lobatchewsky's writings, cf. *Urkunden zur Geschichte der nichteuklidischen Geometrie*, i., *Nikolaj Iwanowitsch Lobatschefsky*, by F. Engel and P. Stäckel (Leipzig, 1898). For John Bolyai's *Appendix*, cf. *Absolute Geometrie nach Johann Bolyai*, by J. Frischauf (Leipzig, 1872), and also the new edition of his father's large work, *Tentamen . . .*, published by the Mathematical Society of Budapest; the second volume contains the appendix. Cf. also J. Frischauf, *Elemente der absoluten Geometrie* (Leipzig, 1876); M. L. Gérard, *Sur la géométrie non-Euclidienne* (thesis for doctorate) (Paris, 1892); de Tilly, *Essai sur les principes fondamentales de la géométrie et de la mécanique* (Bordeaux, 1879); Sir R. S. Ball, "On the Theory of Content," *Trans. Roy. Irish Acad.* vol. xxix. (1889);[2] F. Lindemann, "Mechanik bei projectiver Maasbestimmung," *Math. Annal.* vol. vii;[3] W. K. Clifford, "Preliminary Sketch of Biquaternions,"

1. In the version of the article in ESP, 'section VII' was changed to 'Axioms of Geometry'. ESP did not correct other references to 'section VII' within the article. See p. 299.
2. pp. 123–182.
3. Whitehead appears to have the wrong title, as the only article by Lindemann in the cited journal is: 'Ueber unendlich kleine Bewegungen und über Kraftsysteme bei allgemeiner projectivischer Massbestimmung', *Mathematische Annalen*, vol. 7, no. 1 (March 1873), pp. 56–143.

Proc. of Lond. Math. Soc. (1871),[1] and *Coll. Works*;[2] A. Buchheim, "On the Theory of Screws in Elliptic Space," *Proc. Lond. Math. Soc.* vols. xv., xvi., xvii.;[3] H. Cox, "On the Application of Quaternions and Grassmann's Algebra to different Kinds of Uniform Space," *Trans. Camb. Phil. Soc.* (1882);[4] M. Dehn, "Die Legendarischen Sätze über die Winkelsumme in Dreieck," *Math. Ann.* vol. 53 (1900),[5] and "Über den Rauminhalt," *Math. Annal.* vol. 55 (1901).[6]

For expositions of the whole subject, cf. F. Klein, *Nicht-Euklidische Geometrie* (Göttingen, 1893); R. Bonola, *La Geometria non-Euclidea* (Bologna, 1906); P. Barbarin, *La Géométrie non-Euclidienne* (Paris, 1902); W. Killing, *Die nicht-Euklidischen Raumformen in analytischer Behandlung* (Leipzig, 1885). The last-named work also deals with geometry of more than three dimensions; in this connexion cf. also G. Veronese, *Fondamenti di geometria a più dimensioni ed a più specie* |730a| *di unità rettilinee* ... (Padua, 1891, German translation, Leipzig, 1894);[7] G. Fontené, *L'Hyperespace à (n - 1) dimensions* (Paris, 1892); and A. N. Whitehead, *loc. cit.* Cf. also E. Study, "Über nicht-Euklidische and Liniengeometrie," *Jahr. d. Deutsch. Math. Ver.* vol. xv. (1906);[8] W. Burnside, "On the Kinematics of non-Euclidean Space," *Proc. Lond. Math. Soc.* vol. xxvi. (1894).[9] A bibliography on the subject up to 1878 has been published by G. B. Halsted, *Amer. Journ. of Math.* vols. i. and ii.;[10] and one up to 1900 by R. Bonola, *Index operum ad geometriam absolutam spectantium* . . . (1902, and Leipzig, 1903).

1. Date corrected from 1873. Series 1, vol. 4, no. 1 (November), pp. 381–395.
2. William Kingdon Clifford, *Lectures and Essays*, edited by Leslie Stephen and Frederick Pollock (London: Macmillan, 1879).
3. 'On the Theory of Screws in Elliptic Space', vol. 15 (November 1883), pp. 83–98; 'On the Theory of Screws in Elliptic Space (Supplementary Note)', vol. 16 (November 1884), pp. 15–27; 'On the Theory of Screws in Elliptic Space (Third Note)', vol. 17 (November 1885), pp. 240–254.
4. Vol. 13, pp. 69–143.
5. No. 3 (September), pp. 404–439.
6. Corrected from 1902. No. 3 (September), pp. 465–478.
7. As *Grundzüge der Geometrie*, translated by Adolf Schepp.
8. pp. 476–527.
9. No. 1 (November), pp. 33–58.
10. 'Bibliography of Hyper-Space and Non-Euclidean Geometry', vol. 1, no. 3 (1878), pp. 261–276; 'Addenda to Mr. Halsted's Paper on the Bibliography of Hyper-Space and Non-Euclidean Geometry', vol. 1, no. 4 (1878), pp. 384–385; 'Addenda to Bibliography of Hyper-Space and Non-Euclidean Geometry', vol. 2, no. 1 (March 1879), pp. 65–70.

Axioms of Geometry

1910 **Hugh Chisholm (ed.), *Encyclopaedia Britannica*, 11th edition, vol. 11 (Cambridge: Cambridge University Press), pp. 730–736**

1947 *Essays in Science and Philosophy* (New York: Philosophical Library), pp. 243–268

This Encyclopaedia *entry naturally has a feel that is distinct from that of a research article. Indeed, the topic and its components are worthy of inclusion precisely because they are paramount to the mathematics of the day, and yet the space afforded Whitehead was severely limited. Consequently, such articles in technical areas are typically dense both conceptually and typographically despite our understandable tendency to think that they will be far less demanding. Thus, the density of Whitehead's exposition is par for the course, and so our introductory summary can act only as a roadmap to its contents, with emphasis on its historical context, looking both forward and backward. To understand the vast importance of the nexus of ideas covered in this brief piece, we begin with a concise review of the transformation of geometry underway at the time that Whitehead composed it – and, to judge from his writings on the subject, a transformation of which Whitehead was fully aware.*

In the academic year 1973–1974, Prof. Lipman Bers, one of the top-tier mathematicians of the twentieth century, casually remarked in a lecture to first-year graduate students at Columbia that the reason we did not continue to study geometry in the Euclidean style at university was that it had been subsumed by analytic methods. This was not to say that older, synthetic methods were not valuable and hugely necessary to the advancement of geometry; certainly, the present-day mathematician finds much to lament in this displacement by analysis, as did Whitehead. Prof. Bers was merely remarking on the trajectory of mathematics in the twentieth century. Moreover, to unpack and to interpret this statement, one might say that he was not talking so much about the theorems of Euclidean geometry as about its foundations in its axioms and postulates. These have a well-known, vexing history owing to the difficulty in disentangling the formalisms from the obvious semantics: we know what a circle is long before we can define it, and this intuitive knowledge can sometimes contaminate and invalidate a formal proof. The professor's point was, to our best understanding, that these foundational issues were ultimately resolved with a proper definition of the system of real numbers, from which its more subtle and critical properties could be deduced. This was achieved in two stages, beginning with a seminal idea by Dedekind that matured into an algebraic and analytic construction of the real numbers with a far richer potential for generalisation. (We are speaking, respectively, of the Dedekind cuts mentioned in Whitehead's article and the subsequent notion of equivalence classes of Cauchy sequences.) Once the system of real numbers was well grounded, the business of geometry could be conducted with its foundations secured relative to that system.

Its primary objects – lines, angles, line segments, polygons and conic sections – were then identified with subsets of spaces built directly from the real numbers. With this development, the millennia-long quest for an efficient, comprehensive and coherent set of axioms all but disappears. Of course, something is lost, too, in abandoning the intuitive quest for the meaning of fundamental geometric notions without recourse to so topologically complex an object as the real numbers, and one might wonder if that will someday once again become the accustomed style.

The great charm of Whitehead's entry 'Axioms of Geometry' lies in its capture of the precise period in which the transition described above was taking place. Whitehead combines his vast and profound understanding of the history of geometry and its subdisciplines with his fluent knowledge of current methods to yield a dense summary that moves most of the way from the axiomatisations of the semantics of geometry to the point where such axioms permit us to introduce coordinate systems based on the real numbers. Moreover, he makes clear the essentially decisive topological distinction that the real numbers enjoy over the rational numbers, the notion of completeness *(although he does not use this term, but rather refers to the* Dedekind property*). What is so striking here for later mathematicians is how that crucial property applies in geometry. (Mathematics, like science, tends, as Thomas Kuhn famously noted, to obliterate its technical history with every advance.)*

Whitehead's article moves through an introductory section, followed by more detailed considerations of projective geometry, descriptive geometry, their mutual connection and then what Whitehead calls metrical geometry, *which is founded upon the work of Sophus Lie. The development of projective and descriptive geometries largely follows his books* The Axioms of Projective Geometry *(1906) and* The Axioms of Descriptive Geometry *(1907). The last topic, involving distances, is by far the most demanding mathematically, and the term* metrical geometry *has fallen out of use in the way that Whitehead intends here.*

The concluding paragraphs are somewhat speculative on the relationship between geometry and the physical reality of space, and Whitehead makes two significant assertions. First, he declares the priority of space itself over its subsequent identity with numerical coordinates. Thus, he would resist our present-day commonplace assertion that the set of all ordered pairs of real numbers in fact constitutes the Euclidean plane. Second, he claims that the most natural association of coordinates and distances with physical space is a question subject to interrogation by science, as opposed to an arbitrary convention. The relationship between geometry as a branch of mathematics and some particular species of geometry as a natural descriptor of the real world was certainly in play at the turn of the twentieth century, and Whitehead's position on the philosophical criteria for selection of a plausible geometry would evolve. Nonetheless, coming five years before Einstein's theory of general relativity, these assertions are remarkably noteworthy. In The Principle of Relativity *(1922, see especially chapter III), Whitehead objects to Einstein's mass-driven space-time metric as, in essence, an indefensible violation of what he then considered an essentially metaphysical requirement. Moreover, an actual interrogation of the facts of the world would later show that Whitehead's alternative approach to gravitation fails where Einstein's succeeds. This does not, however, invalidate the cogency of his objection to Einstein's formulation but certainly does leave room for further thought.*

The version reprinted in Essays in Science and Philosophy *removed all the figures from the article and all references to figures, but otherwise left the article virtually unchanged.*

Theories of Space[1]

|730a| Until the discovery of the non-Euclidean geometries (Lobatchewsky, 1826 and 1829;[2] J. Bolyai, 1832;[3] B. Riemann, 1854),[4] geometry was universally considered as being exclusively the science of existent space. (See section VI. *Non-Euclidean Geometry.*) In respect to the science, as thus conceived, two controversies may be noticed. First, there is the controversy respecting the absolute and relational theories of space. According to the absolute theory, which is the traditional view (held explicitly by Newton), space has an existence, in some sense whatever it may be, independent of the bodies which it contains. The bodies occupy space, and it is not intrinsically unmeaning to say that any definite body occupies *this* part of space, and not *that* part of space, without reference to other bodies occupying space. According to the relational theory of space, of which the chief exponent was Leibnitz,[5†] space is nothing but a certain assemblage of the relations between the various particular bodies in space. The idea of space with no bodies in it is absurd. Accordingly there can be no meaning in saying that a body is *here* and not *there*, apart from a reference to the other bodies in the universe. Thus, on this theory, absolute motion

---

1. The original article handled headings oddly; they appeared floating in the outer margin with text wrapping around them, two lines below where they should properly appear. These marginal headings have been moved to the more conventional placement used in ESP.
2. Lobatchewski's (1792–1856) name is now usually rendered as Nikolai *Lobachevsky*. Lobachevsky presented a paper in French titled 'Exposition succincte des principes de la Géométrie, avec une démonstration rigoureuse du théorème des parallèles' ('A Brief Exposition of the Principles of Geometry, with a Rigorous Demonstration of the Theorem of Parallels') to the Section for physico-mathematical sciences at his home university of Kazan in 1826, but this was never published. In 1829 parts of this unpublished manuscript were used in Lobachevsky's article 'О началах геометрии' ('On the Origin of Geometry') in the *Kazan Herald*. See George Bruce Halsted, '*Urkunden zur Geschichte der Nichteuklidischen Geometrie* Von F. Engel und P. Staeckel', *American Mathematical Monthly*, vol. 6, nos 6–7 (June–July 1899), pp. 166–172.
3. The referenced 1832 work of János Bolyai (1802–1860) was actually printed as an appendix to a book by his father: Farkas Bolyai, *Tentamen juventutem studiosam in elementa matheseos purae, elementaris ac sublimioris, methodo intuitiva, evidentiaque huic propria, introducendi* (*'An Attempt to Introduce Studious Youths to the Elements of Pure Mathematics'*). The appendix was titled 'Appendix scientiam spatii absolute veram exhibens; a veritate aut falsitate axiomatis XI Euclidei (a priori haud unquam decidenda)'. Halsted translated the appendix into English and published it as *The Science Absolute of Space: Independent of the Truth or Falsity of Euclid's Axiom XI (Which Can Never Be Decided* A Priori*)* (Austin: The Neomon, 1896).
4. Bernhard Riemann (1826–1866) delivered a lecture at the University of Göttingen entitled 'Ueber die Hypothesen, welche der Geometrie zu Grunde liegen' ('On the Hypotheses Which Lie at the Bases of Geometry') in 1854. It was not published until two years after Riemann's death, by Richard Dedekind (in *Abhandlungen der Königlichen Gesellschaft der Wissenschaften zu Göttingen*, vol. 13 (1868), pp. 133–150). It was translated into English by William Kingdon Clifford in 1873 (*Nature*, vol. 8, nos. 183–184, pp. 14–17, 36–37).
5. † For an analysis of Leibnitz's ideas on space, cf. B. Russell, *A Critical Exposition of the Philosophy of Leibniz*, chs. viii.–x [Cambridge: Cambridge University Press, 1903]. [The original erroneously had the title as '*The Philosophy of Leibnitz*'.]

is intrinsically unmeaning. It is admitted on all hands that in practice only relative motion is directly measurable. Newton, however, maintains in the *Principia* (scholium to the 8th definition) that it is indirectly measurable by means of the effects of "centrifugal force" as it occurs in the phenomena of rotation. This irrelevance of absolute motion (if there be such a thing) to science has led to the general adoption of the relational theory by modern men of science. But no decisive argument for either view has at present been elaborated.[1 †] Kant's view of space as being a form of perception at first sight appears to cut across this controversy. But he, saturated as he was with the spirit of the Newtonian physics, must (at least in both editions of the *Critique*) be classed with the upholders of the absolute theory. The form of perception has a type of existence proper to itself independently of the particular bodies which it contains. For example he writes:[2†] "Space does not represent any quality of objects by themselves, or objects in their relation to one another, *i.e.* space does not represent any determination which is inherent in the objects themselves, and would remain, even if all subjective conditions of intuition were removed."

Axioms

The second controversy is that between the view that the axioms applicable to space are known only from experience, and the view that in some sense these axioms are given *a priori*. Both these views, thus broadly stated, are capable of various subtle modifications, and a discussion of them would merge into a general treatise on epistemology. The cruder forms of the *a priori* view have been made quite untenable by the modern mathematical discoveries. Geometers now profess ignorance in many respects of the exact axioms which apply to existent space, and it seems unlikely that a profound study of the question should thus obliterate *a priori* intuitions.

Another question irrelevant to this article, but with some relevance to the above controversy, is that of the derivation |730b| of our perception of existent space from our various types of sensation. This is a question for psychology.[3†]

Definition of Abstract Geometry.—Existent space is the subject matter of only one of the applications of the modern science of abstract geometry, viewed as a branch of pure mathematics. Geometry has been defined[4†] as "the study of series of two or more dimensions." It has also been defined[5†] as "the science of cross classification." These definitions are founded upon the actual practice of mathematicians in respect to their use of the term "Geometry." Either of them brings out the fact that geometry

1. † Cf. Hon. Bertrand Russell, "Is Position in Time and Space Absolute or Relative?" *Mind*, n.s. vol. 10 (1901) [no. 39, pp. 293–317], and A. N. Whitehead, "Mathematical Concepts of the Material World," *Phil. Trans.* (1906), Vol. 205 [corrected from 'p. 205'; see p. 214].
2. † Cf. *Critique of Pure Reason*, 1st section; "Of Space," conclusion A, Max Müller's translation [London: Macmillan, 1881].
3. † Cf. Ernst Mach, *Erkenntniss und Irrtum* (Leipzig); the relevant chapters are translated by T. J. McCormack, *Space and Geometry* (London, 1906); also A. Meinong, *Über die Stellung der Gegenstandstheorie im System der Wissenschaften* (Leipzig, 1907).
4. † Cf. Russell, *Principles of Mathematics*, § 352 (Cambridge, 1903).
5. † Cf. A. N. Whitehead, *The Axioms of Projective Geometry*, § 3 (Cambridge, 1906).

is not a science with a determinate subject matter. It is concerned with any subject matter to which the formal axioms may apply. Geometry is not peculiar in this respect. All branches of pure mathematics deal merely with types of relations. Thus the fundamental ideas of geometry (*e.g.* those of *points* and of *straight lines*) are not ideas of determinate entities, but of any entities for which the axioms are true. And a set of formal geometrical axioms cannot in themselves be true or false, since they are not determinate propositions, in that they do not refer to a determinate subject matter. The axioms are propositional functions.[1†] When a set of axioms is given, we can ask (1) whether they are consistent, (2) whether their "existence theorem" is proved, (3) whether they are independent. Axioms are consistent when the contradictory of any axiom cannot be deduced from the remaining axioms. Their existence theorem is the proof that they are true when the fundamental ideas are considered as denoting some determinate subject matter, so that the axioms are developed into determinate propositions. It follows from the logical law of contradiction that the proof of the existence theorem proves also the consistency of the axioms. This is the only method of proof of consistency. The axioms of a set are independent of each other when no axiom can be deduced from the remaining axioms of the set. The independence of a given axiom is proved by establishing the consistency of the remaining axioms of the set, together with the contradictory of the given axiom. The enumeration of the axioms is simply the enumeration of the hypotheses[2†] (with respect to the undetermined subject matter) of which some at least occur in each of the subsequent propositions.

Any science is called a "geometry" if it investigates the theory of the classification of a set of entities (the points) into classes (the straight lines), such that (1) there is one and only one class which contains any given pair of entities, and (2) every such class contains more than two members. In the two geometries, important from their relevance to existent space, axioms which secure an order of the points on any line also occur. These geometries will be called "Projective Geometry" and "Descriptive Geometry." In projective geometry any two straight lines in a plane intersect, and the straight lines are closed series which return into themselves, like the circumference of a circle. In descriptive geometry two straight lines in a plane do not necessarily intersect, and a straight line is an open series without beginning or end. Ordinary Euclidean geometry is a descriptive geometry; it becomes a projective geometry when the so-called "points at infinity" are added.

Projective Geometry.

Projective geometry may be developed from two undefined fundamental ideas, namely, that of a "point" and that of a "straight line." These undetermined ideas take different specific meanings for the various specific subject matters to which projective geometry can be applied. The number of the axioms is always to some extent arbitrary, being dependent upon the verbal forms of statement which are adopted.

---

1. †Cf. Russell, *Princ. of Math.*, ch. i.
2. †Cf. Russell, *loc. cit.*, and G. Frege, "Über die Grundlagen der Geometrie," *Jahresber. der Deutsch. Math. Ver.* (1906) [*Jahresbericht Der Deutschen Mathematiker-Vereinigung*, vol. 15, pp. 293–309, 377–402, 423–430].

They will |731a| be presented[1†] here as twelve in number, eight being "axioms of classification," and four being "axioms of order."

Axioms of Classification.—The eight axioms of classification are as follows:

1. Points form a class of entities with at least two members.
2. Any straight line is a class of points containing at least three members.
3. Any two distinct points lie in one and only one straight line.
4. There is at least one straight line which does not contain all the points.
5. If A, B, C are non-collinear points, and A′ is on the straight line BC, and B′ is on the straight line CA, then the straight lines AA′ and BB′ possess a point in common.

Definition.—If A, B, C are any three non-collinear points, the *plane* ABC is the class of points lying on the straight lines joining A with the various points on the straight line BC.

6. There is at least one plane which does not contain all the points.
7. There exists a plane α, and a point A not incident in α, such that any point lies in some straight line which contains both A and a point in α.

Definition.—Harm. (ABCD) symbolizes the following conjoint statements: (1) that the points A, B, C, D are collinear, and (2) that a quadrilateral can be found with one pair of opposite sides intersecting at A, with the other pair intersecting at C, and with its diagonals passing through B and D respectively. Then B and D are said to be "harmonic conjugates" with respect to A and C.

8. Harm. (ABCD) implies that B and D are distinct points.

In the above axioms 4 secures at least two dimensions, axiom 5 is the fundamental axiom of the plane, axiom 6 secures at least three dimensions, and axiom 7 secures at most three dimensions. From axioms 1–5 it can be proved that any two distinct points in a straight line determine that line, that any three non-collinear points in a plane determine that plane, that the straight line containing any two points in a plane lies wholly in that plane, and that any two straight lines in a plane intersect. From axioms 1–6 Desargues's[2] well-known theorem on triangles in perspective can be proved.

The enunciation of this theorem is as follows: If ABC and A′B′C′ are two coplanar triangles such that the lines AA′, BB′, CC′ are concurrent, then the three points of intersection of BC and B′C′,[3] of CA and C′A′, and of AB and A′B′ are collinear; and conversely if the three points of intersection are collinear, the three lines are concurrent. The proof which can be applied is the usual projective proof by which a third triangle A″B″C″ is constructed not coplanar with the other two, but in perspective with each of them.

It has been proved[4†] that Desargues's theorem cannot be deduced from axioms 1–5, that is, if the geometry be confined to two dimensions. All the proofs proceed

1. †This formulation—though not in respect to number—is in all essentials that of M. Pieri, cf. "I principii della Geometria di Posizione," *Accad. R. di Torino* (1899) [vol. 48, no. 2, pp. 1–62; date corrected from 1898]; also cf. Whitehead, *loc. cit.*
2. Girard Desargues (1591–1661) was a French mathematician who was one of the founders of projective geometry.
3. Supplied missing comma.
4. †Cf. G. Peano, "Sui fondamenti della Geometria," p. 73, *Rivista di matematica*, vol. iv. (1894) [articlespans pp. 51–90], and D. Hilbert, *Grundlagen der Geometrie* (Leipzig, 1899) [translated by E.

by the method of producing a specification of "points" and "straight lines" which satisfies axioms 1–5, and such that Desargues's theorem does not hold.

It follows from axioms 1–5 that Harm. (ABCD) implies Harm. (ADCB) and Harm. (CBAD), and that, if A, B, C be any three distinct collinear points, there exists at least one point D such that Harm. (ABCD). But it requires Desargues's theorem, and hence axiom 6, to prove that Harm. (ABCD) and Harm. (ABCD′) imply the identity of D and D′.

The necessity for axiom 8 has been proved by G. Fano,[1†] who has produced a three dimensional geometry of fifteen points, *i.e.* a method of cross classification of fifteen entities, in which each straight line contains three points, and each plane contains seven straight lines. In this geometry axiom 8 does not hold. Also from axioms 1–6 and 8 it follows that Harm. (ABCD) implies Harm. (BCDA).

Definitions.—When two plane figures can be derived from one another by a single projection, they are said to be in *perspective.* When two plane figures can be derived one from the other by a finite series of perspective relations between intermediate figures, they |731b| are said to be *projectively* related. Any property of a plane figure which necessarily also belongs to any projectively related figure, is called a *projective* property.

The following theorem, known from its importance as "the fundamental theorem of projective geometry," cannot be proved[2†] from axioms 1–8. The enunciation is: "A projective correspondence between the points on two straight lines is completely determined when the correspondents of three distinct points on one line are determined on the other." This theorem is equivalent[3†] (assuming axioms 1–8) to another theorem, known as Pappus's Theorem, namely: "If *l* and *l′* are two distinct coplanar lines, and A, B, C are three distinct points on *l*, and A′, B′, C′ are three distinct points on *l′*, then the three points of intersection AA′ and B′C, of A′B and CC′, of BB′ and C′A, are collinear." This theorem is obviously Pascal's well-known theorem respecting a hexagon inscribed in a conic, for the special case when the conic has degenerated into the two lines *l* and *l′*. Another theorem also equivalent (assuming axioms 1–8) to the fundamental theorem is the following:[4†] If the three collinear pairs of points, A and A′, B and B′, C and C′, are such that the three pairs of opposite sides of a complete quadrangle pass respectively through them, *i.e.* one pair through A and

J. Townsend as *The Foundations of Geometry* (Chicago: Open Court, 1902)]; and F. R. Moulton, "A Simple non-Desarguesian Plane Geometry," *Trans. Amer. Math. Soc.*, vol. iii. (1902). [The original citation had 'R. F. Moulton', reversing the author's first and middle names. Full citation: Forest Ray Moulton, 'A Simple Non-Desarguesian Plane Geometry', *Transactions of the American Mathematical Society*, vol. 3, no. 2 (April, 1902), pp. 192–195.]

1. † Cf. "Sui postulati fondamentali della geometria projettiva," *Giorn. di matematica*, vol. xxx. (1891) [The cited paper was actually published in 1892, but the author dated it 'August 1891' at the bottom. Full citation: Gino Fano, 'Sui postulati fondamentali della geometria proiettiva in uno spazio lineare a un numero qualunque di dimensioni', *Giornale di matematiche di Battaglini*, vol. 30 (1892), pp. 106–132]; also of Pieri, *loc. cit.*, and Whitehead, *loc. cit.*
2. † Cf. Hilbert, *loc. cit.*; for a fuller exposition of Hilbert's proof cf. K. T. Vahlen, *Abstrakte Geometrie* (Leipzig, 1905), also Whitehead, *loc. cit.*
3. † Cf. H. Wiener, *Jahresber. der Deutsch. Math. Ver.* vol. i. (1890) ['Ueber Grundlagen und Aufbau der Geometrie', pp. 45–48]; and F. Schur, "Über den Fundamentalsatz der projectiven Geometric," *Math. Ann.* vol. li. (1898) [date corrected from 1899; no. 3 (September), pp. 401–409].
4. † Cf. Hilbert, *loc. cit.*, and Whitehead, *loc. cit.*

A′ respectively, and so on, and if also the three sides of the quadrangle which pass through A, B, and C, are concurrent in one of the corners of the quadrangle, then another quadrangle can be found with the same relation to the three pairs of points, except that its three sides which pass through A, B, and C, are not concurrent.

Thus, if we choose to take any one of these three theorems as an axiom, all the theorems of projective geometry which do not require ordinal or metrical ideas for their enunciation can be proved. Also a conic can be defined as the locus of the points found by the usual construction, based on Pascal's theorem, for points on the conic through five given points. But it is unnecessary to assume here any one of the suggested axioms; for the fundamental theorem can be deduced from the axioms of order together with axioms 1-8.

Axioms of Order.—It is possible to define (cf. Pieri, *loc. cit.*) the property upon which the order of points on a straight line depends. But to secure that this property does in fact range the points in a serial order, some axioms are required. A straight line is to be a closed series; thus, when the points are in order, it requires two points on the line to divide it into two distinct complementary segments, which do not overlap, and together form the whole line. Accordingly the problem of the definition of order reduces itself to the definition of these two segments formed by any two points on the line; and the axioms are stated relatively to these segments.

Definition.—If A, B, C are three collinear points, the points on the *segment ABC* are defined to be those points such as X, for which there exist two points Y and Y′ with the property that Harm. (AYCY′) and Harm. (BYXY′) both hold. The *supplementary segment ABC* is defined to be the rest of the points on the line. This definition is elucidated by noticing that with our ordinary geometrical ideas, if B and X are any two points between A and C, then the two pairs of points, A and C, B and X, define an involution with real double points, namely, the Y and Y′ of the above definition. The property of belonging to a segment ABC is projective, since the harmonic relation is projective.

The first three axioms of order (cf. Pieri, *loc. cit.*) are:

9. If A, B, C are three distinct collinear points, the supplementary segment ABC is contained within the segment BCA.

10. If A, B, C are three distinct collinear points, the common part of the segments BCA and CAB is contained in the supplementary segment ABC.

11. If A, B, C are three distinct collinear points, and D lies in the segment ABC, then the segment ADC is contained within the segment ABC.

From these axioms all the usual properties of a closed order follow. It will be noticed that, if A, B, C are any three collinear points, C is necessarily traversed in passing from A to B by one route along the line, and is not traversed in passing from A to B along the other route. Thus there is no meaning, as referred to closed straight lines, in the simple statement that C lies between A and B. But there may be a relation of separation between two pairs of collinear points, such as A and C, and B and D. The couple B and D is said to separate A and C, if |732a| the four points are collinear and D lies[1] in the segment complementary to the segment ABC. The property

1. The original erroneously omitted the 'D', which left the clause with no subject and rendered the sentence unintelligible.

of the separation of pairs of points by pairs of points is projective. Also it can be proved that Harm. (ABCD) implies that B and D separate A and C.

Definitions.—A series of entities arranged in a serial order, open or closed, is said to be *compact,* if the series contains no immediately consecutive entities, so that in traversing the series from any one entity to any other entity it is necessary to pass through entities distinct from either. It was the merit of R. Dedekind and of G. Cantor explicitly to formulate another fundamental property of series. The Dedekind property[1†] as applied to an open series can be defined thus: An open series possesses the Dedekind property, if, however, it be divided into two mutually exclusive classes *u* and *v*, which (1) contain between them the whole series, and (2) are such that every member of *u* precedes in the serial order every member of *v*, there is always a member of the series, belonging to one of the two, *u* or *v*, which precedes every member of *v* (other than itself if it belong to *v*), and also succeeds every member of *u* (other than itself if it belong to *u*). Accordingly in an open series with the Dedekind property there is always a member of the series marking the junction of two classes such as *u* and *v*. An open series is *continuous* if it is compact and possesses the Dedekind property. A closed series can always be transformed into an open series by taking any arbitrary member as the first term and by taking one of the two ways round as the ascending order of the series. Thus the definitions of compactness and of the Dedekind property can be at once transferred to a closed series.

12. The last axiom of order is that there exists at least one straight line for which the point order possesses the Dedekind property.

It follows from axioms 1–12 by projection that the Dedekind property is true for all lines. Again the *harmonic system ABC*, where A, B, C are collinear points, is defined[2†] thus: take the harmonic conjugates A′, B′, C′ of each point with respect to the other two, again take the harmonic conjugates of each of the six points A, B, C, A′, B′, C′ with respect to each pair of the remaining five, and proceed in this way by an unending series of steps. The set of points thus obtained is called the harmonic system ABC. It can be proved that a harmonic system is compact, and that every segment of the line containing it possesses members of it. Furthermore, it is easy to prove that the fundamental theorem holds for harmonic systems, in the sense that, if A, B, C are three points on a line *l*, and A′, B′, C′ are three points on a line *l′*, and if by any two distinct series of projections, A, B, C are projected into A′, B′, C′, then any point of the harmonic system ABC corresponds to the same point of the harmonic system A′B′C′ according to both the projective relations which are thus established between *l* and *l′*. It now follows immediately that the fundamental theorem must hold for all the points on the lines *l* and *l′*, since (as has been pointed out) harmonic systems are "everywhere dense" on their containing lines. Thus the fundamental theorem follows from the axioms of order.

A system of numerical coordinates can now be introduced, possessing the property that linear equations represent planes and straight lines. The outline of the argument by which this remarkable problem (in that "distance" is as yet undefined) is solved, will now be given. It is first proved that the points on any line can in a

1. †Cf. Dedekind, *Stetigkeit und irrationale Zahlen* (1872) [Braunschweig: Friedrich Vieweg und Sohn].
2. †Cf. v. Staudt, *Geometrie der Lage* (1847) [Nürnberg: F. Corn].

certain way be definitely associated with all the positive and negative real numbers, so as to form with them a one-one correspondence. The arbitrary elements in the establishment of this relation are the points on the line associated with 0, 1 and ∞

This association[1]† is most easily effected by considering a class of projective relations of the line with itself, called by F. Schur (*loc. cit.*) *prospectivities.*

Let *l* (fig. 69)[2] be the given line, *m* and *n* any two lines intersecting at U on *l*, S and S′ two points on *n*. Then a projective relation between *l* and itself is formed by projecting *l* from S on to *m*, and then by projecting *m* from S′ back on to *l*. All such projective |732b| relations, however *m*, *n*, S and S′ be varied, are called "prospectivities," and U is the double point of the prospectivity. If a point O on *l* is related to A by a prospectivity, then all prospectivities, which (1) have the same double point U, and (2) relate O to A, give the same correspondent (Q, in figure) to any point P on the line *l*; in fact they are all the same prospectivity, however *m*, *n*, S, and S′ may have been varied subject to these conditions. Such a prospectivity will be denoted by (OAU^2).

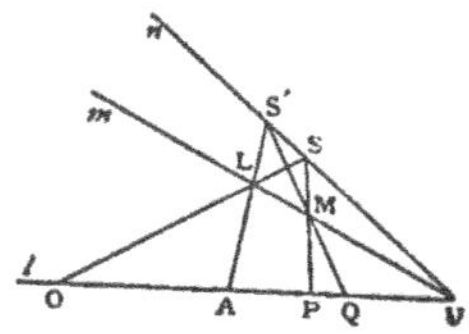

FIG. 69.

The sum of two prospectivities, written $(OAU^2)+(OBU^2)$, is defined to be that transformation of the line *l* into itself which is obtained by first applying the prospectivity (OAU^2) and then applying the prospectivity (OBU^2). Such a transformation, when the two summands have the same double point, is itself a prospectivity with that double point.

With this definition of addition it can be proved that prospectivities with the same double point satisfy all the axioms of magnitude. Accordingly they can be associated in a one-one correspondence with the positive and negative real numbers. Let E (fig. 70) be any point on *l*, distinct from O and U. Then the prospectivity (OEU^2) is associated with unity, the prospectivity (OOU^2) is associated with zero, and (OUU^2) with ∞. The prospectivities of the type (OPU^2), where P is any point on the segment OEU, correspond to the positive numbers; also if P′ is the harmonic conjugate of P with respect to O and U, the prospectivity $(OP'U^2)$ is associated with the corresponding negative number. (The subjoined figure explains this relation of the positive and negative prospectivities.) Then any point P on *l* is associated with the same number as is the prospectivity (OPU^2).

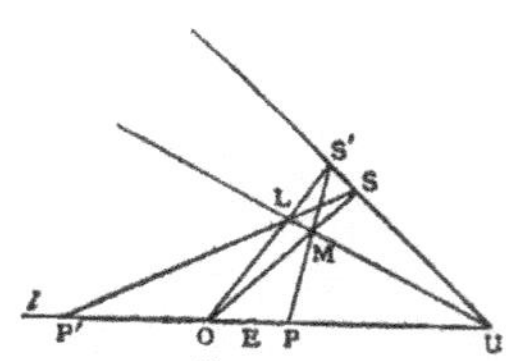

FIG. 70.

It can be proved that the order of the numbers in algebraic order of magnitude agrees with the order on the line of the associated points. Let the numbers, assigned according to the preceding specification, be said to be associated with the

1. †Cf. Pasch, *Vorlesungen über neuere Geometrie* (Leipzig, 1882), a classic work; also Fiedler, *Die darstellende Geometrie* (1st ed., 1871 [Leipzig: Teubner], 3rd ed., 1888 [Leipzig: Teubner]); Clebsch, *Vorlesungen über Geometrie*, vol. iii. [Whitehead likely means '*part* 3'; edited by Ferdinand Lindemann (Leipzig: Teubner, 1891)]; Hilbert, *loc. cit.*; F. Schur, *Math. Ann.* Bd. lv. (1901) [date corrected from 1902; 'Ueber die Grundlagen der Geometrie', vol. 55, no. 2 (June), pp. 265–292]; Vahlen, *loc. cit.*; Whitehead, *loc. cit.*
2. ESP removed all the figures that appeared in the original article and all references to them. Further instances of these deletions are not noted.

points according to the "numeration-system (OEU)." The introduction of a coordinate system for a plane is now managed as follows: Take any triangle OUV in the plane, and on the lines OU and OV establish the numeration systems $(\mathrm{OE_1U})$ and $(\mathrm{OE_2V})$, where E_1 and E_2 are arbitrarily chosen. Then (cf. fig. 71) if M and N are associated with the numbers x and y according to these systems, the coordinates of P are x and y. It then follows that the equation of a straight line is of the form $ax+by+c = 0$. Both coordinates of any point on the line UV are infinite. This can be avoided by introducing homogeneous coordinates X, Y, Z, where $x = \mathrm{X/Z}$, and $y = \mathrm{Y/Z}$, and $\mathrm{Z} = 0$ is the equation of UV.

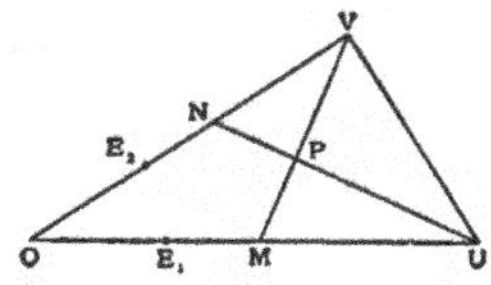

FIG. 71.

The procedure for three dimensions is similar. Let OUVW (fig. 72) be any tetrahedron, and associate points on OU, OV, OW with numbers according to the numeration systems $(\mathrm{OE_1U}), (\mathrm{OE_2V})$, and $(\mathrm{OE_3W})$. Let the planes VWP, WUP, UVP cut OU, OV, OW in L, M, N respectively; and let x, y, z be the numbers associated with L, M, N respectively.[1] Then P is the point (x,y,z). Also homogeneous coordinates can be introduced as before, thus avoiding the infinities on the plane UVW.

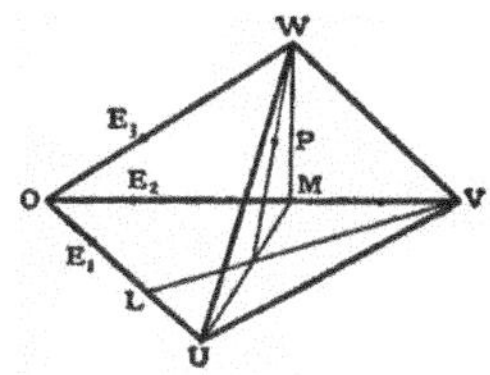

FIG. 72.

The cross ratio of a range of four collinear points can now be defined as a number characteristic of that range. Let the coordinates of any point P_r of the range $\mathrm{P_1P_2P_3P_4}$ be

$$\frac{\lambda_r a+\mu_r+a'}{\lambda_r+\mu_r},\ \frac{\lambda_r b+\mu_r b'}{\lambda_r+\mu_r},\ \frac{\lambda_r c+\mu_r c'}{\lambda_r+\mu_r},\quad (r=1,2,3,4)$$

and let $(\lambda_r\mu_s)$ be written for $\lambda_r\mu_s - \lambda_s\mu_r$. Then the cross ratio $\{\mathrm{P_1P_2P_3P_4}\}$ is defined to be the number $(\lambda_1\mu_2)(\lambda_3\mu_4)/(\lambda_1\mu_4)(\lambda_3\mu_2)$. The equality of the cross ratios of the ranges $(\mathrm{P_1P_2P_3P_4})$ and $(\mathrm{Q_1Q_2Q_3Q_4})$ is proved to be the necessary and sufficient condition for their mutual projectivity. The cross ratios of all harmonic ranges are then easily seen to be all equal to -1, by comparing with the range $(\mathrm{OE_1UE_1'})$ on the axis of x.

Thus all the ordinary propositions of geometry in which distance and angular measure do not enter otherwise than in cross ratios can now be enunciated and proved. Accordingly the greater part of the analytical theory of conics and quadrics belongs to geometry |733a| at this stage. The theory of distance will be considered after the principles of descriptive geometry have been developed.

Descriptive Geometry.

Descriptive geometry is essentially the science of multiple order for open series. The first satisfactory system of axioms was given by M. Pasch.[2†] An improved

1. The point N referenced in the text is not shown in the figure, nor is the plane suggested that would define N; this is to keep the figure relatively uncluttered.
2. †Cf. *loc. cit.*

version is due to G. Peano.[1]† Both these authors treat the idea of the class of points constituting the segment lying *between* two points as an undefined fundamental idea. Thus in fact there are in this system two fundamental ideas, namely, of points and of segments. It is then easy enough to define the prolongations of the segments, so as to form the complete straight lines. D. Hilbert's[2]† formulation of the axioms is in this respect practically based on the same fundamental ideas. His work is justly famous for some of the mathematical investigations contained in it, but his exposition of the axioms is distinctly inferior to that of Peano. Descriptive geometry can also be considered[3]† as the science of a class of relations, each relation being a two-termed serial relation, as considered in the logic of relations, ranging the points between which it holds into a linear open order. Thus the relations are the straight lines, and the terms between which they hold are the points. But a combination of these two points of view yields[4]† the simplest statement of all. Descriptive geometry is then conceived as the investigation of an undefined fundamental relation between three terms (points), and when the relation holds between three points A, B, C, the points are said to be "in the [linear] order ABC."

O. Veblen's axioms and definitions, slightly modified, are as follows:—

1. If the points A, B, C are in the order ABC, they are in the order CBA.

2. If the points A, B, C are in the order ABC, they are not in the order BCA.

3. If the points A, B, C are in the order ABC, A is distinct from C.

4. If A and B are any two distinct points, there exists a point C such that A, B, C are in the order ABC.

Definition.—The *line* AB $(A \neq B)$ consists of A and B, and of all points X in one of the possible orders, ABX, AXB, XAB. The points X in the order AXB constitute the *segment* AB.

5. If points C and D $(C \neq D)$ lie on the line AB, then A lies on the line CD.

6. There exist three distinct points A, B, C not in any of the orders ABC, BCA, CAB.

7. If three distinct points A, B, C (fig. 73) do not lie on the same line, and D and E are two distinct points in the orders BCD and CEA, then a point F exists in the order AFB, and such that D, E, F are collinear.

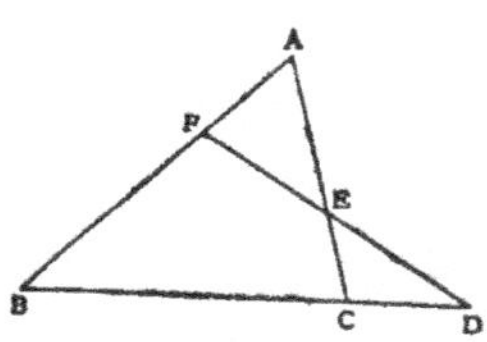

FIG. 73.

Definition.—If A, B, C are three non-collinear points, the *plane* ABC is the class of points which lie on any one of the lines joining any two of the points belonging to the boundary of the triangle ABC, the *boundary* being formed by the segments BC, CA and AB. The *interior* of

1. †Cf. *I Principii di geometria* (Turin, 1889) and "Sui fondamenti della geometria," *Rivista di mat.*, vol. iv. (1894) [pp. 51–90].
2. †Cf. *loc. cit.*
3. †Cf. Vailati, *Rivista di mat.*, vol. iv [Whitehead does not give a title and the reference here is unclear; Vailati published three short pieces in this volume of *Rivista di matematica* (vol. 4, 1894) – two were book reviews and one reviewed the catalogue of the University of Texas for 1893–1894; Whitehead may have meant to reference a different volume of *Rivista di matematica*], and Russell, *loc. cit.* § 376.
4. †Cf. Eliakim Hastings Moore, "On the Projective Axioms of Geometry," *Trans. Amer. Math. Soc.* vol. iii. (1902) [vol. 3, no. 1 (January), pp. 142–158; the original erroneously attributed this article to 'O. Veblen', and in fact Oswald Veblen did publish 'A System of Axioms for Geometry' in the *Transactions of the American Mathematical Society*, vol. 5, no. 3 (July 1904), pp. 343–384].

the triangle ABC is formed by the points in segments such as PQ, where P and Q are points respectively on two of the segments BC, CA, AB.

8. There exists a plane ABC, which does not contain all the points.

Definition.—If A, B, C, D are four non-coplanar points, the space ABCD is the class of points which lie on any of the lines containing two points on the surface of the tetrahedron ABCD, the *surface* being formed by the interiors of the triangles ABC, BCD, DCA, DAB.

9. There exists a space ABCD which contains all the points.

|733b| 10. The Dedekind property holds for the order of the points on any straight line.

It follows from axioms 1–9 that the points on any straight line are arranged in an open serial order. Also all the ordinary theorems respecting a point dividing a straight line into two parts, a straight line dividing a plane into two parts, and a plane dividing space into two parts, follow.

Again, in any plane α consider a line l and a point A (fig. 74).

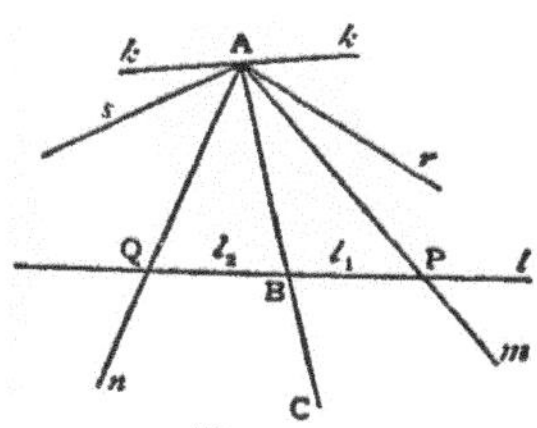

FIG. 74.

Let any point B divide l into two half-lines l_1 and l_2. Then it can be proved that the set of half-lines, emanating from A and intersecting l_1 (such as m), are bounded by two half-lines, of which ABC is one. Let r be the other. Then it can be proved that r does not intersect l_1. Similarly for the half-line, such as n, intersecting l_2. Let s be its bounding half-line. Then two cases are possible. (1) The half-lines r and s are collinear, and together form one complete line. In this case, there is one and only one line (viz. $r + s$) through A and lying in α which does not intersect l. This is the Euclidean case, and the assumption that this case holds is the *Euclidean parallel axiom*. But (2) the half-lines r and s may not be collinear. In this case there will be an infinite number of lines, such as k for instance, containing A and lying in α, which do not intersect l. Then the lines through A in α are divided into two classes by reference to l, namely the *secant* lines which intersect l, and the *non-secant* lines which do not intersect l. The two boundary non-secant lines, of which r and s are respectively halves, may be called the two parallels to l through A.

The perception of the possibility of case 2 constituted the starting-point from which Lobatchewsky constructed the first explicit coherent theory of non-Euclidean geometry, and thus created a revolution in the philosophy of the subject. For many centuries the speculations of mathematicians on the foundations of geometry were almost confined to hopeless attempts to prove the "parallel axiom" without the introduction of some equivalent axiom.[1†]

Associated Projective and Descriptive Spaces.—A region of a projective space, such that one, and only one, of the two supplementary segments between any pair of points within it lies entirely within it, satisfies the above axioms (1–10) of descriptive geometry, where the points of the region are the descriptive points, and the portions of straight lines within the region are the descriptive lines. If the excluded part of the original projective space is a single plane, the Euclidean parallel axiom

1. †Cf. P. Stäckel and F. Engel, *Die Theorie der Parallellinien von Euklid bis auf Gauss* (Leipzig, 1895).

also holds, otherwise it does not hold for the descriptive space of the limited region. Again, conversely, starting from an original descriptive space an associated projective space can be constructed by means of the concept of *ideal points*.[1†] These are also called *projective points*, where it is understood that the simple points are the points of the original descriptive space. An *ideal point* is the class of straight lines which is composed of two coplanar lines *a* and *b*, together with the lines of intersection of all pairs of intersecting planes which respectively contain *a* and *b*, together with the lines of intersection with the plane *ab* of all planes containing any one of the lines (other than *a* or *b*) already specified as belonging to the ideal point. It is evident that, if the two original lines *a* and *b* intersect, the corresponding ideal point is nothing else than the whole class of lines which are concurrent at the point *ab*. But the essence of the definition is that an ideal point has an existence when the lines *a* and *b* do not intersect, so long as they are coplanar. An ideal point is termed *proper*, if the lines composing it intersect; otherwise it is *improper*.

A theorem essential to the whole theory is the following: if any two of the three lines *a*, *b*, *c* are coplanar, but the three lines are not all coplanar, and similarly for the lines *a*, *b*, *d*, then *c* and *d* are coplanar. It follows that any two lines belonging to an ideal point can be used as the pair of guiding lines in the definition. An ideal point is said to be *coherent* with a plane, if any of the lines composing it lie in the plane. An *ideal line is* the class of ideal points each of which is coherent with two given planes. |734a| If the planes intersect, the ideal line is termed *proper*, otherwise it is *improper*. It can be proved that any two planes, with which any two of the ideal points are both coherent, will serve as the guiding planes used in the definition. The ideal planes are defined as in projective geometry, and all the other definitions (for segments, order, &c.) of projective geometry are applied to the ideal elements. If an ideal plane contains some proper ideal points, it is called *proper*, otherwise it is *improper*. Every ideal plane contains some improper ideal points.

It can now be proved that all the axioms of projective geometry hold of the ideal elements as thus obtained; and also that the order of the ideal points as obtained by the projective method agrees with the order of the proper ideal points as obtained from that of the associated points of the descriptive geometry. Thus a projective space has been constructed out of the ideal elements, and the proper ideal elements correspond element by element with the associated descriptive elements. Thus the proper ideal elements form a region in the projective space within which the descriptive axioms hold. Accordingly, by substituting ideal elements, a descriptive space can always be considered as a region within a projective space. This is the justification for the ordinary use of the "points at infinity" in the ordinary Euclidean geometry; the reasoning has been transferred from the original descriptive space to the associated projective space of ideal elements; and with the Euclidean parallel axiom the improper ideal elements reduce to the ideal points on a single improper ideal plane, namely, the plane at infinity.[2†]

1. † Cf. Pasch, *loc. cit.*, and R. Bonola, "Sulla introduzione degli enti improprii in geometria projettive," *Giorn. di mat.* vol. xxxviii. (1900) [pp. 105–116]; and Whitehead, *Axioms of Descriptive Geometry* (Cambridge, 1907).
2. † The original idea (confined to this particular case) of ideal points is due to von Staudt (*loc. cit.*).

Congruence and Measurement.—The property of physical space which is expressed by the term "measurability" has now to be considered. This property has often been considered as essential to the very idea of space. For example, Kant writes,[1†] "Space is represented as an infinite given *quantity.*" This quantitative aspect of space arises from the measurability of distances, of angles, of surfaces and of volumes. These four types of quantity depend upon the two first among them as fundamental. The measurability of space is essentially connected with the idea of *congruence*, of which the simplest examples are to be found in the proofs of equality by the method of superposition, as used in elementary plane geometry. The mere concepts of "part" and of "whole" must of necessity be inadequate as the foundation of measurement, since we require the comparison as to quantity of regions of space which have no portions in common. The idea of congruence, as exemplified by the method of superposition in geometrical reasoning, appears to be founded upon that of the "rigid body," which moves from one position to another with its internal spatial relations unchanged. But unless there is a previous concept of the metrical relations between the parts of the body, there can be no basis from which to deduce that they are unchanged.

It would therefore appear as if the idea of the congruence, or metrical equality, of two portions of space (as empirically suggested by the motion of rigid bodies) must be considered as a fundamental idea incapable of definition in terms of those geometrical concepts which have already been enumerated. This was in effect the point of view of Pasch.[2†] It has, however, been proved by Sophus Lie[3†] that congruence is capable of definition without recourse to a new fundamental idea. This he does by means of his theory of finite continuous groups (see Groups, Theory of), of which the definition is possible in terms of our established geometrical ideas, remembering that coordinates have already been introduced. The displacement of a rigid body is simply a mode of defining to the senses a one-one transformation of all space into itself. For at any point of space a particle may be conceived to be placed, and to be rigidly connected with the rigid body; and thus there is a definite correspondence of any point of space with the new point occupied by the associated particle after displacement. Again two suc-|734b|cessive displacements of a rigid body from position A to position B, and from position B to position C, are the same in effect as one displacement from A to C. But this is the characteristic "group" property. Thus the transformations of space into itself defined by displacements of rigid bodies form a group.

Call this group of transformations a congruence-group. Now according to Lie a congruence-group is defined by the following characteristics:—

1. A congruence-group is a finite continuous group of one-one transformations, containing the identical transformation.

---

1. † Cf. *Critique*, "Trans. Aesth." Sect. 1.
2. † Cf. *loc. cit.*
3. † Cf. Über die Grundlagen der Geometrie (Leipzig, *Ber.*, 1890) [Sophus Lie, 'Über die Grundlagen der Geometrie. Erste Abhandlung. . .' *Königlich Sächsische Gesellschaft der Wissenschaften (Hrsg.): Berichte über die Verhandlungen der Königlich-Sächsischen Gesellschaft der Wissenschaften zu Leipzig, Mathematisch-Physische Klasse*, vol. 42 (1890), pp. 284–321]; and *Theorie der Transformationsgruppen* (Leipzig, 1893), vol. iii.

2. It is a sub-group of the general projective group, *i.e.* of the group of which any transformation converts planes into planes, and straight lines into straight lines.

3. An infinitesimal transformation can always be found satisfying the condition that, at least throughout a certain enclosed region, any definite line and any definite point on the line are latent, *i.e.* correspond to themselves.

4. No infinitesimal transformation of the group exists, such that, at least in the region for which (3) holds, a straight line, a point on it, and a plane through it, shall all be latent.

The property enunciated by conditions (3) and (4), taken together, is named by Lie "Free mobility in the infinitesimal." Lie proves the following theorems for a projective space:—

1. If the above four conditions are only satisfied by a group throughout part of projective space, this part either (α) must be the region enclosed by a real closed quadric, or (β) must be the whole of the projective space with the exception of a single plane. In case (α) the corresponding congruence group is the continuous group for which the enclosing quadric is latent; and in case (β) an imaginary conic (with a real equation) lying in the latent plane is also latent, and the congruence group is the continuous group for which the plane and conic are latent.

2. If the above four conditions are satisfied by a group throughout the whole of projective space, the congruence group is the continuous group for which some imaginary quadric (with a real equation) is latent.

By a proper choice of non-homogeneous co-ordinates the equation of any quadrics of the types considered, either in theorem 1(α), or in theorem 2, can be written in the form $1+c\left(x^2+y^2+z^2\right)=0$, where c is negative for a real closed quadric, and positive for an imaginary quadric. Then the general infinitesimal transformation is defined by the three equations:

$$\left.\begin{aligned} dx/dt &= u - \omega_3 y + \omega_2 z + cx(ux+vy+wz),\\ dy/dt &= v - \omega_1 z + \omega_3 x + cy(ux+vy+wz),\\ dz/dt &= w - \omega_2 x + \omega_1 y + cz(ux+vy+wz).\end{aligned}\right\}\ \text{(A)}$$

In the case considered in theorem 1 (β), with the proper choice of co-ordinates the three equations defining the general infinitesimal transformations are:

$$\left.\begin{aligned} dx/dt &= u - \omega_3 y + \omega_2 z,\\ dy/dt &= v - \omega_1 z + \omega_3 x,\\ dz/dt &= w - \omega_2 x + \omega_1 y.\end{aligned}\right\}\ \text{(B)}$$

In this case the latent plane is the plane for which at least one of x, y, z are infinite, that is, the plane $0\cdot x+0\cdot y+0\cdot z+a=0$; and the latent conic is the conic in which the cone $x^2+y^2+z^2=0$ intersects the latent plane.

It follows from theorems 1 and 2 that there is not one unique congruence group, but an indefinite number of them. There is one congruence-group corresponding to each closed real quadric, one to each imaginary quadric with a real equation, and one to each imaginary conic in a real plane and with a real equation. The quadric thus associated with each congruence-group is called the *absolute* for that group, and in the degenerate case of 1 (β) the absolute is the latent plane together with the latent imaginary conic. If the absolute is real, the congruence-group is *hyperbolic*; if imaginary, it is *elliptic*; if the absolute is a plane and imaginary conic, the group is parabolic. Metrical geometry is simply the theory of the properties of some particular congruence-group selected for study.

The definition of distance is connected with the corresponding congruence-group by two considerations in respect to a range of five points $(A_1, A_2, P_1, P_2, P_3)$, of which A_1 and A_2 are on the absolute.

Let $\{A_1P_1A_2P_2\}$ stand for the cross ratio (as defined above) of the range $(A_1P_1A_2P_2)$, with a similar notation for the other ranges.

Then

(1) $$\log\{A_1P_1A_2P_2\} + \log\{A_1P_2A_2P_3\} = \log\{A_1P_1A_2P_3\},$$

and

(2), if the points A_1, A_2, P_1, P_2 are transformed into A_1', A_2', P_1', P_2' by any transformation of the congruence-group, (α) $\{A_1P_1A_2P_2\} = \{A_1'P_1'A_2'P_2'\}$, since the transformation is projective, and (β) A_1', A_2' are on the absolute since A_1 and A_2 are on it. Thus if we define |735a| the distance P_1P_2 to be $\frac{1}{2}k\log\{A_1P_1A_2P_2\}$, where A_1 and A_2 are the points in which the line P_1P_2 cuts the absolute, and k is some constant, the two characteristic properties of distance, namely, (1) the addition of consecutive lengths on a straight line, and (2) the invariability of distances during a transformation of the congruence-group, are satisfied. This is the well-known Cayley-Klein projective definition[1†] of distance, which was elaborated in view of the addition property alone, previously to Lie's discovery of the theory of congruence-groups. For a hyperbolic group when P_1 and P_2 are in the region enclosed by the absolute, $\log\{A_1P_1A_2P_2\}$ is real, and therefore k must be real. For an elliptic group A_1 and A_2 are conjugate imaginaries, and $\log\{A_1P_1A_2P_2\}$ is a pure imaginary, and k is chosen to be κ/i, where κ is real and[2] $i = \sqrt{-1}$.

Similarly the angle between two planes, p_1 and p_2, is defined to be $(1/2i)$ $\log(t_1p_1t_2p_2)$, where t_1 and t_2 are tangent planes to the absolute through the line p_1p_2. The planes t_1 and t_2 are imaginary for an elliptic group, and also for an hyperbolic group when the planes p_1 and p_2 intersect at points within the region enclosed by the absolute. The development of the consequences of these metrical definitions is the subject[3] of non-Euclidean geometry.

The definitions for the parabolic case can be arrived at as limits of those obtained in either of the other two cases by making k ultimately to vanish. It is also obvious that, if P_1 and P_2 be the points (x_1, y_1, z_1) and (x_2, y_2, z_2), it follows from equations (B) above that[4] $\left\{(x_1-x_2)^2+(y_1-y_2)^2+(z_1-z_2)^2\right\}^{\frac{1}{2}}$ is unaltered by a congruence transformation and also satisfies the addition property for collinear distances. Also the previous definition of an angle can be adapted to this case, by making t_1 and t_2 to be the tangent planes through the line p_1p_2 to the imaginary conic. Similarly if p_1

1. †Cf. A. Cayley, "A Sixth Memoir on Quantics," *Trans. Roy. Soc.*, 1859 [*Philosophical Transactions of the Royal Society*, vol. 149, pp. 61–90], and *Coll. Papers*, vol. ii. [*The Collected Mathematical Papers of Arthur Cayley*, Cambridge: Cambridge University Press, 1889]; and F. Klein, *Math. Ann.* vol. iv., 1871 ['Ueber die sogenannte Nicht-Euklidische Geometrie', no. 4 (December), pp. 573–625].
2. In the original, the following equation was simply 'i = √ –', which is either an error or very non-standard notation.
3. Corrected spelling from 'subjet'.
4. In the original, for the following expression the '$\frac{1}{2}$' was erroneously not printed as a superscript.

and p_2 are intersecting lines, the same definition of an angle holds, where t_1 and t_2 are now the lines from the point p_1p_2 to the two points where the plane p_1p_2 cuts the imaginary conic. These points are in fact the "circular points at infinity" on the plane. The development of the consequences of these definitions for the parabolic case gives the ordinary Euclidean metrical geometry.

Thus the only metrical geometry for the whole of projective space is of the elliptic type. But the actual measure-relations (though not their general properties) differ according to the elliptic congruence-group selected for study. In a descriptive space a congruence-group should possess the four characteristics of such a group throughout the whole of the space. Then form the associated ideal projective space. The associated congruence-group for this ideal space must satisfy the four conditions throughout the region of the proper ideal points. Thus the boundary of this region is the absolute. Accordingly there can be no metrical geometry for the whole of a descriptive space unless its boundary (in the associated ideal space) is a closed quadric or a plane. If the boundary is a closed quadric, there is one possible congruence-group of the hyperbolic type. If the boundary is a plane (the plane at infinity), the possible congruence-groups are parabolic; and there is a congruence-group corresponding to each imaginary conic in this plane, together with a Euclidean metrical geometry corresponding to each such group. Owing to these alternative possibilities, it would appear to be more accurate to say that systems of quantities can be found in a space, rather than that space is a quantity.

Lie has also deduced[1†] the same results with respect to congruence-groups from another set of defining properties, which explicitly assume the existence of a quantitative relation (the distance) between any two points, which is invariant for any transformation of the congruence-group.[2†]

The above results, in respect to congruence and metrical geometry, considered in relation to existent space, have led to the doctrine[3†] that it is intrinsically unmeaning to ask which system of metrical geometry is true of the physical world. Any one of these systems can be applied, and in an indefinite number of ways. The only question before us is one of convenience in respect to simplicity of statement of the physical laws. This point of view seems to neglect the consideration that science is to be relevant to the definite perceiving minds of men; and that (neglecting the ambiguity introduced by the invariable slight inexactness of observation which is not relevant to this special doctrine) |735b| we have, in fact, presented to our senses a definite set of transformations forming a congruence-group, resulting in a set of measure relations which are in no respect arbitrary. Accordingly our scientific laws are to be stated relevantly to that particular congruence-group. Thus the investigation of the type (elliptic, hyperbolic or parabolic) of this special congruence-group is a perfectly definite problem, to be decided by experiment. The consideration of experiments adapted to this object requires some development of non-Euclidean geometry (see section VI., *Non-Euclidean Geometry*). But if the doctrine means that,

---

1. † Cf. *loc. cit.*
2. † For similar deductions from a third set of axioms, suggested in essence by Peano, *Riv. mat.* vol. iv. *loc. cit.* cf. Whitehead, *Desc. Geom. loc. cit.*
3. † Cf. H. Poincaré, *La Science et l'hypothèse*, ch. iii [Paris: Flammarion, 1902].

assuming some sort of objective reality for the material universe, beings can be imagined, to whom *either* all congruence-groups are equally important, *or* some other congruence-group is specially important, the doctrine appears to be an immediate deduction from the mathematical facts. Assuming a definite congruence-group, the investigation of surfaces (or three-dimensional loci in space of four dimensions) with geodesic geometries of the form of metrical geometries of other types of congruence-groups forms an important chapter of non-Euclidean geometry. Arising from this investigation there is a widely-spread fallacy, which has found its way into many philosophic writings, namely, that the possibility of the geometry of existent three-dimensional space being other than Euclidean depends on the physical existence of Euclidean space of four or more dimensions. The foregoing exposition shows the baselessness of this idea.

Bibliography.—For an account of the investigations on the axioms of geometry during the Greek period, see M. Cantor, *Vorlesungen* über *die Geschichte der Mathematik*, Bd. i. and iii.; T. L. Heath, *The Thirteen Books of Euclid's Elements, a New Translation from the Greek, with Introductory Essays and Commentary, Historical, Critical, and Explanatory* (Cambridge, 1908)—this work is the standard source of information; W. B. Frankland, *Euclid, Book I., with a Commentary* (Cambridge, 1905)[1]—the commentary contains copious extracts from the ancient commentators. The next period of really substantive importance is that of the 18th century. The leading authors are: G. Saccheri, S.J., *Euclides ab omni naevo vindicatus* (Milan, 1733). Saccheri was an Italian Jesuit who unconsciously discovered non-Euclidean geometry in the course of his efforts to prove its impossibility. J. H. Lambert, *Theorie der Parallellinien* (1766); A. M. Legendre, *Élements de géométrie* (1794). An adequate account of the above authors is given by P. Stäckel and F. Engel, *Die Theorie der Parallellinien von Euklid bis auf Gauss* (Leipzig, 1895). The next period of time (roughly from 1800 to 1870) contains two streams of thought, both of which are essential to the modern analysis of the subject. The first stream is that which produced the discovery and investigation of non-Euclidean geometries, the second stream is that which has produced the geometry of position, comprising both projective and descriptive geometry not very accurately discriminated. The leading authors on non-Euclidean geometry are K. F. Gauss, in private letters to Schumacher, cf. Stäckel and Engel, *loc. cit.*; N. Lobatchewsky, rector of the university of Kazan, to whom the honour of the effective discovery of non-Euclidean geometry must be assigned. His first publication was at Kazan in 1826.[2] His various memoirs have been re-edited by Engel; cf. *Urkunden zur Geschichte der nichteuklidischen Geometrie* by Stäckel and Engel, vol. i. "Lobatchewsky." J. Bolyai discovered non-Euclidean geometry apparently in independence of Lobatchewsky. His memoir was published in 1832[3] as an appendix to a work by his father W. Bolyai, *Tentamen juventutem. . . .* This memoir has been separately edited by J. Frischauf, *Absolute Geometrie nach J. Bolyai* (Leipzig,

1. William Barrett Frankland, *The First Book of Euclid's Elements, with a Commentary Based Principally Upon That of Proclus Diadochus* (Cambridge: Cambridge University Press, 1905).
2. Actually, while Lobachevsky first presented his ideas in 1826, they would not be published until 1829. See p. 301n2.
3. Corrected from 1831.

1872); B. Riemann, *Über die Hypothesen, welche der Geometric zu Grunde liegen* (1854);[1] cf. *Gesamte Werke*,[2] a translation in *The Collected Papers* of W. K. Clifford.[3] This is a fundamental memoir on the subject and must rank with the work of Lobatchewsky. Riemann discovered elliptic metrical geometry, and Lobatchewsky hyperbolic geometry. A full account of Riemann's ideas, with the subsequent developments due to Clifford, F. Klein and W. Killing, will be found in *The Boston Colloquium for 1903* (New York, 1905), article "Forms of Non-Euclidean Space," by F. S. Woods.[4] A. Cayley, *loc. cit.* (1859), and F. Klein, "Über die sogenannte nichteuklidische Geometrie," *Math. Annal.* vols. iv. and vi. (1871 and 1873),[5] between them elaborated the projective theory of distance; H. Helmholtz, "Über die thatsächlichen Grundlagen der Geometrie" (1866), and "Über die Thatsachen, die der Geometrie zu Grunde liegen" (1868), both in his *Wissenschaftliche Abhandlungen*, vol. ii.,[6] and S. Lie, *loc. cit.* (1890 and 1893), between them elaborated the group theory of congruence.

The numberless works which have been written to suggest equivalent alternatives to Euclid's parallel axioms may be neglected as being of trivial importance, though many of them are marvels of geometric ingenuity.

The second stream of thought confined itself within the circle of ideas of Euclidean geometry. Its origin was mainly due to a |736a| succession of great French mathematicians, for example, G. Monge, *Géométrie descriptive* (1799);[7] J. V. Poncelet, *Traité des proprietés projectives des figures* (1822); M. Chasles, *Aperçu historique sur l'origine et le développement des méthodes en géométrie* (Bruxelles, 1837), and *Traité de géométrie supérieure* (Paris, 1852); and many others. But the works which have been, and are still, of decisive influence on thought as a store-house of ideas relevant to the foundations of geometry are K. G. C. von Staudt's two works, *Geometrie der Lage* (Nürnberg, 1847); and *Beiträge zur Geometrie der Lage* (Nürnberg, 1856, 3rd ed. 1860).

The final period is characterized by the successful production of exact systems of axioms, and by the final solution of problems which have occupied mathematicians for two thousand years. The successful analysis of the ideas involved in serial continuity is due to R. Dedekind, *Stetigkeit and irrationale Zahlen* (1872), and to G. Cantor, *Grundlagen einer allgemeinen Mannigfaltigkeitslehre* (Leipzig, 1883), and *Acta math.* vol. 2.[8]

---

1. Delivered as a lecture in 1854, but not published until 1868. See p. 301n4.
2. Bernhard Riemann, *Gesammelte mathematische Werke und Wissenschaftlicher Nachlass*, edited by Heinrich Weber (Leipzig: Teubner, 1876).
3. Originally published as Bernhard Riemann, 'On the Hypotheses Which Lie at the Bases of Geometry', translated by William Kingdon Clifford, *Nature*, vol. 8, nos 183–184 (1873), pp. 14–17, 36–37).
4. Frederick Shenstone Woods, 'Forms of Non-Euclidean Space', in Edward Burr Van Vleck, Henry Seely White and Frederick Shenstone Woods, *The Boston Colloquium: Lectures on Mathematics Delivered from September 2 to 5, 1903* (New York: Macmillan, 1905), pp. 31–74.
5. For full citation of the volume 4 part, see p. 315n1. The original article erroneously listed the volume 6 issue as published in 1872. Full citation: Felix Klein, 'Ueber die sogenannte Nicht-Euklidische Geometrie', *Mathematische Annalen*, vol. 6, no. 2 (June 1873), pp. 112–145.
6. Herman von Helmholtz, *Wissenschaftliche Abhandlungen, Vol. II* (Leipzig: Barth, 1883).
7. Date corrected from 1800.
8. Cantor published seven papers in a row in *Acta Mathematica*, vol. 2 (December 1883), all between pp. 305–414.

Complete systems of axioms have been stated by M. Pasch, *loc. cit.*; G. Peano, *loc. cit.*; M. Pieri, *loc. cit.*; B. Russell, *Principles of Mathematics*; O. Veblen, *loc. cit.*; and by G. Veronese in his treatise, *Fondamenti di geometria* (Padua, 1891[1]; German transl. by A. Schepp, *Grundzüge der Geometrie*, Leipzig, 1894). Most of the leading memoirs on special questions involved have been cited in the text; in addition there may be mentioned M. Pieri, "Nuovi principii di geometria projettiva complessa," *Trans. Accad. R. d. Sci.* (Turin, 1905);[2] E. H. Moore, "On the Projective Axioms of Geometry," *Trans. Amer. Math. Soc.*, 1902;[3] O. Veblen and W. H. Bussey, "Finite Projective Geometries," *Trans. Amer. Math. Soc.*, 1906;[4] A. B. Kempe, "On the Relation between the Logical Theory of Classes and the Geometrical Theory of Points," *Proc. Lond. Math. Soc.*, 1889;[5] J. Royce, "The Relation of the Principles of Logic to the Foundations of Geometry," *Trans. of Amer. Math. Soc.*, 1905;[6] A. Schoenflies, "Über die Möglichkeit einer projectiven Geometrie bei transfiniter (nichtarchimedischer) Massbestimmung," *Deutsch. M.-V. Jahresb.*, 1906.[7]

For general expositions of the bearings of the above investigations, cf. Hon. Bertrand Russell, *loc. cit.*; L. Couturat, *Les Principes des mathématiques* (Paris, 1905); H. Poincaré, *loc. cit.*; Russell and Whitehead, *Principia mathematica* (Cambridge Univ. Press). The philosophers whose views on space and geometric truth deserve especial study are Descartes, Leibnitz, Hume, Kant and J. S. Mill.

---

1. Available at <https://archive.org/details/FondamentiDiGeometriaVeronese> (accessed 4 April 2025).
2. *Memorie della Reale Accademia della Scienze di Torino*, vol. 55, no. 2, pp. 189–235.
3. Vol. 3, no. 1 (January), pp. 142–158.
4. Date corrected from 1905; vol. 7, no. 2 (April), pp. 241–259.
5. Date corrected from 1890; vol. 21, no. 1 (November), pp. 147–182.
6. Vol. 6, no. 3 (July), pp. 353–415.
7. *Jahresbericht der Deutschen Mathematiker-Vereinigung*, vol. 15 (1906), pp. 26–41.

The Philosophy of Mathematics

1910 *Science Progress in the Twentieth Century*, vol. 5, no. 18 (October), pp. 234–239

In this, the only book review Whitehead is known to have published, he provides a decidedly unfavourable review of Hastings Berkeley's Mysticism in Modern Mathematics. *Just a few months prior (April 1910), Whitehead had resigned his position at Trinity College, Cambridge and subsequently moved his family to London. The review also appears shortly before the publication of the first volume of* Principia Mathematica, *which was released in December of the same year.*

THE most recent work[1†] on this subject suffers from the indeterminateness of its aims and also from its neglect of distinctions vital to the subject. In Part I., which is composed of the first three chapters, the interest is chiefly psychological. The relation of Thought to Language and Language as an instrument of Reason are considered. Then this general subject is narrowed to a consideration of the psychology of mathematicians, and of the reasons why mathematicians tend to lapse into what the author calls mysticism. But in the rest of the book the psychology gradually fades, and the substantial investigation appears to be concerned with the logical correctness of the statements and explanations of various mathematicians, intermingled with the author's own expositions. These two investigations should be kept sharply distinct: logically it is indifferent whether the error of a mathematician be due to mysticism or to indigestion, nor does a medical knowledge of these two diseases help us in determining the truth or falsehood of his statements. But the facile imputation of mysticism, like the invocation of the supernatural to explain events, has, I think, seriously misled the author by preventing him from apprehending the real points which the mathematicians criticised are endeavouring with more or less success to explain.

In Part II. the concepts of Algebra and of Imaginary Loci in Geometry are considered. Part III. is headed "Metageometry" and deals with the principles of Geometry. The book is written from the prevalent standpoint of fifteen years ago, apparently in complete unconsciousness that great discoveries have since been made by which the subject has been revolutionised. The author has read Mr. Bertrand Russell's *Principles of Mathematics*, but has missed the point that the theory of the "variable" is the key which unlocks the whole subject. Indeed, |235| a lack of penetration as to what the authors quoted are really driving at is a defect of the whole work; and even

---

1. †*Mysticism in Modern Mathematics*, by Hastings Berkeley, Oxford University Press, 1910, pp. xii, 264; 8*s*. net.

when the passages quoted are wrong, they are generally at least as satisfactory as the criticisms passed on them.

I agree, however, with the main conclusion which seems to follow from the critical examinations in the book—namely that a thorough reconstruction of the philosophy of mathematics was badly wanted at the end of the last century. The book under review cannot by any stretch of imagination be said to have supplied it. A slight sketch of the problem to be solved will show how the doctrine of the variable has enabled a satisfactory philosophy to be developed.

In any consideration of the principles of mathematics two distinct subjects should be kept separate—namely (1) the nature of mathematical propositions considered in themselves apart from any admixture of particular application and (2) the discussion of the groups of particular facts which are special cases of mathematical truths. These two subjects are respectively the problem of the nature of pure mathematics, and that of the applications of mathematics. That the addition of two and two make four is a mathematical theorem; that the dropping of two apples into an empty basket and then of two more apples, with none taken out, will leave four apples in the basket is a theorem of applied mathematics. This distinction in its grosser forms has been known and recognised for centuries, probably ever since mathematics has been seriously studied. The recent progress in the philosophy of mathematics has shown that it cuts deeper and is of wider import than our predecessors imagined.

The problem of disengaging mathematics from its applications is not so easy as it looks. In popular thought it has been done most completely in the arithmetic of integers, and with a decreasing measure of success in the case of fractions, negative numbers, real numbers and complex numbers. Geometry has been the field of prolonged controversy over this point. The necessity for a solution of the problem becomes urgent when it is found that practically identical types of reasoning occur in connection with widely different material. The mathematician then seeks to disengage the abstract train of reasoning between hypothesis and conclusion from the variety of materials to which it can be applied. A fable will best illustrate the |236| probable course of thought as soon as this process has commenced in any important branch of mathematical knowledge.

A people once existed, civilised and addicted to learning, who, as it happened, had only thought of number in connection with fishes. They would say two-fishes and three-fishes make five-fishes; but never once had they thought of two and three and five, and never had it occurred to them that two stones and three stones make five stones. Their philosophers, erudite men steeped in the classical languages and ancient wisdom, had elaborated subtle theories of the *a priori* nature of fish existence. Also the fish-thinkers, as the mathematicians were termed, were wont to commence their more complete treatises with some consideration of deep-sea dredging. In time other aggregates were considered but at first without reference to fish-arithmetic. Some fish-thinkers elaborated a disconnected theory, rather curious than important as it appeared to them, of deux-stones, trois-stones, quatre-stones, etc.; and, finally, in connection with fish-arithmetic a bold man denied that a fish was an ultimate thought-unit and introduced the notion of fractions—an idea which he utterly failed to make intelligible, as it was conclusively proved by philosophers that any portion of a fish was not a fish, and that it was illegitimate to apply to it the fish-concept.

Unfortunately the narrow specialism of a mathematical education left many mathematicians utterly unable to appreciate the crushing force of this philosophic refutation. Also the confusion was increased by the effort of an eminent mathematician to explain the idea of fractions in some popular lectures. He asked his hearers to imagine a universe in which it was impossible for any being to see at one and the same time the head and the tail of a fish. For more than a generation afterwards all philosophers and many mathematicians imagined that the theory of fractions was inextricably bound up with the existence of such a fantastic world. In short, no sound thinker—of the sort that carries weight—ever treated fractions seriously; and in standard philosophic treatises they were habitually dismissed as mathematical fads without any real import for serious thought.

At this point we may dismiss the fable and return to the central problem, How can mathematical truths be completely disengaged from all adventitious ideas? Twenty years ago, |237| and even now in many cases, mathematicians who considered this question were driven to an extreme formalism. They laid the emphasis on the rules adopted conventionally to produce equivalent collocations of symbols. This formalism was always —and rightly—severely criticised by those who were philosophers first and mathematicians afterwards, if at all. But perhaps such critics did not always understand that an important problem remained which the mathematicians were endeavouring to solve.

But meanwhile a notable discovery has been made by the joint and partially independent work of three men: Frege, Peano and Bertrand Russell—a German, an Italian and an Englishman—by which a flood of light has been let in upon the whole question, so that the problem has received its solution in all essentials. The discovery is that of the generalised conception of the variable and of its essential presence in all mathematical reasoning. This discovery empties mathematics of everything but its logic. For the future mathematics is logic, whereas according to the old formalism mathematics is logic plus conventions as to marks, and according to the older tradition mathematics is logic applied to the domains of number, quantity and space.

A very cursory examination of one of the many ways in which the abstract science of Geometry can be presented will form the best explanation of the position of the doctrine of the variable in the philosophy of mathematics. Instead of thinking of the class of straight lines of actual space, we start by considering *any* class κ whose members are also classes. We may, if we like, call the members of κ the straight lines; but this is a detail of nomenclature and does not alter the fact that κ is *any* class whose members are classes.

Then, in the place of the old axioms of Geometry which were stated as true of physical space, certain propositions involving κ are *considered*, not stated as true but merely enunciated for inspection. In this new sense we will call them the "axioms," as a convenient short title. For example three such "axioms" are:

"If λ is a member of κ, λ is a class with at least three members;"

"If λ and μ are distinct members of κ, λ and μ cannot possess more than one common member, if any;"|238|

"If λ, μ, ρ are distinct members of κ, and any two of them possess a common member but there is no member common to all three, and if the same is true for λ, μ, σ, then ρ and σ possess a common member."

In all about ten or a dozen such "axioms" are required to produce a subject interesting from its apparent relevance to the space of experience.

Then the subject consists of the hypothetical propositions which can be proved concerning κ, *whatever class of classes* κ *may be*, where the hypotheses of the propositions consist of all or some of these "axioms" concerning κ. These hypotheticals are true for any value of κ, whether the "axioms" which form[1] the hypotheses be false or true for that value. And further if for some special determination of κ, we know that the "axioms" are true, then the conclusions of the propositions are also true for this value of κ. These further propositions, arising from a special determination of κ for which the "axioms" are true, belong in general to applied mathematics.

Accordingly the modern mathematician withdraws, or should withdraw, from the apodictic certainty of his results the whole question of fact involved in any application of his reasoning. What then are the properties of space as known to us in our common human experience? The mathematician should leave the answer to the physicist, or to the psychologist, or to the plain man in the street, or to the metaphysician, or to whomever is the proper person to deal with this question. It is true that mathematics forms an essential element in the inductive proof of the properties of space, if the proof is inductive; an interesting chapter could be written upon this point. But the result arrived at has not the shadow of a right to masquerade as possessing the certainty of mathematics; it may be as certain as you like but it is not mathematical certainty. This rapid account, taking geometry as an example, must suffice to show how the discovery of the doctrine of the variable has enabled mathematical philosophy to be reconstructed.

Another distinction, which the book under consideration ignores in practice, is that the problem of the presentation of mathematical ideas to students in the initial stages of knowledge must be entirely separated from the discussion of the true "principia" of the subject. Elementary mathematics and the elements of mathematics are widely different subjects, though |239| a confusion still exists and has survived from bygone times when the psychology of education was not understood so well as it is at present. It seems natural that the learner should begin at the beginning of the subject. The truth is entirely the opposite. The learner's natural beginning is the study of the complex facts of some application of the subject. The simple abstract ideas are very difficult to grasp, because in practical life we never consider them directly; for beginners they require some particular embodiment.

In one instance, however, Mr. Berkeley has not made this error—namely when he criticises parts of the introductory chapter of my *Universal Algebra*. I am glad to be able to close this article on a note of agreement with him. I think the formalist position adopted in that chapter, whilst it has the merit of recognising an important problem, does not give a true solution, which is to be found, as explained above, in the doctrine of the variable. To that extent I agree with Mr. Berkeley's criticisms,

1. The original erroneously had 'from'.

though I hardly think that he always interprets correctly the meaning of my statements. His own solutions of the group of problems presented by the ideas of algebra, which ignore the doctrine of the variable, appear to me to be entirely inadequate, lacking in coherence, clearness and definition.

Mathematics

1911 **Hugh Chisholm (ed.), *Encyclopaedia Britannica*, 11th edition, vol. 17 (Cambridge: Cambridge University Press), pp. 878–883**

1947 *Essays in Science and Philosophy* (New York: Philosophical Library), pp. 269–288

This article is far less technical than Whitehead's contributions on geometry, likely because the topic is so enormous that it simply will not admit the same level of address within the confines of an encyclopaedia entry. Thus, his main points deal with the inextricably linked issues of the characterisation and foundations of mathematics. He begins with the invocation of the conventional notion that mathematics is the 'science of discrete and continuous magnitude' with the suggestion that this will prove inadequate and must be expanded to broader forms of reasoning, a position that in its seminal form he attributes to Leibniz.

To make his case, Whitehead begins by gathering some logical connections among notions of set theory, number and geometry that at first seem to bear out the conventional characterisation. This is not deceptive on his part, but rather necessary to his ultimate conclusion. At the centre of this analysis is the notion of number, which Whitehead addresses in its most ordinary form (the natural numbers – 0, 1, 2, etc.) while expanding the concept in two directions. Working away from foundational issues, he broadens the concept to include integers (the natural numbers and their corresponding negatives), the rational numbers (fractions), real numbers and their higher-dimensional extensions, such as the complex numbers and the worlds of multidimensional geometries. Catching geometry in the net of numbers is a significant unification for Whitehead, but one from which he will retreat after further working out latent foundational issues. Nonetheless, his account of the extension of the natural numbers to multidimension analytic geometry is a good one, and largely consistent with how this would be taught today based on naïve set theory (although we should now rely routinely on the explicit mention of equivalence relations, quotient sets and the completion of a number system by equivalence classes of Cauchy sequences). Indeed, many mathematicians will speak as though the concept of a natural number was given to us by the Peano axioms, explicitly stated in Whitehead's article, while the rest are then constructed with standard set-theoretic methods supplemented with the notion of convergence. The centrality of the Peano axioms is key to Whitehead's larger argument, as we shall soon see.

In the opposite direction, Whitehead considers the foundations of the concept of number from the general perspective of sets and relations. Thus, he shows how cardinal and ordinal numbers both finite and infinite are derivable from some presumably suitable version of set theory. (The distinction between cardinal and ordinal roughly corresponds to the separate but related notions of numbers as the business of counting versus numbering as the business of ordering.) He is overtly aware of the foundational problems of Cantorian set theory – in particular, Russell's paradox – but is not yet in a position to replace it by the full formalism of, say, Zermelo–Fraenkel set theory. Instead, he makes a gesture towards what would later be known in PM as the theory of types (although that, too, ultimately fails). Whatever the ultimate foundational shortcomings of his development, he does give a plausible account of how the Peano axioms arise from sets and relations, and hence mathematics as the study of number in a generalised sense is indeed reduced to set theory and relations. But it is exactly here that Whitehead asserts the insufficiency of number as the cornerstone of mathematics.

Despite his cogent set of connections to the notion of number as a mathematical priority, Whitehead maintains that the concept of number is merely one aspect of the more general theory of sets and relations.[1] *Hence the problem of foundations and the inseparable question of the nature of mathematical inquiry is reduced to the foundations of sets and relations, and in this connection he is prepared to assert that it further reduces in turn to 'the science concerned with the logical deduction of consequences from the general premisses of all reasoning' (p. 332). This, of course, is closely connected to logicism, which he and Russell were working out in PM. Whitehead could not know then that logicism, as a species of formalism, was doomed to fall on the diamond-tipped sword of Gödel's limitative theorems in 1931, and that Russell's theory of types, which Whitehead suggests as the antidote to the poison of Russell's paradox, would prove equally inadequate (and indeed incoherent insofar as the Axiom of Reducibility infirms the entire programme). Nonetheless, Whitehead does now revisit geometry as an example of the quoted characterisation of mathematics as logic, moving away from any analytic foundations with the note that measurement (number as magnitude) comes late to its development. He also mentions group theory – a construct entirely lacking in UA, although abstract groups preceded this book by several decades – as an example of more general reasoning and in this way his characterisation here is quite in accord with the development of the diverse informal axiomatic theories that would characterise twentieth-century mathematics (e.g. groups, rings, fields, vector spaces, topological spaces and topological groups). These would then subsequently provide the classes of objects that would become the basis for the grand generalisation and cultural revolution founded on the mid-century introduction of categories and functors.*

One other noteworthy component of this article remains: the assertion that applied mathematics is characterised (in contrast to pure mathematics) as having no a priori necessity. The focus of the enterprise is the selection of a set of assumptions that are apropos of the domain of inquiry. Whitehead then goes on to say that once these assumptions are fixed, one has essentially returned to pure mathematics. Still, he suggests that the artful articulation of these assumptions gives applied mathematics a distinctive feel and, in this connection, one would strongly suspect that scientific giants such as Einstein and Schrödinger would agree.

In retrospect, one might say that the deepest value of this article lies in the vivid sharing of the transformation of mathematics that was taking place in the search for foundations and, moreover, in the burgeoning growth of informal axiomatic theories.

1. He makes an entirely congruent assertion in his 'Memoir on the Algebra of Symbolic Logic (Part I)' (p. 117).

This remains true even if the direction Whitehead would take in securing the foundations of mathematics was not to succeed. Such was the value of his own encyclopaedic knowledge of the discipline.

MATHEMATICS (Gr. μαθηματική, sc. τέχνη or ἐπιστήμη; from μάθημα, "learning" or "science"), the general term for the various applications of mathematical thought, the traditional field of which is number and quantity. It has been usual to define mathematics as "the science of discrete and continuous magnitude." Even Leibnitz,[1†] who initiated a more modern point of view, follows the tradition in thus confining the scope of mathematics properly so called, while apparently conceiving it as a department of a yet wider science of reasoning. A short |879a| consideration of some leading topics of the science will exemplify both the plausibility and inadequacy of the above definition. Arithmetic, algebra, and the infinitesimal calculus, are sciences directly concerned with integral numbers, rational (or fractional) numbers, and real numbers generally, which include incommensurable numbers. It would seem that "the general theory of discrete and continuous quantity" is the exact description of the topics of these sciences. Furthermore, can we not complete the circle of the mathematical sciences by adding geometry? Now geometry deals with points, lines, planes and cubic contents. Of these all except points are quantities: lines involve lengths, planes involve areas, and cubic contents involve volumes. Also, as the Cartesian geometry shows, all the relations between points are expressible in terms of geometric quantities. Accordingly, at first sight it seems reasonable to define geometry in some such way as "the science of dimensional quantity." Thus every subdivision of mathematical science would appear to deal with quantity, and the definition of mathematics as "the science of quantity" would appear to be justified. We have now to consider the reasons for rejecting this definition as inadequate.

Types of Critical Questions.—What are numbers? We can talk of five apples and ten pears. But what are "five" and "ten" apart from the apples and pears? Also in addition to the cardinal numbers there are the ordinal numbers: the fifth apple and the tenth pear claim thought. What is the relation of "the fifth" and "the tenth" to "five" and "ten"? "The first rose of summer" and "the last rose of summer" are parallel phrases, yet one explicitly introduces an ordinal number and the other does not. Again, "half a foot" and "half a pound" are easily defined. But in what sense is there "a half," which is the same for "half a foot" as "half a pound"? Furthermore, incommensurable numbers are defined as the limits arrived at as the result of certain procedures with rational numbers. But how do we know that there is anything to reach? We must know that $\sqrt{2}$ exists before we can prove that any procedure will reach it. An expedition to the North Pole has nothing to reach unless the earth rotates.

Also in geometry, what is a point? The straightness of a straight line and the planeness of a plane require consideration. Furthermore, "congruence" is a difficulty. For when a triangle "moves," the points do not move with it. So what is it that

1. † Cf. *La Logique de Leibnitz*, ch. vii., by L. Couturat (Paris, 1901).

keeps unaltered in the moving triangle? Thus the whole method of measurement in geometry as described in the elementary textbooks and the older treatises is obscure to the last degree. Lastly, what are "dimensions"? All these topics require thorough discussion before we can rest content with the definition of mathematics as the general science of magnitude; and by the time they are discussed the definition has evaporated. An outline of the modern answers to questions such as the above will now be given. A critical defence of them would require a volume.[1†]

Cardinal Numbers.—A one-one relation between the members of two classes α and β is any method of correlating all the members of α to all the members of β, so that any member of α has one and only one correlate in β, and any member of β has one and only one correlate in α. Two classes between which a one-one relation exists have the same cardinal number and are called cardinally similar; and the cardinal number of the class α is a certain class whose members are themselves classes—namely, it is the class composed of all those classes for which a one-one correlation with α exists. Thus the cardinal number of α is itself a class, and furthermore α is a member of it. For a one-one relation can be established between the members of α and α by the simple process of correlating each member of α with itself. Thus the cardinal number one is the class of unit classes, the cardinal number two is the class of doublets, and so on. Also a unit class is any class with the property that it possesses a member x such that, if y is any member of the class, then x and y are identical. A doublet is any class which possesses a member x such that the modified class formed by all the other members except x is a unit class. And so on for all the finite cardinals, which are thus defined successively. The cardinal number zero is the class of classes with no members; but there is only one such class, namely—the null class. Thus this cardinal |879b| number has only one member. The operations of addition and multiplication of two given cardinal numbers can be defined by taking two classes α and β, satisfying the conditions (1) that their cardinal numbers are respectively the given numbers, and (2) that they contain no member in common, and then by defining by reference to α and β two other suitable classes whose cardinal numbers are defined to be respectively the required sum and product of the cardinal numbers in question. We need not here consider the details of this process.

With these definitions it is now possible to *prove* the following six premisses applying to finite cardinal numbers, from which Peano[2†] has shown that all arithmetic can be deduced:—

i. Cardinal numbers form a class.

ii. Zero is a cardinal number.

iii. If a is a cardinal number, $a + 1$ is a cardinal number.

iv. If s is any class and zero is a member of it, also if when x is a cardinal number and a member of s, also $x + 1$ is a member of s, then the whole class of cardinal numbers is contained in s.

v. If a and b are cardinal numbers, and $a + 1 = b + 1$, then $a = b$.

vi. If a is a cardinal number, then $a + 1 \neq 0$.

1. † Cf. *The Principles of Mathematics*, by Bertrand Russell (Cambridge, 1903).
2. † Cf. *Formulaire mathématique* (Turin, ed. of 1903); earlier formulations of the bases of arithmetic are given by him in the editions of 1898 and of 1901. The variations are only trivial.

It may be noticed that (iv) is the familiar principle of mathematical induction. Peano in an historical note refers its first explicit employment, although without a general enunciation, to Maurolycus in his work, *Arithmeticorum libri duo* (Venice, 1575).

But now the difficulty of confining mathematics to being the science of number and quantity is immediately apparent. For there is no self-contained science of cardinal numbers. The proof of the six premisses requires an elaborate investigation into the general properties of classes and relations which can be deduced by the strictest reasoning from our ultimate logical principles. Also it is purely arbitrary to erect the consequences of these six principles into a separate science. They are excellent principles of the highest value, but they are in no sense the necessary premisses which must be proved before any other propositions of cardinal numbers can be established. On the contrary, the premisses of arithmetic can be put in other forms, and, furthermore, an indefinite number of propositions of arithmetic can be proved directly from logical principles without mentioning them. Thus, while arithmetic may be defined as that branch of deductive reasoning concerning classes and relations which is concerned with the establishment of propositions concerning cardinal numbers, it must be added that the introduction of cardinal numbers makes no great break in this general science. It is no more than an interesting subdivision in a general theory.

Ordinal Numbers.—We must first understand what is meant by "order," that is, by "serial arrangement." An order of a set of things is to be sought in that relation holding between members of the set which constitutes that order. The set viewed as a class has many orders. Thus the telegraph posts along a certain road have a space-order very obvious to our senses; but they have also a time-order according to dates of erection, perhaps more important to the postal authorities who replace them after fixed intervals. A set of cardinal numbers have an order of magnitude, often called *the* order of the set because of its insistent obviousness to us; but, if they are the numbers drawn in a lottery, their time-order of occurrence in that drawing also ranges them in an order of some importance. Thus the order is defined by the "serial" relation. A relation (R) is serial[1†] when (1) it implies diversity, so that, if x has the relation R to y, x is diverse from y; (2) it is transitive, so that if x has the relation R to y, and y to z, then x has the relation R to z; (3) it has the property of connexity, so that if x and y are things to which any things bear the relation R, or which bear the relation R to any things, then *either* x is identical with y, *or* x has the relation R to y, *or* y has the relation R to x. These conditions are necessary and sufficient to secure that our ordinary ideas of "preceding" and "succeeding" hold in respect to the relation R. The "field" of relation R is the class of things ranged in order by it. Two relations R and R′ are said to be ordinally similar, if a one-one relation holds between the members of the two fields of R and R′, such that if x and y are any two members of the field of R, such that x has the relation R to y, and if x' and y' are the correlates in the field of R′ of x and y, then in all such cases x' has the relation R′ to y', and conversely, interchanging the dashes on the letters, *i.e.* R and R′, x and x', &c. It is evident that the ordinal similarity of two relations implies

1. †Cf. Russell, *loc. cit.*, pp. 199–256.

the cardinal similarity of their fields, but not conversely. Also, two relations need not be serial in order to be ordinally similar; but if one is serial, so is the other. The relation-number of a relation is the class whose members are all those relations which are ordinally similar to it. This class will include the original relation itself. The relation-number of a relation should be compared with the cardinal number of a class. When a relation is serial its relation-number is often called its serial type. The addition and multiplication of two relation-numbers is defined by taking two relations R and S, such that (1) their fields have no |880a| terms in common; (2) their relation-numbers are the two relation-numbers in question, and then by defining by reference to R and S two other suitable relations whose relation-numbers are defined to be respectively the sum and product of the relation-numbers in question. We need not consider the details of this process. Now if *n* be any finite cardinal number, it can be proved that the class of those serial relations, which have a field whose cardinal number is *n*, is a relation-number. This relation-number is the ordinal number corresponding to *n*; let it be symbolized by $\dot{n}$. Thus, corresponding to the cardinal numbers 2, 3, 4 . . . there are the ordinal numbers $\dot{2},\dot{3},\dot{4}$ . . . The definition of the ordinal number 1 requires some little ingenuity owing to the fact that no serial relation can have a field whose cardinal number is 1; but we must omit here the explanation of the process. The ordinal number $\dot{0}$ is the class whose sole member is the null relation—that is, the relation which never holds between any pair of entities. The definitions of the finite ordinals can be expressed without use of the corresponding cardinals, so there is no essential priority of cardinals to ordinals. Here also it can be seen that the science of the finite ordinals is a particular subdivision of the general theory of classes and relations. Thus the illusory nature of the traditional definition of mathematics is again illustrated.

Cantor's Infinite Numbers.—Owing to the correspondence between the finite cardinals and the finite ordinals, the propositions of cardinal arithmetic and ordinal arithmetic correspond point by point. But the definition of the cardinal number of a class applies when the class is not finite, and it can be proved that there are different infinite cardinal numbers, and that there is a least infinite cardinal, now usually denoted by $\aleph_0$, where ℵ is the Hebrew letter aleph. Similarly, a class of serial relations, called *well-ordered* serial relations, can be defined, such that their corresponding relation-numbers include the ordinary finite ordinals, but also include relation-numbers which have many properties like those of the finite ordinals, though the fields of the relations belonging to them are not finite. These relation-numbers are the infinite ordinal numbers. The arithmetic of the infinite cardinals does not correspond to that of the infinite ordinals. The theory of these extensions of the ideas of number is dealt with in the article NUMBER. It will suffice to mention here that Peano's fourth premiss of arithmetic does not hold for infinite cardinals or for infinite ordinals. Contrasting the above definitions of number, cardinal and ordinals, with the alternative theory that number is an ultimate idea incapable of definition, we notice that our procedure exacts a greater attention, combined with a smaller credulity; for every idea, assumed as ultimate, demands a separate act of faith.

The Data of Analysis.—Rational numbers and real numbers in general can now be defined according to the same general method. If *m* and *n* are finite cardinal numbers, the rational number *m/n* is the relation which any finite cardinal number

x bears to any finite cardinal number y when $n \times x = m \times y$. Thus the rational number one, which we will denote by 1_r, is not the cardinal number 1; for 1_r is the relation 1/1 as defined above, and is thus a relation holding between certain pairs of cardinals. Similarly, the other rational integers must be distinguished from the corresponding cardinals. The arithmetic of rational numbers is now established by means of appropriate definitions, which indicate the entities meant by the operations of addition and multiplication. But the desire to obtain general enunciations of theorems without exceptional cases has led mathematicians to employ entities of ever-ascending types of elaboration. These entities are not created by mathematicians, they are employed by them, and their definitions should point out the construction of the new entities in terms of those already on hand. The real numbers, which include irrational numbers, have now to be defined. Consider the serial arrangement of the rationals in their order of magnitude. A real number is a class (α, say) of rational numbers which satisfies the condition that it is the same as the class of those rationals each of which precedes at least one member of α. Thus, consider the class of rationals less than 2_r; any member of this class precedes some other members of the class—thus 1/2 precedes 4/3, 3/2 and so on; also the class of predecessors of predecessors of 2_r is itself the class of predecessors of 2_r. Accordingly this class is a real number; it will be called the real number 2_R. Note that the class of rationals less than or equal to 2_r is not a real number. For 2_r is not a predecessor of some member of the class. In the above example 2_R is an integral real number, which is distinct from a rational integer, and from a cardinal number. Similarly, any rational real number is distinct from the corresponding rational number. But now the irrational real numbers have all made their appearance. For example, the class of rationals whose squares are less than 2_r satisfies the definition of a real number; it is the real number $\sqrt{2}$. The arithmetic of real numbers follows from appropriate definitions of the operations of addition and multiplication. Except for the immediate purposes of an explanation, such as the above, it is unnecessary for mathematicians to have separate symbols, such as 2, 2_r and 2_R, or 2/3 and $(2/3)_R$. Real numbers with signs (+ or −) are now defined. If a is a real number, $+a$ is defined to be the relation which any real number of the form $x + a$ bears to the real number x, and $-a$ is the relation which any real number x bears to the real number $x + a$. The addition and multiplication of these "signed" real |880b| numbers is suitably defined, and it is proved that the usual arithmetic of such numbers follows. Finally, we reach a complex number of the nth order. Such a number is a "one-many" relation which relates n signed real numbers (or n algebraic complex numbers when they are already defined by this procedure) to the n cardinal numbers 1, 2 . . . n respectively. If such a complex number is written (as usual) in the form $x_1e_1 + x_2e_2 + \ldots + x_ne_n$, then this particular complex number relates x_1 to 1, x_2 to 2, . . . x_n to n. Also the "unit" e_1 (or e_s) considered as a number of the system is merely a shortened form for the complex number $(+1)\,e_1 + oe_2 + \ldots + oe_n$. This last number exemplifies the fact that one signed real number, such as 0, may be correlated to many of the n cardinals, such as 2 . . . n in the example, but that each cardinal is only correlated with one signed number. Hence the relation has been called above "one-many." The sum of two complex numbers $x_1e_1 + x_2e_2 + \ldots + x_ne_n$

and $y_1e_1 + y_2e_2 + \ldots + y_ne_n$ is always defined to be the complex number $(x_1 + y_1)e_1 + (x_2 + y_2)e_2 + \ldots + (x_n + y_n)e_n$. But an indefinite number of definitions of the product of two complex numbers yield interesting results. Each definition gives rise to a corresponding algebra of higher complex numbers. We will confine ourselves here to algebraic complex numbers—that is, to complex numbers of the second order taken in connexion with that definition of multiplication which leads to ordinary algebra. The product of two complex numbers of the second order—namely, $x_1e_1 + x_2e_2$ and $y_1e_1 + y_2e_2$, is in this case defined to mean the complex $(x_1y_1 - x_2y_2)e_1 + (x_1y_2 + x_2y_1)e_2$. Thus $e_1 \times e_1 = e_1$, $e_2 \times e_2 = -e_1$, $e_1 \times e_2 = e_2 \times e_1 = e_2$. With this definition it is usual to omit the first symbol e_1, and to write i or $\sqrt{-1}$ instead of e_2. Accordingly, the typical form for such a complex number is $x + yi$, and then with this notation the above-mentioned definition of multiplication is invariably adopted. The importance of this algebra arises from the fact that in terms of such complex numbers with this definition of multiplication the utmost generality of expression, to the exclusion of exceptional cases, can be obtained for theorems which occur in analogous forms, but complicated with exceptional cases, in the algebras of real numbers and of signed real numbers. This is exactly the same reason as that which has led mathematicians to work with signed real numbers in preference to real numbers, and with real numbers in preference to rational numbers. The evolution of mathematical thought in the invention of the data of analysis has thus been completely traced in outline.

Definition of Mathematics.— It has now become apparent that the traditional field of mathematics in the province of discrete and continuous number can only be separated from the general abstract theory of classes and relations by a wavering and indeterminate line. Of course a discussion as to the mere application of a word easily degenerates into the most fruitless logomachy. It is open to any one to use any word in any sense. But on the assumption that "mathematics" is to denote a science well marked out by its subject matter and its methods from other topics of thought, and that at least it is to include all topics habitually assigned to it, there is now no option but to employ "mathematics" in the general sense[1†] of the "science concerned with the logical deduction of consequences from the general premisses of all reasoning."

Geometry.—The typical mathematical proposition is: "If $x, y, z \ldots$ satisfy such and such conditions, then such and such other conditions hold with respect to them." By taking fixed conditions for the hypothesis of such a proposition a definite department of mathematics is marked out. For example, geometry is such a department. The "axioms" of geometry are the fixed conditions which occur in the hypotheses of the geometrical propositions. The special nature of the "axioms" which constitute geometry is considered in the article GEOMETRY (*Axioms*). It is sufficient to observe here that they are concerned with special types of classes of classes and of classes

1. †The first unqualified explicit statement of *part* of this definition seems to be by B. Peirce, "Mathematics is the science which draws necessary conclusions" (*Linear Associative Algebra*, § i. (1870), republished in the *Amer. Journ. of Math.*, vol. iv. (1881) [no. 1, pp. 97–229]). But it will be noticed that the second half of the definition in the text—"from the general premisses of all reasoning"—is left unexpressed. The full expression of the idea and its development into a philosophy of mathematics is due to Russell, *loc. cit.*

of relations, and that the connexion of geometry with number and magnitude is in no way an essential part of the foundation of the science. In fact, the whole theory of measurement in geometry arises at a comparatively late stage as the result of a variety of complicated considerations.

Classes and Relations.—The foregoing account of the nature of mathematics necessitates a strict deduction of the general properties |881a| of classes and relations from the ultimate logical premisses. In the course of this process, undertaken for the first time with the rigour of mathematicians, some contradictions have become apparent. That first discovered is known as Burali-Forti's contradiction,[1†] and consists in the proof that there both is and is not a greatest infinite ordinal number. But these contradictions do not depend upon any theory of number, for Russell's contradiction[2†] does not involve number in any form. This contradiction arises from considering the class possessing as members all classes which are not members of themselves. Call this class w; then to say that x is a w is equivalent to saying that x is not an x. Accordingly, to say that w is a w is equivalent to saying that w is not a w. An analogous contradiction can be found for relations. It follows that a careful scrutiny of the very idea of classes and relations is required. Note that classes are here required in extension, so that the class of human beings and the class of rational featherless bipeds are identical; similarly for relations, which are to be determined by the entities related. Now a class in respect to its components is many. In what sense then can it be one? This problem of "the one and the many" has been discussed continuously by the philosophers.[3†] All the contradictions can be avoided, and yet the use of classes and relations can be preserved as required by mathematics, and indeed by common sense, by a theory which denies to a class—or relation—existence or being in any sense in which the entities composing it—or related by it—exist. Thus, to say that a pen is an entity and the class of pens is an entity is merely a play upon the word "entity"; the second sense of "entity" (if any) is indeed derived from the first, but has a more complex signification. Consider an incomplete proposition, incomplete in the sense that some entity which ought to be involved in it is represented by an undetermined x, which may stand for any entity. Call it a propositional function; and, if ϕx be a propositional function, the undetermined variable x is the argument. Two propositional functions ϕx and ψx are "extensionally identical" if any determination of x in ϕx which converts ϕx into a true proposition also converts ψx into a true proposition, and conversely for ψ and ϕ. Now consider a propositional function $F\chi$ in which the variable argument χ is itself a propositional function. If $F\chi$ is true when, and only when, χ is determined to be either ϕ or some other propositional function extensionally equivalent to ϕ, then the proposition $F\phi$ is of the form which is ordinarily recognized as being about the class determined by ϕx taken in extension—that is, the class of entities for which ϕx is a true proposition when x is determined to be any one of them. A similar theory holds for relations which arise from the consideration of propositional functions

1. †"Una questione sui numeri transfiniti," *Rend. del circolo mat. di Palermo*, vol. xi. (1897) [pp. 154–164]; and Russell, *loc. cit.*, ch. xxxviii.
2. †Cf. Russell, *loc. cit.*, ch. x.
3. †Cf. *Pragmatism: a New Name for some Old Ways of Thinking* (1907) [by William James (New York: Longmans, Green, and Co.)].

with two or more variable arguments. It is then possible to define by a parallel elaboration what is meant by classes of classes, classes of relations, relations between classes, and so on. Accordingly, the number of a class of relations can be defined, or of a class of classes, and so on. This theory[1†] is in effect a theory of the *use* of classes and relations, and does not decide the philosophic question as to the sense (if any) in which a class in extension is one entity. It does indeed deny that it is an entity in the sense in which one of its members is an entity. Accordingly, it is a fallacy for any determination of x to consider "x is an x" or "x is not an x" as having the meaning of propositions. Note that for any determination of x, "x is an x" and "x is not an x," are neither of them fallacies but are both meaningless, according to this theory. Thus Russell's contradiction vanishes, and an examination of the other contradictions shows that they vanish also.

Applied Mathematics.—The selection of the topics of mathematical inquiry among the infinite variety open to it has been guided by the useful applications, and indeed the abstract theory has only recently been disentangled from the empirical elements connected with these applications. For example, the application of the theory of cardinal numbers to classes of physical entities involves in practice some process of counting. It is only recently that the *succession* of processes which is involved in any act of counting has been seen to be irrelevant to the idea of number. Indeed, it is only by experience that we can know that any definite process of counting will give the true cardinal number of some class of entities. It is perfectly possible to imagine a universe in which any act of counting by a being in it annihilated some members of the class counted during the time and only during the time of its continuance. A legend of the Council of Nicea[2†] illustrates this point: "When the Bishops took their |881b| places on their thrones, they were 318; when they rose up to be called over, it appeared that they were 319; so that they never could make the number come right, and whenever they approached the last of the series, he immediately turned into the likeness of his next neighbour." Whatever be the historical worth of this story, it may safely be said that it cannot be disproved by deductive reasoning from the premisses of abstract logic. The most we can do is to assert that a universe in which such things are liable to happen on a large scale is unfitted for the practical application of the theory of cardinal numbers. The application of the theory of real numbers to physical quantities involves analogous considerations. In the first place, some physical process of addition is presupposed, involving some inductively inferred law of permanence during that process. Thus in the theory of masses we must know that two pounds of lead when put together will counterbalance in the scales two pounds of sugar, or a pound of lead and a pound of sugar. Furthermore, the sort of continuity of the series (in order of magnitude) of rational numbers is known to be different from that of the series of real numbers. Indeed, mathematicians now reserve "continuity" as the term for the latter kind of continu-

1. † Due to Bertrand Russell, cf. "Mathematical Logic as based on the Theory of Types," *Amer. Journ. of Math.* vol. xxx. (1908) [no. 3 (July), pp. 222–262]. It is more fully explained by him, with later simplifications, in *Principia mathematica* (Cambridge).
2. † Cf. Stanley's *Eastern Church*, Lecture v [Arthur Penrhyn Stanley, *Lectures on the History of the Eastern Church* (London: John Murray, 1861)].

ity; the mere property of having an infinite number of terms between any two terms is called "compactness." The compactness of the series of rational numbers is consistent with quasi-gaps in it—that is, with the possible absence of limits to classes in it. Thus the class of rational numbers whose squares are less than 2 has no upper limit among the rational numbers. But among the real numbers all classes have limits. Now, owing to the necessary inexactness of measurement, it is impossible to discriminate directly whether any kind of continuous physical quantity possesses the compactness of the series of rationals or the continuity of the series of real numbers. In calculations the latter hypothesis is made because of its mathematical simplicity. But, the assumption has certainly no a priori grounds in its favour,[1] and it is not very easy to see how to base it upon experience. For example, if it should turn out that the mass of a body is to be estimated by counting the number of corpuscles (whatever they may be) which go to form it, then a body with an irrational measure of mass is intrinsically impossible. Similarly, the continuity of space apparently rests upon sheer assumption unsupported by any a priori or experimental grounds. Thus the current applications of mathematics to the analysis of phenomena can be justified by no a priori necessity.

In one sense there is no science of applied mathematics. When once the fixed conditions which any hypothetical group of entities are to satisfy have been precisely formulated, the deduction of the further propositions, which also will hold respecting them, can proceed in complete independence of the question as to whether or no any such group of entities can be found in the world of phenomena. Thus rational mechanics, based on the Newtonian Laws, viewed as mathematics is independent of its supposed application, and hydrodynamics remains a coherent and respected science though it is extremely improbable that any perfect fluid exists in the physical world. But this unbendingly logical point of view cannot be the last word upon the matter. For no one can doubt the essential difference between characteristic treatises upon "pure" and "applied" mathematics. The difference is a difference in method. In pure mathematics the hypotheses which a set of entities are to satisfy are given, and a group of interesting deductions are sought. In "applied mathematics" the "deductions" are given in the shape of the experimental evidence of natural science, and the hypotheses from which the "deductions" can be deduced are sought. Accordingly, every treatise on applied mathematics, properly so-called, is directed to the criticism of the "laws" from which the reasoning starts, or to a suggestion of results which experiment may hope to find. Thus if it calculates the result of some experiment, it is not the experimentalist's well-attested results which are on their trial, but the basis of the calculation. Newton's *Hypotheses non fingo*[2] was a proud boast, but it rests upon an entire misconception of the capacities of the mind of man in dealing with external nature.

|882a| *Synopsis of Existing Developments of Pure Mathematics.*—A complete classification of mathematical sciences, as they at present exist, is to be found in the *International Catalogue of Scientific Literature* promoted by the Royal Society. The classification in question was drawn up by an international committee of eminent

1. We have adopted ESP's correction changing a period to a comma.
2. 'I feign/frame/contrive no hypotheses'.

mathematicians, and thus has the highest authority. It would be unfair to criticize it from an exacting philosophical point of view. The practical object of the enterprise required that the proportionate quantity of yearly output in the various branches, and that the liability of various topics as a matter of fact to occur in connexion with each other, should modify the classification.

Section A deals with pure mathematics. Under the general heading "*Fundamental Notions*" occur the subheadings "*Foundations of Arithmetic*," with the topics rational, irrational and transcendental numbers, and aggregates; "*Universal Algebra*," with the topics complex numbers, quaternions, ausdehnungslehre, vector analysis, matrices, and algebra of logic; and "*Theory of Groups*," with the topics finite and continuous groups. For the subjects of this general heading see the articles ALGEBRA, UNIVERSAL; GROUPS, THEORY OF; INFINITESIMAL CALCULUS; NUMBER; QUATERNIONS; VECTOR ANALYSIS. Under the general heading "*Algebra and Theory of Numbers*" occur the subheadings "*Elements of Algebra*," with the topics rational polynomials, permutations, &c., partitions, probabilities; "*Linear Substitutions*," with the topics determinants, &c., linear substitutions, general theory of quantics; "*Theory of Algebraic Equations*," with the topics existence of roots, separation of and approximation to, theory of Galois, &c.; "*Theory of Numbers*," with the topics congruences, quadratic residues, prime numbers, particular irrational and transcendental numbers. For the subjects of this general heading see the articles ALGEBRA; ALGEBRAIC FORMS; ARITHMETIC; COMBINATORIAL ANALYSIS; DETERMINANTS; EQUATION; FRACTION, CONTINUED; INTERPOLATION; LOGARITHMS; MAGIC SQUARE; PROBABILITY. Under the general heading "*Analysis*" occur the subheadings "*Foundations of Analysis*," with the topics theory of functions of real variables, series and other infinite processes, principles and elements of the differential and of the integral calculus, definite integrals, and calculus of variations; "*Theory of Functions of Complex Variables*," with the topics functions of one variable and of several variables; "*Algebraic Functions and their Integrals*," with the topics algebraic functions of one and of several variables, elliptic functions and single theta functions, Abelian integrals; "*Other Special Functions*," with the topics Euler's, Legendre's, Bessel's and automorphic functions; "*Differential Equations*," with the topics existence theorems, methods of solution, general theory; "*Differential Forms and Differential Invariants*," with the topics differential forms, including Pfaffians, transformation of differential forms, including tangential (or contact) transformations, differential invariants; "*Analytical Methods connected with Physical Subjects*," with the topics harmonic analysis, Fourier's series, the differential equations of applied mathematics, Dirichlet's problem; "*Difference Equations and Functional Equations*," with the topics recurring series, solution of equations of finite differences and functional equations. For the subjects of this heading see the articles DIFFERENTIAL EQUATIONS; FOURIER'S SERIES; CONTINUED FRACTIONS; FUNCTION; FUNCTION OF REAL VARIABLES; FUNCTION COMPLEX; GROUPS, THEORY OF; INFINITESIMAL CALCULUS; MAXIMA AND MINIMA; SERIES; SPHERICAL HARMONICS; TRIGONOMETRY; VARIATIONS, CALCULUS OF. Under the general heading "*Geometry*" occur the subheadings "*Foundations*," with the topics principles of geometry, non-Euclidean geometries, hyperspace, methods of analytical geometry; "*Elementary Geometry*," with the topics planimetry, stereometry, trigonometry, descriptive geometry; "*Geometry of Conics and Quadrics*," with the implied topics; "*Algebraic Curves and Surfaces of Degree higher than the Second*," with the implied topics; "*Transformations and General Methods for Algebraic Configurations*," with the topics collineation, duality, transformations, correspondence, groups of points on algebraic curves and surfaces, genus of curves and surfaces, enumerative geometry, connexes, complexes, congruences, higher elements in space, algebraic configurations in hyperspace; "*Infinitesimal Geometry: applications of Differential and Integral Calculus to Geometry*," with the topics kinematic geometry, curvature, rectification and quadrature, special transcendental curves and surfaces; "*Differential Geometry: applications of Differential Equations to Geometry*," with the topics curves on surfaces, minimal surfaces, surfaces determined by differential properties, conformal and other representation of surfaces on others, deformation of surfaces, orthogonal and isothermic surfaces. For the subjects under this heading see the articles CONIC SECTIONS; CIRCLE; CURVE; GEOMETRICAL CONTINUITY; GEOMETRY, *Axioms of*; GEOMETRY, *Euclidean*; GEOMETRY, *Projective*; GEOMETRY, *Analytical*; GEOMETRY, *Line*; KNOTS, MATHEMATICAL THEORY OF; MENSURATION; MODELS; PROJECTION; SURFACE; TRIGONOMETRY.

This survey of the existing developments of pure mathematics confirms the conclusions arrived at from the previous survey of the theoretical principles of the subject. Functions, operations, transformations, substitutions, correspondences, are but names for various types of relations. A group is a class of relations possessing |882b| a special property. Thus the modern ideas, which have so powerfully extended and unified the subject, have loosened its connexion with "number" and "quantity," while bringing ideas of form and structure into increasing prominence. Number must indeed ever remain the great topic of mathematical interest, because it is in reality the great topic of applied mathematics. All the world, including savages who cannot count beyond five, daily "apply" theorems of number. But the complexity of the idea of number is practically illustrated by the fact that it is best studied as a department of a science wider than itself.

Synopsis of Existing Developments of Applied Mathematics.—Section B of the *International Catalogue* deals with mechanics. The heading "*Measurement of Dynamical Quantities*" includes the topics units, measurements, and the constant of gravitation. The topics of the other headings do not require express mention. These headings are: "*Geometry and Kinematics of Particles and Solid Bodies*"; "*Principles of Rational Mechanics*"; "*Statics of Particles, Rigid Bodies, &c.*"; "*Kinetics of Particles, Rigid Bodies, &c.*"; "*General Analytical Mechanics*"; "*Statics and Dynamics of Fluids*"; "*Hydraulics and Fluid Resistances*"; "*Elasticity*." For the subjects of this general heading see the articles MECHANICS; DYNAMICS, ANALYTICAL; GYROSCOPE; HARMONIC ANALYSIS; WAVE; HYDRODYNAMICS; ELASTICITY; MOTION, LAWS OF; ENERGY ENERGETICS; ASTRONOMY (*Celestial Mechanics*); TIDE. Mechanics (including dynamical astronomy) is that subject among those traditionally classed as "applied" which has been most completely transfused by mathematics—that is to say, which is studied with the deductive spirit of the pure mathematician, and not with the covert inductive intention overlaid with the superficial forms of deduction, characteristic of the applied mathematician.

Every branch of physics gives rise to an application of mathematics. A prophecy may be hazarded that in the future these applications will unify themselves into a mathematical theory of a hypothetical substructure of the universe, uniform under all the diverse phenomena. This reflection is suggested by the following articles: AETHER; MOLECULES; CAPILLARY ACTION; DIFFUSION; RADIATION, THEORY OF; and others.

The applications of mathematics to statistics (see STATISTICS and PROBABILITY) should not be lost sight of; the leading fields for these applications are insurance, sociology, variation in zoology and economics.

The History of Mathematics.—The history of mathematics is in the main history of its various branches. A short account of the history of each branch will be found in connexion with the article which deals with it. Viewing the subject as a whole, and apart from remote developments which have not in fact seriously influenced the great structure of the mathematics of the European races, it may be said to have had its origin with the Greeks, working on pre-existing fragmentary lines of thought derived from the Egyptians and Phoenicians. The Greeks created the sciences of geometry and of number as applied to the measurement of continuous quantities. The great abstract ideas (considered directly and not merely in tacit use) which have dominated the science were due to them—namely, ratio, irrationality, continuity, the point, the straight line, the plane. This period lasted[1†] from the time of Thales, *c.* 600 B.C., to the capture of Alexandria by the Mahommedans, A.D. 641. The medieval Arabians invented our system of numeration and developed algebra. The next period of advance stretches from the Renaissance to Newton and Leibnitz at the end of the 17th century. During this period logarithms were invented, trigonometry and algebra developed, analytical geometry invented, dynamics put upon a sound basis, and the period closed with the magnificent invention of (or at least the perfecting of) the differential calculus by Newton and Leibnitz and the discovery of gravitation. The 18th century witnessed a rapid development of analysis, and the period culminated with the genius of Lagrange and Laplace. This period may be conceived as continuing throughout the first quarter of the 19th century. It was remarkable both for the brilliance of its achievements and for the large number of French mathematicians of the first rank who flourished during it. The next period was inaugurated in analysis by K. F. Gauss, N. H. Abel and A. L. Cauchy. Between them the general theory of the complex variable, and of the various "infinite" processes of mathematical analysis, was established, while other mathematicians, such as Poncelet, Steiner, Lobatschewsky[2] and von Staudt, were founding modern geometry, and Gauss inaugurated the differential geometry of surfaces. The applied mathematical sciences of light, electricity and electromagnetism, |883a| and of heat, were now largely developed. This school of mathematical thought lasted beyond the middle of the century, after which a change and further development can be traced. In the next and last period the progress of pure mathematics has been dominated by the critical spirit introduced by the German mathematicians under the guidance of Weierstrass, though foreshadowed by earlier analysts, such as Abel. Also such ideas as those of invariants, groups and of form, have modified the entire science. But the progress

1. †Cf. *A Short Account of Mathematics*, by W. W. R. Ball [The original erroneously had 'History' rather than 'Account'. Walter W. Rouse Ball, *A Short Account of the History of Mathematics* (London: Macmillan, 1888)].
2. Lobatchewski's (1792–1856) name is now usually rendered as Nikolai *Lobachevsky*.

in all directions has been too rapid to admit of any one adequate characterization. During the same period a brilliant group of mathematical physicists, notably Lord Kelvin (W. Thomson), H. V. Helmholtz, J. C. Maxwell, H. Hertz, have transformed applied mathematics by systematically basing their deductions upon the Law of the conservation of energy, and the hypothesis of an ether pervading space.

BIBLIOGRAPHY.—References to the works containing expositions of the various branches of mathematics are given in the appropriate articles. It must suffice here to refer to sources in which the subject is considered as one whole. Most philosophers refer in their works to mathematics more or less cursorily, either in the treatment of the ideas of number and magnitude, or in their consideration of the alleged a priori and necessary truths. A bibliography of such references would be in effect a bibliography of metaphysics, or rather of epistemology. The founder of the modern point of view, explained in this article, was Leibnitz, who, however, was so far in advance of contemporary thought that his ideas remained neglected and undeveloped until recently; cf. *Opuscules et fragments inédits de Leibnitz. Extraits des manuscrits de la bibliothèque royale de Hanovre*, by Louis Couturat (Paris, 1903), especially pp. 356–399, "Generales inquisitiones de analysi notionum et veritatum" (written in 1686); also cf. *La Logique de Leibnitz*, already referred to. For the modern authors who have rediscovered and improved upon the position of Leibnitz, cf. *Grundgesetze der Arithmetik, begriffsschriftlich abgeleitet von Dr. G. Frege, a.o. Professor an der Univ. Jena* (Bd. i., 1893; Bd. ii., 1903, Jena); also cf. Frege's earlier works, *Begriffsschrift, eine der arithmetischen nachgebildete Formelsprache des reinen Denkens* (Halle, 1879), and *Die Grundlagen der Arithmetik* (Breslau, 1884); also cf. Bertrand Russell, *The Principles of Mathematics* (Cambridge, 1903), and his article on "Mathematical Logic" in *Amer. Journ. of Math.* (vol. xxx., 1908).[1] Also the following works are of importance, though not all expressly expounding the Leibnitzian point of view: cf. G. Cantor, "Grundlagen einer allgemeinen Mannigfaltigkeitslehre," *Math. Annal.*, vol. xxi. (1883)[2] and subsequent articles in vols. xlvi.[3] and xlix.;[4] also R. Dedekind, *Stetigkeit und irrationales Zahlen* (1st ed., 1872), and *Was sind und was sollen die Zahlen?* (1st ed., 1887), both tracts translated into English under the title *Essays on the Theory of Numbers* (Chicago, 1901). These works of G. Cantor and Dedekind were of the greatest importance in the progress of the subject. Also cf. G. Peano (with various collaborators of the Italian school), *Formulaire de mathématiques* (Turin, various editions, 1894–1908; the earlier editions are the more interesting philosophically); Felix Klein, *Lectures on Mathematics* (New York, 1894); W. K. Clifford, *The Common Sense of the exact Sciences* (London, 1885); H. Poincaré, *La Science et l'hypothèse* (Paris, 1st ed., 1902), English translation under the title, *Science and Hypothesis* (London, 1905); L. Couturat, *Les Principes des mathématiques* (Paris, 1905); E. Mach,

1. Whitehead's original reference was to the '*Amer. Quart. Journ.*', but there is no 'Quarterly' in the journal's title.
2. The title Whitehead cites was an 1883 book by Cantor; in the cited *Mathematische Annalen*, vol. 21, Cantor published, in two parts, 'Ueber unendliche, lineare Punktmannichfaltigkeiten', nos 1 and 4 (March and December 1883), pp. 51–58 and 545–591.
3. Georg Cantor, 'Beiträge zur Begründung der transfiniten Mengenlehre (1)', *Mathematische Annalen*, vol. 46, no. 4 (November 1895), pp. 481–512.
4. Georg Cantor, 'Beiträge zur Begründung der transfiniten Mengenlehre (2)', *Mathematische Annalen*, vol. 49, no. 2 (June 1897), pp. 207–246.

Die Mechanik in ihrer Entwickelung (Leipzig,[1] 1883), English translation under the title, *The Science of Mechanics* (London, 1893); K. Pearson, *The Grammar of Science* (London, 1st ed., 1892; 2nd ed., 1900, enlarged); A. Cayley, *Presidential Address* (Brit. Assoc., 1883);[2] B. Russell and A. N. Whitehead, *Principia Mathematica* (Cambridge, 1910).[3] For the history of mathematics the one modern and complete source of information is M. Cantor's *Vorlesungen über Geschichte der Mathematik* (Leipzig, 1st. Bd., 1880; 2nd Bd., 1892; 3rd Bd., 1898; 4th Bd., 1908; 1st Bd., *von den ältesten Zeiten bis zum Jahre 1200, n. Chr.;* 2nd Bd. *von 1200–1668*; 3rd Bd., *von 1668–1758*; 4th Bd., *von 1795 bis 1799*); W. W. R. Ball, *A Short Account*[4] *of Mathematics* (London 1st ed., 1888, three subsequent editions, enlarged and revised, and translations into French and Italian).

1. Corrected from 'Prague'.
2. 'Address by Arthur Cayley, President', in *Report of the Fifty-Third Meeting of the British Association for the Advancement of Science* (London: John Murray, 1884), pp. 1–37.
3. PM's three volumes were published in 1910, 1912 and 1913, but the date originally listed here was 1911, which we have changed to 1910. Note that Russell and Whitehead's names are also reversed from how they would appear in the finished publication. It is possible that when Whitehead submitted this *Encyclopaedia* article he anticipated the first volume of PM to be published in 1911, and for Russell's name to be listed first.
4. The original erroneously had 'History'.

The Place of Elementary Mathematics in a Liberal Education

1911 ***Journal of the Association of Teachers of Mathematics for the Southeastern Part of England*, vol. 1, no. 1 (December), pp. 2–12**

1947 *Essays in Science and Philosophy* (New York: Philosophical Library), pp. 175–188

This address – Whitehead's first on education – appeared as an article in the journal of the short-lived[1] *Association of Teachers of Mathematics for the Southeastern Part of England, in December 1911. It was originally delivered as the presidential address at the Association's inaugural meeting on 24 November 1911, five months after his appointment at University College London. The meeting took place at Tonbridge School in Kent, an independent boys' boarding school founded in 1553.*

A driving force of the Association was Tonbridge mathematics teacher George Edward St Lawrence Carson (1873–1934), who served as the Association's Chairman. He stated in an editorial on the first page of this same inaugural journal issue that 'England is almost alone in preserving an unnatural separation between the professed mathematician and the teacher of Elementary Mathematics. Should this Association take the smallest part in bridging this chasm its existence will be more than justified.'[2]

The title in the journal was simply 'The Presidential Address'. We have taken the above title from the short introductory note that followed: 'Dr. A. N. Whitehead delivered his address at the Inaugural Meeting held at Tonbridge on November 24th, the subject being "The Place of Elementary Mathematics in a Liberal Education."'

It is not clear why Whitehead did not choose to have this address reprinted in either The Organisation of Thought *or* The Aims of Education, *both of which contained multiple chapters on education. When it was finally included in* Essays in Science and Philosophy *in 1947, it appeared mostly unchanged. It was retitled as 'Mathematics and Liberal Education: An Address', and the first two paragraphs were excised, likely because they were too specific to the event at which Whitehead delivered the remarks.*

Whitehead was delivering this address at a time when education in general in England was shifting, and mathematics education in particular. The teaching of geometry in England had recently been thrown into confusion with the disestablishment of Euclid's Elements *as the definitive textbook for the subject in 1903 (see p. 348n2),*

1. The Association's Chairman (see below) had planned to publish the three issues of the journal annually, but his departure in 1913 to take up a position in mathematics and education at the University of Liverpool resulted in the journal being abandoned after eight issues. See Michael H. Price, *Mathematics for the Multitude? A History of the Mathematical Association* (London: Mathematical Association, 1994), p. 106.
2. George Edward St Lawrence Carson, 'Editorial', *Journal of the Association of Teachers of Mathematics for the Southeastern Part of England,* vol. 1, no. 1 (December), p. 1.

effectively forcing mathematics teachers throughout the country to rethink the curriculum from the ground up.

Many of Whitehead's best-known ideas about education can already be seen in this address, including his stress on teaching a few things deeply rather than a wide array of subjects shallowly, and on the importance of linking education to immediately demonstrable purposes, rather than merely teaching generalities and abstractions.

The formation of this Association is a welcome reminder of the spirit of enterprise which now informs the teaching of mathematics. No longer are we content to impart a few technical rules in a lifeless way; and it will certainly not be due to any supineness in the teachers of the present day if mathematics does not obtain its full value as an element in a liberal education.

To-day I am not going to speak of mathematics as a necessary technical preparation for some professions, nor shall I consider the training of the few able boys[1] who are going to distinguish themselves at the University. There is really not much to say about such boys. On the whole they are very well taught, and are interested in their subject. Year by year, in scholarship examinations, I have viewed their performances with respectful astonishment. I have only one remark to make, namely, that at the University those boys develop best whose literary education has been properly attended to. By literary education I mean a true, stimulating literary education, and not a thin technical instruction in Latin and Greek grammar.[2]

The subject of my address to-day is the consideration of the part which the elements of mathematics should play in a liberal education for the generality of boys up to the age of nineteen. The boys I mean are, of course, those who are capable of a liberal education. Wealth can do much, but it cannot give brains, and it cannot give character; nor can it give the intellectual interests which come from the union of brains with character. Accordingly, I exclude the residuum of boys, and am thinking of those only with fair brains and decent interests. Happily for England these constitute the great majority of the ordinary students who pass on to our Universities.

|3| It will help us to understand the nature of our problem if we spend a few minutes in examining the ultimate reason for the existing upheaval in the scholastic world. We are, in fact, in the midst of an educational revolution caused by the dying away of the classical impulse which has dominated European thought since the time of the Renaissance. I find it a little difficult to explain my meaning exactly. I am not referring to the mere teaching of a little more or a little less of Latin and Greek. What I mean is the loss of that sustained reference to classical literature for the sake of

---

1. Whitehead was an advocate both for women's education and women's rights – as his 'On Ideals' (p. 38) and 'Liberty and the Enfranchisement of Women' (p. 272) show – and so his continual reference to 'boys' in this address, rather than to 'boys and girls' or to a gender-neutral 'children/students', may be reflective of the fact that he was delivering this address at an all-boys boarding school. Yet it is also a reminder that equality for women in England at this time still had a way to go. No women could vote in Parliamentary elections until 1918, and the two most prestigious universities – Oxford and Cambridge – would not grant equivalent degrees to women until 1920 and 1948, respectively.
2. ESP deleted these first two paragraphs.

finding in it the expression of our best thoughts on all subjects. The Greek masterpieces of literature remain masterpieces of literature, and the labours of a band of brilliant scholars have reinterpreted them to the modern world. But for all that, the scene presented to our view by the human life of to-day is essentially different to that presented either to the Greeks of two thousand years ago, or even to our grandfathers at the beginning of the nineteenth century.

There are three fundamental changes which make an unbridgeable gap. Science now enters into the very texture of our thoughts; its methods and results colour the imaginations of our poets; they modify the conclusions of philosophers and theologians. Again, mechanical inventions, which are the product of science, by altering the material possibilities of life, have transformed our industrial system, and thus have changed the very structure of Society. Finally, the idea of the World now means to us the whole round world of human affairs, from the revolutions of China to those of Peru. Even to our fathers it merely conveyed the idea of the nations of Europe, and, in particular, of the Mediterranean shores. But this provincial phase of thought is rapidly becoming impossible.

The total result of these changes is that the supreme merit of immediate relevance to the full compass of modern life has been lost to classical literature. To take a trivial example: in Athens, a reference to a potter's wheel might recall vivid memories of habitual sights; in London, it requires a footnote. Whether we regret it or no, the absolute dominance of classical ideas in education is necessarily doomed.

But it is not possible to alter the whole basis of our curriculum by a mere change in the time-table. In the first place, there is the difficulty which amateur reformers in education usually forget—that there is no time. Lack of time is the rock upon which the fairest educational schemes are wrecked. It has wrecked that scheme which our fathers constructed to meet the growing demand for the introduction of modern ideas. They simply increased the number of subjects taught. Latin and Greek were to be retained, the time given to mathematics enlarged, modern history and physical science to be added; and, at the same time, geography, music, drawing, and nature studies were not to be neglected. Also, of course, modern languages were regarded as indispensable.

The task of education with such a scheme of studies is frankly impossible. In all modern educational reform the watchword must be "concentration."

|4| But if we examine this older curriculum as it existed in practice, we shall note, unless I am much mistaken, that in general it possessed one very striking peculiarity. It was believed, with some reason, that every cultivated idea had found its best expression in the classical literatures. The result was that the whole of the general training in ideas was annexed, and with some reason, to the study of the classical languages. Other studies were, in fact, pursued as mere technical acquirements. The boys might learn both German and Greek; but it was from Sophocles and not from Goethe that they drew their ideas. Mathematics, for example, was divested of all discussion of ideas, and reduced to the aimless acquirement of formal methods of procedure. In other words, modern thought was not introduced into the educational curriculum, but merely modern technique. If, for the mass of boys and young men, we are to concentrate our education upon modern subjects, we must first transform them into a real vehicle for the inculcation of ideas. We have, in fact, to civilize them.

Now, nothing is more difficult than to transmit to our pupils real general ideas, as distinct from pretentious phrases. Nobody with any sense can confront a class for long without discovering that all sound teaching is concerned with definite, accurate achievements on the part of the pupils—to construe grammatically a Latin sentence, to solve a quadratic equation, or to find, with some precision, the specific heat of lead. Vague generalities are worse than useless, and if we attempt to embody abstractions in short, precise formulæ, the pupils will simply learn them by heart as empty sounds.

In view of this difficulty let us examine briefly how the classical languages achieved their undoubted success as vehicles of a liberal education. The advantage of education based upon them is that at every step definite aims are placed before the learner. He has to construe the author, to know the meaning and grammatical status of each word, and to render the sense in the precise equivalent English. There is nothing vague in all this; it is an accurate achievement which the pupil has to accomplish. It has also the useful property, which every teacher will appreciate, that it is easy to test whether the pupil has in reality tried to accomplish his task. He may not make sense of his translation, but he can at least know the meaning of the various words and their cases or their tenses; and, in addition to all this, the classical languages possess the supreme merit that great ideas are simultaneously presented to the mind. The noblest authors of Greece and Rome can be read. Some of us may still remember construing in our school-days Lucretius' reflections on the nature of the universe, and the account of the battle in the harbour of Syracuse, its triumph and its despair.

I will not consider further the question of the literary side of education. Not because I undervalue its importance, but because we are especially responsible for the logical training. We have really two general aims before us. In the first place, we have to teach what logic is. I do not mean by this that we should indulge in |5| the somewhat futile task of affixing names to elementary logical processes after the manner of primers in formal logic. But we have to make our pupils feel by an acquired instinct what it means to be logical, and to know a precise idea when they see it; or, rather what unfortunately is more often wanted, to know an unprecise idea when they see it. In the second place, we have to make them understand that logic applies to life. This is, in fact, the harder task. Most people agree that there are abstract precise ideas capable of logical treatment, but very few really believe that a sensible man need take any account of them. Such and such ideas, they will say, are all right in theory, but in practice they are useless. It is here that the astounding success of modern science in transforming the world makes an examination of the elements of its logical methods so vital a part of modern education. In this region ancient thought is frankly useless. It is possible that the Greeks were in all respects abler than we are, and that, if here, they would conduct our scientific investigations in a manner superior to anything to which we can attain. But the ancient Greeks are not here, and the fact remains that our modern scientific thought completely overshadows anything of the same sort which existed in the ancient world. In this connection it is a mistake to think that the Greeks discovered the elements of mathematics, and that we have added the advanced parts of the subject. The opposite is more nearly the case;

they were interested in the higher parts of the subject and never discovered its elements. The practical elements, as they are now employed in physical science, and the theoretical elements upon which the whole reposes, were alike unknown to them. Weierstrass' theory of limits and Georg Cantor's theory of sets of points are much more allied to Greek modes of thought than are our modern arithmetic, our modern theory of positive and negative numbers, our modern graphical representation of the functional relation, or our modern idea of the algebraic variable. Elementary mathematics is one of the most characteristic creations of modern thought. It is characteristic of modern thought by virtue of the intimate way in which it correlates theory and practice.

I make this point in no idle spirit, but to enforce a very serious conclusion. In the past, the teaching of elementary mathematics has suffered, not only because its ideas were sucked away from it by the dominant classics, but also because it was treated as a collection of mere uninteresting prolegomena to more advanced parts of the subject. But the mass of pupils never advanced to these further parts, and, in consequence, gained nothing but a set of purposeless dodges.

We must conceive elementary mathematics as a subject complete in itself, to be studied for its own sake. It must be purged of every element which can only be justified by reference to a more prolonged course of study. There can be nothing more destructive of true education than to spend long hours in the acquirement of ideas and methods which lead nowhere. It is fatal to all intellectual vitality. It produces, on the one hand, a sense of incompetence, of lack of grasp, |6| and of inability really to penetrate to the true meaning of things; and, on the other hand, by a natural revolt of the self-respecting intellect, it produces a distaste for ideas, and a suspicion that they are all equally futile. I have had great experience with the average product of our schools as sent up to the Universities. My general conclusion is not that they have been idle at school, or have been taught carelessly. On the contrary, their education has evidently been supervised with a conscientious vigour. But there is a widely-spread sense of boredom with the very idea of learning. I attribute this to the fact that they have been taught too many things merely in the air, things which have no coherence with any train of thought such as would naturally occur to anyone, however intellectual, who has his being in this modern world. The whole apparatus of learning appears to them as nonsense. Of course, any individual schoolmaster is helpless in this matter; he is in the grip of the examination system. It is here that the utility of such associations as the one which celebrates its meeting to-day is apparent. It enables the results of first-hand experience to acquire the authority of a collective demand capable of constraining the nameless Furies who draw up our schedules of examinations.

But to return to elementary mathematics: we conceive it as a group of abstract ideas, and our course is to have a threefold character, namely: (1) the pupil is finally to be left with a precise perception of the nature of the abstractions acquired by constant use of them, illumined by explanations and finally by precise statements; (2) the logical treatment of such ideas is to be exemplified by trains of reasoning which employ them and interconnect them; and (3) the application of these ideas to the course of nature conceived in its widest sense as including human society is to be made familiar.

The subject, as thus broadly sketched out, is limited by the following considerations: there is very little time; only such ideas are to be introduced as are of fundamental importance to all mathematical reasoning, and they are not to be too complicated for the average boy to understand.

Most of these requisites explain themselves, and I need say nothing more about them. It will be seen that the whole spirit of these suggestions is towards a cutting down of the mere quantity of abstract reasoning to be performed, but towards an extension of the time devoted to a consideration of the ideas in themselves especially by the aid of their applications to examples. By examples, I mean important examples. What we want is one hour of the Caliph Omar,[1] to burn up and utterly destroy all the silly mathematical problems which cumber our text-books. I protest against the presentation of mathematics as a silly subject with silly applications.

For example, take the theory of graphs, which, on its theoretical side, should teach the boys the abstract idea of a functional relation between variable quantities. This abstract idea is embodied in a few simple theoretical examples, such as the rectilinear graph of the linear algebraic function of one variable, the parabolic graph of the quadratic function;[2] also the wavy graphs of the sine and cosine illustrate the |7| general nature of periodic functions. In this way the boy grows familiar with the idea of an abstract precise law. If time permits the law of the inverse square can be exhibited in a graph, and also the fundamental law of the geometrical increase by plotting functions such as 2^x, 3^x, etc., and for the abler students by the consideration of the series exp$x$, that is, e^x. But for the mass of boys a few well-chosen examples of precise functional relations would surely be better than a more ambitious course over a wider field.

Now, still keeping for the present to the abstract side of the work, the consideration of the zeros of these functions at once introduces equations as a necessary branch of study. Linear and quadratic equations acquire an important meaning, and so do the zeros of the sine and cosine functions. At this point I suggest the study of abstract algebra might well be stopped. I would utterly sweep away all prolonged multiplications and divisions, and the theories of greatest common measure and least common multiple, and complicated forms of linear and quadratic equations. They lead to nothing important in the boys' minds and consume a vast amount of valuable time. It would be quite sufficient to confine practice in multiplication to cases in which one factor is linear, and practice in division to cases in which the divisor is linear. Similarly, factorization, if admitted, should be rigidly confined to the case of two linear factors with the view of exemplifying the theory of the zeros of quadratic functions. Again, for the mass of boys, the algebraic treatment of fractional expressions consumes uselessly valuable time. Of course, ample practice in algebraical manipulation is necessary, but it should be restricted to a few types of the most necessary operations.

---

1. Whitehead may be referring to the mathematician Omar Khayyam (1048–1131); an 1869 translation of some of his works entitled *Rubáiyát of Omar Khayyám* was very popular throughout the English-speaking world.
2. The original has no punctuation, but extra space between 'function' and 'also' suggests that punctuation of some kind was intended. We have supplied a semicolon.

It will be noticed that while advocating the omission of a large part of the algebra usually taught, I would include the definitions of sine and cosine, and the study of their graphs. I do not suggest that trigonometry, properly so called, be introduced. By this I mean the application of the trigonometrical functions to the theory of the triangle. This would, in general, take up far too much time with very little intellectual result. The true use of these functions in elementary mathematics is their representation of the idea of periodicity. Perhaps, however, I am too sanguine in hoping that their comprehension is within the range of the ordinary boy.

The general outcome of these suggestions is that elementary algebra would be restricted to the consideration of the simplest functions of one variable. The golden rule should be that until the end of their course, students should never see expressions with more than one unknown in them. It is traditional at the very commencement of the study of the science to present functions containing a large number of letters, *a, b, c, d, e, f,* with directions to substitute particular values for them. I am utterly unable to conceive what is the educational value of this inane procedure. All that is wanted to begin with is the calculation of the values of particular cases of the simple functions for particular values of the single variable. Then the idea of any value of the function arises, and, to put the matter technically, we |8| come to the letter "*y*" standing by itself on the left-hand side of the equation, and the function on the right-hand side.

Finally, we *perhaps* reach the idea of algebraic form and introduce the coefficients as parameters with the letters *a, b, c.* But except for the sake of algebraic form more than one variable, as an argument to a function, is never wanted in elementary study.

So much for the theoretical side of the subject. I have treated this first because in view of further suggestions on the practical applications, it was necessary to explain in what way time was to be gained, and what are the theoretical ideas to be led up to and to be applied.

But I wish emphatically to guard myself against suggesting that such abstract ideas as the variable and functionality are best introduced by the consideration of abstract algebraic functions however simple. On the contrary, a preparatory consideration of concrete examples by graphical methods is surely necessary. There we reach one of the chief causes of the weakness of the traditional mathematical training. It is entirely out of relation to the real exhibition of the mathematical spirit in modern thought, with the result that it remained satisfied with examples which were both silly and unsystematic. Now the effect which we want to produce on our pupils is to generate a capacity to apply ideas to the concrete universe. Thus the examples which we choose form the very backbone of our teaching. The study of algebra should commence with a systematic study of the practical application of mathematical ideas of quantity to some important subject. Now what subject can we choose by which to represent the flux of quantities without the necessary intervention of algebraic technique from the very beginning?[1] Many suggestions might be made, and it is obvious that many subjects in competent hands might be equally good. My suggestion in its crudest, and most aggressive, form is that half of

1. We have changed the ending full stop to a question mark.

the teaching of modern history should be handed over to the mathematicians. The phrase "handed over" is not quite accurate; for the half which I mean is the half which, although the true foundation of all knowledge of nations, is hardly taught. Our classical colleagues, excellent fellows as they are, have their limitations; and among them is this one, that they are not very fitted by their mental equipment to appreciate quantitative estimates of the forces which are moulding modern society. But without such estimates modern history as it unfolds itself before us is a meaningless tangle.

Now among other peculiarities of the nineteenth century is this one, that by initiating the systematic collection of statistics it has made the quantitative study of social forces possible. There are to our hands statistics of trade both external and internal, statistics of railway traffic, statistics of harvest, of prices, of health, of population, of education, of crime, of income-tax returns, of national expenditure, of weather, of prices, of pauperism, and of times of sunrise and sunset throughout the year. The reduction of these to graphs, the careful study of the peculiarities of these graphs, the search for correlations among them, |9| and the study of the public events which corresponded in time to peculiarities in graphical form, would teach more mathematics and more knowledge of modern social forces than all our present methods put together. Our relations with our colonies, with France, with Germany, with the United States, could be traced statistically. Problems could be set for solution by the boys—such as to state verbally the effect of war on the social life of a nation as exhibited by the graphs. Also they can be given the statistics and told to exhibit them in graphical form, and to state the general characteristics of the graphs thus[1] obtained. The notion of rates of increase, embodying the essential ideas of the differential calculus, thus emerges.

Finally, theory and practice could be combined by finding graphs which approximately satisfy the simple functional laws which are being simultaneously studied.

I am quite aware that this suggestion of statistical study may seem fantastic, and perhaps be pronounced impossible. Of course, in a rapid first sketch, one cannot hope to have put all the details of the idea in their right relation to one another. But before the whole suggestion is definitely dismissed, I should like to know exactly why it is impossible and why it is fantastic. The information to be imparted is of the utmost importance for the subsequent conduct of life in self-governing communities; it illustrates important abstract ideas, and the means of study lie easily to hand. Also the course of work would involve simple definite efforts on the part of the students. This method of conducting the elementary study of mathematical analysis appears to me to be eminently practical, and at every stage to carry with it its own justification.

The mathematical treatment of our space-ideas is obviously of the first educational importance. I cannot pretend to be very satisfied with the immediate effects of the abolition of Euclid's *Elements* as a text-book.[2] A lamentable deficiency in

1. In the original publication, the 't' in 'thus' was mistakenly printed upside down.
2. Euclid's *Elements* dominated British schools as a geometry textbook through the end of the 1800s. It was only in the mid-1860s that signs of discontent began to appear, with some opponents objecting that Euclid's reasoning was unsuitable for beginners, and that his demonstrations were artificial and so lengthy that they exercised the memory of students rather than their intelligence. An anti-Euclid

logical rigour has crept in, with entirely bad effects on the scholastic value of the subject. My belief is[1] that the science as an educational instrument has been ruined from the time of Euclid downwards by fallacious views of logical method which seem to be both prevalent and traditional. There is an idea that the logical premises of a subject like geometry are propositions which have some peculiar quality of self-evidence, which is not merely one of degree. In fact, it is implied that there are natural premises which have to be used as such because they are self-evident and incapable of proof.

Probably, when the view is thus crudely stated it would be repudiated by everyone. But I believe it to be true that the usual presentation of geometry as a deductive science, based on axioms which the student is simply told to accept, does, in fact, habitually generate this fallacious notion; thereby, the harm done to a sound conception of the relation of logic to induction is nearly as great as is the good received from the training in the art of reasoning. The same error |10| crops up in an even more pestilential form when authors on mechanics imply that that science is based on separate verifications of the various laws of motion. Half our difficulties in the elementary teaching of the deductive sciences arise from the tacit unconscious acceptance of this abominable heresy.

I am told that there are some animals whose centres of intelligence, such as they are, are fairly uniformly distributed throughout their bodies, so that, however you cut them in half, both parts are equally sensible. Something like this is the case in any science. The propositions which, for some reason or other claim our credence, are distributed throughout the whole body of the subject. The function of deductive

association was formed in 1871, the Association for the Improvement of Geometrical Teaching (AIGT); the Association would start the *Mathematical Gazette* in 1894, and changed its name to the Mathematical Association in 1897. Unsuccessful for about thirty years in dethroning the *Elements* as a textbook, the Association's ultimate victory came with unexpected suddenness in the early 1900s, due partly to the influence of Whitehead's friend and colleague Andrew Forsyth (1858–1942). Forsyth had become President of the Mathematical Association in 1903, and was an influential member of a Syndicate appointed by the University of Cambridge to consider examinations in mathematics. On 9 May 1903 the Syndicate recommended the disestablishment of the *Elements* at Cambridge, a recommendation that was adopted by the Senate on 11 June 1903. Confusion followed what might be termed the deregulation of geometry teaching in England, a confusion with which mathematics teachers were still very much struggling at the time of Whitehead's address. See Amirouche Moktefi, 'Geometry: The Euclid Debate', in Raymond Flood, Adrian Rice and Robin Wilson (eds), *Mathematics in Victorian Britain* (Oxford: Oxford University Press, 2011), pp. 321–338.

It is also worth noting here that Forsyth played an indirect role in Whitehead's departure from Trinity College the year prior to the publication of this article. In 1910, Forsyth fell in love and eloped with a married woman. He resigned both his chair and his fellowship at Trinity, but Whitehead hoped that the Trinity Council would allow Forsyth to keep the fellowship, since he was a friend and had served the school well for decades. But a motion to ask Forsyth to reconsider his fellowship resignation got nowhere, and in the end his resignation was accepted by a vote of eight to four. Whitehead was not alone in being angry at the decision, as twenty-seven fellows would subsequently protest the acceptance of Forsyth's resignation. Whitehead responded by resigning his own lectureship about two months later, then moved from Cambridge to London without first lining up a new job. The Forsyth affair was apparently not the only reason for Whitehead's departure; he told Victor Lowe that he had been 'in a groove in Cambridge' and needed a change of scenery. Forsyth would gain a new chair in mathematics at Imperial College in 1913, and end up hiring Whitehead for a position under him a year later. See Victor Lowe, *Alfred North Whitehead: The Man and His Work, Volume I: 1861–1910* (Baltimore: Johns Hopkins University Press, 1985), pp. 315–317.

1. ESP added 'is'; we include it because the original erroneously contained no main verb.

logic is, by the creation of a coherent logical system, to tie them together, so as to enable us to pool their evidence. But often there is more evidence for the more complex propositions than for the premises. The chief requisite for a premise is not obvious truth, but simplicity. There is no obvious truth about the law of gravitation; but the science of attractions, which is founded on it, is verified all along the line in so far as it is applied to ordinary matter.

Thus, in order to pool our evidence for a body of propositions to the utmost extent, it is desirable that the premises assumed should be as few and simple as possible, and, of course, the more fully they claim our credence the better. But none of these requisites are absolutely necessary on pain of logical fallacy. Our selection of premises is arbitrary, and must be guided by the purpose which we have in view. Now these logical considerations have a profound influence on our conceptions of the true mode in which to present geometry. They lead to the conclusion that the old traditional presentation is wrong.

In the first place it is of great importance that students, before considering any logical proofs, should be made thoroughly familiar with the set of ideas and propositions which are to compose the schedule of the subject. They should note that some propositions appear obvious, and that others are capable of experimental verification by the measurement of accurately drawn diagrams.

In this way, a schedule of important propositions should be thoroughly appreciated. Then a selection of some of the simpler and more obvious propositions should be made and treated as logical premises. They are *our* axioms of geometry, not *the* axioms of geometry; and from them, by the most rigid reasoning, justifying every step as we go along, the remainder of the schedule should be proved.

But I would remind you that there is no logical fallacy in retaining a logically redundant set of premises. We can, if we like, point out to our pupils that some of the assumed propositions can be proved from the others, but there is absolutely no necessity to give the proof if we think it too long or too difficult. What is necessary for education is that the pupil should definitely know what propositions have been assumed, that these axioms should carry with them some strong evidence of truth, and that the reasoning should be rigidly accurate and full. But where assumption is so easy, logical fallacies are unpardonable. Also |11| it may be desirable to retain some propositions in the schedule for experimental consideration which are not ultimately subjected to logical proof. For example, the theory of similarity is the foundation of all maps and plans, and it is highly desirable to appreciate its elementary propositions even if they cannot be proved.

But, as in the case of algebra, the schedule should be rigidly purged of all propositions which might appear to the student to be merely curiosities without any important bearings.

But what are the important bearings of geometrical truths? In a sense, the science is its own justification. It is the framework almost instinctively adopted to state our experiences of the universe. In order to explain why we feel tired, we state the number of miles which we have walked; to explain why it took so many days to plough a field, we state its number of acres. In every attempted explanation of the material facts of life we have recourse to geometrical ideas. Geometry is the queen of physical sciences. Accordingly, in a sense, we might bring any geometrical theorem into the schedule. But our time is limited, and we shall do well to concentrate on a few truths of the widest application and of most immediate importance. Whatever we put into the schedule necessarily excludes something else, and this consideration governs our selection.

The treatment of the whole doctrine of similarity makes almost a small subject in itself. It faces us in the selection of the scales of our graphs and other diagrams. Also it naturally coalesces with the doctrine of arithmetical proportion which in its elements receives a simple algebraic treatment. This, again, finds an application in the proportional variation of the entries in statistical tables, corresponding to variations either of population or of other fundamental aggregates. I should like, at this point, to enter a plea for the inclusion of the parallelogram of forces and the polygon of forces as the fundamental example of the application of geometry to science. But already I have been led into a discourse which, against my original intention, has wandered away into a technical discussion.

We have been considering the place of elementary mathematics in a liberal education. What, in a few words, is the final outcome of our thoughts? It is that the elements of mathematics should be treated as the study of a set of fundamental ideas, the importance of which the student can immediately appreciate: that every proposition and method which cannot pass this test, however important for a more advanced study, should be ruthlessly cut out: that with the time thus gained, the fundamental ideas placed before the pupils can be considerably enlarged so as to include what in essence is the method of co-ordinate geometry, the fundamental idea of the differential calculus in relation to rates of increase, and the geometrical notion of similarity. Also, lastly, it has been insisted that important systematic applications of these ideas to the concrete world should be simultaneously studied—for example, some sets of social or scientific statistics and the use of the polygon of forces in the graphical solution of mechanical problems. Again, this rough |12| summary can be further abbreviated into one essential principle, namely, simplify the details and emphasize the important principles and applications.

The suggestions which I have ventured to put forward have been made with unfeigned diffidence. I am emboldened to speak by the conviction that we have now a golden opportunity for reconstituting our scheme of mathematical education. But such opportunities are dangerous. If mathematical teaching is not now revivified by a breath of reality, we cannot hope that it will survive as an important element in the liberal education of the future.

The Principles of Mathematics in Relation to Elementary Teaching

1913 E. W. Hobson and A. E. H. Love (eds), *Proceedings of the Fifth International Congress of Mathematicians, Volume II* (Cambridge: Cambridge University Press), pp. 449–454

1913 *L'Enseignement Mathématique*, vol. 15, pp. 105–111

1917 *The Organisation of Thought* (London: Williams and Norgate), pp. 92–104

The International Congress of Mathematicians is a conference that occurs every four years and continues to this day, but in 1912 it was still relatively new. The brainchild of Felix Klein and Georg Cantor, the first meeting had taken place in 1893, at the Chicago World's Fair.

The first meeting Whitehead attended was the 1900 conference in Paris, along with his wife Evelyn, Bertrand Russell and Alys Russell. He did not present a paper there, but acted as one of five secretaries for the conference, and the only English one. Both Whitehead and Russell were greatly impressed by Giuseppe Peano, and would later adopt parts of his symbolism for the Principia Mathematica, *the second volume of which was published in the same year as the Fifth Congress.*[1]

The 1912 meeting in Cambridge drew 574 members. More than a third (221) were British; the next-best represented were Americans (60) and Germans (53).[2] *Many of the mathematical luminaries of the day were in attendance, including Peano, Ernst Zermelo and Edmund Landau, and Whitehead's friends Russell, William Burnside and Andrew Forsyth. Whitehead would take part with Peano in discussing Edward Vermilye Huntington's paper on postulates for abstract geometry on 23 August,*[3] *before delivering his own paper on 24 August in Section IV, 'Philosophy, History, Didactics'. Karl Friedrich August Gutzmer, a German, chaired the Section; the other paper delivered was by a Slovenian, Rihard Zupančič, who presented in French.*[4]

Whitehead's central pedagogical points in this address might be described as threefold: (1) the teacher must begin by addressing the ideas that already exist in the child's mind, rather than try to simply impose the finished/polished ones, in order to strengthen the power of abstract thought and logical reasoning; (2) logical precision is the faculty at which mathematical education aims, rather than a precondition of its commencement; and (3) teaching advanced mathematics to engineers and physicists

1. Victor Lowe, *Alfred North Whitehead: The Man and His Work, Volume I: 1861–1910* (Baltimore: Johns Hopkins University Press, 1985), pp. 230–234.
2. E. W. Hobson and A. E. H. Love (eds), *Proceedings of the Fifth International Congress of Mathematicians, Volume I* (Cambridge: Cambridge University Press, 1913), p. 28.
3. Hobson and Love, *Proceedings, Vol. 1*, p. 54.
4. Hobson and Love, *Proceedings, Vol. 1*, p. 58.

without attending to the underlying logic or theory is useless because it produces a merely mechanical knowledge lacking the understanding that is needed to arrive at results in applied science.

We are concerned not with the advanced teaching of a few specialist mathematical students, but with the mathematical education of the majority of boys in our secondary schools. Again these boys must be divided into two sections; one section consists of those who desire to restrict their mathematical education, the other section is made up of those who will require some mathematical training for their subsequent professional careers, either in the form of definite mathematical results or in the form of mathematically trained minds.

I shall call the latter of these two sections the mathematical section, and the former the non-mathematical section. But I must repeat that by the mathematical section is meant that large number of boys who desire to learn more than the minimum amount of mathematics. Furthermore most of my remarks about these sections of boys[1] apply also to elementary classes of our university students.

The subject of this paper is the investigation of the place which should be occupied by modern investigations respecting Mathematical Principles[2] in the education of both of these sections of schoolboys.

To find a foothold from which to start the enquiry, let us ask, why the non-mathematical section should be taught any mathematics at all beyond the barest elements of arithmetic. What are the qualities of mind which a mathematical training is designed to produce when it is employed as an element in a liberal education?

My answer, which applies equally to both sections of students, is that there are two allied forms of mental discipline which should be acquired by a well designed mathematical course. These two forms though closely allied are perfectly distinct.

The first form of discipline is not in its essence logical at all. It is the power of clearly grasping abstract ideas, and of relating them to particular circumstances. In other words, the first use of mathematics is to strengthen the power of abstract thought. I repeat that in its essence this has nothing to do with logic, though as a matter of fact a logical discipline is the best method of producing the desired effect. It is not the philosophical theory of abstract ideas which is to be acquired, but the habit and the power of using them. There is one and only one way of acquiring the habit and the power of using anything, that is by the simple common-|450|place method of habitually using it. There is no other short cut. If in education we desire to produce a certain conformation of mind, we must day by day, and year by year, accustom the students' minds to grow into the desired structural shape. Thus to teach

1. The casual sexism in this reference to 'boys' is incongruous with Whitehead's support of participation in the education of students of both sexes. See p. 342n1.
2. The version of the paper that appears in OT standardised the capitalisation of the two prior versions, rendering terms like 'Mathematical Principles', 'Geometry', 'Quantity', 'Number', 'Elementary Algebra', etc. as uniformly lowercase. Both earlier versions were inconsistent in their capitalisation, sometimes capitalising a term and sometimes not.

the power of grasping abstract ideas and the habit of using them, we must select a group of such ideas, which are important and are also easy to think about because they are clear and definite.

The fundamental mathematical truths concerning geometry, ratio, quantity, and number, satisfy these conditions as do no others. Hence the fundamental universal position held by mathematics as an element of a liberal education.

But what are the Fundamental Mathematical truths concerning Geometry, Quantity and Number?

At this point we come to the great question of the relation between the modern science of the Principles of Mathematics and a Mathematical Education.

My answer to the question as to these Fundamental Mathematical truths is that, in any absolute sense, there are none. There is no unique small body of independent primitive unproved propositions which are the necessary starting-points of all mathematical reasoning on these subjects. In mathematical reasoning the only absolute necessary presuppositions are those which make logical deduction possible. Between these absolute logical truths and so-called Fundamental truths concerning Geometry, Quantity and Number, there is a whole new world of mathematical subjects concerning the logic of propositions, of classes, and of relations.

But this subject is too abstract to form an elementary training ground in the difficult art of abstract thought.

It is for this reason that we have to make a compromise and start from such obvious general ideas as naturally occur to all men when they perceive objects with their senses.

In Geometry, the ideas elaborated by the Greeks and presented by Euclid are, roughly speaking, those adapted for our purpose, namely, ideas of volumes, surfaces, lines, of straightness and of curvature, of intersection and of congruence, of greater and less, of similarity, shape, and scale. In fact, we use in education those general ideas of spatial properties which must be habitually present in the mind of anyone who is to observe the world of phenomena with understanding.

Thus we come back to Plato's opinion that for a liberal education, Geometry, as he knew it, is the queen of sciences.

In addition to Geometry, there remain the ideas of quantity, ratio, and of number. This in practice means, Elementary Algebra. Here the prominent ideas are those of 'any number,' in other words, the use of the familiar x, y, z, and of the dependence of variables on each other, or otherwise, the idea of functionality. All this is to be gradually acquired by the continual use of the very simplest functions which we can devise, of linear functions, graphically represented by straight lines, of quadratic functions graphically represented by parabolas, and of those simple implicit functions graphically represented by conic sections. Thence, with good fortune and a willing class, we can advance to the ideas of rates of increase, still confining ourselves to the simplest possible cases.

|451| I wish here emphatically to remind you that both in Geometry and in Algebra a clear grasp of these general ideas is not what the pupil starts from, it is the goal at which he is to arrive. The method of progression is continual practice in the consideration of the simplest particular cases, and the goal is not philosophical analysis but the power of use.

But how is he to practise himself in their use? He cannot simply sit down and think of the relation $y = x + 1$, he must employ it in some simple obvious way.

This brings us to the second power of mind which is to be produced by a mathematical training, namely, the power of logical reasoning. Here again, it is not the knowledge of the philosophy of logic which it is essential to teach, but the habit of thinking logically. By logic, I mean deductive logic.

Deductive logic is the science of certain relations, such as implication, etc., between general ideas. When logic begins, definite particular individual things have been banished. I cannot relate logically this thing to that thing, for example this pen to that pen, except by the indirect way of relating some general idea which applies to this pen to some general idea which applies to that pen. And the individualities of the two pens are quite irrelevant to the logical process. This process is entirely concerned with the two general ideas. Thus the practice of logic is a certain way of employing the mind in the consideration of such ideas; and an elementary mathematical training is in fact nothing else but the logical use of the general ideas respecting Geometry and Algebra which we have enumerated above. It has therefore, as I started this paper by stating, a double advantage. It makes the mind capable of abstract thought, and it achieves this object by training the mind in the most important kind of abstract thought, namely, deductive logic.

I may remind you that other choices of a type of abstract thought might be made. We might train the children to contemplate directly the beauty of abstract moral ideas, in the hope of making them religious mystics. The general practice of education has decided in favour of logic, as exemplified in elementary mathematics.

We have now to answer the further question, what is the *rôle* of Logical Precision in the Teaching of Mathematics? Our general answer to the implied question is obvious, Logical Precision is one of the two objects of the teaching of mathematics, and it is the only weapon by which the teaching of mathematics can achieve its other object. To teach mathematics is to teach logical precision. A mathematical teacher who has not taught that, has taught nothing.

But having stated this thesis in this unqualified way, its meaning must be carefully explained; for otherwise its real bearing on the problem of education will be entirely misunderstood.

Logical precision is the faculty to be acquired. It is the quality of mind which it is the object of the teaching to impart. Thus the habit of reading great literature is the goal at which a literary education aims. But we do not expect a child to start its first lesson by reading for itself Shakespeare. We recognize that reading is impossible till the pupil has learnt its alphabet and can spell, and then we start it with books of one syllable.

In the same way, a mathematical education should grow in logical precision. It |452| is folly to expect the same careful logical analysis at the commencement of the training as would be appropriate at the end. It is an entire misconception of my thesis to construe it as meaning that a mathematical training should assume in the pupil a power of concentrated logical thought. My thesis is in fact the exact opposite, namely that this power cannot be assumed, and has got to be acquired, and that a mathematical training is nothing else than the process of acquiring it. My whole groundwork of assumption is that this power does not initially exist in a fully developed state. Of course like every other power which is acquired, it must be developed gradually.

The various stages of development must be guided by the judgment and the genius of the teacher. But what is essential is, that the teacher should keep clearly in his mind that it is just this power of logical precise reasoning which is the whole object of his efforts. If his pupils have in any measure gained this, they have gained all.

We have not yet however fully considered this part of our subject. Logical precision is the full realization of the steps of the argument. But what are the steps of the argument? The full statement of all the steps is far too elaborate and difficult an operation to be introduced into the mathematical reasoning of an educational curriculum. Such a statement involves the introduction of abstract logical ideas which are very difficult to grasp, because there is so rarely any need to make them explicit in ordinary thought. They are therefore not a fit subject-ground for an elementary education.

I do not think that it is possible to draw any theoretical line between those logical steps which form a theoretically full logical investigation, and those which are full enough for most practical purposes, including that of education. The question is one of psychology, to be solved by a process of experiment. The object to be attained is to gain that amount of logical alertness which will enable its possessors to detect fallacy and to know the types of sound logical deduction. The objects of going further are partly philosophical, and also partly to lay bare abstract ideas whose investigation is in itself important. But both these objects are foreign to education.

My opinion is that, on the whole, the type of logical precision handed down to us by the Greek mathematicians is roughly speaking what we want. In Geometry, this means the sort of precision which we find in Euclid. I do not mean that we should use his famous *Elements* as a textbook,[1] nor that here and there a certain compression in his mode of exposition is not advisable. All this is mere detail. What I do mean is that the sort of logical transition which he made explicit, we should make explicit, and that the sort of transition which he omits, we should omit.

I doubt however whether it is desirable to plunge the student into the full rigour of Euclidean Geometry without some mitigation. It is for this reason that the modern habit, at least in England, of laying great stress, in the initial stages, on training the pupil in simple constructions from numerical data is to be praised. It means that after a slight amount of reasoning on the Euclidean basis of accuracy, the mind of the learner is relieved by doing the things in various special cases, and noting by rough measurements that the desired results are actually attained. |453| It is important however that the measurements be not mistaken for the proofs. Their object is to make the beginner apprehend what the abstract ideas really mean.

Again in algebra, the notation and the practical use of the symbols should be acquired in the simplest cases, and the more theoretical treatment of the symbolism reserved to a suitable later stage. In fact my rule would be initially to learn the meaning of the ideas by a crude practice in simple ways, and to refine the logical procedure in preparation for an advance to greater generality. In fact the thesis of my paper can be put in another way thus, the object of a mathematical education is

1. Regarding the use of Euclid's *Elements* as a textbook, see p. 348n2.

to acquire the powers of analysis, of generalization, and of reasoning. The two processes of analysis and generalization were in my previous statement put together as the power of grasping abstract ideas.

But in order to analyse and to generalize, we must commence with the crude material of ideas which are to be analysed and generalized. Accordingly it is a positive error in education to start with the ultimate products of this process, namely the ideas in their refined analysed and generalized forms. We are thereby skipping an important part of the training, which is to take the ideas as they actually exist in the child's mind, and to exercise the child in the difficult art of civilizing them and clothing them.

The schoolmaster is in fact a missionary, the savages are the ideas in the child's mind; and the missionary shirks his main task if he refuse to risk his body among the cannibals.

At this point I should like to turn your attention to those pupils forming the mathematical section. There is an idea, widely prevalent, that it is possible to teach mathematics of a relatively advanced type—such as differential calculus, for instance—in a way useful to physicists and engineers without any attention to its logic or its theory.

This seems to me to be a profound mistake. It implies that a merely mechanical knowledge without understanding of ways of arriving at mathematical results is useful in applied science. It is[1] of no use whatever. The results themselves can all be found stated in the appropriate pocket-books and in other elementary works of reference. No one when applying a result need bother himself as to why it is true. He accepts it and applies it. What is of supreme importance in physics and in engineering is a mathematically trained mind, and such a mind can only be acquired by a proper mathematical discipline.

I fully admit that the proper way to start such a subject as the Differential Calculus is to plunge quickly into the use of the notation in a few absurdly simple cases, with a crude explanation of the idea of rates of increase. The notation as thus known can then be used by the lecturers in the Physical and Engineering Laboratories. But the mathematical training of the applied scientists consists in making these ideas precise and the proofs accurate.

I hope that the thesis of this paper respecting the position of logical precision in the teaching of mathematics has been rendered plain. The habit of logical precision with its necessary concentration of thought upon abstract ideas is not |454| wholly possible in the initial stages of learning. It is the ideal at which the teacher should aim. Also logical precision, in the sense of logical explicitness, is not an absolute thing: it is an affair of more or less. Accordingly the quantity of explicitness to be introduced at each stage of progress must depend upon the practical judgment of the teacher. Lastly, in a sense, the instructed mind is less explicit; for it travels more quickly over a well-remembered path, and may save the trouble of putting into words trains of thought which are very obvious to it. But on the other hand it atones for this rapidity by a concentration on every subtle point where a fallacy can lurk. The habit of logical precision is the instinct for the subtle difficulty.

---

1. *L'Enseignement Mathématique* changed 'is' to 'seems to me to be'.

Presidential Address to the London Branch of the Mathematical Association

1913 ***Mathematical Gazette*, vol. 7, no. 104 (March), pp. 87–94**

1917 *The Organisation of Thought* (London: Williams and Norgate), pp. 69–91

1929 *The Aims of Education* (New York: Macmillan), pp. 116–135

A prefatory note to the address in the Mathematical Gazette specified that the annual meeting of the London Branch of the Mathematical Association was on 8 February 1913 at what was then the Polytechnic, Regent Street, and is now the University of Westminster. Whitehead was elected President at the meeting – succeeding Micaiah John Muller Hill (1856–1929) – and delivered his address.

Upon republication in both Organisation of Thought *and* Aims of Education, *the address was re-titled 'The Mathematical Curriculum'. It is apparent from the number of changes in common between these two later versions – and particularly a pair of erroneously deleted words*[1] *– that the OT version was used as a basis for the AE version. The latter did, however, correct a few errors, and also removed multiple references to the occasion of the address which had been left intact in OT.*

In this address, Whitehead tackles the idea of mathematics as a 'recondite' subject – that is, obscure, incomprehensible or irrelevant. He begins by cheerfully admitting to the reconditeness of mathematics in current teaching, made so by its lack of apparent relevance to everyday life, reduced to the memorisation of seemingly irrelevant formulas. He advocates instead a shorter curriculum squarely focused on elements of mathematics with obvious applications and implications for the majority of students (rather than simply those seeking to specialise in mathematics), and finishes by outlining five 'stages' of geometry that should be taught – congruence, similarity, trigonometry, analytical geometry and projective geometry – which would 'constitute a training in logical method together with an acquisition of the precise ideas which lie at the base of the scientific and philosophical investigations of the universe' (Whitehead's final paragraph).

The situation in regard to education at the present time cannot find its parallel without going back for some centuries to the break up of the medieval traditions of learning. Then, as now, the traditional intellectual outlook, despite the authority which it had justly acquired from its notable triumphs, had grown to be too narrow for the interests of mankind. The result of this shifting of human interest was a

1. See p. 362n1.

demand for a parallel shifting of the basis of education, so as to fit the pupils for the ideas which later in life would in fact occupy their minds. Any serious fundamental change in the intellectual outlook of human society must necessarily be followed by an educational revolution. It may be delayed for a generation by vested interests or by the passionate attachment of some leaders of thought to the cycle of ideas within which they received their own mental stimulus at an impressionable age. But the law is inexorable that education to be living and effective must be directed to informing pupils with ideas, and to creating for them capacities which will enable them to appreciate the current thought of the epoch.

There is no such thing as a successful system of education in a vacuum, that is to say, a system which is divorced from immediate contact with the existing intellectual atmosphere. Education which is not modern shares the fate of all organic things which are kept too long.

But the blessed word "modern" does not really solve our difficulties. What we mean is, "relevant to modern thought, either in the ideas imparted or in the aptitudes produced."[1] Something found out only yesterday may not really be modern in this sense. It may belong to some bygone system of thought prevalent in a previous age, or, what is very much more likely, it may be too recondite. When we demand that education should be relevant to modern thought, we are referring to thoughts broadly spread throughout cultivated society. It is this question of the unfitness of recondite subjects for use in general education which I wish to make the keynote of my address this afternoon.

It is in fact rather a delicate subject for us mathematicians. Outsiders are apt to accuse our subject of being recondite. Let us grasp the nettle at once and frankly admit that in general opinion it is the very typical example of reconditeness. By this word I do not mean difficulty, but that the ideas involved are of highly special application, and rarely influence thought.

This liability to reconditeness is the characteristic evil which is apt to destroy the utility of mathematics in liberal education. So far as it clings to the educational use of the subject, so far we must acquiesce in a miserably low level of mathematical attainment among cultivated |88| people in general. I yield to no one in my anxiety to increase the educational scope of mathematics. The way to achieve this end is not by a mere blind demand for more mathematics. We must face the real difficulty which obstructs its extended use.

Is the subject recondite? Now, viewed as a whole, I think it is. *Securus judicat orbis terrarum*—the general judgment of mankind is sure.

The subject as it exists in the minds and in the books of students of mathematics *is* recondite. It proceeds by deducing innumerable special results from general ideas, each result more recondite than the preceding. It is not my task this afternoon to defend mathematics as a subject for profound study. It can very well take care of itself. What I want to emphasise is, that the very reasons which make this science a delight to its students are reasons which obstruct its use as an educational instru-

1. Both OT and AE deleted the quotation marks. Likely this was done to dispel confusion about whether Whitehead was quoting a third party (it does not appear that he was), rather than proposing his own definition.

ment—namely, the boundless wealth of deductions from the interplay of general theorems, their complication, their apparent remoteness from the ideas from which the argument started, the variety of methods, and their purely abstract character which brings, as its gift, eternal truth.

Of course, all these characteristics are of priceless value to students; for ages they have fascinated some of the keenest intellects. My only remark is that, except for a highly selected class, they are fatal in education. The pupils are bewildered by a multiplicity of detail, without apparent relevance either to great ideas or to ordinary thoughts. The extension of this sort of training in the direction of acquiring more detail is the last measure to be desired in the interests of education.

The conclusion at which we arrive is that mathematics, if it is to be used in general education, must be subjected to a rigorous process of selection and adaptation. I do not mean, what is of course obvious, that however much time we devote to the subject the average pupil will not get very far. But that, however limited the progress, certain characteristics of the subject, natural at any stage, must be rigorously excluded. The science as presented to young pupils must lose its aspect of reconditeness. It must, on the face of it, deal directly and simply with a few general ideas of far-reaching importance.

Now, in this matter of the reform of mathematical instruction, the present generation of teachers may take a very legitimate pride in its achievements. It has shown immense energy in reform, and has accomplished more than would have been thought possible in so short a time. It is not always recognised how difficult is the task of changing a well-established curriculum entrenched behind public examinations.

But for all that, great progress has been made, and, to put the matter at its lowest, the old dead tradition has been broken up. I want to indicate this afternoon the guiding idea which should direct our efforts at reconstruction. I have already summed it up in a phrase. "We must aim at the elimination of reconditeness from the educational use of the subject."[1]

Our courses of instruction should be planned to illustrate simply a succession of ideas of obvious importance. All pretty divagations should be rigorously excluded. The goal to be aimed at is that the pupil should acquire familiarity with abstract thought, should realise how it applies to particular concrete circumstances, and should know how to apply general methods to its logical investigation. With this educational ideal nothing can be worse than the aimless accretion of theorems in our text-books, which acquire their position merely because the children can be made to learn them and examiners can set neat questions on them. The bookwork to be learnt should all be very important as illustrating ideas. The examples set—and let there be as many examples as teachers find necessary—should be direct illustra-|89|tions of the theorems, either by way of abstract particular cases or by way of application to concrete phenomena. Here it is worth remarking that it is quite useless to simplify the bookwork, if the examples set in examinations in fact

1. Both OT and AE again deleted the quotation marks from this sentence, probably for the same reason as in the third paragraph (see p. 359n1). They then connected it with the previous sentence and the word 'namely': 'I have already summed it up in a phrase, namely, we must aim...'.

require an extended knowledge of recondite details. There is a mistaken idea that problems test ability and genius, and that bookwork tests cram. This is not my experience. Only boys who have been specially crammed for scholarships can ever do a problem paper successfully. Bookwork properly set, not in mere snippets according to the usual bad plan, is a far better test of ability, provided that it is supplemented by direct examples. But this is a digression on the bad influence of examinations on teaching.

The main ideas which lie at the base of mathematics are not at all recondite. They are abstract. But one of the main objects of the inclusion of mathematics in a liberal education is to train the pupils to handle abstract ideas. The science constitutes the first large group of abstract ideas which naturally occur to the mind in any precise form. For the purposes of education, mathematics consists of the relations of number, the relations of quantity, and the relations of space. This is not a general definition of mathematics, which, in my opinion, is a much more general science. But we are now discussing the use of mathematics in education. These three groups of relations, concerning number, quantity, and space, are interconnected.

Now, in education we proceed from the particular to the general. Accordingly, children should be taught the use of these ideas by practice among simple examples. My point is this:— The goal should be, not an aimless accumulation of special mathematical theorems, but the final recognition that the preceding years of work have illustrated those relations of number, and of quantity, and of space, which are of fundamental importance. Such a training should lie at the base of all philosophical thought. In fact elementary mathematics rightly conceived would give just that philosophical discipline of which the ordinary mind is capable. But what at all costs we ought to avoid, is the pointless accumulation of details. As many examples as you like; let the children work at them for terms, or for years. But these examples should be direct illustrations of the main ideas. In this way, and this[1] only, can the fatal reconditeness be avoided.

I am not now speaking in particular of those who are to be professional mathematicians, or of those who for professional reasons require a knowledge of certain mathematical details. We are considering the liberal education of all students, including these two classes. This general use of mathematics should be the simple study of a few general truths, well illustrated by practical examples. This study should be conceived by itself, and completely separated in idea from the professional study mentioned above, for which it would make a most excellent preparation. Its final stage should be the recognition of the general truths which the work done has illustrated. As far as I can make out, at present the final stage is the proof of some property of circles connected with triangles. Such properties are immensely interesting to mathematicians. But are they not rather recondite, and, what is the precise relation of such theorems to the ideal of a liberal education? The end of all the grammatical studies of the student in classics is to read Virgil and Horace—the greatest thoughts of the greatest men. Are we content, when pleading

1. The original had 'that', but this is idiomatically incorrect in English; we have adopted the change made in OT and AE, which both replaced 'that' with 'this'.

for the adequate representation in education of our own science, to say that the end of a mathematical training is that the student should know the properties of the nine-point circle? I ask you frankly, is it not rather a "come down"?

This generation of mathematical teachers has done so much strenuous |90| work in the way of reorganising mathematical instruction that there is no need to despair of its being able to elaborate a curriculum which shall leave in the minds of the pupils something even nobler than "the ambiguous case."

Let us think how this final review, closing the elementary course, might be conducted for the more intelligent pupils. Partly no doubt it requires a general oversight of the whole work done, considered without undue detail so as to emphasise the general ideas used, and their possibilities of importance when subjected to further study. Also the analytical and geometrical ideas find immediate application in the physical laboratory where a course of simple experimental mechanics should have been worked through. Here the point of view is twofold, the physical ideas and the mathematical ideas illustrate each other.

The mathematical ideas are essential to the precise formulation of the mechanical laws. The idea of a precise law of nature, the extent to which such laws are in fact verified in our experience, and the rôle of abstract thought in their formulation, then become practically apparent to the pupil. The whole topic of course requires detailed development with full particular illustration, and is not suggested as requiring merely a few bare abstract statements.

It would, however, be a grave error to put too much emphasis on the mere process of direct explanation of the previous work by way of final review. My point is, that the latter end of the course should be so selected that in fact the general ideas underlying all the previous mathematical work should be brought into prominence. This may well be done by apparently entering on a new subject. For example, the ideas of quantity and the ideas of number are fundamental to all precise thought. In the previous stages they will not have been sharply separated; and children are, rightly enough, pushed on to algebra without too much bother about number[1] and quantity. But the more intelligent among them at the end of their curriculum would gain immensely by a careful consideration of those fundamental properties of quantity in general which lead to the introduction of numerical measurement. This is a topic which also has the advantage that the necessary books are actually to hand. Euclid's fifth book is regarded by those qualified to judge as one of the triumphs of Greek mathematics. It deals with this very point. Nothing can be more characteristic of the hopelessly illiberal character of the traditional mathematical education than the fact that this book has always been omitted. It deals with ideas, and therefore was ostracised. Of course a careful selection of the more important propositions and a careful revision of the argument are required. This also is to hand in the publications of my immediate predecessor in the office of president,

1. Both OT and AE erroneously deleted the words 'about number', without which the sentence does not make sense. That both versions contain this error strongly suggests that the AE version was based on the OT version.

Prof. Hill.[1] Furthermore, in Sir T. L. Heath's complete edition of Euclid,[2] there is a full commentary embodying most of what has been said and thought on the point. Thus it is perfectly easy for teachers to inform themselves generally on the topic.[3] The whole book would not be wanted, but just the few propositions which embody the fundamental ideas. The subject is not fit for backward pupils; but certainly it could be made interesting to the more advanced class. There would be great scope for interesting discussion as to the nature of quantity, and the tests which we should apply to ascertain when we are dealing with quantities. The work would not be at all in the air, but would be illustrated at every stage by reference to actual examples of cases where the quantitative character is absent, or obscure, or doubtful, or evident. Temperature, heat, electricity, pleasure and pain, mass and distance could all be considered.

Another idea which requires illustration is that of functionality. A |91| function in analysis is the counterpart of a law in the physical universe, and of a curve in geometry. Children have studied the relations of functions to curves from the first beginning of their study of algebra, namely in drawing graphs. Of recent years there has been a great reform in respect to graphs. But at its present stage it has either gone too far or not far enough. It is not enough merely to draw a graph. The idea behind the graph—like the man behind the gun—is essential in order to make it effective. At present there is some tendency merely to set the children to draw curves, and there to leave the whole question.

In the study of simple algebraic functions and of trigonometrical functions we are initiating the study of the precise expression of physical laws. Curves are another way of representing these laws. The simple fundamental laws—such as the inverse square and the direct distance—should be passed under review, and the applications of the simple functions to express important concrete cases of physical laws considered. I cannot help thinking that the final review of this topic might well take the form of a study of some of the main ideas of the differential calculus applied to simple curves. There is nothing particularly difficult about the conception of a rate of change; and the differentiation of a few powers of x, such as x^2, x^3, etc., could easily be effected; perhaps by the aid of geometry even $\sin x$ and $\cos x$ could be differentiated. If we once abandon our fatal habit of cramming the children with theorems which they do not understand and will never use, there will be plenty of time to concentrate their attention on really important topics. We can give them familiarity with conceptions which really influence thought.

Before leaving this topic of Physical Laws and mathematical functions, there are other points to be noticed. The fact that the precise law is never really verified by observation in its full precision is capable of easy illustration and of affording excel-

1. Micaiah John Muller Hill published *The Contents of the Fifth and Sixth Books of Euclid* in 1900 (Cambridge: Cambridge University Press, 1900).
2. Thomas Little Heath (1861–1940) had published his new translation of Euclid's *Elements* in 1908, just five years after Cambridge had disestablished it as the definitive geometry textbook (see p. 348n2). Whitehead regularly referenced Heath's translations of Greek classics in his Harvard classroom. See Thomas Little Heath, *The Thirteen Books of Euclid's Elements* (Cambridge: Cambridge University Press, 1908).
3. AE deleted the three preceding sentences, starting from 'This also is to hand...'.

lent examples. Again, statistical laws, namely laws which are only satisfied on the average by large numbers, can easily be studied and illustrated. In fact a slight study of statistical methods and their application to social phenomena affords one of the simplest examples of the application of algebraic ideas.

Another way in which the students' ideas can be generalised is by the use of the History of Mathematics, conceived not as a mere assemblage of the dates and names of men, but as an exposition of the general current of thought which occasioned the subjects to be objects of interest at the time of their first elaboration. The use of the History of Mathematics is to be considered at a later stage of our proceedings this afternoon.[1] Accordingly I merely draw attention to it now, to point out that perhaps it is the very subject which may best obtain the results for which I am pleading.

We have indicated two main topics, namely general ideas of quantity and of Laws of Nature, which should be an object of study in the mathematical curriculum of a liberal education. But there is another side to mathematics which must not be overlooked. It is the chief instrument for discipline in logical method.

Now, what is logical method, and how can anyone be trained in it?

Logical method is more than the mere knowledge of valid types of reasoning and practice in the concentration of mind necessary to follow them. If it were only this, it would still be very important; for the human mind was not evolved in the bygone ages for the sake of reasoning, but merely to enable mankind with more art to hunt between meals for fresh food supplies. Accordingly few people can follow close reasoning without considerable practice.

More than this is wanted to make a good reasoner, or even to enlighten ordinary people with knowledge of what constitutes the essence of the |92| art. The art of reasoning consists in getting hold of the subject at the right end, of seizing on the few general ideas which illuminate the whole, and of persistently marshalling all subsidiary facts round them. Nobody can be a good reasoner unless by constant practice he has realised the importance of getting hold of the big ideas and of hanging on to them like grim death. For this sort of training geometry is, I think, better than algebra. The field of thought of algebra is rather obscure, whereas space is an obvious insistent thing evident to all. Then the process of simplification, or abstraction, by which all irrelevant properties of matter, such as colour, taste, and weight, are put aside is an education in itself. Again, the definitions and the propositions assumed without proof illustrate the necessity of forming clear notions of the fundamental facts of the subject matter[2] and of the relations between them. All this belongs to the mere prolegomena of the subject. When we come to its development, its excellence increases. The learner is not initially confronted with any symbolism which bothers the memory by its rules, however simple they may be. Also, from the very beginning the reasoning, if properly conducted, is dominated by well-marked ideas which guide each stage of development. Accordingly the essence of logical method receives immediate exemplification.

---

1. AE deleted the preceding sentence, and also the word 'Accordingly' from the sentence following. Whitehead was referring to a paper that was to be delivered by J. Katz entitled 'The Use of the History of Mathematics in Teaching Elementary Mathematics', referenced in the prefatory note to his address. It appears that Katz's paper was never published.
2. Both OT and AE hyphenated 'subject matter'.

Let us now put aside for the moment the limitations introduced by the dullness of average pupils and the pressure on time due to other subjects, and consider what geometry has to offer in the way of a liberal education. I will indicate some stages in the subject, without meaning that necessarily they are to be studied in this exclusive order. The first stage is the study of *congruence*. Our perception of congruence is in practice dependent on our judgments of the invariability of the intrinsic properties of bodies when their external circumstances are varying. But however it arises, congruence is in essence the correlation of two regions of space, point by point, so that all homologous distances and all homologous angles are equal. It is to be noticed that the definition of the equality of lengths and angles is their congruence, and all tests of equality, such as the use of the yard measure, are merely devices for making immediate judgments of congruence easy. I make these remarks to suggest that apart from the reasoning connected with it, congruence, both as an example of a larger and very far-reaching idea and also for its own sake, is well worthy of attentive consideration. The propositions concerning it elucidate the elementary properties of the triangle, the parallelogram, and the circle, and of the relations of two planes to each other. It is very desirable to restrict the proved propositions of this part within the narrowest bounds, partly by assuming redundant axiomatic propositions, and partly by introducing only those propositions of absolutely fundamental importance.

The second stage is the study of similarity. This can be reduced to three or four fundamental propositions. Similarity is an enlargement of the idea of congruence, and, like that idea, is another example of a one-to-one correlation of points of spaces. Any extension of study of this subject might well be in the direction of the investigation of one or two simple properties of similar and similarly situated rectilinear figures. The whole subject receives its immediate applications in plans and maps. It is important, however, to remember that trigonometry is really the method by which the main theorems are made available for use.

The third stage is the study of the elements of trigonometry. This is the study of the periodicity introduced by rotation and of properties preserved in a correlation of similar figures. Here for the first time we introduce a slight use of the algebraic analysis founded on the study of number and quantity. The importance of the periodic char-|93|acter of the functions requires full illustration. The simplest properties of the functions are the only ones required for the solution of triangles, and the consequent applications to surveying. The wealth of formulæ, often important in themselves, but entirely useless for this type of study, which crowd our books should be rigorously excluded, except so far as they are capable of being proved by the pupils as direct examples of the bookwork.

This question of the exclusion of formulæ is best illustrated by considering this example of Trigonometry, though of course I may well have hit on an unfortunate case in which my judgment is at fault. A great part of the educational advantage of the subject can be obtained by confining study to Trigonometry of one angle and by exclusion of the addition formulæ for the sine and cosine of the sum of two angles. The functions can be graphed, and the solution of triangles effected. Thus the aspects of the science as (1) embodying analytically the immediate results of some of the theorems deduced from congruence and similarity, (2) as a solution of the main problem of surveying, (3) as a study of the fundamental functions required to express periodicity and wave motion will all be impressed on the pupils' minds both by bookwork and example.

If it be desired to extend this course, the addition formulæ should be added. But great care should be taken to exclude specialising the pupils in the wealth of formulæ which comes in their train. By "exclude" is meant that the pupils should not have spent time or energy in acquiring any facility in their deduction. The teacher may find it interesting to work a few such examples before a class. But such results are not among those which learners need retain. Also, I would exclude the whole subject of circumscribed and inscribed circles both from Trigonometry and from the previous geometrical courses. It is all very pretty, but I do not understand what its function is in an elementary non-professional curriculum.

Accordingly, the actual bookwork of the subject is reduced to very manageable proportions. I was told the other day of an American college where the students are expected to know by heart 90 formulæ or results in Trigonometry alone. We are not quite so bad as that. In fact, in Trigonometry we have nearly approached the ideal here sketched out as far as our elementary courses are concerned.

The fourth stage introduces Analytical Geometry. The study of Graphs in algebra has already employed the fundamental notions, and all that is now required is a rigorously pruned course on the straight line, the circle, and the three types of conic sections, defined by the forms of their equations. At this point there are two remarks to be made. It is often desirable to give our pupils mathematical information which we do not prove. For example, in co-ordinate geometry, the reduction of the general equation of the second degree is probably beyond the capacities of most of the type of students whom we are considering. But that need not prevent us from explaining the fundamental position of conics, as exhausting the possible types of such curves.

The second remark is to advocate the entire sweeping away of geometrical conics as a separate subject. Naturally, on suitable occasions the analysis of analytical geometry will be lightened by the use of direct deduction from some simple figure. But geometrical conics, as developed from the definition of a conic section by the focus and directrix property, suffers from glaring defects. It is hopelessly recondite. The fundamental definition of a conic, $SP = e. PM$, usual in this subject at this stage, is thoroughly bad. It is very recondite, and has no obvious importance. Why should such curves be studied at all, any |94| more than those defined by an indefinite number of other formulæ? But when we have commenced the study of the Cartesian Methods, the equations of the first and second degrees are naturally the first things to think about.

In this ideal course of Geometry, the fifth stage is occupied with the elements of Projective Geometry. The general ideas of cross ratio and of projection are here fundamental. Projection is yet a more general instance of that one-to-one correlation which we have already considered under congruence and similarity. Here again we must avoid the danger of being led into a bewildering wealth of detail.

The intellectual idea which projective geometry is to illustrate is the importance in reasoning of the correlation of all cases which can be proved to possess in common certain identical properties. The preservation of the projective properties in projection is the one important educational idea of the subject. Cross ratio only enters as the fundamental metrical property which is preserved. The few propositions considered are selected to illustrate the two allied processes which are made possible by this procedure. One is proof by simplification. Here the simplification is psychological and not logical—for the general case is logically the simplest. What is meant is:— Proof by considering the case which is in fact the most familiar to us, or the easiest to think about. The other procedure is the deduction of particular cases from general truths[1] as soon as we have a means of discovering such cases or a criterion for testing them.

The projective definition of conic sections and the identity of the results obtained with the curves derived from the general equation of the second degree are capable of simple exposition, but lie on the border-line of the subject. It is the sort of topic on which information can be given, and the proofs suppressed.

The course of geometry as here conceived in its complete ideal—and ideals can never be realised—is not a long one. The actual amount of mathematical deduction at each stage in the form of bookwork is very slight. But much more explanation would be given, the importance of each proposition being illustrated by examples, either worked out or for students to work, so selected as to indicate the fields of thought to which it applies. By such a course the student would gain an analysis of the leading properties of space, and of the chief methods by which they are investigated.

The study of the elements of mathematics, conceived in this spirit, would constitute a training in logical method together with an acquisition of the precise ideas which lie at the base of the scientific and philosophical investigations of the universe. Would it be easy to continue the excellent reforms in mathematical instruction, which this generation has already achieved, so as to include in the curriculum this wider and more philosophic spirit? Frankly, I think that this result would be very hard to achieve as the result of single individual efforts. For reasons which I have already briefly indicated, all reforms in education are very difficult to effect. But the continued pressure of combined effort, provided that the ideal is really present in the minds of the mass of teachers, can do much, and effects in the end surprising modification. Gradually the requisite books get written, still more gradually the examinations are reformed so as to give weight to the less technical aspects of the subject, and then all recent experience has shown that the majority of teachers are only too ready to welcome any practicable means of rescuing the subject from the reproach of being a mechanical discipline.

1. Both OT and AE added a comma after 'general truths' and qualified the phrase with the word 'known': 'known general truths'.

Space, Time, and Relativity

1916 ***Proceedings of the Aristotelian Society*, vol. 16, pp. 104–129**

1917 *The Organisation of Thought* (London: Williams and Norgate), pp. 191–228

1929 *The Aims of Education and Other Essays* (New York: Macmillan), pp. 232–247

This essay was originally delivered on 9 September 1915 to Section A ('Mathematical and Physical Science') of the Manchester meeting of the British Association for the Advancement of Science (now known as the British Science Association, or BSA). This version is not extant, though the title appears in the Association's report as 'Time Space and Relativity'.[1]

Four months later, on 3 January 1916, Whitehead presented the paper to the Aristotelian Society. He had added about 3,000 words of commentary on the original 5,000-word paper. According to the Society's minutes, Whitehead read only the additional commentary aloud, rather than the whole paper, at its Thirty-Seventh Session. Listed as taking part in the ensuing discussion were the Society's President and chair of the meeting, H. Wildon Carr (1857–1931), Ludwik Silberstein (1872–1948), Peter Chalmers Mitchell (1864–1945), H. S. Shelton, Arthington Worsley (1861–1944) and George Dawes Hicks (1862–1941).[2] *The essay – with the additional commentary – was subsequently published in* The Organisation of Thought *with few substantive changes.*

When it was published in The Aims of Education *twelve years later, it was cut down substantially, from about 8,000 words to around 4,000. The cuts mostly consisted of the most technical and mathematically heavy portions. It is impossible to know exactly why these cuts were made, but two reasons suggest themselves: it would make the chapter more digestible for a popular audience, and it would cut down on total length (AE contained ten chapters to OT's eight, and the publisher may have wished the book to be shortened).*

This was Whitehead's first essay delivered to the Aristotelian Society,[3] *and represents something of a transition point from his earlier work in mathematics and logic to work in science (particularly physics) and philosophy. We shall address three aspects of the*

1. *Report of the 85th Meeting of the British Association for the Advancement of Science* (London, John Murray, 1916), p. 366.
2. 'Abstract of the Minutes of the Proceedings of the Aristotelian Society for the Thirty-Seventh Session', *Proceedings of the Aristotelian Society*, New Series, vol. 16, p. 364.
3. The Aristotelian Society was founded in 1880 to further the systematic study of philosophy. Whitehead joined in 1915, when he was fifty-four years old. That he had joined relatively late in life when younger men had been members for decades (e.g. Bertrand Russell had joined the Society in 1896, at the age of twenty-four) may help explain some of his rhetorical reticence in this essay. Whitehead served as President of the Society from 1922 to 1923, and his Presidential Address 'Uniformity and Contingency', delivered on 6 November 1922, was later reprinted in *Essays in Science and Philosophy* (1947). Other prominent members of the Society at the time included G. E. Moore, T. P. Nunn, C. Lloyd Morgan, H. Wildon Carr, C. D. Broad, D'Arcy Thompson, Lord Haldane, Norman Kemp Smith and Samuel Alexander. See Victor Lowe, *Alfred North Whitehead: The Man and His Work, Volume II: 1910–1947* (Baltimore: Johns Hopkins University Press, 1990), pp. 90–92.

work: (1) the geometric framework in which Whitehead develops his derivation of the Lorentz transformation (defined in the next paragraph); (2) the key stages of both the mathematics and the physics of his derivation; (3) the overall philosophical framework (including his supplementary notes) for his exposition insofar as this framework blatantly presages the foundations of process metaphysics. In discussing the first two of these aspects, we shall make some explicit comparisons with Whitehead's treatment and that of Einstein in his seminal paper 'On the Electrodynamics of Moving Bodies' ('Zur Elektrodynamik bewegter Körper', 1905), one of the most profound and significant works in the history of science.

A reference frame, or coordinate system, is called inertial *if Newton's law of inertia holds with respect to it (Whitehead uses the term* Dynamical *rather than inertial). Suppose there are two inertial frames, and one is moving with respect to the other at a uniform velocity. Note that this uniformity is necessary if both frames are to be inertial. Then any physical event can be characterised in both systems by four coordinates, three associated with directions in space and the fourth pertaining to time. The central question is this: for the same event, how are the coordinates in the first system, which can be thought of as the stationary system, related to the coordinates of the second, which can be thought of as the moving system? Nothing essential is lost if the relative motion of the two frames is considered as being no more than the second sliding along the x-axis of the first, with the other two sets of axes parallel, and arrange that both times are set to zero when the origins of the two systems coincide. The change of coordinates from the stationary frame to the moving frame is called the associated* Lorentz transformation.[1] *The amazing conclusion of the theory of special relativity is that the nature of this transformation is such that both time and length differ between observers in these two systems.*

To ease into this technical question, Whitehead begins with a metaphysical discussion of space and time and the idea that all measurements are relations between material things. He notes that in the abstract there are three possible forms of geometry, and that we customarily locate ordinary things in the Euclidean framework, which represents only one of the three choices. Moreover, the notion of distance, or metric, in that geometry is still not uniquely defined by its genre, and thus we have the possibility of two Euclidean observers using different metrics. (Much of this material is reminiscent of his earlier encyclopaedia article 'Axioms of Geometry' (p. 299).) Whitehead begins his mathematical development here (p. 373) and derives some general equations on how the systems must be related. The Lorentz transformation cannot yet be seen, but the table is set to accommodate different metrics for different observers within the same class of geometries.

Whitehead next introduces inertial frames and begins his development of the Lorentz transformation proper (pp. 373–374). His main moves are: (1) an application of symmetry to simplify the analysis, effectively trivialising the transformation of y- *and* z-*coordinates; (2) the development of a framework for the requisite equations using elementary techniques of multivariable calculus; (3) analysis of how the velocity of a moving object must change between frames and its implications for the transformation*

1. Neither Einstein nor Whitehead mentions Lorentz in the works we address here; the relationship between Lorentz and Einstein is quite beyond the scope of this introduction.

time; (4) introduction of a hypothetical velocity that is frame-independent (ultimately realised by the speed of light); (5) revisitation of his prior equations in light of 'this peculiar value' (p. 377) to yield the results given previously by Einstein.

Einstein's paper is distinctly less philosophical and more axiomatic than Whitehead's, although many of the same elements are evident in both expositions, including arguments by symmetry and straightforward applications of multivariable calculus. Einstein grounds the framework of his analysis upon physical measurement but takes no pains to explicate any elements of the underlying geometry. Yet he is much more explicit about the two key principles of special relativity – the independence of physical laws from particular inertial frames and the constancy of the speed of light as measured in any inertial frame – and applies these to the immediate demise of absolute simultaneity. Of particular note, he does not withhold the special property of the speed of light, but begins with it, making his work perhaps more physically and mathematically parsimonious than Whitehead's, but perhaps less insightful in mathematical and philosophical matters, which Whitehead chooses to emphasise.

Finally, it should be noted that the introductory metaphysical framework, the concluding section of the original address and the 'Supplementary Notes' are all distinctly on the way to the mature conception of process theory that Whitehead would present a decade later. Early in the supplement, in opposition to the a priori status that Kant attributes to space and time, he writes (his p. 121), 'I would suggest—rather timidly—that this doctrine should be given a different twist, which in fact turns it in the opposite direction—namely, that in the act of experience we perceive a whole formed of related differentiated parts'. The explicit notion of concrescence of course does not appear, but it is lurking.

Fundamental problems concerning space and time have been considered from the standpoints created by many different sciences. The object of this paper is the humble one of bringing some of these standpoints into relation with each other. This necessitates a very cursory treatment of each point of view.

Mathematical physicists have evolved their theory of relativity to explain the negative results of the Morley-Michelson experiment and of the Trouton experiment.[1] Experimental psychologists have considered the evolution of spatial ideas from the crude sense-data of experience. Metaphysicians have considered the majestic uniformity of space and time, without beginning and without end, without boundaries, and without exception in the truths concerning them; all these qualities the more arresting to our attention, from the confused accidental nature of the empirical universe which is conditioned by them. Mathematicians have studied the axioms of geometry, and can now deduce all that is believed to be universally true of space and of time by the strictest logic from a limited number of assumptions.

1. Regarding the far less well-known Trouton–Noble experiments, see F. T. Trouton and H. R. Noble, 'The Mechanical Forces Acting on a Charged Electric Condenser Moving Through Space', *Philosophical Transactions of the Royal Society, Series A, Containing Papers of a Mathematical or Physical Character*, vol. 202 (1904), pp. 165–181.

These various lines of thought have been evolved with surprisingly little interconnection. Perhaps it is as well. The results of science are never quite true. By a healthy independence of thought perhaps we sometimes avoid adding other people's errors to our own. But there can be no doubt that the normal method of cross-fertilizing thought is by considering the same, or allied problems to our own, in the form which they assume in other sciences.

|105| Here I do not propose to enter into a systematic study of these various chapters of science. I have neither the knowledge, nor the time.

First, let us take the ultimate basis of any theory of relativity. All space measurement is from stuff in space to stuff in space. The geometrical entities of empty space never appear. The only geometrical properties of which we have any direct knowledge are properties of those shifting, changeable appearances which we call things in space. It is the sun which is distant, and the ball which is round, and the lamp-posts which are in linear order. Wherever mankind may have got its idea of an infinite unchangeable space from, it is safe to say that it is not an immediate deliverance of direct observation.

There are two antagonistic philosophical ways of recognising this conclusion.

One is to affirm that space and time are conditions for sensible experience, that without projection into space and time sensible experience would not exist. Thus, although it may be true to say that our knowledge of space and time is given in experience, it is not true to say that it is deduced from experience in the same sense that the Law of Gravitation is so deduced. It is not deduced, because in the act of experiencing we are necessarily made aware of space as an infinite given whole, and of time as an unending uniform succession. This philosophical position is expressed by saying that space and time are *a priori* forms of sensibility.

The opposed philosophical method of dealing with the question is to affirm that our concepts of time and space are deductions from experience, in exactly the same way as the Law of Gravitation is such a deduction. If we form exact concepts of points, lines and surfaces, and of successive instants of time, and assume them to be related as expressed by the axioms of geometry and the axioms for time, then we find that we have framed a concept which, with all the |106| exactness of which our observations are capable, expresses the facts of experience.

These two philosophic positions are each designed to explain a certain difficulty. The *a priori* theory explains the absolute universality ascribed to the laws of space and time, a universality not ascribed to any deduction from experience. The experiential theory explains the derivation of the space-time concepts without introducing any other factors beyond those which are admittedly present in framing the other concepts of physical science.

But we have not yet done with the distinctions which in any discussion of space or time must essentially be kept in mind. Put aside the above question as to how these space-time concepts are related to experience—What are they when they are formed?

We may conceive of the points of space as self-subsistent entities which have the indefinable relation of being occupied by the ultimate stuff (matter, I will call it) which is there. Thus, to say that the sun is *there* (wherever it is) is to affirm the relation of occupation between the set of positive and negative electrons which we call the sun and a certain set of points, the points having an existence essentially independent of the sun. This is the absolute theory of space. The absolute theory is not popular just now, but it has very respectable authority on its side—Newton, for one—so treat it tenderly.

The other theory is associated with Leibniz. Our space concepts are concepts of relations between things in space. Thus there is no such entity as a self-subsistent point. A point is merely the name for some peculiarity of the relations between the matter which is, in common language, said to be in space.

It follows from the relative theory that a point should be definable in terms of the relations between material things. So far as I am aware, this outcome of the theory has escaped the notice of mathematicians, who have invariably assumed |107| the point as the ultimate starting ground of their reasoning. Many years ago I explained some types of ways in which we might achieve such a definition, and more recently have added some others.[1] Similar explanations apply to time. Before the theories of space and time have been carried to a satisfactory conclusion on the relational basis, a long and careful scrutiny of the definitions of points of space and instants of time will have to be undertaken, and many ways of effecting these definitions will have to be tried and compared. This is an unwritten chapter of mathematics,[2] in much the same state as was the theory of parallels in the eighteenth century.

In this connection I should like to draw attention to the analogy between time and space. In analysing our experience we distinguish events, and we also distinguish things whose changing relations form the events. If I had time it would be interesting to consider more closely these concepts of events and of things. It must suffice now to point out that things have certain relations to each other which we consider as relations between the space extensions of the things; for example, one space can contain the other, or exclude it, or overlap it. A point in space is nothing else than a certain set of relations between spatial extensions.

Analogously, there are certain relations between events which we express by saying that they are relations between the temporal durations of these events, that is, between the temporal extensions of the events. [The durations of two events A and B may one precede the other, or may partially overlap, or may one contain the other, giving in all six possibilities.] The properties of the extension of an event in time are largely analogous to the extension of an object in space. Spatial extensions are expressed by relations between objects, temporal extensions by relations between events.

The point in time is a set of relations between temporal extensions. It needs very little reflection to convince us that a point in time is no direct deliverance of experience. We live |108| in durations, and not in points. But what community, beyond the mere name, is there between extension in time and extension in space? In view of the intimate connection between time and space revealed by the modern theory of relativity, this question has taken on a new importance.

1. 'Many years ago . . . ' may refer to Whitehead's 'On Mathematical Concepts of the Material World' (see p. 214), while '. . . and more recently have added some others' is likely referring to his 'La Théorie Relationniste de L'Espace' ('The Relational Theory of Space'), which was published in 1916 in French (see p. 401), but first presented by Whitehead in April 1914, very likely in English.

2. Whitehead is most likely referring to the fourth volume of *Principia Mathematica*, which was planned but never completed.

I have not thought out an answer to this question. I suggest, however, that time and space embody those relations between objects on which depends our judgment of their externality to ourselves. Namely, extension in space and extension[1] in time both embody and perhaps necessitate a judgment of externality. This suggestion is very vague, and I must leave it in this crude form.

Diverse Euclidean Measure Systems

Turning now to the mathematical investigations on the axioms of geometry, the outcome, which is most important for us to remember, is the great separation which it discloses between non-metrical projective geometry, and metrical geometry. Non-metrical projective geometry is by far the more fundamental. Starting with the concepts of points, straight lines, and planes (of which not all three need be taken as indefinable), and with certain very simple non-metrical properties of these entities—such as, for one instance, that two points uniquely determine a straight line—nearly the whole of geometry can be constructed. Even quantitative co-ordinates can be introduced, to facilitate the reasoning. But no mention of distance, area, or volume, need have been introduced. Points will have an order on the line, but order does not imply any settled distance.

When we now enquire what measurements of distance are possible, we find that there are different systems of measurement all equally possible. There are three main types of system: any system of one type gives Euclidean geometry, any system of another type gives Hyperbolic (or Lobatchewskian) geometry, any system of the third type gives Elliptic geometry. |109| Also different beings, or the same being if he chooses, may reckon in different systems of the same type, or in systems of different types. Consider the example which will interest us later. Two beings A and B agree to use the same three intersecting lines as axes of *x*, *y*, *z*. They both employ a system of measurement of the Euclidean type, and (what is not necessarily the case) agree as to the plane at infinity. That is, they agree as to the lines which are parallel. Then with the usual method of rectangular Cartesian axes, they agree that the co-ordinates of P are the lengths ON, NM, MP. So far all is harmony. A fixes on the segment OU_1, on O*x*,[2] as being the unit length, and B on the segment OV_1, on O*x*. A calls his co-ordinates (x, y, z), and B calls them (X,Y,Z).

Then it is found [since both systems are Euclidean] that, whatever point P be taken,[3]

$$X = \beta x, \quad Y = \gamma y, \quad Z = \delta z. \quad [\beta \neq \gamma \neq \delta.]$$

They proceed to adjust their differences, and first take the *x*-co-ordinates. Obviously they have taken different units of length along O*x*. The length OU_1, which A calls one unit, B calls β units. B changes his unit length to OU_1, from its original length OV_1, and obtains X = *x*. But now, as he must use the same unit for all his measurements, his other co-ordinates are altered in the same ratio. Thus we now have

1. Both OT and AE changed the two instances of 'extension' in this sentence to 'location'.
2. The original erroneously had 'OX', which both OT and AE corrected.
3. In the final, bracketed expression below, the original had a period outside the closing bracket rather than inside it. Both OT and AE moved the period to the inside, since it applies only to the information inside the bracket rather than the whole sequence of inequalities.

$$X = x,\ Y = \gamma y/\beta,\ Z = \delta z/\beta.$$

The fundamental divergence is now evident. A and B agree as to their units along Ox. They settled that by taking[1] a given segment OU_1 as having the unit length. But they cannot agree as to what segment along Oy is equal to OU_1. A says it is OU_2, and B that it is OU'_2. Similarly for lengths along OZ.

The result is that A's spheres

$$x^2 + y^2 + z^2 = r^2,$$

|110| are B's ellipsoids,

$$X^2 + \beta^2 Y^2/\gamma^2 + \beta^2 Z^2/\delta^2 = r^2,$$

i.e.

$$X^2/\beta^2 + Y^2/\gamma^2 + Z^2/\delta^2 = r^2/\beta^2.$$

Thus the measurement of angles by the two is hopelessly at variance.

If $\beta \neq \gamma \neq \delta$, there is one and only one set of common rectangular axes at O, namely that from which they started. If $\gamma = \delta$, but $\beta \neq \gamma$, then there are a singly infinite number of common rectangular axes found by rotating the axes round Ox. This is, for us, the interesting case. The same phenomena are reproduced by transferring to any parallel axes.

The root of the difficulty is that A's measuring rod, which for him is a rigid invariable body, appears to B as changing in length when turned in different directions. Similarly all measuring rods, satisfactory to A, violate B's immediate judgment of invariability, and change according to the same law. There is no way out of the difficulty. Two rods ρ and σ coincide whenever laid one on the other; ρ is held still, and both men[2] agree that it does not change. But σ is turned round. A says it is invariable, B says it changes. To test the matter, ρ is turned round to measure it, and exactly fits it. But while A is satisfied, B declares that ρ has changed in exactly the same way as did σ. Meanwhile B has procured two material rods satisfactory to him as invariable, and A makes exactly the same objections.

We shall say that A and B employ diverse Euclidean metrical systems.

The most extraordinary fact of human life is that all beings seem to form their judgments of spatial quantity according to the same metrical system.

Relativity in Modern Physics[3]

Owing to the fact that points of space are incapable of direct recognition, there is a difficulty—apart from any abstract |111| question of the nature of space—in deciding on the motion to be ascribed to any body. Even if there be such a thing as absolute position, it is impossible in practice to decide directly whether a body's absolute position has changed. All spatial measurement is relative to matter.

1. Both OT and AE added the words 'along that axis': 'They settled that by taking *along that axis* a given segment'.
2. 'men' did not appear in the original version, but was added in both OT and AE. We have adopted this change since it clarifies that the otherwise ambiguous verb 'agree' in fact refers to A and B.
3. AE deleted the heading and most of the section. A note appears below where the text in AE picked up.

Newton's laws of motion in their modern dress evade this difficulty by asserting that a framework of axes of co-ordinates can be defined by their relations to matter such that, assuming these axes to be at rest, and all velocities to be measured relatively to them, the laws hold. The same expedient has to be employed for time, namely, the laws hold when the measurement of the flow of time is made by the proper reference to periodic events. Thus the laws assert that the framework and the natural clock adapted for their use have been successfully found.

But, if one framework will do, an infinity of others serve equally well; namely, not only—as is of course the case—all those at rest relatively to the first framework, but also all those which move without relative rotation with uniform velocity relatively to the first. This whole set of frameworks is on a level in respect to Newton's laws. We will call them Dynamical frameworks.

Now, suppose there are two observers, A and B. They agree in their non-metrical projective geometry, *e.g.*, what A calls a straight line so does B. They also both apply a Euclidean metrical system of measurement to this space. Their two metrical systems also agree in having the same plane at infinity, that is, lines which are parallel for A are also parallel for B. Furthermore, they have both successfully applied Newton's laws to the movement of matter, and agree in having the same sets of dynamical axes. But the framework (among these sets) which A chooses to regard as at rest is different from the frame (among the same sets) which B so regards.

Without alteration of their respective judgments of rest, |112| they choose their co-ordinate axes so that the origins (O for A, and O′ for B) are in relative motion along OO′, which is the axis of x for both.

Further, since OO′ is the line of symmetry of their diverse Euclidean systems, we assume that the two measure-systems agree for planes perpendicular to OO′, *i.e.*, we assume a symmetry round OO′. Then if for A at O, the distance OO′ be ξ, the relations at any instant between A's co-ordinates (x,y,z) and B's co-ordinates (x',y',z') for the same point P are given by

$$x' = \beta(x-\xi), \quad y' = y, \quad z' = z.$$

Also, according to A's clock, O′ is moving forward with a uniform velocity v. Also we measure A's time from the instant of the coincidence of O and O′.

Thus $$\xi = vt,$$

and $$x' = \beta(x - vt), \quad y' = y, \quad z' = z.$$

We now consider B's clock, and ask for the most general supposition which is consistent with the fact that their judgments as to the fact of uniform motion are in agreement.

We do not assume that events in various parts of space which A considers to be simultaneous are so considered by B. But we assume that at any point P, co-ordinates (x,y,z) for A, there is a determinate relation between B's time T and x, y, z, t.

Put $$\mathrm{T} = f(x,y,z,t).$$

Write $$\mathrm{P} = \frac{\delta \mathrm{T}}{\delta x}, \quad \mathrm{Q} = \frac{\delta \mathrm{T}}{\delta y}, \quad \mathrm{R} = \frac{\delta \mathrm{T}}{\delta z}, \quad \mathrm{S} = \frac{\delta \mathrm{T}}{\delta t}.$$

Now suppose that the point P is moving, and that (u_1, u_2, u_3) is its set of component velocities along the axes according to A's "space and clock" system, and (U_1, U_2, U_3) is its set of component velocities according to B's "space and clock" system. Then by mere differentiation it follows by a short mathematical deduction that—[1]

|113|

$$U_1 = \left\{\frac{d\beta}{dt}(x - vt) + (\beta u_1 - v)\right\} / \{Pu_1 + Qu_2 + Ru_3 + S\},$$
$$U_2 = u_2 / \{Pu_1 + Qu_2 + Ru_3 + S\},$$
$$U_3 = u_3 / \{Pu_1 + Qu_2 + Ru_3 + S\}.$$

But we have assumed that, whatever the direction of the resultant velocity (u_1, u_2, u_3), the velocities (U_1, U_2, U_3) and (u_1, u_2, u_3) are both uniform when either is uniform.

Hence it is easily proved that β, P, Q, R, S are independent of the co-ordinates (x, y, z) and of the time t. In other words, they are constant.

Hence we obtain

$$U_1 = \beta(u_1 - v) / \{Pu_1 + Qu_2 + Ru_3 + S\},$$

and

$$T = Px + Qy + Rz + St.$$

But we assumed that OO′, *i.e.*, Ox, is an axis of symmetry. It follows from this assumption that

$$Q = R = 0.$$

We thus obtain the simplified results

$$\left.\begin{aligned} T &= Px + St, \\ U_1 &= \beta(u_1 - v)/(Pu_1 + S), \\ U_2 &= u_2/(Pu_1 + S), \\ U_3 &= u_3/(Pu_1 + S). \end{aligned}\right\} \qquad \text{(I)}$$

Here we remember that (u_1, u_2, u_3) are the velocities of any particle according to A's "space and clock" system, and that (U_1, U_2, U_3) are the velocities of the same point according to B's "space and clock" system. We have obtained the most general relations consistent with the facts that (1) they both employ Euclidean systems, related as described above, and (2) they agree in their judgments on the uniformity of velocity.

We now compare their judgments on the magnitudes of velocities.

Let the magnitude of the velocity of P be V according to A's judgment, and V′ according to B's judgment.

Then

$$V^2 = u_1^2 + u_2^2 + u_3^2,$$
$$V'^2 = U_1^2 + U_2^2 + U_3^2.$$

1. In the third equation below, the original erroneously had 'U^3' instead of 'U_3', an error which was fixed in OT.

|114| Also we can put

$$u_1 = l\mathrm{V}, \quad u_2 = m\mathrm{V}, \quad u_3 = n\mathrm{V},$$

where (l,m,n) have nothing to do with the magnitude V, but simply depend on the direction of motion. In fact (l,m,n), are the "direction cosines" of the velocity according to A's judgment. By substituting in the above equation for V^2 we see that

$$l^2 + m^2 + n^2 = 1.$$

Now, substituting for (u_1, u_2, u_3) in the equations (I) above, and squaring and adding, and eliminating $m^2 + n^2$ by the relation just found, we at once find

$$\mathrm{V}'^2 = \left\{ \left(\beta^2 - 1\right)\mathrm{V}^2 l^2 - 2\beta^2 \mathrm{V} v l + \beta^2 v^2 + \mathrm{V}^2 \right\} / \left(\mathrm{PV}l + \mathrm{S}\right)^2.$$

It is thus seen that in general the relation of V′ to V depends on the direction cosine l. Now l is the cosine of the angle which the direction of the velocity V makes with Ox, according to A's judgment.

The meaning of this relation is that if A discharges, from guns at the point P, shells with a given muzzle velocity V, according to his judgment, B will consider that their muzzle velocities are different from each other, except in the case of pairs of guns equally inclined to the axis OO′. Instances of this type of diversity of judgment can be noted any day by anyone who looks out of the window of a railway carriage, and forgets that he is travelling.

Now, suppose the velocity V′ bears a relation to the velocity V, which is independent of l. Then l must disappear from the above formula. There are two conditions to be satisfied:

One condition is

$$\mathrm{V}^2 = \beta^2 v^2 / \left(\beta^2 - 1\right),$$

or in a more convenient form

$$\beta^2 = 1 / \left(1 - v^2 / \mathrm{V}^2\right).$$

|115| The meaning of this condition is that there is one, and only one, muzzle velocity V (according to A's judgment), namely, the muzzle velocity given by the above formula, which can have the property that B will judge that all the guns are firing in their diverse directions with one common muzzle velocity.

Let us now suppose that V has this peculiar value: that is, if we look on this value V as known, we must suppose that β is given by the second of the above formulae.

The other condition allows P and S to be put in the forms

$$\mathrm{P} = -\beta v / \lambda \mathrm{V}^2, \ \mathrm{S} = \beta / \lambda,$$

where

$$\mathrm{V}' = \lambda \mathrm{V}.$$

Thus we have the bundle of formulæ

$$\beta^2 = 1/(1 - v^2/\mathrm{V}^2),$$

$$\mathrm{T} = \beta\{t - vx/\mathrm{V}^2\}/\lambda,$$

$$\mathrm{V}' = \lambda \mathrm{V}.$$

The value which we give to λ is purely a matter for the adjustment of units. If we want A and B to agree in their judgments of the magnitude of this peculiar muzzle-velocity, we put $\lambda = 1$.

We then get the formulæ usually adopted, namely,

$$\left.\begin{aligned} \beta^2 &= 1/(1-v^2/V^2), \\ T &= \beta\{t - vx/V^2\}, \\ V' &= V. \end{aligned}\right\} \qquad \text{(II)}$$

But if we prefer that A and B should reckon (according to A's judgment) in the same units of time, we put $\lambda = \beta$, and obtain

$$\begin{aligned} \beta^2 &= 1/(1-v^2/V^2), \\ T &= t - vx/V^2, \\ V' &= \beta V. \end{aligned}$$

|116| But A and B are in any case in such hopeless difficulties over their comparisons of time judgments that the detail of using the same units does not help them much. Accordingly the formulæ marked (II) are those used. Thus A and B agree in their judgments as to the magnitude of one special velocity V, whatever may be the direction in which the entity possessing it is moving.

In order to reach this measure of agreement, they have to disagree as to their space judgments and their time judgments. The root cause of their disagreement is their diverse judgment as to which axis system is to be taken at rest for the purpose of measuring velocities.

Before discussing the nature of the disagreement disclosed in formula (II), let us ask why we should bring these difficulties on our heads by supposing that two people in relative motion, who both (for the purpose of measuring velocities) assume that they are at rest, should agree in their judgments in respect to this special velocity V.

Such an agreement has no counterpart in any of our obvious judgments made from railway carriages. Surely we can wait till the contingency occurs before discussing the confusion which it creates.

But the contingency has occurred. It occurs when we consider the velocity of light. Perhaps I may venture to remind a philosophical society that light moves so very quickly that it is difficult to consider its velocity at all. So we need not be surprised that this peculiar fact concerning its velocity is not more obvious.

Now V being the velocity of light, unless v is large, v/V (and still more v^2/V^2) will be quite inappreciable. The only velocity ready to hand which is big enough to give v/V an appreciable value is the velocity of the earth in its orbit.

Many diverse experiments have been made, and they all[1] agree in concluding that a man who assumes the earth to be at |117| rest will find by measurement that the velocity of light is the same in all directions. Furthermore, when the same man turns

1. The original article misspelled 'all' as 'al'. This was corrected in OT.

his attention to interstellar or interplanetary phenomena, and assumes the sun to be at rest, he will again find the velocity of light to be the same in all directions. These are well attested experiments made at long intervals of time.

This is the exact contingency contemplated above.

Again the velocity of light *in vacuo* has recently taken on a new dignity. It used to be one among other wave velocities, such as the velocity of sound in air, or in water, or the velocity of surface waves in water. But Clerk Maxwell discovered that all electromagnetic influences are propagated with the velocity of light, and now modern physical science half suspects that electromagnetic influences are the only physical influences which relate the changes in the physical world. Accordingly the velocity of light becomes the fundamental natural velocity, and experiment shows that our judgment of its magnitude is not affected by our choice of the framework at rest, so long as we keep to a set of dynamical axes. These experiments on light have been confirmed by other electromagnetic experiments not involving light.

Thus we are driven to equations (II), where V is the velocity of light.

The first conclusion to be drawn from equations (II) is that two people who make different choices of bodies at rest will disagree as to their measuring rods in the way described above. There is no peculiar difficulty about that. The only wonder is that all people agree so well in their judgments as to metrical systems. A mathematical angel would naturally expect incarnate men to be in violent disagreement on this subject.

But the case of time is different. For simplicity of statement we speak of A as at O, and B as at O′. We remember that O′ is moving relatively to O with velocity v in direction OO′. Suppose A and B are looking in this direction; and |118| they both measure their time from the instant when they met, as O′ passed over O. Then we have

$$\mathrm{T}=\left\{t-vx/\mathrm{V}^2\right\}/\left\{1-v^2/\mathrm{V}^2\right\}.$$

Now, suppose we consider all the events all over space which A considers to have happened simultaneously at the time t. The events of this set which occurred anywhere on a plane perpendicular to OO′ at a distance x in front of O (according to A's reckoning), will have occurred according to B's reckoning at the time T as given above. Let us fix our attention on the fact that B does not consider all these events to be simultaneous. For let T_1 and T_2 be B's times for such events on planes x_1 and x_2. Then

$$\mathrm{T}_1-\mathrm{T}_2=v\left(x_2-x_1\right)/\left(\mathrm{V}^2-v^2\right).$$

Thus if x_2 be greater than x_1, T_2 is less than T_1. Thus B judges the more distant events in front of him to have happened earlier than the nearer events in front of him, and *vice versâ* for the events behind. This disturbance of the judgment of simultaneity is the fundamental fact. Obviously the measurement of time intervals is a detail compared to simultaneity. A may think a sermon long, and B may think it short, but at least they should both agree that it stopped when the clock hand pointed at the hour. The worst of the matter is that so far as any test can be applied there is no method of discriminating between the validities of their judgments.

Thus we are confronted with two distinct concepts of the common world, A's space-time concept, and B's space-time concept. Who is right? It is no use staying for an answer. We must follow the example of the wise old Roman, and pass on to other things.[1]

Thus estimates of quantity in space and time, and, to some extent, even estimates of order, depend on the individual observer. But what are the crude deliverances of sensible |119| experience, apart from that world of imaginative reconstruction which for each of us has the best claim to be called our real world? Here the experimental psychologist steps in. We cannot get away from him. I wish we could, for he is frightfully difficult to understand. Also, sometimes his knowledge of the principles of mathematics is rather weak, and I sometimes suspect—No, I will not say what I sometimes think: probably he, with equal reason, is thinking the same sort of thing of us.

I will, however, venture to summarise conclusions, which are, I believe, in harmony with the experimental evidence, both physical and psychological, and which are certainly suggested by the materials for that unwritten chapter in mathematical logic which I have already commended to your notice. The concepts of space and time and of quantity are capable of analysis into bundles of simpler concepts. In any given sensible experience it is not necessary, or even usual, that the whole complete bundle of such concepts apply. For example, the concept of externality may apply without that of linear order, and the concept of linear order may apply without that of linear distance.

Again, the abstract mathematical concept of a space-relation may confuse together distinct concepts which apply to the given perceptions. For example, linear order in the sense of a linear projection from the observer is distinct from linear order in the sense of a row of objects stretching across the line of sight.

Mathematical physics assumes a given world of definitely related objects, and the various space-time systems are alternative ways of expressing those relations as concepts in a form which also applies to the immediate experience of observers.

Yet there must be one way of expressing the relations between objects in a common external world. Alternative methods can only arise as the result of alternative standpoints; |120| that is to say, as the result of leaving something added by the observer sticking (as it were) in the universe.

But this way of conceiving the world of physical science, as composed of hypothetical objects, leaves it as a mere fairy tale. What is really actual are the immediate experiences. The task of deductive science is to consider the concepts which apply to these data of experience, and then to consider the concepts relating to these concepts, and so on to any necessary degree of refinement. As our concepts become more abstract, their logical relations become more general, and less liable to exception. By this logical construction we finally arrive at conceptions, (i) which have determinate exemplifications in the experience of the individuals, and (ii) whose

1. Whitehead may be referring to Marcus Aurelius here, who said that when there is no distinction in meaning, 'Why then be troubled?' (Marcus Aurelius, *Meditations* 9.39). Marcus Aurelius, *The Meditations of Marcus Aurelius Antoninus*, translated by A. S. L. Farquharson (Oxford: Oxford University Press, 1990).

This marks the end of the long section of text that was deleted in AE. Whitehead added a few sentences of transitional material here in AE: 'This statement, however, is only true within the limits of accuracy obtained by human observation. When we endeavour to frame a self-consistent physical theory we have to admit that diverse spatio-temporal systems of measurement are relevant to the behaviour of things.'

logical relations have a peculiar smoothness. For example, conceptions of mathematical time, of mathematical space, are such smooth conceptions. No one lives in "an infinite given whole," but in a set of fragmentary experiences. The problem is to exhibit the concepts of mathematical space and time as the necessary outcome of these fragments by a process of logical building up. Similarly for the other physical concepts. This process builds a common world of conceptions out of fragmentary worlds of experience. The material pyramids of Egypt are a conception, what is actual are the fragmentary experiences of the races who have gazed on them.

So far as science seeks to rid itself of hypothesis, it cannot go beyond these general logical constructions. For science, as thus conceived, the divergent time orders considered above present no difficulty. The different time systems simply register the different relations of the mathematical construct to those individual experiences (actual or hypothetical) which could exist as the crude material from which the construct is elaborated.

But after all it should be possible so to elaborate the mathematical construct so as to eliminate specific reference to particular experiences. Whatever be the data of experience, there must be something which can be said of them as a whole, |121| and that something is a statement of the general properties of the common world. It is hard to believe that with proper generalisation time and space will not be found among such properties.

Supplementary Notes on the Above Paper[1]

The first four[2] pages of the paper consist of a summary of ideas which ought to be in our minds while considering problems of time and space. The ideas are mostly philosophical, and the summary has been made by an amateur in that science; so there is no reason to ascribe to it any importance except that of a modest reminder. There are only two points in this summary to which I would draw attention.

In the middle of p. 371[3] there occurs:—[4]

"Wherever mankind . . . unending uniform succession."

If I understand Kant rightly—which I admit to be very problematical—he holds that in the act of experience we are aware of space and time as ingredients necessary for the occurrence of experience. I would suggest—rather timidly—that this doctrine should be given a different twist, which in fact turns it in the opposite direction—namely, that in the act of experience we perceive a whole formed of related differentiated parts. The relations between these parts possess certain

1. OT changed this heading to read: '*Commentary added on reading the Paper before the Aristotelian Society*'. AE deleted the heading altogether, though a blank line before the next paragraph was retained. AE also removed the next few paragraphs, and picked up with 'If I understand Kant rightly. . .'.
2. OT changed this number was changed from 'four' in the original to 'six' in OT, and it has been changed again for this edition to accurately reflect the portions of the essay that Whitehead intended to reference.
3. OT changed 'In the middle of p. 105' to 'On pp. 192 and 193'. Page references were, of course, different for the different versions of the essay. These have been changed to reflect proper page referents for this version. Further changes to internal page references have been made silently.
4. This 'dog's bollocks' construction, that is, a colon followed immediately by a dash (primarily a British convention), was intended to indicate a restful pause.

characteristics, and time and space are the expressions of some of the characteristics of these relations. Then the generality and uniformity which are ascribed to time and space express what may be termed the uniformity of the texture of experience.

The success of mankind—modest though it is—in deducing uniform laws of nature is, so far as it goes, a testimony that this uniformity of texture goes beyond those characteristics of the data of experience which are expressed as time and space. |122| Time and space are necessary to experience in the sense that they are characteristics of our experience; and, of course, no one can have our experience without running into them. I cannot see that Kant's deduction amounts to much more than saying that "what is, is"—true enough, but not very helpful.

But I admit that what I have termed the "uniformity of the texture of experience" is a most curious and arresting fact. I am quite ready to believe that it is a mere illusion: and later on in the paper I suggest that this uniformity does not belong to the immediate relations of the crude data of experience, but is the result of substituting for them more refined logical entities, such as relations between relations, or classes of relations, or classes of classes of relations. By this means it can be demonstrated—I think—that the uniformity which must be ascribed to experience is of a much more abstract attenuated character than is usually allowed. This process of lifting the uniform time and space of the physical world into the status of logical abstractions has also the advantage of recognising another fact, namely, the extremely fragmentary nature of all direct individual experience.[1]

My point in this respect is that fragmentary individual experiences are all that we know, and that all speculation must start from these disjecta membra as its sole datum. It is not true that we are directly aware of a smooth running world, which in our speculations we are to conceive as given. In my view the creation of the world is the first unconscious act of speculative thought; and the first task of a self-conscious philosophy is to explain how it has been done.

There are roughly two rival explanations. One is to assert the world as a postulate. The other way is to obtain it as a deduction, not a deduction through a chain of reasoning, but a deduction through a chain of definitions which, in fact, lifts thought on to a more abstract level in which the logical ideas are more complex, and their relations are more universal. In this way the broken limited experiences sustain that con-|123|nected infinite world in which in our thoughts we live. There are three more remarks while on this point I wish to make:—

(i) The fact that immediate experience is capable of this deductive superstructive must mean that it itself has a certain uniformity of texture. So this great fact still remains.

(ii) I do not wish to deny the world as a postulate. Speaking without prejudice, I do not see how in our present elementary state of philosophical advance we can get on without middle axioms, which, in fact, we habitually assume.

My position is that by careful scrutiny we should extrude such postulates from every part of our organised knowledge in which it is possible to do without them.

Now, physical science organises our knowledge of the relations between the deliverances of our various senses. I hold that in this department of knowledge such

1. AE added the important qualifier 'conscious': 'individual *conscious* experience'.

postulates, though not entirely to be extruded, can be reduced to a minimum in the way which I have described.[1]

I have not the slightest knowledge of theories respecting our emotions, affections, and moral sentiments, and I can well believe that in dealing with them further postulates are required. And in practice I recognise that we all make such postulates, uncritically.

(iii) The next paragraphs on p. 371 are as follows:—

"The opposed philosophical method . . . physical science."

It will be noted that, in the light of what has just been stated, the first of these paragraphs (which, I hope, faithfully expresses the experiential way of approaching the problem) really obscures the point which I have been endeavouring to make. The phrase, "If we form the exact concepts of points, etc.," is fatally ambiguous as between the method of postulating entities with assigned relations, and the method of forming logical constructions, and thus reaching points, etc., as the result of a chain of definitions.

|124| Turning now to p. 372, we come to the following paragraphs:—

"The other theory . . . eighteenth century."

We note again that the relational theory of space from another point of view brings us back to the idea of the fundamental space entities as being logical constructs from the relations between things. The difference is that this paragraph is written from a more developed point of view, as it implicitly assumes the things in space, and conceives space as an expression of certain of their relations. Combining this paragraph with what has gone before, we see that the suggested procedure is first to define "things" in terms of the data of experience, and then to define space in terms of the relations between things.[2]

This procedure is explicitly assumed in the next short paragraph: "In this connection . . . from the events."

The gist of the remaining paragraphs of this section is contained in the paragraph at the bottom of p. 372: "The point in time . . . new importance."

The sentence, "We live in durations, and not in points," can be amplified by the addition, "We live in space-extensions and not in space-points."

It must be noted that "whole and part" as applied to extensions in space or time must be different from the "all and some" of logic, unless we admit points to be the fundamental entities. For "spatial whole and spatial part" can only mean "all and some" if they really mean "all the points and some of the points." But if extensions and their relations are more fundamental than points, this interpretation is precluded. I suggest that "spatial whole and spatial part" is intimately connected with the fundamental relation between things from which our space ideas spring.

The relation of space whole to space part has many formal properties which are identical with the properties of "all and some." Also when points have been defined, we can replace it |125| by the conception of "all the points and some of the points." But the confusion between the two relations is fatal to sound views on the subject.

1. AE deleted the next half-dozen paragraphs, picking up with 'We note again the relational theory . . .'.
2. AE deleted the rest of the essay save for the last four sentences.

Diverse Euclidean Measure Systems

The next section deals with the measure systems applicable to space.

A measure system is a group of congruent transformations of space into itself. Consider a rigid body occupying all space. Let this body be moved in any way so that the particles of the body which occupied points P_1, P_2, P_3, etc., now occupy points Q_1, Q_2, Q_3, etc. Then any point P_1 in space is uniquely related to the corresponding point Q_1 in space by a one-to-one transformation with certain characteristics. By the aid of these transformations we can achieve the definition of distance in a way which definitely determines the distance between any two points, provided that we can define what we mean by a congruent transformation without introducing the idea of distance. If we introduced the idea of distance, we should simply say that a congruent transformation is one which leaves all distances unchanged, *i.e.*, if P_1, P_2, are transformed into Q_1, Q_2, then the distance P_1P_2 is equal to the distance Q_1Q_2.

But mathematicians have succeeded in defining congruent transformations without any reference to distance.

There are alternative groups of such congruent transformations, and each group gives a different measure system for space. The distance P_1P_2 may equal the distance Q_1Q_2 for one measure system, and will not equal it for another measure system. All these different measure systems are on the same level, equally applicable. A being with a strong enough head could think of them all at once as applying to space. The result so far as it interests us in respect to the theory of relativity is explained on pp. 372–374, ending with "The root of the difficulty[1] . . . same metrical system." This final sentence bears on Poincaré's assertion that the measure system adopted is purely |126| "conventional." I presume that by "conventional" a certain arbitrariness of choice is meant; and in that case, I must express entire dissent. It is true that within the circle of geometrical ideas there is no means of giving any preference to any one measure system, and any one is as good as any other. But it is not true that if we look at a normal carriage wheel, and at an oval curve one foot broad and ten feet long, we experience any arbitrariness of judgment in deciding which has approximately the form of a circle. Accordingly to Poincaré the choice between them, as representing a circle, is entirely conventional.

Again we equally form immediate judgments as to whether a body is approximately rigid. We know that a paving stone is rigid, and that a concertina is not rigid. This again necessitates a determinate measure system, selected from among the others.

Accordingly we conclude that (i) each being does in fact employ a determinate measure system, which remains the same, except possibly for very small variations, and (ii) the measure systems of different human beings agree, to within the limits of our observations. These conclusions are not the less extraordinary because no plain man has ever doubted them.

It is an interesting subject to investigate exactly what are the fundamental uniformities of experience which necessitate this conclusion. It is not so easy as it looks,

1. OT changed 'root of the difficulty' to 'most extraordinary fact'. The changed passage reference is a little easier to follow since it refers only to the final paragraph of the cited discussion, whereas it formerly referred to several paragraphs.

since we have to divest ourselves of all aid of scientific hypothesis if our conclusions are to be demonstrative.

Relativity in Modern Physics

Pp. 374–375, "Owing to the fact...[1] which B so regards."

The fundamental formulæ for the theory of relativity are the relations between diverse co-ordinate systems given on p. 375, and formulæ II at the top of p. 378. The general explanation of one method in which these formulæ arise—namely, Einstein's method—is given on pp. 375–379. Namely, we seek the condition that for all dynamical axes the velocity of light |127| should be the same, and the same in all directions. It should be noted that the experiments which, so far as they go, confirm these formulæ, can also be explained in another way which makes the theory of relativity unnecessary. We need only ascribe to the ether a certain property of contraction in the direction of motion, and the thing is done. So no one need be bludgeoned into accepting the rather bizarre doctrine of relativity, nor indeed any other scientific generalisation. The good old homely ether, which we all know, can in this case serve the purpose. Just as an author of genius, if he lives long enough, survives the inevitable accusation of immorality, so the ether by dint of persistence has outlived all reputation of extravagance. But if we detach ourselves from the glib phraseology concerning it, the scientific ether is uncommonly like the primitive explanation of the soul, as a little man inside us, which can sometimes be caught escaping in the form of a butterfly. As soon as the ether has to be patched up with special properties to explain special experiments, its scientific use is problematical, and its philosophic use is nil.

Philosophically the ether seems to me to be an ambitious attempt to give a complete explanation of the physical universe by making an elephant stand on a tortoise. Scientifically it has a perfectly adequate use by veiling the extremely abstract character of scientific generalisations under a myth, which enables our imaginations to work more freely. I am not advocating the extrusion of ether from our scientific phraseology, even though at special points we have to abandon it.

But the key to the reasons why it is worth while to consider seriously the doctrine of relativity is to be found on p. 379. "Again the velocity of light . . . not involving light." Namely, we have begun to suspect that all physical influences require time for their propagation in space. This generalisation is a long way from being proved. Gravitation stands like a lion in the path. But if it is the case, then all idea of an immediate |128| presentation to us of an aspect of the world as it in fact is, must necessarily be abandoned. What we perceive at any instant is already ancient history, with the dates of the various parts hopelessly mixed.

We must add to this the difficulty of determining what is at rest and what is in motion, and the further difficulty of determining a definite uniform flow of time. It is no use discussing this matter as though, but for the silly extravagant doctrine

1. The original contained an extra, fourth period in the ellipsis, which was deleted in OT.

of relativity, everything would be plain sailing. It isn't. You may be quite sure that when, after prolonged study, you endeavour to give the simplest explanation of a grave difficulty, you will be accused of extravagance. I have no responsibility for the doctrine of relativity, and hold no brief for it, but it has some claim to be considered as a comparatively simple way out of a scientific maze.

In the first place, we use the Newtonian dynamical sets of axes, and the Newtonian clock to extricate ourselves partially from the difficulties of rest, motion, and time. These have proved capable of scientific determination within the limits of our experimental accuracy. Thus the only thing left over is the choice of the axes at rest, which is a completely indeterminate problem on Newtonian principles.

Again, so far as we can at present guess by adopting the theory that all metrical influence is electromagnetic, all influences are propagated with the velocity of light *in vacuo*. This electromagnetic hypothesis is by no means established, but it gives the simplest of all possible results in respect to the propagation of influence, which we therefore adopt.

But what dynamical axes are we taking as at rest? Now our practical choice gives a range of relative velocities small compared to that of light. So except for certain refined experiments it does not matter. There are two possibilities:—

(i) We may assume that one set of axes are at rest, and that the others will show traces of motion in respect to the velocity of light; or

|129| (ii) That the velocity of light is the same in all directions whichever be the dynamical axes assumed.

The first supposition is negatived by experiment, and hence we are driven to the second supposition; which immediately lands us in the whole theory of relativity.

But if we will not have this theory we must reject the earlier supposition that the velocity of light *in vacuo* is the same in all directions. This we do, in fact, by assuming an ether, and assuming a certain law for its modification. Then we, in fact, adopt the first supposition so far as to hold that there are dynamical axes specially at rest, namely, at rest relatively to the undisturbed ether. Then an assumed law for the modification of the ether so alters the velocity of light that we explain why no dynamical axes show traces of motion.

I wish now to go back to the point which I made a few minutes ago, that what we perceive at any instant is ancient history with its dates hopelessly mixed. In the earlier part of my comments I emphasised[1] the point that our only data[2] as to the physical world are our sensible perceptions. We must not slip into the fallacy of assuming that we are comparing a given world with given perceptions of it. The physical world is in some general sense of the term a deduced concept.

Our problem is, in fact, to fit the world to our perceptions, and not our perceptions to the world.

---

1. AE changed to 'emphasise' and deleted 'In the earlier part of my comments'. AE picked up with this sentence after deleting a large portion of the preceding content.
2. AE added the qualifier 'exact': 'our only *exact* data'.

The Aims of Education – A Plea for Reform

1916 ***Mathematical Gazette*, vol. 8, no. 121 (January), pp. 191–203**

1917 *The Organisation of Thought* (London: Williams and Norgate), pp. 1–28

1929 *The Aims of Education* (New York: Macmillan), pp. 1–23

1929 *The New Republic*, vol. 58, no. 750 (17 April), pp. 244–246

1947 Claude M. Fuess and Emory S. Basford (eds), *Unseen Harvests: A Treasuring of Teaching* (New York: Macmillan), pp. 92–102

This address was delivered at the annual meeting of the Mathematical Association on 5 January 1916. The meeting took place at 2 p.m., with Whitehead's address immediately following a short report of the Council and Treasurer's report.

The meeting took place at London Day Training College (LDTC) at Southampton Row, which trained teachers in London universities of both sexes. Founded in 1902, it had become a school of the University of London in 1909 and been officially renamed the University of London, Institute of Education (IOE), but continued to be known informally as the LDTC for some time.

The text of this address appeared as the first chapter of both The Organisation of Thought *and* The Aims of Education. *It seems clear that Whitehead regarded it as particularly important. His main argument is that education needed to divest itself of teaching 'inert ideas', that is, ideas that are not utilised in the pupil's life, that have no practical application. Related themes are the evils of standardised examinations and teaching 'too many subjects', as these lend themselves to the memorisation of disconnected and useless facts.*

As with the other essays that appeared in Organisation of Thought *and* Aims of Education, *the AE version deleted portions that identified the specific occasion of its delivery as a meeting of the Mathematics Association, or that focused too specifically on English education to be considered relevant to an American audience.*

The essay was also printed in The New Republic, *an American magazine, and* Unseen Harvests, *an edited book of essays on education.* The New Republic *version (abbreviated to NR in the notes) was considerably abridged; it restored one of the five paragraphs at the top of the essay that AE had deleted, but then cut the entire second half (about 3,400 words out of the original 6,400).* The Unseen Harvests *version (abbreviated UH in the notes) was explicitly reprinted from AE, but added a short biography of Whitehead at the top* *(see p. 389n4)**.*

When I had the honour of being made President of the Mathematical Association, I did not foresee the unusual responsibility which it entailed. It was my intention to take as the theme of a presidential address the consideration of some aspect of those special subjects to which my own researches have principally been directed. Events have forced me to abandon that intention.[1] It is useless to discuss abstract questions in the midst of dominant practical preoccupation. We cannot disregard the present crisis in European civilization. It affects every function of life. In the harder struggle for existence which lies before the nation, all departments of national effort will be reviewed for judgment. The mere necessity for economy in resources will provoke this reformation.

We are concerned with education. This Association, so rich in its membership of educationalists, with the conception of reform as the very reason of its being,[2] is among those bodies which must take the lead in guiding that educational reconstruction which by a sociological law follows every social revolution. We do not want impracticable ideals, only to be realized beyond the clouds in

"Some wild, weird clime,
Out of Space, and out of Time."[3]

We require to know what is possible now in England, a nation conscious of its high achievements, and of great failures, shaken to its foundations, distrustful of the old ways, and dreading fantastic novelties.

1. Whitehead is undoubtedly referring to the First World War.
2. The Mathematical Association was founded in 1871 as the Association for the Improvement of Geometrical Teaching in 1871 (it was renamed in 1897). Interested in mathematical education reform generally, its explicit goal was to overthrow Euclid's *Elements* as the sole geometry textbook in England, a goal which was achieved in 1903 (see p. 348n2). In an article that appeared in the same issue of the *Mathematical Gazette* as Whitehead's address, the recently deceased William Henry Hoar Hudson (1838–1915) wrote that 'Euclid has been dethroned, but it may still remain a question whether all the Improvement that was hoped for has or has not taken place. It was admitted in the arguments that were used that the main fault was that of learning by heart the exact words of a particular translation of Euclid's book. If this fault has not been corrected, the mere substitution of another text-book can avail but little. The worship of the letter killeth.' W. H. H. Hudson, 'The Work of a Local Chapter', *Mathematical Gazette*, vol. 8, no. 121 (January 1916), p. 216.
3. Whitehead appears to here be quoting (though not quite exactly) from Edgar Allan Poe's poem *Dreamland*, lines 7 and 8 of which read:

   From a wild weird clime that lieth, sublime,
   Out of SPACE—Out of TIME.

   It is no stretch at all to think that Whitehead may have first heard these lines from Walter Raleigh (1861–1922). Both men were born in the same year, and both were Cambridge Apostles. Raleigh had quoted the same lines in 1890 in an inaugural address at University College, Liverpool. See Walter Raleigh, *Poetry and Fact* (Liverpool: Henry Young and Sons, 1890). In his biography of Whitehead, Victor Lowe wrote that 'Until Miss Evelyn Wade came into Whitehead's life, Walter Raleigh was the most vivid person of his own generation that he knew. . . . His humor and gaiety led [the Cambridge Apostle] meetings away from the Victorianism of [Henry] Sidgwick and his period. Even Whitehead, basically a Victorian, escaped to some extent; to that extent, Raleigh was a permanent influence on him.' Victor Lowe, *Alfred North Whitehead: The Man and His Work, Volume I: 1861–1910* (Baltimore: Johns Hopkins University Press, 1985), p. 122. It is perhaps also worth noting that Raleigh wrote a book on William Wordsworth in 1903; Whitehead would later make much of a comparison of Wordsworth and Percy Bysshe Shelley in his *Science and the Modern World* in 1925.

I will take my courage in both hands, and put before you an outline of educational principles. What I am going to say is of course entirely without your authority, and does not pledge or prejudge any action of the Association. We are primarily concerned only with the intellectual side of education, and, as mathematicians, are naturally concerned to illustrate details more particularly by reference to mathematics. Thus much to explain deliberate omissions in what follows.

Consider now the general and special education of two types of boys,[1] namely those in secondary schools who in after life must form the professional and directing classes in commerce, industry, and public administration, and again those in junior technical schools, and later in advanced continuation classes, who are going to form the class of skilled artisans and foremen of workshops. These two sets compose the educated strength of the nation. We must form no ideals which include less than these entire classes within their scope. What I shall say, will in phraseology apply |192| more directly to the secondary schools, but with unessential changes it will apply equally to the other group.[2]

What is the first commandment to be obeyed in any educational theme?[3] It is this: Do not teach too many subjects. The second command is this: What you teach, teach thoroughly. The devil in the scholastic world has assumed the form of a general education consisting of scraps of a large number of disconnected subjects; and, with the artfulness of the serpent, he has entrenched himself behind the matriculation examination of the University of London, with a wire entanglement formed by the Oxford and Cambridge schools' examination.[4]

Culture is activity of thought, and receptiveness to beauty, and humane feeling. Scraps of information have nothing to do with it. A merely well-informed man is the most useless bore on God's earth. What we should aim at producing is men who possess both culture and expert knowledge in some special direction. Their expert knowledge will give them the ground to start from, and their culture will lead them as deep as philosophy and as high as art. We have to remember that the valuable intellectual development is self-development, and that it mostly takes place between

1. The casual sexism in this reference to 'boys' is incongruous with Whitehead's support of participation in the education of students of both sexes. See p. 342n1.
2. NR began with the following paragraph, deleting all preceding material.
3. OT and NR changed 'theme' to 'scheme'.
4. NR deleted 'matriculation examination of the University of London, with a wire entanglement formed by the Oxford and Cambridge schools' examination' and replaced it with the more general 'entrance examinations of universities'.

AE and UH began with the following paragraph, deleting all preceding material. As with many other AE deletions from OT essays, this seems to have been done because AE was a book published in America for an American audience, and the forgoing was likely considered to be too specific for a non-English context.

The UH version – which was explicitly based on the AE version – added the following short biography of Whitehead to the top of the essay: '*ALFRED NORTH WHITEHEAD, mathematician and philosopher, was born in England in 1861. He was educated at Sherburne School and at Trinity College, Cambridge. He taught mathematics in his native country until 1924, when he became Professor of Philosophy at Harvard. He retired in 1937 and now lives quietly in Cambridge, Massachusetts. His is a deeply religious nature. A mystic, he regards religious experience as the rational path to knowledge.*' It seems unlikely that Whitehead ever saw this text prior to the book's publication. If nothing else, he probably would have corrected the misspelling of Sherborne School, and may have objected to the editors' characterisation of him in the last two sentences.

the ages of 16 and 30. As to training, the most important part is given by mothers before the age of 12. A saying due to Archbishop Temple illustrates my meaning. Surprise was expressed at the success in after-life of a man, who as a boy at Rugby had been somewhat undistinguished. He answered, "It is not what they are at eighteen, it is what they become afterwards that matters."[1]

In training a child to activity of thought, above all things we must beware of what I will call "inert ideas"—that is to say, ideas that are merely received into the mind without being utilized, or tested, or thrown into fresh combinations.

In the history of education, the most striking phenomenon is that schools of learning, which at one epoch are alive with a ferment of genius, in a succeeding generation exhibit merely pedantry and routine. The reason is, that they are overladen with inert ideas. Education with inert ideas is not only useless: it is, above all things, harmful—*Corruptio optimi, pessima.*[2] Except at rare intervals of intellectual ferment, education in the past has been radically infected with inert ideas. That is the reason why uneducated clever women, who have seen much of the world, are in middle life so much the most cultured part of the community. They have been saved from this horrible burden of inert ideas. Every intellectual revolution which has ever stirred humanity into greatness has been a passionate protest against inert ideas. Then, alas, with pathetic ignorance of human psychology, it has proceeded by some educational scheme to bind humanity afresh with inert ideas of its own fashioning.

|193| Let us now ask how in our system of education we are to guard against this mental dry rot. We recur to our[3] two educational commandments, "Do not teach too many subjects," and again, "What you teach, teach thoroughly."

The result of teaching small parts of a large number of subjects is the passive reception of disconnected ideas, not illumined with any spark of vitality. Let the main ideas which are introduced into a child's education be few and important, and let them be thrown into every combination possible. The child should make them his own, and should understand their application here and now in the circumstances of his actual life. From the very beginning of his education, the child should experience the joy of discovery. The discovery which he has to make, is that general ideas give an understanding of that stream of events which pours through his life, which is his life. By understanding I mean more than a mere logical analysis, though that is included. I mean "understanding" in the sense in which it is used in the French proverb, "To understand all, is to forgive all." Pedants sneer at an education which is useful. But if education is not useful, what is it? Is it a talent, to be hidden away in a napkin? Of course, education should be useful, whatever your aim in life. It was useful to Saint Augustine, and it was useful to Napoleon. It is useful, because understanding is useful.

1. The cited Archbishop is Frederick Temple (1821–1902), Archbishop of Canterbury from 1896 to 1902. The reference to 'Rugby' is to Rugby School, a grammar school for boys founded in 1567 and the birthplace of rugby football; Temple was Head Master from 1858 to 1869.
2. 'Corruption of the best is the worst of all'.
3. AE and UH changed 'recur to our' to 'enunciate'.

I pass lightly over that understanding which should be given by the literary side of education.[1] It is not peculiarly the function of this Association to consider it.[2] Nor do I wish to be supposed to pronounce on the relative merits of a classical or a modern curriculum. I would only remark that the understanding which we want is an understanding of an insistent present. The only use of a knowledge of the past is to equip us for the present. No more deadly harm can be done to young minds than by depreciation of the present. The present contains all that there is. It is holy ground; for it is the past, and it is the future. At the same time it must be observed that an age is no less past if it existed two hundred years ago than if it existed two thousand years ago. Do not be deceived by the pedantry of dates. The ages of Shakespeare and of Molière are no less past than are the ages of Sophocles and of Vergil. The communion of saints is a great and inspiring assemblage, but it has only one possible hall of meeting, and that is, the present; and the mere lapse of time through which any particular group of saints must travel to reach that meeting-place, makes very little difference.

Passing now to the scientific and logical side of education, we remember that here also ideas which are not utilized are positively harmful. By utilizing an idea, I mean relating it to that stream, compounded of sense perceptions, feelings, hopes, desires, and of mental activities relating[3] thought to thought, which forms our life. I can imagine a set of beings which might fortify their souls |194| by passively reviewing disconnected ideas. Humanity is not built that way—except perhaps some editors of newspapers.

In scientific training, the first thing to do with an idea is to prove it. But allow me for one moment to extend the meaning of 'prove'; I mean—to prove its worth. Now an idea is not worth much unless the propositions in which it is embodied are true. Accordingly an essential part of the proof of an idea is the proof, either by experiment or by logic, of the truth of the propositions. But it is not essential that this proof of the truth should constitute the first introduction to the idea. After all, its assertion by the authority of respectable teachers is sufficient evidence to begin with. In our first contact with a set of propositions, we commence by appreciating their importance. That is what we all do in after-life. We do not attempt, in the strict sense, to prove or to disprove anything, unless its importance makes it worthy of that honour. These two processes of proof, in the narrow sense, and of appreciation do not require a rigid separation in time. Both can be proceeded with nearly concurrently. But in so far as either process must have the priority, it should be that of appreciation by use.

Furthermore, we should not endeavour to use propositions in isolation. Emphatically I do not mean, a neat little set of experiments to illustrate proposition I. and then the proof of proposition I., a neat little set of experiments to illustrate proposition II. and then the proof of proposition II.,[4] and so on to the end of the

1. Cf. 'Technical Education and Its Relation to Science and Literature', EA2, p. 3.
2. AE, NR, and UH deleted this sentence.
3. AE and UH changed 'relating' to 'adjusting'.
4. OT, AE, NR and UH capitalised both instances of 'proposition I' and both instances of 'proposition II' and removed the periods following the Roman numerals. Oddly, this original version capitalised this second instance of 'proposition II' only; we have rendered it as lower case for the sake of consistency.

book. Nothing could be more boring. Inter-related truths are utilized *en bloc*, and the various propositions are employed in any order, and with any reiteration. Choose some important applications of your theoretical subject; and study them concurrently with the systematic theoretical exposition. Keep the theoretical exposition short and simple, but let it be strict and rigid so far as it goes. It should not be too long for it easily to be known with thoroughness and accuracy. The consequences of a plethora of half-digested theoretical knowledge are deplorable. Also the theory should not be muddled up with the practice. The child should have no doubt when it is proving and when it is utilizing. My point is that what is proved should be utilized, and that what is utilized should—so far as is practicable—be proved. I am far from asserting that proof and utilization are the same thing.

At this point of my discourse, I can most directly carry forward my argument in the outward form of a digression. We are only just realizing that the art and science of education require a genius and a study of their own; and that this genius and this science are more than a bare knowledge of some branch of science or of literature. This truth was partially perceived in the past generation; and headmasters, somewhat crudely, were apt to supersede learning in their colleagues by requiring left-hand bowling and a taste for |195| football. But culture is more than cricket, and more than football, and more than extent of knowledge.

Education is the acquisition of the art of the utilization of knowledge. This is an art very difficult to impart. Whenever a text-book is written of real educational worth, you may be quite certain that some reviewer will say that it will be difficult to teach from it. Of course, it will be difficult to teach from it. If it were easy, the book ought to be burned; for it cannot be educational. In education, as elsewhere, the broad primrose path leads to a nasty place. This evil path is represented by a book or a set of lectures which will practically enable the student to learn by heart all the questions likely to be asked at the next external examination. And I may say in passing that no educational system is possible unless every question directly asked of a pupil at any examination is either framed or modified by the actual teacher of that pupil in that subject. The external assessor may report on the curriculum or on the performance of the pupils, but never should be allowed to ask the pupil a question which has not been strictly supervised by the actual teacher, or at least inspired by a long conference with him. There are a few exceptions to this rule, but they are exceptions, and could easily be allowed for under the general rule.

We now return to my previous point, that theoretical ideas should always find important applications within the pupil's curriculum. This is not an easy doctrine to apply, but a very hard one. It contains within itself the problem of keeping knowledge alive, of preventing it from becoming inert, which is the central problem of all education.

The best procedure will depend on several factors, none of which can be neglected, namely, the genius of the teacher, the intellectual type of the pupils, their prospects in life, the opportunities offered by the immediate surroundings of the school, and allied factors of this sort. It is for this reason that the uniform external examination is so deadly. We do not denounce it because we are cranks, and like denouncing established things. We are not so childish. Also, of course, such examinations have their use in testing slackness. Our reason of dislike is very definite and very practical. It kills the best part of culture. When you analyse in the light of experience the central task of education, you find that its successful accomplishment depends on a delicate adjustment of many variable factors. The reason is that we are dealing with human minds, and not with dead matter. The evocation of curiosity, of judgment, of the power of mastering a complicated tangle of circumstances, the use of theory in giving foresight in special cases—all these powers are not to be imparted by a set rule embodied in one schedule of examination subjects.

I appeal to you, as practical teachers.[1] With good discipline, it is |196| always possible to pump into the minds of a class a certain quantity of inert knowledge. You take a text-book and make them learn it. So far, so good. The child then knows how to solve a quadratic equation. But what is the point of teaching a child to solve a quadratic equation? There is a traditional answer to this question. It runs thus: The mind is an instrument, you first sharpen it, and then use it; the acquisition of the power of solving a quadratic equation is part of the process of sharpening the mind. Now there is just enough truth in this answer to have made it live through the ages. But for all its half-truth, it embodies a radical error which bids fair to stifle the genius of the modern world. I do not know who was first responsible for this analogy of the mind to a dead instrument. For aught I know, it may have been one of the seven wise men of Greece, or a committee of the whole lot of them. Whoever was the originator, there can be no doubt of the authority which it has acquired by the continuous approval which it has received from[2] eminent persons. But whatever its weight of authority, whatever the high approval which it can quote, I have no hesitation in denouncing it as one of the most fatal, erroneous, and dangerous conceptions ever introduced into the theory of education. The mind is never passive; it is a perpetual activity, delicate, receptive, responsive to stimulus. You cannot postpone its life until you have sharpened it. Whatever interest attaches to your subject-matter, must be evoked here and now; whatever powers you are strengthening in the pupil, must be exercised here and now; whatever possibilities of mental life your teaching should impart, must be exhibited here and now. That is the golden rule of education, and a very difficult rule to follow.

The difficulty is just this: the apprehension of general ideas, intellectual habits of mind, and pleasurable interest in mental achievement can be evoked by no form of words, however accurately adjusted. All practical teachers know that education is a patient process of the mastery of details, minute by minute, hour by hour, day by day. There is no royal road to learning through an airy path of brilliant generalizations. There is a proverb about the difficulty of seeing the wood because of the trees. That difficulty is exactly the point which I am enforcing. The problem of education is to make the pupil see the wood by means of the trees.

The solution which I am urging, is to eradicate the fatal disconnection of subjects which kills the vitality of our modern curriculum. There is only one subject matter for education, and that is Life in all its manifestations. Instead of this single unity, we offer children—Algebra, from which nothing follows; Geometry, from which nothing follows; Science, from which nothing follows; History, from which nothing follows; a

1. NR deleted this sentence.
2. AE and UH changed 'which it has received from' to 'bestowed upon it by'.

Couple of Languages, never mastered; and lastly, most dreary of all, Literature, represented by plays of Shakespeare, with philological notes and short |197| analyses of plot and character to be in substance committed to memory. Can such a list be said to represent Life, as it is known in the midst of the living of it? The best that can be said of it, is that it is a rapid table of contents which a deity might run over in his mind while he was thinking of creating a world, and had not yet determined how to put it together.[1]

Let us now return to quadratic equations. We still have on hand the unanswered question, Why should children be taught their solution? Unless quadratic equations fit into a connected curriculum, of course there is no reason to teach anything about them. Furthermore, extensive as should be the place of mathematics in a complete culture, I am a little doubtful whether for many types of boys algebraic solutions of quadratic equations do not lie on the specialist side of mathematics. I may here remind you that as yet I have not said anything of the psychology or the content of the specialism, which is so necessary a part of an ideal education. But all that is an evasion of our real question, and I merely state it in order to avoid being misunderstood in my answer.

Quadratic equations are part of algebra, and algebra is the intellectual instrument which has been created for rendering clear the quantitative aspects of the world. There is no getting out of it. Through and through the world is infected with quantity. To talk sense, is to talk in quantities. It is no use saying that the nation is large,—How large? It is no use saying that radium is scarce,—How scarce? You cannot evade quantity. You may fly to poetry and to music, and quantity and number will face you in your rhythms and your octaves. Elegant intellects which despise the theory of quantity, are but half developed. They are more to be pitied than blamed. The scraps of gibberish, which in their schooldays were taught to them in the name of algebra, deserve some contempt.

This question of the degeneration of algebra into gibberish, both in word and in fact, affords a pathetic instance of the uselessness of reforming educational schedules without a clear conception of the attributes which you wish to evoke in the living minds of the children. A few years ago there was an outcry that school algebra was in need of reform, but there was a general agreement that graphs would put everything right. So all sorts of things were extruded, and graphs were introduced. So far as I can see with no sort of idea behind them, but just graphs. Now every examination paper has one or two questions on graphs. Personally, I am an enthusiastic adherent of graphs. But I wonder whether as yet we have gained very much. You cannot put life into any schedule of general education unless you succeed in exhibiting its relation to some essential characteristic of all intelligent or emotional perception. It is a hard saying, but it is true; and I do not see how to make it any easier. In making these little formal altera-|198|tions you are beaten by the very nature of things. You are pitted against too skilful an adversary, who will see to it that the pea is always under the other thimble.

1. NR ends here, deleting the remainder of the address.

Reformation must begin at the other end. First, you must make up your mind as to those quantitative aspects of the world which are simple enough to be introduced into general education; then a schedule of algebra should be framed which will about find its exemplification in these applications. We need not fear for our pet graphs, they will be there in plenty when we once begin to treat algebra as a serious means of studying the world. Some of the simplest applications will be found in the quantities which occur in the simplest study of society. The curves of history are more vivid and more informing than the dry catalogues of names and dates which comprise the greater part of that arid school study. What purpose is effected by a catalogue of undistinguished kings and queens? Tom, Dick, or Harry, they are all dead. General resurrections are failures, and are better postponed. The quantitative flux of the forces of modern society is[1] capable of very simple exhibition. Meanwhile the idea of the variable, of the function, of rate of change, of equations and their solution, of elimination, are being studied as an abstract science for their own sake. Not of course in the pompous phrases with which I am alluding to them here, but with that iteration of simple special cases proper to teaching.

If this course be followed, the route from Chaucer to the Black Death, from the Black Death to modern Labour troubles, will connect the tales of the medieval pilgrims with the abstract science of algebra, both yielding diverse aspects of that single theme, Life. I know what most of you are thinking at this point. It is that the exact course which I have sketched out is not the particular one which you would have chosen, or even see how to work. I quite agree. I am not claiming that I could do it myself. But your objection is the precise reason why a common external examination system is fatal to education. The process of exhibiting the applications of knowledge must, for its success, essentially depend on the character of the pupils and the genius of the teacher. Of course I have left out the easiest applications with which most of us are more at home. I mean the quantitative sides of sciences, such as mechanics and physics.

My meaning can be illustrated by looking more closely into a special case of this type of application. In my rough catalogue of the sort of subjects which should form the schedule for Algebra, I mentioned Elimination. It was not put there by accident, for it covers a very important body of thought.

In the first place, there is the abstract process of algebraic elimination for suitable simple cases. The pupil acquires a firm grasp of this by the process, inevitable in education, of working an adequate |199| number of examples. Again, there are the graphical solutions of the same problem. Then we consider the significance in the external world. We consider the velocity, time, space, acceleration diagrams. We take uniform acceleration; we eliminate "t" between

$$v = u + ft, \text{ and } s = ut + \tfrac{1}{2} ft^2,$$

and eliminate "s" between

$$v^2 = u^2 + 2fs, \text{ and } s = ut + \tfrac{1}{2} ft^2.$$

1. The original and OT both had 'are', but we have adopted AE's correction to 'is', since the subject ('flux') is singular.

Then we remember that constant acceleration is a very special case, and we consider graphical solutions for empirically given variations of v or of f. In preference, we use those empirical formulae which occur in the pupil's experimental work. We compare the strong and weak points of the algebraic and graphical solutions.[1]

Again, in the same connection we plot the statistics of social phenomena against the time. We then eliminate the time between suitable pairs. We can speculate how far we have exhibited a real causal connection, or how far a mere temporal coincidence. We notice that we might have plotted against the time one set of statistics for one country and another set for another country, and thus, with suitable choice of subjects, have obtained graphs which certainly exhibited mere coincidence. Also other graphs exhibit obvious causal connections. We wonder how to discriminate. And so are drawn on as far as we will.

But in considering this description, I must beg you to remember what I have been insisting on above. In the first place, one train of thought will not suit all groups of children. For example, I should expect that artisan children will want something more concrete and, in a sense, swifter than I have set down here. Perhaps I am wrong, but that is what I should guess. In the second place, I am not contemplating one beautiful lecture stimulating, once and for all, an admiring class. That is not the way in which education proceeds. No; all the time the pupils are hard at work solving examples, drawing graphs, and making experiments, until they have a thorough hold on the whole subject. I am describing the interspersed explanations, the directions which should be given to their thoughts. The pupils have got to be made to feel that they are studying something, and are not merely executing intellectual minuets.

In this connection the excellence of some of the most recent text-books on elementary algebra emanating from members of this Association, should create an epoch in the teaching of the subject.[2]

Finally, if you are teaching pupils for some general examination, the problem of sound teaching is greatly complicated. Have you ever noticed the zig-zag moulding round a Norman arch? The ancient work is beautiful, the modern work is hideous. The reason is that the modern work is done to exact measure, the ancient work is varied according to the idiosyncrasy of the workman. |200| Here it is crowded, and there it is expanded. Now the essence of getting pupils through examinations is to give equal weight to all parts of the schedule. But mankind is naturally specialist. One man sees a whole subject, where another can find only a few detached examples. I know that it seems contradictory to allow for specialism in a curriculum especially designed for a broad culture. Without contradictions the world would be simpler, and perhaps duller. But I am certain that in education wherever you exclude specialism you destroy life.

We now come to the other great branch of a general mathematical education, namely Geometry. The same principles apply. The theoretical part should be clear cut, rigid, short, and important. Every proposition not absolutely necessary to

1. AE and UH deleted the two preceding paragraphs, starting with 'My meaning can be illustrated...'.
2. AE and UH deleted this paragraph.

exhibit the main connection of ideas should be cut out, but the great fundamental ideas should all be there. No omission of concepts, such as those of Similarity and Proportion. We must remember that, owing to the aid rendered by the visual presence of a figure, Geometry is a field of unequalled excellence for the exercise of the deductive faculties of reasoning. Then of course there follows Geometrical Drawing, with its training for the hand and eye.

But, like Algebra, Geometry and Geometrical Drawing must be extended beyond the mere circle of geometrical ideas. In an industrial neighbourhood, machinery and workshop practice form the appropriate extension. For example, in the London Polytechnics this has been achieved with conspicuous success. For many secondary schools I suggest that surveying and maps are the natural applications. In particular, plane-table surveying should lead pupils to a vivid apprehension of the immediate application of geometric truths. Simple drawing apparatus, a surveyor's chain, and a surveyor's[1] compass, should enable the pupils to rise from the survey and mensuration of a field to the construction of the map of a small district. The best education is to be found in gaining the utmost information from the simplest apparatus. The provision of elaborate instruments is greatly to be deprecated. To have constructed the map of a small district, to have considered its roads, its contours, its geology, its climate, its relation to other districts, the effects on the status of its inhabitants, will teach more history and geography than any knowledge of Perkin Warbeck or of Behren's Straits.[2] I mean not a nebulous lecture on the subject, but a serious investigation in which the real facts are definitely ascertained by the aid of accurate theoretical knowledge. A typical mathematical problem should be: Survey such and such a field, draw a plan of it to such and such a scale, and on the same linear scale construct a square of equal area.[3] It would be quite a good procedure to impart the necessary geometrical propositions without their proofs. Then, concurrently in the same term, the proofs of the propositions would be learnt while the survey was being made.

|201| Fortunately, the specialist side of education presents an easier problem than does the provision of a general culture. For this there are many reasons. One is that many of the principles of procedure to be observed are the same in both cases, and it is unnecessary to recapitulate. Another reason is that specialist training takes place—or should take place—at a more advanced stage of the pupil's course, and thus there is easier material to work upon. But undoubtedly the chief reason is that the specialist study is normally a study of peculiar interest to the student. He is studying it because, for some reason, he wants to know it. This makes all the difference. The general culture is designed to foster an activity of mind; the specialist

---

1. The original had 'prismatic', which OT, AE and UH changed to 'surveyor's'.
2. Whitehead's examples here do not appear to have any special significance aside from their general irrelevance to the typical English pupil. Perkin Warbeck was a pretender to the English throne who claimed to be Richard of Shrewsbury, Duke of York, rightful heir to the throne. In 1497 he surrendered to Henry VII and confessed that he was actually a Fleming born in Tournai around 1474; he was executed in 1499. 'Behren's Straits' appears to be an old alternative spelling of what is today known as the Bering Strait, between the Pacific and Arctic oceans.
3. OT, AE and UH changed 'on the same linear scale construct a square of equal area' to simply 'find the area'.

course utilizes this activity. But it does not do to lay too much stress on these neat antitheses. As we have already seen, in the general course foci of special interest will arise; and similarly in the special study, the external connections of the subject drag thought outwards.

Again, there is not one course of study which merely gives general culture, and another which gives special knowledge. The subjects pursued for the sake of a general education are special subjects specially studied; and, on the other hand, one of the ways of encouraging general mental activity is to foster a special devotion. You may not divide the seamless coat of learning. What education has to impart is an intimate sense for the power of ideas, for the beauty of ideas, and for the structure of ideas, together with a particular body of knowledge which has peculiar reference to the life of the being possessing it.

The appreciation of the structure of ideas is that side of a cultured mind which can only grow under the influence of a special study. I mean that eye for the whole chess-board, for the bearing of one set of ideas on another. Nothing but a special study can give any appreciation for the exact formulation of general ideas, for their relations when formulated, for their service in the comprehension of life. A mind so disciplined should be both more abstract and more concrete. It has been trained in the comprehension of abstract thought and in the analysis of facts.

Finally, there should grow the most austere of all mental qualities; I mean the sense for style. It is an aesthetic sense, based on admiration for the direct attainment of a foreseen end, simply and without waste. Style in art, style in literature, style in science, style in logic, style in practical execution have fundamentally the same aesthetic qualities, namely, attainment and restraint. The love of a subject in itself and for itself, where it is not the sleepy pleasure of pacing a mental quarter-deck, is the love of style as manifested in that study.

Here we are brought back to the position from which we started, the utility of education. Style, in its finest sense, is the last acquirement of the educated mind; it is also the most useful. It pervades the whole being. The administrator with a sense for style, hates |202| waste; the engineer with a sense for style, economises his material; the artisan with a sense for style, prefers good work. Style is the ultimate morality of mind.

But above style, and above knowledge, there is something, a vague shape like fate above the Greek gods. That something is Power. Style is the fashioning of power, the restraining of power. But, after all, the power of attainment of the desired end is fundamental. The first thing is to get there. Do not bother about your style, but solve your problem, justify the ways of God to man, administer your province, or do whatever else is set before you.

Where then does style help? In this, with style the end is attained without side issues, without raising undesirable inflammations. With style you attain your end and nothing but your end. With style the effect of your activity is calculable, and foresight is the last gift of gods to men. With style your power is increased, for your mind is not distracted with irrelevancies, and you are more likely to attain your object. Now style is the exclusive privilege of the expert. Whoever heard of the style of an amateur painter, of the style of an amateur poet? Style is always the product of specialist study, the peculiar contribution of specialism to culture.

English education in its present phase suffers from a lack of definite aim, and from an external machinery which kills its vitality. Hitherto in this address I have been considering the aims which should govern education. In this respect England halts between two opinions. It has not decided whether to produce amateurs or experts. The profound change in the world which the nineteenth century has produced is that the growth of knowledge has given foresight. The amateur is essentially a man with appreciation and with immense versatility in mastering a given routine. But he lacks the foresight which comes from special knowledge. The object of this address is to suggest how to produce the expert without loss of the essential virtues of the amateur. The machinery of our secondary education is rigid where it should be yielding, and lax where it should be rigid. Every school is bound on pain of extinction to train its boys for a small set of definite examinations. No headmaster has a free hand to develop his general education or his specialist studies in accordance with the opportunities of his school, which are created by its staff, its environment, its class of boys, and its endowments. I suggest that no system of external tests which aims primarily at examining individual scholars can result in anything but educational waste.

Primarily it is the schools and not the scholars which should be inspected. Each school should grant its own leaving certificates, based on its own curriculum. The standards of these schools should be sampled and corrected. But the first requisite for educational |203| reform is the school as a unit, with its approved curriculum based on its own needs, and evolved by its own staff. If we fail to secure that, we simply fall from one formalism into another, from one dunghill of inert ideas into another.

In stating that the school is the true educational unit in any national system for the safeguarding of efficiency, I have conceived the alternative system as being the external examination of the individual scholar. But every Scylla is faced by its Charybdis—or, in more homely language, there is a ditch on both sides of the road. It will be equally fatal to education if we fall into the hands of a supervising department which is under the impression that it can divide all schools into two or three rigid categories, each type being forced to adopt a rigid curriculum. When I say that the school is the educational unit, I mean exactly what I say, no larger unit, no smaller unit. Each school must have the claim to be considered in relation to its special circumstances. The classifying of schools for some purposes is necessary. But no absolutely rigid curriculum, not modified by its own staff, should be permissible. Exactly the same principles apply, with the proper modifications, to universities and to technical colleges.

When one considers in its length and in its breadth the importance of this question of the education of a nation's young, the broken lives, the defeated hopes, the national failures, which result from the frivolous inertia with which it is treated, it is difficult to restrain within oneself a savage rage. In the conditions of modern life the rule is absolute, the race which does not value trained intelligence is doomed. Not all your heroism, not all your social charm, not all your wit, not all your victories on land or at sea, can move back the finger of fate. To-day we maintain ourselves. To-morrow science will have moved forward yet one more step, and there will be no appeal from the judgment which will then be pronounced on the uneducated.

We can be content with no less than the old summary of educational ideal which has been current at any time from the dawn of our civilization. The essence of education is that it be religious.

Pray, what is religious education?

A religious education is an education which inculcates duty and reverence. Duty arises from our potential control over the course of events. Where attainable knowledge could have changed the issue, ignorance has the guilt of vice. And the foundation of reverence is this perception, that the present holds within itself the complete sum of existence, backwards and forwards, that whole amplitude of time, which is eternity.[1]

1. A note at the bottom of the original read: 'A vote of thanks to the President for his Address was proposed by Mr. A. Lodge and carried unanimously'. Alfred Lodge (1854–1937) was the first President of the Mathematical Association after its name change in 1897.

La Théorie Relationniste de L'Espace (The Relational Theory of Space)

1916 *Revue de Métaphysique et de Morale*, vol. 23, no. 3 (May), pp. 423–454

For Whitehead, the first decade of the twentieth century was a period of close collaboration with Russell to demonstrate that all of mathematics is part of the logic of relations. It resulted in the publication of the Principia Mathematica, *in three volumes, 1910–1913. However, with this collaboration and publication the aim was not yet reached, and Whitehead was to write a fourth volume, on geometry, a major branch of mathematics not dealt with in the previous volumes. On 1 October 1913, Whitehead wrote to Russell that he had done a lot of writing for this fourth volume, and also that he wanted to attend the congress of mathematical logicians in Paris in April 1914 to air some of his work with respect to the concept of space.*[1] *On 10 January 1914, he wrote to Russell to say that he had completed his paper for the congress and that, provided it survived Russell's criticisms, it would go nearly without change into volume four of* Principia Mathematica.[2] *Whitehead added that he would submit it for reading at the congress as well as for publication in the* Revue de Métaphysique et de Morale.

The paper was read at the congress early in April 1914, but its publication in the Revue de Métaphysique et de Morale *was delayed two years – it was published in May 1916. A report of the congress given in* L'Enseignement mathématique[3] *records a delegate's remark that Whitehead's opinions did not agree with Russell's conclusions in* The Principles of Mathematics; *Whitehead replied that his collaboration with Russell did not preclude divergences and that, besides, Russell's ideas on space had since developed.*[4]

What is the status of this paper – in which the logic of relations is applied to the concept of space – in the development of Whitehead's own thought, independent of his Principia Mathematica collaboration with Russell? It can best be seen as marking a transition with respect to Whitehead's thoughts on space and time between his memoir 'On Mathematical Concepts of the Material World' read to the Royal Society in 1905 (published 1906; p. 214), and his 1919 book An Enquiry Concerning the Principles of Natural Knowledge.

In his 1905 memoir, Whitehead considered five alternative mathematical concepts of the material world in terms of the logic of relations. In a classical concept, the ultimate existents of the material world are points of space, instants of time and particles of matter; in an ether concept, they are instants of time and particles of ether; and in

1. Alfred North Whitehead, 'Letter from Alfred North Whitehead to Bertrand Russell, 1 October 1913', Bertrand Russell Archives, Box 5.54, 710.057467, McMaster University Library.
2. Victor Lowe, *Alfred North Whitehead: The Man and His Work, Volume II: 1910–1947* (Baltimore: Johns Hopkins University Press, 1990), p. 15.
3. Arnold Reymond, 'Le Premier Congrès de Philosophie mathématique', *L'Enseignement mathématique*, vol. 16 (1914), pp. 370–378.
4. Cf. Victor Lowe, *Understanding Whitehead* (Baltimore: Johns Hopkins University Press, 1966), p. 178.

Whitehead's implicitly preferred concept – the fifth one – they are instants of time and 'linear objective reals', which Whitehead specified by saying that 'they are the lines of force of the modern physicist' (p. 482). In all alternatives, instants of time are ultimate, and hence Whitehead did not yet arrive at a relational theory of time in this memoir. But the case is different for points of space. They are ultimate in a classical concept, and – given Whitehead's identification of particles of ether with 'moving points' – only trivially relational in an ether concept. But points of space are truly non-ultimate and relational in the fifth concept. They are 'defined complex entities' in his theory, 'being certain classes of linear objective reals' (p. 482) arising from an essential relation in the field of linear objective reals. One can compare the exact definition of points of space in Whitehead's fifth concept of the material world with the definition of projective points in projective geometry: 'the projective point is a class of straight lines' and, likewise, Whitehead's points of space are classes of linear objective reals. These classes arise from an essential 'polyadic (i.e., many-termed) relation R' in the field of linear objective reals and the '"axioms" satisfied by R' imply 'the theorems of Euclidean geometry'.

Between 1905 and 1919, Whitehead gradually took into account two major insights. He realised that – in order to avoid the bifurcation of nature in an abstract world of science and a concrete world of perception – he had to replace his definition of points in terms of abstract entities of science (lines of force) with a definition in terms of concrete entities of perception (perceptual objects). And he realised that – in order to take into account Einstein's special theory of relativity – he had to transcend his space talk (perceptual objects) with spacetime talk (perceptual events).

In his 1914 paper 'The Relational Theory of Space', Whitehead took into account the first insight, but not the second. Not that he was not already involved in generalising his approach from classical space to special relativistic spacetime. In fact, on 3 September 1911, even prior to getting acquainted with special relativity, Whitehead wrote to Russell: 'Last night . . . the idea suddenly flashed on me that time could be treated in exactly the same way as I have now got space (which is a picture of beauty, by the bye). So till the small hours of the morning, I was employed in making notes of the various ramifications. The result is a relational theory of time, exactly on four legs with that of space. As far as I can see, it gets over all the old difficulties, and above all abolishes the instants of time . . .'.[1] *However, Whitehead always let his ideas mature slowly before making them public, and in 1914 he limited himself to a treatment of space to address the mathematical logicians in Paris.*

Whitehead adopted a 'relational theory of space', and held that points are 'complex entities, logical functions of those relations between objects which constitute space' (p. 429). More specifically, but drastically simplifying things, in Whitehead's 1914 paper the essential relation is a mereological (part–whole) relation of extension in the field of perceived object-volumes, and the points of apparent space (as well as the points of the closely related scientific space) are classes of converging volumes. This approach of mereologically converging extensive entities, used to define the geometrical abstractions of Euclidean space – points, but also straight lines and planes – is a blueprint for what Whitehead later called his method of extensive abstraction. In 1919, this method was

1. Alfred North Whitehead, 'Letter from Alfred North Whitehead to Bertrand Russell, 3 September 1911', Bertrand Russell Archives, Box 5.54, 710.057464, McMaster University Library.

used by Whitehead to define the geometrical abstractions of Minkowskian spacetime as diverse classes of mereologically converging events in the idealised field of perception. But the 1914 blueprint for the method of extensive abstraction remains worth studying as a chapter in mathematical logic and mereology in particular, and to better understand Whitehead's later applications of the method of extensive abstraction as well as his conviction that the business of philosophy is 'to explain the more abstract things from the more concrete things',[1] *and not the other way around. In fact, Whitehead's 1914 comments, surrounding and justifying the mathematico-logical kernel of his paper, foreshadow several aspects of his later philosophy of nature, and even of his process philosophy.*

Given Whitehead's limited knowledge of French, it is most likely that Whitehead originally wrote the paper in English and that someone else translated it into French to be read and published as 'La Théorie Relationniste de L'Espace'. If so, the English original has never been found. As such, the English version of Whitehead's article found in this volume is a translation from the published French.

Patrick Hurley undertook the first English translation in 1978 under the title 'The Relational Theory of Space'.[2] *Independently, Janet Fitzgerald did her own English translation of the non-mathematical parts of the French paper and published it as an appendix to her 1979 book* Alfred North Whitehead's Early Philosophy of Space and Time,[3] *which contains one of the best discussions of Whitehead's 1914 paper (pp. 41–49).*

In order to produce the English text which appears in this volume, we began with Hurley's translation, which was then evaluated for accuracy and revised by two scholars: Benjamin Semple, Professor of French and Modern Languages and Literature at Gonzaga University, checked the accuracy of the translation and provided translation notes and alternative translations; Hurley's translation and Semple's suggested revisions were then reviewed by Michael Halewood, Professor of Sociology at the University of Essex, who finalised revisions to Hurley's translation. We consider the result of this process, presented here, to be the best translation of the original French text.

It should also be noted that the mathematical symbolism in the original French article had a number of unfortunate global inadequacies and outright errors, very possibly stemming from typographical limitations. For instance, what should have been rendered as a curved opening single quotation mark was instead rendered throughout as a superscripted 'c'. It also used a greater-than sign ('>') in place of an implication sign ('⊃'), a less-than sign ('<') in place of the inclusion symbol for classes ('⊂'), *and '<.' instead of the inclusion symbol for relations* ('⪽'). *Such symbolism has been changed to match what is found in* Principia Mathematica, *which Whitehead references heavily.*[4]

---

1. Alfred North Whitehead, *Process and Reality: An Essay in Cosmology*, corrected edition, edited by David Ray Griffin and Donald W. Sherburne (New York: Free Press, 1978 [1929]), p. 20.
2. Patrick J. Hurley, 'Whitehead's *Relational Theory of Space*: Text, Translation, and Commentary', *Philosophy Research Archives*, vol. 5 (1979), pp. 676–777, available at <https://www.pdcnet.org/pra/content/pra_1979_0005_0676_0777> (accessed 10 February 2025). Cf. Lowe, *Alfred North Whitehead, Vol. II*, pp. 14–15.
3. Janet Fitzgerald, *Alfred North Whitehead's Early Philosophy of Space and Time* (Washington, DC: University Press of America, 1979), pp. 167–178.
4. This introduction was written by Ronny Desmet. We are also grateful to Landon D. C. Elkind for his help in updating the article's symbolism.

I. — LES DIFFÉRENTS SENS DU MOT « ESPACE ».

Dans l'usage du mot « espace » quatre sens différents sont enveloppés, qu'il importe de distinguer. Ces distinctions de sens s'étendent au mot de géométrie dans la mesure où celle-ci est définie comme la science des propriétés de l'espace.

Il faut faire d'abord une première distinction entre « espace apparent » et « espace physique ». L'espace apparent est le lieu des objets, en tant qu'ils nous apparaissent. C'est l'espace dans lequel sont perçus des arbres verts, des sons, des odeurs. Si nous acceptons cette idée que l'espace n'est autre qu'un complexe de certaines relations entre les objets, alors, l'espace apparent est un complexe de certaines relations entre objets perçus.

Nous appellerons les objets perçus objets apparents. L'espace apparent est donc l'espace des objets apparents. Un objet apparent est nécessairement fonction de celui qui perçoit. L'espace apparent de A est, naturellement, différent de l'espace apparent de B. En effet, l'espace apparent de A ne consiste que de certaines relations entre les choses qui apparaissent à A; l'espace apparent de B ne consiste que de certaines relations entre les choses qui apparaissent à B.

Deux sens doivent être distingués dans le mot « espace apparent ». « L'espace apparent immédiat » est ce qui apparaît immédiatement à celui qui perçoit ; c'est exactement ce qui apparaît comme directement perçu. Ce peut être une partie d'une pièce, ou le versant d'une montagne avec quelques objets intermédiaires. C'est l'apparence immédiate et fragmentée du monde pour celui qui perçoit.

« L'espace apparent complet » est l'idée de l'espace total d'un monde complet d'objets apparents dans lequel on ne se réfère plus exclusivement à un sujet percevant. Cette manière de « compléter » présente un double caractère. En premier lieu, les espaces apparents |424| immédiats des différents individus sont ajustés les uns aux autres ; le côté or et le côté argent du bouclier sont combinés en un objet unique. En second lieu, on se forme, quoique grossièrement, une idée d'un monde physique et le monde apparent est alors complété par l'adjonction de l'idée de toutes les perceptions qui pourraient être apparues à des sujets hypothétiques, suivant les lois et l'état du monde physique. Cette conception, ainsi complétée en un double sens, comprend un monde d'objets apparents dans un espace apparent illimité. L'espace apparent complet est l'espace du monde perçu, auquel on se réfère ordinairement dans la conversation des hommes. Il est d'ordinaire conçu comme unique. Par exemple, dans un tribunal, un témoignage concernant la présence d'un tapis rouge dans une pièce vide, ne sera jamais rejeté comme sans intérêt, sous le prétexte qu'il pourrait être également possible de « compléter » l'univers avec ou sans tapis. L'espace apparent complet suppose la conception du « monde physique » que nous devons maintenant examiner.

L'espace physique est l'espace d'un monde hypothétique, le même pour tous, d'objets dont les relations seraient exactement correspondantes à nos sensations. Ce

I. The Different Meanings of the Word 'Space'

In the usage of the word 'space' four different meanings are involved which it is important to distinguish. These distinctions of meaning extend to the subject of geometry insofar as the latter is defined as the science of the properties of space.

First of all, it is necessary to make an initial distinction between 'apparent space' and 'physical space'. Apparent space is the place of objects inasmuch as they appear to us. It is the space in which are perceived green trees, sounds and odours. If we accept this idea that space is none other than a complex of certain relations between objects, then apparent space is a complex of certain relations between perceived objects.

We will call perceived objects apparent objects. Apparent space is then the space of apparent objects. An apparent object is necessarily a function of the one who perceives. The apparent space of A is naturally different from the apparent space of B. Indeed, the apparent space of A consists only of certain relations between things which appear to A; the apparent space of B consists only of certain relations between things which appear to B.

Two meanings must be distinguished in the term 'apparent space.' 'Immediate apparent space' is that which appears immediately to the one who perceives; it is precisely that which appears as directly perceived. It could be part of a room or the side of a mountain with some intermediate objects. It is the immediate and fragmented appearance of the world for the one who perceives.

'Complete apparent space' is the idea of the total space of a complete world of apparent objects in which reference is no longer made exclusively to a perceiving subject. This sense of 'complete' presents a double character. In the first place, the immediate apparent spaces of different individuals are adjusted to one another; the gold side and the silver side of a buckle[1] are combined into a single object. In the second place, we form, although roughly, an idea of a physical world, the idea of all the perceptions which could have appeared to hypothetical subjects according to the laws and the state of the physical world. This conception is thus completed in a double sense and comprises a world of apparent objects in an unlimited apparent space. Complete apparent space is the space of the perceived world to which we ordinarily refer in conversation. It is ordinarily conceived as unique. For example, in a court of law, testimony concerning the presence of a red carpet in an empty room would never be rejected as without interest on the pretext that it would be equally possible to 'complete' the universe with or without the carpet. Complete apparent space presupposes the concept of the 'physical world', which we must examine now.

Physical space is the space of a hypothetical world, the same for everyone, of objects whose relations would exactly correspond to our sensations. That which

1. Patrick Hurley, when asked for permission to use his translation, had a revision of his own: 'Somewhere in the middle of the article Whitehead speaks about the gold and the silver side of something that I translated as "shield". Please change this word to "buckle". I didn't know what Whitehead was talking about then, but today it's clear to me that he was talking about those reversible belts that have a buckle that's gold on one side and silver on the other' (personal correspondence, 29 June 2022).

qui apparaît immédiatement à un sujet est, dans ce cas, rapporté à un complexe de relations entre objets physiques. Par exemple, une sensation visuelle est rapportée à une excitation de cerveau, qui vient d'une excitation des nerfs de la rétine de l'œil, qui vient elle-même du choc des rayons lumineux, etc.

L'analyse exacte du procédé logique essentiel qui est enveloppé dans ce parallélisme et des idées fondamentales qui y ont conduit le genre humain, ne rentre pas dans le domaine de ce mémoire. Nous le mentionnons comme un problème scientifique fondamental qui est ici laissé de côté.

Ce parallélisme est habituellement présenté comme une « causation » des sensations par les états du monde physique. Mais la seule caractéristique essentielle de ce monde physique est le parallélisme de ses événements aux perceptions de tous les sujets connus. Il est évident qu'un tel monde est une construction logique hypothétique.

Les objets de ce monde physique seront appelés « objets physiques ». Selon la physique usuelle, les perceptions résultent des relations changeantes entre objets physiques, survenant dans un certain laps de temps. Par exemple, la sensation visuelle résulte |425| du choc d'une foule d'ondulations de l'éther sur l'œil, le son vient du choc d'une foule d'ondulations de l'air sur le tympan et il en est de même des autres sensations. Donc, les objets apparents du monde apparent sont en corrélation directe avec les *événements* du monde physique et non pas avec les objets du monde physique. La conséquence de ceci est que l'on remplace par la pensée les objets apparents fluides et se désagrégeant par des objets physiques plus permanents ; tout progrès dans l'analyse du monde physique consiste à remplacer les objets instables par des objets permanents. Par exemple, les corps apparents sont remplacés par les molécules physiques, les molécules par les atomes, les atomes par les électrons. Et chaque fois qu'un tel remplacement a lieu, les propriétés du corps relativement complexe sont conçues, comme les propriétés d'*événements*, survenant dans un ensemble de composants plus simples, en interaction.

Le quatrième usage du mot « espace » est celui « d'espace abstrait » et la science qui y correspond est la géométrie abstraite.

La géométrie, comme science mathématique, est nécessairement la géométrie abstraite. En effet, dans toute recherche mathématique, bien qu'une application particulière puisse être l'unique but en raison de son importance, la recherche est essentiellement générale. Les conclusions des propositions sont vraies pour n'importe quel ensemble d'entités pour lesquelles les hypothèses sont vraies. La géométrie appliquée, comme toute mathématique appliquée, est simplement la géométrie pure dont une application dominante nous préoccupe.

Ainsi, les quatre sens du mot espace sont distingués par les termes « espace apparent immédiat », « espace apparent complet », « espace physique », « espace abstrait ».

Il existe de nombreux espaces apparents immédiats et de nombreux espaces abstraits. Il est d'usage de supposer qu'il n'y a qu'un espace apparent complet et qu'un espace physique.

La croyance à l'unité d'un espace apparent complet semble justifiable. Mais la croyance à l'unité d'un espace physique appelle des réserves, dans la mesure où l'espace est considéré comme consistant simplement de certaines propriétés d'un

appears immediately to a subject is, in this case, referred back to a complex of relations between physical objects. For example, a visual sensation is referred back to an excitation of the brain, which comes from an excitation of the nerves of the retina of the eye, which in turn comes from the impact of light rays, etc.

The exact analysis of the essential logical procedure which is involved in this parallelism, and of the fundamental ideas which have led humans to adopt such a position, fall outside the scope of this treatise. We only mention it as a fundamental scientific problem which is hereafter put aside.

This parallelism is customarily presented as a 'cause' of the sensations by the states of the physical world. But the only essential characteristic of this physical world is the parallelism of its events to the perceptions of all of the known subjects. It is evident that such a world is a hypothetical logical construction.

The objects of this physical world will be called 'physical objects'. According to ordinary physics, perceptions result from the changing relations between physical objects and occur during a certain lapse of time. For example, visual sensation results from the impact upon the eye of a mass of vibrations of the ether, sound comes from the impact on the eardrum of a mass of vibrations of the air, and the same goes for the other sensations. Hence, the apparent objects of the apparent world are in direct correlation with the *events* of the physical world, and not with the objects of the physical world. The consequence of this is that through thought we replace the fluid and disintegrating apparent objects with physical objects that are more permanent; all progress in the analysis of the physical world consists in replacing unstable objects with permanent objects. For example, apparent bodies are replaced by physical molecules, the molecules by atoms, the atoms by electrons. And each time that such a replacement occurs the properties of the relatively complex body are conceived as the properties of *events*, arising from a group of simpler components in interaction.

The fourth usage of the word 'space' is that of 'abstract space', and the science which corresponds to it is abstract geometry.

Geometry, as a mathematical science, is necessarily abstract geometry. In fact, in all mathematical research, although a particular application may be the unique objective by reason of its importance, the research is essentially general. The conclusions of the propositions are true for any group of entities for which the hypotheses are true. Applied geometry, like all applied mathematics, is simply pure geometry of which one dominant application holds our attention.

Thus, the four meanings of the word 'space' are distinguished by the terms 'immediate apparent space', 'complete apparent space', 'physical space' and 'abstract space'.

There exist numerous immediate apparent spaces and numerous abstract spaces. It is customary to suppose that there is but one complete apparent space and but one physical space.

Belief in the unity of complete apparent space seems justifiable. But belief in the unity of physical space calls for reservations to the extent that the space is considered as consisting simply in certain properties of a complex of relations between

complexe de relations entre des objets physiques. A partir des relations simples entre objets physiques, il sera certainement possible de construire différents complexes de relations, qui donneront lieu à différentes définitions |426| des points, lignes, plans, etc., avec le type logique des propriétés qui sont définies comme spatiales.

Mais pour la physique, il se trouve sûrement un intérêt dominant dans un certain complexe de relations, qui formera alors l'espace physique, en fonction de quoi toute action physique s'exprime. Ainsi, dans un sens pratique, il y a seulement un espace physique; dans un sens théorique, il y en a un nombre infini.

II.—LES OBJETS PHYSIQUES ET L'ESPACE PHYSIQUE.

L'espace physique est l'espace de la science physique, l'espace dans lequel électrons et molécules se meuvent et agissent l'un sur l'autre, ou par l'intermédiaire de l'éther.

Le monde physique existe dans le temps et le temps du monde physique est identique au temps du monde apparent complet et des divers mondes apparents immédiats des différents sujets percevants. Un temps commun est le lieu dans lequel le parallélisme des mondes différents est sauvegardé et rendu possible. Il est évident que sur ce point le problème réel du temps est la formation du temps commun pour le monde apparent complet, en dehors des temps différents des mondes immédiats apparents des divers sujets percevants.

Les relations entre objets physiques peuvent être divisées en deux classes : relations directes et relations indirectes. Si A est rapporté à B et B à C, alors le fait même de ces deux relations constitue une relation indirecte de A à C. Mais les relations indirectes supposent des relations directes, qui ne sont plus elles-mêmes décomposables. Le premier problème de la détermination du monde physique est de déterminer certaines caractéristiques générales de ces relations directes et des objets qu'elles mettent en rapport.

Les relations directes entre objets physiques sont, par hypothèse, déterminées par leurs relations avec quelque partie antécédente du temps. Donc, considérées dynamiquement en connexion avec le temps ultérieur, elles sont les « causes » des événements futurs. A ce dernier point de vue, elles sont conçues comme les actions des objets les uns sur les autres, causant les relations directes des moments ultérieurs. Mais les actions entre objets physiques ne sont pas des relations différentes des relations causées. Il n'y a qu'un seul genre de fait à considérer, à savoir que l'état de l'univers physique |427| à travers un certain laps de temps détermine les états ultérieurs. En conséquence, les relations directes des objets physiques pendant un certain[1] laps de temps sont conçues comme des agents actifs dans l'apparition des états ultérieurs, et, ainsi conçues, sont appelées « actions des objets les uns sur les autres ».

Pour faire apparaitre la nécessité d'une analyse minutieuse de nos idées sur les relations spatiales entre les objets et les lois qui régissent le changement dans ces relations, considérons les trois axiomes primaires qui, souvent, implicitement ou explicitement, gouvernent la pensée à ce sujet. Ces trois axiomes sont :

1. Corrected 'certains' to 'certain'.

physical objects. By beginning with the simple relations between physical objects it certainly will be possible to construct different complexes of relations which will give rise to different definitions of points, lines, planes, etc., having the logical type of properties which are defined as spatial.

But for physics there is certainly a dominant interest in a certain complex of relations which will then constitute physical space, in terms of which all physical action is expressed. Thus, in a practical sense, there is only one physical space; in a theoretical sense, there are an infinite number.

II. Physical Objects and Physical Space

Physical space is the space of physical science, the space in which electrons and molecules move and act on one another (directly) or though the mediation of the ether.

The physical world exists in time, and the time of the physical world is identical to the time of the complete apparent world and of the various immediate apparent worlds of the different perceiving subjects. A common time is the locus in which the parallelism of the different worlds is safeguarded and rendered possible. It is evident that on this point the real problem of time is the formation of a common time for the complete apparent world which is independent of the different times of the immediate apparent worlds of the various perceiving subjects.

The relations between physical objects can be divided into two classes: direct relations and indirect relations. If A is related to B, and B to C, then the very fact of these two relations constitutes an indirect relation from A to C. But the indirect relations presuppose direct relations which are themselves not further resolvable. The first problem concerning the determination of the physical world is to determine certain general characteristics of these direct relations and of the objects which they relate to each other.

Direct relations between physical objects are, by hypothesis, determined by their relations with some antecedent part of time. Thus, considered dynamically in connection with future time, they are the 'causes' of future events. From this last point of view, they are conceived as the action of objects on one another causing the direct relations of future moments. But the actions between physical objects are not relations different from the caused relations. There is only one kind of fact to consider; namely, that the state of the physical universe during a certain lapse of time determines the future states. Accordingly, the direct relations of physical objects during a certain lapse of time are conceived as active agents in the emergence of future states and, thus conceived, are called 'actions of objects on one another'.

To make evident the necessity of a meticulous analysis of our ideas concerning the spatial relations between objects and the laws which govern the change in these relations, let us consider the three primary axioms which often, implicitly or explicitly, govern the thought on this subject. These three axioms are:

(1) Un objet ne peut pas être entièrement à deux places à la fois; de manière à être tout entier à chacune des deux places.

(2) Deux objets ne peuvent pas être dans la même place à la fois.

(3) Deux objets à distance ne peuvent pas agir l'un sur l'autre.

Si les deux derniers axiomes sont maintenus, il s'ensuit qu'aucune action n'est possible entre deux objets différents. Car, s'ils sont au même point, ils sont le même objet ; et s'ils sont en des points différents, ils sont à distance et ne peuvent donc agir l'un sur l'autre.

Bien plus, si le premier et le troisième axiome sont maintenus et le second écarté, alors l'action entre des objets distincts est possible, à savoir entre les objets qui sont tous au même point. Mais, en vertu de l'argument précédent, aucune action n'est possible entre des objets à des points différents. Ainsi chaque point contiendrait un monde complètement indépendant du monde des autres points.

Le précédent raisonnement n'est pas admis en général. On pose que l'hypothèse de la continuité de la matière à travers l'espace permet à l'action d'être transmise sans qu'on renonce à nier l'action à distance.

Dans cette alternative, l'action est transmise par les parties contiguës du milieu continu. Mais il n'y a pas de parties contiguës dans un milieu continu. Deux points quelconques sont séparés par une distance et par conséquent, au nom du troisième axiome, les parties du milieu ne peuvent agir l'une sur l'autre, en aucun couple de points.

On peut dire encore qu'à la limite, comme la distance entre deux points devient infiniment petite, l'action commence. Mais ceci repose sur une conception erronée de la théorie des limites, condamnée par Weierstrass. Il n'y a pas de distances infiniment petites. Toute |428| distance est finie et si les matières de deux points séparés agissent l'une sur l'autre, elles agissent à travers une distance finie.

On dira encore que l'action n'est pas entre des matières, réduites à des points, mais entre des matières, occupant des volumes, et qu'il y a des volumes contigus.

Cependant considérons la limite commune aux deux volumes contigus. L'action se produit seulement à travers cette limite. Mais il n'y a pas de volumes infiniment petits. Nous avons donc deux volumes finis agissant l'un sur l'autre à travers leur limite commune. Or, deux volumes finis peuvent être partagés en deux parts. Aucune position de l'un ou l'autre volume, qui peut être incluse dans une position non contiguë à la limite, ne contribue à l'action : car elle est à distance de la limite. Mais alors, toute matière, entourant un point quelconque qui n'est pas sur la limite, ne contribue pas à l'action. Il n'y a donc aucune matière, occupant un volume, qui contribue en quelque façon à l'action, L'action est donc due à la matière des points situés sur la limite. Mais cette action a été démontrée impossible.

Cette objection peut être évitée en affirmant que, quand les volumes sont réduits jusqu'à un certain degré, l'action entre volumes contigus doit être rapportée à chaque volume pris comme un tout et ne doit pas être calculée comme la somme des actions séparées des parties du volume.

Si l'on se tient à cette position, on doit envisager deux sortes de considérations :

(α) Puisqu'il n'y a pas de volumes infiniment petits, la limite d'étendue doit être finie. Donc on doit admettre une structure atomique du milieu continu, de telle sorte que ce milieu est composé d'éléments finis ultimes qui agissent, comme des touts, sur

(1) An object cannot be entirely in two places at the same time, in such a way that it is entirely in each of the two places.

(2) Two objects cannot be in the same place at the same time.

(3) Two objects at a distance cannot act on one another.

If the last two axioms are maintained it follows that no action is possible between two different objects. For, if they are at the same point they are the same object; and if they are at different points they are at a distance and thus cannot act on one another.

Furthermore, if the first and the third axiom are maintained and the second discarded, then action between distinct objects is possible; namely, between objects which are all at the same point. But in virtue of the preceding argument, no action is possible between objects at different points. Thus, each point would contain a world completely independent of the world of other points.

The preceding argument is not generally acknowledged. It can be argued that the hypothesis of the continuity of matter through space permits action to be transmitted without admitting the existence of action at a distance.

In this alternative, action is transmitted by contiguous parts of the continuous medium. But there are no contiguous parts in a continuous medium. Any two points are separated by a distance and consequently, by the third axiom, the parts of the medium cannot act on one another through any pair of points.

Still, one may say that, ultimately, as this distance between two points becomes infinitely small, the action begins. But this rests upon an erroneous concept of the theory of limits, condemned by Weierstrass. There are no infinitely small distances. Every distance is finite, and if the masses of two separate points act on one another, they act across a finite distance.

Again, one will say that the action is not between masses reduced to points, but between masses occupying volumes, and that there are contiguous volumes.

However, let us consider the boundary common to two contiguous volumes. Action occurs only across this boundary. But there are no infinitely small volumes. Thus, we have two volumes acting on one another across their common boundary. Now two finite volumes can be divided into two parts. No position of either one or the other volume, which can be included in a position not contiguous with the boundary, contributes to the action; for it is at a distance from the boundary. But then every mass surrounding any point which is not on the boundary does not contribute to the action. Thus, there is no mass occupying a volume which contributes in any way to the action. The action is thus due to the mass of the points situated on the boundary. But this action has been demonstrated to be impossible.

This objection can be avoided by affirming that when volumes are reduced in size up to a certain degree the action between contiguous volumes must be attributed to each volume taken as a whole, and must not be calculated as the sum of the separate actions of the parts of the volume.

If we hold to this position, we must face two kinds of considerations:

(α) Since there are no infinitely small volumes, the limit of extension must be finite. Thus, we must admit an atomic structure to the continuous medium, of such a kind that this medium is composed of ultimate finite elements which act, as wholes,

les éléments contigus. Ainsi les éléments, quoiqu'ils soient idéalement, géométriquement divisibles, sont dans la mesure où il est question d'action physique, indivisibles.

(β) Cette structure atomique du milieu étant admise, deux volumes atomiques ne peuvent être contigus à moins que l'un des deux soit sans surface. Car il n'y a pas de points contigus et, par conséquent, il n'y a pas de points contigus sur deux limites distinctes.

Donc, d'après (α) et (β), le seul milieu continu qui puisse transmettre l'action, en accord avec les trois axiomes précités, c'est le milieu composé d'éléments finis atomiques, de telle sorte que tous |429| les éléments atomiques à surfaces soient entremêlés d'éléments atomiques sans surfaces.

Cette conception est logiquement possible, elle est très improbable et n'a jamais en fait été adoptée.

L'objection réelle n'est pourtant pas l'improbabilité de la conception à quoi mènent les trois axiomes, mais plutôt les conceptions non analysées et non critiquées de l'espace et des objets, dont elle procède.

Nier l'action à distance est en fait nier les relations directes entre objets physiques n'occupant pas les mêmes points ; mais cela implique la négation de la théorie de l'espace-relation. Car il est difficile de voir comment l'espace peut être constitué par des relations entre objets qui ne sont pas en relation. Dénier l'action à distance peut seulement convenir à la théorie relationniste à condition que les relations entre les objets physiques dont procèdent les relations spatiales ne fassent pas partie de ces relations qui sont les antécédents des changements des relations spatiales à des temps ultérieurs. Cela peut être vrai, mais il est téméraire d'ériger ce principe en axiome *a priori* de la science.

III. — LA THÉORIE RELATIONNISTE DE L'ESPACE.

La géométrie comme théorie mathématique, a coutume de prendre pour point de départ tout ou partie des entités spatiales fondamentales, points, lignes courbes ou droites, surfaces et volumes[1†]. Elle les prend comme idées simples primitives, c'est-à-dire, en langage abstrait, comme des « variables » qui ne sont pas logiquement fonctions de variables plus simples. Mais si la théorie relationniste de l'espace est adoptée, soit pour le monde apparent, soit pour le monde physique, cela ne peut pas être la première étape de la recherche géométrique. Pour la théorie relationniste[2] de l'espace, il est essentiel que les points, par exemple, soient des entités complexes, fonctions logiques de ces relations entre objets qui constituent l'espace. Car, si un point est une chose simple, incapable d'être |430| logiquement définie au moyen de

1. †Cf. par exemple: Staudt, *Geometrie de Lage*, §1 et §2. « Die Geometrie geht von der Vorstellung eines unbegrenzten Raümes aus; ein vollständig begrenztes Stück desselben heisst ein Körper. Körper und uberhaupt körperliche Raüme werden dürch Flächen, Flächen durch Linien, Linien durch Punkte getheilt und begrenzt...Ein Punkt ist untheilbar. » [Added closing quotation mark.]
2. The French original had 'relativiste' here, which we have corrected to 'relationniste' and translated as 'relational'. 'Relativistic' and 'absolutist' theories of space would equally want points to be complex entities, so there is no reason why 'relativistic' would be chosen here. What Whitehead goes on to say supports that 'relational' was his original meaning: he is discussing the needs of geometry, not the needs of relativistic physics.

upon the contiguous elements. Thus, although the elements may be geometrically divisible, they are, to the extent that there is a question of physical action, indivisible.

(β) This atomic structure of the medium being accepted, two atomic volumes cannot be contiguous unless one of the two is without a surface. For there are no contiguous points and, consequently, there are no contiguous points on two distinct boundaries.

Thus, according to (α) and (β), the only continuous medium that can transmit action, in accordance with the three previously cited axioms, is the medium composed of finite atomic elements in such a way that all of the atomic elements with surfaces are intermingled with atomic elements without surfaces.

This conception is logically possible, but is very improbable and has never, in fact, been adopted.

The real objection, however, is not the improbability of the conception to which the three axioms lead, but rather the unanalysed and uncriticised notions of space and of objects from which it proceeds.

To deny action at a distance is in fact to deny direct relations between physical objects not occupying the same points; but this implies the negation of the theory of the space-relation. For it is difficult to see how space can be constituted by relations between objects which are not in relation. To deny action at a distance can be in accord with the relational theory only on the condition that the relations between physical objects, from which proceed spatial relations, do not form a part of those relations which are the antecedents of changes of spatial relations at future times. This may be true, but it is rash to set up this principle as an *a priori* axiom of science.

III. The Relational Theory of Space

Geometry, as mathematical theory, customarily takes for its point of departure all or part of the fundamental spatial entities: points, straight or curved lines, surfaces and volumes.[1†] It takes them as simple, primitive ideas; that is, in abstract language, as 'variables' which are not logically functions of simpler variables. But if the relational theory of space is adopted, whether for the apparent world or for the physical world, this cannot be the first stage of geometrical research. For the relational theory of space it is essential that points, for example, be complex entities, logical functions of those relations between objects which constitute space. For, if a point is a simple thing, incapable of being logically defined by means of relations between objects, then points are, in fact, absolute positions. Thus, the relation of

1. †Cf. For example: Karl Georg Christian von Staudt, *Geometrie de Lage* [*Geometry of Location*] (Nürnberg: F. Corn, 1847), §1 and §2. 'Geometry is based on the idea of an unlimited space; a completely limited piece of it is called a body. Bodies and physical spaces in general are divided and limited by surfaces, surfaces by lines, lines by points . . . A point is indivisible.'

relations entre objets, alors les points sont, en fait, des positions absolues. Donc la relation d'« être en un point » doit être une relation primitive incapable de définition, et ainsi, on doit prendre comme seul fait ultime de la géométrie, les relations primitives des objets à leurs positions absolues. Mais ceci n'est autre que la théorie absolue de l'espace qui, nominalement du moins, a été presque universellement abandonnée. Donc, la première occupation des géomètres, recherchant les fondements de leur science, est de définir les points en fonction des relations entre objets.

L'ordre fondamental des idées est d'abord un monde de choses en relation, puis l'espace dont les entités fondamentales sont définies au moyen de ces relations et dont les propriétés se déduisent de la nature de ces relations. Les éléments ultimes du monde, ainsi mis en relation, n'ont pas besoin nécessairement d' « occuper » une position d'espace, d'avoir des positions *uniques* dans l'espace. Du moins cette hypothèse n'est nullement nécessaire dans la théorie relationniste. Elle est nécessaire jusqu'à un certain point pour l'espace apparent, puisque quelques-unes de ces « choses » sont des objets apparents et doivent par là même être perçues dans l'espace apparent. Mais les sujets percevants ne sont déjà assujettis à aucune position assignable. Et quant à l'espace physique il est possible que les molécules, les électrons, l'éther de la physique soient très bien conçus comme objets complexes dérivés, tandis que les objets physiques fondamentaux, c'est-à-dire les termes reliés par les rapports les plus simples, peuvent n'avoir aucune connexion avec des « positions » définies dans l'espace. Quelle que soit la théorie physique, finalement adoptée par la science, des objets physiques « dans l'espace », simples ou dérivés, doivent être posés par elle. Dans la théorie relationniste de l'espace, nous ne devons pas considérer ces corps physiques comme existant d'abord dans l'espace, puis agissant les uns sur les autres, directement ou indirectement. Ils sont dans l'espace parce qu'ils agissent les uns sur les autres, et l'espace n'est rien d'autre que l'expression de certaines propriétés de leur interaction. Un livre de géométrie, en tant qu'elle est traitée comme une science applicable à l'espace physique, n'est rien d'autre que la première partie d'un traité de physique. Son sujet n'est pas « les prolégomènes à la Physique », c'est de la Physique.

L'idée fondamentale dont part la théorie relationniste,[1] dans sa |431| construction du concept d'un monde existant dans l'espace, est celle d'une classe de relations (σ). Partons donc d'une classe quelconque (σ) de relations et cherchons quelles sont les définitions possibles de quelques concepts spatiaux fondamentaux et quelles propriétés doit avoir σ pour que les propositions usuelles puissent être vraies des concepts ainsi définis. Un monde, ainsi fondé sur une classe σ, sera appelé un «monde-σ».

Si R est une relation quelconque, telle que xRy signifie « x a la relation R à y », alors la classe des entités telles que x, qui se tiennent à gauche de R est appelée (Cf. Whitehead et Russell, *Principia mathematica*, vol. I, *33) le « domaine de R » et est représentée par D'R. [2] La classe des entités à droite de R est appelée le « domaine

1. Corrected from 'relativiste'; see p. 412n2.
2. The original article used a superscripted 'c' instead of a curved opening quotation mark, which was likely a typographical limitation of the French printer. We have rendered these marks as curved opening quotation marks throughout to match PM notation.

'being at a point' must be a primitive relation incapable of definition, and thus, we must take as the sole ultimate fact of geometry the primitive relations of objects to their absolute position. But this is none other than the absolute theory of space which, nominally at least, has been all but universally abandoned. Thus, the primary occupation of geometricians investigating the foundations of their science is to define points as a function of relations between objects.

The fundamental order of ideas is first a world of things in relation; then the space whose fundamental entities are defined by means of these relations and whose properties are deduced from the nature of these relations. The ultimate elements of the world, thus placed in relation, do not necessarily have to 'occupy' a position in space, or to have *unique* positions in space. At least this hypothesis is not at all necessary in the relational theory. It is necessary up to a certain point for apparent space since some of those 'things' are apparent objects and must for that very reason be perceived in apparent space. But the perceiving subjects are not previously constrained to any one assignable position. And in regard to physical space, it is possible that the molecules, electrons and ether of physics be conceived as derived complex objects, while the fundamental physical objects, that is, the terms connected by the simplest relations, may have no connection with the 'positions' defined in space. Whatever be the physical theory finally adopted by science, the physical objects 'in space', simple or derived, must be posited by it. In the relational theory of space, we must not consider these physical bodies as existing first in space, then acting on one another, directly or indirectly. They are in space because they act on one another, and space is none other than the expression of certain properties of their interaction. A book of geometry, insofar as it is treated as a science applicable to physical space, is none other than the first part of a treatise of physics. Its subject is not 'the prolegomena to physics'; it is part of physics.

The fundamental idea from which the relational theory proceeds, in its construction of the concept of a world existing in space, is that of a class of relations (σ). Let us begin, then, with any class (σ) of relations and let us investigate the possible definitions of a few fundamental spatial concepts and what properties σ must have in order that the usual propositions can be true of concepts so defined. A universe thus based on a class σ will be called a 'σ-universe'.

If R is any relation such that xRy signifies 'x has the relation R to y', then the class of entities such as x, which stand to the left of R, is called (Cf. Whitehead and Russell, *Principia Mathematica*, Vol. I, *33) the 'domain of R' and is represented by D'R. The class of entities to the right of R is called the 'converse domain of R' and

converse de R » et désignée par Ɑ 'R. Ces deux classes ne sont pas nécessairement mutuellement exclusives. Considérons la classe formée en constituant en une seule classe tous les domaines de toutes les relations appartenant à σ. Nous l'appellerons « domaine-σ ». De même, la classe formée en constituant en une seule classe tous les domaines converses de toutes les relations appartenant à σ sera appelée le « domaine-σ converse ». En symboles (Cf. *Principia mathematica*, *33, *41, *44) le domaine-σ peut être écrit D'$\dot{s}$'σ ou s'D"σ et le domaine-σ converse: Ɑ '$\dot{s}$'σ ou s' Ɑ "σ.

La classe σ sera appelée la « classe-base » du monde-σ. Les entités appartenant au domaine-σ converse seront appelées les « objets-σ » du « monde-σ », de telle sorte que le domaine-σ converse est la classe des « objets-σ ».

Par exemple, considérons le monde complet apparent. Supposons que σ représente la classe des relations entre un sujet percevant possible et l'objet apparent perçu et étendu. Alors, un membre quelconque R de σ est un mode de perception d'un objet par le sens, de telle sorte que pRx signifie que le sujet possible p perçoit l'objet apparent étendu x suivant le mode sensible R. Ainsi un objet apparent est nécessairement membre du domaine converse d'au moins un membre de σ; en d'autres termes, les objets apparents sont les objets-σ définis précédemment. Avec cette interprétation de σ, le domaine-σ est la classe des sujets actuels et possibles.

Considérons maintenant le monde physique. Soit σ la classe des relations directes entre objets physiques. Alors, un membre quelconque R de cette classe est une certaine relation directe entre objets physiques, c'est-à-dire quelque mode d'action physique entre |432| ces objets. Donc, xRy signifie que l'objet physique x a la relation directe R à l'objet physique y. Ainsi tout objet physique sera un membre du domaine converse d'un membre déterminé de σ ; en d'autres termes, les objets physiques sont les objets-σ.[1] Il est évident que, dans ce cas, le domaine-σ est identique au domaine-σ converse.

Proposons-nous maintenant de définir un moyen par lequel le concept géométrique de « point » peut être défini pour un monde-σ. Le premier point de cette définition est la définition d'une relation dérivée E_σ qui sera appelée « inclusion-σ » et qui est analogue, dans ses propriétés formelles, à la relation de tout à partie.

Dans l'espace apparent un point est (pratiquement) une aire ou un volume suffisamment petit pour que le sujet soit incapable d'y introduire une division exacte en parties. De tels « minima sensibilia » ne manquent ni de surface ni de volume, mais de la stabilité nécessaire pour la division. De même, dans l'espace physique, un point est (pratiquement) une aire ou un volume suffisamment petit pour qu'une division soit inutile dans l'état actuel de la science.

Mais ce sont là des usages approximatifs du concept de «point». La science déductive exacte a progressé en substituant des concepts exacts aux concepts approximatifs. Par exemple, la Géométrie a inauguré ses brillants progrès en employant dans son sens vrai l'idée de point : ce qui n'a essentiellement et exactement ni parties ni grandeur. La science du nombre, appliquée à la mesure, réclame aussi le concept exact de nombre irrationnel, et le calcul infinitésimal requiert le concept exact de limite.

1. We have deleted an extra, double 'σ'.

is designated by Ɑ‘R. These two classes are not necessarily mutually exclusive. Let us consider the class formed by including in a single class all the domains of all the relations belonging to σ. We will call it a 'σ-domain'. In the same way, the class formed by including in a single class all the converse domains of all the relations belonging to σ will be called the 'converse σ-domain'. In symbols (Cf. *Principia Mathematica* *33, *41, *44) the σ-domain can be written D‘ṡ‘σ or *s*‘D“σ and the converse σ-domain: Ɑ‘ṡ‘σ or *s*‘Ɑ“σ.

The class σ will be called the 'base class' of the σ-universe. The entities belonging to the converse σ-domain will be called the 'σ-objects' of the 'σ-universe,' in such a way that the converse σ-domain is the class of 'σ-objects'.

For example, let us consider the complete apparent world. Let us suppose that σ represents the class of relations between a possible perceiving subject and the perceived and extended apparent object. Then, any member R of σ is a mode of perception of an object by the senses in such a way that pRx signifies that the possible subject p perceives the extended apparent object x according to the sensible mode R. Thus, an apparent object is necessarily a member of the converse domain of at least one member of σ; in other words, the apparent objects are the σ-objects defined previously. With this interpretation of σ, the σ-domain is the class of actual and possible subjects.

Let us now consider the physical world. Let σ be the class of direct relations between physical objects. Then, any member of R of this class is a certain direct relation between physical objects; that is, some mode of physical action between these objects. Accordingly, xRy signifies that the physical object x has the direct relation R to the physical object y. Thus every physical object will be a member of the converse domain of a determinate member of σ; in other words, the physical objects are the σ-objects. It is evident that, in this case, the σ-domain is identical to the converse σ-domain.

Let us now propose to define a means by which the geometrical concept of 'point' can be defined for a σ-universe. The first point of this definition is the definition of a derived relation E_σ which will be called 'σ-inclusion' and which is analogous, in its formal properties, to the relation of whole to part.

In apparent space a point is (practically speaking) an area or a volume sufficiently small so that the subject is incapable of introducing an exact division into parts. Such 'minima sensibilia' lack neither surface nor volume but, rather, the stability necessary for the division. Likewise, in physical space, a point is (practically speaking) an area or a volume sufficiently small so that a division is useless in the actual state of the science.

But these are approximate usages of the concept of 'point'. Exact deductive science has progressed by substituting exact concepts for approximate concepts. For example, geometry inaugurated its brilliant progress by employing the idea of point in its true sense: that which has essentially and exactly neither parts nor size. The science of number, applied to measurement, likewise requires the exact concept of irrational number, and infinitesimal calculus requires the exact concept of limit.

La simplicité du raisonnement exact ne dépend pas de la simplicité logique des concepts qu'il emploie. Elle dépend du fait que les lois qui gouvernent les combinaisons logiques et ses concepts sont simples et en particulier, qu'elles sont parfaitement générales. De la sorte, les propositions sont établies pour toutes les entités concevables, sans exceptions. Par exemple, les nombres irrationnels sont des « segments » de la série des nombres[1] rationnels; et le concept de nombre réel est logiquement plus complexe que celui de nombre rationnel. Mais dans le raisonnement, son emploi est plus simple en raison de la généralité supérieure des propositions. Le concept de « nombre complexe » est logiquement plus complexe que celui de « nombre réel » mais, dans l'analyse algébrique, son usage est plus simple en raison de la généralité supérieure des |433| propositions qui le concernent. Ainsi, dans la science déductive, on ne cherche pas la simplicité logique dans le concept des entités qui forment le champ de la pensée, mais la simplicité et la généralité dans les relations par quoi ces entités sont rapprochées. Mieux encore, pour que la science exacte ainsi obtenue soit réellement importante dans la sphère de pensée dont elle est dérivée, il faut qu'une interprétation approximative de ses théorèmes exacts soit possible dans le domaine primitif des concepts.

En appliquant ces remarques à la géométrie, nous voyons que nous n'avons pas besoin d'une définition logiquement simple des points, lignes et surfaces, mais de définitions qui conservent les propriétés générales et simples qui leur sont attribuées en géométrie, et qui permettent aussi de leur substituer les concepts approximatifs de point (par exemple) dans les propositions approximatives des domaines plus simples de pensée où les concepts fondamentaux apparaissent.

IV.—LA RELATION DE TOUT A PARTIE ET L'« INCLUSION- σ ».

Un objet peut avoir des parties en trois sens:

(*i*) Le « tout » peut être l'ensemble d'une certaine classe et la partie peut être une certaine fraction, mais non l'ensemble, de cette classe. Dans ce sens « tout et partie » signifient le « tout et quelque » de la logique. L'objet et ses parties sont à ce point de vue deux classes de choses et la classe qui est la partie est logiquement contenue dans la classe qui est le tout.

(*ii*) Un objet complexe est formé par un ensemble d'objets simples (ou simples) en connexion. Ces objets plus simples sont donc en un sens parties de l'objet complexe. Mais ils sont ce qu'on appelle « des parties hétérogènes ». Car l'objet plus simple est différent par essence de l'objet complexe. Ainsi le sucre est, en ce sens, partie hétérogène d'un pudding, la note partie hétérogène d'un accord. Nous appellerons un objet plus simple un « composant » d'un objet plus complexe quand il lui est ainsi relié et nous n'emploierons jamais le mot « partie » dans ce sens.

(*iii*) « Tout et partie » peut signifier « tout spatial et partie spatiale ». Dans ce sens, la tête est partie du corps du cheval, la province partie du territoire de la nation, le mètre partie du kilomètre.

1. Corrected 'membres' to 'nombres'.

The simplicity of exact reasoning does not depend on the logical simplicity of the concepts which it employs. It depends on the fact that the laws which govern the logical combinations and its concepts are simple and, in particular, that they are perfectly general. In this way, propositions are established for all entities conceivable, without exception. For example, irrational numbers are 'segments' of the series of rational numbers; and the concept of real number is logically more complex than that of rational number. But in reasoning, its use is simpler by reason of the greater generality of the propositions. The concept of 'complex number' is logically more complex than that of 'real number' but, in algebraic analysis, its use is simpler by reason of the greater generality of the propositions which concern it. Thus, in deductive science, we do not seek logical simplicity in the concept of the entities which form the field of thought, but simplicity and generality in the relations by which these entities are brought together. Furthermore, in order that the exact science thus obtained be really important in the sphere of thought from which it is derived, it is necessary that an approximate interpretation of its exact theorems be possible in the original domain of the concepts.

In applying these remarks to geometry, we see that we do not have need for a logically simple definition of points, lines and surfaces, but for definitions which preserve the general and simple properties which are attributed to them in geometry, and which permit also the substitution of approximate concepts of point (for example) in the approximate propositions of simpler domains of thought where the fundamental concepts appear.

IV. The Relation of Whole to Part and 'σ -Inclusion'

An object can have parts in three senses:

(i) The 'whole' can be the whole of a certain class, and the part can be a certain fraction, but not the whole, of this class. In this sense 'whole and part' signify the 'all and some' of logic. The object and its parts are, from this point of view, classes of things, and the class which is the part is logically contained in the class which is the whole.

(ii) A complex object is formed by a whole of simple objects (or simples) in connection. These simpler objects are then, in a sense, parts of the complex object. But they are what we call 'the heterogeneous parts', for the simpler object is different by its very nature from the complex object. Thus sugar is, in this sense, a heterogeneous part of a pudding, and the note is a heterogeneous part of a chord. We will call a simpler object a 'component' of a more complex object when it is thus related to it, and we will never employ the word 'part' in this sense.

(iii) 'Whole and part' can signify 'spatial whole and spatial part'. In this sense, the head is part of the body of a horse, the province is part of the territory of a nation, and the metre is part of the kilometre.

La partie spatiale est un objet du même genre que le tout dont |434| elle fait partie et elle se distingue ainsi essentiellement du « composant », considéré au deuxième sens de « tout et partie ». Un objet (apparent ou physique) s'étend dans l'espace, où il occupe un volume (aire ou ligne) et tout volume (aire ou ligne) qui est partie de l'espace où l'objet s'étend contient une partie de cet objet, au sens de « partie homogène ». L'expression « partie spatiale » sera toujours employée dans ce sens de partie spatialement homogène d'un objet étendu.

Il n'est pas si évident que ce sens spatial de « tout et partie » n'est pas un cas particulier du premier sens. S'il en est ainsi, les deux corps spatiaux sont deux collections de choses, et la plus petite collection (ou classe) est formée d'une partie de la plus grande. Mais quelles choses contribuent à former la collection qui constitue un corps ? Ce ne peuvent être des corps du même genre et plus petits ; car alors ces corps devraient être donnés comme parties spatiales du corps original. Mais cette idée de « partie spatialement homogène » est justement la notion qu'il faut expliquer par l'idée de corps conçu comme collection de choses. Nous tournons donc dans un cercle si nous espérons expliquer le tout et la partie dans l'espace par la notion de corps conçu comme collection de parties plus petites du même genre.

Ainsi, si l'identification du « tout et partie » spatial avec le « tout et quelque » logique doit être maintenue, nous sommes conduits à considérer une portion étendue d'espace comme une collection de points, et un corps étendu comme une collection de choses occupant des points. Ces choses-points, dont la collection forme un corps étendu, sont donc les composants du corps au sens où composant a été défini précédemment. Il n'y a certainement aucune contradiction dans cette conception soit du monde apparent soit du monde physique. La seule objection est le doute très grave sur le fait de savoir si elle comprend tous les faits de la question.

Considérons d'abord le monde apparent. Des objets apparents étendus sont perçus comme unités et non pas comme collections de points. Ce qui est donné dans les perceptions consiste en objets étendus en relations spatiales les uns avec les autres et avec leurs parties homogènes. Mais tout objet immédiatement donné est une chose unique et non pas une collection de points, séparément perçus. La connexion du « tout et partie » spatial et du « tout et quelque » logique est expliquée de cette façon. Quand l'attention se |435| fixe sur une ou plusieurs parties, la perception immédiate du tout spatial est perdu. Elle est alors mentalement reconstruite comme le « tout » logique de ses parties perçues. Un objet de grandes dimensions peut n'être jamais donné comme une chose unique, immédiatement perçu, mais il reste toujours comme construction mentale à savoir comme étant « toutes » ses parties perçues.

Mais ceci suppose la relation immédiatement donnée d'un objet comme partie spatialement homogène d'un autre, à moins que nous n'admettions que nous percevons un objet étendu comme une collection d'entités-points.

Arrivons maintenant au monde physique. Nous avons à considérer (dans le présent état de la science) une question non de fait, mais de convenance. Nous avons une classe de relations σ et le domaine-σ et le domaine-σ converse forment ensemble un monde-σ. Or, nos objets-σ ultimes doivent nécessairement être conçus chacun comme une unité ultime. S'il en était autrement, les relations de la classe σ ne seraient pas les relations ultimes de notre monde-σ. De même qu'en

The spatial part is an object of the same kind as the whole of which it is a part, and it is thus essentially distinguished from the 'component' considered in the second sense of 'whole and part'. An object (apparent or physical) is extended in space when it occupies a volume (area or line), and every volume (area or line) which is a part of the space where the object is extended contains a part of this object in the sense of 'homogeneous part'. The expression 'spatial part' will always be employed in this sense of a spatially homogeneous part of an extended object.

It is not so obvious that this spatial sense of 'whole and part' is not a particular case of the first sense. If it is, the two spatial bodies are each two collections of things, and the smaller collection (or class) comprises a part of the larger. But what things contribute to form the collection which constitutes a body? They cannot be smaller bodies of the same kind; for then these bodies would be given as spatial parts of the original body. But this idea of a spatially homogeneous part is precisely the notion which must be explained through the idea of a body conceived as a collection of things. We therefore turn in a circle if we expect to explain the whole and part in space through the notion of a body conceived as a collection of smaller parts of the same kind.

Thus, if the identification of the spatial 'whole and part' with the logical 'all and some' is to be maintained, we are led to consider an extended portion of space as a collection of points and an extended body as a collection of things occupying these points. These thing-points, the collection of which forms an extended body, are then the components of the body in the sense in which component has been defined previously. There is certainly no contradiction in this conception of either the apparent world or of the physical world. The only objection is the very serious doubt concerning the matter of knowing whether it deals with all the facts of the question.

Let us first consider the apparent world. Extended apparent objects are perceived as unities and not as collections of points. That which is given in perceptions consists of extended objects in spatial relations with one another and with their homogeneous parts. But every immediately given object is a unique thing and not a collection of points perceived separately. The connection of the spatial 'whole and part' with the logical 'all and some' is explained in this fashion. When we fix our attention on one or several parts, the immediate perception of the spatial whole is lost. It is then mentally reconstructed as the logical 'whole' of its perceived parts. An object of large dimensions may never be given as a unique thing immediately perceived, but it always remains as a mental construction, namely as being a 'whole' of its perceived parts.

But this presupposes the immediately given relation of an object as a spatially homogeneous part of another, unless we concede that we perceive an extended object as a collection of entity-points.

Let us turn now to the physical world. We must consider (in the present state of science) a question not of fact but of convenience. We have a class of relations σ, and the σ-domain and the converse σ-domain form together a σ-universe. Now, our ultimate σ-objects must necessarily be each conceived as an ultimate unity. If it were otherwise, the relations of the class σ would not be the ultimate relations of our σ-universe. Just as in geometry we are brought back to points 'without parts or

géométrie nous sommes ramenés aux points « sans parties ni grandeur », de même dans le monde-σ, nous devons partir d'unités indécomposables. Donc, le « tout et quelque » logique ne peut pas plus s'appliquer à la décomposition d'un objet-σ, qu'il ne peut s'appliquer à un point. Mais nous pouvons nous demander si: 1^0 au moyen des relations de la classe σ, on peut définir une relation qui aurait les propriétés logiques que nous assignons à « tout et partie » ou si; 2^0 l'une des relations de la classe σ ne possède pas ces propriétés logiques.

La question de mot, à savoir s'il est désirable d'appeler une telle relation du nom de « tout et partie », est sans intérêt. Nous l'éviterons ici en nommant cette relation « la relation d'inclusion-σ ». Ce nom sera donné à une certaine relation dérivée, définie au moyen des membres de la classe σ.

La relation d'inclusion-σ, une fois définie, peut être employée pour définir les points. Elle nous rend capables d'atteindre un élément de la définition, à savoir de définir ce que nous appellerons « les points-σ matériels ». Ce n'est pas d'ailleurs le seul moyen par lequel les points peuvent être définis, au moins pour l'espace physique[1]†.

|436| Nous définissons une relation dérivée E_σ, que nous appellerons « la relation-inclusion dérivée de σ » ou, plus brièvement, la « relation-inclusion-σ ». Cette relation existe entre deux objets-σ quelconques a et b, c'est-à-dire: $aE_\sigma b$ a lieu quand et seulement quand:

1° Si R est un membre de σ et x une entité quelconque ayant la relation R à b, x a toujours aussi la relation R à a.

2° b est un objet-σ, c'est-à-dire un membre du domaine-σ converse.

La condition 2° est ajoutée pour éviter les cas oiseux, c'est-à-dire pour réduire les entités a et b au domaine d'entités à quoi nous entendons appliquer la relation, c'est-à-dire aux objets-σ.

En symboles (Cf. W. et R., *Princ. math.*, vol. I, *10, 21) la définition de E_σ prend la forme:[2]

$$E_\sigma = \hat{a}\hat{b}\{\mathrm{R} \in \sigma \,.\, x\mathrm{R}b \,.\supset_{x,\mathrm{R}} .\, x\mathrm{R}a : b \in \text{Ɑ}`\dot{s}`\sigma\} \quad \text{Df.}$$

Par exemple, soit σ la classe de toutes les relations possibles de la perception sensible entre un sujet et un objet apparent dans le monde apparent complet. Alors (b étant un objet apparent):

$$\mathrm{R} \in \sigma \,.\, x\mathrm{R}b \,.\supset_{x,\mathrm{R}} .\, x\mathrm{R}a$$

affirme que tout sujet, qui perçoit b en quelque mode, perçoit aussi a suivant le même mode.

Quand ces conditions sont remplies, nous disons que a inclut b au regard de σ, ou:

b est une partie-σ de a.

Par exemple, voir la tête d'un chien c'est voir le chien ; toucher la tête d'un chien, c'est toucher le chien ; et caresser la tête d'un chien, c'est caresser le chien, etc.

---

1. † Cf. Mémoire de l'auteur sur « Mathematical concepts of the material World ». *Phil. Trans.*, série A, vol. CCV, 1906.
2. In the formula below, the original article used a greater-than sign ('>') instead of an implication sign ('⊃'); this was a global error that we have fixed throughout the document.

size', in the same way in the σ-universe, we must start out with non-decomposable unities. Thus, the 'all and some' of logic can no more be applied to the decomposition of a σ-object than it can to a point. But we can ask ourselves first, by means of the relations of the class σ, whether we can define a relation which would have the logical properties which we assign to 'whole and part', or, second, whether one of the relations of the class σ does not possess these logical properties.

The semantic question as to whether it is desirable to call such a relation by the name of 'whole or part' is without interest. We will avoid it here by calling this relation 'the relation of σ-inclusion'. This name will be given to a certain derived relation, defined by means of members of the class σ.

The relation of σ-inclusion, once defined, can be used to define points. It enables us to arrive at an element of the definition, namely, of defining what we will call 'material σ-points'. Moreover, this is not the only means by which points can be defined, at least for physical space.[1†]

We define a derived relation E_σ, which we will call 'the derived inclusion relation of σ,' or, more briefly, the 'σ-inclusion-relation'. This relation exists between any two σ-objects such as a and b, that is, $a E_\sigma b$ takes place, when and only when:

(1) If R is a member of σ and x is any entity having the relation R to b, then x also always has the relation R to a.

(2) b is a σ-object, that is, a member of the converse σ-domain.

The second condition is added to avoid the trivial case; that is, to restrict the entities a and b to the domain of entities to which we intend to apply the relation, namely, to the σ-objects.

In symbols (Cf. W. and R., *Princ. Math.*, Vol. I, *10, *21) the definition of E_σ takes the form:

$$E_\sigma = \hat{a}\hat{b}\{R \in \sigma . xRb . \supset_{x,R} . xRa : b \in \text{Cl}\,`\dot{s}`\sigma\} \quad \text{Df.}$$

For example, let σ be the class of all the possible relations of sensible perception between a subject and an apparent object in the complete apparent world. Then (b being an apparent object):

$$R \in \sigma . xRb . \supset_{x,R} . xRa$$

affirms that every subject which perceives b in some way also perceives a in the same way.

When these conditions are fulfilled, we say that a includes b with respect to σ, or:

b is a σ-part of a.

For example, to see the head of a dog is to see the dog; to touch the head of a dog is to touch the dog, and to pet the head of a dog is to pet the dog, etc.

1. †Cf. Author's memoir on 'Mathematical concepts of the material World', *Phil. Trans.*, series A, vol. 205, 1906. [See p. 214.]

Pour l'espace physique, σ doit être identifié à la classe des relations directes entre objets physiques. Ainsi b étant un objet physique:

$$\mathrm{R} \in \sigma \,.\, x\mathrm{R}b \,.\supset_{x,\mathrm{R}} .\, x\mathrm{R}a$$

affirme que tout objet physique x qui est une relation directe avec b a la même relation avec a. Quand ces conditions sont remplies, c'est-à-dire, quand agir sur b c'est nécessairement agir sur a, l'objet physique b est appelé une partie-σ de l'objet physique a. Ainsi la définition de l' « inclusion-σ » peut être appliquée à la fois au monde apparent complet ou au monde physique. En conséquence, la forme logique est le seul élément qui nous intéresse ici. Comme |437| il a été dit plus haut, dans notre discussion logique, σ est une classe quelconque de relations, E_σ est une relation dérivée de σ par une construction logique qui conserve son sens (mais non pas nécessairement son importance) pour toute particularisation de la classe.

Considérons maintenant quelles hypothèses sur la classe σ assureront que la relation E_σ, précédemment définie, a les propriétés logiques caractéristiques du « tout et partie » spatial dans l'espace apparent.

(1) Le « domaine » de E_σ, à savoir la classe des objets qui sont soumis à cette relation, est l'ensemble total des objets-σ. En symboles

$$\mathrm{D}'\mathrm{E}_\sigma = \text{Ɑ}'\mathrm{E}_\sigma = \text{Ɑ}'\dot{s}'\sigma.$$

La preuve de ceci se tire de la preuve de la proposition suivante. Nous ne pouvons pas, sans faire d'hypothèses sur σ, prouver que, si x est quelque objet-σ, alors x ou est inclus-σ par,[1] ou inclut-σ quelque objet-σ autre que lui-même.

(2) La relation de « tout et partie » spatial ou bien est réflexive, ou bien enveloppe une diversité. C'est une pure affaire de mots et de convenance technique que nous admettions comme un cas extrême qu'un objet spatial doit toujours être considéré comme partie de lui-même, ou que nous considérions que « partie » signifie « partie à proprement parler », de telle sorte qu'aucun objet spatial ne soit partie de lui-même. Ce qui est exclus c'est le cas où quelques objets spatiaux seraient parties d'eux- mêmes et où quelques autres ne seraient pas parties d'eux- mêmes.

On prouve facilement que E_σ est réflexive. En symboles

$$a \in \mathrm{C}'\mathrm{E}_\sigma \,.\supset_a .\, a\mathrm{E}_\sigma a.$$

(3) La relation de « tout et partie » spatial est transitive: si b est partie de a, et c partie de b, c est partie de a. En symboles:

$$a\mathrm{E}_\sigma b \,.\, b\mathrm{E}_\sigma c \,.\supset_{a,b,c} .\, a\mathrm{E}_\sigma c.$$

Cette propriété peut être facilement prouvée pour E_σ, sans aucune hypothèse sur σ. Plus simplement la proposition s'exprime:[2]

$$\mathrm{E}_\sigma^2 \mathrel{\dot\subset} \mathrm{E}_\sigma.$$

(4) La relation de « tout et partie » spatial,[3] appliquée à des objets divers est asymétrique; si b est partie de a et est différent de a, |438| alors a n'est pas partie

1. An extraneous 'α' has been removed
2. Corrected '$\mathrm{E}_\sigma^\alpha \mathrel{\dot\subset} \mathrm{E}_\sigma$' to '$\mathrm{E}_\sigma^2 \mathrel{\dot\subset} \mathrm{E}_\sigma$'. Also, the original article used a less-than sign ('$<$') instead of the inclusion symbol for classes ('$\subset$') and ('$<.$') instead of the inclusion symbol for relations ('$\dot\subset$'); these were a global errors that we have fixed throughout the document.
3. Corrected from 'spatiale'.

For physical space, σ must be identified with the class of direct relations between physical objects. Thus, b being a physical object:

$$\mathrm{R} \in \sigma \,.\, x\mathrm{R}b \,.\supset_{x,\mathrm{R}} .\, x\mathrm{R}a$$

affirms that any physical object x which has a direct relation with b has the same relation with a. When these conditions are fulfilled, that is, when to act upon b is necessarily to act upon a, the physical object b is called a σ-part of the physical object a. Thus, the definition of 'σ-inclusion' can be applied both to the apparent world and the physical world. Consequently, the logical form is the only element that interests us here. As was said above, in our logical discussion, σ is any class of relations, E_σ is a relation derived from σ by a logical construction which conserves its meaning (but not necessarily its importance) for every particularisation of the class.

Let us now consider which hypotheses on the class σ will ensure that the relation E_σ, defined previously, has the logical properties characteristic of spatial 'whole and part' in apparent space.

(1) The 'domain' of E_σ, that is, the class of objects which are subject to this relation, is the total group of σ-objects. In symbols:

$$\mathrm{D}'\mathrm{E}_\sigma = \mathrm{a}'\mathrm{E}_\sigma = \mathrm{a}'\dot{s}'\sigma.$$

The proof of this is taken from the proof of the next proposition. We cannot, without making some hypotheses on σ, prove that if x is some σ-object, then x either is σ-included by, or σ-includes, some σ-object other than itself.

(2) The relation of spatial 'whole and part' is either reflexive or else it involves a diversity. It is purely a matter of words and of technical convenience that we allow as an extreme case that a spatial object must always be considered as part of itself, or that we consider that 'part' signifies 'part properly speaking' in such a way that no spatial object can be part of itself. What is excluded is the case where some spatial objects would be parts of themselves and where other spatial objects would not be parts of themselves.

We can easily prove that E_σ is reflexive. In symbols:

$$a \in \mathrm{C}'\mathrm{E}_\sigma \,.\supset_a .\, a\mathrm{E}_\sigma a.$$

(3) The relation of spatial 'whole and part' is transitive: if b is part of a, and c part of b, then c is part of a. In symbols:

$$a\mathrm{E}_\sigma b \,.\, b\mathrm{E}_\sigma c \,.\supset_{a,b,c} .\, a\mathrm{E}_\sigma c.$$

This property can easily be proved for E_σ without any hypothesis on σ. More simply, the proposition is expressed:

$$\mathrm{E}_\sigma^2 \Subset \mathrm{E}_\sigma.$$

(4) The relation of spatial 'whole and part' applied to diverse objects is asymmetrical; if b is part of a and is different from a, then a is not part of b or, in other words, if a is part of b and b part of a, then a and b are identical.

de b ou, en d'autres termes, si a est partie de b et b partie de a, alors a et b sont identiques.

Jusqu'à présent nous n'avons pas eu à faire de suppositions au regard de la classe σ afin d'identifier les propriétés formelles de E_σ avec celles du « tout et partie » spatial. Mais cette propriété, à savoir

$$aE_\sigma b \,.\; a \neq b \,.\supset_{a,b} . \sim (bE_\sigma a)$$
$$aE_\sigma b \,.\, bE_\sigma a \,.\supset_{a,b} . \; a = b$$

ne peut être prouvée pour E_σ sans hypothèse concernant la nature de σ. Cette hypothèse est en fait l'affirmation de la propriété requise pour la classe σ. Mais il est instructif de rechercher ce que cette affirmation devient en fait lorsque nous substituons à $aE_\sigma b$ et $bE_\sigma a$ l'énoncé complet des propositions pour lesquelles elles valent.

Représentons par $\overrightarrow{R}\text{‘}a$ la classe des choses qui ont la relation R à a. Alors:

$$xRb \,.\supset_x . \, xRa$$

peut être remplacé par:

$$\overrightarrow{R}\text{‘}b \subset \overrightarrow{R}\text{‘}a.$$

Par cette substitution, la proposition $aE_\sigma b$ peut s'écrire, sous forme plus commode:

$$R \in \sigma \,.\supset_R . \, \overrightarrow{R}\text{‘}b \subset \overrightarrow{R}\text{‘}a : b \in \text{C}\text{‘}\dot{s}\text{‘}\sigma.$$

conséquemment les propositions conjointes « $aE_\sigma b$ et $bE_\sigma a$ » peuvent s'écrire:

$$R \in \sigma \,.\supset_R . \, \overrightarrow{R}\text{‘}a = \overrightarrow{R}\text{‘}b : a, b \in \text{C}\text{‘}\dot{s}\text{‘}\sigma.$$

Ainsi notre hypothèse devient:

$$R \in \sigma \,.\supset_R . \, \overrightarrow{R}\text{‘}a = \overrightarrow{R}\text{‘}b : a, b \in \text{C}\text{‘}\dot{s}\text{‘}\sigma : \supset_{a,b} . \; a = b.$$

ce qui s'énonce: si a et b sont objets-σ, et si pour tout membre (R) de σ, la classe des choses qui ont la relation R à a est la même que la classe des choses qui ont la relation R à b, alors a est identique à b.

Appliquée au monde apparent complet, cette hypothèse devient « si les relations des objets apparents a et b à tout sujet possible sont les mêmes, alors a est identique à b ». Mais ceci est pratiquement la doctrine de Leibniz sur l'identité des indiscernables, et con-|439|séquemment, c'est une hypothèse par laquelle nous voulons limiter les classes σ pour lesquelles nous pouvons rechercher utilement les propriétés de E_σ.

Appliquée au monde physique, l'hypothèse devient: « deux objets physiques qui ont les mêmes relations directes à tout objet physique sont identiques ». Ceci reproduit encore la doctrine de Leibniz dans la forme où elle est applicable au monde physique. En conséquence, nous ferons sur σ, quand il sera nécessaire, l'hypothèse que:

$$R \in \sigma \,.\supset_R . \, \overrightarrow{R}\text{‘}a = \overrightarrow{R}\text{‘}b : a, b \in \text{C}\text{‘}\dot{s}\text{‘}\sigma : \supset_{a,b} . \; a = b$$

et, pour abréger, nous la représenterons par:

« Indiscern. Hyp. σ ».

Donc :

$$\text{Indiscern. Hyp. } \sigma . \equiv : aE_\sigma b \,.\, bE_\sigma a \,.\supset_{a,b} . \; a = b.$$

Until now we have not had to make suppositions with respect to the class σ in order to identify the formal properties of E_σ with those of spatial 'whole and part'. But this property, namely:

$$aE_\sigma b \,.\, a \neq b \,.\supset_{a,b} . \sim(bE_\sigma a)$$
$$aE_\sigma b \,.\, bE_\sigma a \,.\supset_{a,b} . a = b$$

cannot be proved for E_σ without a hypothesis concerning the nature of σ. This hypothesis is in fact the affirmation of the required property for the class σ. But it is instructive to investigate what this affirmation becomes in fact when we substitute for $aE_\sigma b$ and $bE_\sigma a$ the complete statement of propositions for which they are valid.

Let us represent by $\overrightarrow{R}`a$ the class of things which have the relation R to a. Then:

$$xRb \,.\supset_x . xRa$$

can be replaced by:

$$\overrightarrow{R}`b \subset \overrightarrow{R}`a.$$

By this substitution, the proposition $aE_\sigma b$ can be written under the more convenient form:

$$R \in \sigma \,.\supset_R . \overrightarrow{R}`b \subset \overrightarrow{R}`a : b \in \text{Cl}`\dot{s}`\sigma.$$

Consequently, the conjoined propositions '$aE_\sigma b$ and $bE_\sigma a$' can be written:

$$R \in \sigma \,.\supset_R . \overrightarrow{R}`a = \overrightarrow{R}`b : a, b \in \text{Cl}`\dot{s}`\sigma.$$

Thus our hypothesis becomes

$$R \in \sigma \,.\supset_R . \overrightarrow{R}`a = \overrightarrow{R}`b : a, b \in \text{Cl}`\dot{s}`\sigma : \supset_{a,b} . a = b.$$

which states: if *a* and *b* are σ-objects and if, for any member (R) of σ, the class of things which have the relation R to *a* is the same as the class of things which have the relation R to *b*, then *a* is identical to *b*.

Applied to the complete apparent world this hypothesis becomes 'if the relations of apparent objects *a* and *b* to any possible subject are the same, then *a* is identical to *b*'. But this is practically the doctrine of Leibniz on the identity of indiscernibles and, consequently, it is a hypothesis by which we want to limit the classes σ to those for which we can usefully investigate the properties of E_σ.

Applied to the physical world, the hypothesis becomes: 'two physical objects which have the same direct relation to any physical object are identical'. This reproduces once again the doctrine of Leibniz in the form in which it is applicable to the physical world. Accordingly, we will make upon σ, when it will be necessary, the hypothesis that:

$$R \in \sigma \,.\supset_R . \overrightarrow{R}`a = \overrightarrow{R}`b : a, b \in \text{Cl}`\dot{s}`\sigma : \supset_{a,b} . a = b$$

and, to abbreviate, we will represent it by:

$$\text{Indiscern. Hyp. } \sigma$$

Thus:

$$\text{Indiscern. Hyp. } \sigma . \equiv : aE_\sigma b \,.\, bE_\sigma a \,.\supset_{a,b} . a = b.$$

(5) Tout objet spatial a des parties spatiales autres que lui-même. Cette propriété du « tout et partie » spatial est l'hypothèse de la divisibilité infinie ; c'est le fondement de la continuité de l'espace. La vérité de la proposition n'est pas évidente. Dans l'espace apparent, l'existence de minima visibilia paraît la contredire. Dans le monde physique, l'hypothèse de la continuité rend les déductions mathématiques possibles dans l'état présent de la connaissance mathématique, mais il n'y a pas de raison bien définie pour l'adopter si ce n'est sa commodité.

La propriété que la relation E_σ devrait avoir, pour satisfaire à cette condition, est que « si y est un objet quelconque qui est une partie-σ de quelque objet x, alors un objet z existe, différent de y, qui est une partie-σ de y ». En symboles (*Princ. math.*, vol. I, *23, 50).[1]

$$\text{Ɑ}'E_\sigma \subset D'(j \dot{\cap} E_\sigma).$$

Cette propriété ne peut être prouvée pour E_σ où σ est une classe quelconque de relations. Elle doit donc, quand cela est repris dans le raisonnement, être assumée comme une hypothèse sur la classe σ.

Il est important d'écrire tout au long le sens de E_σ, de manière à analyser complètement la nature exacte de cette hypothèse pour σ. A l'aide d'une légère transformation de symboles, nous avons:

$$x \in \text{Ɑ}'\dot{s}'\sigma . \supset_x :. (\exists y) : y \in \text{Ɑ}'\dot{s}'\sigma . y \neq x : R \in \sigma . \supset_R . \overrightarrow{R}'y \subset \overrightarrow{R}'x.$$

|440| Ce qui s'énonce: si x est un objet-σ, il y a un objet-σ y, différent de x tel que si R est une relation quelconque appartenant à la classe σ, les choses qui ont la relation R à y sont comprises dans les choses qui ont la relation R à x.

Cette hypothèse concernant σ s'appellera « hypothèse-divisibilité concernant σ ».

La théorie des relations spatiales, telle qu'elle est développée ici, est indépendante de la vérité ou de la fausseté de cette hypothèse. La portée de notre science — en dehors de sa justification logique — dépendra de l'hypothèse qu'il peut exister des séries d'objets-σ, successivement inclus-σ l'un dans l'autre, qui n'ont aucun objet-σ, soit divisible-σ, soit indivisible-σ, inclus-σ dans tous les membres de la série.

Il nous faut un symbole pour la classe des « objets-σ indivisibles-σ », c'est-à-dire pour la classe des objets-σ dont chacun inclut-σ aucun autre objet-σ que lui-même. Il nous est donné par (*Princ. Math.*, I, *23, 50)

$$C'E_\sigma - D'(j \dot{\cap} E_\sigma).$$

Mais il sera plus commode de substituer le symbole spécial: indiv. (σ). Donc:

$$\text{indiv.}\ (\sigma) = C'E_\sigma - D'(j \dot{\cap} E_\sigma)\ \text{Df}.$$

1. In the formula below, the original article used a lowercase 'j'; here and throughout we have substituted an uppercase 'J' to match PM notation.

(5) Every spatial object has some spatial parts other than itself. This property of spatial 'whole and part' is the hypothesis of infinite divisibility; it is the foundation of the continuity of space. The truth of this proposition is not obvious. In apparent space the existence of minima visibilia appears to contradict it. In the physical world the hypothesis of continuity renders mathematical deductions possible in the present state of mathematical knowledge, but there is no convincing reason to adopt it other than convenience.

The property which the relation E_σ ought to have to satisfy this condition is that 'if *y* is any object which is a σ-part of some object *x*, then an object *z* exists, different from *y*, which is a σ-part of *y*'. In symbols (*Princ. Math.*, Vol. I, *23, *50):

$$\text{Cl}\,{}^{\prime}\text{E}_\sigma \subset \text{D}{}^{\prime}(j \dot{\frown} \text{E}_\sigma).$$

This property cannot be proved for E_σ when σ is any class of relations. It must then, when this is taken up again in the argument, be assumed as a hypothesis on the class σ.

It is important to write out at length the meaning of E_σ in order to analyse completely the exact nature of this hypothesis for σ. With the aid of a slight transformation of symbols, we have:

$$x \in \text{Cl}\,{}^{\prime}\dot{s}{}^{\prime}\sigma . \supset_x :. (\exists y) : y \in \text{Cl}\,{}^{\prime}\dot{s}{}^{\prime}\sigma . y \neq x : \text{R} \in \sigma . \supset_{\text{R}} . \overrightarrow{\text{R}}{}^{\prime}y \subset \overrightarrow{\text{R}}{}^{\prime}x.$$

Which states: if *x* is a σ-object, there is a σ-object *y*, different from *x*, such that if R is any relation belonging to the class σ, the things which have the relation R to *y* are included in the things which have the relation R to *x*.

This hypothesis concerning σ will be called 'the divisibility hypothesis concerning σ'.

The theory of spatial relations, as it is developed here, is independent of the truth or the falsity of this hypothesis. The scope of our science – apart from its logical justification – will depend on the hypothesis that there can exist some series of σ-objects successively σ-included in each other, which have no σ-object, either σ-divisible or σ-indivisible, σ-included in all the members of the series.

We need a symbol for the class of 'σ-indivisible σ-objects,' that is, for the class of σ-objects each of which σ-includes no σ-object other than itself. It is given to us by (*Princ. Math.*, I, *23, *50):

$$\text{C}{}^{\prime}\text{E}_\sigma - \text{D}{}^{\prime}(j \dot{\frown} \text{E}_\sigma).$$

But it will be more convenient to substitute the special symbol: indiv. (σ). Thus:

$$\text{indiv.}\,(\sigma) = \text{C}{}^{\prime}\text{E}_\sigma - \text{D}{}^{\prime}(j \dot{\frown} \text{E}_\sigma) \text{ Df.}$$

V. — LA DÉFINITION RELATIONNISTE DES CONCEPTS-SPATIAUX. (Considérations générales.)

Nous pouvons maintenant procéder à la définition des concepts spatiaux fondamentaux à l'aide de la relation d' « Inclusion ». Il n'est pas nécessaire que nous affirmions que cette relation est E_σ, telle qu'elle a été définie dans la section précédente. Toute relation T, définie en quelque manière, se référant à la classe fondamentale σ, ou étant en elle-même un membre de σ, servira pourvu qu'elle possède les propriétés logiques requises.

Ces propriétés sont:

(i) T est une relation entre objets-σ, c'est-à-dire:

$$C\text{'}T \subset \text{Cl}\text{'}\dot{s}\text{'}\sigma.$$

|441| Cette propriété ne sera pas requise comme hypothèse des propositions; elle est posée ici comme une définition de sens.

(ii) Si x est un objet-σ, alors $x\mathrm{T}x$.

(iii)[1] T est transitive; $\mathrm{T}^2 \mathbin{\dot\subset} \mathrm{T}$.

(iv) Si $a\mathrm{T}b$ et $b\mathrm{T}a$, alors $a=b$,

c'est-à-dire $a\mathrm{T}b.b\mathrm{T}a. \supset_{a,b} a=b$.

L'hypothèse (ii) sera appelée: « Ident Hyp T », l'hypothèse (iii): « Trans Hyp T », l'hypothèse (iv): « Indiscern Hyp T ».

Notre méthode consiste d'abord à définir les entités géométriques, soit points, lignes, surfaces, de manière à obtenir ce que nous appellerons « Points-T matériels », « Lignes-T matérielles », « surfaces-T matérielles ».

Celles-ci correspondent aux points, lignes, surfaces « occupés ». Si nous acceptons l'hypothèse que toutes les positions sont actuellement occupées par des objets-σ, alors les définitions sont complètes. Mais si nous admettons un espace inoccupé, c'est-à-dire des points inoccupés par des objets-σ, alors nous avons à faire une théorie généralisée des points idéaux (que j'espère expliquer dans un mémoire ultérieur). Ces points idéaux sont les points de l'espace complet de la géométrie. Un point idéal est occupé quand il y a un « point-T matériel » qui lui correspond; sinon il est inoccupé.

Ainsi, les constructions requises, essentielles à la théorie relationniste de l'espace, sont présentées sur la base de la relation d'inclusion T. L'espace ainsi construit sera l'espace-T.

Un « objet matériel de l'espace-T » est défini comme une collection (classe) de points-T matériels. Le concept de « tout et partie », appliqué à un « objet matériel de l'espace-T » est simplement le « tout et quelque » logique, rapporté à des collections de points-T matériels. Si (suivant le mode de définition) chaque objet-σ est rapporté uniquement à une collection de points-T matériels, alors il est souvent indifférent de spécifier si, dans l'espace-T, nous entendons par objets l'objet-σ ou « l'objet matériel de l'espace-T ». Ainsi le « tout et quelque » logique (rapporté aux points matériels-T) et l'inclusion-T (rapportée aux objets-σ) se confondent.

---

1. Corrected 'T$^\alpha$' to 'T^2'.

V. The Relational Definition of Spatial Concepts (General Considerations)

We can now proceed to the definition of the fundamental spatial concepts with the help of the relation of 'Inclusion'. It is not necessary that we state that this relation is E_σ, such as it has been defined in the previous section. Every relation T, defined in some manner, referring to the fundamental class σ, or, being in itself a member of σ, will serve provided that it possesses the requisite logical properties.

These properties are:

(i) T is a relation between σ-objects, that is:

$$C`T \subset \complement`\dot{s}`\sigma.$$

This property will not be required as a hypothesis for the propositions; it is posited here as a definition of meaning.

(ii) If x is a σ-object, then $x\mathrm{T}x$.

(iii) T is transitive; $\mathrm{T}^2 \in \mathrm{T}$.

(iv) If $a\mathrm{T}b$ and $b\mathrm{T}a$, then $a{=}b$,

that is, $a\mathrm{T}b.b\mathrm{T}a. \supset_{a,b} a = b$.

Hypothesis (ii) will be called: 'Ident Hyp T', hypothesis (iii): 'Trans Hyp T', hypothesis (iv): 'Indiscern Hyp T'.

Our method consists first of all in defining the geometrical entities, that is, points, lines and surfaces, so as to obtain what we will call 'material T-points', 'material T-lines' and 'material T-surfaces.'

These entities correspond to 'occupied' points, lines and surfaces. If we accept the hypothesis that all positions are actually occupied by σ-objects, then the definitions are complete. But if we admit an unoccupied space, that is to say points unoccupied by σ-objects, then we have to develop a generalised theory of ideal points (which I hope to present in a later treatise). These ideal points are the points of the complete space of geometry. An ideal point is occupied when there is a 'material T-point' which corresponds to it; if not, it is unoccupied.

Thus, the required constructions, essential to the relational theory of space, are presented on the basis of the inclusion relation T. The space thus constructed will be T-space.

A 'material object of T-space' is defined as a collection (class) of material T-points. The concept of 'whole and part', applied to a 'material object of T-space', is simply the 'all and some' of logic related to collections of material T-points. If (according to the mode of definition) each σ-object is related uniquely to a collection of material T-points, then it is often unnecessary to specify if, in T-space, we understand by objects the σ-object or 'the material object of T-space'. Thus, the 'all and some' of logic (related to material T-points) and T-inclusion (related to σ-objects) merge together.

Dans le mémoire précédemment cité, *Mathematical concepts*[1] *of the material World*, une méthode de définition des entités géométriques, indépendante de cette relation, était donnée. Il est certain que de multiples modes de définition sont possibles et il est très essentiel qu'ils soient recherchés, de manière que les méthodes qui |442| s'accordent le mieux avec les faits perceptifs (pour l'espace apparent) ou avec les hypothèses scientifiques (pour l'espace physique) puissent être développées.

On peut remarquer que plus d'une méthode se présente à nous si nous n'acceptons pas l'hypothèse de la divisibilité (appliquée à T). Il peut exister une classe d'« indivisibles-T » qui, par hypothèse, peut être identifiée avec tous les points-T matériels. Ainsi notre méthode géométrique est simplifiée si nous assujettissons la science à identifier l'ultime objet-σ unité avec la propriété d'occuper un point unique. Ceci peut, en dernière analyse, être tenu pour la meilleure méthode, soit pour l'espace apparent, soit pour l'espace physique. Mais c'est l'ambition de la logique de se libérer de l'imagination scientifique; conséquemment, notre méthode ne justifie ni la divisibilité pour T, ni le contraire.

La conception générale que nous avons maintenant à préciser est celle d'un objet, coupé progressivement en parties de plus en plus petites jusqu'à ce que ses dimensions aient disparu et qu'il ne reste plus qu'un point. Suivant cette conception; un point est quelquefois appelé une « limite conceptuelle » obtenue par le procédé précédent. Il est facile d'établir la nature générale de ce concept d'un point au moyen d'objets-σ et de la relation E_σ.

Considérons une série x_1, x_2, x_3,...[2] d'un nombre infini d'objets-σ, tels que $x_1 E_\sigma x_2$, $x_2 E_\sigma x_3$, etc., de telle sorte que finalement x_1, x_2, x_3,... *convergent* vers une *limite conceptuelle* qui n'a pas de parties. Il est évident que les deux mots critiques, dont dépend la signification du procédé ici décrit, n'ont encore aucune signification déterminée. Le sens de « convergence » d'une série infinie de nombres est précis et défini. Mais les objets-σ ne sont pas des nombres, et le sens mathématique de « convergence » ne s'applique pas simplement.

De plus, que signifie ici une *limite*? La « limite » d'une fonction en Analyse a un sens précis qui ne s'applique pas ici. Il y a un autre sens du mot « limite », plus général, à savoir le sens dans lequel G. Cantor emploie le terme de « points-limites » d'une série (Cf. *Princ. math.*, *207). Mais cette conception n'est pas utile pour nous ici, à moins que nous n'acceptions l'existence d'indivisibles-T (pour toutes ces séries). Mais si nous faisons cette hypothèse, il n'est pas nécessaire (comme nous l'avons noté précédemment) de compliquer notre définition des points par l'emploi des séries d'indivisibles-T formés par la relation d'inclusion-T.

|443| Conséquemment (en dehors d'une hypothèse spéciale) l'idée de « limite conceptuelle », en tant qu'appliquée aux séries d'objets-σ, reste encore sans signification précise.

1. Corrected 'concept' to 'concepts'.
2. Supplied missing ellipsis.

In the memoir previously cited, *Mathematical Concepts of the Material World,* a method of defining geometrical entities independently of this relation was given. It is certain that many modes of definition are possible, and it is very essential that they be investigated in such a way that the methods which best agree with the perceptual facts (for apparent space) or with the scientific hypotheses (for physical space) may be developed.

We can remark that more than one method presents itself to us if we do not accept the hypothesis of divisibility (applied to T). There can exist a class of 'T-indivisibles' which, by hypothesis, can be identified with all of the material T-points. Thus, our geometrical method is simplified if we make the requirement that science identify the ultimate σ-object unit with the property of occupying a unique point. This may, in the final analysis, be considered as the best method either for apparent space or for physical space. But it is the ambition of logic to free itself from scientific imagination; consequently, our method justifies neither divisibility for T nor the contrary.

The general conception which we now have to render precise is that of an object progressively cut into smaller and smaller parts until its dimensions have disappeared and nothing remains but a point. In accord with this concept, a point is sometimes called a 'conceptual limit' obtained by the above procedure. It is easy to establish the general nature of the concept of a point by means of σ-objects and the relation E_σ.

Let us consider a series $x_1, x_2, x_3,...$ of an infinite number of σ-objects, such that $x_1 E_\sigma x_2$, $x_2 E_\sigma x_3$, etc., of such a kind that finally $x_1, x_2, x_3,...$ *converge* toward a *conceptual limit* which has no parts. It is evident that the two critical words, upon which depends the meaning of the procedure here described, have as yet no determinate meaning. The meaning of 'convergence' of an infinite series of numbers is precise and definite. But σ-objects are not numbers and the mathematical meaning of 'convergence' simply does not apply.

Moreover, what does *limit* signify here? The 'limit' of a function in Analysis has a precise meaning which does not apply here. There is another meaning of the word 'limit', more general, namely the meaning with which G. Cantor employed the term 'limit-points' of a series (Cf. *Princ. Math.*, *207). But this concept is not useful to us here unless we accept the existence of T-indivisibles (for all of these series). But if we make this hypothesis, it is not necessary (as we have noted previously) to complicate our definition of points by the use of the series of T-indivisibles formed by the relation of T-inclusion.

Consequently (except as a special hypothesis) the idea of 'conceptual limit', insofar as it applied to the series of σ-objects, remains still without a precise meaning.

VI.—CLASSES GÉOMÉTRIQUES SÉRIELLES-T ET ÉGALITÉ-T.

L'ensemble des classes sérielles-T—symbolisées par ser'T—est composé de classes d'objets-σ qui satisfont aux deux conditions suivantes: (i) une classe qui est un membre de ser'T n'est pas la classe nulle (ii) si x et y sont deux membres d'une classe qui est un membre de ser'T,[1] ou x a la relation T à y, ou y a la relation T à x.

En symboles

$$\left\{\mathrm{ser}`\mathrm{T} = \hat{\alpha}\exists!\,\alpha :.\, x, y \in \alpha \,.\supset_{x,y}: x\mathrm{T}y . \vee . y\mathrm{T}x\right\} \text{ Df.}$$

Ici la première condition est insérée simplement pour exclure le cas inopportun de la classe nulle (Λ).

Prop. 1—Il est facile de voir que toute relation *transitive* T transforme une classe sérielle-T en une série. En symboles (Cf. *Princ. math.*, vol. I, *36[2] et vol. II, *201, *204).

$$\mathrm{T} \in \mathrm{trans.}\ \alpha \in \mathrm{ser}`\mathrm{T} \,.\supset.\, \left(j \dot{\frown} \mathrm{T}\right) \upharpoonleft \alpha \in \mathrm{ser.}$$

L'ensemble des « classes géométriques sérielles-T » est symbolisé par « géom T » et tout membre (σ) de l'ensemble, outre qu'il est sériel-T, est caractérisé par la propriété que le seul objet-σ (*s'il y en a un*) à quoi tout membre de σ peut avoir la relation T doit être un indivisible-T, où un « indivisible-T » est un membre de la classe

$$\mathrm{C}`\mathrm{T} - \mathrm{D}`\left(\mathrm{T} \dot{\frown} j\right)$$

ce qui sera dénoté par « indiv (T) ». En symboles

$$\text{géom}`\mathrm{T} = \hat{\alpha}\left\{\alpha \in \mathrm{ser}`\mathrm{T} \,.\, \mathrm{p}`\overleftarrow{\mathrm{T}}``\alpha \subset \mathrm{indiv}\ (\mathrm{T})\right\} \text{ Df.}$$

L'importance de notre procédure ultérieure dépend entièrement de ce fait: est-il désirable d'admettre l'existence de cas où $p`\overleftarrow{\mathrm{T}}``\alpha = \Lambda$. Ce cas est compris dans la définition précédente du géom'T, car Λ est contenu dans toute classe.

Par exemple dans l'espace apparent considérons l'ensemble formé par des cylindres suivant le même axe, mais sur des longueurs |444| différentes, et qui ont la même section plane constante, de telle sorte que les volumes des cylindres forment une série, dont chaque terme contient le suivant. Nous avons une « classe-inclusion-sérielle ». C'est une « classe géométrique inclusion-sérielle » s'il n'y a aucune limite inférieure (autre que 0) à la longueur des cylindres. Alors, il n'y a aucun volume commun à tous les cylindres et le seul élément géométrique commun à tous les nombres de la série est la section plane qui est une surface et non un volume.

Soit encore un ensemble de cylindres finis, tous de même longueur et de même axe, mais dont les sections planes ont des rayons différents; ils forment une « classe-inclusion-sérielle » et s'il n'y a aucune limite inférieure (autre que 0) au rayon, la série est géométrique. Le seul élément spatial commun est l'axe, qui est un segment d'une ligne droite et non un volume.

De même pour une « classe géométrique inclusion-sérielle » de sphères concentriques dont le seul élément spatial commun est le centre. D'autres exemples peuvent être tirés des séries-inclusion d'aires sur la même surface.

---

1. Corrected 'ser T' to 'ser'T'.
2. The reference has been corrected from '*35' to '*36', and in the below formula, the symbol '$\downharpoonleft$' has been corrected to '$\upharpoonleft$'.

VI. Geometric T-Serial and T-Equality Classes

The group of T-serial classes - symbolised by ser'T - is composed of classes of σ-objects which satisfy the two following conditions: (i) a class which is a member of ser'T is not the null class, and (ii) if x and y are two members of a class which is a member of ser'T, then either x has the relation T to y, or y has the relation T to x.

In symbols:

$$\left\{\text{ser'T} = \hat{\alpha}\exists\,!\,\alpha:.\,x,y\in\alpha\,.\supset_{x,y}: x\text{T}y.\vee.y\text{T}x\right\}\ \text{Df.}$$

Here the first condition is inserted simply to exclude the inopportune case of the null class (Λ).

Prop. 1 - It is easy to see that every *transitive* relation T transforms a T-serial class into a series. In symbols (cf. *Princ. Math.*, Vol. I, *36 and Vol. II, *201, *204):

$$\text{T}\in\text{trans.}\ \alpha\in\text{ser'T}\,.\supset.\ (j\dot{\frown}\text{T})\restriction\alpha\in\text{ser.}$$

The group of 'geometric T-serial classes' is symbolised by 'geom T' and every member (σ) of the group, in addition to being T-serial, is characterised by the property that the only σ-object (*if there is one*) to which every member of σ can have the relation T must be a T-indivisible, where a 'T-indivisible' is a member of the class

$$\text{C'T}-\text{D'}(\text{T}\dot{\frown}j)$$

which will be denoted by 'indiv (T)'. In symbols

$$\text{geom'T} = \hat{\alpha}\left\{\alpha\in\text{ser'T}\,.\,\text{p'}\overleftarrow{\text{T}}\text{''}\alpha\subset\text{indiv}\,(\text{T})\right\}\ \text{Df.}$$

The importance of our subsequent procedure depends entirely on this fact: It is desirable to admit the existence of the case where $p'\overleftarrow{\text{T}}''\alpha=\Lambda$. This case is included in the preceding definition of geom'T, for Λ is contained in every class.

For example, in apparent space let us consider the group formed by cylinders having the same axis but different lengths, and which have the same constant plane section, of such a kind in such a way that the volumes of the cylinders form a series of which each term contains the subsequent one. We have a 'serial-inclusion-class'. It is a 'geometric serial-inclusion class' if there is no lower limit (other than 0) to the length of the cylinders. Then, there is no volume common to all of the cylinders, and the only geometrical element common to all of the members of the series is the plane section, which is a surface and not a volume.

Again, if we have a group of finite cylinders, all of the same length and with the same axis, but whose plane section have different radii, they form a 'serial-inclusion-class', and if there is no lower limit (other than 0) to the radius, the series is geometric. The only common spatial element is the axis, which is a segment of a straight line and not a volume.

It is the same for a 'geometric serial-Inclusion class' of concentric spheres of which the only common spatial element is the centre. Other examples can be drawn of inclusion-series of areas on the same surface.

D'importantes modifications dans notre procédure logique surviennent du fait que surfaces, lignes, points, ainsi déterminés par les « classes géométriques inclusion-sérielles » ne sont pas nécessairement situés à l'intérieur les uns des autres. Par exemple, si dans le premier cas, nous ne prenons que les moitiés des cylindres d'un seul côté de la section plane centrale, la nouvelle série géométrique définit encore cette section commune qui est à la surface de tous les membres de la série. De même, dans le second cas, nous pouvons couper par moitié tous les cylindres par un plan passant par l'axe commun et coupant tous les cylindres longitudinalement. Prenant les moitiés d'un seul côté du plan, nous obtenons une nouvelle série géométrique qui détermine la même ligne que celle qui est définie par les cylindres tout entiers. De même nous pouvons partager en deux toutes les sphères concentriques par un plan diamétral commun.

Prop. 2.—Il suit immédiatement des définitions de indiv (T) et de géom'T, qu'il ne peut y avoir deux indivisibles-T différents appartenant à la même classe sérielle-T. Conséquemment, une classe géométrique sérielle-T *ou bien* a un *seul* indivisible-T formant son dernier terme, *ou bien* est telle qu'il n'y a aucun objet-σ inclus-T parmi ses termes.

|445| Nous désirons maintenant pouvoir diviser l'ensemble géom'T en classes, suivant que la convergence est vers un « point-T matériel », ou une « ligne-T matérielle » ou une « surface-T matérielle ». En d'autres termes, nous désirons différencier les classes géométriques sérielles-T ponctuelles, linéaires, superficielles, les unes des autres. La seule classe essentielle pour les développements géométriques ultérieurs est l'ensemble convergent vers des points.

Si deux classes sérielles-T α et β sont telles que $\alpha \subset \mathrm{T}``\beta$, alors α est dit « couvrir β par rapport à T ». Ou, plus brièvement, « couvrir-T β ». Soit T_c représentant la relation de couvrir-T, de telle sorte que « $\alpha \mathrm{T}_c \beta$ » signifie que α et β soient des classes sérielles-T et que tout membre de α ait la relation T à un membre déterminé de β. En symboles, la définition devient:

$$\mathrm{T}_c = \hat{\alpha}\hat{\beta}\,\{\alpha,\beta \in \mathrm{ser}`\mathrm{T} \,.\, \alpha \subset \mathrm{T}``\alpha\}\ \mathrm{Df.}$$

Si T est transitive et β une classe sérielle-T, alors, si x inclut-T tout membre y de β, il inclut-T tout membre de β qui suit y dans l'ordre-T. Donc, quand $\alpha \mathrm{T}_c \beta$, l'« extrême fin » de β dans l'ordre-T est inclus-T par les membres de α.

Prop. 3.—Si T est transitive et si α et β sont classes sérielles-T, si α est géométrique et $\alpha \mathrm{T}_c \beta$, alors β est géométrique. La preuve est évidente.

Deux classes sérielles-T α et β sont dites être « égales-T » quand chacune couvre-T l'autre, c'est-à-dire quand[1] $\alpha \mathrm{T}_c \beta$ et $\beta \mathrm{T}_c \alpha$. Cette relation est dénotée par T_{eq}. En symboles

$$\mathrm{T}_{eq} = \mathrm{T}_c \dot{\cap} \breve{\mathrm{T}}_c .$$

Il est évident que toute classe sérielle-T se couvre-T elle-même.

La classe des classes sérielles-T qui sont égales-T à une classe α est appelée la classe égale-T de α. En symboles la « classe égale-T de » est dénotée par

$$\overrightarrow{\mathrm{T}_{\mathrm{eq}}}`\alpha.$$

1. In the following expression, the original erroneously had '$\alpha \mathrm{T}_c \alpha$' instead of '$\alpha \mathrm{T}_c \beta$'.

Some important modifications in our logical procedure arise from the fact that surfaces, lines and points thus determined by 'geometric serial-inclusion classes' are not necessarily situated inside one another. For example, if, in the first case, we take only the halves of the cylinders on one side of the central plane section, the new geometric series still defines this common section which is on the surface of all the members of the series. Likewise, in the second case, we can cut all of the cylinders in half by passing a plane through the common axis and cutting all of the cylinders longitudinally. Taking the halves on one side of the plane, we obtain a new geometric series which determines the same line as that which is defined by the complete cylinders. In the same way, we can divide in two all of the concentric spheres by a common diametral plane.

Prop. 2 – It immediately follows from the definitions of indiv (T) and geom'T, that there cannot be two different T-indivisibles belonging to the same T-serial class. Consequently, a geometric T-serial class *either* has a *single* T-indivisible comprising its last term, *or else* is such that there is no σ-object T-included among its terms.

We now want to be able to divide the group geom'T into classes, depending on whether the convergence is toward a 'material T-point', a 'material T-line', or a 'material T-surface'. In other words, we want to differentiate the point, line, and surface type of geometric T-serial classes from one another. The only class essential for the final geometrical developments is the group converging toward points.

If two T-serial classes α and β are such that $\alpha \subset \mathrm{T}``\beta$, then α is said 'to cover β in respect to T'. Or, more briefly, 'to T-cover β'. Let T_c represent the relation of T-covering in such a way that '$\alpha \mathrm{T}_c \beta$' signifies that α and β are T-serial classes and that every member of α has the relation T to a determinate member of β. In symbols the definition becomes:

$$\mathrm{T}_c = \hat{\alpha}\hat{\beta}\ \{\alpha, \beta \in \mathrm{ser}`\mathrm{T} . \alpha \subset \mathrm{T}``\alpha\}\ \text{Df.}$$

If T is transitive and β is a T-serial class, then, if x T-includes every member y of β, it T-includes every member of β which follows y in the T-order. Therefore, when $\alpha \mathrm{T}_c \beta$, the 'extreme end' of β in the T-order is T-included by the members of α.

Prop. 3 – If T is transitive and if α and β are T-serial classes, if α is geometric and $\alpha \mathrm{T}_c \beta$, then β is geometric. The proof is evident.

Two T-serial classes α and β and are said to be 'T-equal' when each T-covers the other, that is, when $\alpha \mathrm{T}_c \beta$ and $\beta \mathrm{T}_c \alpha$. This relation is denoted by T_{eq}. In symbols

$$\mathrm{T}_{eq} = \mathrm{T}_c \dot{\cap} \breve{\mathrm{T}}_c.$$

It is evident that every T-serial class T-covers itself.

The class of T-serial classes which are T-equal to a class α is called the T-equal class of α. In symbols the 'T-equality class of' is denoted by

$$\overrightarrow{\mathrm{T}_{eq}}`\alpha.$$

Prop. 4.—Il est évident que si T est transitif et si on a « Ident Hyp T »,[1] T_{eq} est réflexive, transitive et symétrique.

Prop. 5.—Si T est transitive et si α est un membre de géom'T, alors si $\beta T_{eq}\alpha$, β est aussi un membre de géom'T.

Prop. 6.—Si T est transitive et $\alpha T_{eq}\beta$ et $\beta T_c \gamma$, alors $\alpha T_c \gamma$.

|446| *Prop. 7.*—Si T est transitive et $\alpha T_{eq}\beta$, et $\gamma T_{eq}\delta$ et $\alpha T_c \gamma$, alors $\beta T_c \delta$.[2]

Prop. 8.—Si T est transitive et qu'on a Indiscern Hyp T et si α et β sont membres de géom'T. Si α et β possède le même indivisible-T, alors $\alpha T_{eq}\beta$. Donc, conversement, si $\alpha T_{eq}\beta$ et α converge vers un indivisible-T, alors β converge vers le même indivisible-T.

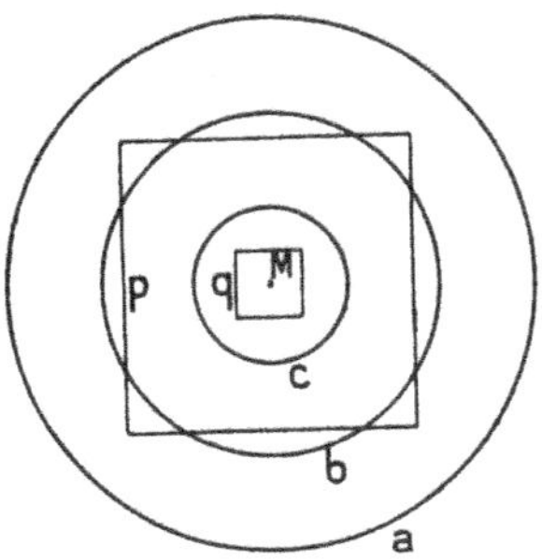

Fig. 1.

Le rapport de la théorie des classes égales-T géométriques sérielles-T à la théorie de l'espace apparent s'éclaire en considérant des aires apparentes tracées sur une même surface. Considérons (fig. 1) la classe géométrique inclusion sérielle formée par l'ensemble des cercles concentriques *a*, *b*, *c*... de centre *M*, que nous supposerons n'être pas un élément apparent inclusion-indivisible. Considérons aussi la classe géométrique inclusion-sérielle des carrés concentriques *p*, *q*,... de même centre *M* et de côtés parallèles. Alors naturellement les deux classes géométriques sérielles sont égales-inclusion. De même, des exemples peuvent être trouvés de deux classes géométriques inclusion-sérielles d'aires apparentes convergentes vers le même segment de ligne comme limite.

De même, il est facile de voir qu'une « classe géométrique inclusion-sérielle » d'aires qui « convergent vers une ligne » comme limite peut couvrir-inclusion une classe géométrique inclusion-sérielle d'aires qui « convergent vers un point » comme limite, mais non inversement.

Mieux encore, deux classes géométriques inclusion-sérielles d'aires tracées sur la même surface qui convergent vers le même point à l'*intérieur* des aires sont nécessairement égales-inclusion. Mais si le point est sur une ligne limite commune à toutes les aires de l'une (ou des deux) des classes géométriques, alors cela n'est pas nécessairement le cas.

Il sera commode de construire la théorie générale du développement d'un espace-T de manière qu'il soit immédiatement applicable à un concept particulier d'objets apparents. De légères variations dans la procédure peuvent être facilement introduites pour adapter la théorie à des vues analogues.

|447| Quelques remarques seront donc encore faites sur les objets apparents qui expliqueront nos intentions dans les définitions ultérieures de l'espace-T.

Des objets apparents sont perçus comme des *aires*, non pas nécessairement comme plans. L'intérieur de ces objets est conçu, mais l'appréhension immédiate

1. Corrected a semicolon to a comma.
2. Corrected '$\beta T_{eq}\delta$' to '$\beta T_c \delta$'.

Prop. 4 – It is evident that if T is transitive and if we have 'Ident Hyp T', T_{eq} is reflexive, transitive and symmetrical.

Prop. 5 – If T is transitive and if α is a member of geom'T, then if $\beta T_{eq}\alpha$, β is also a member of geom'T.

Prop. 6 – If T is transitive and $\alpha T_{eq} \beta$ and $\beta T_c \gamma$, then $\alpha T_c \gamma$.

Prop. 7 – If T is transitive and $\alpha T_{eq} \beta$, and $\gamma T_{eq} \delta$ and $\alpha T_c \gamma$, then $\beta T_c \delta$.

Prop. 8 – If T is transitive and if we have Indiscern Hyp T and if α and β are members of geom'T, then if α and β possess the same T-indivisible, then $\alpha T_{eq}\beta$. Therefore, conversely, if $\alpha T_{eq} \beta$ and α converges toward a T-indivisible, then β converges toward the same T-indivisible.

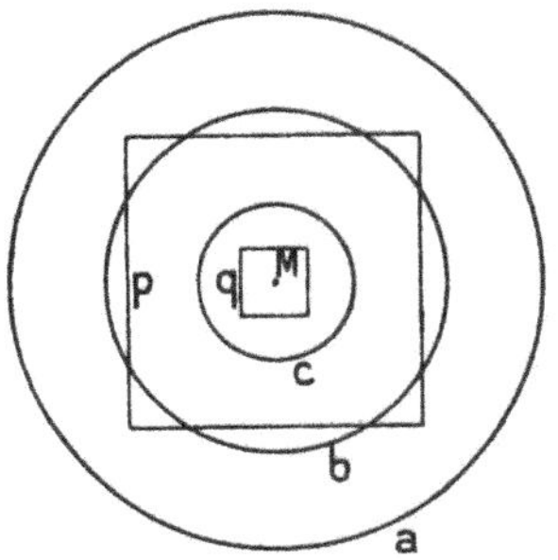

Fig. 1.

The relation of the theory of geometric T-equal T-serial classes to the theory of apparent space is illuminated through the consideration of apparent areas traced on the same surface. Let us consider (fig. 1) the geometric serial inclusion class formed by the group of concentric circles *a, b, c,* ... with centre *M,* which we will suppose not to be an apparent indivisible-inclusion element. Let us also consider the geometric serial-inclusion class of concentric squares *p, q,* . . . with the same centre *M* and with parallel sides. Then naturally the two geometric serial classes are inclusion-equal. In the same way, examples can be found of two geometric serial-inclusion classes of apparent areas converging toward the same line segment as a limit.

In the same way, it is easy to see that a 'geometric serial-inclusion class' of areas which 'converge toward a line' as a limit can inclusion-cover a geometric serial-inclusion class of areas which 'converge toward a point' as a limit, but not vice versa.

Furthermore, two serial-inclusion geometric classes of areas traced on the same surface which converge toward the same point at the *interior* of the areas are necessarily inclusion-equal. But if the point is on a boundary line common to all of the areas of one (or both) of the geometric classes, then that is not necessarily the case.

It will be convenient to construct the general theory of the development of a T-space in a way that will be immediately applicable to a particular concept of apparent objects. Minor variations in the procedure can be easily introduced to adapt the theory to analogous views.

A few remarks will therefore again be made concerning apparent objects which will explain our intentions in the final definitions of the T-space.

Some apparent objects are perceived as *areas,* and not necessarily as planes. The interior of these objects is conceived, but the immediate apprehension is always the

est toujours la perception d'un objet-surface. Dans la mesure où les points sont directement perçus, ce doit être par la perception d'un point-aire indivisible. Donc, si les relations de deux objets-surfaces sont *perçues* comme exactement définies par un point géométrique, alors un point-aire indivisible est perçu comme le point qui définit cette relation.

Dans certains cas, des lignes de division (par exemple, où la couleur change brusquement) sont perçues. Si un espace apparent unique en inclut un autre, alors les deux sont nécessairement sur la surface du même objet-volume (conçu). Une série-inclusion géométrique d'aires peut « converger » vers un point ou ligne, ou bien (i) à l'intérieur de tout membre de la série, ou bien (ii) sur la limite de tout membre qui suit un membre donné de la série, pourvu que la limite commune soit l'une des lignes particulières perçues.

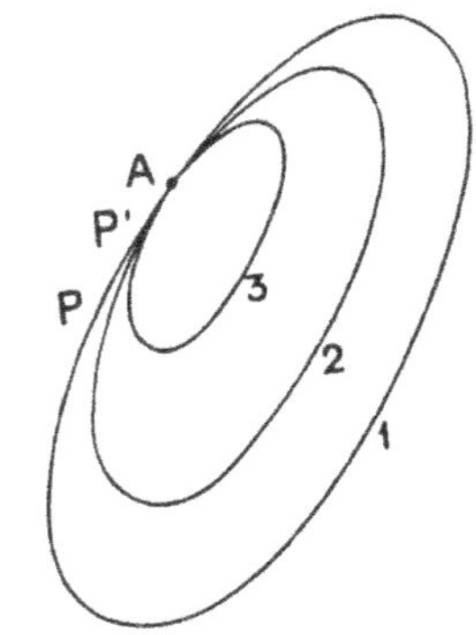

Fig. 2.

Dans le second cas, si les limites successives (1, 2, 3...) (Cf. fig. 2) se terminent toutes au même point, alors, de par la règle énoncée précédemment, la détermination exacte par le point A sera supposée être due à un point-aire indivisible, perçu en A. Ainsi la série, si elle est « géométrique » doit se terminer au point-aire indivisible A et de par Prop. 8, être égale-T à toute autre série géométrique dont tout membre inclut A.

Il est évident par conséquent qu'une série géométrique ne peut pas avoir sa limite commune terminée en deux points fixes à l'un et l'autre bout. Car la série posséderait alors deux indivisibles.

Mais le second cas est compatible avec l'absence de point-indivisible, car (Cf. fig. 3) les parties communes des limites peuvent se resserrer continuellement, s'approchant comme limite soit d'un segment de ligne, soit d'un point. Cette série géométrique peut ne posséder aucun point-indivisible. Le fondement de ce raisonnement est |448| que les points géométriques ou lignes ne sont pas donnés en perception mais sont le résultat du *concept* de série géométrique. Les aires, se contenant l'une l'autre, ne sont pas exactement définies en perception par les parties homogènes des surfaces ; elles existent seulement comme *possibilités* de subdivision. Il y a (ou il peut y avoir selon la théorie logique ici développée) des points et lignes exceptionnels qui sont objets de perception. De là la possibilité de communes limites définies pour toutes les aires d'une série. Mais quand la loi de convergence requiert dans sa définition un point défini ou une ligne définie, ce point ou cette ligne doit être un point perçu ou une ligne perçue; par conséquent, le point perçu doit être un indivisible point-aire perçu. En dehors de ces cas exceptionnels, qu'inclut notre raisonnement, la convergence se fait essentiellement vers un point *déterminé* ou ligne *déterminée* à l'intérieur de chaque membre de la série. En se référant à la figure (4), il faut noter que le même point P

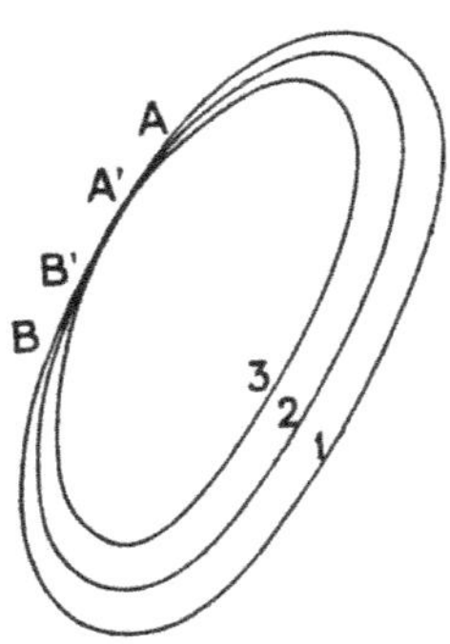

Fig. 3.

perception of a surface-object. To the extent that points are directly perceived, this must be through the perception of an indivisible area-point. Thus, if the relations between two surface objects are *perceived* as exactly defined by a geometric point, then an indivisible area-point is perceived as the point which defines this relation.

In certain cases, some lines of division (for example, where colour changes abruptly) are perceived. If a unique apparent space includes another, then the two are necessarily on the surface of the same volume-object (conceived). A geometric inclusion-series of areas can 'converge' toward a point or a line either (1) at the interior of every member of the series or else (2) on the boundary of every member which follows a given member of the series, provided that the common boundary is one of the particular lines perceived.

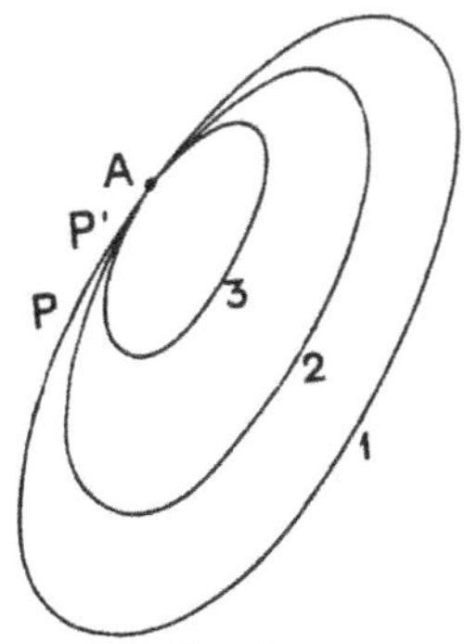

Fig. 2.

In the second case, if the successive boundaries (1, 2, 3, . . .) (Cf. fig. 2) terminate all at the same point then, by the rule stated previously, the exact determination by the point A will be supposed to be due to an indivisible area-point, perceived in A. Thus, the series, if it is 'geometric', must terminate at the indivisible area-point A, and by Prop. 8, be T-equal to every other geometric series of which every member includes A.

It is evident as a consequence that a geometric series cannot have its common boundary terminate at two fixed points at each end. For the series would then possess two indivisibles.

But the second case is compatible with the absence of the indivisible point, for (Cf. fig. 3) the common parts of the boundaries can continually contract, approaching as a limit either a line segment, or a point. This geometric series cannot possess any indivisible point. The basis of this argument is that the geometric points or lines are not given in perception but are the result of the *concept* of the geometric series. The areas, containing one another, are not exactly defined in perception by the homogeneous parts of the surfaces; they exist only as *possibilities* of subdivision. There are (or there can be according to the logical theory here developed) some exceptional points and lines which are objects of perception. From thence follows the possibility of common boundaries defined for all of the areas of a series. But when the law of convergence requires in its definition a definite point or a definite line, that point or that line must be a perceived point or a perceived line; as a consequence, the perceived point must be an indivisible perceived area-point. Apart from these exceptional cases, which our argument includes, the convergence is made essentially toward a *determined* point or a *determined* line from the interior of each member of the series. In reference to figure 4 it is necessary to note that the same point P (or line) can be the locus of

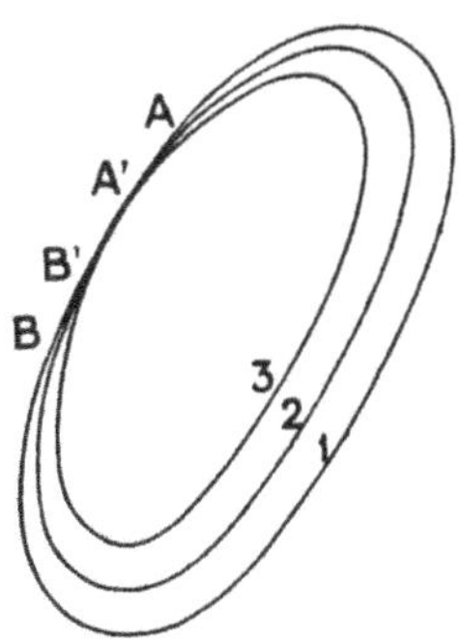

Fig. 3.

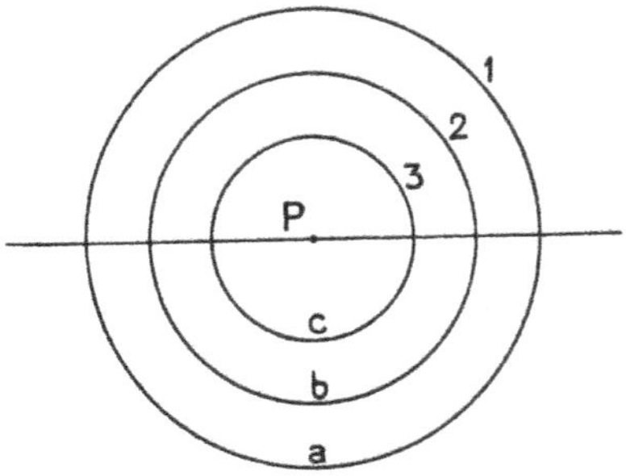

Fig. 4.

(ou ligne) peut être le lieu de convergence de deux séries-inclusion géométriques dont l'une ne « couvre » pas l'autre. Le point P est le lieu de convergence des semi-cercles 1, 2, 3... et a, b, c...et des cercles complets. Notre raisonnement général doit admettre cette possibilité.

Le résultat général de la théorie est que, quoiqu'un point soit (dans des cas exceptionnels) perçu directement, il est en général simplement le concept de la possibilité de séries de subdivisions. Il en est de même pour une ligne ou un segment de ligne.

VII.—POINTS-T MATÉRIELS ET SEGMENTS-T MATÉRIELS.

Nous sommes maintenant en mesure de définir la méthode pour discerner parmi les membres de géom'T ceux qui « convergent vers |449| des points » et ceux qui « convergent vers des segments de ligne ». Le premier ensemble sera dénoté par géom_1'T et le second par géom_2'T. Les définitions doivent être faites seulement au moyen de définitions logiques portant sur T et sans l'aide d'aucune nouvelle idée fondamentale,—c'est-à-dire que notre géométrie doit être fondée sur la notion fondamentale de inclusion-T. Nous distinguons dans géom'T trois degrés successifs de simplicité décroissante, que nous dénommerons l'ensemble des « premiers éléments de géom'T » l'ensemble des « seconds éléments de géom'T » et le « résidu de géom'T ». Une méthode plus générale, impliquant moins de postulats est exposée plus loin.

L'ensemble des « premiers éléments géom'T », symbolisé par prm_1'T est l'ensemble des classes géométriques sérielles-T qui sont égales-T à toute classe géométrique sérielle-T qu'elles couvrent-T. En symboles:

$$\text{prm}_1\text{'T} = \hat{\alpha}\left\{\alpha \in \text{géom'T} : \alpha \, \text{T}_c \, \beta \, . \supset_\beta . \, \beta \, \text{T}_c \, \alpha\right\} \text{ Df.}$$

Dans l'application à l'espace apparent précédemment indiquée (où $x\text{T}y$ signifie que x et y sont des aires apparentes et que x inclut y) prm_1'T comprendra toute classe géométrique sérielle-T d'aires qui inclut un point-aire indivisible, et aussi toute classe géométrique sérielle-T d'aires qui converge vers un point P sur un côté d'une ligne donnée passant par P. Il est évident sur la figure que deux classes géométriques sérielles-T α et β d'un type tel que l'une couvre-T l'autre, doivent être égales-T. L'ensemble des « seconds éléments de géom'T »,[1] symbolisé

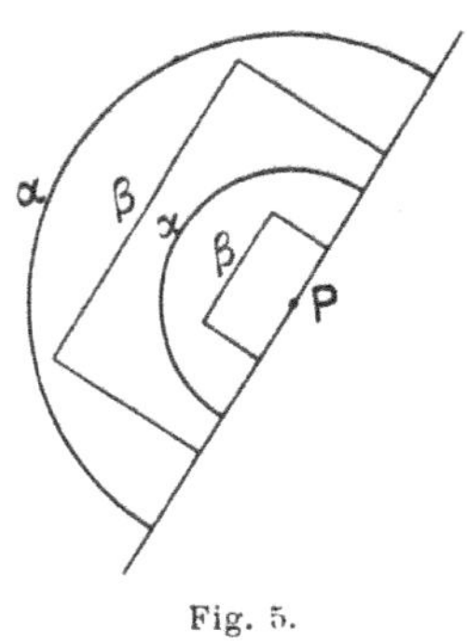

Fig. 5.

1. Closing quotation mark added.

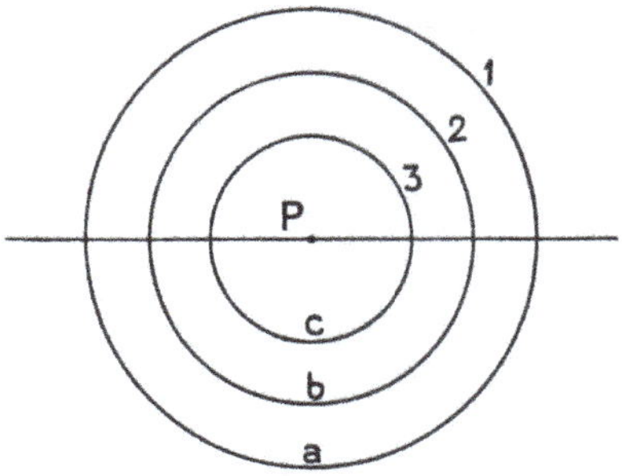

Fig. 4.

convergence of two geometric inclusion series of which one does not cover the other. The point P is the locus of convergence of the semicircles 1, 2, 3 ... and a, b, c ... and of the complete circles. Our general argument must allow for this possibility.

The general result of the theory is that, although a point may be (in some exceptional cases) perceived directly, it is in general simply the concept of the possibility of a series of subdivisions. It is the same for a line or a segment of a line.

VII. Material T-Points and Material T-Segments

We are now able to define the method for discerning among the members of geom'T those which 'converge toward points' and those which 'converge toward line segments'. The first group will be denoted by $\text{geom}_1\text{'T}$ and the second by $\text{geom}_2\text{'T}$. The definitions must be made solely by means of logical definitions pertaining to T and with the aid of no new fundamental idea – that is, our geometry must be based on the fundamental notion of T-inclusion. We distinguish in geom'T three successive degrees of decreasing simplicity, which we will designate the group of 'first elements of geom'T', the group of 'second elements of geom'T', and the 'residue of geom'T'. A more general method, implying fewer postulates, is set forth later on.

The group of 'first elements of geom'T', symbolised by $\text{prm}_1\text{'T}$, is the group of geometric T-serial classes which are T-equal to every geometric T-serial class which they T-cover. In symbols:

$$\text{prm}_1\text{'T} = \hat{\alpha}\left\{\alpha \in \text{geom'T} : \alpha \text{T}_c \beta \, . \supset_\beta . \, \beta \text{T}_c \alpha\right\} \text{ Df.}$$

In the application to apparent space previously indicated (when *x*T*y* signifies that *x* and *y* are apparent areas and that *x* includes *y*) $\text{prm}_1\text{'T}$ will contain every geometric T-serial class of areas which includes an indivisible area-point, and also every geometric T-serial class of areas which converges toward a point P on one side of a given line passing through P. It is evident from the figure that two geometric T-serial classes α and β of a type such that one T-covers the other, must be T-equal. The group of 'second elements of geom'T', symbolised by $\text{prm}_2\text{'T}$, is the group of geometric T-serial classes

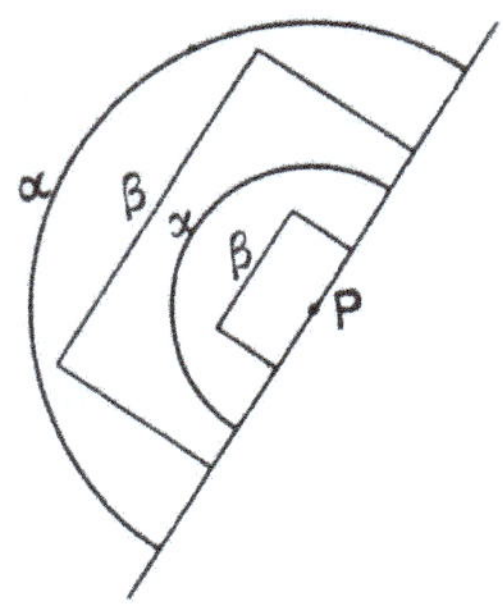

Fig. 5.

par $\text{prm}_2\text{'T}$ est l'ensemble des classes géométrique sérielles-T qui sont égales-T à tout membre de géom'T, en dehors de $\text{prm}_1\text{'T}$, qu'elles couvrent-T. En symboles:

$$\text{prm}_2\text{'T} = \hat{\alpha}\left\{\alpha \in \text{géom'T} : \alpha \mathrm{T}_c \beta \,.\, \beta \sim\in \text{prm}_1\text{'T} \,.\, \supset_\beta .\, \beta \mathrm{T}_c \alpha\right\} \quad \text{Df.}$$

Si l'on applique cette définition à l'espace apparent, la classe $\text{prm}_2\text{'T}$ comprendra toute classe géométrique sérielle-T d'aires qui converge vers un point dans l'intérieur des aires. |450| Le résidu de géom'T comprendra le reste de géom'T. Dans l'application précédente le résidu consiste de classes géométriques sérielles-T d'aires, convergentes vers des segments de lignes. Ces classes sérielles-T d'aires apparentes peuvent n'avoir aucune qualité des « éléments » définis précédemment. En effet, toute classe convergente vers AB peut couvrir-T toute classe convergente vers CD (cf. la figure ci-contre). Mais la relation n'est pas réciproque.

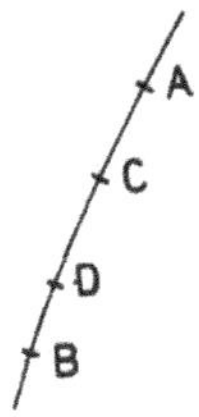

Fig. 6.

Ainsi $\text{géom}_1\text{'T}$ est la somme de deux ensembles $\text{prm}_1\text{'T}$; et $\text{prm}_2\text{'T}$; et $\text{géom}_2\text{'T}$ est le reste de géom'T. En symboles:

$$\text{géom}_1\text{'T} = \text{prm}_1\text{'T} \cup \text{prm}_2\text{'T} \quad \text{Df.}$$
$$\text{géom}_2\text{'T} = \text{géom'T} - \text{géom}_1\text{'T} \quad \text{Df.}$$

Nous avons maintenant à grouper ces membres de $\text{géom}_1\text{'T}$ qui « convergent » vers le même point. Puisque le concept d'un point ne s'est pas présenté comme une idée fondamentale dans la définition de $\text{géom}_1\text{'T}$ (mais seulement lorsque nous avons expliqué verbalement les possibilités de sens), un tel groupe peut être pris comme étant lui-même le point géométrique. Aucune autre entité n'est nécessaire. Aussi un point-T matériel est un groupe de membres de $\text{géom}_1\text{'T}$ qui convergent vers le même point et nous montrons comment cette idée est capable d'une définition purement logique.

Deux membres de $\text{géom}_1\text{'T}$ (α et β) seront appelés connexes-T, si x étant un membre quelconque de α et y de β, un ou plusieurs objets-σ existent à qui x et y ont la relation *T*. La relation de connection-*T* est dénotée par T_{cn}. En symboles:

$$\mathrm{T}_{cn} = \hat{\alpha}\hat{\beta}\left\{\alpha, \beta \in \text{géom}_1\text{'T} : x \in \alpha \,.\, y \in \beta \,.\, \supset_{x,y} .\, \exists !\, \overleftarrow{\mathrm{T}}\text{'}x \cap \overleftarrow{\mathrm{T}}\text{'}y\right\} \quad \text{Df.}$$

Il est évident que « $\alpha \mathrm{T}_{cn} \beta$ » et « $\beta \mathrm{T}_{cn} \alpha$ » sont des propositions équivalentes : c'est-à-dire $\mathrm{T}_{cn} = \breve{\mathrm{T}}_{cn}$. Maintenant en accord avec la définition de Frege de la « relation ancestrale » correspondante à une relation *R* (cf. *Princ. math.*, vol. I, *90)[1] si R_* représente cette relation, alors « $u\, \mathrm{R}_* v$ » signifie que quelque chaine d'éléments $z_1\, z_2 \ldots z_n$ peut être trouvée telle que

$$u \mathrm{R} z_1 \,.\, z_1 \mathrm{R} z_2 \ldots\ldots z_n \mathrm{R} v.$$

|451| Donc suivant la définition ci-dessus, si u est un membre quelconque du champ de *R*, nous avons $u\, R_* u$. Appliquons au cas de T_{cn} nous voyons que:[2]

1. We have corrected '*20' to '*90'.
2. In the below expression, the French original used a five-pointed star to indicate the ancestral relation; here and in the four equations below, we have changed this to an asterisk to reflect the notation used in *Principia Mathematica*.

which are T-equal to every member of geom'T, apart from prm_1'T, which they T-cover. In symbols:

$$\text{prm}_2\text{'T} = \hat{\alpha}\{\alpha \in \text{geom'T} : \alpha \text{T}_c \beta . \beta \sim\in \text{prm}_1\text{'T} . \supset_\beta . \beta \text{T}_c \alpha\} \text{ Df.}$$

If we apply this definition to apparent space, the class prm_2'T will contain every geometric T-serial class of areas which converges toward a point in the interior of the areas. The residue of geom'T will include the remainder of geom'T. In the preceding application the residue consists of geometric T-serial classes of areas converging toward line segments. These T-serial classes of apparent areas may have no quality of the 'elements' defined previously. Indeed, every class converging toward AB can T-cover every class converging toward CD (Cf. the figure opposite). But the relation is not reciprocal.

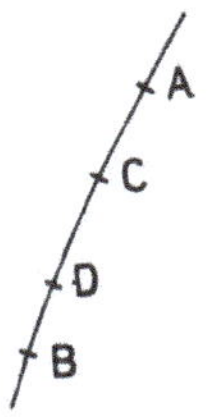

Fig. 6.

Thus geom_1'T is the sum of the two groups prm_1'T and prm_2'T; and geom_2'T is the remainder of geom'T. In symbols:

$$\text{geom}_1\text{'T} = \text{prm}_1\text{'T} \cup \text{prm}_2\text{'T} \text{ Df.}$$
$$\text{geom}_2\text{'T} = \text{geom'T} - \text{geom}_1\text{'T} \text{ Df.}$$

We now have to group these members of geom_1'T which 'converge' toward the same point. Since the concept of a point did not present itself as a fundamental idea in the definition of geom_1'T (but only when we verbally explained the possibilities of meaning) one such group can be taken as being itself the geometric point. No other entity is necessary. Also, a material T-point is a group of members of geom_1'T which converge toward the same point, and we will show how this idea is susceptible of a purely logical definition.

Two members of geom_1'T (α and β) will be called T-connected if, x being any member of α and y of β, one or more σ-objects exist to which x and y have the same relation T. The relation of T-connection is denoted by T_{cn}. In symbols:

$$\text{T}_{cn} = \hat{\alpha}\hat{\beta}\{\alpha, \beta \in \text{geom}_1\text{'T} : x \in \alpha . y \in \beta . \supset_{x,y} . \exists ! \overleftarrow{\text{T}}\text{'}x \cap \overleftarrow{\text{T}}\text{'}y\} \text{ Df.}$$

It is evident that '$\alpha \text{T}_{cn} \beta$' and '$\beta \text{T}_{cn} \alpha$' are equivalent propositions: that is $\text{T}_{cn} = \breve{\text{T}}_{cn}$. Now in accord with Frege's definition of the 'ancestral relation' corresponding to a relation R (Cf. *Princ. Math.* Vol. I, *90) if R_* represents this relation, then "$u \text{R}_* v$' signifies that some chain of elements $z_1 z_2 ... z_n$ can be found such that:

$$u \text{R} z_1 . z_1 \text{R} z_2 ..... z_n \text{R} v.$$

Thus according to the definition given above, if u is any member of the field of T, we have $u R_* u$. Applied to the case of T_{cn} we see that:

$$\alpha(\mathrm{T}_{cn})_* \beta$$

signifie que α et β sont reliés par une chaîne de membres de géom_1'T en connection-T et qui sont eux-mêmes membres de géom_1'T.

Un « point-T matériel »[1] sera défini comme l'ensemble des membres de géom_1'T qui sont reliés à un membre déterminé de géom_1'T par une chaîne de membres de géom_1'T en connexion-T. La classe des « points-T matériels » sera définie par $Pt(\mathrm{T})$. En symboles:

$$Pt(\mathrm{T}) = \hat{\pi}\left\{(\exists\alpha) \,.\, \alpha \in \text{géom}_1\text{'T} \,.\, \pi = \left(\overrightarrow{\mathrm{T}_{cn}}\right)_* \text{'}\alpha\right\} \; Df.$$

Ceci peut être écrit en abrégé:

$$Pt(T) = \left(\overrightarrow{\mathrm{T}_{cn}}\right)_* \text{''géom}_1\text{'T}.$$

Pour expliquer cette définition, notons que:

$$\pi = \left(\overrightarrow{\mathrm{T}_{cn}}\right)_* \text{'}\alpha$$

est une forme abrégée de:

$$\pi = \hat{\beta}\left\{\beta(\mathrm{T}_{cn})_* \alpha\right\}.$$

Appliquant cette définition aux aires apparentes, il est facile de voir que deux classes géométriques sérielles-T d'aires qui convergent chacune vers un point et sont connexes-T, doivent converger vers le même point. De même deux classes géométriques sérielles-T qui convergent vers le même point doivent être reliées par une chaîne de classes géométriques sérielles-T connexes-T.

Un point-T matériel sera appelé, incident-T dans un membre de géom_2'T, quand le membre de géom_2'T couvre-T un membre du point.

Deux membres de géom_2'T convergent vers le même segment-T matériel quand le même ensemble de points-T matériels sont incidents dans les deux. La classe des « segments-T matériels » est dénotée par $\text{segm}(\mathrm{T})$, et l'un quelconque de ses membres consiste de l'ensemble des membres de géom_2'T qui ont les mêmes points-T matériels incidents en eux qu'un membre donné de géom_2'T.

|452| Dès lors, un point-T matériel (π) sera appelé « incident-T dans un segment-T matériel (ρ) »[2] quand $\rho \subset \mathrm{T}_c\text{''}\pi$.

Il sera noté que la propriété d'être sans parties et grandeur, qui est la définition euclidienne du point est ici représentée par la qualité d'« élémentarité » qui s'attache à toute classe géométrique sérielle-T qui converge vers un point, c'est-à-dire qui est membre d'un point.

Si notre objet est de construire un espace-T qui peut être identifié à l'espace du monde physique, les objets-σ seront naturellement conçus suivant trois dimensions. De même, dans ce monde hypothétique, il ne semble pas probable qu'il y aura avantage à retenir l'idée d'objets physiques à surface commune. Il s'ensuit qu'une classe géométrique inclusion-sérielle d'objets physiques peut toujours être conçue comme convergente vers un point ou un segment de ligne, ou une aire qui est entièrement située à l'intérieur de chaque objet de la classe. La classe $(\text{prm}_1\text{'T})$ précédemment définie, peut donc être identifiée à toutes les classes d'objets

1. Quotation marks added and an opening quotation mark to the left of 'sera' removed..
2. Added closing quotation mark.

$$\alpha(\mathrm{T}_{cn})_*\beta$$

signifies that α and β are connected by a chain of members of geom_1'T in T-connection and that they are themselves members of geom_1'T.

A 'material T-point' will be defined as the group of members of geom_1'T which are connected to a determinate member of geom_1'T by a chain of members of geom_1'T in T-connection. The class of 'material T-points' will be defined by $Pt(\mathrm{T})$. In symbols:

$$Pt(\mathrm{T}) = \hat{\pi}\left\{(\exists\alpha) \,.\, \alpha \in \mathrm{geom}_1\text{'T} \,.\, \pi = \left(\overrightarrow{\mathrm{T}_{cn}}\right)_*\text{'}\alpha\right\} \; Df.$$

This may be written more briefly as:

$$Pt(\mathrm{T}) = \left(\overrightarrow{\mathrm{T}_{cn}}\right)_*\text{''}\mathrm{geom}_1\text{'T}.$$

In order to explain the definition, let us note that:

$$\pi = \left(\overrightarrow{\mathrm{T}_{cn}}\right)_*\text{'}\alpha$$

is an abbreviated form of:

$$\pi = \hat{\beta}\left\{\beta(\mathrm{T}_{cn})_*\alpha\right\}.$$

Applying this definition to apparent areas, it is easy to see that two geometric T-serial classes of areas, each of which converges toward a point and is T-connected, must converge toward the same point. Likewise, two geometric T-serial classes which converge toward the same point must be connected by a chain of geometric T-connected T-serial classes.

A material T-point will be called 'T-incident in a member of geom_2'T' when the member of geom_2'T T-covers a member of the point.

Two members of geom_2'T converge toward the same material T-segment when the same group of material T-points occurs in both. The class of 'material T-segments' is denoted by $\mathrm{segm}(\mathrm{T})$, and any one of its members consists of the group of members of geom_2'T which have the same material T-points occurring in them as a given member of geom_2'T.

From here on a material T-point (π) will be called 'T-incident in a material T-segment (ρ)' when $\rho \subset \mathrm{T}_c\text{''}\pi$.

It will be noted that the property of being without parts and size, which is the Euclidian definition of a point, is here represented by the quality of 'elementarity' which belongs to every geometric T-serial class which converges toward a point, that is, which is a member of a point.

If our object is to construct a T-space which can be identified with the space of the physical world, σ-objects will naturally be conceived according to three dimensions. Likewise, in this hypothetical world, it does not seem probable that there will be any advantage in retaining the idea of physical objects with a common surface. It follows that a geometric serial-inclusion class of physical objects can always be conceived as converging toward a point, or a line segment, or an area which is entirely situated at the interior or each object of the class. The class $(\mathrm{prm}_1\text{'T})$, defined

physiques qui convergent vers des points. La classe $\text{prm}_2\text{'T}$ sera nécessairement nulle.[1]

La classe $\text{géom}_1\text{'T}$ est donc la classe $\text{prm}_1\text{'T}$. Donc la classe des points-T matériels, à savoir la classe $Pt(\text{T})$, est définie comme précédemment.

Il n'y a en réalité aucune différence entre ces théories des points pour les aires apparentes ou pour les objets physiques conçus comme volumes. Dans les deux cas :

$$\text{géom}_1\text{'T} = \text{prm}_1\text{'T} \cup \text{prm}_2\text{'T}.$$

Mais, dans le monde physique, comme il a été conçu précédemment, $\text{prm}_2\text{'T} = \Lambda$, et dans le monde apparent; $\exists!\ \text{prm}_2\text{'T}$. Donc, dans les deux cas, $\text{prm}_3\text{'T} = \Lambda$.

En dehors des autres axiomes géométriques concernant les relations rectilignes entre points, les hypothèses sur la relation T requises pour cette théorie relationniste[2] des points-T matériels, appliquée soit au monde apparent, soit au monde physique sont:

(i) $C\text{'T} \subset \mathrm{C\!\!\!I}\text{'}\dot{s}\text{'}\sigma$

(ii) Ident Hyp T

(iii) Trans Hyp T

(iv) Indiscern Hyp T

(v) $\exists!\ \text{géom}_1\text{'T}$

(vi) $\alpha \in \text{géom'T} - \text{géom}_1\text{'T} \,.\supset_\alpha.\, (\exists\beta)\,.\,\alpha \text{T}_c \beta \,.\sim(\beta \text{T}_c \alpha)$.

|453| Si nous substituons E_σ à *T*, alors les conditions (i), (ii), (iii), sont nécessairement remplies. En outre, nous pouvons nous débarrasser de (vi), par un réarrangement de nos définitions qui s'affranchit entièrement des suppositions assez spéciales relatives aux limites des objets apparents en physique, introduites pour la simplicité de l'explication.

Supposons que le reste-T d'un set (ρ) de classes sérielles-T soit composé de tout membre de ρ qui couvre-T un membre déterminé de ρ, non égal-T à lui-même. Dénotons le reste-T de ρ par $\text{Res}\,(\text{T})\text{'}\rho$.

En symboles

$$\text{Res}\,(\text{T}) = \hat{\tau}\hat{\rho}\left\{\alpha \in \tau \,.\, \equiv_\alpha .\, \alpha \in \rho \cap \text{T}_c\text{''}\left(\rho - \overrightarrow{\text{T}_{eq}}\text{'}\alpha\right)\right\} \quad \text{Df.}$$

Il suit de là (*Princ. math.*, vol. I, $*30$) que :

$$\text{Res}\,(\text{T})\text{'}\rho = \hat{\alpha}\left\{\alpha \in \rho \cap \text{T}_c\text{''}\left(\rho - \overrightarrow{\text{T}_{eq}}\text{'}\alpha\right)\right\} \quad \text{Df.}$$

Dès lors « $\xi\left\{\text{Res}\,(\text{T})\right\}_* \rho$ » affirme que ξ est ou ρ ou le reste-T de ρ, ou bien est le reste-T du reste-T de ρ, ou le reste-T du reste-T, du reste-T de ρ, etc.

Donc, ([3]Cf. *Princ. math.*, vol. I $*40$) la classe:

$$p\text{'}\overrightarrow{\left\{\text{Res}\,(\text{T})\right\}_*}\text{'}\rho$$

est la partie de ρ qui est contenue dans tout reste-T qui sont un reste-T.

Donc:

$$\rho - p\text{'}\overrightarrow{\left\{\text{Res}\,(\text{T})\right\}_*}\text{'}\rho$$

1. We have corrected 'prm$_c$T' to 'prm$_2$'T'.
2. Corrected from 'relativiste'; see p. 412n2.
3. We have supplied a missing opening parenthesis.

previously, can therefore be identified with all the classes of physical objects which converge toward points. The class prm_2'T will necessarily be null.

The class geom_1'T is thus the class prm_1'T. Therefore, the class of material T-points, namely the class $Pt(\mathrm{T})$, is defined as previously stated.

In reality there is no difference between these theories of points for apparent areas or for physical objects conceived as volumes. In both cases:

$$\mathrm{geom}_1\text{'T} = \mathrm{prm}_1\text{'T} \cup \mathrm{prm}_2\text{'T}.$$

But in the physical world, as it has been previously conceived, $\mathrm{prm}_2\text{'T} = \Lambda$, and in the apparent world $\exists!\ \mathrm{prm}_2$'T. Therefore, in both cases, $\mathrm{prm}_3\text{'T} = \Lambda$.

Apart from the other geometric axioms concerning the rectilinear relations between points, the hypotheses required on the T-relation for this relational theory of material T-points, applied either to the apparent world or to the physical world, are:

(i) $C\text{'T} \subset \mathrm{Cl}\text{'}\dot{s}\text{'}\sigma$

(ii) Indent Hyp T

(iii) Trans Hyp T

(iv) Indiscern Hyp T

(v) $\exists!$ géom$_1$'T

(vi) $\alpha \in \mathrm{geom}\text{'T} - \mathrm{geom}_1\text{'T} . \supset_\alpha . (\exists\beta) . \alpha \mathrm{T}_c \beta . \sim(\beta \mathrm{T}_c \alpha)$.

If we substitute E_σ for T, then conditions (i), (ii), and (iii) are necessarily fulfilled. Furthermore, we can disencumber ourselves from (vi) by a rearrangement of our definitions, which frees us altogether from the quite special suppositions relative to the boundaries of apparent objects in physics, introduced for simplicity of explication.

Let us suppose that the T-residue of a set (ρ) of T-serial classes is composed of every member of ρ which T-covers a determinate member of ρ, not T-equal to itself. Let us denote the T-residue of ρ by Res $(\mathrm{T})\text{'}\rho$.

In symbols:

$$\mathrm{Res}\ (\mathrm{T}) = \hat{\tau}\hat{\rho}\left\{\alpha \in \tau\ .\ \equiv_\alpha .\ \alpha \in \rho \cap \mathrm{T}_c\text{''}\left(\rho - \overrightarrow{\mathrm{T}_{eq}}\text{'}\alpha\right)\right\} \quad \mathrm{Df.}$$

It follows from this (*Princ. Math.*, Vol. I, *30) that:

$$\mathrm{Res}\ (\mathrm{T})\text{'}\rho = \hat{\alpha}\left\{\alpha \in \rho \cap \mathrm{T}_c\text{''}\left(\rho - \overrightarrow{\mathrm{T}_{eq}}\text{'}\alpha\right)\right\} \quad \mathrm{Df.}$$

Thence '$\xi\{\mathrm{Res}\ (\mathrm{T})\}_* \rho$' affirms that ξ is either ρ, or the T-residue of ρ, or else it is the T-residue of the T-residue of ρ, or the T-residue of the T-residue of the T-residue of ρ, etc.

Therefore (Cf. *Princ. Math.*, Vol. I, *40) the class:

$$p\text{'}\overrightarrow{\{\mathrm{Res}\ (\mathrm{T})\}_*}\text{'}\rho$$

is the part of ρ which is contained in every T-residue which is a T-residue.

Therefore:

$$\rho - p\text{'}\overrightarrow{\{\mathrm{Res}\ (\mathrm{T})\}_*}\text{'}\rho$$

est la partie de ρ qui disparaît aux degrés successifs et par conséquent a[1] la qualité d'élémentarité.

$$\text{géom}_1\text{'T} = \text{géom'T} - p'\left\{\overrightarrow{\text{Res (T)}}\right\}_*\text{'}\rho \quad \text{Df.}$$

Et nous procédons comme précédemment à la définition de[2] $Pt(\mathrm{T})$.

L'hypothèse (vi), précédemment énoncée, peut maintenant être prouvée. Donc, pour E_σ, s'il est substitué à *T*, nous faisons seulement les deux hypothèses:

$$\text{Indiscern Hyp } E_\sigma \quad \text{et} \quad \exists!\, \text{géom}_1\text{'}E_\sigma.$$

|454| La théorie de la définition des points-T idéaux et des lignes droites fait l'objet d'un mémoire distinct que j'espère publier bientôt.

La relation d'« occupation » est maintenant définissable. Un objet-σ x « occupe » un point-T idéal quand il y a un point-T matériel en corrélation avec le point-T idéal et x est un membre de l'une des classes géométriques sérielles-T qui constitue le point-T matériel.

1. Corrected 'à' to 'a'.
2. The original erroneously failed to italicise the 't' in the following expression.

is the part of ρ which disappears in successive degrees and which consequently has the quality of elementarity.

$$\text{geom}_1\text{‘T} = \text{geom‘T} - p‘\left\{\overrightarrow{\text{Res (T)}}\right\}_*\text{‘}\rho \quad \text{Df.}$$

And we proceed as previously to the definition of $\text{P}t(\text{T})$.

Hypothesis (vi), previously stated, can now be proved. Thus, for E_σ, if it is substituted for *T*, we make only the two hypotheses:

$$\text{Indiscern Hyp E}_\sigma \quad \text{and} \quad \exists\,!\,\text{geom}_1\text{‘E}_\sigma.$$

The theory of the definition of ideal T-points and straight lines comprises the object of a distinct treatise which I hope to publish soon.

The relation of 'occupation' is now definable. A σ-object x 'occupies' an ideal T-point when there is a material T-point in correlation with the ideal T-point and x is a member of one of the geometric T-serial classes which constitutes the material T-point.

The Organisation of Thought

1916 *Chemical News*, vol. 114, no. 2965 (21 September), pp. 139–143

1916 *Science*, vol. 44, no. 1134 (22 September), pp. 409–419

1916 *Nature*, vol. 98, no. 2448 (28 September), pp. 80–81

1917 *Report of the 86th Meeting of the British Association for the Advancement of Science* (London: John Murray), pp. 355–363

1917 *Scientific American Supplement*, vol. 83, no. 2161 (2 June), pp. 338–339; no. 2162 (9 June), p. 366

1917 *Proceedings of the Aristotelian Society*, vol. 17, pp. 58–76

1917 *The Organisation of Thought* (London: Williams and Norgate), pp. 105–133

1929 *The Aims of Education* (New York: Macmillan), pp. 153–179

As with 'Space, Time, and Relativity' in 1915, Whitehead first read 'The Organisation of Thought' to Section A ('Mathematical and Physical Science') of the British Association for the Advancement of Science (now known as the British Science Association, or BSA), specifically on 6 September 1916. This year, however, he was President of his section, and his full address was published in the Association's report. Three and a half months later, on 18 December 1916, he delivered it again, to the Aristotelian Society, with no major changes. Listed as taking part in the discussion were the President and Chair, H. Wildon Carr (1857–1931), Ludwik Silberstein (1872–1948), Abraham Wolf (1876–1948), Bertrand Russell (1872–1970), Arthington Worsley (1861–1944), T. Percy Nunn (1870–1944), Frederick William Thomas (1867–1956) and George Dawes Hicks (1862–1941).[1]

In the meantime, however, versions of the address proliferated. It first appeared in print in the Chemical News *on 21 September 1916, just fifteen days after Whitehead first delivered it. On 22 September it appeared in* Science. *On 28 September an abridged version appeared in* Nature. *In June 1917, it appeared in the* Scientific American Supplement *with a new subtitle ('The Basis of Organized Action'), split into two pieces published a week apart. It then appeared in the Aristotelian Society's* Proceedings *and was published in Whitehead's books* The Organisation of Thought *and* The Aims of Education.

Although it appears that the Chemical News *was first to print the address for wide public distribution (by a hair, since* Science *published it a day later), we use the British Association version as the source text, partly because it was the organisation at which Whitehead originally presented. It might therefore be regarded as the 'most original' version of the essay, but this is not completely clear. Interestingly, the British Association version is the only one which uses single quotation marks, while all other versions use double quotation marks.*

1. Aristotelian Society, 'Abstract of the Minutes of the Proceedings of the Aristotelian Society for the Thirty-Eighth Session', *Proceedings of the Aristotelian Society*, vol. 17, p. 481.

A few other things can be said with confidence about the many different versions of this address. One is that, aside from the heavily abridged version in Nature, *there are no large changes from one version to another. Differences are mostly confined to grammar, spelling and the odd word change here and there, mostly for stylistic reasons; AE's deletion of half a sentence was otherwise the largest change. It also seems clear that the Aristotelian Society version and the ones that appeared in Whitehead's books were revised slightly later, owing to the expanded footnote that is common to the three of them. In most cases it is difficult to tell where each publication sourced the text, but it is clear that the* Scientific American Supplement *based its text on the* Chemical News *(and credits it in a footnote) and that, as usual, AE based its text on OT.*

Whitehead notes that 'Science is the organisation of thought', and the address is primarily concerned with an inquiry into the 'nature of the texture of science'. While asserting the equal value and importance of the 'practical' and 'theoretical' sources of science – which are now typically called applied and pure science – Whitehead focuses on the latter here. In particular, a significant portion of the essay is focused on logic as the organising principle of science.

Perhaps of particular interest in this essay is Whitehead's proffered division of Principia *into four main sections, the last of which includes '[t]he whole of mathematics' as 'construction' from 'forms of propositions' and proof 'by reference to' the 'algebraic section of logic'.*

The subject of this address is the organisation of thought, a topic evidently capable of many diverse modes of treatment. I intend more particularly to give some account of that department of logical science with which some of my own studies have been connected. But I am anxious, if I can succeed in so doing, to handle this account so as to exhibit the relation with certain considerations which underlie general scientific activities.

It is no accident that an age of science has developed into an age of organisation. Organised thought is the basis of organised action. Organisation is the adjustment of diverse elements so that their mutual relations may exhibit some predetermined quality. An epic poem is a triumph of organisation, that is to say, it is a triumph in the unlikely event of it being a good epic poem. It is the successful organisation of multitudinous sounds of words, associations of words, pictorial memories of diverse events and feelings ordinarily occurring in life, combined with a special narrative of great events: the whole so disposed as to excite emotions which, as defined by Milton, are simple, sensuous, and passionate. The number of successful epic poems is commensurate, or, rather, is inversely commensurate with the obvious difficulty of the task of organisation.

Science is the organisation of thought. But the example of the epic poem warns us that science is not any organisation of thought. It is an organisation of a certain definite type which we will endeavour to determine.

Science is a river with two sources, the practical source and the theoretical source. The practical source is the desire to direct our actions to achieve predetermined ends. For example, the British nation, fighting for justice, turns to science, which teaches it the importance of compounds of nitrogen. The theoretical source is the desire to understand. Now I am going to emphasise the importance of theory in science. But to avoid misconception I most emphatically state that I do not consider one source as in any sense nobler than the other, or intrinsically more interesting.[1] I cannot see why it is nobler to strive to understand than to busy oneself with the right ordering of one's actions. Both have their bad sides; there are evil ends directing actions, and there are ignoble curiosities of the understanding.

The importance, even in practice, of the theoretical side of science arises from the fact that action must be immediate, and takes place under circumstances which are excessively complicated. If we wait for the necessities of action before we commence to arrange our ideas, in peace we shall have lost our trade, and in war we shall have lost the battle.

Success in practice depends on theorists who, led by other motives of exploration, have been there before, and by some good chance have hit upon |356| the relevant ideas. By a theorist I do not mean a man who is up in the clouds, but a man whose motive for thought is the desire to formulate correctly the rules according to which events occur. A successful theorist should be excessively interested in immediate events, otherwise he is not at all likely to formulate correctly anything about them. Of course, both sources of science exist in all men.

Now, what is this thought organisation which we call science? The first aspect of modern science which struck thoughtful observers was its inductive character. The nature of induction, its importance, and the rules of inductive logic have been considered by a long series of thinkers, especially English thinkers, Bacon, Herschel,[2] J. S. Mill, Venn, Jevons,[3] and others. I am not going to plunge into an analysis of the process of induction. Induction is the machinery and not the product, and it is the product which I want to consider. When we understand the product we shall be in a stronger position to improve the machinery.

First, there is one point which it is necessary to emphasise. There is a tendency in analysing scientific processes to assume a given assemblage of concepts applying to nature, and to imagine that the discovery of laws of nature consists in selecting by means of inductive logic some one out of a definite set of possible alternative relations which may hold between the things in nature answering to these obvious concepts. In a sense this assumption is fairly correct, especially in regard to the earlier stages of science. Mankind found itself in possession of certain concepts respecting nature—for example, the concept of fairly permanent material bodies—and proceeded to determine laws which related the corresponding percepts in nature. But the formulation of laws changed the concepts, sometimes gently by an added precision, sometimes violently. At first this process was not much noticed, or at least was felt to be a process curbed within narrow bounds, not touching fundamental ideas. At the stage where we now are, the formulation of the concepts can be seen to be as important as the formulation of the empirical laws connecting the

1. The *Nature* version deleted the rest of this paragraph.
2. John Frederick William Herschel (1792–1871).
3. William Stanley Jevons (1835–1882).

events in the universe as thus conceived by us. For example, the concepts of life, of heredity, of a material body, of a molecule, of an atom, of an electron, of energy, of space, of time, of quantity, and of number.[1] I am not dogmatising about the best way of getting such ideas straight. Certainly it will only be done by those who have devoted themselves to a special study of the facts in question. Success is never absolute, and progress in the right direction is the result of a slow, gradual process of continual comparison of ideas with facts. The criterion of success is that we should be able to formulate empirical laws, that is, statements of relations, connecting the various parts of the universe as thus conceived, laws with the property that we can interpret the actual events of our lives as being our fragmentary knowledge of this conceived interrelated whole.

But, for the purposes of science, what is the actual world? Has science to wait for the termination of the metaphysical debate till it can determine its own subject-matter? I suggest that science has a much more homely starting-ground. Its task is the discovery of the relations which exist within that flux of perceptions, sensations, and emotions which forms our experience of life. The panorama yielded by sight, sound, taste, smell, touch, and by more inchoate sensible feelings, is the sole field of its[2] activity. It is in this way that science is the thought organisation of experience. The most obvious aspect of this field of actual experience is its disorderly character. It is for each person a *continuum*, fragmentary, and with elements not clearly differentiated. The comparison of the sensible experiences of diverse people brings its own difficulties. I insist on the radically untidy, ill-adjusted character of the fields of actual experience from which science starts. To grasp this fundamental truth is the first step in wisdom, when constructing a philosophy of science. This fact is concealed by the influence of language, moulded by science, which foists on us exact concepts as though they represented the immediate deliverances of experience. The result is that we imagine that we have immediate experience of a world of perfectly defined objects implicated in perfectly defined events which, as known to us by the direct deliverance of our senses, happen at exact instants of time, in a space formed by exact points, without parts and without |357| magnitude: the neat, trim, tidy, exact world which is the goal of scientific thought.

My contention is that this world is a world of ideas, and that its internal relations are relations between abstract concepts, and that the elucidation of the precise connection between this world and the feelings of actual experience is the fundamental question of scientific philosophy. The question which I am inviting you to consider is this: How does exact thought apply to the fragmentary, vague *continua* of experience? I am not saying that it does not apply, quite the contrary. But I want to know how it applies. The solution I am asking for is not a phrase however brilliant, but a solid branch of science, constructed with slow patience, showing in detail how the correspondence is effected.

The first great steps in the organisation of thought were due exclusively to the practical source of scientific activity, without any admixture of theoretical impulse. Their slow accomplishment was the cause and also the effect of the gradual

1. The *Nature* version deleted the rest of this paragraph.
2. *Proceedings of the Aristotelian Society*, OT and AE deleted 'its'.

evolution of moderately rational beings. I mean the formation of the concepts of definite material objects, of the determinate lapse of time, of simultaneity, of recurrence, of definite relative position, and of analogous fundamental ideas, according to which the flux of our experiences is mentally arranged for handy reference: in fact, the whole apparatus of common-sense thought. Consider in your mind some definite chair. The concept of that chair is simply the concept of all the interrelated experiences connected with that chair—namely, of the experiences of the folk who made it, of the folk who sold it, of the folk who have seen it or used it, of the man who is now experiencing a comfortable sense of support, combined with our expectations of an analogous future, terminated finally by a different set of experiences when the chair collapses and becomes fire-wood. The formation of that type of concept was a tremendous job, and zoologists and geologists tell us that it took many tens of millions of years. I can well believe it.

I now emphasise two points. In the first place, science is rooted in what I have just called the whole apparatus of common-sense thought. That is the *datum* from which it starts, and to which it must recur. We may speculate, if it amuses us, of other beings in other planets who have arranged analogous experiences according to an entirely different conceptual code—namely, who have directed their chief attention to different relations between their various experiences. But the task is too complex, too gigantic, to be revised in its main outlines. You may polish up common sense, you may contradict it in detail, you may surprise it. But ultimately your whole task is to satisfy it.

In the second place, neither common sense nor science can proceed with its[1] task of thought organisation without departing in some respect from the strict consideration of what is actual in experience. Think again of the chair. Among the experiences upon which its concept is based, I included our expectations of its future history. I should have gone further and included our imagination of all the possible experiences which in ordinary language we should call perceptions of the chair which might have occurred. This is a difficult question, and I do not see my way through it. But at present in the construction of a theory of space and of time, there seem insuperable difficulties if we refuse to admit ideal experiences.

This imaginative perception of experiences, which, if they occurred, would be coherent with our actual experiences, seems fundamental in our lives. It is neither wholly arbitrary, nor yet fully determined. It is a vague background which is only made in part definite by isolated activities of thought. Consider, for example, our thoughts of the unseen flora of Brazil.

Ideal experiences are closely connected with our imaginative reproduction of the actual experiences of other people, and also with our almost inevitable conception of ourselves as receiving our impressions from an external complex reality beyond ourselves. It may be that an adequate analysis of every source and every type of experience yields demonstrative proof of such a reality and of its nature. Indeed, it is hardly to be doubted that this is the case. The precise elucidation of this question is the problem of metaphysics. One of the points which I am urging in this address is that the basis of science does not depend on the assumption of any of the conclu-

1. The original erroneously had 'their', which only *Nature* corrected to 'its'.

sions of metaphysics; but that |358| both science and metaphysics start from the same given groundwork of immediate experience, and in the main proceed in opposite directions on their diverse tasks.

For example, metaphysics inquires how our perceptions of the chair relate us to some true reality. Science gathers up these perceptions into a determinate class, adds to them ideal perceptions of analogous sort, which under assignable circumstances would be obtained, and this single concept of that set of perceptions is all that science needs; unless indeed you prefer that thought find its origin in some legend of those great twin brethren, the Cock and Bull.[1]

My immediate problem is to inquire into the nature of the texture of science. Science is essentially logical. The nexus between its concepts is a logical nexus, and the grounds for its detailed assertions are logical grounds. King James said, 'No bishops, no king.'[2] With greater confidence we can say, 'No logic, no science.' The reason for the instinctive dislike which most men of science feel towards the recognition of this truth is, I think, the barren failure of logical theory during the past three or four centuries. We may trace this failure back to the worship of authority which in some respects increased in the learned world at the time of the Renaissance. Mankind then changed its authority, and this fact temporally acted as an emancipation. But the main fact, and we can find complaints[3†] of it at the very commencement of the modern movement, was the establishment of a reverential attitude towards any statement made by a classical author. Scholars became commentators on truths too fragile to bear translation. A science which hesitates to forget its founders is lost. To this hesitation I ascribe the barrenness of logic.[4] Another reason for distrust of logical theory and of mathematics is the belief that deductive reasoning can give you nothing new. Your conclusions are contained in your premises, which by hypothesis are known to you.

In the first place this last condemnation of logic neglects the fragmentary, disconnected character of human knowledge. To know one premise on Monday, and another premise on Tuesday, is useless to you on Wednesday. Science is a permanent record of premises, deductions, and conclusions, verified all along the line by its correspondence with facts. Secondly, it is untrue that when we know the premises we also know the conclusions. In arithmetic, for example, mankind are not

1. AE deleted the portion of the sentence after the semicolon, starting with 'unless indeed'.
2. During his reign, King James I was faced with a Puritan demand for the elimination or at least reduction in power of Anglican bishops, but James saw the authority of bishops as key to his royal power; if he ruled by divine right, he must have bishops to bestow it upon him.
3. † *e.g.*, in 1551 by Italian schoolmen. *Cf.* Sarpi's *History of the Council of Trent,* under that date. [The 1551 date likely refers to the second convocation of the Council of Trent, 1551–1552. The second sentence of this note – Whitehead's only footnote in the address – did not appear in the original version or in the *Chemical News*, *Science*, *Nature* or *Scientific American Supplement*, but did appear in *Proceedings of the Aristotelian Society*, OT or AE. Paolo Sarpi (1552–1623) was a Venetian historian. His 1619 *Istoria del Concilio Tridentino* argued that the Catholic Church's doctrinal decrees at Trent were politically motivated rather than divinely directed. It was widely read, particularly by Protestants, so much so that the Vatican later commissioned a three-volume rebuttal of it. Whitehead references the same book in his *Science and the Modern World*, chapter 1.]
4. The *Nature* version deleted most of the rest of the address from this point. It kept the line 'It will be necessary to sketch in broad outline some relevant features of modern logic' (from a few paragraphs down) followed by an ellipsis as a placeholder for all the discussion which it cut out. It then picked up with the three concluding paragraphs.

calculating boys. Any theory which proves that they are conversant with the consequences of their assumptions must be wrong. We can imagine beings who possess such insight. But we are not such creatures. Both these answers are, I think, true and relevant. But they are not satisfactory. They are too much in the nature of bludgeons, too external. We want something more explanatory of the very real difficulty which the question suggests. In fact, the true answer is embedded in the discussion of our main problem of the relation of logic to natural science.

It will be necessary to sketch in broad outline some relevant features of modern logic. In doing so I shall try to avoid the profound general discussions and the minute technical classifications which occupy the main part of traditional logic. It is characteristic of a science in its earlier stages—and logic has become fossilised in such a stage—to be both ambitiously profound in its aims and trivial in its handling of details. We can discern four departments of logical theory. By an analogy which is not so very remote I will call these departments or sections the arithmetic section, the algebraic section, the section of general-function theory, the analytical section. I do not mean that arithmetic arises in the first section, algebra in the second section, and so on; but the names are suggestive of certain qualities of thought in each section which are reminiscent of analogous qualities in arithmetic, in algebra, in the general theory of a mathematical function, and in the[1] analysis of the properties of particular functions.

The first section—namely, the arithmetic stage—deals with the relations of definite propositions to each other, just as arithmetic deals with definite numbers. Consider any definite proposition; call it '*p*.' We conceive that there is always another proposition which is the direct contradictory to '*p*'; call it 'not-*p*.' When we have got two propositions, *p* and *q*, we can form derivative |359| propositions from them, and from their contradictories. We can say, 'At least one of *p* or *q* is true, and perhaps both.' Let us call this proposition '*p* or *q*.' I may mention as an aside that one of the greatest living philosophers has stated that this use of the word 'or'—namely, '*p* or *q*' in the sense that either or both may be true—makes him despair of exact expression. We must brave his wrath, which is unintelligible to me.[2]

We have thus got hold of four new propositions, namely, '*p* or *q*,' and 'not-*p* or *q*,' and '*p* or not-*q*,' and 'not-*p* or not-*q*.' Call these the set of disjunctive derivatives. There are, so far, in all eight propositions, *p*, not-*p*, *q*, not-*q*, and the four disjunctive derivatives. Any pair of these eight propositions can be taken, and substituted for *p* and *q* in the foregoing treatment. Thus each pair yields eight propositions, some of which may have been obtained before. By proceeding in this way we arrive at an unending set of propositions of growing complexity, ultimately derived from the two original propositions *p* or *q*. Of course, only a few are important. Similarly we can start from three propositions, *p, q, r*, or from four propositions, *p, q, r, s*, and so on. Any one of the propositions of these aggregates may be true or false. It has

1. *Proceedings of the Aristotelian Society*, OT and AE added the word 'mathematical': 'in the *mathematical* analysis'.
2. Whitehead seems likely to be referring to F. H. Bradley here, who wrote in his *The Principles of Logic* that 'I confess I should despair of human language, if such distinctions as separate "and" from "or" could be broken down' (F. H. Bradley, *The Principles of Logic, Volume I*, 2nd edition (London: Oxford University Press, 1928), p. 134).

no other alternative. Whichever it is, true or false, call it the 'truth-value' of the proposition.

The first section of logical inquiry is to settle what we know of the truth-values of these propositions, when we know the truth-values of some of them. The inquiry, so far as it is worth while carrying it, is not very abstruse, and the best way of expressing its results is a detail which I will not now consider. This inquiry forms the arithmetic stage.

The next section of logic is the algebraic stage. Now, the difference between arithmetic and algebra is that in arithmetic definite numbers are considered, and in algebra symbols—namely, letters—are introduced which stand for any numbers. The idea of a number is also enlarged. These letters, standing for any numbers, are called sometimes variables and sometimes parameters. Their essential characteristic is that they are undetermined, unless, indeed, the algebraic conditions which they satisfy implicitly determine them. Then they are sometimes called unknowns. An algebraic formula with letters is a blank form. It becomes a determinate arithmetic statement when definite numbers are substituted for the letters. The importance of algebra is a tribute to the study of form. Consider now the following proposition,

The specific heat of mercury is 0·033.

This is a definite proposition which, with certain limitations, is true. But the truth-value of the proposition does not immediately concern us. Instead of mercury put a mere letter which is the name of some undetermined thing: we get,

The specific heat of x is 0·033.

This is not a proposition; it has been called by Russell a propositional function. It is the logical analogy of an algebraic expression. Let us write $f(x)$ for any propositional function.

We could also generalise still further, and say,

The specific heat of x is y.

We thus get another propositional function, $F(x, y)$ of two arguments x and y, and so on for any number of arguments.

Now, consider $f(x)$. There is the range of values of x, for which $f(x)$ is a proposition, true or false. For values of x outside this range, $f(x)$ is not a proposition at all, and is neither true nor false. It may have vague suggestions for us, but it has no unit meaning of definite assertion. For example,

The specific heat of water is 0·033

is a proposition which is false; and

The specific heat of virtue is 0·033

is, I should imagine, not a proposition at all; so that it is neither true nor false, though its component parts raise various associations in our minds. This |360| range of values, for which $f(x)$ has sense, is called the 'type' of the argument x.

But there is also a range of values of x for which $f(x)$ is a true proposition. This is the class of those values of the argument which *satisfy* $f(x)$. This class may have no members, or, in the other extreme, the class may be the whole type of the arguments.

We thus conceive two general propositions respecting the indefinite number of propositions which share in the same logical form, that is, which are values of the same propositional function. One of these propositions is,

$f(x)$ yields a true proposition for each value of x of the proper type;

the other proposition is,

There is a value of x for which $f(x)$ is true.

Given two, or more, propositional functions $f(x)$ and $\varphi(x)$ with the same argument x, we form derivative propositional functions, namely,

$f(x)$ or $\varphi(x)$, $f(x)$ or not-$\varphi(x)$,

and so on with the contradictories, obtaining, as in the arithmetical stage, an unending aggregate of propositional functions. Also each propositional function yields two general propositions. The theory of the interconnection between the truth-values of the general propositions arising from any such aggregate of propositional functions forms a simple and elegant chapter of mathematical logic.

In this algebraic section of logic the theory of types crops up, as we have already noted. It cannot be neglected without the introduction of error. Its theory has to be settled at least by some safe hypothesis, even if it does not go to the philosophic basis of the question. This part of the subject is obscure and difficult, and has not been finally elucidated, though Russell's brilliant work has opened out the subject.

The final impulse to modern logic comes from the independent discovery of the importance of the logical variable by Frege and Peano. Frege went further than Peano, but by an unfortunate symbolism rendered his work so obscure that no one fully recognised his meaning who had not found it out for himself. But the movement has a large history reaching back to Leibniz and even to Aristotle. Among English contributors are De Morgan, Boole, and Sir Alfred Kempe; their work is of the first rank.

The third logical section is the stage of general-function theory. In logical language, we perform in this stage the transition from intension to extension, and investigate the theory of denotation. Take the propositional function $f(x)$. There is the class, or range of values for x, whose members satisfy $f(x)$. But the same range may be the class whose members satisfy another propositional function $\varphi(x)$. It is necessary to investigate how to indicate the class by a way which is indifferent as between the various propositional functions which are satisfied by any member of it, and of it only. What has to be done is to analyse the nature of propositions about a class—namely, those propositions whose truth-values depend on the class itself and not on the particular meaning by which the class is indicated.

Furthermore, there are propositions about alleged individuals indicated by descriptive phrases: for example, propositions about 'the present King of England,' who does exist, and 'the present Emperor of Brazil,' who does not exist. More complicated, but analogous, questions involving propositional functions of two variables involve the notion of 'correlation,' just as functions of one argument involve classes. Similarly functions of three arguments yield three-cornered correlations, and so on. This logical section is one which Russell has made peculiarly his own by work which must always remain fundamental. I have called this the section of functional theory, because its ideas are essential to the construction of logical denoting functions which include as a special case ordinary mathematical functions such as sine, logarithm, &c. In each of these three stages it will be necessary gradually to introduce an appropriate symbolism, if we are to pass on to the fourth stage.

The fourth logical section, the analytic stage, is concerned with the investigation of the properties of special logical constructions, that is, of classes and |361| correlations of special sorts. The whole of mathematics is included here. So the section is a large one. In fact, it is mathematics, neither more nor less. But it includes an analysis of mathematical ideas not hitherto included in the scope of that science, nor, indeed, contemplated at all. The essence of this stage is construction. It is by means of suitable constructions that the great framework of applied mathematics, comprising the theories of number, quantity, time, and space, is elaborated.

It is impossible even in brief outline to explain how mathematics is developed from the concepts of class and correlation, including many-cornered correlations, which are established in the third section. I can only allude to the headings of the process which is fully developed in the work, *Principia Mathematica*,[1] by Mr. Russell and myself. There are in this process of development seven special sorts of correlations which are of peculiar interest. The first sort comprises one-to-many, many-to-one, and one-to-one correlations. The second sort comprises serial relations, that is, correlations by which the members of some field are arranged in a serial order, so that, in the sense defined by the relation, any member of the field is either before or after any other member. The third class comprises inductive relations, that is, correlations on which the theory of mathematical induction depends. The fourth class comprises selective relations, which are required for the general theory of arithmetic operations, and elsewhere. It is in connection with such relations that the famous multiplicative axiom arises for consideration. The fifth class comprises vector relations, from which the theory of quantity arises. The sixth class comprises ratio relations, which interconnect number and quantity. The seventh class comprises three-cornered and four-cornered relations which occur in Geometry.

A bare enumeration of technical names, such as the above, is not very illuminating, though it may help to a comprehension of the demarcations of the subject. Please remember that the names are technical names, meant, no doubt, to be suggestive, but used in strictly defined senses. We have suffered much from critics who consider it sufficient to criticise our procedure on the slender basis of a knowledge of the dictionary meanings of such terms. For example, a one-to-one correlation depends on the notion of a class with only one member, and this notion is defined without appeal to the concept of the number one. The notion of diversity is all that

1. The original had 'Mathematica Principia' (not italicised and in single quote marks). *Proceedings of the Aristotelian Society*, OT and AE changed this to '*Principia Mathematica*' (italicised and without quote marks), which we have adopted.

is wanted. Thus the class a has only one member, if (1) the class of values of x which satisfies the propositional function,

x is not a member of a,

is not the whole type of relevant values of x, and (2) the propositional function,

x and y are members of a, and x is diverse from y,

is false, whatever be the values of x and y in the relevant type.

Analogous procedures are obviously possible for higher finite cardinal members. Thus, step by step, the whole cycle of current mathematical ideas is capable of logical definition. The process is detailed and laborious, and, like all science, knows nothing of a royal road of airy phrases. The essence of the process is, first to construct the notion in terms of the forms of propositions, that is, in terms of the relevant propositional functions, and secondly to prove the fundamental truths which hold about the notion by reference to the results obtained in the algebraic section of logic.

It will be seen that in this process the whole apparatus of special indefinable mathematical concepts, and special *a priori* mathematical premises, respecting number, quantity, and space, has vanished. Mathematics is merely an apparatus for analysing the deductions which can be drawn from any particular premises, supplied by common sense, or by more refined scientific observation, so far as these deductions depend on the forms of the propositions. Propositions of certain forms are continually occurring in thought. Our existing mathematics is the analysis of deductions, which concern those forms and in some way are important, either from practical utility or theoretical interest. Here I am speaking of the science as it in fact exists. A theoretical definition of mathematics must include in its scope any deductions depending on the mere forms |362| of propositions. But, of course, no one would wish to develop that part of mathematics which in no sense is of importance.

This hasty summary of logical ideas suggests some reflections. The question arises, How many forms of propositions are there? The answer is, an unending number. The reason for the supposed sterility of logical science can thus be discerned. Aristotle founded the science by conceiving the idea of the form of a proposition, and by conceiving deduction as taking place in virtue of the forms. But he confined propositions to four forms, now named A, I, E, O. So long as logicians were obsessed by this unfortunate restriction, real progress was impossible. Again, in their theory of form, both Aristotle and subsequent logicians came very near to the theory of the logical variable. But to come very near to a true theory, and to grasp its precise application, are two very different things, as the history of science teaches us. Everything of importance has been said before by somebody who did not discover it.

Again, one reason why logical deductions are not obvious is that logical form is not a subject which ordinarily enters into thought. Common-sense deduction probably moves by blind instinct from concrete proposition to concrete proposition, guided by some habitual association of ideas. Thus common sense fails in the presence of a wealth of material.

A more important question is the relation of induction, based on observation, to deductive logic. There is a tradition of opposition between adherents of induction and of deduction. In my view, it would be just as sensible for the two ends of a worm to quarrel. Both observation and deduction are necessary for any knowledge worth having. We cannot get at an inductive law without having recourse to a propositional function. For example, take the statement of observed fact,

This body is mercury, and its specific heat is 0·033.

The propositional function is formed,

Either *x* is not mercury, or its specific heat is 0·033.

The inductive law is the assumption of the truth of the general proposition, that the above propositional function is true for every value of *x* in the relevant type.

But it is objected that this process and its consequences are so simple that an elaborate science is out of place. In the same way, a British sailor knows the salt sea when he sails over it. What, then, is the use of an elaborate chemical analysis of sea-water? There is the general answer, that you cannot know too much of methods which you always employ; and there is the special answer, that logical forms and logical implications are not so very simple, and that the whole of mathematics is evidence to this effect.

One great use of the study of logical method is not in the region of elaborate deduction, but to guide us in the study of the formation of the main concepts of science. Consider Geometry, for example. What are the points which compose space? Euclid tells us that they are without parts and without magnitude. But how is the notion of a point derived from the sense-perceptions from which science starts? Certainly points are not direct deliverances of the senses. Here and there we may see or unpleasantly feel something suggestive of a point. But this is a rare phenomenon, and certainly does not warrant the conception of space as composed of points. Our knowledge of space properties is not based on any observations of relations between points. It arises from experience of relations between bodies. Now a fundamental space relation between bodies is that one body may be part of another. We are tempted to define the 'whole and part' relation by saying that the points occupied by the part are some of the points occupied by the whole. But 'whole and part' being more fundamental than the notion of 'point,' this definition is really circular and vicious.

We accordingly ask whether any other definition of 'spatial whole and part' can be given. I think that it can be done in this way, though, if I be mistaken, it is unessential to my general argument. We have come to the conclusion that an extended body is nothing else than the class of perceptions of it by all its percipients, actual or ideal. Of course, it is not any class of perceptions, but a certain definite sort of class which I have not defined here, except by |363| the vicious method of saying that they are perceptions of a body. Now, the perceptions of a part of a body are among the perceptions which compose the whole body. Thus two bodies *a* and *b* are both classes of perceptions; and *b* is part of *a* when the class which is *b* is contained in the class which is *a*. It immediately follows from the logical form of this definition that if *b* is part of *a*, and *c* is part of *b*, then *c* is part of *a*. Thus the relation 'whole to part' is transitive. Again, it will be convenient to allow that a body is part of itself. This is a mere question of how you draw the definition. With this understanding, the relation is reflexive. Finally, if *a* is part of *b*, and *b* is part of *a*, then *a* and *b* must be identical. These properties of 'whole and part' are not fresh assumptions, they follow from the logical form of our definition.

One assumption has to be made if we assume the ideal infinite divisibility of space. Namely, we assume that every class of perceptions which is an extended body contains other classes of perceptions which are extended bodies diverse from itself. This assumption makes rather a large draft on the theory of ideal perceptions. Geometry vanishes unless in some form you make it. The assumption is not peculiar to my exposition.

It is then possible to define what we mean by a point. A point is the class of extended objects which, in ordinary language, contain that point. The definition, without presupposing the idea of a point, is rather elaborate, and I have not now time for its statement.

The advantage of introducing points into Geometry is the simplicity of the logical expression of their mutual relations. For science, simplicity of definition is of slight importance, but simplicity of mutual relations is essential. Another example of this law is the way physicists and chemists have dissolved the simple idea of an extended body, say of a chair, which a child understands, into a bewildering notion of a complex dance of molecules and atoms and electrons and waves of light. They have thereby gained notions with simpler logical relations.

Space as thus conceived is the exact formulation of the properties of the apparent space of the common-sense world of experience. It is not necessarily the best mode of conceiving the space of the physicist. The one essential requisite is that the correspondence between the common-sense world in its space and the physicists' world in its space should be definite and reciprocal.[1]

I will now break off the exposition of the function of logic in connection with the science of natural phenomena. I have endeavoured to exhibit it as the organising principle, analysing the derivation of the concepts from the immediate phenomena, examining the structure of the general propositions which are the assumed laws of nature, establishing their relations to each other in respect to reciprocal implications, deducing the phenomena we may expect under given circumstances.

Logic, properly used, does not shackle thought. It gives freedom and, above all, boldness. Illogical thought hesitates to draw conclusions, because it never knows either what it means, or what it assumes, or how far it trusts its own assumptions, or what will be the effect of any modification of assumptions. Also the mind untrained in that part of constructive logic which is relevant to the subject in hand will be ignorant of the sort of conclusions which follow from various sorts of assumptions, and will be correspondingly dull in divining the inductive laws. The fundamental training in this relevant logic is, undoubtedly, to ponder with an active mind over the known facts of the case, directly observed. But where elaborate de-

1. This marks the end of the long section that the *Nature* version had abridged. It printed the final three paragraphs in full.

ductions are possible, this mental activity requires for its full exercise the direct study of the abstract logical relations. This is applied mathematics.

Neither logic without observation, nor observation without logic, can move one step in the formation of science. We may conceive humanity as engaged in an internecine conflict between youth and age. Youth is not defined by years, but by the creative impulse to make something. The aged are those who, before all things, desire not to make a mistake. Logic is the olive branch from the old to the young, the wand which in the hands of youth has the magic property of creating science.

Textual Notes

These notes identify (by page and line number of the present text) minor changes between different versions of a text. If only one alternative version of a text exists, the notes will not identify it. Otherwise, the most common abbreviations are to Whitehead's three books of collected essays:

OT = *The Organisation of Thought Educational and Scientific* (1917)
AE = *The Aims of Education & Other Essays* (1929)
ESP = *Essays in Science and Philosophy* (1947)

Abbreviations for other alternative versions of essays/articles appear at the top of each piece's respective notes.

Sets of Operations in Relation to Groups of Finite Order

90:15 type] *deleted struck* dealt with
90:15 of algebra] *deleted struck* with
90:21 either as a] *deleted struck false start*
90:23 entity.] *deleted struck* Similarly
90:24 on] *deleted struck repeat word*
90:24 some] *deleted struck* sp
90:26 in] *deleted struck* qua
90:27 distinct] *deleted struck* of
90:27 these objects] *deleted struck* symbols
90:28 third] *deleted struck* symbol
90:31 Groups.] *deleted struck* But
90:33 called] *deleted struck* ob
91:1 *N*] *deleted struck* sym
91:1 operations,] *deleted struck* is called
91:2 special] *deleted struck* ⟨?⟩ it is called a group
91:3 whole] *deleted struck false start*
91:4 other] *deleted struck* sums are its su
91:7 subject.] *altered from* subjects
91:8 the] *deleted struck false start*
91:9 slight.] *deleted struck* T
91:9 the] *deleted struck indecipherable word*
91:9 no] *deleted struck* interest
91:13 an] *deleted struck* independent
91:13 algebra] *deleted struck comma*
91:13 any] *deleted struck false start*
91:25 *H*] *deleted struck comma*
91:28 other] *moved* to each other *from end of sentence as indicated by author's arrow*
91:31 where] *deleted struck* such
91:33 |7|] *deleted struck* |4|
91:34 them] *deleted struck indecipherable word*
91:34 called,] *deleted struck indecipherable word*
91:38 (ii)] *deleted struck* (2)
91:39 (iii)] *deleted struck* (3)
91:39 is] *deleted struck* always
91:40 and] *deleted struck comma*
92:1 |8|] *deleted struck* |5|
92:1 (iv)] *deleted struck* (4)
92:1 property] *altered from* properties
92:6 (v)] *deleted struck* (5)
92:6 corresponds] *deleted struck* another
92:9 Hence] *deleted struck indecipherable word*
92:12 |9|] *deleted struck* |6|
92:15 imply] implies *had been written over* imply *in pencil, but original is correct*
92:20 $S_p = S_0$.] *deleted struck* These are deduced by multiplying by S_p and using postulate (4) (iv). Thus if S_q and S_r are distinct, then S_pS_q and S_pS_r are distinct operations.
92:21 powers,] *The remainder of the* S *characters on manuscript pages 9 and 10 had subscripts that were scratched out, save that for a few a subscript* 0 *was retained*
92:26 |10|] *deleted struck* |7|
92:29 that] *deleted struck* S_q
92:35 hypothesis,] *deleted struck* . Then ⟨?⟩
93:1 |11|] *deleted struck* |8|
93:3 logical] *deleted struck repeat word*
93:3 + S_r] *deleted struck* by the
93:13 |12|] *deleted struck* |9|
93:16 the] *deleted struck* operations
93:22 nomenclature.] *deleted struck* Multiplication is defined to be distributive in regard to addi *from reverse side of page*
93:23 |13|] *deleted struck* |10|
93:27 then] *deleted struck* S
93:32 not] *altered from* note
94:5 meaning] *deleted struck indecipherable word (possibly* unless*)*
94:5–6 operation] *deleted struck* it will be ⟨?⟩
94:11 |15|] *deleted struck* |12|
94:11 ***Multiplicity.***] *Period supplied for consistency with other headings*
94:12 of] *deleted struck* oper
94:15 Again] *deleted struck indecipherable word (possibly* the*)*
94:16 which] *deleted struck* comprises
94:18 be any] *deleted struck* operation

94:22 |16|] *deleted struck* |13|
94:23 But] *deleted struck* if
94:23 may be] *deleted struck* to
94:35 |17|] *deleted struck* |14|
94:37 then] *deleted struck* if
95:1 have] *deleted struck* fu
95:1 the] *deleted struck indecipherable word(s)*
95:2 .] *Period supplied*
95:5 hold] *deleted struck* in the ca
95:5 In the] *deleted struck false start*
95:5 two] *deleted struck* sets of
95:9 |18|] *deleted struck* |15|
95:10 contains every] *deleted struck* symbol and
95:10 must] *deleted struck* t
95:19 |19|] *deleted struck* |16|
95:20 The] *altered from* Then
95:29 taken] *deleted struck* sym
95:29 simultaneously.] *deleted struck* But if $H+K=H$, then $H+K=H+K'$ implies that $K=K'$
95:32 Similarly] *deleted struck* S
96:1 |20|] *deleted struck* |17|
96:8 we] *deleted struck indecipherable word immediately prior*
96:9 be] *deleted struck* any
96:18 zero,] *deleted struck indecipherable word*
96:21 form] *deleted struck* K
96:23 |22|] *deleted struck* |18|
96:23 unequal] *deleted struck* symbols
96:27 2,] *deleted struck indecipherable fraction*
97:1 |23|] *deleted struck* |19|
97:2 when] *deleted struck indecipherable word*
97:2 the] *deleted struck* proved
97:5 Then] *deleted struck indecipherable word(s)*
97:10 *g*,] *deleted struck* it follows
97:11 be] *deleted struck* true
97:11 satisfied.] *deleted struck* If a group contains S it contains S^{-1}. For multiply the equation $Sg=g$ by S^{-1}. Then we obtain $g=S^{-1}g$.
97:12 |24|] *deleted struck* |20|
97:32 necessarily] *deleted struck* in
98:4 |26|] *deleted struck* |22|
98:4 *Sequence*] *altered from* Sequences
98:4 *Set*.] *deleted struck* Let the series of m consecuti
98:5 Let any] *deleted struck* set
98:6 a power] *deleted struck* series
98:6 *q*] *deleted struck* members
98:8 H^p,] *deleted struck indeciperhable letter*
98:11 *H*,] *deleted struck indeciperhable letter*
98:14 power] *deleted struck* series
98:15 |27|] *deleted struck* |23|
98:23 if] *deleted struck* the multiplicities
98:27 |28|] *deleted struck* |24|
98:27 *H*] *deleted struck* ⟨?⟩ contains all succeeding powers of H of the
98:27 the] *deleted struck* seque
98:32 Hence] *deleted struck* g + H
99:1 |29|] *deleted struck* |25|
99:1 §8.] *altered from* §9. *Whitehead accidentally skipped §8. The following sections have all been renumbered accordingly.*
99:4 first] *deleted struck indecipherable word*
99:4 which] *deleted struck* repr
99:4 former] *deleted struck* member
99:7 H^n.] *deleted struck* .....(1) *and* Now multiply this equation by H, H^2.... Then

$$\left.\begin{aligned} H^{m+n+1} &= H^{n+1} \\ H^{m+n+2} &= H^{n+2} \\ H^{m+n+r} &= H^{n+r} \end{aligned}\right\} \text{.....(1)}$$

99:8 |30|] *deleted struck* |26|
99:8 multiply] *altered from* multiplying by
99:16 Call] *deleted struck* H
99:17 |31|] *deleted struck* |27|
99:18 part] *altered from* parts
99:18 of the] *deleted struck repeated words*
99:19 *H*.] *deleted struck* Let the $(m+n-1)$ distinct terms of the
99:22 two] *deleted struck* ⟨?⟩ equal
99:24 non-recurring] *deleted struck* sel collection
99:24 part.] *deleted struck* It is obvious that the multiplicities of all the recurring digits are equal. Let *p* be V invariably used to denote V this multiplicity. Then $p \geq r$, where *r* is, as usual, the multiplicity of *H*.
99:24 |32|] *deleted struck indecipherable word*
99:26 period] *altered from* periodic
99:33 Accordingly] *deleted struck false start immediately prior*
99:38 |33|] *deleted struck* |28|
99:40 For] *deleted struck repeated word*
99:40 this] *deleted struck indecipherable word*
99:40 be] *deleted struck* written
100:5 [*H*]] *deleted struck* for
100:11 Also] *deleted struck false start*
100:11 the] *deleted struck* pow
100:13 |34|] *deleted struck* |29|
100:13 group] *deleted struck indecipherable word*
100:14 to] *deleted struck* the
100:14 another] *deleted struck* succeeding
100:20 Hence,] *deleted struck indecipherable word*
100:20 of the product] *deleted struck indicipherable character*
100:21 into] *deleted struck indecipherable words*
100:23 |35|] *deleted struck* |30|
100:28 H^p]. *deleted struck* .....(2)
100:32 2*p* =] *deleted struck indicipherable character*
101:4 *m*,] *altered from* n
101:5 $(2s+1)$.] *deleted struck indecipherable symbol(s)*
101:11 (I)] *The parentheses had been struck; we restored them for consistency*
101:11 least] *deleted struck indcipherable word*
101:14 Then] *deleted struck* p =
101:16 |37|] *deleted struck* |32|
101:18 $tm \geq n$,] *deleted struck indecipherable character*
101:22 can] *deleted struck* will

101:26 [*H*].] *deleted struck* Corollary. It follows that the only ⟨?⟩

101:26 |38|] *deleted struck* |33|

101:41 |39|] *deleted struck* |34|

102:3 represents] *deleted struck indecipherable word*

102:5 there] *deleted struck indecipherable word*

102:12 |40|] *deleted struck* |35|

102:12 ***Period.***] *deleted struck* Proposition The Period

102:16 + *S*.] *deleted struck* Hence $H^{tm}=(H+S)^{tm}=H^{tm}+...+H^{tm-1}S$. Hence *H*

102:27 Hence] *deleted struck* p

102:28 |41|] *deleted struck* |36|

102:28 *Corollary*] *deleted struck false starts*

102:29 any] *deleted struck indecipherable word, possibly* sub

102:34 |42|] *deleted struck* |37|

102:34 operation] *deleted struck* occurring in ⟨?⟩

102:34 digit] *deleted struck* S

102:34 as] *deleted struck* the

102:36 $p\alpha$,] *altered from* α

103:1 $=s'm$] *deleted struck indecipherable symbol*

103:4 orders] *deleted struck* prime

103:4 |43|] *deleted struck* |38|

103:5 contained,] *deleted struck* in

103:8 *Corollary* III.] *deleted struck* If m be greater than unity

103:11 |44|] *deleted struck* |39| *and* If two digits of H of the forms

103:11 as a] *deleted struck* set

103:12 $(\alpha-\beta)$.] altered from $(\alpha\sim\beta)$.

103:13 H^{β}] *deleted struck* contain in com

103:25 $H^{tm}=$] *deleted struck* $H^{tm}+...+$

103:27 |45|] *deleted struck* |40|

103:34 *Corollary* I.] *period supplied*

103:34 two] *deleted struck* terms

103:34 of] *deleted struck indecipherable word*

103:36 *Corollary* II.] *period supplied*

104:2 of the] *deleted struck* factor

104:5 |47|] *deleted struck* |42|

104:6 of] *deleted struck* ea

104:8 by] *deleted struck* th

104:9 *mp*.] *deleted struck* Corollary. Since *p* is greater than not less than *r*, the multiplicity of *H*, we have it follows that the order of [*H*] is not less than *mr*.

104:14 For] *deleted struck indecipherable word*

104:15 = *p*] *deleted struck indecipherable symbol*

104:31 then the] *deleted struck* there is one

104:31 the recurring digit] *deleted struck false start*

104:31 $H^{(s+t)m-\alpha}$] *deleted struck* which

104:35 such] *deleted struck* digits

104:35 inverse] *deleted struck* pairs of

104:40 If a] *deleted struck* dig

105:3 H^{tm}] *deleted struck comma*

105:5 For] *deleted struck* it

105:5 2.] *deleted struck* A Now if *p* be the period

105:6 order,] *deleted struck* have an odd period

105:6 in] *deleted struck* such a

105:8 |51|] *deleted struck* |43|

105:9 Each] *deleted struck* The digits

105:9 digit] *altered from* digits

105:9 is a] *deleted struck* are

105:9 set] *altered from* sets

105:10 any] *deleted struck* re

105:11 digit,] *deleted struck false start*

105:14 |52|] *deleted struck* |44|

105:16 *K*] *deleted struck* K$^{(?)}$ *immediately prior*

105:26 $n'\leq n$,] *deleted struck* m′

105:33 |53|] *deleted struck* |45|

105:33 ***Groups.***] *period supplied*

105:37 of] *deleted struck* the

106:11 |54|] *deleted struck* |55|

106:11 α] *deleted struck indecipherable word*

106:11 m_{α}] *deleted struck* can

106:12 every] *deleted struck* divo

106:14 that] *deleted struck* every

106:18 μ.] *deleted struck* and therefore to μ

106:23 |55|] *deleted struck* |47|

106:26 $p_2 y$.] *altered from* $y\beta$

106:28 integers.] *period supplied*

106:28 two] *deleted struck* power sequ

106:28 *H*] *deleted struck indecipherable word*

106:30 $[H^{\alpha}]=[H^{\beta}]$.] *deleted struck* Thus the ⟨?⟩ number

106:32 as] *deleted struck* the first

106:37 |56|] *deleted struck* |48|

106:41 digital] *deleted struck* sub

107:1 |57|] *deleted struck* |49|

107:2 When] *deleted struck* If *at beginning of line*

107:4 by] *deleted struck* H

107:6 called] *deleted struck* unilatera

107:8 the] *deleted struck* equation

107:10 *H*.] *deleted struck indecipherable superscript*

107:14 |58|] *deleted struck* |50|

107:26 |59|] *deleted struck* |51|

107:27 S_p and S_q] *deleted struck indecipherable superscript*

107:29 its] *deleted struck* right

107:30 its] *deleted struck* left

107:31 H^1, H^2,] *altered from* H_1, H_2,

107:36 so on.] *deleted struck* But since every group is absorbed by a

107:37 |60|] *deleted struck* |52|

107:37 *Corollary* I.] *period supplied*

107:39 *Corollary* II.] *period supplied*

107:42 |61|] *deleted struck* |53|

108:1 every] *deleted struck* group

108:1 group] *deleted struck* is absorbed

108:2 to] *deleted struck* each ⟨?⟩

108:6 satisfied] *deleted struck* by

108:9 |62|] *deleted struck* |54|

108:9 *Corollary* I.] *period supplied*

108:10 This] *deleted struck* gr

108:10 the] *deleted struck* corollary

108:10 is] *deleted struck* Corollary II If *H* contains g_1 or g_1', the period is unity. For them H^{tm} con

108:12 |63|] *deleted struck* |55|
108:12 its] *deleted struck* contained
108:23 making] *altered from* taking
108:23 enough,] *deleted struck* this
108:24 digits.] *deleted struck* But
108:26 digits] *deleted struck* f
108:29 absorbed groups] *removed extraneous dots*
108:31 |64|] *deleted struck* |56|
108:33 are] *deleted struck* absorbe
108:33 absorbed] *deleted struck false start*
108:36 expressed] *deleted struck* thus in this a
108:37 *k* be any] *deleted struck* self
108:38 Let] *deleted struck* pq
108:38 is] *deleted struck* known tha
109:29 equation] *deleted struck* set expression
109:31 kS_2] *deleted struck* which can be written
109:37 must] *deleted struck indeciperable character*
110:3 In] *deleted struck* The Now *at beginning of line*
110:7 as] *deleted struck indecipherable character*
110:7 a] *deleted struck* set of *q* fundamental operations is
110:14 *g/k*] *this and other instances altered from* $\frac{g}{k}$ *for ease of reading*
110:15 (kS_{q-1}).] *deleted struck* Now every theorem which is true of
110:21 *g*.] *deleted struck* Thus if k_1
110:34 $[H]/H^{tm}$] *this and other instances altered from* $\frac{[H]}{H^{tm}}$ *for ease of reading*
111:1 |71|] *deleted struck* |57| *and* |65|
111:1 ***§16***] *deleted struck repeated number*
111:6 characteristic] *deleted struck indecipherable word*
111:10 recurring.] *period supplied*
111:13 Hence] *deleted struck* the chara
111:15 |72|] *deleted struck* |58| *and* |66|
111:18 one] *deleted struck* g
111:27 power] *deleted struck* series
111:28 digit] *altered from* digits
111:29 |73|] *deleted struck* |59| *and* |67|
111:29 exist] *deleted struck indecipherable word*
111:37 *p*–1] *altered from* p
111:38 |74|] *deleted struck* |60| *and* |68|
112:2 *r*, the] *deleted struck* order
112:4 |75|] *deleted struck indecipherable characters*
112:11 Also] *deleted struck false start*
112:15 |76|] *deleted struck* |62| *and* |70|
112:15 (vi)] *deleted struck false start*
112:15 taking] *deleted struck* Let
112:23 H^{p-2}.] *deleted struck* Also since *p* is even, H^{p-2} contains
112:24 any] *deleted struck* digit
112:24 *H*] *deleted struck* which is an even power
112:28 *p*–2] *altered from* p
112:30 H^{p-2},] *deleted struck* and also it is evident th
112:32 |77|] *deleted struck* |63| *and* |71|
112:35 odd power.] *deleted struck* All these properties are sufficiently of S+S(?) one suffi
112:36 |78|] *deleted struck* |64| *and* |72|
112:37–38 addition,] *deleted struck* that
113:4 here the] *deleted struck* likeness
113:4 For the] *deleted struck* indi
113:5 in the] *deleted struck* al
113:6 so.] *deleted struck* Both algebras are linear in the sense that the product of two things produces another thing of the same general type.
113:7 |79|] *deleted struck* |73|
113:7 Logic] *deleted struck indecipherable word*
113:10 is in] *deleted struck* every
113:11 representing] *deleted struck* in
113:13 *i*.] *deleted struck* ix = x
113:14 *i*.] *deleted struck indecipherable word*
113:21 |80|] *deleted struck* |74|
113:21 The] *deleted struck indecipherable character(s)*
113:21 inverse] *deleted struck* S^{-1}
113:25 properties] *altered from* property
113:29 |81|] *deleted struck* |75|
113:29 The] *deleted struck indecipherable character*
113:31 of a] *deleted struck* cl
113:32 containing] *deleted struck* numb
113:34 Algebra of] *deleted struck* the Theory
113:34 Groups of] *deleted struck* indecipherable character
113:34 number] *deleted struck* does not
113:35 with a] *deleted struck* thoug
113:38 individuals] *deleted struck* are ⟨?⟩
113:39 |82|] *deleted struck* |76|
113:39 set.] *deleted struck* Since every individual is conceived as
113:40 sets of] *deleted struck repeat word*
114:4 no] *deleted struck* to
114:4 thought] *deleted struck* to recognize that one individual is not another individual.

Non-Euclidean Geometry

Change notes refer to Russell's 1902 version unless otherwise specified

283:36 *History*.] *added* History
283:36 Savile] *deleted comma*
284:3 Moderns,"] *moved comma from outside quotation mark*
284:6 geometry] *altered from* Geometry
284:19 Christoph Clavius] *added* Christoph
284:22 Girolamo Saccheri] *added* Girolamo
285:4 geometry] *altered from* Geometry
285:6 from] *altered from* to
285:11 geometry] *altered from* Geometry
285:15 published] *deleted comma*
285:16 geometry] *altered from* Geometry
285:20 Bernhard Riemann,] *added* Bernhard
285:24 geometry] *altered from* Geometry
285:24 Karl Friedrich Gauss] *added* Karl Friedrich
285:28 geometry] *altered from* Geometry
285:28 Felix Klein] *added* Felix
286:15 Arthur Cayley] *added* Arthur
286:17 imaginary] *deleted comma*

286:20 Sophus Lie] *added* Sophus
286:26 geometry] *altered from* Geometry
286:29 geometry] *altered from* Geometry
286:30 geometry] *altered from* Geometry
287:4 geometry] *altered from* Geometry
287:8 geometry] *altered from* Geometry
287:8 Nicholas Lobatchewsky] *added* Nicholas
287:8–9 professor] *altered from* Professor
287:9 mathematics] *altered from* Mathematics
287:16 geometry] *altered from* Geometry
288:11 geometry] *altered from* Geometry
290:6 origin)] *deleted comma*
291:3 Hermann von Helmholtz,] *added* Hermann von
291:6 (see VII. Axioms of Geometry).] *altered from* (v. infra)
291:10 Eugenio Beltrami] *added* Eugenio
291:11 connexion] *altered from* connection
291:13 geometry] *altered from* Geometry
291:14 geometry] *altered from* Geometry
292:7 R^2w^{-4}] *altered from* $\frac{R^2}{w^{-4}}$ (Russell was in error)
293:2 connexion] *altered from* connection
293:2 limit-lines] *altered from* limit lines
293:14 imaginary] *deleted comma*
293:22 R^2x^{-2}] *altered from* $\frac{R^2}{x^{-2}}$ (Russell was in error)
293:27 identical] ESP misspelled as indentical
294:16 &c.] *altered from* etc.
294:26 geometry] *altered from* Geometry
294:28 geometry] *altered from* Geometry
295:6 Laguerre] *reference added*
295:12 The projective] *altered from* This
296:15 geometry] *altered from* Geometry
297:5 &c.] *altered from* etc.
297:11 geometry] *altered from* Geometry
297:17 geometry] *altered from* Geometry

Axioms of Geometry

301:5 Euclidean] *altered to* Euclidian *throughout; further instances are not noted*
304:11 lines] *altered to* line
310:15 (points),] *altered comma to semicolon*
313:13 superpostion] *altered to* super-position
314:24 z^2)] *deleted the closing parenthesis*
315:13 congruence-group,] *deleted comma*
315:14 (β) A_1', A_2'] *added comma*
317:17 *Critical,*] *deleted comma*

Mathematics

328:32 premisses] *altered to* premises
330:38 NUMBER] *deleted* NUMBER

The Place of Elementary Mathematics in a Liberal Education

344:22 in] *altered to* of
346:38–39 algebraical] *altered to* algebraic
347:30 as the] *deleted* the
349:14 reasoning.] *altered to* reason

The Principles of Mathematics in Relation to Elementary Teaching

EM = L'Enseignement Mathématique
353:6 sections;] *OT altered semicolon to colon*
353:7 education,] *OT altered comma to semicolon*
353:14 Furthermore] *OT added comma*
353:15 university] *OT altered to* University
353:17 Mathematical Principles] *OT altered to* mathematical principles
353:19 enquiry] *OT altered to* inquiry
353:24 well designed] *OT altered to* well-designed
353:32 There] *EM altered to* Now there
353:32 only one way] *OT deleted* one
354:4–5 geometry, ratio, quantity, and number] *EM altered to* Geometry, Ratio, Quantity, and Number
354:5 Hence] *OT added comma*
354:7 Fundamental Mathematical] *OT altered to* fundamental mathematical
354:7–8 Geometry, Quantity and Number] *OT altered to* geometry, quantity and number
354:8 Number?] *EM removed paragraph break*
354:10 Principles of Mathematics and a Mathematical Education] *OT altered* to principles of pathematics and a mathematical education
354:11 Fundamental Mathematical] *OT altered to* fundamental mathematical
354:11–12 is that, in any absolute sense, there] *OT altered to* is, that in any absolute sense there
354:15 presuppositions] *OT altered to* pre-suppositions
354:16 Fundamental] *OT altered to* fundamental
354:17 Geometry, Quantity and Number] *OT altered to* geometry, quantity and number
354:18 relations.] *EM removed paragraph break*
354:24 Geometry] *OT altered to* geometry
354:28 anyone] *OT altered to* any one
354:30 Geometry] *OT altered to* geometry
354:32 Geometry] *OT altered to* geometry
354:33 means,] *EM deleted comma*
354:33 Elementary Algebra] *OT altered to* elementary algebra
354:34 'any number,'] *OT and EM altered to* "any number,"
354:37 devise,] *OT altered comma to semicolon*
354:37 lines,] *OT altered comma to semicolon*
354:37–38 quadratic functions] *OT added comma*
354:38 parabolas,] *OT altered comma to semicolon*
354:38–39 implicit functions] *OT added semicolon*
354:42 Geometry] *OT altered to* geometry
354:43 Algebra] *OT altered to* algebra
355:1 use?] *EM altered question mark to period*
355:12 are] *EM altered to* is
355:16 Geometry and Algebra] *OT altered to* geometry and algebra
355:24 *rôle*] *OT removed italics*

355:24–25 Logical Precision in the Teaching of Mathematics] *OT altered to* logical precision in the teaching of mathematics
355:26 obvious,] *OT altered comma to colon*
355:26 Logical Precision] *OT altered to* logical precision
355:29 that,] *OT and EM deleted comma*
355:31 problem of education] *OT deleted* of education
355:36 recognize] *OT altered to* recognise
355:44 namely] *OT added comma*
356:8 yet however fully] *OT altered to* yet, however, fully
356:9 realization] *OT altered to* realisation
356:13 grasp,] *EM deleted comma*
356:24 My opinion] *EM altered to* My own opinion
356:24 is] *OT added comma*
356:25 is roughly speaking what] *OT altered to* is, roughly speaking, what
356:27 textbook] *OT altered to* text-book
356:28 advisable.] *EM added paragraph break*
356:29 is] *OT added comma*
356:31 doubt however that] *OT altered to* doubt, however, that
356:32 Euclidean Geometry] *OT altered to* euclidean geometry
356:33 stress, in the initial stages, on] *OT altered to* stress in the initial stages on
356:35 Euclidean] *OT altered to* euclidean
356:37–38 important however that] *OT altered to* important, however, that
356:42 In fact my] *OT altered to* My
356:45 is] *OT added comma*
357:1 generalization] *OT altered to* generalisation
357:2 generalization] *OT altered to* generalisation
357:4 generalize] *OT altered to* generalise
357:5 generalized] *OT altered to* generalised
357:7 generalized] *OT altered to* generalised
357:9 civilizing] *OT altered to* civilising
357:22 pocket-books] *OT altered to* pocketbooks
357:27–28 Differential Calculus] *OT altered to* differential calculus

Presidential Address to the London Branch of the Mathematical Association

358:30 medieval] *OT and AE altered to* mediæval
359:8 with ideas] *OT and AE altered to* with those ideas
359:8 them capacities] *OT and AE altered to* them those capacities
359:23 us] *AE deleted* us
360:11 arrive is] *OT and AE added comma*
361:19 this:—] *OT and AE deleted em dash*
361:40 and,] *OT and AE deleted comma*
362:18 rôle] *OT and AE altered to* role
363:22 and] *OT altered to* and and
363:33 understand] *OT and AE added comma*
363:36 Physical Laws] *OT and AE altered to* physical laws
364:12–13 Laws of Nature] *OT and AE altered to* laws of nature
364:16 anyone] *OT and AE altered to* any one
364:33–34 definitions and the propositions assumed without proof illustrate] *OT and AE altered to* definitions, and the propositions assumed without proof, illustrate
365:43 cosine of] *AE altered to* cosine and
366:2 motion] *OT and AE added* comma
366:15 90] *OT and AE altered to* ninety
366:18 Graphs] *OT and AE altered to* graphs
366:27 geometrical conics] *AE altered to* "geometrical conics"
366:36 Methods] *OT and AE altered to* methods
367:5 is:—] *OT and AE deleted em dash*
367:25 instruction,] *OT and AE deleted comma*

Space, Time, and Relativity

370:35 attention,] *AE deleted comma*
371:5 cross-fertilizing] *OT and AE altered to* cross-fertilising
371:8 knowledge] *OT and AE deleted comma*
372:29 extensions] *AE altered to* extension
373:13 one] *OT and AE deleted* one
373:18 enquire] *OT and AE altered* to inquire
373:24–25 beings A and B agree] *OT and AE altered to* beings, A and B, agree
373:29 co-ordinates] *OT and AE altered to* coordinates
373:31 co-ordinates] *OT and AE altered to* coordinates
373:35 co-ordinates] *OT and AE altered to* coordinates
373:39 co-ordinates] *OT and AE altered to* coordinates
374:3 OU_1] *AE altered to* OU,
374:12 one and only one set] *AE altered to* one, and only one, set
374:17 difficulty] *OT and AE added comma*
374:23 changes.] *OT deleted period*
374:25 way] *AE altered to* day
375:2 co-ordinates] *OT altered to* coordinates
375:21 judgments] *OT altered to* judgment
375:22 co-ordinate] *OT altered to* coordinate
375:27 A's co-ordinates] *OT altered to* A's coordinates
375:27 B's co-ordinates] *OT altered to* B's coordinates
375:30 *v*. Also] *OT altered to* v, and
375:38–39 co-ordinates] *OT altered to* coordinates
376:4 follows by] *OT altered to follows* after
376:5 that—] *OT deleted dash*
376:34 B's] OT altered to B's′
377:13 is] *OT added comma*
377:14 V,] *OT deleted comma*
377:17 anyone] *OT altered to* any one
377:21 satisfied:] *OT deleted colon*
377:26 condition is,] *OT added comma*
378:4 namely,] *OT deleted comma*

378:14 time judgments] *OT altered to* time-judgments
379:3 well attested] *OT altered to* well-attested
379:6 velocities] *OT deleted comma*
380:9 suspect—No] *OT and AE altered to suspect*— No
381:11 considered] *AE altered to* mentioned
381:15 possible so] *AE deleted* so
382:12–13 illusion:] *OT and AE altered colon to semicolon*
382:34 make:—] *OT and AE deleted the colon*
382:41 is] *OT and AE added comma*
383:7 p. 371] *OT altered* pp. 105 and 106 *to* pp. 193 and 194
383:7 follows:—] *OT deleted colon*
383:10 experiential] *OT altered to* experiental
383:16 p. 372] *OT altered* pp. 105–6 *to* pp. 193 and 194
383:16 paragraphs:—] *OT deleted colon*
383:19 space entities] *OT and AE altered to* space-entities
383:20 difference is] *OT and AE added comma*
383:29 p. 372] *OT altered* p. 107 *to* p. 196
384:6 is] *OT altered to* in
384:13 Q_2, then] *OT deleted comma*
384:22 372–4] *OT altered* 107–10 *to* p. 197–200
384:32 Again] *OT added comma*
384:36 does in fact employ] *OT altered to* does, in fact, employ
385:4 Pp. 374–5] *OT altered* Pp. 110–11 *to* Pp. 201–202
385:6 co-ordinate] *OT altered to* coordinate
385:6 p. 375] *OT altered* p. 112 *to* p. 203
385:6 top of p. 378] *OT altered* bottom of p. 115 *to* bottom of p. 207
385:8 pp. 375–9] *OT altered* pp. 111–118 *to* pp. 201–211
385:23 nil] *OT italicized* nil
385:31 p. 379] *OT altered* p. 117 *to* pp. 209, 210
385:34 is] *OT altered to* be
386:18 possibilities:—] *OT deleted colon*
386:20 or] *OT added em dash*
386:37 is in some general sense of the term a] *AE altered to* is, in some general sense of the term, a

The Aims of Education—A Plea for Reform

NR = *The New Republic*
UH = *Unseen Harvests: A Treasuring of Teaching*
388:7 civilization] *OT altered to* civilisation
388:15 realized] *OT altered to* realised
389:10 schools] *OT deleted comma*
390:1 16 and 30] *OT, AE, NR, and UH altered to* sixteen and thirty
390:2 12] *OT, AE, NR, and UH altered to* twelve
390:8 utilized] *OT, AE, and UH altered to* utilised
390:11 is,] *NR deleted comma*
390:16 most] *NR altered to* more
390:22 dry rot] *AE and UH altered to* dryrot
390:30 make,] *NR deleted comma*
390:34 all,] *NR deleted comma*
390:37 Augustine,] *OT, AE, NR, and UH deleted comma*
391:11 Molière] *UH altered to* Moliere
391:11 Vergil] *OT, AE, NR, and UH altered to* Virgil
391:14 meeting-place,] *NR deleted comma*
391:17 utilized] *OT, AE, and UH altered to* utilised
391:17 utilizing] *OT, AE, and UH altered to* utilising
391:24 'prove'] *OT, AE, NR, and UH altered to* "prove"
391:33 honour] *NR altered to* honor
391:33 processes of proof, in] *NR altered to* processes, of proof in
391:33 appreciation] *AH and UH added comma*
391:37 endeavour] *NR altered to* endeavor
391:38 mean,] *NR deleted comma*
392:1 Inter-related] *AE and NR altered to* Interrelated
392:1 utilized] *OT, AE, and UH altered to* utilised
392:6 easily to be] *AE and UH altered to* to be easily
392:9 utilizing] *OT, AE, and UH altered to* utilising
392:9 utilized] *OT, AE, and UH altered to* utilised
392:10 utilized] *OT, AE, and UH altered to* utilised
392:11 utilization] *OT, AE, and UH altered to* utilisation
392:12 of my discourse] *NR deleted* of my discourse
392:13 realizing] *OT, AE, and UH altered to* realising
392:20 utilization] *OT, AE, and UH altered to* utilisation
392:21 text-book] *NR altered to* textbook
392:23 Of course,] *OT, AE, NR, and UH deleted comma*
393:1 analyse] *NR altered to* analyze
393:10 text-book] *NR altered to* textbook
393:26 subject-matter,] *AE, NR, and UH* deleted comma
393:27 whatever powers you are strengthening in the pupil,] *NR deleted* whatever powers you are strengthening in the pupil,
393:29 impart,] *NR deleted comma*
393:35–6 generalizations] *OT, AE, and UH altered to* generalisations
393:39 urging,] *NR deleted comma*
393:40 subject matter] *OT, AE, NR, and UH altered to* subject-matter
394:1 Couple of Languages] *NR altered to* couple of languages
394:5 it,] *OT, AE, NR, and UH deleted comma*
394:5 it, is] *OT, AE, and UH altered to* it is,
394:9 question,] *OT, AE, and UH altered comma to period*

394:26 schooldays] *OT, AE, and UH altered to* school-days
394:34 see] *OT, AE, and UH added comma*
395:13 Meanwhile] *AE and UH added comma*
395:16 Not of course in] *OT, AE, and UH altered to* Not, of course, in
395:19 medieval] *OT and AE altered to* mediæval; *UH altered to* mediaeval
395:32 Algebra] *OT altered to* algebra
396:2 solutions for] *OT altered to* solutions or
396:3 formulae] *OT altered to* formulæ
396:7 causal] *OT altered to* casual
396:11 causal] *OT altered to* casual
396:29 zig-zag] *UH altered to* zigzag
396:31 reason is] *OT, AE, and UH added comma*
396:43–44 clear cut] *OT, AE, and UH altered to* clear-cut
397:2 all be] *OT, AE, and UH altered to* be all
397:5 Then of course there] *OT, AE, and UH altered to* Then, of course, there
398:1 utilizes] *OT, AE, and UH altered to* utilises
398:21 aesthetic] OT and AE altered to æsthetic
398:24 aesthetic] OT and AE altered to æsthetic
398:30 style, ... style,] *AE and UH deleted commas*
398:31 style,] *AE and UH deleted comma*
398:39 Where then does] *OT, AE, and UH altered to* Where, then, does
399:23 dunghill] *OT, AE, and UH altered to* dung-hill
400:2 civilization] *OT, AE, and UH altered to* civilisation

The Organisation of Thought
ChN = *The Chemical News*
Sc = *Science*
N = *Nature*
SAS = *Scientific American Supplement*
AS = *Proceedings of the Aristotelian Society*
453:21 address] *AS altered to* paper
453:21 organisation] *Sc and SAS altered to* organization
453:24 But] *AS added comma*
453:27 organisation] *Sc and SAS altered to* organization
453:28 Organised] *Sc and SAS altered to* Organized
453:28 organised] *Sc and SAS altered to* organized
453:28 Organisation] *Sc and SAS altered to* Organization
453:30 organisation,] *Sc and SAS altered to* organization; *ChN and SAS changed comma to semicolon; N changed comma to dash; AS added dash*
453:31 it] *AS, OT, and AE altered to in*
453:31 organisation] *Sc and SAS altered to* organization
453:35 sensuous,] *Sc deleted comma*
453:36 or,] *Sc, AS, OT, and AE deleted comma*
453:36 inversely commensurate] *N, AS, OT, and AE* added comma
453:37 organisation] *Sc and SAS altered to* organization
453:38 organisation] *Sc and SAS altered to* organization
453:39 organisation ... organisation] *Sc and SAS altered to* organization
453:40 endeavour] *Sc and SAS altered to* endeavor
454:3 Now] *AS added comma*
454:3 emphasise] *Sc and SAS altered to* emphasize
454:9 practice] *Sc altered to* practise
454:10 under] *N altered to* in
454:13 battle.] *AS, OT, and AE removed paragraph break*
454:14 practice] *Sc altered to* practise
454:16 clouds,] *SAS deleted comma*
454:20 course,] *ChN deleted comma*
454:21 organisation] *Sc and SAS altered to* organization
454:24 thinkers,] *ChN, AS, OT, and AE altered comma to colon; SAS altered comma to dash*
454:25 Jevons,] *Sc deleted comma*
454:29 emphasise] *Sc and SAS altered to* emphasize
454:29–30 tendency in analysing scientific processes to] *AS altered to* tendency, in analysing scientific processes, to
454:30 analysing] *Sc and SAS altered to* analyzing
454:31–32 selecting by means of inductive logic some] *AS altered to* selecting, by means of inductive logic, some
454:37 percepts] *N altered to* precepts
454:38 concepts] *AS added dash*
455:1 us. For] *N altered to* us—for
455:1 life,] SAS deleted comma
455:3 dogmatising] *Sc and SAS altered to* dogmatizing
455:9 conceived,] *AS added dash*
455:12 purposes] *OT and AE altered to* purpose
455:13 its] *ChN and SAS altered to* it
455:15 sensations,] *Sc deleted comma*
455:16 emotions] *AS added comma*
455:17 touch,] *Sc deleted comma*
455:18 organisation] *Sc and SAS altered to* organization
455:19 is] *AE altered to* in
455:24 wisdom] *ChN and SAS deleted comma*
455:27 is] *OT and AE added comma*
455:30 magnitude] *AE altered to* mganitude
455:32 contention is] *OT and AE added comma*
455:37 apply,] *N altered comma to semicolon; AS, OT, and AE altered comma to colon*
455:38 phrase] *Sc, AS, OT, and AE added comma*
455:42 organisation] *Sc and SAS altered to* organization
456:4 experiences] *ChN, SAS, AS, OT, and AE altered to* experience
456:4 reference:] *ChN and SAS altered colon to semicolon*

456:5 common-sense] *AE altered to* commonsense*; OT probably would have done the same (based on other instances in the essay) had the word not been hyphenated across two lines*
456:7 the experiences] *AS, OT, and AE altered to* the experience
456:8 seen it] *Sc added comma*
456:11 fire-wood] *OT and AE altered to* firewood
456:14 emphasise] *Sc and SAS altered to* emphasize
456:15 common-sense] *OT and AE altered to* commonsense
456:16 of] *AE altered to* on
456:18 namely,] *AS deleted* namely,
456:18–19 attention to] *AS altered to* attention, namely, to
456:20–21 common sense] *OT and AE altered to* commonsense
456:21 But] *AS altered to* Yet
456:23 common sense] *OT and AE altered to* commonsense
456:24 organisation] *Sc and SAS altered to* organization
456:26 based,] *ChN, N, and SAS deleted comma*
456:30 But at present in] *AS, OT, and AE altered to* But, at present, in
456:30 time,] *ChN, N, AS, SAS, OT, and AE* deleted comma
456:31 experiences.] *AS removed paragraph break*
456:34 arbitrary,] *ChN, N, and SAS deleted comma*
456:42 hardly] *N altered to* scarcely
456:43 address] *AS altered to* paper
456:43 is] *OT and AE added comma*
457:6 of analogous] *N altered to* of an analogous
457:6 under] *N altered to* in
457:9 Cock and Bull] *Sc altered to* cock and bull
457:13 'No bishops, no king.'] *Sc, N, AS, OT, and AE altered to* "No bishops, no king."; *ChN and SAS altered to* "No bishops, no King."
457:13 'No logic, no science.'] *ChN, Sc, N, SAS, AS, OT, and AE altered to* "No logic, no science."
457:16–17 authority] *AS, OT, and AE added comma*
457:18 temporally] *ChN, Sc, and SAS altered to* temporarily
457:19 complaints[3†]] *ChN and SAS had the note as a parenthetical instead. Sc, N, AS, OT, and AE capitalized the 'e' in 'e.g.'. OT and AE used a semicolon rather than a period before the new material. AE erroneously had the author's named as 'Scarpi'.*
457:27 place] *AS added comma*
457:30 deductions,] *Sc deleted comma*
458:2 assumptions] *AS altered to* assumption
458:9 so] *AS added comma*
458:11 fossilised] *Sc and SAS altered to* fossilized
458:13 details.] *AS, OT, and AE added paragraph break*
458:15 analytical] *ChN, Sc, and SAS altered to* analytic
458:23 '*p*.'] *ChN, Sc, SAS, AS, OT, and AE altered single quote marks to double quote marks throughout this paragraph*
458:25 them,] *ChN and SAS deleted comma*
458:26 least] *OT and AE altered to* last
458:28 'or'] *SAS altered to* 'or
458:31 '*p* or *q*'] *ChN, Sc, SAS, AS, OT, and AE altered single quote marks to double quote marks throughout this sentence*
458:33 not-*q*] *SAS failed to italicize* q
458:34 taken,] *ChN and SAS deleted comma*
458:36 way] *AS added comma*
458:38 course,] *ChN and SAS deleted comma*
459:1 'truth-value'] *ChN, Sc, SAS, AS, OT, and AE altered to* "truth-value"
459:3 inquiry] *ChN altered to* enquiry
459:4 truth-values] *SAS altered to* truth values
459:4 inquiry] *ChN altered to* enquiry
459:6 inquiry] *ChN altered to* enquiry
459:9 is] *OT and AE added comma*
459:17 proposition,] *AS altered comma to colon; OT and AE altered comma to dash*
459:18 0·033.] *Sc and SAS altered middle dot to period; SAS also deleted period*
459:21 thing:] *ChN and SAS altered colon to semicolon*
459:21 get,] *Sc and AS deleted comma; OT and AE altered comma to dash*
459:22 0·033.] *Sc and SAS altered middle dot to period; SAS also deleted period; OT and AE indented the line following*
459:25 function.] *AS removed paragraph break; AE removed indent from the line following*
459:26 generalise] *Sc and SAS altered to* generalize
459:26 say,] *Sc deleted comma*
459:28 F(*x*, *y*)] *AS, OT, and AE added comma*
459:31 true] *SAS altered to* ture
459:32 suggestions] *AE altered to* suggestion
459:34 0·033] *Sc and SAS altered middle dot to period*
459:35 and] *OT and AE added comma*
459:36 0·033] *Sc and SAS altered middle dot to period*
459:39 'type'] *ChN, Sc, SAS, AS, OT, and AE altered to* "type"
460:1 members,] *AS altered comma to semicolon*
460:4 form,] *AS added dash*
460:5 is,] *Sc and AS deleted comma*
460:7 is,] *Sc deleted comma*
460:8 true.] *SAS deleted period*
460:18 cannot] *Sc altered to* can not
460:23 logical] *AS, OT, and AE altered to* logic
460:25 recognised] *Sc and SAS altered to* recognized
460:25 himself] *AE altered to* himeslf
460:26 Leibniz] ChN and SAS altered to Liebniz
460:27 Boole,] Sc deleted comma
460:35 which] *AS altered to* that

460:36 analyse] *Sc and SAS altered to* analyze
460:40 'the present King of England,'] *ChN, Sc, SAS, AS, OT, and AE altered single quote marks to double quote marks throughout this paragraph*
461:7 &c] *Sc, SAS, AS, OT, and AE altered to* etc
461:11 less. But] *OT and AE altered to* less, but
461:15 time,] *Sc deleted comma*
461:16 impossible even in brief outline to] *AS, OT, and AE altered to* impossible, even in brief outline, to
461:17 many-cornered] *SAS altered to* many cornered
461:18–19 process] *AS, OT and AE added comma*
461:19 *Principia Mathematica*] *ChN, Sc, and SAS the original single quote marks to double quote marks*
461:23 a] *AS, OT, and AE deleted* a
461:32 three-cornered and four-cornered] *SAS altered to* three cornered and four cornered
461:32 Geometry] *Sc, SAS, AS, OT, and AE altered to* geometry
461:37 criticise] Sc and SAS altered to criticize
462:4 and (2)] *AS, OT, and AE altered to* and if (2)
462:6 false,] *Sc deleted comma*
462:10 first] *AS, OT, and AE added comma*
462:12 secondly] *AS, OT, and AE added comma*
462:16 *a priori*] *ChN and SAS altered to* à priori
462:17 quantity,] *Sc deleted comma*
462:18 analysing] *Sc and SAS altered to* analyzing
462:22 deductions,] *AS, OT, and AE deleted comma*
462:25 course,] *OT and AE deleted comma*
462:28 is,] *Sc altered comma to colon; AS deleted comma*
462:28 an] *Sc, OT, and AE altered to* An
462:35 true theory,] *AS deleted comma*
462:36 application,] *AS deleted comma*
462:40 Common-sense] *SAS altered to* Common sense; *AE altered to* Commonsense *(OT probably would have done the same based on other instances in the essay had the word not been hyphenated across two lines)*
462:42 common sense] *OT and AE altered to* commonsense
463:5 cannot] *Sc altered to* can not
463:5 at] *Sc deleted* at
463:6 fact,] *SAS altered comma to colon*
463:7 0·033] *ChN* 3·033; *Sc altered to* 0.033; *SAS altered to* 3.033
463:9 0·033] *Sc and SAS altered middle dot to period*
463:15 cannot] *Sc altered to* can not
463:21 Geometry] *Sc, SAS, OT, and AE altered to* geometry
463:25 see or unpleasantly feel something] *AS altered to* see, or unpleasantly feel, something
463:25 But] *AS altered to* Yet
463:28 Now] *AS added comma*
463:29 space relation] *OT and AE altered to* space-relation
463:30 'whole and part'] *ChN, Sc, SAS, AS, OT, and AE altered single quote marks to double quote marks throughout this paragraph*
463:34 'spatial whole and part'] *ChN, Sc, SAS, AS, OT, and AE altered single quote marks to double quote marks throughout this paragraph*
463:37 perceptions] *AS, OT, and AE altered to* perception
463:40 of a body] *OT and AE deleted* a
464:5 assumptions, they] *AS altered to* assumptions. They
464:6 definition.] *AS removed paragraph break*
464:8 Namely, we assume] *AS altered to* We assume, namely,
464:12 exposition.] *AS removed paragraph break*
464:17 Geometry] *Sc, SAS, OT, and AE altered to* geometry
464:25 common-sense] *OT and AE altered to* commonsense
464:27 common-sense] *SAS altered to* common sense; *AE altered to* commonsense *(OT probably would have done the same based on other instances in the essay had the word not been hyphenated across two lines)*
464:30 endeavoured] *Sc and SAS altered to* endeavored
464:30 organising] *Sc and SAS altered to* organizing
464:31 analysing] *Sc and SAS altered to* analyzing
464:34 under] *N altered to* in
464:35 freedom and, above] *AS, OT, and AE altered to* freedom, and above
465:5 years,] *OT and AE deleted comma*

Bibliography

Aristotelian Society, 'Abstract of the Minutes of the Proceedings of the Aristotelian Society for the Thirty-Eighth Session', *Proceedings of the Aristotelian Society*, New Series, vol. 17 (1917), pp. 481–484.

Aristotelian Society, 'Abstract of the Minutes of the Proceedings of the Aristotelian Society for the Thirty-Seventh Session', *Proceedings of the Aristotelian Society*, New Series, vol. 16 (1916), pp. 364–366.

Aurelius, Marcus, *The Meditations of Marcus Aurelius Antoninus*, translated by A. S. L. Farquharson (Oxford: Oxford University Press, 1990).

Ball, Robert Stawell, 'On the Theory of the Content', *Transactions of the Royal Irish Academy*, vol. 29 (1887–1892), pp. 123–182.

Ball, Walter W. Rouse, *A Short Account of the History of Mathematics* (London: Macmillan, 1888).

Barbarin, Paul Jean Joseph, *La Géométrie non-Euclidienne* (Paris: C. Nand, 1902).

Beltrami, Eugenio, 'Essai d'interprétation de la géométrie noneuclidéenne. Trad. par J. Hoüel', *Annales scientifiques de l'École Normale Supérieure*, Series 1, vol. 6 (1869), pp. 251–288.

Beltrami, Eugenio, 'Saggio di interpretazione della geometria non-Euclidea', *Giornale di matematiche*, vol. 6 (1868), pp. 284–312.

Beltrami, Eugenio, 'Teoria fondamentale degli spazii di curvatura costante', *Annali di matematica*, vol. 2 (1868), pp. 232–255.

Beltrami, Eugenio, 'Théorie fondamentale des espaces de courbure constante. Trad. par. J. Hoüel', *Annales scientifiques de l'École Normale Supérieure*, Series 1, vol. 6 (1869), pp. 347–375.

Berkeley, Hastings, *Mysticism in Modern Mathematics* (Oxford: Oxford University Press, 1910).

Bolyai, János, 'Appendix scientiam spatii absolute veram exhibens; a veritate aut falsitate axiomatis XI Euclidei (a priori haud unquam decidenda)', in Farkas Bolyai, *Tentamen juventutem studiosam in elementa matheseos purae, elementaris ac sublimioris, methodo intuitiva, evidentiaque huic propria, introducendi* (Maros Vásárhelyini: Typis Collegii Reformatorum per Josephum et Simeonem Kali de felso Vist., 1832).

Bolyai, János, *The Science Absolute of Space: Independent of the Truth or Falsity of Euclid's Axiom XI (Which Can Never Be Decided* A Priori*)*, translated by George Bruce Halsted (Austin: The Neomon, 1896).

Bonola, Roberto, 'Index operum ad geometriam absolutam spectantium', in *Ioannis Bolyai in memoriam* (Klausenburg, 1902), pp. 81–154. First published as *Index operum ad geometriam absolutam spectantium* (Italy: Societatis franklinianae budapestinensis, 1867).

Bonola, Roberto, *La Geometria non-Euclidea* (Bologna: Nicola Zanichelli, 1906).

Bonola, Roberto, 'Sulla Introduzione degli Enti improprii in Geometria projettiva', *Giornale di matematiche di Battaglini*, vol. 38 (1900), pp. 105–116.

Borel, Emile, *Leçons sur la Théorie des Fonctions* (Paris: Gauthier-Villars et fils, 1898).

Boutroux, Pierre Léon, 'Correspondance Mathématique: Et Relation Logique', *Revue de Métaphysique et de Morale*, vol. 13, no. 4 (July 1905), pp. 620–637.

Bradley, F. H., *The Principles of Logic, Volume I*, 2nd edition (London: Oxford University Press, 1928).

British Association for the Advancement of Science, *Report of the 85th Meeting of the British Association for the Advancement of Science* (London, John Murray, 1916).

Buchheim, Arthur, 'On the Theory of Screws in Elliptic Space', *Proceedings of the London Mathematical Society*, vol. 15 (November 1883), pp. 83–98.

Buchheim, Arthur, 'On the Theory of Screws in Elliptic Space (Supplementary Note)', *Proceedings of the London Mathematical Society*, vol. 16 (November 1884), pp. 15–27.

Buchheim, Arthur, 'On the Theory of Screws in Elliptic Space (Third Note)', *Proceedings of the London Mathematical Society*, vol. 17 (November 1885), pp. 240–254.

Burali-Forti, Cesare, 'Una questione sui numeri transfiniti', *Rendiconti del Circolo Matematico di Palermo*, vol. 11 (December 1897), pp. 154–164.

Burnside, William, 'On the Kinematics of Non-Euclidean Space', *Proceedings of the London Mathematical Society*, vol. 26, no. 1 (November 1894), pp. 33–58.

Burnside, William, Referee's report by William Burnside on paper 'Sets of Operations in Relation to Groups of Finite Order' by Alfred North Whitehead', Royal Society Archives, RR/14/269.

Burnside, William, *Theory of Groups of Finite Order* (Cambridge: Cambridge University Press, 1897).

Cantor, Georg, 'Beiträge zur Begründung der transfiniten Mengenlehre (1)', *Mathematische Annalen*, vol. 46, no. 4 (November 1895), pp. 481–512.

Cantor, Georg, 'Beiträge zur Begründung der transfiniten Mengenlehre (2)', *Mathematische Annalen*, vol. 49, no. 2 (June 1897), pp. 207–246.

Cantor, Georg, *Grundlagen einer allgemeinen Mannigfaltigkeitslehre* (Leipzig: Teubner, 1883).

Cantor, Georg, *Sur les fondements de la théorie des ensembles transfinis*, traduction de F. Marotte (Paris: A. Hermann, 1899).

Cantor, Georg, 'Ueber unendliche, lineare Punktmannichfaltigkeiten (1)', *Mathematische Annalen*, vol. 21, no. 1 (March 1883), pp. 51–58.

Cantor, Georg, 'Ueber unendliche, lineare Punktmannichfaltigkeiten (2)', *Mathematische Annalen*, vol. 21, no. 4 (December 1883), pp. 545–591.

Cantor, Moritz Benedikt, *Vorlesungen über Geschichte der Mathematik* (Leipzig: Teubner, 1880–1920).

Carson, George Edward St Lawrence, 'Editorial', *Journal of the Association of Teachers of Mathematics for the Southeastern Part of England*, vol. 1, no. 1 (December), p. 1.

Cayley, Arthur, 'A Sixth Memoir on Quantics', *Philosophical Transactions of the Royal Society*, vol. 149 (1859), pp. 61–90.

Cayley, Arthur, 'Address by Arthur Cayley, President', in *Report of the Fifty-Third Meeting of the British Association for the Advancement of Science* (London: John Murray, 1884), pp. 1–37.

Cayley, Arthur, *The Collected Mathematical Papers of Arthur Cayley, Volume II* (Cambridge: Cambridge University Press, 1889).

Chasels, Michel, *Aperçu historique sur l'origine et le développement des méthodes en géométrie* (Brussels: M. Hayez, 1837).

Chasles, Michel, *Traité de géométrie supérieure* (Paris: Bachelier, 1852).

Chisholm, Hugh, 'Letter from Hugh Chisholm to Bertrand Russell, 30 October 1905', Victor Lowe Papers, MS-0284, Series 2, Box 2.1, Special Collections, Sheridan Libraries, Johns Hopkins University.

Chisholm, Hugh, 'Letter from Hugh Chisholm to Bertrand Russell, 20 November 1905', Victor Lowe Papers, MS-0284, Series 2, Box 2.1, Special Collections, Sheridan Libraries, Johns Hopkins University.

Clebsch, Alfred, *Vorlesungen über geometrie*, edited by Ferdinand Lindemann (Leipzig: Teubner, 1891).

Clifford, William Kingdon, *The Common Sense of the Exact Sciences* (London: Kegan Paul Trench, and Co., 1885).

Clifford, William Kingdon, *Lectures and Essays*, edited by Leslie Stephen and Frederick Pollock (London: Macmillan, 1879).

Clifford, William Kingdon, 'Preliminary Sketch of Biquaternions', *Proceedings of the London Mathematical Society*, series 1, vol. 4, no. 1 (November 1871), pp. 381–395.

Couturat, Louis, *La Logique de Leibniz* (Paris: F. Alcan, 1901).

Couturat, Louis, 'La philosophie des Mathématiques de Kant', *Revue de Métaphysique et de Morale*, vol. 12, no. 3 (May 1904), pp. 321–383.

Couturat, Louis, *Les principes des Mathématiques* (Paris: F. Alcan, 1905).

Couturat, Louis, 'Les principes des Mathématiques — VI. La géometrie', *Revue de Métaphysique et de Morale*, vol. 12, no. 5 (September 1904), pp. 810–844.

Couturat, Louis, *Opuscules et fragments inédits de Leibniz. Extraits des manuscrits de la bibliothèque royale de Hanovre* (Paris: F. Alcan, 1903).
Cox, Homersham, 'On the Application of Quaternions and Grassmann's Algebra to Different Kinds of Uniform Space', *Transactions of the Cambridge Philosophical Society*, vol. 13 (1881–1883), pp. 69–143.
Cox, Peter, 'My Ouse Washes' history and management research project, <http://www.ousewashes.info/>.
Darboux, Jean-Gaston, *Leçons sur la théorie générale des surfaces et les applications géométriques du calcul infinitésimal, deuxième partie* [*Lessons on the general theory of surfaces and geometric applications of infinitesimal calculus, second part*] (Paris: Gauthier-Villars, 1889).
Darboux, Jean-Gaston, *Leçons sur la théorie générale des surfaces et les applications géométriques du calcul infinitésimal, troisième partie* [*Lessons on the general theory of surfaces and geometric applications of infinitesimal calculus, third part*] (Paris: Gauthier-Villars, 1894).
Davidson, Thomas, *Chambers's Twentieth Century Dictionary of the English Language* (London: W. and R. Chambers, 1907).
Dedekind, Richard, *Essays on the Theory of Numbers* (Chicago: Open Court, 1901).
Dedekind, Richard, *Stetigkeit und irrationale Zahlen* (Braunschweig: Friedrich Vieweg und Sohn 1872).
Dedekind, Richard, *Was sind und was sollen die Zahlen?* (Braunschweig: Friedrich Vieweg und Sohn, 1887).
Dehn, Max, 'Die Legendarischen Sätze über die Winkelsumme in Dreieck', *Mathematische Annalen*, vol. 53, no. 3 (September 1900), pp. 404–439.
Dehn, Max, 'Ueber den Rauminhalt', *Mathematische Annalen*, vol. 55, no. 3 (September 1901), pp. 465–478.
Elliott, Edwin Bailey, Corrections for paper 'Sets of Operations in Relation to Groups of Finite Order' by Alfred North Whitehead, Royal Society Archives, RR/14/271.
Elliott, Edwin Bailey, Letter from Edwin Bailey to Andrew Forsyth, 11 February 1899, Royal Society Archives, RR/14/272.
Engel, Friedrich, and Paul Stäckel, *Urkunden zur Geschichte der nichteuklidischen Geometrie, Erster Theil, Nikolaj Iwanowitsch Lobatschefskij* (Leipzig: Teubner, 1898).
Fano, Gino, 'Sui postulati fondamentali della geometria proiettiva in uno spazio lineare a un numero qualunque di dimensioni', *Giornale di matematiche di Battaglini*, vol. 30 (1892), pp. 106–132
Fiedler, Wilhelm, *Die darstellende Geometrie*, 1st edition (Leipzig: Teubner, 1871).
Fiedler, Wilhelm, *Die darstellende Geometrie*, 3rd edition (Leipzig: Teubner, 1888).
Fitzgerald, Janet, *Alfred North Whitehead's Early Philosophy of Space and Time* (Washington, DC: University Press of America, 1979).
Fontené, Georges, *L'Hyperespace à (n – 1) dimensions* (Paris: Gauthier-Villars et Fils, 1892).
Frankland, William Barrett, *The First Book of Euclid's Elements, with a Commentary Based Principally Upon That of Proclus Diadochus* (Cambridge: Cambridge University Press, 1905).
Frege, Gottlob, *Begriffsschrift: eine der arithmetischen nachgebildete Formelsprache des reinen Denkens* (Halle an der Saale: Verlag von Louis Nebert, 1879).
Frege, Gottlob, *Die Grundlagen der Arithmetik* (Breslau: Wilhelm Koebner, 1884).
Frege, Gottlob, *Grundgesetze der Arithmetik, Volume I* (Jena: Hermann Pohle, 1893).
Frege, Gottlob, *Grundgesetze der Arithmetik, Volume II* (Jena: Hermann Pohle, 1903).
Frege, Gottlob, 'Über die Grundlagen der Geometrie', *Jahresbericht Der Deutschen Mathematiker-Vereinigung*, vol. 15 (1906), pp. 293–309, 377–402, 423–430.
Frischauf, Johann, *Absolute Geometrie nach J. Bolyai* (Leipzig: Teubner, 1872).
Frischauf, Johann, *Elemente der absoluten Geometrie* (Leipzig: Teubner,, 1876).
Gérard, Louis, *Sur la géométrie non-Euclidienne* (Paris: Gauthier-Villars et fils, 1892).
Grassmann, Hermann, *Hermann Grassmann's Gesammelte Mathematische und Physikalische Werke*, edited by Friedrich Engel (Leipzig: Teubner, 1894).
Hall, Basil, *Fragments of Voyages and Travels Including Anecdotes of a Naval Life: Chiefly for the Use of Young Persons, Volume III* (Edinburgh: Robert Cadell, 1831).

Halsted, George Bruce, 'Addenda to Bibliography of Hyper-Space and Non-Euclidean Geometry', *American Journal of Mathematics*, vol. 2, no. 1 (March 1879), pp. 65–70.

Halsted, George Bruce, 'Addenda to Mr. Halsted's Paper on the Bibliography of Hyper-Space and Non-Euclidean Geometry', *American Journal of Mathematics*, vol. 1, no. 4 (1878), pp. 384–385.

Halsted, George Bruce, 'Bibliography of Hyper-Space and Non-Euclidean Geometry', *American Journal of Mathematics*, vol. 1, no. 3 (1878), pp. 261–276.

Halsted, George Bruce, '*Urkunden zur Geschichte der Nichteuklidischen Geometrie* Von F. Engel und P. Staeckel', *American Mathematical Monthly*, vol. 6, Nos 6–7 (June–July 1899), pp. 166–172.

Hamilton, William Rowan, *Lectures on Quaternions* (Dublin: Hodges and Smith, 1853).

Heath, Thomas Little, *The Thirteen Books of Euclid's Elements* (Cambridge: Cambridge University Press, 1908).

Helmholtz, Herman von, 'The Origin and Meaning of Geometrical Axioms', *Mind*, vol. 1, no. 3 (July 1876), pp. 301–321.

Helmholtz, Herman von, 'The Origin and Meaning of Geometrical Axioms (II)', *Mind*, vol. 3, no. 10 (April 1878), pp. 212–225.

Helmholtz, Herman von, *Vorträge und Reden, Zweiter Band* (Braunschweig: Friedrich Vieweg und Sohn, 1903).

Helmholtz, Herman von, *Wissenschaftliche Abhandlungen, Zweiter Band* (Leipzig: Barth, 1883).

Henning, Brian G., 'On the Recently Discovered Whitehead Papers', Whitehead Research Project, 14 January 2019, <http://whiteheadresearch.org/2019/01/14/on-the-recently-discovered-whitehead-papers/>.

Henning, Brian G., 'Preface: A Brief History of the Critical Edition of Whitehead', in Brian G. Henning and Joseph Petek (eds), *Whitehead at Harvard, 1924–1925* (Edinburgh: Edinburgh University Press, 2020).

Henning, Brian G., 'Revisiting Lucien Price's Dialogues', Whitehead Research Project, 3 January 2020, <https://whiteheadresearch.org/2020/01/03/revisiting-lucien-prices-dialogues/>.

Hilbert, David, *The Foundations of Geometry*, translated by E. J. Townsend (Chicago: Open Court, 1902).

Hilbert, David, *Grundlagen der Geometrie* (Leipzig: Teubner, 1899).

Hill, Micaiah John Muller, *The Contents of the Fifth and Sixth Books of Euclid* (Cambridge: Cambridge University Press, 1900).

Hobson, E. W., and A. E. H. Love (eds), *Proceedings of the Fifth International Congress of Mathematicians, Volume I* (Cambridge: Cambridge University Press, 1913), p. 28.

Hoüel, Jules, 'Etudes géométriques sur la théorie des parallèles', *Mémoires de la Société des Sciences Physiques et Naturelles de Bordeaux, Tome IV* (Paris: Gauthier-Villars, 1867), pp. 83–128.

Hoüel, Jules, 'La Science absolue de l'espace', *Mémoires de la Société des Sciences Physiques et Naturelles de Bordeaux, Tome V* (Paris: Gauthier-Villars, 1867), pp. 189–248.

Hudson, W. H. H., 'The Work of a Local Chapter', *Mathematical Gazette*, vol. 8, no. 121 (January 1916), pp. 215–219.

Hurley, Patrick J., 'Whitehead's *Relational Theory of Space*: Text, Translation, and Commentary', *Philosophy Research Archives*, vol. 5 (1979), pp. 676–777, available online at <https://www.pdcnet.org/pra/content/pra_1979_0005_0676_0777>.

James, William, *Pragmatism: A New Name for some Old Ways of Thinking* (New York: Longmans, Green, and Co., 1907).

Johnson, William Ernest, 'Sur la théorie des equations logiques', in *Bibliothèque du Congrès International de Philosophie, Tome III: Logique et Histoire des Sciences* (Paris: Armand Colin, 1901), pp. 185–199.

Kant, Immanuel, *Critique of Pure Reason*, translated by F. Max Müller (London: Macmillan, 1881).

Kempe, A. B., 'On the Relation between the Logical Theory of Classes and the Geometrical Theory of Points', *Proceedings of the London Mathematical Society*, vol. 21, no. 1 (November 1889), pp. 147–182.

Killing, Wilhelm Karl Joseph, *Die nicht-Euklidischen Raumformen in analytischer Behandlung* (Leipzig: Teubner, 1885).
Klein, Felix, *Lectures on Mathematics* (New York: Macmillan, 1894).
Klein, Felix, *Nicht-Euklidische Geometrie* (Göttingen, 1893).
Klein, Felix, 'Ueber die sogenannte Nicht-Euklidische Geometrie', *Mathematische Annalen*, vol. 4, no. 4 (December 1871), pp. 573–625.
Klein, Felix, 'Ueber die sogenannte Nicht-Euklidische Geometrie', *Mathematische Annalen*, vol. 6, no. 2 (June 1873), pp. 112–145.
Klein, Felix, 'Zur nicht-Euklidischen Geometrie', *Mathematische Annalen*, vol. 37, no. 4 (December 1890), pp. 544–572.
Kline, Mary-Jo, and Susan Holbrook Perdue, *A Guide to Documentary Editing*, 3rd edition (Charlottesville: University of Virginia Press, 2008).
Laguerre-Verly, Edmon, 'Sur la théorie des foyers', *Nouvelles annales de mathématiques*, Series 1, vol. 12 (1853), pp. 57–66.
Lamb, Horace, *A Treatise on the Mathematical Theory of the Motion of Fluids* (Cambridge: Cambridge University Press, 1879).
Lambert, Johann Heinrich, *Theorie der Parallellinien* (Leipzig: 1786).
Legendre, Adrien-Marie, *Élements de géométrie* (Paris: Firmin Didot, 1794).
Lie, Sophus, *Theorie der Transformationsgruppen, dritter und letzter abschnitt* (Leipzig, 1893).
Lie, Sophus, 'Über die Grundlagen der Geometrie. Erste Abhandlung . . .', *Königlich Sächsische Gesellschaft der Wissenschaften (Hrsg.): Berichte über die Verhandlungen der Königlich-Sächsischen Gesellschaft der Wissenschaften zu Leipzig, Mathematisch-Physische Klasse*, vol. 42 (1890), pp. 284–321.
Lindemann, Ferdinand von, 'Über die Zahl π', *Mathematische Annalen*, vol. 20, no. 2 (1882), pp. 213–225.
Lindemann, Ferdinand von, 'Ueber unendlich kleine Bewegungen und über Kraftsysteme bei allgemeiner projectivischer Massbestimmung', *Mathematische Annalen*, vol. 7, no. 1 (March 1873), pp. 56–143.
Lobachevsky, Nikolai, *Geometrical Researches on the Theory of Parallels*, translated by George Bruce Halsted (Austin: University of Texas, 1891).
Lobachevsky, Nikolai, 'Géométrie imaginaire', *Journal für die reine und angewandte Mathematik (Crelle's Journal)*, vol. 17 (1837), pp. 295–320.
Lobachevsky, Nikolai, *Geometrische Untersuchungen zur Theorie der Parallellinien* (Berlin: Fincke, 1840).
Lobachevsky, Nikolai, *Geometrische Untersuchungen zur Theorie der Parallellinien*, 2nd edition (Berlin: Mayer & Müller, 1887).
Lowe, Victor, *Alfred North Whitehead: The Man and His Work, Volume I: 1861–1910* (Baltimore: Johns Hopkins University Press, 1985).
Lowe, Victor, *Alfred North Whitehead: The Man and His Work, Volume II: 1910–1947* (Baltimore: Johns Hopkins University Press, 1990).
Lowe, Victor, *Understanding Whitehead* (Baltimore: Johns Hopkins University Press, 1966).
Macaulay, W. H., 'Newton's theory of kinetics', *Bulletin of the American Mathematical Society*, vol. 3, no. 10 (1897), pp. 363–371.
Mach, Ernst, *Die Mechanik in ihrer Entwickelung* (Leipzig: F. A. Brockhaus, 1883).
Mach, Ernst, *Erkenntniss und Irrtum* (Leipzig: Barth, 1906).
Mach, Ernst, *Space and Geometry*, translated by Thomas J. McCormack (La Salle: Open Court, 1906).
Mach, Ernst, *The Science of Mechanics*, translated by Thomas J. McCormack (Chicago: Open Court, 1893).
Mackintosh, Alexander, *Joseph Chamberlain: An Honest Biography* (London: Hodder and Stoughton, 1914).
Maurolico, Francesco, *Arithmeticorum libri duo* (Venice: Franciscum Franciscium Senensem, 1575).
Meinong, Alexius, *Über die Stellung der Gegenstandstheorie im System der Wissenschaften* (Leipzig: R. Voigtländer, 1907).

Moktefi, Amirouche, 'Geometry: The Euclid Debate', in Raymond Flood, Adrian Rice and Robin Wilson (eds), *Mathematics in Victorian Britain* (Oxford: Oxford University Press, 2011), pp. 321–338.
Monge, Gaspard, *Géométrie descriptive* (Paris: Baudouin, 1799).
Moore, E. H., 'On the Projective Axioms of Geometry', *Transactions of the American Mathematical Society*, vol. 3, no. 1 (January 1902), pp. 142–158.
Moulton, Forest Ray, 'A Simple Non-Desarguesian Plane Geometry', *Transactions of the American Mathematical Society*, vol. 3, no. 2 (April 1902), pp. 192–195.
Mummert, Haley, 'The History of the Plimsoll Line', traderiskguaranty.com, 6 May 2020, <https://traderiskguaranty.com/trgpeak/history-plimsoll-line/>.
Murphy, Arthur E., 'Review: *Essays in Science and Philosophy*', *Philosophical Review*, vol. 56 (1947), pp. 709–711.
Oxford English Dictionary, s.v. 'Clerk of the Weather (*n.*)', June 2024, <https://doi.org/10.1093/OED/5518823350>.
Pasch, Moritz, *Vorlesungen über neuere Geometrie* (Leipzig: Teubner, 1882).
Peano, Giuseppe, *Arithmetices Principia: Nova Methodo Exposita* (Rome: Fratres Bocca, 1889).
Peano, Giuseppe, *Formulaire de Mathématiques, Tome I* (Turin: Bocca frères, Ch. Clausen, 1895).
Peano, Giuseppe, *Formulaire de Mathématiques, Tome II, No. 1: Logique mathématique* (Turin: Bocca frères, Ch. Clausen, 1897).
Peano, Giuseppe, *Formulaire de Mathématiques, Tome II, No. 2* (Turin: Bocca frères, Ch. Clausen, 1898).
Peano, Giuseppe, *Formulaire de Mathématiques, Tome II, No. 3* (Turin: Bocca frères, Ch. Clausen, 1899).
Peano, Giuseppe, *Formulaire de Mathématiques, Tome IV* (Turin: Bocca frères, Ch. Clausen, 1903).
Peano, Giuseppe, *I Principii di Geometria Logicamente Esposti* (Turin: Fratelli Bocca, 1889).
Peano, Giuseppe, *Notations de Logique Mathématique* (Turin: Guadagnini, 1894).
Peano, Giuseppe, *Revue de Mathématiques, Tome VII, No. 1* (Turin: Bocca frères, Ch. Clausen, 1900).
Peano, Giuseppe, 'Sui fondamenti della Geometria', *Rivista Di Matematica*, vol. 4 (1894), pp. 51–90.
Pearson, Karl, *The Grammar of Science* (London: Walter Scott, 1892).
Pearson, Karl, *The Grammar of Science*, 2nd edition, revised and enlarged (London: Adam and Charles Black, 1900).
Peirce, Benjamin, *Linear Associative Algebra* (Washington, DC: 1870).
Peirce, Benjamin, 'Linear Associative Algebra', *American Journal of Mathematics*, vol. 4, no. 1 (1881), pp. 97–229.
Peirce, C. S., 'On a New List of Categories', *Proceedings of the American Academy of Arts and Sciences*, vol. 7 (May 1865 – May 1868), pp. 287–298.
Peirce, C. S., 'On an Improvement in Boole's Calculus of Logic', *Proceedings of the American Academy of Arts and Sciences*, vol. 7 (May 1865 – May 1868), pp. 250–261.
Peirce, C. S., 'On the Algebra of Logic', *American Journal of Mathematics*, vol. 3, no. 1 (March 1880), pp. 15–57.
Peirce, C. S., 'On the Logic of Number', *American Journal of Mathematics*, vol. 4, no. 1 (1881), pp. 85–95.
Peirce, C. S., 'On the Natural Classification of Arguments', *Proceedings of the American Academy of Arts and Sciences*, vol. 7 (May 1865 – May 1868), pp. 261–287.
Petek, Joseph, *Unearthing the Unknown Whitehead* (Lanham: Lexington Press, 2022).
Pieri, Mario, 'I principii delia geometria di posizione composti in sistema logico deduttivo', *Memoria della Reale Accademia della Scienze di Torino*, vol. 48, no. 2 (1899), pp. 1–62.
Pieri, Mario, 'Nuovi principii di geometria projettiva complessa', *Memorie della Reale Accademia della Scienze di Torino*, vol. 55, no. 2 (1905), pp. 189–235.
Poincaré, Henri, *La Science et l'Hypothèse* (Paris: Flammarion, 1902).
Poincaré, Henri, *Science and Hypothesis*, translated by W. J. Greenstreet (London: Walter Scott, 1905).

Poncelet, Jean-Victor, *Traité des proprietés projectives des figures* (Paris: Gauthier-Villars, 1822).
Poretsky, Platon, 'La loi des racines en Logique', in *Revue de Mathématiques, Tome VI* (Turin: Bocca Frères Libraires, 1899), pp. 5–8.
Price, Lucien, *Dialogues of Alfred North Whitehead* (Boston: Little, Brown, 1954).
Price, Michael H., *Mathematics for the Multitude? A History of the Mathematical Association* (London: Mathematical Association, 1994).
Raleigh, Walter, *Poetry and Fact* (Liverpool: Henry Young and Sons, 1890).
Reymond, Arnold, 'Le Premier Congrès de Philosophie mathématique', *L'Enseignement mathématique*, vol. 16 (1914), pp. 370–378.
Riemann, Bernhard, *Gesammelte mathematische Werke und Wissenschaftlicher Nachlass*, edited by Heinrich Weber (Leipzig: Teubner, 1876).
Riemann, Bernhard, 'On the Hypotheses Which Lie at the Bases of Geometry', translated by William Kingdon Clifford, *Nature*, vol. 8, nos 183–184 (1873), pp. 14–17, 36–37.
Riemann, Bernhard, 'Ueber die Hypothesen, welche der Geometrie zu Grunde liegen', *Abhandlungen der Königlichen Gesellschaft der Wissenschaften zu Göttingen*, vol. 13 (1868), pp. 133–150.
Roberts, Stuart, 'The Rising Tide: Women at Cambridge', University of Cambridge, 2019, <https://www.cam.ac.uk/stories/the-rising-tide>.
Ross, David, 'The Cambridgeshire Fens', Britain Express, <https://www.britainexpress.com/counties/cambridgeshire/fens.htm>.
Rowley, Martin, 'Weather in History 1850 to 1899 AD', Weatherwebdotnet, <https://premium.weatherweb.net/weather-in-history-1850-to-1899-ad/>.
Royce, Josiah, 'The Relation of the Principles of Logic to the Foundations of Geometry', *Transactions of the American Mathematical Society*, vol. 6, no. 3 (July 1905), pp. 353–415.
Russell, Bertrand, *A Critical Exposition of the Philosophy of Leibniz* (Cambridge: Cambridge University Press, 1900).
Russell, Bertrand, *An Essay on the Foundations of Geometry* (Cambridge: Cambridge University Press, 1897).
Russell, Bertrand, 'Geometry, Non-Euclidean', in Bertrand Russell, *The Collected Papers of Bertrand Russell, Volume III: Toward the 'Principles of Mathematics': 1900–1902*, edited by Gregory H. Moore (London: Routledge, 1993), pp. 470–504.
Russell, Bertrand, 'Geometry, Non-Euclidean', in Hugh Chisholm (ed), *Encyclopaedia Britannica*, 10th edition, vol. 28 (Cambridge: Cambridge University Press, 1902), pp. 664–674.
Russell, Bertrand, 'Is Position in Time and Space Absolute or Relative?', *Mind*, vol. 10, no. 39 (1901), pp. 293–317.
Russell, Bertrand, 'Mathematical Logic as Based on the Theory of Types', *American Journal of Mathematics*, vol. 30, no. 3 (July 1908), pp. 222–262.
Russell, Bertrand, 'On Denoting', *Mind*, vol. 14, no. 56 (1905), pp. 479–493.
Russell, Bertrand, *The Principles of Mathematics* (Cambridge: Cambridge University Press, 1903).
Russell, Bertrand, 'Sur la logique des relations avec des applications à la théorie des series', *Revue de Mathématiques*, vol. 7, nos 2–3 (1901), pp. 115–136, 137–148.
Saccheri, Giovanni, *Euclides ab omni naevo vindicatus* (Milan: Paolo Antonio Montani, 1733).
Savile, Henry, *Praelectiones tresdecim in principium Elemetorum Euclidis: Oxonii habitae M.DC. XX.* (Oxford: Iohannes Lichfield and Iacobus Short, 1621).
Schilpp, Paul (ed.), *The Philosophy of Alfred North Whitehead* (Evanston: Northwestern University Press, 1941).
Schoenflies, Arthur, 'Die Entwickelung der Lehre von den Punktmannigfaltigkeiten', *Jahresbericht der Deutschen Mathematiker-Vereinigung*, vol. 8 (1900), pp. 1–250.
Schoenflies, Arthur, Über die Möglichkeit einer projectiven Geometrie bei transfiniter (nichtarchimedischer) Massbestimmung', *Jahresbericht der Deutschen Mathematiker-Vereinigung*, vol. 15 (1906), pp. 26–41.
Schröder, Ernest, *Der Operationskreis des Logikkalkülus* (Leibzig: Teubner, 1877).
Schröder, Ernest, *Vorlesungen über die Algebra der Logik, Volume I* (Leibzig: Teubner, 1890).
Schröder, Ernest, *Vorlesungen über die Algebra der Logik, Volume II* (Leibzig: Teubner, 1891).

Schur, Friedrich, 'Über den Fundamentalsatz der projectiven Geometric', *Mathematische Annalen*, vol. 51, no. 3 (September 1898), pp. 401–409.
Schur, Friedrich, 'Ueber die Einführung der sogenannten idealen Elemente in die projective Geometrie', *Mathematische Annalen*, vol. 38 (1891), pp. 113–124.
Schur, Friedrich, 'Ueber die Grundlagen der Geometrie', *Mathematische Annalen*, vol. 55, no. 2 (June 1901), pp. 265–292.
Siemens, C. William, *On the Conservation of Solar Energy: A Collection of Papers and Discussions* (London: Macmillan, 1883).
Slater, John G., Letter from John G. Slater to Victor Lowe, 5 December 1976, MS-0284, Victor Lowe Papers, Box 2.3, 'Women's Suffrage, W on' folder, Special Collections, Sheridan Libraries, Johns Hopkins University.
Stäckel, Paul, and Friedrich Engel, *Die Theorie der Parallellinien von Euklid bis auf Gauss* (Leipzig: Teubner, 1895).
Stäckel, Paul, and Friedrich Engel, 'Gauss, die beiden Bolyai und die nichteuklidische Geometrie', *Mathematische Annalen*, vol. 49, no. 2 (June 1897), pp. 149–167.
Stanley, Arthur Penrhyn, *Lectures on the History of the Eastern Church* (London: John Murray, 1861).
Staudt, Karl Georg Christian von, *Beiträge zur Geometrie der Lage* (Nürnberg: F. Corn, 1856).
Staudt, Karl Georg Christian von, *Geometrie de Lage* (Nürnberg: F. Corn, 1847).
Stokes, George Gabriel, 'On the Theories of the Internal Friction of Fluids in Motion, and of the Equilibrium and Motion of Elastic Solids', in George Gabriel Stokes, *Mathematical and Physical Papers, Volume I* (Cambridge: Cambridge University Press, 1880).
Study, E., 'Über nicht-Euklidische and Liniengeometrie', *Jahresbericht Der Deutschen Mathematiker-Vereinigung*, vol. 15 (1906), pp. 476–527.
Tilly, Joseph Marie, *Essai sur les principes fondamentales de la géométrie et de la mécanique* (Brussels: G. Mayolez, 1879).
Trouton, F. T., and H. R. Noble, 'The Mechanical Forces Acting on a Charged Electric Condenser Moving Through Space', *Philosophical Transactions of the Royal Society, Series A, Containing Papers of a Mathematical or Physical Character*, vol. 202 (1904), pp. 165–181.
Vahlen, Karl Theodor, *Abstrakte Geometrie* (Leipzig: Teubner, 1905).
Veblen, Oswald, 'A System of Axioms for Geometry', *Transactions of the American Mathematical Society*, vol. 5, no. 3 (July 1904), pp. 343–384.
Veblen, Oswald, and William Henry Bussey, 'Finite Projective Geometries', *Transactions of the American Mathematical Society*, vol. 7, no. 2 (April 1906), pp. 241–259.
Venn, John, *Symbolic Logic*, 1st edition (London: Macmillan, 1881).
Venn, John, *Symbolic Logic*, 2nd edition (London: Macmillan, 1894).
Veronese, Giuseppe, *Fondamenti di geometria* (Padua, 1891).
Veronese, Giuseppe, *Grundzüge der Geometrie*, translated by Adolf Schepp (Leipzig: Teubner, 1894).
Virgil, *The Aeneid*, translated by Robert Fagles (London: Penguin Books, 2010).
Whitehead, Alfred North, *The Aims of Education* (New York: Macmillan, 1929).
Whitehead, Alfred North, 'The Aims of Education', *New Republic*, vol. 58, no. 750 (17 April 1929), pp. 244–246.
Whitehead, Alfred North, 'The Aims of Education', in Alfred North Whitehead, *The Aims of Education* (New York: Macmillan, 1929), pp. 1–23.
Whitehead, Alfred North, 'The Aims of Education', in Claude M. Fuess and Emory S. Basford (eds), *Unseen Harvests: A Treasuring of Teaching* (New York: Macmillan, 1947), pp. 92–102.
Whitehead, Alfred North, 'The Aims of Education – A Plea for Reform', *Mathematical Gazette*, vol. 8, no. 121 (January 1916), pp. 191–203.
Whitehead, Alfred North, 'The Aims of Education – A Plea for Reform', in Alfred North Whitehead, *The Organisation of Thought* (London: Williams and Norgate, 1917), pp. 1–28.
Whitehead, Alfred North, *The Axioms of Descriptive Geometry* (Cambridge: Cambridge University Press, 1907).
Whitehead, Alfred North, 'Axioms of Geometry', in Hugh Chisholm (ed.), *Encyclopaedia Britannica*, 11th edition, vol. 11 (Cambridge: Cambridge University Press, 1910), pp. 730–736.

Whitehead, Alfred North, 'Axioms of Geometry', in Alfred North Whitehead, *Essays in Science and Philosophy* (New York: Philosophical Library, 1947), pp. 243–268.
Whitehead, Alfred North, *The Axioms of Projective Geometry* (Cambridge: Cambridge University Press, 1906).
Whitehead, Alfred North, 'A Celebrity at Home: The Clerk of the Weather', *Cambridge Review*, vol. 7, no. 10 (February 1886), pp. 202–203.
Whitehead, Alfred North, 'Davy Jones', *Cambridge Review*, vol. 7, no. 12 (May 1886), pp. 311–312.
Whitehead, Alfred North, *Essays in Science and Philosophy* (New York: Philosophical Library, 1947).
Whitehead, Alfred North, 'The Fens as Seen from Skates', *Cambridge Review*, vol. 12, no. 20 (February 1891), pp. 212–213.
Whitehead, Alfred North, 'The Geodesic Geometry of Surfaces in Non-Euclidean Space', *Proceedings of the London Mathematical Society*, vol. 29 (1898), pp. 275–324.
Whitehead, Alfred North, *The Harvard Lectures of Alfred North Whitehead, 1924–1925: Philosophical Presuppositions of Science*, edited by Paul A. Bogaard and Jason Bell (Edinburgh: Edinburgh University Press, 2017).
Whitehead, Alfred North, *The Harvard Lectures of Alfred North Whitehead, 1925–1927: General Metaphysical Problems of Science*, edited by Brian G. Henning, Joseph Petek and George Lucas (Edinburgh: Edinburgh University Press, 2021).
Whitehead, Alfred North, 'La Théorie Relationniste de L'Espace (The Relational Theory of Space)', *Revue de Métaphysique et de Morale*, vol. 23, no. 3 (May 1916), pp. 423–454.
Whitehead, Alfred North, 'Letter from Alfred North Whitehead to Bertrand Russell, 3 September 1911', Bertrand Russell Archives, Box 5.54, 710.057464, McMaster University Library.
Whitehead, Alfred North, 'Letter from Alfred North Whitehead to Bertrand Russell, 1 October 1913', Bertrand Russell Archives, Box 5.54, 710.057467, McMaster University Library.
Whitehead, Alfred North, 'Letter from Alfred North Whitehead to Victor Lowe, 21 August 1942', Alfred North Whitehead Collection, MS-0282, Box 2, Special Collections, Sheridan Libraries, Johns Hopkins University.
Whitehead, Alfred North, 'Liberty and the Enfranchisement of Women' (pamphlet), Cambridge Women's Suffrage Association (Cambridge: W. Seaton & Son, Mill Rd., 5 November 1906).
Whitehead, Alfred North, 'Liberty and the Enfranchisement of Women', *Process Studies*, vol. 7, no. 1 (spring 1977), pp. 37–39.
Whitehead, Alfred North, 'The Logic of Relations, Logical Substitution Groups, and Cardinal Numbers', *American Journal of Mathematics*, vol. 25, no. 2 (April 1903), pp. 157–178.
Whitehead, Alfred North, 'The Mathematical Curriculum', in Alfred North Whitehead, *The Organisation of Thought* (London: Williams and Norgate, 1917), pp. 69–91.
Whitehead, Alfred North, 'The Mathematical Curriculum', in Alfred North Whitehead, *The Aims of Education* (New York: Macmillan, 1929), pp. 116–135.
Whitehead, Alfred North, 'Mathematics', in Hugh Chisholm (ed), *Encyclopaedia Britannica*, 11th edition, vol. 17 (Cambridge: Cambridge University Press, 1911), pp. 878–883.
Whitehead, Alfred North, 'Mathematics', in Alfred North Whitehead, *Essays in Science and Philosophy* (New York: Philosophical Library, 1947), pp. 269–288.
Whitehead, Alfred North, 'Mathematics and Liberal Education: An Address', in Alfred North Whitehead, *Essays in Science and Philosophy* (New York: Philosophical Library, 1947), pp. 175–188.
Whitehead, Alfred North, 'Memoir on the Algebra of Symbolic Logic (Part I)', *American Journal of Mathematics*, vol. 23, no. 2 (April 1901), pp. 139–165.
Whitehead, Alfred North, 'Memoir on the Algebra of Symbolic Logic (Part II)', *American Journal of Mathematics*, vol. 23, no. 4 (October 1901), pp. 297–316.
Whitehead, Alfred North, 'Note', *Revue de Métaphysique et de Morale*, vol. 13, no. 6 (November 1905), pp. 916–917.

Whitehead, Alfred North, 'On Cardinal Numbers', *American Journal of Mathematics*, vol. 24, no. 4 (October 1902), pp. 367–394.
Whitehead, Alfred North, 'On Ideals: With Reference to the Controversy Concerning the Admission of Women to Degrees in the University', *Cambridge Review*, vol. 17 (14 May 1896), pp. 310–311.
Whitehead, Alfred North, 'On Mathematical Concepts of the Material World' (abstract), *Proceedings of the Royal Society of London, Series A, Containing Papers of a Mathematical and Physical Character*, vol. 77, no. 517 (29 March 1906), pp. 290–291.
Whitehead, Alfred North, 'On Mathematical Concepts of the Material World', *Philosophical Transactions of the Royal Society of London*, vol. 205 (May 1906), pp. 465–525.
Whitehead, Alfred North, 'On the Motion of Viscous Incompressible Fluids: A Method of Approximation', *Quarterly Journal of Pure and Applied Mathematics*, vol. 23, no. 89 (February 1888), pp. 78–93.
Whitehead, Alfred North, *The Organisation of Thought* (London: Williams and Norgate, 1917).
Whitehead, Alfred North, 'The Organisation of Thought', *The Chemical News*, vol. 114, no. 2965 (21 September 1916), pp. 139–143.
Whitehead, Alfred North, 'The Organisation of Thought (abridged)', *Nature*, vol. 98, no. 2448 (28 September 1916), pp. 80–81.
Whitehead, Alfred North, 'The Organisation of Thought', *Proceedings of the Aristotelian Society*, vol. 17 (1916–1917), pp. 58–76.
Whitehead, Alfred North, 'The Organization of Thought', *Science*, vol. 44, no. 1134 (22 September 1916), pp. 409–419.
Whitehead, Alfred North, 'The Organisation of Thought', in *Report of the 86th Meeting of the British Association for the Advancement of Science* (London: John Murray, 1917), pp. 355–363.
Whitehead, Alfred North, 'The Organisation of Thought', in Alfred North Whitehead, *The Organisation of Thought* (London: Williams and Norgate, 1917), pp. 105–133.
Whitehead, Alfred North, 'The Organisation of Thought', in Alfred North Whitehead, *The Aims of Education* (New York: Macmillan, 1929), pp. 153–179.
Whitehead, Alfred North, 'The Organization of Thought – I', *Scientific American Supplement*, vol. 83, no. 2161 (2 June 1917), pp. 338–339.
Whitehead, Alfred North, 'The Organization of Thought – II', *Scientific American Supplement*, vol. 83, no. 2162 (9 June 1917), p. 366.
Whitehead, Alfred North, 'The Philosophy of Mathematics', *Science Progress in the Twentieth Century*, vol. 5, no. 18 (October 1910), pp. 234–239.
Whitehead, Alfred North, 'The Place of Elementary Mathematics in a Liberal Education', *Journal of the Association of Teachers of Mathematics for the Southeastern Part of England*, vol. 1, no. 1 (December 1911), pp. 2–12.
Whitehead, Alfred North, 'Presidential Address to the London Branch of the Mathematical Association', *Mathematical Gazette*, vol. 7, no. 104 (March 1913), pp. 87–94.
Whitehead, Alfred North, *Process and Reality: An Essay in Cosmology*, corrected edition, edited by David Ray Griffin and Donald W. Sherburne (New York: Free Press, 1978 [1929]).
Whitehead, Alfred North, *The Principle of Relativity, with Applications to Physical Science* (Cambridge: Cambridge University Press, 1922).
Whitehead, Alfred North, 'The Principles of Mathematics in Relation to Elementary Teaching', *L'Enseignement Mathématique*, vol. 15 (1913), pp. 105–111.
Whitehead, Alfred North, 'Second Approximations to Viscous Fluid Motion', *Quarterly Journal of Pure and Applied Mathematics*, vol. 23, no. 90 (June 1888), pp. 143–152.
Whitehead, Alfred North, 'Sets of Operations in Relation to Groups of Finite Order', Unpublished, Royal Society Archives (1899), AP/74/4.
Whitehead, Alfred North, 'Sets of Operations in Relation to Groups of Finite Order' (abstract), *Proceedings of the Royal Society of London*, vol. 64 (1899), pp. 319–320.
Whitehead, Alfred North, 'Space, Time, and Relativity', *Proceedings of the Aristotelian Society, New Series*, vol. 16 (1916), pp. 104–129.
Whitehead, Alfred North, 'Space, Time, and Relativity', in Alfred North Whitehead, *The Organisation of Thought* (London: Williams and Norgate, 1917), pp. 191–228.

Whitehead, Alfred North, 'Space, Time, and Relativity', in Alfred North Whitehead, *The Aims of Education and Other Essays* (New York: MacMillan, 1929), pp. 232–247.

Whitehead, Alfred North, 'Theorems on Cardinal Numbers', *American Journal of Mathematics*, vol. 26, no. 1 (January 1904), pp. 31–32.

Whitehead, Alfred North, *A Treatise on Universal Algebra* (Cambridge: Cambridge University Press, 1898).

Whitehead, Alfred North, *A Treatise on Universal Algebra, with Applications, Volume I* (Cambridge: Cambridge University Press, 1898).

Whitehead, Alfred North, 'A Visitation', *Cambridge Fortnightly*, vol. 1, no. 4 (6 March 1888), pp. 81–83.

Whitehead, Alfred North, and Bertrand Russell, 'Non-Euclidean Geometry', in Alfred North Whitehead, *Essays in Science and Philosophy* (New York: Philosophical Library, 1947), pp. 289–312.

Whitehead, Alfred North, and Bertrand Russell, 'Non-Euclidean Geometry', in Hugh Chisholm (ed.), *Encyclopaedia Britannica*, 11th edition, vol. 11 (Cambridge: Cambridge University Press, 1910), pp. 724–730.

Whitehead, Alfred North, and Bertrand Russell, *Principia Mathematica, Volume I* (Cambridge: Cambridge University Press, 1910).

Whitehead, Alfred North, and Bertrand Russell, *Principia Mathematica, Volume II* (Cambridge: Cambridge University Press, 1912).

Whitehead, Alfred North, and Bertrand Russell, *Principia Mathematica, Volume III* (Cambridge: Cambridge University Press, 1913).

Wiener, H., 'Ueber Grundlagen und Aufbau der Geometrie', *Jahresbericht der Deutschen Mathematiker-Vereinigung*, vol. 1 (1890), pp. 45–48.

Woman Citizen Publishing Society, 'Notes and Comments', *Women's Franchise*, 20 February 1908, p. 394.

Woodbridge, Barry A., *Alfred North Whitehead: A Primary–Secondary Bibliography* (Bowling Green: Philosophy Documentation Center, 1977).

Woods, Frederick Shenstone, 'Forms of Non-Euclidean Space', in Edward Burr Van Vleck, Henry Seely White and Frederick Shenstone Woods, *The Boston Colloquium: Lectures on Mathematics Delivered from September 2 to 5, 1903* (New York: Macmillan, 1905), pp. 31–74.

Zermelo, Ernst, 'Über die Addition transfiniter Cardinalzahlen', *Nachrichten von der Königl. Gesellschaft der Wissenschaften zu Göttingen. Mathematisch-physikalische Klasse aus dem Jahre 1901* (1902), pp. 34–38.

Index

Note: Page numbers in italics refer to the editorial introductions. Shortened paper titles in parentheses indicate the context for terms used differently in different contexts. 'Fluid Motion' refers to both papers on fluid motion; 'Memoir' refers to both papers on algebra of symbolic logic.